Handbook of Experimental Pharmacology

Volume 163/I

Editor-in-Chief

K. Starke, Freiburg i. Br.

Editorial Board

G.V.R. Born, London
M. Eichelbaum, Stuttgart
D. Ganten, Berlin
F. Hofmann, München
B. Kobilka, Stanford, CA
W. Rosenthal, Berlin
G. Rubanyi, Richmond, CA

Springer
*Berlin
Heidelberg
New York
Hong Kong
London
Milan
Paris
Tokyo*

Angiotensin
Vol. I

Contributors

E.M. Abdel-Rahman, A.L. Albiston, D.B. Averill, M. Bader,
J.C. Balt, P. Beaucage, D. de Blois, C. Boden, M. Brede,
K.B. Brosnihan, S.Y. Chai, E. Chamoux, J. de Champlain,
M.C. Chappell, E. Clauser, S. Conchon, A.H.J. Danser,
S. Dimmeler, D.I. Diz, D. Felix, C.M. Ferrario, P.E. Gallagher,
N. Gallo-Payet, D. Ganten, L. Gendron, J. Haendeler, P. Hamet,
L. Hein, H. Imboden, X. Jeunemaitre, B. Kimura, V. Kren,
M.-A. Laplante, J. Lee, C. Maric, S.G. McDowall,
F.A.O. Mendelsohn, J.-P. Montani, P. Moreau, T. Mustafa,
C. Nahmias, G. Nickenig, H. Nishimura, S.N. Orlov,
M.D. Payet, M. Pfaffendorf, M.I. Phillips, K. Sandberg,
H.M. Siragy, M. Stoll, Y. Sun, E.A. Tallant, J. Tremblay,
T. Unger, P. Vanderheyden, B.N. Van Vliet, G. Vauquelin,
S. Wassmann, C. Wruck

Editors

Thomas Unger and Bernward A. Schölkens

Springer

Professor
Thomas Unger
Center for Cardiovascular Research (CCR)
Institut für Pharmakologie und Toxikologie
Charité Campus Mitte
Charité – Universitätsmedizin Berlin
Hessische Strasse 3-4
10115 Berlin, Germany
e-mail: Thomas.Unger@charite.de

Professor
Bernward A. Schölkens
Aventis Pharma Deutschland GmbH
Drug Innovation & Approval
Bldg. H 831, Room C 320
65926 Frankfurt/Main, Germany
e-mail: Bernward.Schoelkens@aventis.com

With 94 Figures and 19 Tables

ISSN 0171-2004
ISBN 3-540-40640-9 Springer-Verlag Berlin Heidelberg New York

Library of Congress Cataloging-in-Publication Data
Angiotensin / contributors, E.M. Abdel-Rahman ... [et al.] ; editors, Thomas Unger and Bernward A. Schölkens.   p. ; cm. – (Handbook of experimental pharmacology ; v. 163) Includes bibliographical references and indexes.
ISBN 3-540-40640-9 (hard : alk. paper) ISBN 3-540-40641-7 (hard : v. 2 : alk. paper)
1. Angiotensins. I. Abdel-Rahman, E. M. II. Unger, Th. III. Schölkens, Bernward A., 1943-   . IV. Series.
[DNLM: 1. Angiotensins–physiology. 2. Receptors, Angiotensin–physiology. 3. Renin-Angiotensin System–genetics. 4. Renin-Angiotensin System–physiology. 5. Angiotensin-Converting Enzyme Inhibitors–pharmacokinetics. W1 HASIL v. 163 2004 / WG 106 A58778 2004]
OP905.H3 vol. 163   [QP572.A54]   615'.1s–dc22   [616.1'32]   2003060602

This work is subject to copyright. All rights are reserved, whether the whole or part of the material is concerned, specifically the rights of translation, reprinting, re-use of illustrations, recitation, broadcasting, reproduction on microfilm or in any other way, and storage in data banks. Duplication of this publication or parts thereof is permitted only under the provisions of the German Copyright Law of September 9, 1965, in its current version, and permission for use must always be obtained from Springer-Verlag. Violations are liable for Prosecution under the German Copyright Law.

Springer-Verlag is a part of Springer Science+Business Media
springeronline.com

© Springer-Verlag Berlin Heidelberg 2004
Printed in Germany

The use of general descriptive names, registered names, etc. in this publication does not imply, even in the absence of a specific statement, that such names are exempt from the relevant protective laws and regulations and free for general use.

Product liability: The publishers cannot guarantee the accuracy of any information about dosage and application contained in this book. In every individual case the user must check such information by consulting the relevant literature.

Cover design: design & production GmbH, Heidelberg
Typesetting: Stürtz AG, 97080 Würzburg

Printed on acid-free paper   27/3150 hs – 5 4 3 2 1 0

Preface

The 1974 volume on angiotensin edited by Irvine H. Page and F. Merlin Bumpus expanded the *Handbook of Experimental Pharmacology* series. In the preface, the editors of the first edition commented on their subject matters as follows:

"...Initially, it seemed that the action of angiotensin was relatively simple but this proved grossly misleading... Even after two decades [since angiotensin was identified as the major effector peptide of the renin–angiotensin system (editors' note)] the multiplicity of its actions appears not to have been fully discovered. To call attention to its many functions is one of the purposes of this book."

Thirty years later, this statement still holds true. Nevertheless, this new edition of the volume on angiotensin attempts to provide an updated account of the knowledge and findings accumulated since the complexity of angiotensin was so accurately recognized.

Certainly, the editors of the first volume on angiotensin would have been gratified by the wealth of new data on their subject flooding the literature since 1974, adding to the complexity of actions of angiotensin peptides they had predicted. This, of course, does not make our present-day task of understanding the multiple facets of the renin–angiotensin system any easier. However, it justifies our current endeavor to take a new look at this venerable system that still offers so many hidden miracles to be unveiled and described and, after all, has proved in the last 30 years to be of utmost clinical importance.

It is indeed the advent of the inhibitors of the renin–angiotensin system, notably the angiotensin-converting enzyme inhibitors and the angiotensin AT_1 receptor antagonists, which has helped enormously in gaining deeper insights into the system, notwithstanding the fact that these compounds have become some of the most successful drugs ever developed, not only to control hypertension but also to protect target organs like the kidney, heart, blood vessels, and brain, and, most importantly, to reduce cardiovascular mortality.

Naturally, the focus of any published work on medical science changes over the years and between editions. In 1974, much emphasis was laid on the biochemistry of the renin–angiotensin system, since the major discoveries had been made in this area. The interaction of renin and converting enzyme with their substrates, newly developed assays to analyze the various components of the system, and angiotensin analogs and their structure–function relationships were in the center of interest.

In the years that followed, attention gradually shifted to other areas, such as the genetics of the renin–angiotensin system, angiotensin receptors, their regulation, signaling pathways and various functions, and the inhibitors of the renin–angiotensin system with their mechanisms of action and their clinical use.

Moreover, whereas 30 years ago angiotensin II was still predominantly seen as a regulator of blood pressure and body volume, this peptide and its active fragments together with another major effector molecule of the system, the adrenal steroid aldosterone, are now considered to play an important role in a variety of (patho-) physiological functions that may be as diversified as vascular growth and atheroma formation, renal protein handling and glomerulosclerosis, cardiac left ventricular hypertrophy, fibrosis and postinfarction remodeling, or central osmoregulation and neuroregeneration. And along with the immense progress in biological sciences that we have witnessed during the last decades, the renin–angiotensin–aldosterone system has been connected to a number of biological phenomena such as cellular differentiation, neuroplasticity, and apoptosis.

This shift in scientific interest and research activities is reflected in the present volume, although we strongly felt that the fundamental knowledge on the system accumulated and substantiated over the last 100 years should never be omitted but be present as an undercurrent to help our understanding and, even more importantly, to put our temporary knowledge of today into a historical perspective.

The editors of the 1974 volume finished their preface stating, "...Books today are expensive and time-consuming to read..."

Again, their statement holds true for the two volumes of this new edition. Today, electronic media provide us with virtually any information including, of course, what has been written on angiotensin to date, and a host of review articles has been published on almost every aspect of the renin–angiotensin–aldosterone system. However, as with Page and Bumpus in their day, we are convinced that even in our time, there is still a place for books of this kind which invite the scholar to in-depth reading of what acknowledged experts have compiled as the essentials in their field.

When we asked the authors if they were willing to contribute to this edition, we did so with a certain degree of apprehension for the above-mentioned reasons but were overwhelmed by their unanimous positive response to our request. We would like to thank all authors for their efforts to make this volume a solid source of comprehensive information. We would also like to express our gratitude to our secretaries, Miranda Schröder and Undine Schelle, as well as to Sibylle Melzer and Ellen Scheibe and also to Susanne Dathe, our partner at Springer-Verlag, for their continuous, invaluable support, which enabled us to achieve our task.

Berlin and Frankfurt, January 2004   Thomas Unger
                                                                          Bernward A. Schölkens

List of Contributors

(Addresses stated at the beginning of respective chapters.)

Abdel-Rahman, E.M.   423
Albiston, A.L.   519
Averill, D.B.   477

Bader, M.   229
Balt, J.C.   351
Beaucage, P.   149
Boden, C.   375
Blois, D. de   71
Brede, M.   207
Brosnihan, K.B.   477

Chai, S.Y.   519
Chamoux, E.   399
Champlain, J. de   149
Chappell, M.C.   477
Clauser, E.   269
Conchon, S.   269

Danser, A.H.J.   129
Dimmeler, S.   99
Diz, D.I.   477

Felix, D.   111
Ferrario, C.M.   477

Gallagher, P.E.   477
Gallo-Payet, N.   399
Ganten, D.   229
Gendron, L.   399

Haendeler, J.   99

Hamet, P.   71
Hein, L.   207

Imboden, H.   111

Jeunemaitre, X.   173

Kimura, B.   251
Kren, V.   71

Laplante, M.-A.   149
Lee, J.   519

Maric, C.   335
McDowall, S.G.   519
Mendelsohn, F.A.O.   519
Montani, J.-P.   3
Moreau, P.   149
Mustafa, T.   519

Nahmias, C.   375
Nickenig, G.   317
Nishimura, H.   31

Orlov, S.N.   71

Payet, M.D.   399
Pfaffendorf, M.   351
Phillips, M.I.   251

Sandberg, K.   335
Siragy, H.M.   423

Stoll, M.   449
Sun, Y.   71

Tallant, E. A.   477
Tremblay, J.   71

Unger, T.   449

Vanderheyden, P.   297
Van Vliet, B. N.   3
Vauquelin, G.   297

Wassmann, S.   317
Wruck, C.   449

List of Contents

Part 1. General Aspects

General Physiology and Pathophysiology
of the Renin–Angiotensin System 3
    J.-P. Montani, B. N. Van Vliet

Phylogeny and Ontogeny of the Renin–Angiotensin System 31
    H. Nishimura

Angiotensin as a Cytokine Implicated in Accelerated Cellular Turnover 71
    P. Hamet, S. N. Orlov, D. de Blois, Y. Sun, V. Kren, J. Tremblay

Regulation of Angiogenesis by Angiotensin II 99
    J. Haendeler, S. Dimmeler

Immunohistochemistry of Angiotensin 111
    H. Imboden, D. Felix

Renin–Angiotensin System: Plasma Versus Tissues 129
    A. H. J. Danser

Angiotensin–Endothelin Interactions 149
    P. Moreau, M.-A. Laplante, P. Beaucage, J. de Champlain

Part 2. Genetics

Genetics of the Human Renin–Angiotensin System 173
    X. Jeunemaitre

Knockout Models of the Renin–Angiotensin System 207
    M. Brede, L. Hein

Transgenics of the RAS 229
    M. Bader, D. Ganten

Gene Therapy and the Renin–Angiotensin System 251
    M. I. Phillips, B. Kimura

Part 3. ANG Receptors

AT$_1$

AT$_1$ Receptor Molecular Aspects 269
  S. Conchon, E. Clauser

AT$_1$ Receptor Interactions 297
  G. Vauquelin, P. Vanderheyden

AT$_1$ Receptor Regulation 317
  S. Wassmann, G. Nickenig

Angiotensin AT$_1$ Receptor Signal Transduction 335
  C. Maric, K. Sandberg

Sympathetic Interactions of AT$_1$ Receptors 351
  J.C. Balt, M. Pfaffendorf

AT$_2$

Molecular Aspects of AT$_2$ Receptor 375
  C. Nahmias, C. Boden

AT$_2$ Receptor of Angiotensin II and Cellular Differentiation 399
  N. Gallo-Payet, L. Gendron, E. Chamoux, M.D. Payet

AT$_2$ Renal Aspects 423
  E.M. Abdel-Rahman, H.M. Siragy

AT$_2$ Function and Target Genes 449
  C. Wruck, M. Stoll, T. Unger

Angiotensin-(1-7). Its Contribution to Arterial Pressure
Control Mechanisms 477
  C.M. Ferrario, D.B. Averill, K.B. Brosnihan, M.C. Chappell,
  D.I. Diz, P.E. Gallagher, E.A. Tallant

Angiotensin AT$_4$ Receptor 519
  S.Y. Chai, F.A.O. Mendelsohn, J. Lee, T. Mustafa, S.G. McDowall,
  A.L. Albiston

Subject Index 539

Contents of Companion Volume 163/II

Angiotensin

Part 4. Tissues

Vascular

Angiotensin II and Oxidative Stress 3
    N. Tsilimingas, A. Warnholtz, M. Wendt, T. Münzel

Angiotensin II and Atherosclerosis.................................. 21
    H. Drexler, B. Schieffer

Angiotensin II and Vascular Extracellular Matrix 39
    J. W. Fischer

Renin-Angiotensin Inhibitors and Vascular Effects..................... 65
    E. L. Schiffrin

Brain/Nervous System

Angiotensin Pathways and Brain Function............................ 81
    F. Qadri

Involvement of the Renin Angiotensin System
in the Regulation of the Hypothalamic Pituitary Adrenal Axis............ 101
    G. Aguilera

Angiotensin Actions on the Brain Influencing Salt and Water Balance 115
    M. J. McKinley, D. A. Denton, M. L. Mathai, B. J. Oldfield, R. S. Weisinger

Angiotensin Receptor Signaling in the Brain: Ionic Currents
and Neuronal Activity 141
    C. Sumners, E. M. Richards

Angiotensin, Neuroplasticity and Stroke 163
    A. Blume, J. Culman

Heart

Role of Angiotensin II in Cardiac Remodeling........................ 193
 J. Díez

Pathophysiology of Cardiac AT_1 and AT_2 Receptors 209
 J. Fielitz, V. Regitz-Zagrosek

Activation of the Renin-Angiotensin System After Myocardial Infarction... 237
 W.M. Aartsen, J.F.M. Smits, M.J.A.P. Daemen

Kidney and Adrenal Gland

Angiotensin II, the Kidney and Hypertension 255
 O. Grisk, R. Rettig

Angiotensin and Aldosterone Biosynthesis 285
 A.M. Capponi, M.F. Rossier

Aldosterone: Clinical Aspects 343
 T.L. Goodfriend

Part 5. Inhibition of the Renin-Angiotensin System

ACE Inhibitors

Interactions Between the Renin-Angiotensin
and the Kallikrein-Kinin System..................................... 359
 P. Wohlfart, G. Wiemer

ACE Inhibitors: Pharmacology 375
 P. Gohlke, B.A. Schölkens

AT_1 Antagonists

AT_1 Receptor Antagonists: Pharmacology 417
 M. de Gasparo

Clinical Pharmacology of Angiotensin II Receptor Antagonists........... 453
 M. Maillard, M. Burnier

Combined Blockade of the Renin Angiotensin System with ACE Inhibitors
and AT_1 Receptor Antagonists...................................... 485
 M. Azizi, J. Ménard

NEP/ACE Inhibitors

NEP/ACE Inhibitors: Experimental and Clinical Aspects 519
R. Corti, F. Ruschitzka, T. F. Lüscher

Part 6. Clinical Qutlook

Inhibitors of the RAS: Evidence-Based Medicine 545
W. Schulz

Subject Index 593

Part 1
General Aspects

General Physiology and Pathophysiology of the Renin–Angiotensin System

J.-P. Montani[1] · B. N. Van Vliet[2]

[1] Department of Medicine/Division of Physiology, University of Fribourg,
Rue du Musée 5, 1700 Fribourg, Switzerland
e-mail: Jean-Pierre.Montani@unifr.ch

[2] Division of Basic Medical Sciences, Faculty of Medicine,
Memorial University of Newfoundland, St. John's, NL, A1B 3V6, Canada

1	The Major Players of the Renin–Angiotensin System	5
1.1	History of the Discovery of the RAS	5
1.2	The Renin–Angiotensin Cascade	5
1.3	Angiotensin II Acts on Specific Receptors	8
2	Regulation of Angiotensin II Formation	8
2.1	Synthesis of Circulating Renin	8
2.2	Regulation of Renin Release	9
2.3	Modulation of Angiotensin II Production	10
3	The RAS Is an Important Physiological Control System	11
3.1	The Rapid Actions of Angiotensin Prevent Life-Threatening Hypovolaemia and Hypotension	12
3.1.1	Angiotensin Increases Total Peripheral Resistance	12
3.1.2	Angiotensin Preserves Extracellular Fluid Volume	13
3.1.3	Other Physiological Actions of Angiotensin Contribute to Corporal Integrity	13
3.1.4	Activation of the RAS Is a Useful Response in Many Demanding Situations	14
3.2	The Slower Actions of Angiotensin Contribute to Sodium Balance	15
3.2.1	The Cardiovascular System Is More Than a Simple Closed Circuit	15
3.2.2	Fluid Volume Equilibrium Is Reached When Salt Excretion Is Equal to Salt Intake	16
3.2.3	The Pressure Natriuresis Curve Is Modulated by Changes in Salt Intake	17
3.2.4	The Inability to Modulate the RAS Leads to Salt Sensitivity	19
3.3	The Slowest Actions of Angiotensin Increase the Efficiency of the Cardiovascular System	20
3.3.1	Angiotensin Promotes Vascular Growth and Cardiac Hypertrophy	20
3.3.2	Angiotensin Stimulates Superoxide Anion Formation	20
4	Pathophysiology of the RAS	21
4.1	The RAS May Contribute to the Higher Cardiovascular Risks of Males	21
4.2	The RAS Contributes to Many Forms of Hypertension	22
4.2.1	Role of the RAS in Renovascular Hypertension	22
4.2.2	Role of the RAS in Essential Hypertension	23
4.3	Angiotensin-Induced Cardiac and Vascular Hypertrophies Are Risks Factors	24
4.4	Activation of the RAS Worsens Congestive Heart Failure	24
4.5	Systemic and Local Angiotensin May Initiate and Amplify Vascular Disease	25

| 5 | Conclusions | 26 |

References 26

Abstract The renin–angiotensin system (RAS), one of the oldest hormone systems, is a complex regulatory system with many identifiable actions. However, it may primarily be viewed as a powerful regulatory system for the conservation of salt and blood volume, and the preservation of an adequate blood pressure (BP). To circumvent the major threats of low blood volume and low BP, animals and our ancestors, with a diet relatively poor in sodium, needed powerful mechanisms for salt and water conservation, and their organisms relied heavily on the RAS. Many of the diverse actions of angiotensin II, the major end product of the RAS, can be viewed in a single conceptual framework, as serving to prevent life-threatening shrinkage of intravascular volume (rapid actions of angiotensin, in combination with the sympathetic nervous system), to help maintain volume homeostasis by minimizing the changes in arterial pressure and fluid volumes required to achieve sodium balance (prevention of salt-sensitivity), and to increase the efficiency of cardiovascular dynamics by promoting the growth of the heart and vessels, and sensitizing blood vessels to vasoconstrictor agents (slowest actions of angiotensin). Activation of the RAS is therefore a useful response in many demanding situations. However, an increased activity of the RAS, especially in combination with other cardiovascular risks factors, may lead to a cascade of deleterious effects. Many of these pathophysiological actions of angiotensin II may still be viewed as being homeostatic in principle, but harmful if carried to excess.

Keywords Renin release · Blood volume · Blood pressure · Sodium balance · Salt-sensitivity · Pressure natriuresis · Hypertension · Cardiovascular hypertrophy

Abbreviations

ACE	Angiotensin-converting enzyme
AGT	Angiotensinogen
Ang II	Angiotensin II
AT_1-R	AT_1-receptor
BP	Blood pressure
GFR	Glomerular filtration rate
NO	Nitric oxide
PNC	pressure-natriuresis curve
RAS	Renin–angiotensin system
TGF	Tubuloglomerular feedbacks

The renin–angiotensin system (RAS), one of the oldest hormone systems, is a strong control system for salt conservation, blood volume and blood pressure (BP) preservation. To circumvent the major threats of low blood volume and low BP, animals and our ancestors, with a diet relatively poor in sodium, needed powerful mechanisms for salt and water conservation, and their organisms relied heavily on the RAS. The purpose of this chapter is to give a general overview of the RAS, to stress its usefulness in daily homeostasis, but also to show how its effectiveness can be detrimental in certain circumstances.

1
The Major Players of the Renin–Angiotensin System

1.1
History of the Discovery of the RAS

The story of the discovery of the RAS began more than 100 years ago, on 8 November 1896, when the Finnish physiologist Robert A. Tigerstedt (1853–1923), who was at that time Professor of Physiology at the Karolinska Institute in Stockholm, and Per Gustav Bergman, a medical student, set up to do a crucial experiment. Inspired by the French physiologist Charles-Édouard Brown-Séquard, who started a trend for discovering "inner secretions" in organs by injecting extracts from donor organs into animals, Tigerstedt and Bergman injected cold extracts of donor rabbit kidneys into the jugular vein of recipient rabbits and showed that those extracts consistently increased BP. They concluded that the kidney contained a pressor substance they named renin. In further experiments, they showed that this pressor substance was located in the renal cortex and that the pressure response did not require an intact nervous system (Tigerstedt and Bergman 1898). Intriguingly, these observations were forgotten for many years.

The possibility that the kidney may release a pressor substance was revived by Harry Goldblatt, who could induce experimental hypertension in dogs by clipping one or both renal arteries (Goldblatt et al. 1934). These observations led to a renewed interest in a renal pressor substance, and in 1939 two independent laboratories, Page, Helmer and Kohlstaedt in Indianapolis, and Braun-Menendez, Fasciola and Leloir in Argentina, showed that renin was not itself a pressor substance, but an enzyme acting upon a protein to release a peptide vasoconstrictor that became known as angiotensin (Braun-Menendez and Page 1958).

1.2
The Renin–Angiotensin Cascade

Slowly over the years, the RAS was elucidated, as illustrated in Fig. 1. In response to certain stimuli, renin, a proteolytic enzyme produced by the kidney, is released into the circulation and acts on angiotensinogen (AGT), a circulating

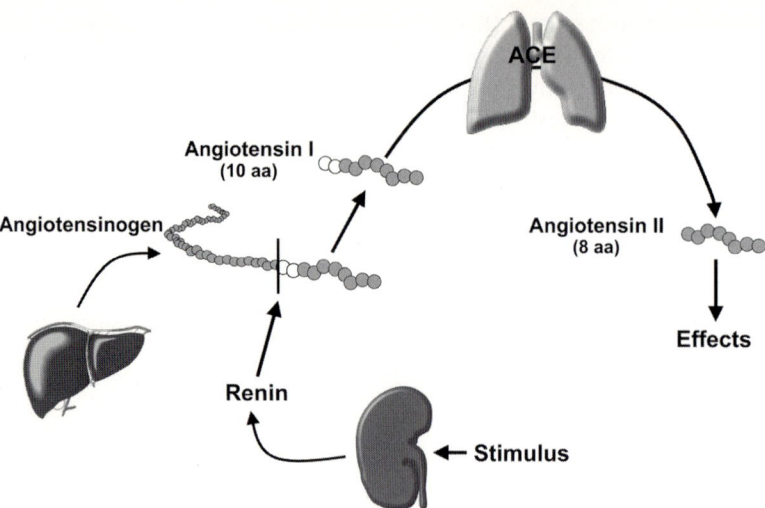

Fig. 1 Simple diagram of the RAS pathway

protein (α_2-globulin) produced by the liver. Renin cleaves AGT to produce angiotensin I (Ang I) a small fragment of only 10 amino acids. Ang I has no biological action in itself, but is converted to angiotensin II (Ang II), an active octapeptide, by angiotensin-converting enzyme (ACE), an enzyme present on the cell surface of many cells and particularly on vascular endothelial cells. As all blood leaving the kidneys and liver eventually flows through the lung, the pulmonary vascular endothelium plays a major role in the rapid conversion of Ang I into Ang II. Finally, Ang II will bind to specific cell surface angiotensin-receptors to elicit multiple actions.

At this stage, for the sake of completeness, four important remarks should be made:

1. Ang I and Ang II can be generated by alternate enzymatic pathways (Fig. 2). Enzymes other than renin, such as tonin and cathepsin D, can promote the formation of Ang I. Similarly, enzymes other than ACE, such as trypsin, cathepsin G or heart chymase, can facilitate the conversion of Ang I into Ang II. However, the contribution of these alternative pathways in Ang II production in humans is still unclear.
2. ACE can act on substrates other than Ang I. Particularly, ACE can promote the degradation of bradykinin, substance P and other small peptides. Although the physiological role of this enzymatic conversion is unclear, pharmacological blockade of ACE with specific inhibitors leads to an accumulation of bradykinin and substance P, which may be responsible for some of the beneficial effects (antihypertensive), but also some of the adverse effects (angioedema, cough) of ACE inhibitors.

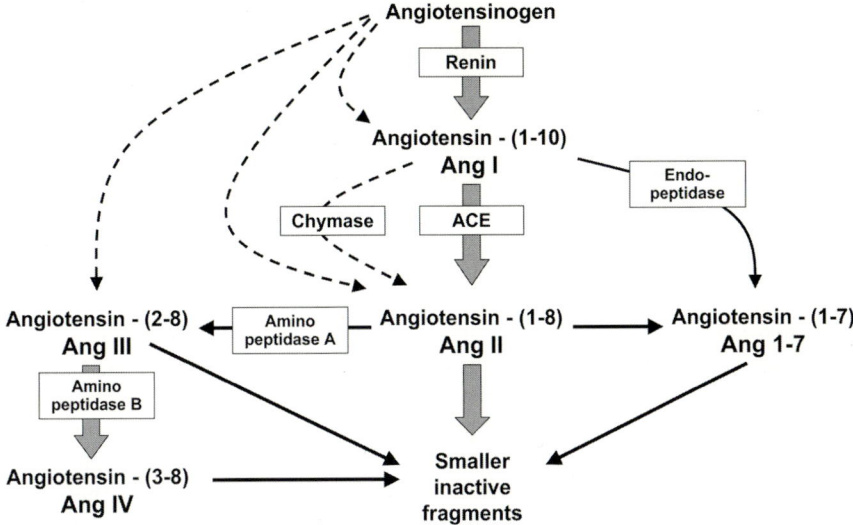

Fig. 2 Alternate pathways of angiotensin II formation and other angiotensin peptides

3. There are several angiotensin peptides with biological effects (Fig. 2). Although Ang II [angiotensin-(1-8)] is the major end-product of the system, the action of other enzymes on Ang II may cleave a further one or two amino acids from the amino end, to yield Ang III [angiotensin-(2-8)] and Ang IV [angiotensin-(3-8)], respectively. Cleavage from the carboxyl end yields angiotensin-(1-7). Ang III and IV may play an important role in the brain, whereas angiotensin-(1-7) has vasodepressor properties and may contribute to the antihypertensive actions of ACE inhibitors; angiotensin-(1-7) may be formed directly from Ang I. These angiotensin peptides will not be discussed further in this Chapter, as they are the major focus of final two chapters of this volume.

4. Components of the RAS reside within many tissues. Although most of the circulating renin comes from the kidney and most of the circulating AGT comes from the liver, components of the RAS (renin, AGT, ACE) may also be expressed locally within tissues. Circulating renin and AGT constitute the systemic RAS, and local components of the RAS constitute the tissue RAS. Local RAS have been described in many organs, such as the brain, heart, vascular wall, kidneys (interstitium), fat tissues, gut, pancreas, reproductive organs and the adrenals, and may play an important local role. For example: brain and intrarenal RAS are thought to contribute to salt balance and BP control; heart and vascular RAS are involved in cardiovascular pathology. The various tissue RAS will be discussed in detail in Part 4 of this book. In this chapter, we will focus on the physiological role of the systemic RAS.

1.3
Angiotensin II Acts on Specific Receptors

Two main cell surface receptors to Ang II have been identified: AT_1 and AT_2. In rodents, there are two isoforms of the AT_1 receptor, designated AT_{1a} and AT_{1b}. This distinction is, however, not relevant to humans, in which a single AT_1 receptor type is found. Other AT receptors have been described: AT_4R and $AT_{1-7}R$ that mediate the effects of other angiotensin peptides and intracellular receptors.

Both the AT_1 and the AT_2 receptors have been cloned and belong to the superfamily of G protein-coupled receptors that contain seven transmembrane regions. They share about 34% homology and have distinct signal transduction pathways. The AT_1 receptor mediates all of the classical actions of Ang II (vasoconstriction, sodium retention, cell growth and proliferation), and can be selectively blocked with pharmacological agents known as sartans. AT_2 receptors are mainly expressed in fetal tissues and their number decreases in the postnatal period; however, their number increases again in tissue injury. AT_2 receptors promote vasodilatation, cell differentiation, inhibition of cell growth and apoptosis, and may play a counterbalancing role to the effects of Ang II on AT_1 receptors.

2
Regulation of Angiotensin II Formation

2.1
Synthesis of Circulating Renin

The human genome contains only one renin gene (Ren-1^c), whereas certain strains of mice, such as 129, have two distinct renin genes (Ren-1^d and Ren-2). Ren-1^c renin gene expression varies in different tissues, but the kidneys are the only organs that can contain substantial amounts of readily releasable active renin. Indeed, bilateral nephrectomy leads to practically undetectable levels of renin in the plasma. Renin is produced by the juxtaglomerular (JG) cells, specialized cells derived from vascular smooth muscle cells located at the end of the afferent arteriole. During sustained stimulation (such as with a low-salt diet), there is not only an increased expression in those cells, but also an expansion in expression to vascular cells situated upstream.

The initial step in renin synthesis is the formation of preprorenin by renin messenger RNA, which is transported into the rough endoplasmic reticulum. The "pre" sequence is then cleaved, leaving prorenin, a likely inactive form of renin. Subsequently, prorenin is transported through the Golgi apparatus, glycosylated with mannose-6-phosphate residues, and deposited in granules where the "pro" sequence is cleaved to form renin the active 40,000-Da single-chain polypeptide enzyme. Renin can then be released luminally into the circulation (or abluminally into the renal interstitium) by exocytosis in a regulated response to specific mediators.

2.2
Regulation of Renin Release

The juxtaglomerular apparatus plays a central role in the regulation of renin release. It comprises the afferent arteriole, the glomerular mesangium and the macula densa cells of the distal tubule of the same nephron. Three classical stimuli, all elicited by a decrease in BP or blood volume, are known to increase renin synthesis and release:

1. Decreased stretch of the afferent arteriole. The smooth muscle cells of the afferent arteriole are very sensitive to stretch. An increase in intravascular pressure raises intracellular calcium, leading to both a contraction of vascular smooth muscle cells (myogenic vasoconstriction) and an inhibition of renin release. Conversely, a low intravascular pressure in the afferent arteriole stimulates renin release. This is a local effect, which does not require any neural input.
2. Decreased delivery of salt (sodium chloride) to the macula densa. Macula densa cells are modified tubular cells at the end of the loop of Henle that ensure a steady input of salt to the distal tubular cells by controlling both the tone of the afferent arteriole (tubuloglomerular feedback, TGF) and the release of renin. Such a mechanism helps maintain glomerular filtration rate (GFR) at a relatively constant level. Decreases in distal tubular salt delivery are sensed by macula densa cells (probably via the amount of salt which is transported through the luminal Na-K-2Cl cotransporter). This leads to a decreased release of some chemical mediators (ATP, adenosine, NO) by macula densa cells and to an increased release of other mediators (prostaglandin PGE_2). In turn, this chemical modulation of JG cells dilates the afferent arteriole and stimulates renin release.
3. Adrenergic stimulation. The JG cells are directly innervated by sympathetic nerve endings, which act on β_1-adrenergic receptors expressed on cell surface. This results in an increased formation of cyclic adenosine monophosphate (cAMP) that stimulates renin release. Renal sympathetic nerves have a very potent effect on renin release, occurring at levels of sympathetic activity much lower than those required for acute sodium retention or renal vasoconstriction (DiBona and Kopp 1997).

All three stimuli operate simultaneously and are usually stimulated by the same conditions: a decrease in blood pressure (BP). When BP falls, there is decreased stretch of JG cells and salt delivery to the macula densa (due to a decrease in GFR and an increase in proximal tubular reabsorption). In addition, the low arterial pressure unloads carotid and aortic baroreceptors, leading to renal sympathetic nerve activation and thus to β_1-receptor stimulation.

In addition to these three classical stimuli, some circulating hormones or substances can directly stimulate or inhibit renin release by the juxtaglomerular apparatus. For the purpose of this chapter, two substances that inhibit renin release are particularly worth mentioning: atrial natriuretic peptide (ANP), a hormone released by the atria in response to an increased blood volume, and Ang II. The inhibitory effect of ANP contributes to renin suppression at high salt intake. The inhibitory effect of Ang II constitutes a "short-loop" negative feedback that allows a rapid suppression of renin release, in contrast to a "long-loop" negative feedback (suppression of renin release by hypertension and hypervolemia), which would require many hours or days.

2.3
Modulation of Angiotensin II Production

Under normal circumstances, renin is the rate-limiting step in the formation of Ang II. However, modulation of Ang II formation by other components of the RAS may come into consideration in certain situations.

Modulation by Angiotensinogen. AGT is the only known precursor protein to the family of angiotensin peptides. Systemic AGT originates primarily from hepatocytes where it is constitutively secreted, and is present in the plasma in stable concentrations (half-life of 16 h, in contrast to 20 min for renin). AGT secretion can be modulated by various compounds, such as glucocorticoids, estrogens, thyroid hormones, insulin, selected cytokines, and Ang II itself (Brasier and Li 1996).

Because the normal concentration of AGT is near the K_m for its reaction with renin (Gould and Green 1971), one would expect any change in AGT levels to be accompanied by parallel changes in the formation and actions of Ang II. On the other hand, an AGT-mediated increase in Ang II levels (and action) should lead to a suppression of renin release via both the short and long negative feedback loops, and thereby a return to normal levels of plasma Ang II concentration. Yet there is indirect evidence that AGT may play a role in human hypertension (Jeunemaitre et al. 1992).

To understand this paradox, two explanations can be presented: (1) A hypertensive effect of AGT is expected in situations when renin release is poorly modulated, such as during renal damage or in 129 mice which have two distinct renin genes, one of which is submaxillary and is not subject to the usual negative feedback control (Wang et al. 2002). Indeed, 129 mice engineered to carry one to four copies of the AGT gene have AGT concentrations and BP levels that correlate to the number of AGT gene copies (Kim et al. 1995). (2) High levels of systemic AGT may also promote hypertension by increasing local formation of Ang II in various tissues, where it is not subjected to systemic feedback control.

Modulation by ACE. Various substances, such as NO and Ang II itself, have been shown to downregulate the activity of ACE in endothelium. However, the physiological role of these modulations remains unclear. Experimental data from animals and computer simulations have indicated that modest changes in ACE activity in either direction have little effect on the production of Ang II itself

(Smithies et al. 2000). As a matter of fact, mice engineered to carry one to three copies of the ACE gene do not show any variation in blood pressure (Krege et al. 1997).

On the other hand, ACE is present not only on vascular endothelial cells, but also on the cell membranes of many different cells. High levels of ACE in the microvilli of proximal tubular cells may produce high local levels of Ang II, promoting sodium reabsorption. The presence of ACE in inflammatory cells could also contribute to vascular disease (Dzau 2001). Furthermore, high levels of ACE may decrease bradykinin levels and thus contribute to the diabetic proteinuria observed in diabetic mice with genetically higher ACE levels (Huang et al. 2001), or in humans with the insertion/deletion ACE polymorphism.

In addition to the various factors that regulate or modulate Ang II formation, there are also a number of mechanisms by which Ang II action may be regulated: e.g. variations in AT receptor density, interactions between AT_1 and AT_2 receptors, or post-receptor modulation. Those complex interactions will be dealt in subsequent chapters of this book.

3
The RAS Is an Important Physiological Control System

The RAS is a complex regulatory system with many identifiable actions. However, it may primarily be viewed as a powerful regulatory system for salt conservation, blood volume and BP. A minimal intake of salt is required to compensate for obligatory salt losses by urine, sweat, faeces and epithelial desquamation. When animals are put on an extremely poor sodium diet, they exhibit hypovolaemia with possibly impaired exercise performance, thus becoming easier prays for predators. Salt depletion also endangers species survival due to poor reproductive functions such as decreased fertility, decreased number of pups in the litter and decreased pup size (McBurnie et al. 1999).

In the case of man, our ancestors routinely consumed a poor sodium diet (10–30 mmol/day) (Eaton and Konner 1985; MacGregor and de Wardener 1998). Stringent mechanisms for salt conservation were thus required to regulate the amount of fluid in our bodies. Without an efficient RAS, our ancestors would have never survived the additional stresses associated with starvation or haemorrhage, and would not have the required hemodynamic reserve for fight-or-flight reactions.

The RAS is a complex system, with more than 60 Ang II actions. In this chapter, we will show how many of these diverse actions of Ang II can be viewed in a single conceptual framework, as serving to prevent life-threatening shrinkage of intravascular volume (rapid actions of angiotensin), to help achieve sodium balance without large alterations in BP (slower actions of angiotensin) and to increase the efficiency of cardiovascular dynamics by promoting the growth of the heart and vessels, and sensitizing blood vessels to vasoconstrictor agents (slowest actions of angiotensin).

3.1
The Rapid Actions of Angiotensin Prevent Life-Threatening Hypovolaemia and Hypotension

Most of the rapid actions of Ang II can be viewed as a concerted response that supports the circulation when it is threatened by intravascular volume shrinkage and/or hypotension. Indeed, the main physiological stimuli for RAS activation are low salt intake, blood volume and BP. In turn, Ang II acts to help raise blood volume and BP via combined actions illustrated in Fig. 3: all are exerted via the AT_1 receptor.

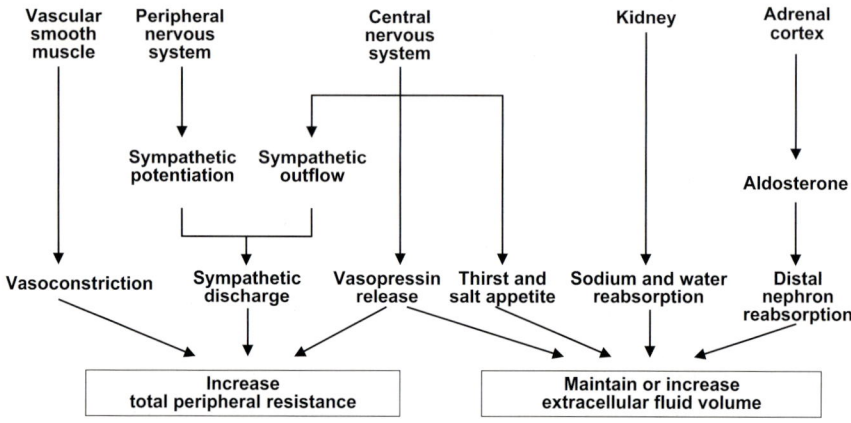

Fig. 3 Schematic diagram, showing the major effects of Ang II on total peripheral resistance and extracellular fluid volume preservation

3.1.1
Angiotensin Increases Total Peripheral Resistance

Ang II is a potent vasoconstrictor agent that elevates vascular tone by both direct and indirect mechanisms. Binding of Ang II to AT_1 receptors located on the surface of vascular smooth muscle cells leads to an immediate contraction. It is interesting to note that Ang II does not exert identical vasoconstrictor effects on all vessels. For example, renal post-glomerular (efferent) arterioles are exquisitely sensitive to Ang II, whereas pre-glomerular (afferent) arterioles show very little direct sensitivity to Ang II (Edwards 1983).

Ang II may also increase vascular tone by indirect mechanisms. Ang II increases sympathetic discharge via direct action at various brain structures that lack a blood–brain barrier, and can also potentiate the release of norepinephrine from adrenergic varicosities within peripheral tissues. This sympathetic effect is normally blunted or even suppressed in vivo by the vasoconstriction-induced rise in arterial pressure, which loads baroreceptors and results in a reflex decrease in sympathetic nerve activity (Lohmeier et al. 2000b). Situations associated with baroreflex impairment (such as heart failure or vasculopathies with aortic and carotid stiffness) may thus unmask the sympathoexcitatory actions of Ang II.

Another central action of Ang II is the stimulation of vasopressin release by the posterior pituitary gland. The quantitative contribution of this effect is not well established. However, very low levels of circulating vasopressin not only causes antidiuresis (via the V_2-receptor), but can also increase total peripheral resistance (Montani et al. 1980) and favour a more efficient renal countercurrent system by vasoconstriction of renal vasa recta (Cowley 2000), thus facilitating renal retention of sodium and water.

3.1.2
Angiotensin Preserves Extracellular Fluid Volume

Ang II also acts to maintain or increase extracellular fluid volume (ECFV), both by promoting water and sodium intakes, and by decreasing water and sodium excretions. Intracerebral infusions of Ang II in experimental animals increase both thirst and salt appetite, leading to increased water drinking, and preferential drinking of a saline solution when the animal is offered both saline and water solutions. This effect is also seen in response to moderate elevation of circulating Ang II levels, due to its actions on various regions of the brain involved in thirst and salt appetite (Fitzsimons 1998).

Ang II acts on many other tissues with the same general goal of sodium and ECFV preservation. It enhances epithelial sodium transport in the gut. In the kidney, it promotes Na^+/H^+ exchange in the apical membrane of proximal tubular cells, augmenting sodium reabsorption. Renal vasoconstriction with predominantly post-glomerular constriction leads to a decrease in peritubular capillary hydrostatic pressure, and to an increase in filtration fraction that concentrates post-glomerular plasma protein concentration, further boosting sodium reabsorption (Hall 1986a). Constriction of efferent arterioles of juxtamedullary nephrons and/or a direct action on vasa recta lowers renal papillary blood flow, enhancing urine-concentrating capability. Finally, Ang II acts on the adrenal glands to promote secretion of aldosterone, a sodium-retaining hormone acting on the distal parts of the nephron.

3.1.3
Other Physiological Actions of Angiotensin Contribute to Corporal Integrity

Other actions of Ang II that may seem unusual or even harmful at first glance, do fit well in the general scheme of homeostatic functions. Some of these actions include:

a. Preservation of glomerular filtration rate at low perfusion pressures. When arterial pressure increases suddenly, myogenic and TGF-induced vasocon-

strictions of the afferent arteriole protect the glomerulus, thereby preventing large increases in GFR. When arterial pressure decreases, these mechanisms are reversed; i.e. decreases afferent arteriolar tone. However, they are not very effective in maintaining GFR in situations of low perfusion pressures, and thus constriction of efferent arteriole by angiotensin then becomes crucial to preserve GFR. In situations when the RAS is activated (low salt intake, volume depletion), blockade of the RAS does indeed impair autoregulation of GFR at low perfusion pressures (Hall et al. 1977). This situation is well known to clinicians prescribing inhibitors of the RAS in hypertensive patients with renal artery stenosis (Hricik et al. 1983). Mechanisms for preservation of GFR during low salt intake and volume depletion are important as they allow the kidney to continue its filtrating and detoxifying functions.
b. Haematopoiesis. ACE inhibitors often lead to a reduction in haematocrit, and RAS activation leads to erythropoiesis. Ang II stimulates the renal production of erythropoietin via an AT_1 receptor-dependent pathway (Gossmann et al. 2001) by decreasing oxygen concentration around peritubular fibroblasts. This action is due to the combined action of Ang II in decreasing renal blood flow and thus oxygen delivery, and in increasing tubular sodium reabsorption and thus oxygen consumption. Ang II may also act directly on AT_1 receptors present on erythroid precursor cells of the bone marrow (Rodgers et al. 2000). In situations of chronic volume depletion (e.g. during extreme salt deprivation or following haemorrhage), an increased haematocrit would compensate for the hypovolaemia-induced decrease in cardiac output.
c. Procoagulatory effects. The mild stimulatory effects of Ang II on the coagulation cascade and on platelet activation (Larsson et al. 2000) can be viewed as a volume-conserving reaction during haemorrhage.
d. Stimulation of liver glycogenolysis. Ang II infusion leads to an elevation in blood glucose levels (Machado et al. 1998) as a consequence of an AT_1-receptor dependent action on hepatic glycogen phosphorylase (Keppens et al. 1993). Hyperglycaemia becomes useful in fight-or-flight situations, as this condition provides energetic fuel to the skeletal muscles.
e. Inotropic actions. Ang II increases cardiac contractility (Mattiazzi 1997). The exact mechanism of this action remains poorly understood, but it may be related both to a potentiation of norepinephrine release at adrenergic endings, and to a direct effect of Ang II on the myocardium. In any case, this represents a useful response during situations such as acute hypovolaemia (e.g. following haemorrhage).

3.1.4
Activation of the RAS Is a Useful Response in Many Demanding Situations

a. An effective RAS attenuates orthostatic hypotension. Standing upright leads to an accumulation of blood in the legs, a decrease in venous return and thereby a decrease in cardiac output. In turn, arterial pressure decreases, which stimulates renin release in a matter of minutes, helping to restore BP.

Complementary systems (sympathetic nervous system and vasopressin) work concurrently to oppose acute drops in BP, and thus compensate for a poorly reactive RAS. Blockade of the RAS may lead to severe orthostatic hypotension if autonomic function or vasopressin release is impaired, or in situations of pre-existing hypovolaemia. Hypertensive patients with restricted sodium intake (10 mmol/day for 5 days) suffered from circulatory collapse during an orthostatic tilt, when renin release was prevented from rising by pretreatment with propranolol (Morganti et al. 1979).

b. An effective RAS prevents hypotension during low salt intake or during dehydration. A poor sodium diet or a state of dehydration leads to a decrease in circulatory filling pressures and blood volume. However, activation of the RAS in these circumstances leads to a restoration of filling pressures and blood volume to near normal values, raising BP back to normal. The importance of RAS activation during low salt intake is illustrated by the fact that blockade of the RAS in dogs maintained on sodium intake of 5 mmol/day for 1 week lowered mean arterial pressure to about 68 mmHg (Hall et al. 1980)—a low value perhaps tolerated at rest, but probably not very well during exercise or orthostasis. Marked hypotension has also been described in patients treated with ACE inhibitors who experience gastrointestinal fluid loss or other types of volume depletion (McMurray and Matthews 1987).

3.2
The Slower Actions of Angiotensin Contribute to Sodium Balance

All the aforementioned renal actions of Ang II, not only help maintain ECFV and BP during prolonged periods of low salt intake, but play also an important role in defending ECFV and BP in the face of high salt intake. The beauty of the RAS is that it works in both directions. In one direction, the system is stimulated by low BP and blood volume induced by a low salt intake, and acts to oppose these perturbations by way of a classical negative feedback system. In the other direction, the system is suppressed in situations of increased BP and blood volume such as may be induced by high salt intake. One of the major roles of a normal RAS regulation is to prevent volume-dependent salt sensitivity. To illustrate this concept, a few words on the role of the kidney on long-term BP control are needed.

3.2.1
The Cardiovascular System Is More Than a Simple Closed Circuit

The cardiovascular system is often viewed as a simple closed circuit consisting of a pump (the heart) and a series of tubes with varying resistance (the vasculature). In this model, all that counts for BP control is the strength of the heart and the resistance of the peripheral vasculature. A more complete representation of circulatory function and BP control would be a cardiovascular system with an input from the outside (fluid and salt intakes) and an output to the out-

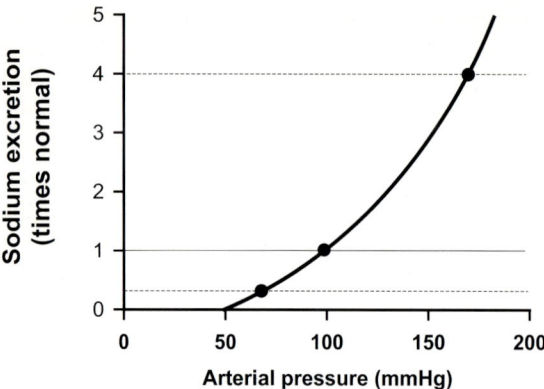

Fig. 4 The basic concept of pressure natriuresis to achieve sodium balance. Three levels of salt intake are depicted (normal, 0.2×normal and 4×normal). Equilibrium is reached at the intersection between the pressure natriuresis curve and the corresponding level of salt intake

side (urinary excretion). Any change in the input would alter blood volume and thereby BP, which in turn leads to changes in the output via pressure-natriuresis. In fact, changes in renal perfusion pressure, regardless of whether the kidney is studied in vivo or in vitro, lead to profound changes in sodium excretion, as illustrated by the solid curve in Fig. 4. The intrarenal mechanisms for this phenomenon have been detailed elsewhere (Granger et al. 2002).

3.2.2
Fluid Volume Equilibrium Is Reached When Salt Excretion Is Equal to Salt Intake

The pressure-natriuresis curve (PNC) in Fig. 4 is at the centre of blood volume and BP control. If the body gains too much fluid (e.g. acute volume load), BP increases. This leads to increased excretion of salt and water via the pressure-natriuresis mechanism, bringing blood volume and BP back towards normal. Conversely, if one looses fluids (e.g. haemorrhage), BP decreases and the kidneys retain salt and water, which helps return blood volume and BP to normal. Equilibrium is thus reached at the intersection point of the PNC with the corresponding salt intake level, as shown on Fig. 4.

Based on this concept, one can understand the general renal body fluid feedback mechanism (Hall et al. 1986b). Any imbalance between salt intake and output will lead to a cascade of events that oppose the initial disturbance; i.e. a classical negative feedback loop. For example, if salt intake is greater than output, ECFV increases, thereby increasing blood volume and thus the mean filling pressure of the circulation; this then increases cardiac output, and thus BP. In turn, the higher BP increases salt output, which opposes the effects of the initial increase in salt intake by way of the pressure–natriuresis mechanism. The system acts very slowly (hours or days), but it has an infinite gain and corrects completely any error in salt balance.

This simplified feedback loop may not explain the whole story. According to this analysis, a fourfold increase in salt intake would lead to volume retention until BP increases to well over 150 mmHg (Fig. 4). Sodium balance would be reached, but at the expense of profound volume retention and tremendous hypertension. Similarly, a poor sodium diet (for example, 1/5 of normal) would require a drop in BP by 30 or 40 mmHg to achieve a state of sodium balance. Yet, this sensitivity to salt does not fit with the small variations in BP that are normally observed when animals (Hall et al. 1980) and humans (Luft et al. 1979) are subjected to large variations of salt intake. Clearly, additional mechanisms must come into play.

3.2.3
The Pressure Natriuresis Curve Is Modulated by Changes in Salt Intake

The PNC depicted in Fig. 4 is not immovable. In fact, it becomes steeper and is shifted to the left during high salt intake. This allows the body to achieve sodium balance with minimal increases in blood pressure. Conversely, during low salt intake, the PNC becomes flatter and is shifted to the right. Joining the equilibrium points at the various salt intakes reveals a very steep "chronic" relationship with little changes in BP (Fig. 5). That is, the chronic relationship between salt intake and blood pressure has become relatively salt-*in*sensitive.

Various neurohormonal mechanisms contribute to the adjustment of the PNC. RAS modulation, above all, plays a crucial role in the adaptation to changes in salt intake. RAS suppression at high salt intake facilitates sodium excretion, and RAS stimulation at low salt intake contributes to sodium conservation. The importance of this modulation is illustrated by the dramatic salt-induced changes in BP that occur when the RAS is blocked with an ACE inhibitor, or when circulating Ang II levels are fixed with an intravenous infusion of angio-

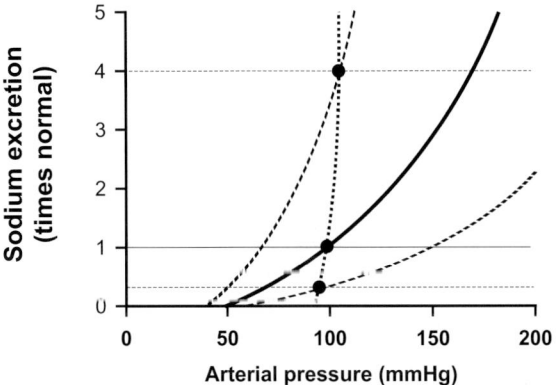

Fig. 5 The modulation of the pressure natriuresis curve during alterations in salt intake allows the body to achieve sodium balance with minimal changes in arterial pressure (*steep dotted line*)

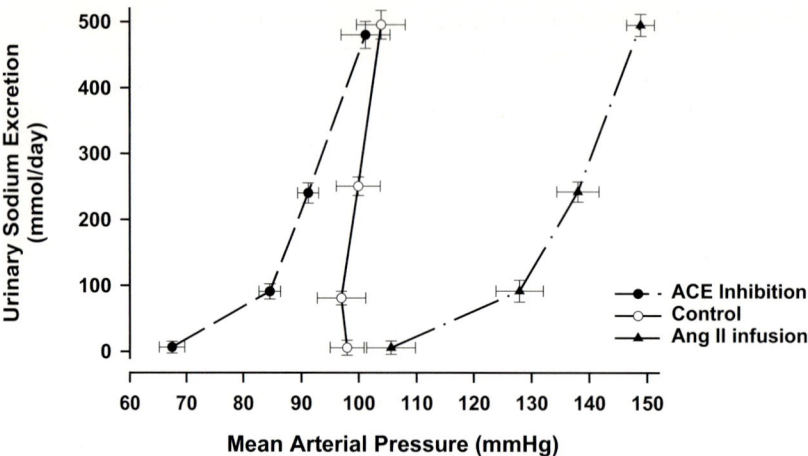

Fig. 6 Steady-state relationships between mean arterial pressure and urinary sodium excretion in dogs subjected to varying salt intakes from 5 mmol/day to 500 mmol/day, lasting 1 week at every level. The dogs were studied under three conditions: (1) with an intact RAS; (2) during chronic blockade of the RAS with captopril; (3) during fixed elevated circulating levels of Ang II via an intravenous infusion of Ang II at 5 ng/kg/min. (From Hall et al. 1980, by permission)

tensin (Hall et al. 1980) (Fig. 6). Lack of RAS activation during low salt intake leads to a dramatic decrease in mean arterial pressure to less than 70 mmHg.

But by which mechanisms do Ang II levels vary with changes in salt intake? The sequence of events is presented in Fig. 7. The initial increase in salt intake (salt with little water) leads to increased plasma osmolality, resulting in thirst

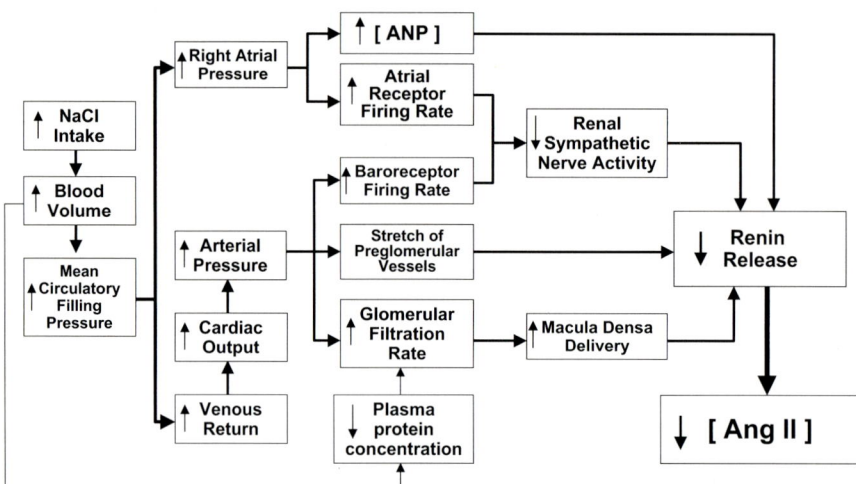

Fig. 7 Block diagram illustrating the mechanisms whereby an increased in salt intake leads to a decrease in Ang II formation

and drinking and thus an increase in extracellular fluid volume. This then increases blood volume and decreases, at least acutely, plasma protein concentration. Subsequently, a complex but logical sequence of events takes place. The decrease in plasma protein concentration favours an increased fluid filtration across the glomerular capillary membrane. The greater blood volume increases mean circulating filling pressure (greater content for the same container), resulting in both an increase in right atrial pressure and venous return.

The increased right atrial pressure stretches the right atrium, and loads low pressure receptors that reflexly decrease vasopressin secretion and renal nerve sympathetic activity. Atrial stretch leads also to a direct increased release of atrial natriuretic peptide (ANP), a hormone that has a direct inhibitory action on renin release (and aldosterone secretion). By its vasodilatory action on preglomerular vessels, ANP also promotes an increase in GFR.

The greater venous return increases cardiac output and thus arterial pressure, which in turn leads to three events: (1) loading of arterial carotid and aortic baroreceptors, and resulting in a decreased renal sympathetic nerve activity; (2) mechanical stretch of preglomerular vessels; (3) increase in delivery of fluid and salt to the macula densa, mediated by the small increase in GFR (favoured by physical forces and accentuated by the vasodilatory ability of ANP on the afferent arteriole). Altogether, these three events promote a decrease in renin release, and thus in Ang II levels.

A similar flowchart can be applied, but in the reverse direction, to explain the increased Ang II levels during low salt intake. Volume depletion leads to a decrease in filling pressures and arterial pressure. Unloading of atrial and arterial baroreceptors, decreased ANP concentration, decreased preglomerular stretch and decreased salt-delivery to the macula densa all contribute to the stimulation of renin release.

3.2.4
The Inability to Modulate the RAS Leads to Salt Sensitivity

As shown in Fig. 6, the ability to suppress renin release during high salt intake and to stimulate renin release during low salt intake is the cornerstone of having very little salt sensitivity. If the RAS is blocked with an ACE inhibitor or if Ang II levels are not allowed to fluctuate naturally in response to varying salt intakes, rapid, volume-dependent salt sensitivity ensues.

Thus one of the major roles of the RAS is to prevent a large drop in BP (and ECFV) during low salt intakes, and a large increase in BP (and ECFV) during high salt intakes. In other words, when the ability to suppress renin at high salt intakes is lost, volume-dependent salt sensitivity develops. This particularly may occur in the following two situations:

a. Ageing. Circulating renin levels decrease steadily with age (Weidmann et al. 1975), possibly related to the observed decrease in glomerular number and size that occur with ageing (Nyengaard and Bendtsen 1992). The response of renin in older individuals is also blunted when the RAS is either stimulated (volume contraction) or inhibited (volume expansion) (Luft et al. 1992). The lower basal levels of plasma renin activity and poor reactivity of the RAS may help explain the higher prevalence of salt sensitivity in older subjects.

b. Low-renin essential hypertension. About one quarter of all essential hypertensive patients have low renin levels that are poorly stimulated by a low salt intake (Fisher et al. 2002). Because renin levels are low to start with, the incapacity to further suppress renin at high salt intake may explain the salt sensitivity frequently observed in low-renin essential hypertension.

On the other end of the PNC, the concept of RAS modulation is particularly useful to understand the increased effectiveness of ACE inhibitors or angiotensin-receptor blockers in lowering blood pressure, if treatment is combined with a reduction in salt intake or with the use of diuretics.

3.3
The Slowest Actions of Angiotensin Increase the Efficiency of the Cardiovascular System

3.3.1
Angiotensin Promotes Vascular Growth and Cardiac Hypertrophy

By way of its effect on the AT_1 receptor, Ang II is also a growth factor, acting on vascular smooth muscle cells and cardiac myocytes. The trophic response to Ang II leads to a slow structural remodelling that helps maintain a higher BP. With a greater vascular muscle mass that increases both the strength of vascular contraction and the sensitivity to chronic vasoconstrictors, the vascular system becomes more effective in maintaining a high vascular tone. With a larger myocardial mass that increases cardiac strength, the heart becomes more capable in maintaining a high blood pressure. In both cases, activation of the local tissue RAS may play an important role in this trophic response.

3.3.2
Angiotensin Stimulates Superoxide Anion Formation

The superoxide radical (O_2^-) is produced endogenously during normal mitochondrial respiration and by various oxidases, especially NADH ((nicotinamide adenine dinucleotide, reduced) and NADPH (nicotinamide adenine dinucleotide phosphate, reduced) oxidases. The superoxide anion may play a physiological role in various tissues as a signalling molecule and may contribute to the regulation of vascular tone (Touyz 2000), particularly in the renal microvasculature (Schnackenberg 2002). In normotensive anaesthetized rats, the administration of tempol, a mimetic of the enzyme superoxide dismutase that decreases superoxide levels, causes an increase in medullary blood flow, urine flow and sodium excretion (Zou et al. 2001), but has no effect on basal afferent arterioles. This suggests that the superoxide anion may participate in maintaining the basal tone of the renal medullary microcirculation. In normal conditions, superoxide may also scavenge the NO formed in macula densa cells by neuronal NO synthase, thereby increasing the gain of the TGF (Ren et al. 2002). Both effects tend to increase the renal ability to retain salt.

Ang II can, via activation of the AT_1 receptor, stimulate NAD(P)H-oxidase and thereby the production of the superoxide anion. Some of the effects of Ang II on vasoconstriction and renal sodium retention may thus be mediated by oxygen radicals. For example, infusions of low doses of Ang II do not produce immediate hypertension, but a slow, progressive elevation of BP over hours or days. Furthermore, the administration of tempol attenuates or prevents Ang II-induced hypertension (Ortiz et al. 2001). The stimulation of oxygen radicals by Ang II may thus be viewed as a slow physiological response to enhance the long-term vasoconstrictor and sodium-retaining effects of Ang II.

4
Pathophysiology of the RAS

So far, we have reviewed the actions of the RAS from the general point of view of cardiovascular homeostasis. However, increased activity of the RAS, especially in combination with other cardiovascular risks factors, may lead to a cascade of deleterious effects such as hypertension, cardiovascular hypertrophy, oxidative stress with endothelial dysfunction, atherosclerosis and tissue inflammation. Many of these pathophysiological actions of Ang II will be reviewed in detail in other chapters of this book. At this stage, it is interesting to consider that many of these Ang II actions may still be viewed as being homeostatic in principle, but harmful if carried to excess.

4.1
The RAS May Contribute to the Higher Cardiovascular Risks of Males

Although being male is not exactly a pathological situation, men before the age of 50 show a higher prevalence of hypertension and a greater cardiovascular morbidity than premenopausal women. Blood pressure in a normotensive population is also higher in men than in women. Interestingly, men have higher plasma renin activities than women, and there is indirect evidence that increased levels of renin may contribute to increased cardiovascular risks. Indeed, hypertensive patients with high levels of plasma renin activity are at higher risk of developing stroke or myocardial infarction than those with low plasma renin activity (Brunner et al. 1972).

Plasma renin levels are higher in male spontaneous hypertensive rats (SHR) and their BP is 25–30 mmHg higher than in female SHR (Reckelhoff et al. 2000). On the other hand, castrated male SHR, female SHR and ovariectomized female SHR all show about the same level of BP. In contrast, female ovariectomized

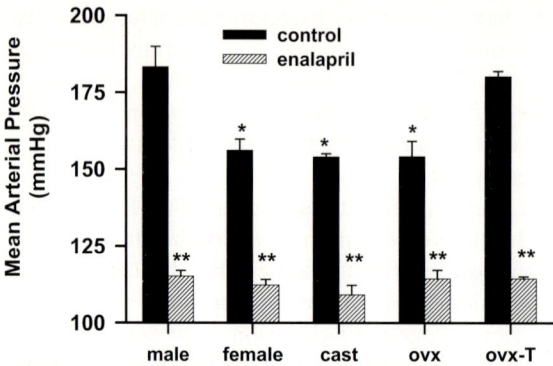

Fig. 8 Mean arterial pressure in five groups of SHR: male, female, castrated males (*cast*), ovariectomized females (*ovx*) and ovariectomized females receiving testosterone (*ovx-T*). Animals were studied with an intact RAS (*black bars*) or after chronic blockade of the RAS with enalapril (*shaded bars*). (From Reckelhoff et al. 2000, by permission)

SHR, given testosterone, show a degree of hypertension approaching the level observed in male SHR, pointing to the role of testosterone in the more severe hypertension of male SHR (Fig. 8). When the RAS is blocked with enalapril, all five groups of animals show remarkably similar BP levels, suggesting that the pressure difference between male and female SHR is due entirely to a more active RAS in male animals, and testosterone may contribute to the higher renin levels in males. This is consistent with the observation that normotensive castrated male Sprague Dawley rats, with undetectable serum testosterone levels, have low renin levels, and that implantation of testosterone pellets of increasing concentrations not only raise testosterone levels in blood, but also plasma renin activity with a significant linear correlation ($r=0.904$) between the two variables (Reckelhoff et al. 2001). Whether this animal observation can be extrapolated to humans remains to be further investigated.

4.2
The RAS Contributes to Many Forms of Hypertension

In humans, known causes of arterial hypertension account for less than 10% of all cases of hypertension (Kaplan 1998). Most often, a precise cause cannot be found and the hypertension is said to be "essential". Since Ang II elevates BP, it is appealing to implicate an overreactivity of the RAS in the pathogenesis of certain forms of hypertension.

4.2.1
Role of the RAS in Renovascular Hypertension

Since the classical experiments of Goldblatt, many studies have appeared, inducing hypertension in animal models by clipping one or both renal arteries. However, the contribution of the RAS in renovascular hypertension depends on the type of stenosis.

a. One kidney-one clip (1K1C) hypertension. Experimentally, a stenotic clip is placed on the renal artery of one kidney whereas the contralateral kidney is removed. The renal artery stenosis reduces renal perfusion pressure, which may explain many of the initial events, including sodium retention and stimulation of the RAS. However, as the animal retains volume over time and becomes hypertensive, the glomerular pressure tends to return towards normal and there is no longer a strong stimulus for renin release. At this stage, administration of an ACE inhibitor has little effect on BP. The hypertension is volume-dependent but no longer renin–dependent. The clinical equivalent of 1K1C is renal artery stenosis in a patient with a solitary kidney, or bilateral renal artery stenoses (2K2C) or stenosis of the aorta above the origin of the renal arteries. All three situations are characterized by a low renal perfusion pressure.
b. Two kidney-one clip (2K1C) hypertension. The pathogenesis of hypertension in this model is more complex. The stenotic kidney is underperfused and thus secretes large amounts of renin. The resulting elevation in plasma Ang II acts on the intact contralateral kidney, both by a direct effect and via stimulation of aldosterone secretion to promote enhanced sodium reabsorption. Initially, both kidneys may retain salt, but the stenotic kidney with its lower distal renal artery pressure and its locally stimulated RAS retains much more salt than the contralateral kidney. As BP rises due to the volume expansion, systemic BP increases ultimately high enough to achieve sodium balance by the pressure natriuresis mechanism, sodium excretion being now slightly elevated in the intact kidney and slightly decreased in the stenotic kidney (Mizelle et al. 1993). Because there is a continuing stimulus for renin release from the stenotic kidney and possible accumulation of intrarenal Ang II in the nonstenotic kidney (Navar et al. 1998), this hypertension is highly angiotensin-dependent and responds well to blockers of the RAS.

4.2.2
Role of the RAS in Essential Hypertension

Low renin levels would be expected in essential hypertension because of the higher renal perfusion pressure. However, the observation that most hypertensive patients have either normal or high renin levels has lead to the view that renin may play a critical role in the pathogenesis of many forms of essential hypertension (Laragh 1992). Several factors have been presented to explain these "inappropriate" high levels of renin, such as a state of high sympathetic drive found in many hypertensive patients (Julius 1988), nephron heterogeneity with a subpopulation of ischaemic nephrons responsible for increased tonic renin release (Sealey et al. 1988) or a defective feedback regulation with nonmodulation of the RAS (Williams et al. 1992). As expected, patients with normal or high renin levels respond better to β-adrenergic blockers and ACE inhibitors, whereas low-renin hypertensive patients respond better to diuretic treatment.

4.3
Angiotensin-Induced Cardiac and Vascular Hypertrophies Are Risks Factors

As mentioned above, the growth-promoting actions of Ang II may be viewed as an appropriate response in conditions in which increased heart strength and prolonged vasoconstriction are required. A certain degree of vascular hypertrophy is also useful in volume-loading hypertension, as it minimizes the amount of volume retention needed within the vascular system to maintain an elevated blood pressure. Indeed, the various forms of volume-dependent hypertension (mineralocorticoid-induced hypertension, high salt intake with a reduced renal mass) are characterized experimentally by an initial increase in cardiac output followed by a secondary autoregulatory vasoconstriction that returns blood volume and cardiac output towards normal (Guyton 1980). Were it not for autoregulation and vascular hypertrophy, reestablishment of sodium balance would be accompanied by much larger changes in fluid volumes.

Although cardiac and vascular hypertrophies may be considered adaptive from the point of view of enhancing short-term survival, they are clearly detrimental when allowed to continue to progress over prolonged periods of time. The increased stiffness of the hypertrophied heart impairs ventricular relaxation and filling. In fact, left ventricular hypertrophy is considered an independent risk factor for cardiovascular events. Vascular hypertrophy makes arteries stiffer, leading to increased pulse pressure and increased pulse wave velocity. An increased pulse pressure for a given mean arterial pressure is in itself a cardiovascular risk factor: the higher systolic pressure constitutes an elevated left ventricular afterload while the lower diastolic pressure reduces the driving pressure for coronary blood flow. Increased pulse wave velocity results in a rapid return of reflection waves to the heart, increasing systolic pressure and decreasing diastolic pressure further. In this context, one can understand the beneficial health effects of blockers of the RAS in reversing cardiac and vascular hypertrophy.

4.4
Activation of the RAS Worsens Congestive Heart Failure

An activation of the RAS may be a natural response to the initial insult of heart failure. Various models of experimental heart failure (rapid ventricular pacing, pulmonary artery occlusion) in which the same animal could be studied before and after induction of heart failure, are characterized by a decrease in arterial pressure (Mizelle et al. 1989; Lohmeier et al. 1995, 2000a), which is expected to stimulate the RAS. Although there is only a modest activation of the RAS in the early compensated phase of heart failure (Lohmeier 2002), even small increments in Ang II concentration may favour fluid retention. The resulting increase in fluid volumes could be beneficial, allowing cardiac output and arterial pressure to return towards normal.

However, excessive activation of the RAS in heart failure is clearly harmful. When exogenous Ang II was administered for 4 days to dogs in compensated heart failure, decompensation occurred with profound sodium retention and marked increases in plasma norepinephrine (Lohmeier et al. 2000a). Cardiopulmonary baroreflex suppression of sympathetic nerve activity, which is impaired in heart failure, could play a critical role in the transition from compensated to decompensated heart failure. Impaired sympathoinhibition would unmask the sympathoexcitatory actions of Ang II, resulting in a positive feedback. Ang II would increase sympathetic nerve activity, which stimulates renin secretion further. In turn, higher plasma levels of Ang II would further stimulate sympathetic activity. As a result, there would be a progressive fluid retention and progressive cardiac dysfunction. Consistent with this hypothesis is the observation that ACE inhibitors have been shown to delay the progression of heart failure and to improve symptoms and prolong survival in patients with ventricular dysfunction.

4.5
Systemic and Local Angiotensin May Initiate and Amplify Vascular Disease

Ang II has direct effect on endothelial and vascular smooth muscles cells, and may play a key role in initiating and amplifying vascular disease. In the normal state, there is a homeostatic balance between locally produced NO and oxygen radicals, such as the superoxide anion and hydrogen peroxide. Under these conditions, NO can exert all of its protective functions as vasodilator, inhibitor of platelet aggregation, inhibitor of vascular smooth muscle growth and migration, and inhibitor of the expression of proinflammatory molecules.

As aforementioned, Ang II by its action on AT_1 receptors can stimulate NAD(P)H oxidases, leading to the production of the oxygen radical superoxide. Quenching of NO by the superoxide anion not only reduces the bioavailability of NO (and thereby of all its protective functions), but also forms peroxynitrite ($ONOO^-$), a powerful oxidant. The combined action of excess Ang II and oxidative stress may unleash a cascade of harmful effects, such as increased vasoconstriction, increased expression of chemoattractant proteins and leukocyte adhesion molecules, stimulation of thrombosis and vascular remodelling. The local inflammatory response promotes an accumulation of various inflammatory cells that can release enzymes that generate Ang II. For example, macrophages are rich in ACE, neutrophils in cathepsin G, mast cells in chymase. The increased local production of Ang II may further promote oxidative stress, leading to a vicious cycle of inflammation and subsequent increase in tissue Ang II (Dzau 2001).

If endothelial function is preserved, this positive feedback can easily be dampened by the actions of NO and antioxidants. However, in the presence of cardiovascular risk factors, the homeostatic balance between pro-oxidants and antioxidants is perturbed. Dyslipidaemia, diabetes and cigarette smoking can all initiate endothelial dysfunction and promote oxidative stress. Excessive activity of the RAS potentiates the vicious cycle described above, inducing vascular remodelling, promoting atherosclerosis and upsetting the balance between the fibrinolytic and coagulation systems. The observations may explain why ACE inhibitors and angiotensin-receptor blockers have beneficial effects on cardiovascular events far beyond blood pressure reduction.

5
Conclusions

Without efficient mechanisms for conserving salt, our ancestors living on a diet relatively poor in sodium would have never survived, as they would not be able to respond to even moderate haemorrhage, and not have the required haemodynamic reserve for fight-or-flight reactions. Ang II, the major end-product of the RAS, has multiple actions that work in a concerted manner to maintain cardiovascular integrity and efficiency. Most of the rapid actions of Ang II can be viewed together, in combination with the sympathetic nervous system, to support the circulation when it is threatened by acute disturbances such as hypovolaemia and hypotension. Slower actions of Ang II help maintain volume homeostasis by minimizing the changes in arterial pressure and ECFV required to achieve sodium balance (prevention of salt sensitivity). Some of the very slow actions of Ang II, such as cardiac and vascular hypertrophy and oxidative stress, although potentially harmful, may be viewed as adaptive responses to improve the efficiency of fluid volume and BP preservation. However, such responses are clearly detrimental if carried to excess.

Acknowledgements. The authors thank Vladan Antic for the preparation of the figures.

References

Brasier AR, Li J (1996) Mechanisms for inducible control of angiotensinogen gene transcription. Hypertension 27:465–475
Braun-Menendez E, Page IH (1958) Suggested revision of nomenclature: angiotensin. Science 127:242
Brunner HR, Laragh JH, Baer L, Newton MA, Goodwin FT, Krakoff LR, Bard RH, Buhler FR (1972) Essential hypertension: renin and aldosterone, heart attack and stroke. N Engl J Med 286:441–449
Cowley AW Jr (2000) Control of the renal medullary circulation by vasopressin V1 and V2 receptors in the rat. Exp Physiol 85(Spec No):223S–231S
DiBona GF, Kopp UC (1997) Neural control of renal function. Physiol Rev 77:75–197
Dzau VJ (2001) Theodore Cooper lecture: tissue angiotensin and pathobiology of vascular disease: a unifying hypothesis. Hypertension 37:1047–1052
Eaton SB, Konner M (1985) Paleolithic nutrition: a consideration of its nature and current implications. N Engl J Med 312:283–289
Edwards RM (1983) Segmental effects of norepinephrine and angiotensin II on isolated renal microvessels. Am J Physiol 244:F526–F534

Fisher ND, Hurwitz S, Jeunemaitre X, Hopkins PN, Hollenberg NK, Williams GH (2002) Familial aggregation of low-renin hypertension. Hypertension 39:914–918

Fitzsimons JT (1998) Angiotensin, thirst, and sodium appetite. Physiol Rev 78:583–686

Goldblatt H, Lynch J, Hanzal RF, Summerville WW (1934) Studies of elevation of systolic blood pressure by means of renal ischaemia. J Exp Med 59:347–379

Gossmann J, Burkhardt R, Harder S, Lenz T, Sedlmeyer A, Klinkhardt U, Geiger H, Scheuermann EH (2001) Angiotensin II infusion increases plasma erythropoietin levels via an angiotensin II type 1 receptor-dependent pathway. Kidney Int 60:83–86

Gould AB, Green D (1971) Kinetics of the human renin and human substrate reaction. Cardiovasc Res 5:86–89

Granger JP, Alexander BT, Llinas M (2002) Mechanisms of pressure natriuresis. Curr Hypertens Rep 4:152–159

Guyton AC (1980) Arterial pressure and hypertension. WB Saunders, Philadelphia, pp 139–155

Hall JE, Guyton AC, Jackson TE, Coleman TG, Lohmeier TE, Trippodo NC (1977) Control of glomerular filtration rate by renin–angiotensin system. Am J Physiol 233:F366–F372

Hall JE, Guyton AC, Smith MJ Jr, Coleman TG (1980) Blood pressure and renal function during chronic changes in sodium intake: role of angiotensin. Am J Physiol 239:F271–F280

Hall JE (1986a) Control of sodium excretion by angiotensin II: intrarenal mechanisms and blood pressure regulation. Am J Physiol 250:R960–R972

Hall JE, Granger JP, Hester RL, Montani JP (1986b) Mechanisms of sodium balance in hypertension: role of pressure natriuresis. J Hypertens 4 (Suppl 4):S57–S65

Hricik DE, Browning PJ, Kopelman R, Goorno WE, Madias NE, Dzau VJ (1983) Captopril-induced functional renal insufficiency in patients with bilateral renal-artery stenoses or renal-artery stenosis in a solitary kidney. N Engl J Med 308:373–376

Huang W, Gallois Y, Bouby N, Bruneval P, Heudes D, Belair MF, Krege JH, Meneton P, Marre M, Smithies O, Alhenc-Gelas F (2001) Genetically increased angiotensin I-converting enzyme level and renal complications in the diabetic mouse. Proc Natl Acad Sci USA 98:13330–13334

Jeunemaitre X, Soubrier F, Kotelevtsev YV, Lifton RP, Williams CS, Charru A, Hunt SC, Hopkins PN, Williams RR, Lalouel JM, Corvol P (1992) Molecular basis of human hypertension: role of angiotensinogen. Cell 71:169–180

Julius S (1988) Interaction between renin and the autonomic nervous system in hypertension. Am Heart J 116:600–606

Kaplan NM (1998) Clinical hypertension, 7th edition. Williams and Wilkins, Baltimore, p 12

Keppens S, Vandekerckhove A, Moshage H, Yap SH, Aerts R, De Wulf H (1993) Regulation of glycogen phosphorylase activity in isolated human hepatocytes. Hepatology 17:610–614

Kim HS, Krege JH, Kluckman KD, Hagaman JR, Hodgin JB, Best CF, Jennette JC, Coffman TM, Maeda N, Smithies O (1995) Genetic control of blood pressure and the angiotensinogen locus. Proc Natl Acad Sci USA 92:2735–2739

Krege JH, Kim HS, Moyer JS, Jennette JC, Peng L, Hiller SK, Smithies O (1997) Angiotensin-converting enzyme gene mutations, blood pressures, and cardiovascular homeostasis. Hypertension 29:150–157

Laragh JH (1992) Lewis K. Dahl Memorial Lecture. The renin system and four lines of hypertension research. Nephron heterogeneity, the calcium connection, the prorenin vasodilator limb, and plasma renin and heart attack. Hypertension 20:267–279

Larsson PT, Schwieler JH, Wallen NH (2000) Platelet activation during angiotensin II infusion in healthy volunteers. Blood Coagul Fibrinolysis 11:61–69

Lohmeier TE, Reinhart GA, Mizelle HL, Montani JP, Hester R, Hord CE Jr, Hildebrandt DA (1995) Influence of the renal nerves on sodium excretion during progressive reductions in cardiac output. Am J Physiol 269:R678–R690

Lohmeier TE, Mizelle HL, Reinhart GA, Montani JP (2000a) Influence of angiotensin on the early progression of heart failure. Am J Physiol 278:R74–R86

Lohmeier TE, Lohmeier JR, Haque A, Hildebrandt DA (2000b) Baroreflexes prevent neurally induced sodium retention in angiotensin hypertension. Am J Physiol 279:R1437–R1448

Lohmeier TE (2002) Neurohumoral regulation of arterial pressure in hemorrhage and heart failure. Am J Physiol 283:R810–R814

Luft FC, Rankin LI, Bloch R, Weyman AE, Willis LR, Murray RH, Weinberger MH (1979) Cardiovascular and humoral responses to extremes of sodium intake in normal white and black men. Circulation 60:697–706

Luft FC, Fineberg NS, Weinberger MH (1992) The influence of age on renal function and renin and aldosterone responses to sodium-volume expansion and contraction in normotensive and mildly hypertensive humans. Am J Hypertens 5:520–528

MacGregor GA, de Wardener HE (1998) Salt, diet and health: Neptune's poisoned chalice: the origins of high blood pressure. Cambridge University Press, Cambridge, pp 100–105

Machado LJ, Marubayashi U, Reis AM, Coimbra CC (1998) The hyperglycemia induced by angiotensin II in rats is mediated by AT1 receptors. Braz J Med Biol Res 31:1349–1352

Mattiazzi A (1997) Positive inotropic effect of angiotensin II. Increases in intracellular Ca2+ or changes in myofilament Ca2+ responsiveness? J Pharmacol Toxicol Methods 37:205–214

McBurnie MI, Blair-West JR, Denton DA, Weisinger RS (1999) Sodium intake and reproduction in BALB/C mice. Physiol Behav 66:873–879

McMurray J, Matthews DM (1987) Consequences of fluid loss in patients treated with ACE inhibitors. Postgrad Med J 63:385–387

Mizelle HL, Hall JE, Montani JP (1989) Role of renal nerves in control of sodium excretion in chronic congestive heart failure. Am J Physiol 256:F1084–F1093

Mizelle HL, Montani JP, Hester RL, Didlake RH, Hall JE (1993) Role of pressure natriuresis in long-term control of renal electrolyte excretion. Hypertension 22:102–110

Montani JP, Liard JF, Schoun J, Mohring J (1980) Hemodynamic effects of exogenous and endogenous vasopressin at low plasma concentrations in conscious dogs. Circ Res 47:346–355

Morganti A, Lopez-Ovejero JA, Pickering TG, Laragh JH (1979) Role of the sympathetic nervous system in mediating the renin response to head-up tilt. Their possible synergism in defending blood pressure against postural changes during sodium deprivation. Am J Cardiol 43:600–604

Navar LG, Zou L, Von Thun A, Tarng Wang C, Imig JD, Mitchell KD (1998) Unraveling the Mystery of Goldblatt Hypertension. News Physiol Sci 13:170–176

Nyengaard JR, Bendtsen TF (1992) Glomerular number and size in relation to age, kidney weight, and body surface in normal man. Anat Rec 232:194–201

Ortiz MC, Manriquez MC, Romero JC, Juncos LA (2001) Antioxidants block angiotensin II-induced increases in blood pressure and endothelin. Hypertension 38:655–659

Reckelhoff JF, Zhang H, Srivastava K (2000) Gender differences in development of hypertension in spontaneously hypertensive rats: role of the renin–angiotensin system. Hypertension 35:480–483

Reckelhoff JF (2001) Gender differences in the regulation of blood pressure. Hypertension 37:1199–1208

Ren Y, Carretero OA, Garvin JL (2002) Mechanism by which superoxide potentiates tubuloglomerular feedback. Hypertension 39:624–628

Rodgers KE, Xiong S, Steer R, diZerega GS (2000) Effect of angiotensin II on hematopoietic progenitor cell proliferation. Stem Cells 18:287–294

Schnackenberg CG (2002) Physiological and pathophysiologal roles of oxygen radicals in the renal microvasculature. Am J Physiol 282:R335–R342

Sealey JE, Blumenfeld JD, Bell GM, Pecker MS, Sommers SC, Laragh JH (1988) On the renal basis for essential hypertension: nephron heterogeneity with discordant renin secretion and sodium excretion causing a hypertensive vasoconstriction-volume relationship. J Hypertens 6:763–777

Smithies O, Kim HS, Takahashi N, Edgell MH (2000) Importance of quantitative genetic variations in the etiology of hypertension. Kidney Int 58:2265–2280

Tigerstedt RA, Bergman PG (1898) Niere und Kreislauf (kidney and circulation). Skand Arch Physiol 8:223–238

Touyz RM (2000) Oxidative stress and vascular damage in hypertension. Curr Hypertens Rep 2:98–105

Wang Q, Hummler E, Nussberger J, Clement S, Gabbiani G, Brunner HR, Burnier M (2002) Blood pressure, cardiac, and renal responses to salt and deoxycorticosterone acetate in mice: role of Renin genes. J Am Soc Nephrol 13:1509–1516

Weidmann P, De Myttenaere-Bursztein S, Maxwell MH, de Lima J (1975) Effect of aging on plasma renin and aldosterone in normal man. Kidney Int 8:325–333

Williams GH, Dluhy RG, Lifton RP, Moore TJ, Gleason R, Williams R, Hunt SC, Hopkins PN, Hollenberg NK (1992) Non-modulation as an intermediate phenotype in essential hypertension. Hypertension 20:788–796

Zou AP, Li N, Cowley AW Jr (2001) Production and actions of superoxide in the renal medulla. Hypertension 37:547–553

Phylogeny and Ontogeny of the Renin-Angiotensin System

H. Nishimura

Department of Physiology, University of Tennessee Health Science Center,
894 Union Avenue, Memphis, TN 38163, USA
e-mail: nishimur@physio1.utmem.edu

1	Introduction	33
2	Phylogeny of the Renin-Angiotensin System	34
2.1	Morphology, Biochemistry, and Molecular Biology	34
2.1.1	Juxtaglomerular Apparatus in Primitive Vertebrates	34
2.1.2	Angiotensinogen and Renin	36
2.1.3	Angiotensin	36
2.1.4	Angiotensin Receptor Subtypes and Signaling	38
2.2	Function and Regulation	43
2.2.1	Control of Renin Release	43
2.2.2	Biological Action of Angiotensin	45
3	Ontogeny of the Renin-Angiotensin System	48
3.1	Biochemistry and Molecular Biology	48
3.1.1	Renin Substrate and Renin	48
3.1.2	Angiotensin and Angiotensin-Converting Enzymes	50
3.1.3	Angiotensin Receptors	50
3.2	Function and Regulation	52
3.2.1	Angiotensin Action	52
3.2.2	Regulation of the Renin-Angiotensin System During Development	56
4	Integration of Phylogeny and Ontogeny	56
4.1	Phylogeny and Ontogeny	56
4.2	Biochemistry of the Renin-Angiotensin System	57
4.3	Function and Regulation	58
4.4	Perspectives	59
	References	61

Abstract Renin substrate in plasma, biological renin activity, and/or granulated cells in the kidney evolved at an early stage of vertebrate phylogeny. Angiotensin (Ang) I and II molecules have been biochemically identified in representative species of all vertebrate classes, and Ang II structure is well preserved throughout the phylogenetical scale. Ang receptors have also been identified pharmacologically and, in limited species, molecularly characterized. The renin–angiotensin system (RAS) is important in maintaining blood pressure/blood volume homeostasis and ion-fluid balance. Recently, the regulatory role of Ang in cell growth and vascularization, possibly via paracrine action, has been an intensive focus of research. The RAS has been detected during early fetal development, and genetic or pharmacological blockade of Ang signaling results in striking developmental abnormalities of the kidney. This chapter will provide: (1) a brief overview of current knowledge of RAS phylogeny and ontogeny; (2) a comparison of these two systems in terms of structural, biochemical, and functional properties of each component of the RAS; and (3) perspectives for future study. Discussions are focused on: (1) the most fundamental functions of the RAS that have been conserved throughout phylogenetical advancement as well as developmental maturation; (2) the physiological significance of the shifting of expression of renin-producing cells in the renal arterial trees with vertebrate advancement or ontogenic maturation; and (3) the roles of environments, or microenvironments, in modulating the expression of component members of the RAS during phylogenetic and ontogenic processes. Although ontogeny does not directly recapitulate phylogeny, comparison of these two time-dependent processes should provide new insight into the molecular and functional evolution and significance of the RAS.

Keywords Renin–angiotensin system · Native angiotensin · Renal growth and development · Blood pressure regulation · Juxtaglomerular apparatus · Angiotensin-converting enzyme · Tissue renin–angiotensin · Evolution and development · Renin-secretory cells · Angiotensin receptor subtypes

Abbreviations

ACE	Angiotensin-converting enzyme
AGT	Angiotensinogen
AL	Ascending limb
Ang	Angiotensin
AQP	Aquaporin
BP	Blood pressure
BW	Body weight
cAT_1	Chicken angiotensin type-1 receptor
CD	Collecting duct
D7–D15	Newborn days 7–15
DCS	Distal convoluted tubule
DHES/s	Dehydroepiandrosterone-sulfate
E18	Embryonic day 18
EDRF	Endothelium-dependent relaxation factor
GFP	Green fluorescent protein
GFR	Glomerular filtration rate
JG	Juxtaglomerular
JGG	Juxtaglomerular granules
NS	Neck segment
PRA	Plasma renin activity
RAS	Renin–angiotensin system

SHR	Spontaneously hypertensive rats
tAT$_1$	Turkey angiotensin type 1 receptor
UT	Urea transporters
VEGF	Vascular endothelium growth factor
VSM	Vascular smooth muscle
WKY	Wistar-Kyoto
ZCatD	Zebrafish cDNA

1
Introduction

The renin–angiotensin system (RAS) appears to have evolved during an early stage of vertebrate evolution. Although the molecular identification of each component of the RAS has not been completed in nonmammalian species, biochemical and pharmacological evidence suggests that the RAS and angiotensin (Ang) receptors are present in all vertebrate classes (Fig. 1). Increasing evidence also indicates that the RAS plays an important role during fetal life in promoting growth and development and that lack or impairment of a component of the RAS induces pathological structure and function at birth and/or in adult life. This chapter will provide a brief overview of current knowledge of RAS phylogeny and ontogeny and perspectives on their integration. Following Haeckel's hypothesis, a number of investigators have tried, but failed, to elaborate more complete taxonomies of the relationship between ontogeny and phylogeny. Although it is not my intent to discuss whether ontogeny recapitulates phylogeny, both are time-dependent processes and show certain similarities. Details of mo-

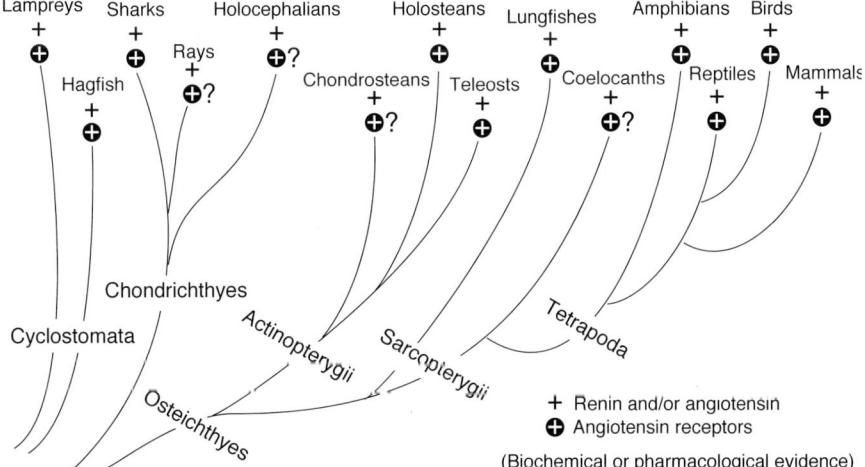

Fig. 1 Diagrammatic presentation of vertebrate phylogenetic tree and presence of biochemical activities of renin and/or angiotensin (+) and molecular, biochemical, or pharmacological evidence for the presence of angiotensin receptors (+ in closed circle). A question mark (?) indicates that the presence of an angiotensin receptor is presumed but not confirmed. (Reproduced from Nishimura 2001)

lecular, biochemical, or physiological properties of each component of the RAS and mechanisms behind them are beyond the scope of this chapter. It is not possible to include all organs and tissues that express the RAS; this review is focused on the kidney and cardiovascular system and their functions.

2
Phylogeny of the Renin–Angiotensin System

2.1
Morphology, Biochemistry, and Molecular Biology

2.1.1
Juxtaglomerular Apparatus in Primitive Vertebrates

The juxtaglomerular (JG) apparatus of the mammalian kidney is composed of JG cells in the afferent arteriole with sympathetic innervation, efferent arterioles, the convoluted early distal tubule (macula densa), and the extracellular

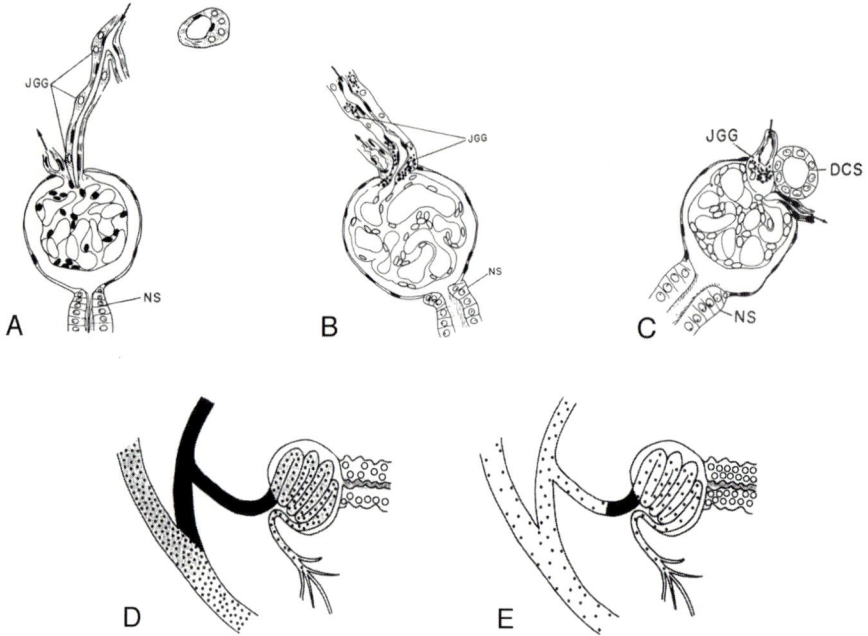

Fig. 2A–E Diagrams of nephron and juxtaglomerular (*JG*) apparatus showing distribution of JG granules (*JGG*) in teleost fish (**A**), amphibian (**B**), and bird (**C**) examined histochemically and by electronmicrography. (Reproduced with permission from Sokabe et al. 1969). *DCS*, distal convoluted tubule; *NS*, neck segment. Diagrams **D** (newborn rat) and **E** (adult rat) indicate the expression of angiotensinogen (*open circles*), renin (*black area*), angiotensin-converting enzyme (*gray area*) and AT_1 receptor (*black dots*) (Reproduced with permission from Harris and Gomez 1997). Both in primitive vertebrates, such as teleosts, and in newborn rat, expressions of renin (JG granules, renin mRNA and protein) are seen at upstream of renal arteries

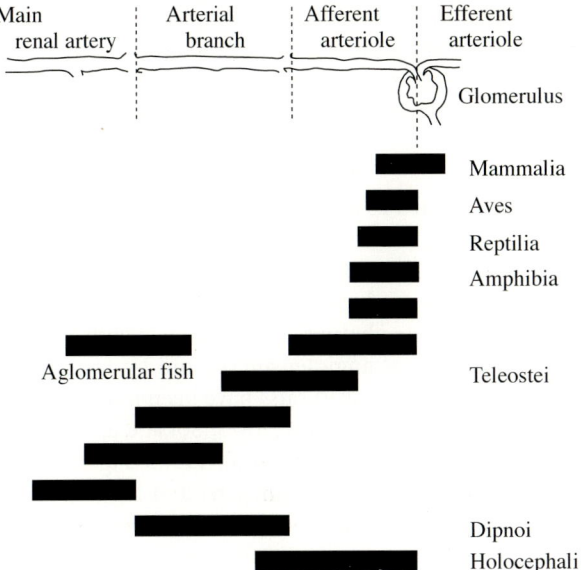

Fig. 3 Distribution of granulated cells along renal arteries and arterioles in various vertebrate animals. Granulated cells are more localized at the juxtaglomerular area with phylogenetic advancement of vertebrates. (Reproduced with permission from Sokabe and Ogawa 1974)

mesangium (Barajas and Latta 1967). The JG apparatus in nonmammalian vertebrates is incomplete (for review, Sokabe and Ogawa 1974; Nishimura 1980). Primitive bony fish, such as holosteans (for example, longnose gar), teleosts, lungfish, amphibians, and reptiles have granulated epithelioid cells resembling mammalian JG cells (Fig. 2). Granulated cells are widely seen in the media of large and small renal arteries and arterioles (Sokabe et al. 1969; Nishimura et al. 1973) (Fig. 3). Kidneys from aglomerular teleosts, in which glomeruli and a part of the proximal tubules have degenerated, also have renin secretory granulated cells. Granulated cells either have not been found (cyclostomes) or are present but negative for Bowie's staining (elasmobranchs) (Lacy and Reale 1990) or the appearance and distribution pattern of the granules differ (holocephalians, such as ratfish) in the more primitive vertebrates (Nishimura et al. 1973; Nishimura 1980).

In the majority of the nonmammalian species examined, either the distal tubules do not attach to the vascular pole of their parent glomeruli (hence, they lack a macula densa) or they return to their own glomeruli and attach to the afferent arterioles as seen in elasmobranchs, holocephalians, and in some amphibians, but no cells resembling mammalian macula densa cells are present (Sokabe et al. 1969; Nishimura et al. 1973, 1980). Avian kidneys possess macula densa cells that have some characteristics of a mammalian macula densa (Sokabe et al. 1969). Therefore, the JG apparatus in birds represents a transitional JG apparatus between the primitive and mammalian types. It therefore appears that the vascular component of the JG apparatus evolved phylogenetically earlier than the macula densa, the tubular component. This is important when we consider the function of the RAS in nonmammalian vertebrates.

2.1.2
Angiotensinogen and Renin

Angiotensinogen (AGT) levels, determined as maximally yielded Ang I from homologous plasma, appear to be higher in taxonomically more advanced animals than in lower vertebrates (Nishimura et al. 1973). The molecular structure of zebrafish AGT has been identified (Vallon et al. 2000). Renal renin activities (determined as Ang I-forming activities by incubating kidney extracts with homologous plasma under adequate inhibition of angiotensinases and by testing the vasopressor activity of the product in anesthetized rats) have been found in representative species of all vertebrates (Nishimura et al. 1973; Sokabe and Ogawa 1974; Nishimura 1980; Henderson et al. 1993; Hazon et al. 1999) (Fig. 1). This suggests that the RAS evolved at an early stage of phylogeny and is widely distributed among vertebrates. The renal renin activity agrees in general with the occurrence of granulated cells except in cartilaginous fish and primitive bony fishes, which show some discrepancy, suggesting that the properties of renin granules may differ in these fish. To date, no renin has been molecularly identified in nonmammalian vertebrates. Zebrafish cDNA (zCatD) that has high homology to human cathepsin D was cloned from a zebrafish kidney cDNA library using human renin cDNA as a probe (Chen et al. 2000); zCatD generates Ang I from both porcine AGT and zebrafish peptide substrate at an acidic pH (Chen et al. 2000; Vallon et al. 2000). Since renin, pepsin, and cathepsin D appear to have been derived by gene duplication from an ancestral gene encoding for an aspartyl protease, it is possible that aspartyl protease acts as an Ang-forming enzyme in primitive vertebrates. Renin from teleost fish, however, has a neutral optimum pH and substrate specificity and clearly differs from cathepsin (Nishimura et al. 1977). The cleavage of homologous AGT by zCatD should be performed. Zebrafish renin, distinct from cathepsins, has been identified recently (K. Gross, personal communication), but the details have not yet been reported. Plasma renin activity (PRA) has also been measured in several species (for review see Nishimura 1987; Henderson et al. 1993).

2.1.3
Angiotensin

Native Ang ligands appear to be stable molecules throughout the phylogenetic scale, with variation occurring in the amino acid at positions 1 (Asn, Asp, or additional side chain of tri-peptide), 3 (Val, Ile, or Pro), 5 (Ile or Val), and 9 (His, Ser, Tyr, Gly, Thr, etc.) of Ang I (Table 1) (for review, Kobayashi and Takei 1996; Nishimura 2001). Dogfish Ang I (Pro^3) (Takei et al. 1993a) and *Crinia* skin Ang II (Ile^3) (Erspamer et al. 1979) show unique amino acid variations in position 3. Structure–function relationship studies indicate that changes in the amino acid

Table 1 Native angiotensins of various vertebrate species (reproduced from Nishimura 2001)

Cyclostomes		
Sea lamprey, *Petromyzon marinus*	[Asn^1,Val^5,Thr^9]Ang I	Takei et al. 2003
Lamprey, *Lampetra fluviatilis*	[Asn^1,Val^5,Thr^9]Ang I	Takei 1999
Elasmobranchs		
Dogfish, *Triakis scyllia*	[Asn^1,Pro^3,Ile^5,Glu^9]Ang I	Takei et al. 1993a
Bony fish		
Holosteans		
Bowfin, *Amia calva*	[Asp^1,Val^5,Asn^9]Ang I	Takei et al. 1998
Teleosts		
Japanese goosefish, *Lophius litulon*	[Asn^1,Val^5,His^9]Ang I	Hayashi et al. 1978
Chum salmon, *Oncorhynchus keta*	[Asn^1,Val^5,Asn^9]Ang I	Takemoto et al. 1983
Japanese eel, *Anguilla japonica*	[Asp^1,Val^5,Gly^9]Ang I	Hasegawa et al. 1983a
American eel, *Anguilla rostrata*	[Asn^1,Val^5,Gly^9]Ang I	Khosla et al. 1985
Trout, *Oncorhynchus mykiss*	[Asn^1, or Asp^1,Val^5]Ang II	Conlon et al. 1996
Flounder, *Platichthys flesus*	[Asn^1,Ile^5,Thr^9]Ang I	Balment et al. 2003
Lungfish		
Australian lungfish, *Neoceratodus forsteri*	[Asn^1,Val^5,Thr^9]Ang I	Joss et al. 1999
Amphibians		
Axolotl, *Ambystoma mexicanum*	[Asp^1,Val^5,His^9]Ang I	Takei et al. 2003
Three-toed amphiuma, *Amphiuma tridactylum*	[Asp^1,Val^5]Ang II	Takei et al. 2003
Bullfrog, *Rana catesbeiana*	[Asn^1,Val^5,Asn^9]Ang I	Hasegawa et al. 1983b
Australian frog, *Crinia georgiana* (skin)	Ala-Pro-Gly-[Asp^1,Ile^3,Val^5]Ang II	Erspamer et al. 1979
Reptiles		
Snake, *Elaphe climacophora*	[Asp^1,Val^5,Tyr^9]Ang I	Nakayama et al. 1977
Turtle, *Pseudemys scripta*	[Asp^1,Val^5,His^9]Ang I	Hasegawa et al. 1984a
Alligator, *Alligator mississippiensis*	[Asp^1,Val^5,Ala^9]Ang I	Takei et al. 1993b
Birds		
Chicken, *Gallus gallus*	[Asp^1,Val^5,Ser^9]Ang I	Nakayama et al. 1973
Emu, *Dromiceus novaehollandiae*	[Asp^1,Val^5,Asn^9]Ang I	Takei et al. 2003
Quail, *Coturnix coturnix*	[Asp^1,Val^5,Ser^9]Ang I	Takei and Hasegawa 1990
Mammals		
Human, horse, dog, rat	[Asp^1,Ile^5,His^9]Ang I	Khosla et al. 1974, 1983
Cow	[Asp^1,Val^5,His^9]Ang I	

at position 1 (Asn or Asp) or position 5 (Ile or Val) cause relatively minor effects with respect to function and binding (Khosla et al. 1974). Removal of the first amino acid, or the change from Asp^1 to Asn^1, reduces vasopressor activity in the rat, whereas native Ang II appears to exert more potent vasopressor/catecholamine-releasing actions in fishes (Nishimura 1985a; Takei et al. 1993a).

Position 9 of the Ang I is variable and is likely to show different specificity to Ang-converting enzyme. Furthermore, antibodies raised against specific Ang I molecules often fail to bind to other Ang I ligands that have a different amino acid in position 9 (Nishimura et al. 1977, 1981a). All mammalian species studied to date have histidine at the 9th position, but variation in the 9th amino acid may exist in mammals (Best et al. 1974). Change in the first amino acid from Asp to Asn or in the 5th from Val to Ile requires a single nucleotide change. Conversion of the amino acid at the 9th position from alanine (alligator Ang I) (Takei et al. 1993b) to serine (avian Ang I) (Nakayama et al. 1973) or from tyrosine (snake Ang I) (Nakayama et al. 1977) to histidine (turtle and mammalian Ang I) requires a one-point mutation of the triplet sequence of the DNA code (Takei et al. 1993b), whereas a two-nucleotide difference exists between alanine and tyrosine (or histidine) (Takei et al. 1993b) and between serine (avian Ang I) and histidine. This structural difference in Ang I agrees with the notion that the evolutionary line leading to the mammals departed from the reptilian-avian line at a very early stage of tetrapod evolution and that the crocodilians are phylogenetically closer to avians. The differences at position 3 (Pro, Ile, Val) may induce conformational changes and thus significantly modulate ligand binding or signal transduction. Indeed, [Pro^3]Ang II shows much lower vasopressor action in the rat compared to other Ang II (Takei et al. 1993a). Furthermore, Ang receptors possess dynamic structures that may undergo conformational changes by shifting between active and inactive forms in response to ligand binding (Pérodin et al. 1996).

2.1.4
Angiotensin Receptor Subtypes and Signaling

AT$_1$-Homolog Receptors and Signaling. The recent development of novel nonpeptide Ang receptor antagonists has led to the discovery of Ang receptor subtype 1 (AT$_1$) and subtype 2 (AT$_2$) (Murphy et al. 1991; Sasaki et al. 1991; Kambayashi et al. 1993; Mukoyama et al. 1993). The biochemical properties and signal transduction pathways (Griendling et al. 1997) of AT$_1$ and AT$_2$ will be discussed in other sections of this book. Ang receptors sharing part of the AT$_1$ receptor protein/nucleotide sequences (referred to as AT$_1$-homolog receptors) have been identified in several nonmammalian species (Table 2) (Nishimura 2001). Cobb and Brown (1993), Brown and coworkers (1997), and Parkyn and coworkers (1997) have molecularly identified Ang receptors in the rainbow trout. A partial clone indicates relatively low homology (47%) to the mammalian AT$_1$ receptor. Teleost Ang receptors show considerable affinity to losartan (trout; Brown et al. 1997) or [Sar^1, Ile^8]Ang II (toadfish; Qin et al. 1999) and stimulate cytosolic Ca^{2+} (Cobb et al. 1999; Qin et al. 1999). Tran Van Chuoi and coworkers (1999) have cloned a cDNA encoding the eel Ang receptor from the intestinal brush border membrane that has 50% homology to the rat AT$_{1A}$ receptor and 35% to rat AT$_2$ receptor subtypes (Fig. 4; H. Ji and K. Sandberg, personal communication).

The AT$_1$-homolog receptor has been identified in amphibians (Ji et al. 1993; Bergsma et al. 1993) (Table 2). Myocardial AT receptors from *Xenopus laevis* consisting of 362 or 363 amino acid residues show ~60% amino acid identity and 65% nucleotide homology with the coding region of the mammalian AT$_1$ receptor. AT$_1$-homolog *Xenopus* receptors are coupled to GTP binding protein

Table 2 AT_1 homologous receptors

	AT_1 homology (%)	Amino acid	Binding affinity (nM)					Structure	Signaling	Function/Action	Reference
			Ang II K_d or IC_{50}	$[Sar^1, Ile^8]$ Ang II	Losartan	PD	CGP				
Teleost fish											
Rainbow trout Glomerulus			0.38		19.3	–			$[Ca^{2+}]_i\uparrow$	Control of GFR	Cobb and Brown 1993; Brown et al. 1997; Cobb et al. 1999; Parkyn et al. 1997
Liver	47 (partial clone)										
Toadfish VSM			1.0	15.8	2500				$[Ca^{2+}]_i\uparrow$	Vasopressor	Qin et al. 1999
Eel Intestine, BBM	50		3.4								Marsigliante et al. 1994; Tran van Chuoi et al. 1999
Amphibians											
Xenopus laevis Myocardium	63	362	1.0	0.40	–	–	540	7 transmembrane domain, G protein coupled	$[Ca^{2+}]_i\uparrow$		Bergsma et al. 1993; Aiyar et al. 1994
	60–65	363	2.0	0.91	–	–	1200	7 transmembrane domain, G protein coupled	$[Ca^{2+}]_i\uparrow$		Sandberg et al. 1991; Ji et al. 1993

Table 2 (continued)

	AT₁ homology (%)	Amino acid	Binding affinity (nM)					Structure	Signaling	Function/Action	Reference
			Ang II K_d or IC_{50}	[Sar¹, Ile⁸] Ang II	Losartan	PD	CGP				
Avians											
Turkey Adrenal	75	359	0.17*	0.45*	5224*	–	29*	7 transmembrane domain, G protein coupled	$[Ca^{2+}]_i$*↑, IP_3*↑	Mineralocorticoid secretion?	Murphy et al. 1993
Chicken Adrenal	75	359	1.39*	185*	–	–	3400*	7 transmembrane domain, G protein coupled	IP_3*↑	Mineralocorticoid secretion?	Kempf et al. 1996
Endothelium	Expressed		~10						EDRF/cGMP↑	Vasorelaxation	Yamaguchi and Nishimura 1988; Nishimura et al. 1994; Nishimura et al. 2003

BBM, brush border membrane; EDRF, endothelium-dependent relaxing factor; GFR, glomerular filtration rate; VSM, vascular smooth muscle; *, Transfected cells; –, no selective binding.

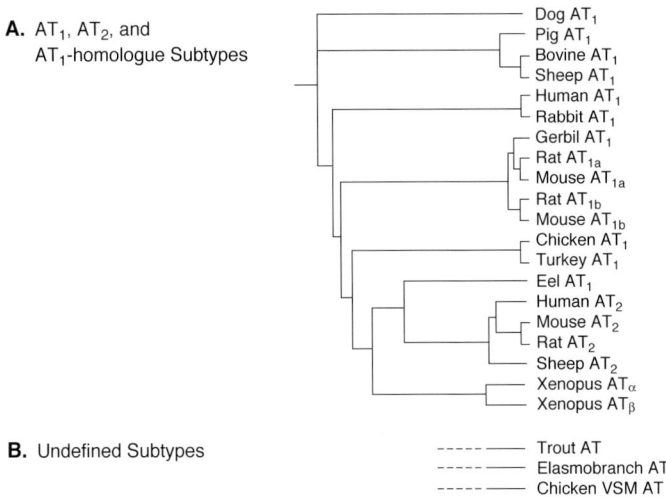

Fig. 4 A Phylogenetic dendrogram of cloned mammalian and nonmammalian angiotensin II receptors. The length of the horizontal connecting bars is inversely proportional to the pairwise similarity scores calculated from the alignment of sequences with use of the Clustal algorithm of PC/GENE. Courtesy of Ji and Sandberg, modified from Sandberg (1994). **B** Unidentified AT subtypes in nonmammalian vertebrates. *AT*, angiotensin receptor; *VSM*, vascular smooth muscle. (Reproduced from Nishimura 2001)

and stimulate the hydrolysis of phosphatidylinositol 4,5-bisphosphate by phospholipase C, leading to the formation of inositol triphosphate (IP_3) and to cytosolic Ca^{2+} release. The *Xenopus* AT receptors show high affinity to Ang II and [Sar^1, Ile^8]Ang II and low affinity to CGP-42112A (Ciba-Geigy), but show no binding to AT_1-selective (losartan) or AT_2-selective PD (Parke-Davis) compounds.

Furthermore, AT_1-homolog receptors (359 amino acid residues) have been cloned in adrenal glands of two avian species, turkeys (tAT_1) (Murphy et al. 1993) and chickens (cAT_1) (Kempf et al. 1996). cAT_1 and tAT_1 are nearly identical, show 75% amino acid identity with the mammalian AT_1 receptor, and involve the inositol phosphate-IP_3 pathway. cAT_1 and tAT_1 transfected to Cos cells exhibit high affinity to Ang II and considerable affinity to [Ile^8]Ang II analogs and CGP (in tAT_1), whereas both show no specific binding to AT_1- or AT_2-selective nonpeptide antagonists. cAT_1 mRNA is expressed in the endothelia (Kempf et al. 1996) of arteries and arterioles in the kidney, to a lesser extent in the epithelial cells of small arteries of the adrenals, and abundantly in the glomeruli and subcapsular region of the adrenals of chick embryos (Nishimura et al. 2003).

Unidentified Angiotensin Receptors and Signaling. Pharmacological and molecular studies suggest the presence of another unidentified receptor(s) in vertebrates (Smith 1996) (Table 3). The dogfish, an elasmobranch, has a CGP-sensitive receptor in the rectal gland and an interrenal receptor that has considerable

Table 3 Unidentified angiotensin receptors

	Binding affinity (nM)					Signaling	Function/Action	Reference
	Ang II K_d or IC_{50}	[Sar^1, Ile^8] Ang II	Losartan	PD	CGP			
Elasmobranchs								
Japanese dogfish								
Rectal gland	0.41–0.75	~100		2–5				Tierney et al. 1997
Interrenal	0.25	~10			~50	$[Ca^{2+}]_i\downarrow$	1α-hydroxycorticosterone secretion?	Armour et al. 1993; Hazon et al. 1997
Avians								
Chicken								
VSM	0.15	1710	–	41000		No $[Ca^{2+}]_i\uparrow$	?	Takei et al. 1988; Stallone et al. 1989; Nishimura et al. 1994
Liver	30	55000	–	49000				Walker et al. 1993
Liver	8.2	–	–				Glycogenolysis?	Bouley et al. 1997

VSM, vascular smooth muscle.

affinity to both losartan and CGP (Tierney et al. 1997; Hazon et al. 1997). Chicken vascular smooth muscle (VSM) has a high-affinity Ang receptor (Takei et al. 1988; Stallone et al. 1989) that shows low binding to 8th amino acid-replaced peptide antagonists (Stallone et al. 1989) and no binding to nonpeptide AT_1 or AT_2 antagonists (Nishimura et al. 1994). The chicken VSM Ang receptor does not mediate cytosolic Ca^{2+} signaling (Walker et al. 1993; H. Nishimura, unpublished). When the endothelium is removed, fowl Ang II ([Asp^1, Val^5]Ang II) causes neither relaxation nor contraction (Hasegawa et al. 1993) of fowl abdominal aortae. Fowl Ang II does not stimulate thymidine incorporation into cultured adult fowl VSM cells (Shimada et al. 1998). Furthermore, cAT_1 mRNA is not expressed in VSM layers of aortae or in small arteries and arterioles of various organs from adult fowl (Kempf et al. 1996; Nishimura et al. 2003), indicating that the Ang receptor in fowl VSM is distinct from the cAT_1 cloned from the fowl adrenal gland. A similar receptor that has no selectivity to nonpeptide antagonists is also present in the liver (Walker et al. 1993; Bouley et al. 1997).

2.2
Function and Regulation

2.2.1
Control of Renin Release

Effects of Blood Volume and Blood Pressure. Since the kidneys from primitive vertebrates lack a macula densa, the possible baroreceptor function in JG cells may be important in the regulation of renin release in these animals. Cumulative hemorrhage or a single massive one increased PRA more than 15 times in the conscious aglomerular toadfish (*Opsanus tau*; Nishimura et al. 1979), and also increased plasma Ang II levels in conscious Japanese eels (*Anguilla japonica*; Kobayashi et al. 1980). Hemorrhage increased PRA in the turtle (*Pseudemys scripta*; Stephens and Creekmore 1984), pigeon (Chan and Holmes 1971), and chicken (Nishimura and Bailey 1982). Furthermore, intraarterial injection of papaverine (10 mg/kg) in the toadfish increased PRA (5- to 20-fold) with a concomitant decrease in blood pressure (BP). Likewise, reduction in renal perfusion pressure caused renin release in the rainbow trout (Bailey and Randall 1981) and in chickens (Wideman et al. 1993). It appears, therefore, that primitive vertebrates, in spite of the absence or incomplete form of a macula densa, have a functional renal baroreceptor mechanism for controlling renin release. In contrast, isosmotic volume expansion in the Australian lungfish (*Neoceretodus forsteri*) decreased PRA and increased the mean dorsal aortic pressure by 2 mmHg (Blair-West et al. 1977).

Furthermore, injection or infusion of Ang-converting enzyme (ACE) inhibitor decreases resting BP levels and increases PRA in conscious bony fish (Nishimura et al. 1978; Nishimura and Bailey 1982), suggesting that the RAS may have a role as a physiological regulator of BP. Since kidneys of lower vertebrates are not, or are only poorly, autoregulated, the changes in aortic pressure readily influence glomerular filtration rate (GFR) and renal blood flow, which appear to be a mechanism for controlling fluid and electrolyte excretion.

Involvement of Neural Mechanisms and Macula Densa. A question rises as to whether the renal nerves are also involved in stimulation of renin release induced by hemorrhage or hypotension. It appears that the large arteries of teleosts are supplied by both "constrictor adrenergic" and "constrictor cholinergic" nerves (Kirby and Burnstock 1969). Cholinergic vasoconstrictor nerves comprise both sympathetic and parasympathetic fibers. Adrenergic receptors and adrenergic innervation in the blood vessels are present, though less marked, in teleost fish (Nakamura et al. 1992); and adrenergic innervation increases, whereas cholinergic innervation diminishes, as phylogeny advances (Burnstock 1969).

In mammals, β-adrenoceptors are localized on the JG cells; and intrarenal infusion of isoproterenol or norepinephrine, or electrical stimulation of renal nerves causes renin release (Davis and Freeman 1976). In aglomerular teleost fish, isoproterenol evoked an increase in PRA with a concomitant reduction in BP, both of which are inhibitable by propranolol (Nakamura et al. 1992). Neither propranolol alone nor depletion of catecholamines by repeated treatment with 6-hydroxydopamine, however, altered the basal level of PRA, suggesting that isoproterenol-induced renin release is likely due to activation of baroreceptors via decreased renal arterial perfusion pressure. Indeed, there is no specific relationship between renin-containing granulated cells and adrenergic nerve endings in renal arterioles (Nakamura et al. 1992).

The role of the macula densa in the control of renin release and the existence of a tubuloglomerular feedback mechanism have not been much explored in nonmammalian vertebrates. Although a close topographical relationship between the distal tubule and the glomerular vascular pole exists in elasmobranch kidneys, communication and signaling between these structures has not been examined. We observed that slow infusion of hypertonic saline into the renal portal system of fowl in which shunt pathways bypassing the kidney were occluded caused diuresis and natriuresis on the infused side without altering GFR, whereas infusion of the same dose into the systemic circulation showed no effect (Bailey and Nishimura 1984). PRA was decreased, suggesting that, in fowl, some functional link may exist between NaCl concentration in tubular fluid and renin secretion via a macula densa mechanism. In mammalian kidneys, a reduced chloride transport across macula densa cells or an application of bumetanide to the tubular lumen stimulates renin secretion from JG cells (Lorenz et al. 1991), and this Cl-dependent renin release is inhibited by blockade of cyclooxygenase 2 (Traynor et al. 1999).

Cellular Mechanism of Control of Renin Release. In aglomerular toadfish, potassium-induced depolarization increases cytosolic Ca^{2+} in VSM (Qin et al. 1999) and decreases renin release from renal slices superfused in vitro (Nishimura and Madey 1989). This depolarization-induced reduction of renin release was inhibited by blockade of Ca^{2+} influx by nifedipine and restored by the Ca^{2+} channel agonist Bay K8644 or by phorbol ester (Nishimura and Madey 1989). Likewise, removal of extracellular Ca^{2+} stimulates renin release. Since renin-secretory cells are modified VSM cells, mechanical stretch induced by increased renal perfusion pressure is likely to cause membrane depolarization and activation of a voltage-gated calcium channel. It appears, therefore, that cytosolic Ca^{2+} levels may be a cellular regulator for control of renin release. In contrast, isoproterenol, dibutyryl, chlorophenylthio-cAMP, 8-bromo-cGMP, and acetylcholine each failed to stimulate renin release or to restore potassium-induced renin suppression. This supports the concept that β-adrenoceptor-mediated control of renin release has not yet evolved in teleost fish. Similarly, prostaglandin E_2, phenylephrine, Ang II, and atrial natriuretic peptide showed no effect on renin levels (H. Nishimura, unpublished).

In mammalian JG cells, intracellular regulation of renin secretion appears to occur at three levels, including vesicle volume, the state of contraction of the actin-myocin complex, and the docking process of the vesicles with the cell membrane. The swelling of renin storage granules evoked by the proton pump-induced influx of KCl may be an important cellular process for the exocytosis of renin secretory granules (Kurtz and Wagner 1999). While an increase in cytosolic Ca^{2+} via a release from the internal store or via stimulation of Ca^{2+} influx stabilizes exocytosis, cAMP protein kinase stimulates this process by weakening the actinomycin shield (Kurtz and Wagner 1999).

2.2.2
Biological Action of Angiotensin

Cardiovascular Effects and Release of Catecholamines. Ang increases BP in selected species of all vertebrate classes and also stimulates cardiac contractility in some species (Nishimura 1980, 1987). In nonmammalian vertebrates, Ang II causes catecholamine release (Nishimura et al. 1981b; Carroll and Opdyke 1982; Bernier et al. 1999b). [Asn^1, Val^5]Ang II (Opdyke and Holcombe 1976; Carroll and Opdyke 1982) and dogfish Ang II ([Asn^1, Pro^3, Ile^5]Ang II) (Bernier et al. 1999a) dose-relatedly increase plasma catecholamines and arterial pressure of the dogfish (*Squalus acanthias*). Treatment with α-adrenoceptor antagonists, phentolamine (Opdyke and Holcombe 1976) and yohimbine (Bernier et al. 1999a), completely abolished Ang-induced vasopressor action but not the catecholamine-releasing effect (Bernier et al. 1999a), suggesting that the vasopressor effect of Ang II is indirectly mediated by catecholamine release via Ang receptors at the adrenergic nerve endings and in the adrenal medulla. Ang releases catecholamines from chromaffin tissues in rainbow trout via a specific Ang II binding site (Bernier and Perry 1997). In teleost fish (Carroll 1981), amphibians (Corwin et al. 1984), and reptiles (Stephens 1984), Ang II contracts isolated arterial preparations. These findings suggest that the receptors mediating catecholamine release may have evolved phylogenetically earlier than the vascular Ang receptors that mediate the direct vasocontractile action of Ang II.

In domestic fowl, Ang II-induced vasopressor action is solely ascribable to catecholamine release from the adrenal medulla and adrenergic nerve endings (Nishimura et al. 1982; Nakamura et al. 1982). Furthermore, Ang II specifically binds to the aortic endothelium and causes endothelium-dependent relaxation (Yamaguchi and Nishimura 1988; Stallone et al. 1990; Nishimura et al. 1994), accompanied by a sharp rise in aortic cGMP levels. Hemoglobin completely inhibits and N_w-nitro-L-arginine methyl ester (L-NAME) partially inhibits Ang II-induced vasorelaxation in fowl aortic rings, suggesting that part of the signal pathway of the endothelial Ang receptor may be via nitric oxide and cGMP (Hasegawa et al. 1993). Ang II-induced depressor and/or vasorelaxing effects of Ang II are inhibitable by 8th amino acid-replaced Ang antagonists but not by inhibitors for cyclooxygenase or monooxygenase, β adrenoceptor antagonists or cholinergic blockers, or by inhibitors for serotonin, vasotocin, histamine, or phospholipases (Nakamura et al. 1982; Yamaguchi and Nishimura 1988). Ang II-induced relaxation appears partly ascribable to a K^+-channel and endothelium-dependent hyperpolarization factor (Nishimura et al. 2003; H. Nishimura, unpublished).

Ang II induces neither inotropic nor chronotropic action in perfused rainbow trout heart preparations (Olson et al. 1994). In intact rainbow trout, Ang II increases the heart rate, and the increase can be attenuated by [Sar^1, Ala^8]Ang II (Fuentes and Eddy 1998). In fresh water-adapted conscious eels, Ang II increases the heart rate and cardiac contractility (Oudit and Butler 1995).

Dipsogenic Action. Intravenous or intracranial application of Ang II stimulates water intake in various classes of vertebrates (extensively reviewed in Nishimura 1987; Kobayashi and Takei 1996; Takei 2000), suggesting that, as in mammals, Ang receptors in the brain mediate dipsogenic action. Furthermore, stimulation of the endogenous RAS by hemorrhage or papaverine-induced hypotension enhances water intake in elasmobranchs (Anderson et al. 2001), freshwater-adapted flounder (Balment and Carrick 1985), and freshwater- and seawater-adapted eels (Tierney et al. 1995) with a concomitant increase in plasma Ang II (Tierney et al. 1995). Captopril decreases both the plasma Ang II level and drinking rate in seawater-adapted eels (Tierney et al. 1995), suggesting that endogenous RAS plays a role in maintaining fluid intake in fish. It remains to be determined, however, whether systemic or brain Ang physiologically maintains extracellular and intracellular fluid volume homeostasis via its dipsogenic effect. The role of the subfornical organ in Ang II-induced drinking has been reported in Japanese quail (Takei 1977). Increases in water intake and plasma Ang II levels following hemorrhage have been reported in the conscious Japanese eels adapted to fresh water (Kobayashi et al. 1980). The evolution of brain Ang receptors responsible for drink stimulation, and their precise localization and properties, however, remain to be elucidated in nonmammalian vertebrates.

Stimulation of Adrenal Steroids. In mammals, Ang stimulates aldosterone production in early and late stages of steroidogenic pathways (Fraser and Lantos 1978; Fraser et al. 1979), respectively, by stimulating conversion of cholesterol to pregnenolone and by stimulating 18-hydroxydehydrogenase activity. It has been reported that Ang II stimulates 1-α hydroxycorticosterone and cortisol that act as mineralocorticoids in, respectively, elasmobranchs and teleost fish (for review, Nishimura 1987; Henderson et al. 1993; Hazon et al. 1999). In ducks, [Asp^1, Val^5]Ang II (native fowl Ang II) stimulates production of aldosterone and 18-hydroxycorticosterone and corticosterone from adrenal cell suspensions (Vinson et al. 1979). Although the available information is limited, a linkage between endogenous RAS activity and adrenal steroids apparently evolved early and may be conserved. For example, in flounder, injection of the vasodilator papaverine evoked hypotension and an increase in plasma cortisol, whereas both the elevation of cortisol and BP restoration after papaverine-induced hypotension were prevented by captopril (Perrott and Balment 1990). Likewise, the transfer of freshwater-adapted eels to seawater increased the plasma levels of both renin and cortisol, which presumably play a role in the stimulation of drinking and water absorption (Henderson et al. 1976). Although aldosterone has been identified in bony fish, amphibians, reptiles, and birds, the role of Ang

II in the control of aldosterone is not as clear as in mammals; rather, ACTH appears a more potent stimulator for both glucocorticoids and mineralocorticoids (Klingbeil 1985; Holmes et al. 1991; for review, Nishimura 1987). Furthermore, in nonmammalian animals, the role of aldosterone or other adrenal steroids in the control of renal Na^+ and K^+ transport has not been established, whereas mineralocorticoid functions of adrenocortical hormones have been clearly demonstrated in extrarenal transport epithelia. Thus, the manifestation of Ang function in hydromineral balance would vary depending on species and target organs. In fowl, the mRNA encoding AT_1-homolog receptor cDNA has been extensively expressed in fowl adrenocortical tissues (Kempf et al. 1996), particularly the subcapsular region of embryonic adrenals (Nishimura et al. 2003). The role of Ang in steroidogenesis has not been investigated in embryonic adrenals.

Renal Action and Role in Osmoregulation. The RAS could influence renal function (1) through its effect on the systemic renal perfusion pressure and flow, (2) by renal action of systemically delivered renin and Ang, and (3) by local action of intrarenally formed renin and Ang (for review, Nishimura 1985b, 1987; Brown et al. 1993; Russell et al. 2001). Furthermore, the renal actions of Ang include its effects on the renal vasculature and renal tubules. In general, high pressor doses of Ang II produce glomerular diuresis via an increase in GFR in bony fish and amphibians (pressure diuresis); whereas Ang II reduces GFR, urine flow, and renal blood flow by constriction of preglomerular arterioles in rainbow trout (glomerular antidiuresis) (Brown et al. 1980). Inhibition of ACE activity by captopril causes profound diuresis in the in situ perfused trout kidney and a reduction of vascular resistance of the trunk (Brown et al. 2000). Thus the effect of Ang on renal hemodynamics and function differs depending on the location of Ang receptors. High cAT_1 mRNA levels are detected by RT-PCR and in situ hybridization in glomeruli of mesonephric kidneys of chick embryos and, to a lesser degree, in metanephric kidneys (Nishimura et al. 2003; H. Nishimura, unpublished). Strong expression in glomerular tufts, presumably mesangial cells, during development may indicate that the cAT_1 receptor plays a role in the growth/maturation of renal glomeruli.

Direct action of Ang II in renal tubule epithelial cells remains unclear. Ang altered neither urine flow nor Na excretion in aglomerular teleosts (Zucker and Nishimura 1981) in which urine is formed via excretion and reabsorption across the renal tubules. In contrast, native Ang II infused into the renal portal system of fowl, after the shunt between the systemic and renal portal circulation was closed, produced marked natriuresis without significantly altering GFR (Nishimura 1987), suggesting that Ang directly influences Na transport in renal tubules. In addition to the effect on the kidney, Ang II influences NaCl and water transport across epithelial tissues such as the urinary bladder, gastrointestinal tract, and nasal salt glands in some species (for review, Nishimura 1987; Kobayashi and Takei 1996). In general, however, the Na-retaining effect of Ang II, either via stimulation of mineralocorticoid secretion or renal tubule transport, is not established in nonmammalian species; rather, it varies depending on the osmoregulatory function of the target organs. In euryhaline teleosts, Ang II receptor concentration and Na^+/K^+ ATPase activity of gill chloride cells increase during adaptation to seawater (Marsigliante et al. 1997), reflecting an enhanced dipsogenic effect of Ang II and possibly NaCl extrusion from the gill in seawater fish.

3
Ontogeny of the Renin–Angiotensin System

Embryonic and fetal development is a result of integrated programs of growth and differentiation leading to organogenesis. A number of reports, primarily on rodent or ovine models, as well as on human subjects, suggest that the RAS plays a critical role in fetal growth and development and serves as a modulator of the ontogenic process. Defect or deletion of a component of the RAS results in abnormal development in architecture and function. All components of the RAS are detected during the fetal period; and gene expressions of AGT, renin, ACE, and Ang receptors are developmentally regulated in a tissue- and time-specific manner. Furthermore, impairment of the RAS during the fetal or perinatal period appears to induce organ dysfunction in maturity via so-called perinatal programming, or the fetal origin, of adult disease. Recently emerging gene-targeting techniques provide useful tools for elucidating the roles of specific genes and their physiological manifestation.

3.1
Biochemistry and Molecular Biology

3.1.1
Renin Substrate and Renin

All components of the RAS are present and functioning early in fetal life (Gomez et al. 1991; Tufro-McReddie and Gomez 1993). In the mesonephric kidney of the human fetus, AGT mRNA determined by in situ hybridization is first detected at the proximal portion of the renal tubules (Corvol et al. 1998). AGT mRNA is also present in the proximal tubules, mesangial cells of the glomerulus, and renal arterial system of the rat kidney at embryonic day 18 (E18); it is most intensely expressed during the neonatal period and is reduced to less than half in mature adults (higher in females than in males) (Darby and Sernia 1995) (Table 4). AGT mRNA from the fetal rat liver (total RNA or poly A+ enriched RNA) is very low and increases by 20- to 50-fold after birth; it gradually reaches the adult level by three postnatal months (Gomez et al. 1988b; Tufro-McReddie and Gomez 1993). In contrast, AGT mRNA is found abundantly in the brain, brown fat, and kidneys, suggesting that the liver may not be the primary source of AGT in fetal life (Gomez et al. 1988b; Harris and Gomez 1997). Differences in AGT mRNA were more clearly seen in spontaneously hypertensive rats (SHR) than in control Wistar-Kyoto (WKY) rats (Gomez et al. 1988b).

Table 4 Time-dependent expression of the renin–angiotensin system in the developing rat kidney

Component (mRNA or protein)	Embryonic day			Newborn/ perinatal	Adult
	E11~15	E17~18	E20~21		
Angiotensinogen	±	+	++	+++++[a]	++
Renin	±	+	++++++++~[b]	+++++++~[c]	+~++
Angiotensin-converting enzymes[d]		±	++	+++	++++++
AT$_1$	±	++	+++	+++++	++++
AT$_2$	++	++++	++++++	+++++	±
AT$_1$/AT$_2$		AT$_1$<<<AT$_2$	AT$_1$<<AT$_2$	AT$_1 \geq$AT$_2$	AT$_1$>>AT$_2$
Angiotensin II				+++++	+

Data (relative levels within each component) are summarized from Darby and Sernia 1995; Gomez et al. 1988a,b, 1989, 1991; Harris and Gomez 1997; Norwood et al. 1997; Shanmugam et al. 1994, 1995; Tufro-McReddie and Gomez 1993; Tufro-McReddie et al. 1993, 1994; Wallace et al. 1978, 1979; Yosipiv and El-Dahr 1996.

[a] Fifty-fold increase in liver.
[b] Twenty-fold of adult level.
[c] Ten-fold of adult level.
[d] Higher ACE activity in fetal lung.

In human mesonephros at 5–6 weeks, detectable renin is expressed in the glomerular capillaries and in the walls of renal arteries from arterioles up to aortae (Corvol et al. 1998). Renin mRNA and protein (histochemical manifestation of JG cells) increase with gestation, whereas expression in the JG apparatus does not increase when normal fetal growth and the growth of the renal cortex are restricted (Kingdom et al. 1999). In metanephric kidneys, the distribution of renin immunoreactive cells (Celio et al. 1985) and renin mRNA is confined to the JG area and arterioles adjacent to the vascular pole (Corvol et al. 1998) and, to a lesser extent, proximal tubules (Celio et al. 1985).

During ontogeny of metanephric rat kidneys, renin mRNA levels and distribution change as maturation progresses. Renin protein and mRNA signals are detectable at E17 in the renal artery, and localization of renin synthesis (mRNA) and storage (protein) shift from arcuate and interlobular arteries to the classical JG apparatus vascular pole and afferent arteries (Gomez et al. 1989) towards term. Renin mRNA and renin protein levels (renin-containing JG cells) are higher in newborns and decrease progressively as animals mature (Gomez et al. 1988b, 1991) (Table 4). Although the reasons for the high and widely distributed renin expression in fetal kidney are unclear, it may be the result of developmentally regulated transacting factors that interact with the 5′ flanking region of the renin gene (Harris and Gomez 1997). Likewise, the transgene of human renin promotor with a *LacZ* reporter gene to mouse embryo at E15 mimics endogenous renin expression and is detected in the interlobular arteries of the kidney, as well as in large arteries outside the kidney (Fuchs et al. 2002).

Pan and coworkers have identified the renin promoter element as a *Hox* protein binding site and verified that *Hox* genes and their cofactors determine the sites of renin expression (Pan et al. 2001). Since *Hox* genes are members of a homeobox family of transcription factors and control cell morphogenesis and differentiation, strong renin expression may coincide with growing structures during development (Pan et al. 2001). Interestingly, in adult rats, chronic inhibition of ACE enhances intrarenal renin synthesis and renin mRNA upstream from the glomeruli in cells that do not express renin in their basal state (Gomez et al. 1988a; Jones et al. 2000). The distribution of renin-expressing cells in ACE-treated adult kidneys resembles that in fetal kidneys, suggesting that the adult vasculature is capable of recruiting JG cells and reproducing the fetal pattern of renin and renin gene expression.

3.1.2
Angiotensin and Angiotensin-Converting Enzymes

Immunoreactive ACE activity is detected in apical and basolateral membranes of differentiating human proximal nephrons and in glomerular endothelial cells of developing capillaries (Mounier et al. 1987). During renal development and postnatally, ACE protein is consistently found on the brush border of proximal tubule cells (Mounier et al. 1987; Corvol et al. 1998). In fetal rat lungs, ACE activity is first detected at E18 and increases two- to fourfold by the first day of birth (Wallace et al. 1978, 1979; Tufro-McReddie and Gomez 1993). ACE activity further increases during postnatal maturation and is stabilized in adult life (Table 4).

Reflecting high levels of AGT and renin levels, intrarenal Ang I and Ang II levels are three- to sixfold higher in newborn than in adult rat kidneys (Yosipiv and El-Dahr 1996) (Table 4). It is not known, however, whether any Ang other than [Asp^1, Ile^5]Ang II or [Asp^1, Val^5]Ang II is expressed during mammalian fetal development, as in the case of neurohypophysial hormones. It will be interesting to examine whether nonmammalian Ang I and Ang II (Table 1) are expressed during mammalian ontogeny. ACE activity is detected in fetal lambs and increases with gestation parallel to the increases in plasma cortisol levels and BP; levels are high in newborns and decrease with time (Forhead et al. 1998).

3.1.3
Angiotensin Receptors

In human mesonephric and metanephric kidneys, AT_1 mRNA is detected in the glomeruli, perhaps in mesangial cells, beginning at stage 12, while AT_2 mRNA is also expressed in undifferentiated mesenchymal tissues at stage 11 (Corvol et al.

1998). AT_2 mRNA is highest at 8–9 weeks of gestation and declines after 20 weeks of gestation, although it is still detectable at birth.

In rat kidneys (Table 4), AT_1 mRNA is minimum at E14, highly expressed at E20, and persists, although becoming lower with maturation, into adulthood (Norwood et al. 1997). The AT_1 receptor is found in the undifferentiated mesenchymal tissues at E14, whereas it is localized in nephron segments (Norwood et al. 1997) at E17. During maturation, localization of AT_1 mRNA in the rat kidney shifts from a diffuse distribution in the nephrogenic cortex to more selective areas such as the glomeruli, adjacent arterioles (media/adventitia; weak signal), and vascular bundles (Tufro-McReddie et al. 1993). Also, the evolution and distribution of two AT_1 receptor subtypes, AT_{1A} and AT_{1B}, slightly differ: from E15–E19 gestation, both subtype mRNAs are present in the kidney and adrenal glands; whereas only AT_{1A} mRNA is expressed in the liver, lung, heart, aorta (media and adventitia), and undifferentiated mesenchymal tissues (Shanmugam et al. 1994). AT_{1A} mRNA is found in mesenchymal cells associated with differentiating glomeruli that later mature to mesangial cells, interstitial tissues, maturing tubules, vasa recta, and JG apparatus in the postnatal kidney; whereas AT_{1B} expression is limited to preglomerular structures and mature glomeruli (Shanmugam et al. 1994).

In contrast, AT_2 expression in the rat cardiopulmonary system is easily detectable from E14–15 through postnatal days 7–15 (D7–D15) and rapidly declines (Shanmugam et al. 1995, 1996). AT_2 mRNA is abundantly found in undifferentiated nephrogenic tissues, but not in glomeruli (Shanmugam et al. 1995). During the fetal period, AT_2 represents the majority of Ang receptor subtypes; whereas AT_1 is the dominant functioning receptor after birth, although AT_2 expression remains in some organs/tissues. Rat kidneys continue to mature after birth; and AT_1 mRNA, but not AT_2, is expressed in immature glomeruli of neonatal day 2 (D2) kidneys (Aguilera et al. 1994; Shanmugam et al. 1995) and in the medulla at D7. The dissociation constants of AT_1 and AT_2 are similar to those of adult receptors (Norwood et al.1997). The time-dependent expression of AT_1 and AT_2 and selective localization suggest that these receptors may have different roles in nephrogenesis. In situ radioligand binding autoradiography indicates that, at E14, specific binding to AT_1 and AT_2 is, respectively, 24% and 76% (Norwood et al. 1997); but the ratio of AT_1 and AT_2 receptors (as much as 1:10) during fetal life differs, depending on the time and method of measurements.

Adrenal glands contain cells from two origins: (1) undifferentiated mesenchymal tissue (cortex, 3 zones) and (2) a cluster of neural crest cells by migration (medulla). In human adrenal glands, the AT_2 subtype represents the dominant specific Ang II binding site in the fetal zone at gestation week 14, whereas the AT_1 is expressed only at the periphery of the gland (Breault et al. 1996). No detectable AT_2 mRNA (in situ hybridization) is present, however, in the zona glomerulosa (Schütz et al. 1996). At E17–21 of rat fetal life, AT_1 mRNA is detected in the zona glomerulosa, medulla (less extensively), and capsule regions (Shanmugam et al. 1994). After birth, AT_1 expression remains high in the glomerulosa. In rat adrenals, strong AT_2 mRNA signals are also detected in the zona glomerulosa, whereas the zona fasciculata and zona reticularis do not express AT_2 mRNA (Shanmugam et al. 1995). The adrenal medulla contains clusters of cells with positive signals. After birth until D15, the zona glomerulosa continues to express AT_2 mRNA (Shanmugam et al. 1995).

Likewise, AT_1 is expressed predominantly in the zona glomerulosa of fetal lambs and to a lesser extent in the zona fasciculata (Coulter et al. 2000), or equally in both zones but not in the medulla (Wintour et al. 1999). AT_1 mRNA can be detected as early as 60 days of gestation and remains high throughout adult life, whereas AT_2 mRNA dramatically decreases at term and after birth (Wintour et al. 1999). AT_1 mRNA has also been detected in the forebrain of the E19 rat, which contains an area important for fluid homeostasis; whereas AT_1 mRNA is nearly absent in medullary nuclei until after birth (Nuyt et al. 2001). AT_2 mRNA expression is found only in medullary zone in 8-day-old mice (Hubert et al. 1999).

In normal adults, the AT_2 receptor is expressed in only limited tissues, including adrenal glands, ovary, and brain (Pucell et al. 1991; for review see de Gasparo et al. 1994; Horiuchi et al. 1999). The AT_2 receptor is also expressed during pathophysiological conditions such as skin wound healing and vascular remodeling. The signal pathways of AT_1 and AT_2 receptor subtypes have been studied in adult tissues (discussed in detail in other chapters of this book), but information on fetal tissues/organs is scant.

3.2
Function and Regulation

3.2.1
Angiotensin Action

The RAS appears to play important roles in cardiovascular and fluid–mineral homeostasis during both fetal and mature mammalian life. The role of the RAS in regulation of BP and fluid–mineral balance evolved early, whereas its role in the control of renal hemodynamic forces may have developed at a later stage of fetal life (Harris and Gomez 1997). Increasing evidence suggests that Ang regulates fetal growth and development by initiating a growth-related signal pathway involving proto-oncogenes and various growth factors and cytokines (Harris and Gomez 1997). The effect of the RAS on body growth and organogenesis has been studied by (1) blockade of the RAS using pharmacological tools, (2) use of gene-targeting approaches, and (3) examining the effects of an impaired RAS during fetal life that result in abnormal manifestation at adulthood (McMillen et al. 2001; Battista et al. 2002). It has been shown that deletion or blockade of any component of the RAS results in striking maldevelopment of kidney architecture and function, affecting both the nephrons and the vascular system (Gomez 1998).

Role of Angiotensin in Body Growth and Renal Development. As in cultured cells or tissues from mature animals, Ang II (10^{-11} M) stimulates both somatic and yolk sac/allantoic development and growth of rat embryo cultured in vitro, and this effect is inhibited by PD123319 (Tebbs et al. 1999). The time and localization of expressions suggest that AT_2 is important for early stages of organogenesis, stimulating differentiation and apoptosis; whereas AT_1 plays a role in renal vascularization and trophic actions, including cell hypertrophy, synthesis of cell matrix, and development of glomeruli and mesangial cells (for review, Harris and Gomez 1997; Ardaillou 1999). The signal pathway involving fetal growth via AT_2 is not well understood. In adult tissues, AT_2 receptor over-expression induced by transfecting a vector into balloon injury-induced carotid neointima causes a reduction in neointima formation; whereas in cultured cells it reduces mitogen-activated protein (MAP) kinase and Ang II-induced (losartan-inhibitable) cell proliferation, suggesting that AT_2 mediates antiproliferative action that counteracts the growth-promoting action of the AT_1 receptor (Nakajima et al. 1995). It thus appears that both AT_1 and AT_2 are needed for normal cell differentiation and organogenesis.

Homozygous AGT gene null-mutant mice, five postnatal weeks old, exhibit structural defects of the kidney, including hypoplastic papilla and prominent medial hypertrophy of interlobular arteries and afferent arterioles (Okubo et al. 1998). AGT-null mice have labile GFR after different amounts of Na intake, while urine osmolality does not increase in response to the vasopressin agonist 1-deamino-D-Arg-8 vasopressin (DDAVP) (Okubo et al. 1998). Animals are hypotensive and exhibit lesions in the cortex resembling nephrosclerosis in which platelet-derived growth factor (PDGF)-B and transforming growth factor (TGF)-β-1 mRNA were upregulated (Niimura et al. 1995).

Pharmacological blockade of Ang I to Ang II conversion by ACE inhibitors induces fetopathy in humans (Barr 1994; Sedman et al. 1995). Maternal intake of ACE inhibitor during the second and third trimester of pregnancy often evokes profound fetal growth restriction, hypotension, renal tubule dysplasia, hypocalvaria, and death (Sedman et al. 1995). Likewise, chronic ACE inhibition in neonatal rats arrests renal growth and nephrovascular maturation (Tufro-McReddie et al. 1995) and induces maldeveloped renal structure similar to that evoked by AGT gene deletion, such as dilated Bowman's capsule space and irregular size/shape of glomeruli (Barr 1994). Similarly, treatment of newborn SHR and WKY rats with captopril or enalapril induces renal dysfunction and structure, including (1) reduced urine osmolality, (2) increased urine production and increased water intake, and (3) renal papilla atrophy and tubulo-interstitial inflammation (Friberg et al. 1994). Mice lacking ACE (mice in which modified ACE allele was used for targeted homologous recombination in embryonic stem cells) exhibit renal papilla atrophy, vascular hyperplasia, and defects in urine-concentrating ability (Esther et al. 1996). Thus, the various investigative approaches indicate that a lack of functioning ACE activities exerts a significant influence on the development of renal architectural and functional defects—ones similar to those observed in the AGT gene deletion model (AGT knockout mice).

Blockade of Ang receptors, either pharmacologically, using AT_1- or AT_2-selective antagonist, or with gene-targeting approaches, also causes maldevelopment of the kidney (Barr 1994; Sedman et al. 1995; Woods and Rasch 1998). Treatment of weaning rats with losartan for 3 weeks reduces the DNA content of the kidney, whereas renin-secreting cells are seen more widely on arteries and arterioles (Tufro-McReddie et al. 1994). Deletion of both AT_{1A} and AT_{1B} genes evokes severe abnormal phenotypes resembling those of homozygous AGT null mutant mice (Tsuchida et al. 1998; Okubo et al. 1998), including low ex utero survival rate, growth retardation, hypotension, and abnormal kidney development (impaired glomerular maturity, plastic papilla, hypertrophy of renal arteries). In contrast, targeted deletion of the AT_{1B} receptor gene did not evoke any distinctive abnormal morphological phenotype (Chen et al. 1997). These findings indicate that AT_1 receptor is responsible for somatic renal growth. AT_{1A} receptors, which are predominantly expressed in mouse kidneys, may compensate for the loss of the functioning AT_{1B} subtype (Chen et al. 1997).

AT_2 gene-null mutant mice also display structural defects of the kidney and urinary tract (Nishimura et al. 1999). Of particular interest is the delayed apoptosis of undifferentiated mesenchyme. Since timely programmed apoptosis of mesenchymal tissues that surround the differentiating Wolffian duct and ureters is required for the normal development of the kidney and urinary tract, the lack of the AT_2 receptor that plays a role in differentiation and time-dependent apoptosis may lead to structural defects of the kidney and urinary tracts. In summary, currently available studies indicate that during fetal and neonatal development, AT_1 mediates proliferation, whereas AT_2 mediates differentiation and apoptosis, both of which are essential for normal development/maturation of kidney structures and functions.

Role in Perinatal Programming of Growth. Increasing evidence suggests that adult diseases originate in abnormal events during the fetal period. It has been postulated that maternal dietary protein restriction suppresses the fetal RAS, leading to impaired fetal growth and renal development (Woods 2000; Langley-Evans 2001). This in turn results in permanently altered renal structure and function in adult life, such as a reduction in the number and size of glomeruli, followed by a reduction in GFR and impaired tubule function, which may further alter the RAS function (Woods 2000). Furthermore, although phenotypes at birth may be normal, abnormal fetal experience determines in adulthood whether an individual is resistant or prone to stress and a variety of health problems, including hypertension, cardiovascular diseases, and renal dysfunction (Ingelfinger and Woods 2002). The RAS, in interaction with various other hormonal systems, appears to play an important role in this perinatal programming, while suppression of Ang action during the perinatal period may trigger abnormal renal and cardiovascular function in adulthood. For example, newborn rat pups treated with losartan for 12 days of the postnatal period demonstrated retarded growth at weaning day (D22) compared to controls. Both groups had similar body weight (BW) and kidney–BW ratios at 22 weeks (adult);

but, in the losartan-treated rats, BP was higher while GFR, effective renal plasma flow, urine-concentrating ability, and the number of glomeruli were significantly lower than in the untreated group (Woods and Rasch 1998). Similarly, intrauterine growth restriction induced by treating the mother rats with a low-Na diet for the last 7 days before term produced pups with retarded growth and with an age-dependent increase in BP (Battista et al. 2002). At 12 weeks of age, renin and aldosterone levels were elevated. Since kidney weight is consistently lower during development/maturation of pups of Na-restricted mothers (although the size of glomeruli and the number of glomeruli/area are similar to those in control rats), the high renin and BP may be a direct consequence of reduction of renal mass or impaired vascular development. Furthermore, captopril reduced BP in adult sheep exposed to growth restriction during the fetal period, but not in control groups (McMillen et al. 2001), suggesting that the RAS plays a greater role in the control of BP in growth-restricted animals. The cellular and molecular mechanisms for this fetal adaptation to insufficient substrate supply remain to be determined.

Role of Angiotensin in Cardiovascular Function. The possible role of the RAS in controlling cardiovascular–renal function during development has been investigated using molecular and pharmacological approaches. The transduction of the mouse Ren-2 gene to the rat increased plasma prorenin and increased BP, whereas renal renin and Ang I levels were low (Mullins et al. 1990). Captopril treatment of fetal lambs at 120–130 days' gestation markedly increased PRA (Gomez and Robillard 1984); and hemorrhage increased PRA in both control and captopril-treated fetal lambs (Gomez and Robillard 1984), suggesting that feedback control between BP/blood volume (BV) and renin activity exists in the fetus. Interestingly, renal blood flow decreased while renal vascular resistance increased in response to hemorrhage in captopril-treated fetuses (Gomez and Robillard 1984), indicating that the RAS is not the essential factor for regulating renal hemodynamic forces in the fetus. Hence, it appears that the RAS plays an important role in control of BP/BV during fetal life, whereas its effect on renal hemodynamic function may have evolved at a later stage of development (Robillard and Nakamura 1988).

Role of Angiotensin in Water and Ion Balance. As in adults, Ang II increased aldosterone levels in fetal lambs in the third trimester (Robillard et al. 1982; Harris and Gomez 1997). Exogenous administration of aldosterone to fetal lambs of the same gestational period decreased PRA, suggesting that, as above, negative feedback control of renin via volume/Na regulation exists in the fetus. Human adrenal glands in midgestation express abundant AT_2 in the fetal zone that secretes dehydroepiandrosterone-sulfate (DHES/s) (Breault et al. 1998). DHES/s is, however, primarily regulated by adrenocorticotropic hormone (ACTH), and Ang II exerts no effect on basal steroid secretion. When human fetal adrenal cells were cultured in serum-free medium, Ang II potentiated the ACTH-induced secretion of cortisol, aldosterone, and DHES secretion (Breault et al. 1998). Central administration of L-NAME into fetal lambs near term reduced Ang II-induced swallowing, suggesting that Ang II and nitric oxide interact in facilitating swallowing (Ei-Haddad et al. 2000).

3.2.2
Regulation of the Renin–Angiotensin System During Development

As discussed above, all components of the RAS exist in the fetus and are developmentally regulated. The factors controlling the time-dependent expression of the RAS, however, are not clear. In rats, renin levels are higher in the fetus than at other times of life, possibly because of increased renin synthesis/secretion and decreased metabolism (Harris and Gomez 1997). In contrast, the sensitivity of the Ang receptor is lower in newborns than in adults. Although the mechanism is unclear, this relative insensitivity may compensate for high Ang II levels (Yosipiv and El-Dahr 1996) in the newborn. The Ang receptor K_d appears similar in fetuses and adults (Norwood et al. 1997).

It has been reported that several factors shown to regulate renin gene expressions and/or renin release in mature animals also control them during fetal life. First, stimulation of the renal nerve in fetal sheep evoked renin release via the β-adrenergic system and increased renal vascular resistance with a concomitant reduction in renal blood flow (Robillard et al. 1987; Robillard and Nakamura 1988). The magnitude of these changes, however, was smaller than in adult kidneys. Bilateral renal denervation in fetal sheep (125 days gestation) decreased the basal level of active PRA and reduced isoproterenol-induced increases in prorenin and active renin levels in vivo, as well as renin mRNA in renal cortical cells (Draper et al. 2000). Second, PRA is higher in fetal lambs (mid- to end gestation) than in mothers and, as discussed above, is increased by hemorrhage (Smith et al. 1974). Ang II infusion (1 µg/h for 3 days) into the ovine fetus at midgestation increased BP and urine flow, whereas PRA and the levels of AT_1, AT_2, and renin mRNA decreased, suggesting that Ang II exerts a negative feedback control on the RAS (Mullins et al. 1990; Moritz et al. 2000). Likewise, Ang II reduced the number of renin-secretory cells equally in both newborn and adult rats.

4
Integration of Phylogeny and Ontogeny

4.1
Phylogeny and Ontogeny

The doctrine of Ernst Heinrich Haeckel that ontogeny recapitulates phylogeny and his interpretation of the gill slits of human embryos as a re-expression of ancestral adult fish features are the subject of arguments among evolutionary biologists (Gould 1977). Attempts to elucidate an intimate relationship between ontogeny and phylogeny have not had much success. In this chapter, I have described the RAS during phylogeny and ontogeny in a parallel fashion so that readers may find similarities and differences between these two categories. Although it is not my intent to discuss whether some characteristics of the RAS in the mammalian fetus/embryo may be a repeat of the adult stage of ancestral vertebrates or whether they merely represent a common feature of the early ontogeny of all vertebrate animals, comparison of phylogeny and ontogeny may provide new insight into the interpretation of existing findings and provide a perspective for future investigations. Furthermore, molecular and physiological changes observed during phylogenetic or ontogenic processes may be related to, and thus have an impact on disease-/tissue-specific gene expressions and phenotypic modulations during pathogenesis (Kintscher and Unger 2003).

4.2
Biochemistry of the Renin–Angiotensin System

At present, molecular information on renin substrate and renin in primitive vertebrates is scarce. Since Ang I and Ang II show variation in the amino acid in positions 1, 3, 5, and 9, it is expected that multiple AGT genes exists among vertebrates. Although mammalian fetuses express native AGT mRNA and/or proteins, this does not exclude the possibility that heterologous AGT genes are expressed during the early ontogeny process. As mentioned in Sect. 2.1.2, the renin-like enzyme identified in zebrafish has high homology to cathepsin D. It is possible that aspartyl proteases other than renin act as Ang I-forming enzymes in primitive vertebrates. Likewise, the RAS in some mammalian tissues differs from the RAS in the kidney and systemic circulation, bypassing the traditional cascade to form Ang (Danser 1996).

In primitive vertebrates renin-secretory cells are diffusely distributed along small arteries and arterioles in the kidney (Figs. 2 and 3). With the phylogenetic advancement of vertebrates, renin-secretory cells are localized in the JG area. A similar time-dependent change in renin expression has been recognized during ontogeny of mammalian kidneys. In the vascular system of fetal kidneys, renin-secretory cells (Gomez et al. 1989), renin transcripts (Jones et al. 1990, 2000), and renin/AGT gene expression (Gomez et al. 1988b; Fig. 2) are widely distributed. As the renal arterial tree develops, renin mRNA-containing cells are progressively localized to more distal blood vessels and finally to JG apparatus (Jones et al. 1990). A reported construct with a renin 5′-flanking sequence fused to green fluorescent protein (GFP) cDNA accurately recapitulates endogenous renin expression profiles (Jones et al. 2000). This renin/GFP transgene expression was also detected in developing VSM cells of renal arteries in mouse embryos, and the expression was localized at the JG area in adults (Jones et al. 2000). Furthermore, a diffuse distribution of renin mRNA and protein in adult rats (Gomez et al. 1988a) and the renin/GFP transgene expression in adult mice (Jones et al. 2000) similar to the fetal type have been reproduced during ACE inhibition, suggesting that adult renal arteries have the ability to express renin or recruit renin-secretory cells in response to physiological stimulation. As shown in Table 1, native Ang in various steps of mutation has been identified in nonmammalian vertebrates. It remains to be determined whether some of these nonmammalian Ang are recapitulated during mammalian fetal life.

Likewise, Ang receptors with considerable homology to the AT_1 subtype have been identified in nonmammalian vertebrates (Nishimura 2001; Table 2; Fig. 4). The degree of homology increases with vertebrate advancement, suggesting that all AT_1-homolog receptors may have evolved from the same progenitor receptor. AT_1 mRNA and protein have been detected around E14 of embryonic life (Table 4) by using AT_1 probes constructed on the basis of AT_1 structures identified in adults. It has not been investigated whether the AT_1-homolog receptors in nonmammalian vertebrates may be expressed during the early embryonic stages of mammals.

4.3
Function and Regulation

The comparison of phylogeny and ontogeny in terms of function and regulation of the RAS has several interesting aspects. First, in both the mammalian fetus and nonmammalian vertebrates, the vasopressor/vasoconstricting action of Ang evolved early. Inhibition of ACE activity decreases the basal level of BP in conscious teleost fish (Nishimura and Bailey 1982) and in human and other mammalian fetuses (Sedman et al. 1995), but usually not in mature mammals. Furthermore, circulating Ang or renin levels increase in response to the reduction in BP/BV in both primitive vertebrates and mammalian fetuses. It thus appears that the role of the RAS as a physiological regulator of BP/BV and the feedback control of renin secretion via the baroreceptor function of JG cells may be a fundamental mechanism of the RAS. In contrast, evidence suggests that, in both fetal life and primitive vertebrates, the tubular action of Ang in the control of ion/water transport and the involvement of the adrenergic nervous system in the control of renin secretion may be later additions during the ontogenic (Harris and Gomez 1997) or phylogenic (Nakamura et al. 1992) processes. This concept is supported by the facts that (1) a close spatial relationship between renin-containing cells and nerve fibers has been detected at E17 in rat fetal kidneys and (2) the density of nerve fibers around renin-secretory cells increases as renin is localized in the JG area (Pupilli et al. 1991). In teleost fish (Nishimura and Madey 1989), cytosolic Ca^{2+} may be a fundamental cellular mechanism controlling renin secretion. Also of interest is the question of whether fetal kidneys contain a Ca^{2+}-dependent mechanism similar to the one seen in adult renin-secretory cells (Fray et al. 1987). Stimulation of adrenocorticosteroid hormones by Ang II has been reported in primitive vertebrates (see Sect. 2.2.2.3). ACTH appears to be a more potent regulator for mineralocorticoid, at least in some species. It will be interesting to determine the relative potencies of ACTH and Ang II in the regulation of mineralocorticoid synthesis/secretion during the ontogeny of mammalian adrenal glands.

As discussed above, a major physiological role of Ang in developing mammals appears to be stimulation of growth and differentiation via AT_1 and AT_2 receptors (Tufro-McReddie and Gomez 1993; Hilgers et al.1997). In nonmammalian vertebrates, the growth-promoting action of Ang has been studied in only a few species. Le Noble and coworkers (1993, 1997) have demonstrated that Ang II causes pronounced angiogenesis of pre- and postcapillary vessels in the chorioallantoic membrane of chick embryos, suggesting that the Ang receptor plays a role in vascular growth and organization during avian development; the response is inhibitable by CGP42112A, but not losartan or PD123319 (Le Noble et al. 1993). Vascular endothelium growth factor (VEGF) appears to play an important role in the development of the vascular system in the avian embryo (Eichmann et al. 1998), but the effect of Ang II on VEGF gene expression or on VEGF receptors has not been determined. Ang II stimulates thymidine incorporation into cultured VSM cells from chicks; but this effect decreases with maturation/age, and no growth-promoting action of Ang is seen in adult birds (Shimada et al. 1998). The paracrine action of Ang as a growth-stimulating factor in nonmammalian species remains to be examined.

4.4
Perspectives

During development, the expression level of the RAS shows time-dependent changes (Table 4). In rat embryos, AGT is first detected at E11 of gestation. AGT mRNA or protein shows 2- to 3-fold (kidney) and 50-fold (liver) increases within 24 h of birth (Darby and Sernia 1995; Gomez et al. 1988b), remaining elevated for 3 weeks, and then it reaches a plateau at the adult level (Tufro-McReddie and Gomez 1993). Renin is first detected in rats at E15–E17 in the aorta and its major branches, extends to the renal vasculature by the end of gestation, and is localized at the vascular pole of the JG area and afferent arterioles after birth (Gomez et al. 1989). Ang receptors appear at E11 of gestation and are broadly expressed in developing tissues; their abundance increases as gestation progresses, with maximum binding near term (Tufro-McReddie and Gomez 1993). During fetal life, AT_2 receptor expression is about tenfold higher than that of AT_1; but it rapidly decreases at birth. ACE activity also shows time-dependent changes (Table 4). ACE activity shows the same substrate affinity at all ages examined, while the V_{max} of the enzyme is two- to threefold lower in fetal than adult kidneys and lungs, reflecting low ACE activity (Wallace et al. 1978, 1979).

It is interesting to note that the evolution of AGT and Ang receptors in rat kidneys occurs a few embryonic days earlier than that of renin. Does this indicate that an aspartyl protease other than renin acts as renin like activity at early gestation? Does this agree with the fact that, in teleost fish, cathepsin D appears to exert renin-like activity? The physiological significance of the progressive elevation of the RAS near term and during the neonatal period is not completely understood. The role of Ang II in differentiation and apoptosis via AT_2 receptors and its growth-promoting effect via AT_1 receptors may be more crucial for development and maturation near term and continue to play a maturation role during the neonatal period. Deletion or impairment of any component of the RAS results in serious abnormality in organogenesis (see Sect. 3.2.1.1). It has been suggested that the RAS is partly responsible for tonic vasoconstriction of renal vessels and mesangium, resulting in limited glomerular blood flow and GFR in developing animals (Robillard and Nakamura 1988). cAT_1 mRNA is also abundantly expressed (PCR and in situ hybridization) at E19 (term, 21 days), primarily on mesangial cells of glomeruli (Nishimura et al. 2003); and its expression decreases in chicks and pullets. Furthermore, cAT_1 mRNA has been detected in the adventitia and outer layer of the media of the abdominal aorta at E19 and in 2- to 3-week-old chicks, but not in adult birds. The time and site of cAT_1 mRNA expression suggest that Ang plays a role in embryonic growth and maturation.

An interesting example of the similarities of ontogeny and phylogeny can be found during the development/maturation of renal tubule transport in the rat and chick kidneys. Liu and coworkers (Liu et al. 2001) compared medullary nephrons from the kidneys of (1) quail (looped nephron; for review, Nishimura and Fan 2002), (2) neonatal rats (long-looped nephron), and (3) mature adult rats (long-looped nephron) and found the following structural and functional similarities between the first two. First, both quail and neonatal rat kidneys lack the thin ascending limb, and the entire thick limb shows characteristics of diluting segments. During maturation of rat kidneys, apoptosis occurs in the lower segment of the ascending limb (AL), where the Na^+-K^+-$2Cl^-$ cotransporter is replaced by a thin AL-specific Cl channel. Second, the descending limb of both kidneys has low water permeability and no aquaporin (AQP)1 water channel expression. Third, AQP2 is present in the collecting duct (CD) of neonatal rats and quail, but the AQP2 responses (increases in water permeability, AQP2 protein level, etc.) to vasopressin/AVT are only modest. Fourth, the thin AL and CD of neonatal rats (and presumably quail kidney) lack urea permeability and urea transporters (UT), whereas UTA_1 mRNA is expressed in mature rat kidneys. Urea recirculation utilizing exchange mechanisms between the vasa recta and tubule segments contributes to the development of an osmotic gradient in mammalian kidneys, whereas uric acid is a primary nitrogen metabolite in birds that does not contribute to enhancement of the medullary osmotic gradient. Similarly, an effective mechanism for maintaining high papillary osmolality utilizing urea recirculation has not yet developed in neonatal rats. It will be interesting to determine whether Ang is involved in these growth/differentiation processes of the renal medulla.

It is difficult to assess whether these time-dependent changes in the RAS in developing mammals are a recapitulation of the changes seen during phylogeny because the time scales involving these two biological processes are entirely different. Moreover, simple comparisons of the structural, functional, or molecular properties of mammalian fetuses and existing nonmammalian vertebrates have little meaning because most of the currently available species represent the end-products of various taxonomical branches. Hence, the collection of information from existing nonmammalian vertebrates does not necessarily reflect the true evolutionary process of modulation of the RAS. It is therefore important to select species taxonomically close to the origin of evolution and the trunk of phylogenetic advancement.

References

Aguilera G, Kapur S, Feuillan P, Sunar-Akbasak B, Bathia AJ (1994) Developmental changes in angiotensin II receptor subtypes and AT_1 receptor mRNA in rat kidney. Kidney Int 46:973–979

Aiyar N, Baker E, Pullen M, Nuthulaganti P, Bergsma DJ, Kumar C, Nambi P (1994) Characterization of a functional angiotensin II receptor in *Xenopus laevis* heart. Pharmacology 48:242–249

Anderson WG, Takei Y, Hazon N (2001) The dipsogenic effect of the renin-angiotensin system in elasmobranch fish. Gen Comp Endocrinol 124:300–307

Ardaillou R (1999) Angiotensin II receptors. J Am Soc Nephrol 10(Suppl 11):S30–S39

Armour KJ, O'Toole LB, Hazon N (1993) Mechanisms of ACTH- and angiotensin II-stimulated 1α-hydroxycorticosterone secretion in the dogfish, *Scyliorhinus canicula*. J Mol Endocrinol 10:235–244

Bailey JR, Randall DJ (1981) Renal perfusion pressure and renin secretion in the rainbow trout, *Salmo gairdneri*. Can J Zool 59:1220–1226

Bailey JR, Nishimura H (1984) Renal response of fowl to hypertonic saline infusion into the renal portal system. Am J Physiol 246 (Reg Integ Comp Physiol 15):R624–R632

Balment RJ, Carrick S (1985) Endogenous renin-angiotensin system and drinking behavior in flounder. Am J Physiol 248 (Reg Integ Comp Physiol 17):R157–R160

Balment RJ, Warne JM, Takei Y (2003) Isolation, synthesis, and biological activity of flounder [Asn^1, Ile^5, Thr^9] angiotensin I. Gen Comp Endocrinol 130:92–98

Barajas L, Latta H (1967) Structure of the juxtaglomerular apparatus. Circ Res 21(Suppl II):15–28

Barr M (1994) Teratogen update: angiotensin-converting enzyme inhibitors. Teratology 50:399–409

Battista MC, Oligny LL, St-Louis J, Brochu M (2002) Intrauterine growth restriction in rats is associated with hypertension and renal dysfunction in adulthood. Am J Physiol 283 (Endocrinol Metab):E124–E131

Bergsma DJ, Ellis C, Nuthulaganti PR, Nambi P, Scaife K, Kumar C, Aiyar N (1993) Isolation and expression of a novel angiotensin II receptor from *Xenopus laevis* heart. Mol Pharmacol 44:277–284

Bernier NJ, Gilmour KM, Takei Y, Perry SF (1999a) Cardiovascular control via angiotensin II and circulating catecholamines in the spiny dogfish, *Squalus acanthias*. J Comp Physiol B 169:237–248

Bernier NJ, McKendry JE, Perry SF (1999b) Blood pressure regulation during hypotension in two teleost species: differential involvement of the renin–angiotensin and adrenergic systems. J Exp Biol 202:1677–1690

Bernier NJ, Perry SF (1997) Angiotensins stimulate catecholamine release from the chromaffin tissue of the rainbow trout. Am J Physiol 273 (Reg Integ Comp Physiol 42):R49–R57

Best JB, Blair-West JR, Coghlan JP, Cran EJ, Fernley RT, Simpson PA (1974) A novel sequence in kangaroo angiotensin I. Clin Exper Pharm Physiol 1:171–174

Blair-West JR, Coghlan JP, Denton DA, Gibson AP, Oddie CJ, Sawyer WH, Scoggins BA (1977) Plasma renin activity and blood corticosteroids in the Australian lungfish, *Neoceratodus forsteri*. J Endocrinol 74:137–142

Bouley R, Gosselin M, Plante H, Servant G, Pérodin J, Arcand M, Guillemette G, Escher E (1997) Characterization of a specific binding site for angiotensin II in chicken liver. Can J Physiol Pharmacol 75:568–575

Breault L, Chamoux E, Lehoux JG, Gallo-Payet N (1998) The role of angiotensin II in human adrenal gland development. Endocrine Res 24(3&4):953–954

Breault L, Lehoux JG, Gallo-Payet N (1996) Angiotensin II receptors in the human adrenal gland. Endocrine Res 22(4):355–361

Brown JA, Oliver JA, Henderson IW, Jackson BA (1980) Angiotensin and single nephron glomerular function in the trout *Salmo gairdneri*. Am J Physiol 239 (Reg Integ Comp Physiol 8):R509–R514

Brown JA, Paley RK, Amer S, Aves SJ (2000) Evidence for an intrarenal renin–angiotensin system in the rainbow trout, *Oncorhynchus mykiss*. Am J Physiol 278 (Reg Integ Comp Physiol):R1685–R1691

Brown JA, Pope SK, Amer S, Cobb CS, Williamson R, Parkyn G, Aves SJ (1997) Angiotensin receptors in teleost fish glomeruli. In: Kawashima S, Kikuyama S (eds) Advances in comparative endocrinology. Proceedings of the XIIIth International Congress of Comparative Endocrinology. Monduzzi Editore, Bologna, Italy, pp 1313–1319

Brown JA, Rankin JC, Yokota SD (1993) Glomerular haemodynamics and filtration in single nephrons of nonmammalian vertebrates. In: Brown JA, Balment RJ, Rankin JC (eds) New insights in vertebrate kidney function. Society for Experimental Biology. Cambridge University Press, New York, pp 1–44

Burnstock G (1969) Evolution of the autonomic innervation of visceral and cardiovascular systems in vertebrates. Pharmacol Rev 21(4):247–324

Carroll RG (1981) Vascular response of the dogfish and sculpin to angiotensin II. Am J Physiol 240 (Reg Integ Comp Physiol 9):R139–R143

Carroll RG, Opdyke DF (1982) Evolution of angiotensin II-induced catecholamine release. Am J Physiol 243 (Reg Integ Comp Physiol 12):R65–R69

Celio MR, Groscurth P, Inagami T (1985) Ontogeny of renin immunoreactive cells in the human kidney. Anat Embryol 173:149–155

Chan MY, Holmes WN (1971) Studies on a "renin–angiotensin" system in the normal and hypophysectomized pigeon (*Columba livia*). Gen Comp Endocrinol 16:304–311

Chen X, Li W, Yoshida H, Tsuchida S, Nishimura H, Takemoto F, Okubo S, Fogo A, Matsusaka T, Ichikawa I (1997) Targeting deletion of angiotensin type 1B receptor gene in the mouse. Am J Physiol 272 (Renal Physiol 41):F299–F304

Chen M, Vallon V, Smart A, Endo Y, Schnemann J, Briggs JP (2000) Studies of the zebrafish (*Danio berio*) renin–angiotensin system. Bulletin Mt Desert Island Biol Lab 39:35–36

Cobb CS, Brown JA (1993) Characterization of putative glomerular receptors for angiotensin II in the rainbow trout *Oncorhynchus mykiss* using the antagonists losartan, PD 123177, and saralasin. Gen Comp Endocrinol 92:123–131

Cobb CS, Williamson R, Brown JA (1999) Angiotensin II-induced calcium signalling in isolated glomeruli from fish kidney (*Oncorhynchus mykiss*) and effects of losartan. Gen Comp Endocrinol 113:312–321

Conlon JM, Yano K, Olson KR (1996) Production of [Asn^1, Val^5]angiotensin II and [Asp^1, Val^5]angiotensin II in kallikrein-treated trout plasma (T60 K). Peptides 17:527–530

Corvol P, Schütz S, Gasc JM (1998) Early expression of all components of the renin–angiotensin system in human development. Adv Nephrol Necker Hosp 28:195–212

Corwin E, Malvin GM, Katz S, Malvin RL (1984) Temperature sensitivity of the renin–angiotensin system in *Ambystoma tigrinum*. Am J Physiol 246 (Reg Integ Comp Physiol 15):R510–R515

Coulter CL, Myers DA, Nathanielsz PW, Bird IM (2000) Ontogeny of angiotensin II type 1 receptor and cytochrome P450$_{c11}$ in the sheep adrenal gland. Biol Reprod 62:714–719

Danser AHJ (1996) Local renin–angiotensin systems. Mol Cell Biochem 157:211–216

Darby IA, Sernia C (1995) In situ hybridization and immunohistochemistry of renal angiotensinogen in neonatal and adult rat kidneys. Cell Tissue Res 281:197–206

Davis JO, Freeman RH (1976) Mechanisms regulating renin release. Am Physiol Society 56:1–56

de Gasparo M, Levens NR, Kamber B, Furet P, Whitebread S, Brechler V, Bottari SP (1994) The angiotensin II AT_2 receptor subtype. In: Saavedra JM, Timmermans PBMWM (eds) Angiotensin receptors. Plenum Press, New York, pp 95–117

Draper ML, Wang J, Valego N, Block WA, Rose JC (2000) Effect of renal denervation on renin gene expression, concentration, and secretion in mature ovine fetus. Am J Physiol 279 (Reg Integ Comp Physiol):R263–R270

Eichmann A, Corbel C, Le Douarin NM (1998) Segregation of the embryonic vascular and hemopoietic systems. Biochem Cell Biol 76:939–946

Ei-Haddad MA, Chao CR, Ma SX, Ross MG (2000) Nitric oxide modulates angiotensin II-induced drinking behavior in the near-term ovine fetus. Am J Obstet Gynecol 182:713–719

Erspamer V, Melchiorri P, Nakajima T, Yasuhara T, Endean R (1979) Amino acid composition and sequence of Crinia-angiotensin, an angiotensin II-like endecapeptide from the skin of the Australian frog *Crinia georgiana*. Experientia 35:1132–1133

Esther CR, Howard TE, Marino EM, Goddard JM, Capecchi MR, Bernstein KE (1996) Mice lacking angiotensin-converting enzyme have low blood pressure, renal pathology, and reduced male fertility. Lab Invest 74:953–965

Forhead AJ, Melvin R, Balouzet V, Fowden AL (1998) Developmental changes in plasma angiotensin-converting enzyme concentration in fetal and neonatal lambs. Reprod Fertil Dev 10:393–398

Fraser R, Brown JJ, Lever AF, Mason PA, Robertson JIS (1979) Control of aldosterone secretion. Clin Sci 56:389–399

Fraser R, Lantos CP (1978) 18-Hydroxycorticosterone: a review. J Steroid Biochem 9:273–286

Fray JCS, Park CS, Valentine AND (1987) Calcium and the control of renin secretion. Endocrine Rev 8:53–93

Friberg P, Sundelin B, Bohman S-O, Bobik A, Nilsson H, Wickman A, Gustafsson H, Petersen J, Adams MA (1994) Renin-angiotensin system in neonatal rats: Induction of a renal abnormality in response to ACE inhibition or angiotensin II antagonism. Kidney Int 45:485–492

Fuchs S, Germain S, Philippe J, Corvol P, Pinet F (2002). Expression of renin in large arteries outside the kidney revealed by human renin promoter/LacZ transgenic mouse. Am J Pathol 161:717–725

Fuentes J, Eddy FB (1998) Cardiovascular responses in vivo to angiotensin II and the peptide antagonist saralasin in rainbow trout Oncorhynchus mykiss. J Exp Biol 201:267–272

Gomez RA (1998) Role of angiotensin in renal vascular development. Kidney Int 54 (Suppl 67):S12–S16

Gomez RA, Lynch KR, Chevalier RL, Everett AD, Johns DW, Wilfong N, Peach MJ, Carey RM (1988a) Renin and angiotensinogen gene expression and intrarenal renin distribution during ACE inhibition. Am J Physiol 254 (Renal Fluid Electrolyte Physiol 23):F900–F906

Gomez RA, Lynch KR, Chevalier RL, Wilfong N, Everett A, Carey RM, Peach MJ (1988b) Renin and angiotensinogen gene expression in maturing rat kidney. Am J Physiol 254 (Renal Fluid Electrolyte Physiol 23):F582–F587

Gomez RA, Lynch KR, Sturgill BC, Elwood JP, Chevalier RL, Carey RM, Peach MJ (1989) Distribution of renin mRNA and its protein in the developing kidney. Am J Physiol 257 (Renal Fluid Electrolyte Physiol 26):F850–F858

Gomez RA, Pupilli C, Everett AD (1991) Molecular and cellular aspects of renin during kidney ontogeny. Pediatr Nephrol 5:80–87

Gomez RA, Robillard JE (1984) Developmental aspects of the renal responses to hemorrhage during converting-enzyme inhibition in fetal lambs. Circ Res 54:301–312

Gould SJ (1977) Prospectus. In: Gould SJ (ed) Ontogeny and phylogeny. Cambridge, London, pp 1–9

Griendling KK, Ushio-Fukai M, Lassègue B, Alexander RW (1997) Angiotensin II signaling in vascular smooth muscle: new concepts. Hypertension 29:366–373

Harris JM, Gomez RA (1997) Renin–angiotensin system genes in kidney development. Microsc Res Tech 39(3):211–221

Hasegawa Y, Cipolle M, Watanabe TX, Nakajima T, Sokabe H, Zehr JE (1984) Chemical structure of angiotensin in the turtle, *Pseudemys scripta*. Gen Comp Endocrinol 53:159–162

Hasegawa Y, Nakajima T, Sokabe H (1983a) Chemical structure of angiotensin formed with kidney renin in the Japanese eel, *Anguilla japonica*. Biomed Res 4:417–420

Hasegawa K, Nishimura H, Khosla MC (1993) Angiotensin II-induced endothelium-dependent relaxation of fowl aorta. Am J Physiol 264 (Reg Integ Comp Physiol 33):R903–R911

Hasegawa Y, Watanabe TX, Sokabe H, Nakajima T (1983b) Chemical structure of angiotensin in the bullfrog *Rana catesbeiana*. Gen Comp Endocrinol 50:75–80

Hayashi T, Nakayama T, Nakajima T, Sokabe H (1978) Comparative studies on angiotensins. V. Structure of angiotensin formed by the kidney of Japanese goosefish and its identification by Dansyl method. Chem Pharm Bull 26:215–219

Hazon N, Cerra MC, Tierney ML, Tota B, Takei Y (1997) Elasmobranch renin angiotensin system and the angiotensin receptor. In: Kawashima S, Kikuyama S (eds) Advances in comparative endocrinology. Proceedings of the XIIIth International Congress of Comparative Endocrinology. Monduzzi Editore, Bologna, Italy, pp 1307–1312

Hazon N, Tierney ML, Takei Y(1999) Renin–angiotensin system in elasmobranch fish: a review. J Exp Zool 284:526–534

Henderson IW, Jotisankasa V, Mosley W, Oguri M (1976) Endocrine and environmental influences upon plasma cortisol concentrations and plasma renin activity of the eel, *Anguilla anguilla* L. J Endocrinol 70:81–95

Henderson IW, Brown JA, Balment RJ (1993) The renin–angiotensin system and volume homeostasis. In: Brown JA, Balment RJ, Rankin JC (eds) New insights in vertebrate kidney function. Society for Experimental Biology seminar series 52. Cambridge University Press, Cambridge, pp 311–350

Hilgers KF, Norwood VF, Gomez RA (1997) Angiotensin's role in renal development. Sem Nephrol 17(5):492–501

Holmes WN, Cronshaw J, Rohde KE (1991) The steroidogenic responsiveness in vitro of adrenal gland tissue from the neonatal mallard duck (*Anas platyrhynchos*). Gen Comp Endocrinol 82:217

Horiuchi M, Akishita M, Dzau VJ (1999) Recent progress in angiotensin II type 2 receptor research in the cardiovascular system. Hypertension 33:613–621

Hubert C, Gasc JM, Berger S, Schütz G, Corvol P (1999) Effects of mineralocorticoid receptor gene disruption on the components of the renin-angiotensin system in 8-day old mice. Molecular Endocrinol 13:297–306

Ingelfinger JR, Woods LL (2002) Perinatal programming, renal development, and adult renal function. AJH 15:46s–49s

Ji H, Sandberg K, Zhang Y, Catt KJ (1993) Molecular cloning, sequencing and functional expression of an amphibian angiotensin II receptor. Biochem Biophys Res Commun 194:756–762

Jones CA, Hurley MI, Black TA, Kane CM, Pan L, Pruitt SC, Gross KW (2000) Expression of a renin/GFP transgene in mouse embryonic, extra-embryonic, and adult tissues. Physiol Genomics 4:75–81

Jones CA, Sigmund CD, McGowan RA, Kane-Haas CM, Gross KW (1990) Expression of murine renin genes during fetal development. Molecular Endocrinol 4:375–383

Joss JMP, Itahara Y, Watanabe TX, Nakajima K, Takei Y (1999) Teleost-type angiotensin is present in Australian lungfish, Neoceratodus forsteri. Gen Comp Endocrinol 114: 206–212

Kambayashi Y, Bardhan S, Takahashi K, Tsuzuki S, Inui H, Hamakubo T, Inagami T (1993) Molecular cloning of a novel angiotensin II receptor isoform involved in phosphotyrosine phosphatase inhibition. J Biol Chem 268:24543–24546

Kempf H, Le Moullec JM, Corvol P, Gasc, JM (1996) Molecular cloning, expression and tissue distribution of a chicken angiotensin II receptor. FEBS Lett 399:198–202

Khosla MC, Bumpus FM, Nishimura H, Opdyke DF, Coviello A (1983) Synthesis of nonmammalian angiotensins and their comparative pressor properties in dogfish shark, domestic chicken, and rat. Hypertension 5 (Suppl 5):V22–V28

Khosla MC, Nishimura H, Hasegawa Y, Bumpus FM (1985) Identification and synthesis of [1-asparagine, 5-valine, 9-glycine]-angiotensin I produced from plasma of American eel, Anguilla rostrata. Gen Comp Endocrinol 57:223–233

Khosla MC, Smeby RR, Bumpus FM (1974) Structure-activity relationship in angiotensin II analogs. In: Page IH, Bumpus FM (eds) Angiotensin, vol 37. Handbuch der Experimentellen Pharmakologie (Handbook of experimental pharmacology). Springer-Verlag, New York, pp 126–161

Kingdom JCP, Hayes M, McQueen J, Howatson AG, and Lindop GBM (1999) Intrauterine growth restriction is associated with persistent juxtamedullary expression of renin in the fetal kidney. Kidney Int 55:424–429

Kintscher U, Unger T (2003) Angiotensin II receptor expression:from maturation to pathogenesis. Am J Physiol (Reg Integ Comp Physiol) 285:R26–R27

Kirby S, Burnstock G (1969) Comparative pharmacological studies of isolated spiral strips of large arteries from lower vertebrates. Comp Biochem Physiol 28:307–319

Klingbeil CK (1985) Corticosterone and aldosterone dose-dependent responses to ACTH and angiotensin II in the duck (Anas platyrhynchos). Gen Comp Endocrinol 59:382–390

Kobayashi H, Takei Y (1996) The renin angiotensin system. Comparative aspects. In: Bradshaw SD, Burggren W, Heller HC, Ishii S, Langer H, Neuweiler G, Randall DJ (eds) Zoophysiology, vol 35. Springer Verlag, Berlin/Heidelberg, pp 1–245

Kobayashi H, Uemura H, Takei Y (1980) Physiological role of the renin–angiotensin system during dehydration. In: Epple A, Stetson MH (eds) Avian endocrinology. Academic Press, New York, pp 319–330

Kurtz A, Wagner C (1999) Cellular control of renin secretion. J Exp Biol 202:219–225

Lacy ER, Reale E (1990) The presence of a juxtaglomerular apparatus in elasmobranch fish. Anat Embryol 182:249–262

Langley-Evans, SC (2001) Fetal programming of cardiovascular function through exposure to maternal undernutrition. Proc Nutrition Soc 60:505–513

Le Noble FAC, Schreurs NHJS, Van Straaten HWM, Slaaf DW, Smits JFM, Rogg H, Struijker-Boudier HAJ (1993) Evidence for a novel angiotensin II receptor involved in angiogenesis in chick embryo chorioallantoic membrane. Am J Physiol 264 (Reg Integ Comp Physiol 33):R460–R465

Le Noble F, Smits J, Struijker-Boudier H (1997) Hypertension, the renin–angiotensin system and vascular development. In: Kawashima S, Kikuyama S (eds) Advances in comparative endocrinology. Proceedings of the XIIIth International Congress of Comparative Endocrinology. Monduzzi Editore, Bologna, Italy, pp 1321–1327

Liu W, Morimoto T, Kondo Y, Iinuma K, Uchida S, Imai M (2001) "Avian-type" renal medullary tubule organization causes immaturity of urine-concentrating ability in neonates. Kidney Int 60:680–693

Lorenz JN, Weihprecht H, Schnermann J, Skøtt O, Briggs JP (1991) Renin release from isolated juxtaglomerular apparatus depends on macula densa chloride transport. Am J Physiol 260 (Renal Fluid Electrolyte Physiol):F486–F493

Marsigliante S, Muscella A, Vinson GP, Storelli C (1997) Angiotensin II receptors in the gill of sea water- and freshwater-adapted eel. J Molecular Endocrinol 18:67–76

Marsigliante S, Verri T, Barker S, Jimenez E, Vinson GP, Storelli C (1994) Angiotensin II receptor subtypes in eel (*Anguilla anguilla*). J Mol Endocrinol 12:61–69

McMillen IC, Adams MB, Ross JT, Coulter CL, Simonetta G, Owens JA, Robinson JS, Edwards LJ (2001) Fetal growth restriction: adaptations and consequences. Reproduction 122:195–204

Moritz K, Koukoulas I, Albiston A, Wintour EM (2000) Angiotensin II infusion to the midgestation ovine fetus: effects on the fetal kidney. Am J Physiol 279 (Reg Integ Comp Physiol):R1290–R1297

Mounier F, Hinglais N, Sich M, Gros F, Lacoste M, Deris Y, Alhenc-Gelas F, Gubler MC (1987) Ontogenesis of angiotensin-I converting enzyme in human kidney. Kidney Int 32:684–690

Mukoyama M, Nakajima M, Horiuchi M, Sasamura H, Pratt RE, Dzau VJ (1993) Expression cloning of type 2 angiotensin II receptor reveals a unique class of seven-transmembrane receptors. J Biol Chem 268:24539–24542

Mullins JJ, Peters J, Ganten D (1990) Fulminant hypertension in transgenic rats harbouring the mouse *Ren*-2 gene. Nature 344:541–544

Murphy TJ, Alexander RW, Griendling KK, Runge MS, Bernstein KE (1991) Isolation of a cDNA encoding the vascular type-1 angiotensin II receptor. Nature 351:233–236

Murphy TJ, Nakamura Y, Takeuchi K, Alexander RW (1993) Accelerated Communication: a cloned angiotensin receptor isoform from the turkey adrenal gland is pharmacologically distinct from mammalian angiotensin receptors. Mol Pharmacol 44:1–7

Nakajima M, Hutchinson HG, Fujinaga M, Hayashida W, Morishita R, Zhang L, Horiuchi M, Pratt RE, Dzau VJ (1995) The angiotensin II type 2 (AT_2) receptor antagonizes the growth effects of the AT_1 receptor: gain of function study using gene transfer. Proc Natl Acad Sci USA 92:10663–10667

Nakamura Y, Madey MA, Nishimura H, Quach D, Barajas L (1992) Lack of control of renin release by adrenergic nervous system in the aglomerular toadfish. Gen Comp Endocrinol 88:62–75

Nakamura Y, Nishimura H, Khosla MC (1982) Vasodepressor action of angiotensin in conscious chickens. Am J Physiol 243 (Heart Circ Physiol 12):H456–H462

Nakayama T, Nakajima T, Sokabe H (1973) Comparative studies on angiotensins. III. Structure of fowl angiotensin and its identification by DNS-method. Chem Pharm Bull 21:2085–2087

Nakayama T, Nakajima T, Sokabe H (1977) Comparative studies on angiotensin IV. Structure of snake (*Elaphe climocophora*) angiotensin. Chem Pharm Bull 25:3255–3260

Niimura F, Labosky PA, Kakuchi J, Okubo S, Yoshida H, Oikawa T, Ichiki T, Naftilan AJ, Fogo A, Inagami T, Hogan BLM, Ichikawa I (1995) Gene targeting in mice reveals a requirement for angiotensin in the development and maintenance of kidney morphology and growth factor regulation. J Clin Invest 96:2947–2954

Nishimura H (1980) Comparative endocrinology of renin and angiotensin. In: Johnson JA, Anderson RR (eds) The renin–angiotensin system. Plenum Press, New York, pp 29–77

Nishimura H (1985a) Evolution of the renin–angiotensin system and its role in control of cardiovascular function in fishes. In: Foreman RE, Gorbman A, Dodd JM, Olsson R (eds) Evolutionary biology of primitive fishes. Plenum Press, New York, pp 275–293

Nishimura H (1985b) Endocrine control of renal handling of solutes and water in vertebrates. Renal Physiol (Basel) 8:279–300

Nishimura H (1987) Role of the renin–angiotensin system in osmoregulation. In: Pang PKT, Schreibman MP (eds) Vertebrate endocrinology: fundamentals and biomedical implications. Academic Press, New York, pp 157–187

Nishimura H (2001) Angiotensin receptors—evolutionary overview and perspectives. Comp Biochem Physiol (Part A) 128:11–30

Nishimura H, Bailey JR (1982) Intrarenal renin–angiotensin system in primitive vertebrates. Kidney Int 22 (Suppl 12):S185–S192

Nishimura H, Crofton JT, Norton VM, Share L (1977) Angiotensin generation in teleost fish determined by radioimmunoassay and bioassay. Gen Comp Endocrinol 32:236–247

Nishimura H, Fan Z (2002) Sodium and water transport and urine concentration in avian kidney. In: Hazon N, Flik G (eds) Osmoregulation and drinking in vertebrates. Experimental Biology Reviews. Bios Scientific, Oxford, pp 129–151

Nishimura H, Lunde LG, Zucker A (1979) Renin response to hemorrhage and hypotension in the aglomerular toadfish Opsanus tau. Am J Physiol 237 (Heart Circ Physiol 6):H105–H111

Nishimura H, Madey MA (1989) Signals controlling renin release in aglomerular toadfish. Fish Physiol Biochem 7:323–329

Nishimura H, Madey MA, Mugaas JN, Khosla MC, Crofton JT (1981a) Radioimmunoassay of fowl angiotensin I. Gen Comp Endocrinol 45:262–272

Nishimura H, Nakamura Y, Sumner RP, Khosla MC (1982) Vasopressor and depressor actions of angiotensin in the anesthetized fowl. Am J Physiol 242 (Heart Circ Physiol 11):H314–H324

Nishimura H, Nakamura Y, Taylor AA, Madey MA (1981b) Renin–angiotensin and adrenergic mechanisms in control of blood pressure in fowl. Hypertension 3 (Suppl 1):I-41–I-49

Nishimura H, Norton VM, Bumpus FM (1978) Lack of specific inhibition of angiotensin II in eels by angiotensin antagonists. Am J Physiol 235 (Heart Circ Physiol 4):H95–H103

Nishimura H, Ogawa M, Sawyer WH (1973) Renin–angiotensin system in primitive bony fishes and a holocephalian. Am J Physiol 224:950–956

Nishimura H, Walker OE, Patton CM, Madison AB, Chiu AT, Keiser J (1994) Novel angiotensin receptor subtypes in fowl. Am J Physiol 267 (Reg Integ Comp Physiol 36):R1174–R1181

Nishimura H, Yerkos E, Hohenfollner K, Miyazaki Y, Ma J, Hunley TE, Yoshida H, Ichiki T, Threadgill D, Phillips JA III, Hogan BML, Fogo A, Brock JW III, Inagami T, Ichikawa I (1999) Role of the angiotensin type 2 receptor gene in congenital anomalies of the kidney and urinary tract, CAKUT, of mice and men. Mol Cell 3:1–10

Nishimura H, Yang Y, Hubert C, Gasc JM, Ruijtenbeek K, De Mey J, Struijiker Boudier HAJ, Corvol P (2003) Maturation-dependent changes of angiotensin receptor expression in fowl. Am J Physiol 285 (Reg Integ Comp Physiol):R231–R242

Norwood VF, Craig MR, Harris JM, Gomez RA (1997) Differential expression of angiotensin II receptors during early renal morphogenesis. Am J Physiol 272 (Reg Integ Comp Physiol 41):R662–R668

Nuyt AM, Lenkei Z, Corvol P. Palkovits M, Llorens-Cortés C (2001) Ontogeny of angiotensin II type 1 receptor mRNAs in fetal and neonatal rat brain. J Comp Neurol 440:192–203

Okubo S, Niimura F, Matsusaka T, Fogo A, Hogan BLM, Ichikawa I (1998) Angiotensinogen gene null-mutant mice lack homeostatic regulation of glomerular filtration and tubular reabsorption. Kidney Int 53:617–625

Olson KR, Chavez A, Conklin DJ, Cousins KL, Farrell AP, Ferlic R, Keen JE, Kne T, Kowalski KA, Veldman T (1994) Localization of angiotensin II responses in the trout cardiovascular system. J Exp Biol 194:117–138

Opdyke DF, Holcombe R (1976) Response to angiotensins I and II and to AI-converting-enzyme inhibitor in a shark. Am J Physiol 231:1750–1753

Oudit GY, Butler DG (1995) Angiotensin II and cardiovascular regulation in a freshwater teleost, Anguilla rostrata LeSueur. Am J Physiol 269 (Reg Integ Comp Physiol 38):R726–R735

Parkyn G, Aves SJ, Brown JA (1997) Cloning and characterisation of the rainbow trout angiotensin receptor. In: Kawashima S, Kikuyama S (eds) Advances in comparative endocrinology. Proceedings of the XIIIth International Congress of Comparative Endocrinology. Monduzzi Editore, Bologna, pp 1329–1332

Pan L, Xie Y, Black TA, Jones CA, Pruitt SC, Gross KW (2001) An Abd-B class HOX·PBX recognition sequence is required for expression from the mouse $Ren-1^c$ gene. J Biol Chem 276:32489–32494

Pérodin J, Bossé R, Gagnon S, Zhou LM, Bouley R, Leduc R, Escher E (1996) Structure-activity relationship of the agonist-antagonist transition on the type 1 angiotensin II receptor; the search for inverse agonists. In: Raizada MK, Philips MI, Sumners C (eds) Recent advances in cellular and molecular aspects of angiotensin receptors, vol 396. Advances in experimental medicine and biology. Plenum Press, New York, pp 131–143

Perrott MN, Balment RJ (1990) The renin-angiotensin system and the regulation of plasma cortisol in the flounder, *Platichthys flesus*. Gen Comp Endocrinol 78:414–420

Pucell AG, Hodges JC, Sen I, Bumpus FM, Husain A (1991) Biochemical properties of the ovarian granulosa cell type 2-angiotensin II receptor. Endocrinology 128:1947–1959

Pupilli C, Gomez RA, Tuttle JB, Peach MJ, Carey RM (1991) Spatial association of renin containing cells and nerve fibers in developing rat kidney. Pediatr Nephrol 5:690–695

Qin ZL, Yan HQ, Nishimura H (1999) Vascular angiotensin II receptor and calcium signaling in toadfish. Gen Comp Endocrinol 115:122–131

Robillard JE, Gomez RA, Vanorden D, Smith FG (1982) Comparison of the adrenal and renal responses to angiotensin II in fetal lambs and adult sheep. Circ Res 50:140–147

Robillard JE, Nakamura KT (1988) Neurohormonal regulation of renal function during development. Am J Physiol 254 (Renal Fluid Electrolyte Physiol 23):F771–F779

Robillard JE, Nakamura KT, Wilkin MK, McWeeny OJ, Dibona GF (1987) Ontogeny of renal hemodynamic response to renal nerve stimulation in sheep. Am J Physiol 252 (Renal Fluid Electrolyte Physiol 21):F605–F612

Russell MJ, Klemmer AM, Olson KR (2001) Angiotensin signaling and receptor types in teleost fish. Comp Biochem Physiol Part A 128:41–51

Sandberg K (1994) Structural analysis and regulation of angiotensin II receptors. Trends Endocrinol Metab 5:28–35

Sandberg K, Ji H, Millan MA, Catt KJ (1991) Amphibian myocardial angiotensin II receptors are distinct from mammalian AT_1 and AT_2 receptor subtypes. FEBS Lett 284:281–284

Sasaki K, Yamano Y, Bardhan S, Iwai N, Murray JJ, Hasegawa M, Matsuda Y, Inagami T (1991) Cloning and expression of a complementary DNA encoding a bovine adrenal angiotensin II type 1 receptor. Nature 351:230–232

Sedman AB, Kershaw DB, Bunchman TE (1995) Recognition and management of angiotensin converting enzyme inhibitor fetopathy. Pediatr Nephrol 9:382–385

Schütz S, Le Moullec JM, Corvol P, Gasc JM (1996) Early expression of all components of the renin-angiotensin-system in human development. Am J Pathol 149:2067–2079

Shanmugam S, Corvol P, Gasc JM (1994) Ontogeny of the two angiotensin II type 1 receptor subtypes in rats. Am J Physiol 267 (Endocrinol Metab 30):E828–E836

Shanmugam S, Corvol P, Gasc JM (1996) Angiotensin II type 2 receptor mRNA expression in the developing cardiopulmonary system of the rat. Hypertension 28:91–97

Shanmugam S, Llorens-Cortes C, Clauser E, Corvol P, Gasc JM (1995) Expression of angiotensin II AT_2 receptor mRNA during development of rat kidney and adrenal gland. Am J Physiol 268 (Renal Fluid Electrolyte Physiol 37):F922–F930

Shimada T, Fabian M, Yan HQ, Nishimura H (1998) Control of vascular smooth muscle cell growth in fowl. Gen Comp Endocrinol 112:115–128

Smith RD (1996) Atypical (non-AT_1, non-AT_2) angiotensin receptors. In: Raizada MK, Phillips MI, Sumners C (eds) Recent advances in cellular and molecular aspects of angiotensin receptors. Advances in experimental medicine and biology, vol 396. Plenum Press, New York, pp 237–245

Smith FG, Lupu AN, Barajas L, Bauer R, Bashore RA (1974) The renin-angiotensin system in the fetal lamb. Pediatr Res 8:611–620

Sokabe H, Ogawa M (1974) Comparative studies of the juxtaglomerular apparatus. Int Rev Cytol 37:271–327

Sokabe H, Ogawa M, Oguri M, Nishimura H (1969) Evolution of the juxtaglomerular apparatus in the vertebrate kidneys. Texas Rept Biol Med 27:867–885

Stallone JN, Nishimura H, Khosla MC (1989) Angiotensin II vascular receptors in fowl aorta: binding specificity and modulation by divalent cations and guanine nucleotides. J Pharmacol Exp Ther 251:1076–1082

Stallone JN, Nishimura H, Nasjletti A (1990) Angiotensin II binding sites in aortic endothelium of domestic fowl. Am J Physiol 258 (Reg Integ Comp Physiol 27):R777–R782

Stephens GA (1984) Angiotensin and norepinephrine effects on isolated vascular strips from a reptile. Gen Comp Endocrinol 54:175–180

Stephens GA, Creekmore JS (1984) Response of plasma renin activity to hypotension and angiotensin converting enzyme inhibitor in the turtle. J Comp Physiol B 154:287–294

Takei Y (1977) The role of the subfornical organ in drinking induced by angiotensin in the Japanese quail, *Coturnix coturnix japonica*. Cell Tissue Res 185:175–181

Takei Y (1999) Novel angiotensin and angiotensin receptors in cartilaginous fish. In: Symposium; Angiotensin receptors and signaling: evolution and perspectives, Experimental Biology, Washington, DC, USA

Takei Y (2000) Comparative physiology of body fluid regulation in vertebrates with special reference to thirst regulation. Jap J Physiol 50:171–186

Takei Y, Hasegawa Y (1990) Vasopressor and depressor effects of native angiotensins and inhibition of these effects in the Japanese quail. Gen Comp Endocrinol 79:12–22

Takei Y, Hasegawa Y, Watanabe TX, Nakajima K, Hazon N (1993a) A novel angiotensin I isolated from an elasmobranch fish. J Endocrinol 139:281–285

Takei Y, Itahara Y, Butler DG, Watanabe TX, Oudit GY (1998) Tetrapod-type [Asp^1] angiotensin is present in a holostean fish, *Amia calva*. Gen Comp Endocrinol 110:140–146

Takei Y, Silldorff EP, Hasegawa Y, Watanabe TX, Nakajima K, Stephens GA, Sakakibara S (1993b) New angiotensin I isolated from a reptile, *Alligator mississippiensis*. Gen Comp Endocrinol 90:214–219

Takei Y, Stallone JN, Nishimura H, Campanile CP (1988) Angiotensin II receptors in the fowl aorta. Gen Comp Endocrinol 69:205–216

Takei Y, Joss JMP, Kloas W, Rankin JC (2003) Isolation and sequencing of angiotensin I from evolutionarily unique vertebrate species. Gen Comp Endocrinol (in press)

Takemoto Y, Nakajima T, Hasegawa Y, Watanabe TX, Sokabe H, Kumagae S, Sakakibara S (1983) Chemical structures of angiotensins formed by incubating plasma with the kidney and the corpuscles of Stannius in the chum salmon, *Oncorhynchus keta*. Gen Comp Endocrinol 51:219–227

Tebbs C, Pratten MK, Pipkin FB (1999) Angiotensin II is a growth factor in the peri-implantation rat embryo. J Anat 195:75–86

Tierney ML, Luke G, Cramb G, Hazon N (1995) The role of the renin-angiotensin system in the control of blood pressure and drinking in the European eel, *Anguilla anguilla*. Gen Comp Endocrinol 100:39–48

Tierney M, Takei Y, Hazon N (1997) The presence of angiotensin II receptors in elasmobranchs. Gen Comp Endocrinol 105:9–17

Tran van Chuoi M, Dolphin CT, Barker S, Clark AJ, Vinson GP (1999) Molecular cloning and characterization of the cDNA encoding the angiotensin II receptor of European eel *Anguilla anguilla*. GenBank Data Base AJ005132

Traynor TR, Smart A, Briggs JP, Schnermann J (1999) Inhibition of macula densa-stimulated renin secretion by pharmacological blockade of cyclooxygenase-2. Am J Physiol 277 (Renal Physiol 46):F706–F710

Tsuchida S, Matsusaka T, Chen X, Okubo S, Niimura F, Nishimura H, Fogo A, Utsunomiya H, Inagami T, Ichikawa I (1998) Murine double nullizygotes of the angiotensin type 1A and 1B receptor genes duplicate severe abnormal phenotypes of angiotensinogen nullizygotes. J Clin Invest 101:755–760

Tufro-McReddie A, Gomez RA (1993) Ontogeny of the renin–angiotensin system. Sem Nephrol 13:519–530

Tufro-McReddie A, Harrison JK, Everett AD, Gomez RA (1993) Ontogeny of type 1 angiotensin II receptor gene expression in the rat. J Clin Invest 91:530–537

Tufro-McReddie A, Johns DW, Geary KM, Dagli H, Everett AD, Chevalier RL, Carey RM, Gomez RA (1994) Angiotensin II type 1 receptor: role in renal growth and gene expression during normal development. Am J Physiol 266 (Renal Fluid Electrolyte Physiol 35):F911–F918

Tufro-McReddie A, Romano LM, Harris JM, Ferder L, Gomez RA (1995) Angiotensin II regulates nephrogenesis and renal vascular development. Am J Physiol 269 (Renal Fluid Electrolyte Physiol 38):F110–F115

Vallon V, Chen M, Schnermann J, Briggs J (2000) Cloning and expression of an angiotensinogen-like gene in zebrafish. FASEB J 14:A140

Vinson GP, Whitehouse BJ, Goddard C, Sibley CP (1979) Comparative and evolutionary aspects of aldosterone secretion and zona glomerulosa function. J Endocrinol 81:5–24

Walker OE, Nishimura H, Taylor A, Lester B (1993) Novel angiotensin receptor subtypes in domestic fowl. FASEB J 7:A406

Wallace KB, Bailie MD, Hook JB (1978) Angiotensin-converting enzyme in developing lung and kidney. Am J Physiol 234 (Reg Integ Comp Physiol 3):R141–R145

Wallace KB, Bailie MD, Hook JB (1979) Development of angiotensin-converting enzyme in fetal rat lungs. Am J Physiol 236 (Reg Integ Comp Physiol 5):R57–R60

Wideman RF, Nishimura H, Bottje WG, Glahn RP (1993) Reduced renal arterial perfusion pressure stimulates renin release from domestic fowl kidneys. Gen Comp Endocrinol 89:405–414

Wintour EM, Moritz K, Butkus A, Baird R, Albiston A, Tenis N (1999) Ontogeny and regulation of the AT_1 and AT_2 receptors in the ovine fetal adrenal gland. Mol Cell Endocrinol 157:161–170

Woods LL (2000) Fetal origins of adult hypertension: a renal mechanism? Nephrol Hypertension 9:419–425

Woods LL, Rasch R (1998) Perinatal Ang II programs adult blood pressure, glomerular number, and renal function in rats. Am J Physiol 275 (Reg Integ Comp Physiol 44):R1593–R1599

Yamaguchi K, Nishimura H (1988) Angiotensin II-induced relaxation of fowl aorta. Am J Physiol 255 (Reg Integ Comp Physiol 24):R591–R599

Yosipiv IV, El-Dahr SS (1996) Activation of angiotensin-generating systems in the developing rat kidney. Hypertension 27:281–286

Zucker A, Nishimura H (1981) Renal responses to vasoactive hormones in the aglomerular toadfish, *Opsanus tau*. Gen Comp Endocrinol 43:1–9

Angiotensin as a Cytokine Implicated in Accelerated Cellular Turnover

P. Hamet[1] · S. N. Orlov[1] · D. deBlois[1] · Y. Sun[1] · V. Kren[2] · J. Tremblay[1]

[1] Centre hospitalier de l'Université de Montréal (CHUM), 3850 St. Urbain Street, Montreal, QC, H2W 1T7, Canada
e-mail: pavel.hamet@umontreal.ca

[2] Institute of Biology and Medical Genetics, First Faculty of Medicine, Charles University, Prague, Czech Republic

1	Introduction	73
2	Abnormal Proliferation/Apoptosis in Hypertensive Neonates	74
3	Role of Cell Cycle Genes and Chaperones	77
4	Genetics of Growth-Related Phenotypes	79
5	Replicative Senescence Versus Accelerated Aging	79
6	Biomarkers of Senescence	81
7	Role of Ion Transporters and Intracellular Monovalent Cations	81
8	Antihypertensive Therapy, Cell Growth, and Apoptosis	86
References		89

Abstract Several models of essential hypertension have revealed abnormalities in pathways regulating cell proliferation and programmed cell death (apoptosis). The increased proliferative phenotype found as early as birth in hypertensives is accompanied by age-dependent alterations in apoptosis, contributing to neonatal hyperplasia of the heart, aorta, and kidneys. During the course of life, accelerated cell turnover occurs and is modifiable by antihypertensive therapy, notably by inhibitors of the renin-angiotensin system. We consider the hypothesis that hypertension may be a case of accelerated aging. Part of this process may involve the defective regulation of cell proliferation in cardiovascular target organs via a putative specific senescence pathway. Candidates include abnormalities in cell cycle control genes, the renin angiotensin pathway, and regulation of the telomerase pathway. Abnormal activity of angiotensin II regulated Na^+ transporters and augmented production of endogenous ouabain-like substances have been detected in experimental models of primary hypertension. Recent data show that both ouabain and intracellular Na^+ are involved in the regulation of gene expression and apoptosis. The relevance of neonatal and early life development as a predictor of cardiovascular disease outcomes later in life is an intriguing issue that remains to be better defined. In this regard, understanding the complex genetic and epigenetic influences contributing to aging and age-related diseases will be a major goal. Because phenotype development can be analyzed longitudinally during the course of life in recombinant inbred rat strains, these models will allow a systematic approach to the molecular analysis of senescence pathway regulation, their determinants early in life, and their control by hereditary and epigenetic factors, including pharmacotherapy.

Keywords Hypertension · Apoptosis · Proliferation · Cellular senescence · Renin-angiotensin system · Telomeres · Recombinant inbred strains · Genetics · Epigenetics · Hypertrophy · Ion transport · Mortalin

Abbreviations

AII	Angiotensin II
ACE	Angiotensin-converting enzyme
BP	Blood pressure
cGMP	Cyclic guanosine monophosphate
chr	Chromosome
ERG	Early response genes
ESPI	Endogenous Na^+,K^+ pump inhibitors
HSP	Heat stress protein(s)
MAPK	Mitogen-activated protein kinase
MOT-2	Mortalin
NHE1	Na^+-H^+ exchanger
NKCC1	Na^+,K^+,Cl^- cotransport
NO	Nitric oxide
PI3 K	Polyphosphoinositide-3-kinase
PKC	Protein kinase C
PLC-γ	Phospholipase C
QTL	Quantitative trait locus
RAS	Renin-angiotensin system
REC	Renal epithelial cells
RIS	Recombinant inbred strains
SHR	Spontaneously hypertensive rats
SMC	Smooth muscle cells
VSMC	Vascular smooth muscle cell(s)
VSMC-E1A	E1A-adenoviral protein

1
Introduction

Angiotensin II (AII) was initially considered a main blood pressure (BP) regulatory substance due to its vasoconstricting effects. Relatively recently, its growth-promoting effects have been recognized, mainly as being mediated by AT_1 receptors, but it was only when AT_2 receptor-mediated inhibition of cell proliferation (and apoptosis) became known (Stoll et al. 1995; Stoll and Unger 2001) that the extent of AII's impact on the balance of proliferation/apoptosis was realized. This has major implications for cardiovascular disorders of generalized proliferation imbalance, such as hypertension (Hamet 1995), or localized abnormalities, such as occurring in atherosclerosis (Choy et al. 2001).

In this chapter, we will review the evidence of accelerated cellular turnover associated with defective proliferation and apoptosis in target organs of hypertension with specific attention to the putative role of AII in this process and in genetic determinants of early development, ion transport, and the impact of remodeling via the implication of apoptosis.

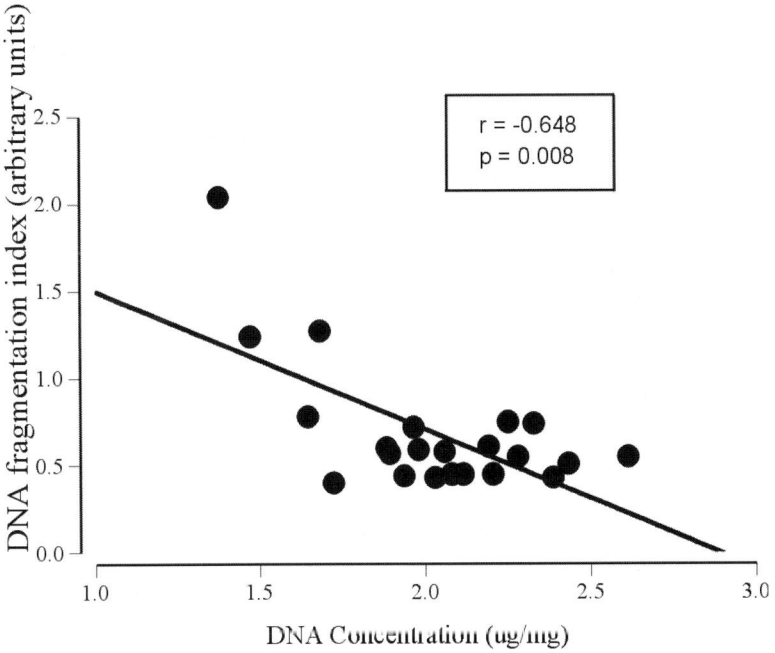

Fig. 1 Correlation between neonatal cardiac DNA fragmentation indicative of apoptosis and DNA concentration (μg/mg) in 20 recombinant inbred strains and 2 progenitor strains. There was negative correlation between newborn cardiac DNA concentration and apoptosis in the heart (r=-0.64, p<0.001). Data shown are mean values of 5 rats per strain. (Reproduced from Hamet et al. (2001) with permission from *Hypertension*)

2
Abnormal Proliferation/Apoptosis in Hypertensive Neonates

In hypertension, the heart shows both increased cell size (mainly cardiomyocytes) and cell number (mainly fibroblasts) (Anversa et al. 1997; Walsh and Dorn 1998), while the aorta presents augmented smooth muscle cell (SMC) DNA due to SMC polyploidy and hyperplasia. Several groups, including our own, have reported hyperplasia in the "prehypertensive period" in the heart (Tanase et al. 1982; Walter and Hamet 1986) and kidney of neonates from diverse strains with genetic hypertension (Pang et al. 1986). An altered balance between cell replication and apoptosis is present in the newborn heart and kidney of spontaneously hypertensive rats (SHR), leading to greater relative organ size and even DNA content, while apoptosis is significantly suppressed (Moreau et al. 1997). DNA accumulation is actually significantly negatively correlated with DNA fragmentation (Fig. 1). In this neonatal period, the ratio of DNA synthesis over apoptosis nevertheless favors proliferation in the heart, kidney, and aorta (Hamet et al. 1996a, 2001; Orlov et al. 2002).

Table 1 summarizes the major findings by our group and several others, supporting the concept of proliferation/apoptotic disorders and accelerated senes-

Table 1 Major phenotypes supporting evidence of proliferative/apoptotic disorder and accelerated senescence in hypertension

Phenotype	Year	Reference(s)
Cardiac and renal hyperplasia in neonates from SHR	1982, 1986	Tanase et al. 1982; Pang et al. 1986
Increased thymidine incorporation into DNA of SHR neonates (heart, kidney, aorta)	1986	Walter and Hamet 1986
Persistence of increased VSMC proliferation in culture	1989, 1992	See Table 2
Shortening of the cell cycle (G_1-S, G_2-M)	1992	See Table 2
Apoptosis in target organs of hypertension	1995	Hamet et al. 1995
Time window of apoptosis	1996	Hamet et al. 1996a
Apoptosis in regression of hyperplasia/hypertrophy	1997–2000	deBlois et al. 1997; Diez et al. 1997; Tea et al. 1999; Intengan and Schiffrin 2000
Suppression of apoptosis in neonatal SHR	1997	Moreau et al. 1997
QTL of elevated heart/ and kidney/body weight ratios, proliferation and apoptosis in hypertension	1995–2001	Hamet et al. 1998
First proposition of accelerated aging with increased cell turnover	1997	Hamet 1997
Altered in vivo vascular cell turnover in SHR	2001	Thorin-Trescases et al. 2001
Inverse relationship between telomere length and pulse pressure	2000	Jeanclos et al. 2000
Excess growth and apoptosis as accelerated aging with increased telomere restriction fragments	2001	Hamet et al. 2001
Telomerase activation in VSMC as a cause of proliferation in hypertension	2001	Cao et al. 2002

Table 2 Markers of proliferative activity in cultured cells from hypertensive rats

Rat strains/cell type	Parameter measured	Data	Reference(s)
SHR vs WKY, VSMC	Rate of cell growth, DNA and protein synthesis	Increased in SHR	Yamori et al. 1981; Clegg and Sambhi 1989; Hadrava et al. 1989; Paquet et al. 1989; Scott-Burden et al. 1989; Bukoski 1990; Hamada et al. 1990; Hadrava et al. 1992; Saltis and Bobik 1992; LaPointe et al. 1995; Nakayama et al. 1999
GH vs WKY, VSMC	DNA synthesis	Increased in GH	Harris et al. 1990
SHR vs WKY, VSMC	Cell cycle analysis	Shortening of the G_0/G_1 phase in SHR	Uehara et al. 1991; Hadrava et al. 1992
SHR vs WKY, VSMC	Cell cycle analysis	Increased in SHR, accumulation of cells in the S phase	Hamada et al. 1990
Rats with DOCA-salt hypertension, VSMC	DNA synthesis, cell cycle analysis	Not altered	Hamada et al. 1990
SHR vs BN.lx, VSMC	DNA synthesis	Increased in SHR	Champagne et al. 1999
SHR-SP vs WKY, VSMC	Rate of cell growth, DNA synthesis	Increased in SHR-SP	Devlin et al. 1995, 2000
SHR vs WKY, skin fibroblasts	Rate of cell growth, DNA synthesis	Increased in SHR	Guicheney et al. 1991
SHR vs WKY, adventitial fibroblasts	Rate of cell growth, DNA synthesis	Increased in SHR	Zhu et al. 1991
SHR vs WKY, adventitial fibroblasts	Cell cycle analysis	Faster exit from G_0 in SHR	Venance et al. 1993
SHR-SP vs WKY, endothelial cells	Rate of cell growth, DNA synthesis	Increased in SHR-SP	Ito et al. 1995
SHR-SP vs WKY, astrocytes	Rate of cell growth, DNA synthesis	Increased in SHR-SP	Yamagata et al. 1995

GH, New Zealand strain of spontaneously hypertensive rats; SHR-SP, stroke-prone SHR. For other abbreviations see text.

cence in hypertension, from enlarged organ size in genetically hypertensive neonates to the recent description of heightened telomerase activity in vascular SMC (VSMC). The phenotype increase of proliferation persists in culture, as initially observed by Yamori and colleagues (1981). The markers of heightened proliferative activity reported in cultured cells from hypertensive rats are summarized in Table 2. Finally, quantitative trait locus (QTL) mapping of elevated heart and kidney weight/body weight ratio, DNA accumulation and apoptosis is summarized in Table 3.

QTLs of enlarged organ size map to the same chromosome (chr)—for example, relative heart weight on rat chr 2—in different genetic models of hypertension, but some studies point to different loci. The availability of "permanent ho-

Table 3 QTLs of elevated heart- and kidney-body weight ratio, proliferation, and apoptosis in hypertension

Chromosome (QTL)	Phenotype	Crosses	Significance	Reference
1(a)	HW	SS/JR×LEW	4.0*	Gu et al. 1996
1(b)	HW	SS/JR×LEW	2.8*	Gu et al. 1996
1(c)	AKW	SHR×BN.lx	(−)1.2*	Hamet et al. 1998
2	HW	SS/JR×MNS	3.0*	Deng et al. 1997
	HW	SS/JR×WKY	4.0*	Deng et al. 1997
	HW	GH×BN	1.3*	Harris et al. 1995
3	NKW	SHR×BN.lx	(−)2.0*	Hamet et al. 1998
	AKW	SHR×BN.lx	(−)2.0*	Hamet et al. 1998
5(a)	HW	SS/JR×LEW	4.0*	Gu et al. 1996
5(b)	AKW	SHR×BN.lx	2.0*	Hamet et al. 1998
8	HW	SHR×BN.lx	2.0*	Kren et al. 1997
10	HW	GH×BN	1.3*	Harris et al. 1995
12	HW	SHR×BN.lx	2.0*	Hamet et al. 1996b
13	HW	GH×BN	3.0*	Harris et al. 1995
17(a)	HW	SHR×BN.lx	3.4	Pravenec et al. 1995
17(b)	HW	SS/JR×LEW	3.7	Gu et al. 1996
18(a)	AHN	SHR×BN.lx	(−)1.3*	Hamet et al. 2001
18(a)	PHN	SHR×BN.lx	(−)1.3*	Hamet et al. 2001
18(b)	AHN	SHR×BN.lx	1.3*	Hamet et al. 2001
18(c)	AHN	SHR×BN.lx	(−)2.0*	Hamet et al. 2001
18(d)	AHN	SHR×BN.lx	(−)1.3*	Hamet et al. 2001
X	HW	GH×BN	2.0*	Harris et al. 1995

Note: statistical significance is presented as the value of a logarithm of odds ratio (LOD score) or as a log of the statistical values (asterisks).
AHN, apoptosis in the heart of newborn rats (a relative amount of fragmented DNA was used to estimate this parameter); AKW, kidney-body weight ratio of adult rats; HW, heart-body weight ratio; LEW, Lewis normotensive rats; NKW, kidney-body weight ratio of newborn rats; PHN, proliferative activity in the heart of newborn rats (DNA content per wet tissue weight was used to estimate this parameter); SS/JR, salt-sensitive strain of Dahl rats; −, negative impact of the SHR allele.

mozygous replicas" of the F_2 generation, obtained by 25-generations of inbreeding after intercrossing between BN.lx and SHR (Pravenec et al. 1989), has allowed us to relate neonatal to adult phenotypes in recombinant inbred strains (RIS); while neonatal kidney weight at least partially determines the phenotype in adulthood, the latter is negatively influenced by BP increases with age (Hamet et al. 1998) (Fig.2).

The renin-angiotensin system (RAS) is known to be excessively stimulated in the neonatal period. It is therefore relevant if one of the significant QTLs of apoptosis inhibition (phenotype of excess shorter fragments over long ones) (Moreau et al. 1997) includes the D17Mit2 locus ($p=0.004$) on rat chr 17, which contains the AT_{1A} receptor gene as the main positional candidate (Moreau P, Sun Y, Kren V, and Hamet P, unpublished data).

To obtain continuous "recordings" of cellular death, we recently developed a novel approach of in utero prelabeling with prolonged follow-up of DNA-specific activity (Hamet et al. 2001; Thorin-Trescases et al. 2001). Despite its limita-

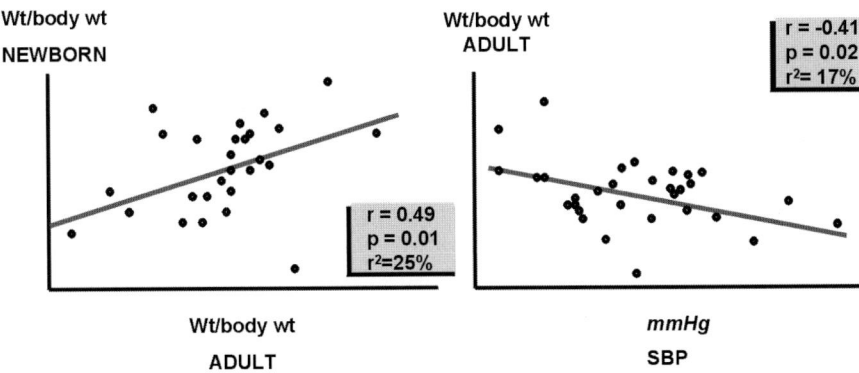

Fig. 2 Correlation between relative kidney weight in newborn and adult male rats of the RIS panel (*left side*). Correlation between adult relative kidney weights and systolic blood pressure in the panel of RIS (*right side*). (Reproduced from Hamet et al. (1998) with permission from *Kidney International*)

tions, this technique showed that increased aorta, kidney, and heart size in the neonatal stage of hypertension reverses by the age of 5-8 weeks in rats, followed by the appearance of secondary hypertrophy (heart, aorta) or atrophy (kidney) with time (Hamet et al. 2001; Thorin-Trescases et al. 2001). Cellular half-life was shorter in the SHR aorta: $t_{1/2}$ was 20 weeks in normotensive Wistar-Kyoto rats, but only 7 weeks in SHR (these values in the kidney were 14 and 8 weeks, respectively). Our most recent studies have demonstrated the presence of increased telomere restriction fragments, particularly of smaller size (Hamet et al. 2001), in the neonatal kidney. Clearly, telomerase activity should also be monitored (Aisner et al. 2002). Taken together, these investigations led us to propose that heightened cell turnover in SHR is related to initial apoptosis inhibition in neonates, resulting in heart, kidney, and aortic hyperplasia (Hamet et al. 2001), shortening of the cell cycle (Uehara et al. 1991; Hadrava et al. 1992), and increased apoptosis later in life (Hamet et al. 1995). While the AT_1 receptor is an attractive target to modulate proliferation/apoptosis imbalance, its direct involvement remains to be tested.

3
Role of Cell Cycle Genes and Chaperones

The above events could result from abnormalities within cell cycle control genes, including p53, MDM 2, $p21^{cip1}$ and $p16^{INK4a}$ as well as genes controlling their proteolytic degradation (Okorokov et al. 1997; Kubbutat and Vousden 1997; Ariyoshi et al. 1998; Pochampally et al. 1999; Nakayama et al. 2001; Heessen et al. 2002) and chaperoning at G_1-S and G_2-M checkpoints. The interaction of chaperones (represented mainly by heat stress proteins, HSP) with the p53 molecule is now better understood (King et al. 2001). In this context, it is relevant to point out that the targeted disruption of mouse AT_2 receptor gene impacts on stress-induced hyperthermia (Watanabe et al. 1999). We have previously demonstrated that HSP27 polymorphism is in linkage with left ventricular hypertrophy and that HSP70 expression is higher in SHR VSMC (Tremblay et al. 1992; Hamet et al. 1996b). Prior HSP expression, on the other hand, has been shown to partially protect against hyperthermic cell cycle arrest (Champagne et al. 1999).

Human aging, according to the gene profiling studies of Ly et al. (2000), is mainly characterized by a cumulative slowdown in G_2-M. It is therefore relevant that a newly identified hypertension-related calcium-regulated gene with heightened expression in SHR kidneys (Solban et al. 2000) induced G_2-M arrest of renal epithelial cells (REC) by increasing p21$^{cip1/WAF-1}$ and decreasing p27^{kip1} (Devlin et al. 2003). AII is involved in cardiomyocyte hypertrophy and cell death via its impact on extracellular signal-regulated kinase 1/2 (MEK-ERK1/2 pathway and JAK-STAT signaling) (Booz et al. 2002; Bueno and Molkentin 2002). Part of the pathway downstream of AT_1 and AT_2 receptors is distinct but a part is shared, mainly by activation of nuclear factor-$\kappa\beta$ transcription factors (Ruiz-Ortega et al. 2000, 2001). In addition, the AT_2 receptor involves activation of protein phosphatases, the nitric oxide (NO)-cyclic guanosine monophosphate (cGMP) system and phospholipase A_2 (Nouet and Nahmias 2000), and its antiproliferative effect appears to include downregulation of SM-20 (Wolf et al. 2002), a mitochondrial protein implicated in apoptosis and expressed in vascular smooth muscle (Wax et al. 1996). It has been reported that apoptosis is heterogeneously distributed in VSMC, depending in part on the cellular phenotype (Bascands et al. 2001). AT_2 receptor expression is not obligatorily related to apoptosis, such as, for instance, apoptosis induced during ischemia/reperfusion-dependent cardioprotection (Moudgil et al. 2002). AII induces p53 expression (Bonnet et al. 2001); via the same downstream pathway implicating p21$^{cip1/WAF-1}$, p27^{kip1} and p53, it also evokes proliferation via superoxide production (Mueller et al. 2002).

Our studies using proteomics have identified another HSP, mortalin (MOT-2 or HSPA9), as a major mediator of the $[Na^+]_i$-induced delay in VSMC apoptosis (for more details, see below). MOT-2 was initially cloned as a highly homologous member of the HSP70 family, not inducible by heat stress, but interacting with other chaperones (Domanico et al. 1993). Injection of anti-MOT-2 antibodies led to transient division of senescent cells (Wadhwa et al. 1993a), whereas transfection with MOT-2 cDNA resulted in induction of senescence in NIH 3T3 fibroblasts (Wadhwa et al. 1993b). It has now been demonstrated to inactivate p53 and extend fibroblast lifespan (Kaul et al. 2000). This action is exercised in concert with HSP90 under control of the co-chaperones Bag-1, Hop and HSP40 (King et al. 2001). The regulation of p53 degradation (by ubiquitination) is the consequence of a complex interplay of these HSPs with wild and mutated p53 (King et al. 2001). Actually, HSP90 inhibits the MDM 2 stabilization of p53 (Peng et al. 2001).

4
Genetics of Growth-Related Phenotypes

MOT-2 is localized on rat chr 18 in a region syntenic to human 5q31-3. This region is highly relevant to hypertension in both rats and humans (Jacob et al. 1991; Krushkal et al. 1998). It has been known since publication of the first total genome scan in SHR (Jacob et al. 1991), and confirmed by many others (Cowley et al. 2000; Garrett et al. 1998), that the locus on chr 18 bears significant components of salt sensitivity (Fig. 3). This is of interest as $[Na^+]_i$ elevation induces MOT-2 expression (Taurin et al. 2002a).

In the largest set of genotypic/phenotypic profiling studies published to date by the Medical College of Wisconsin and accessible on http://brc.mcw.edu/phyprof (Stoll et al. 2001), a significant cluster of loci was identified on rat chr 18. BP, renal and anthropometric phenotypes correlated only on chr 18, suggesting the presence of a "gene cassette" (Fig. 3).

To avoid an untargeted and tedious search for candidate genes determining neonatal growth-related phenotypes, we performed total genome scan in RIS as a unique tool to investigate neonatal and adult phenotype/genotype concordance (Hamet et al. 1998). This approach has demonstrated a strain distribution pattern of RIS for the DNA fragmentation index and DNA concentration in the kidney, allowing their chromosomal mapping (Hamet et al. 2001). The most significant QTLs were found on chrs 1 and 18. Of interest here is the locus on chr 18 at Mit4R525 which is the locus of MOT-2 (Fig. 3). Direct testing of this candidate locus/gene and its putative interaction with components of the RAS is under way.

5
Replicative Senescence Versus Accelerated Aging

Aging-related phenotypes, including disease states, are both genetically and epigenetically defined. The aging process is biologically complex, and the distinction between replicative senescence, old age, and accelerated aging (such as progeria) must be made (Slagboom et al. 2000). Thus, for instance, fibroblasts from the elderly share more molecular determinants with progeria than with replicative senescence. Some markers, such as p53 and p21^{WAF-1}, are augmented in all three conditions, while cdc42 and CAMK2D increase only in progeric and elderly fibroblasts (Park et al. 2001). Clustering demonstrated hierarchical similarities between progeria and old age but not with replicative senescence (Eisen et al. 1998). A detailed analysis of middle age, old age, and progeria was provided by Ly et al. (2000) who related aging with mitotic misregulation, particularly affecting G_2-M transition; of 6,000 genes analyzed, 61 (1%) were modified, 25% of them in cell cycle progression, and 31% in extracellular matrix remodeling. Progeria, while extremely rare, is characterized by the presence of severe hypertension (Ogihara et al. 1986) and VSMC depletion (Stehbens et al. 2001) leading to cardiovascular death (Rosman et al. 2001). Its mechanism is unknown, but

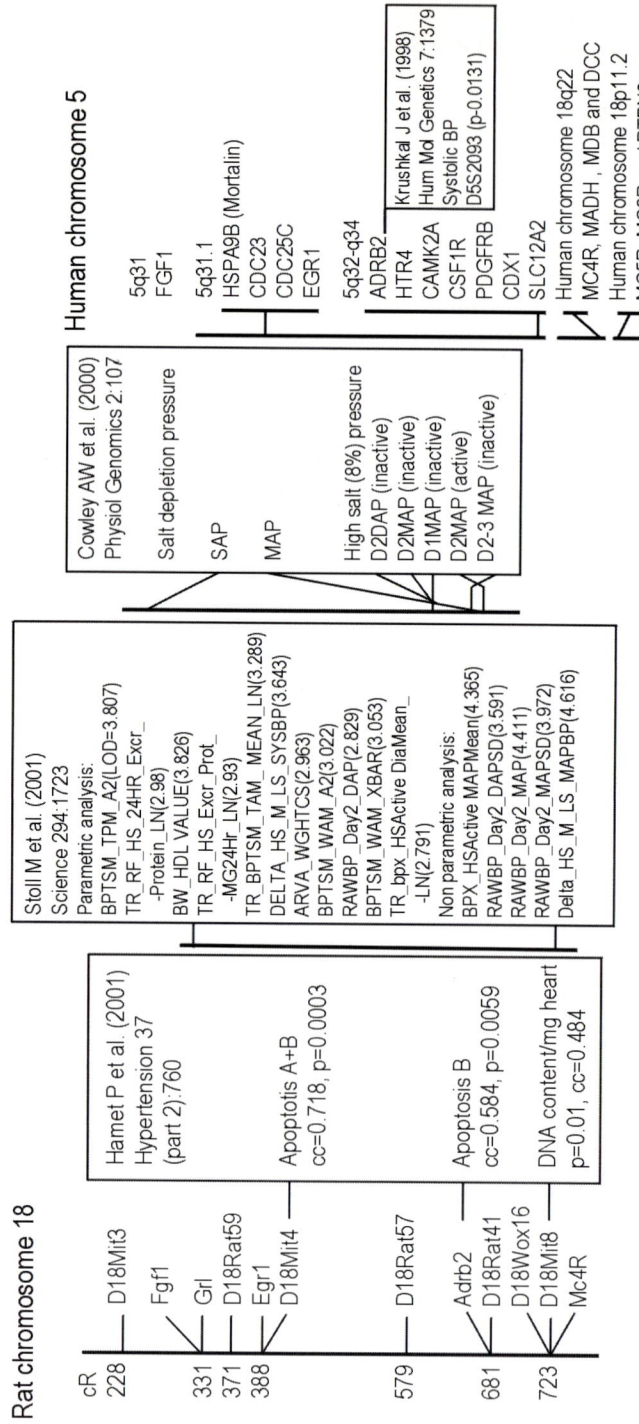

Fig. 3 Quantitative trait loci (QTL) clusters mapped on the long arm of rat chr 18 and its syntenic region on human chr 5q31-q32

sporadic dominant mutation is suspected (Sarkar and Shinton 2001). Telomerase activity is putatively involved, but normal telomere length has been observed in neonatal progeroid (Wiedemann-Rautenstrauch) syndrome (Korniszewski et al. 2001).

6
Biomarkers of Senescence

Senescence-associated β-galactosidase is widely used as a distinct biomarker of senescence as it is not expressed in pre-senescent cells and is absent in "immortal cells" (Dimri et al. 1995). Increased p53-induced gene 3 expression, in response to environmental stimuli such as reactive oxygen species, is long-lasting (Flatt et al. 2000). In human atherosclerotic plaque, Rb and p53 have been shown to act cooperatively in the control of proliferation, cell senescence, and apoptosis, leading to a high level of proliferation and suppression of apoptosis (Bennett et al. 1998). Disappearance of angiotensin-converting enzyme (ACE) activity was identified as a phenotype associated with several karyotypic changes induced by the aging of human endothelial cells in culture (Johnson et al. 1992). More recently, in a seminal study, Tyner and colleagues (2002) have demonstrated that p53 mutant mice display aging-associated phenotypes, including cutaneous atrophy, osteoporosis, impaired stress tolerance, and reduced longevity. Noticeably, several of these phenotypes are also linked with cardiovascular diseases, such as diminished stress tolerance (Dumas et al. 2000a,b) and even osteoporosis, SHR being used as a "model of osteoporosis" (Tsuruoka et al. 2001).

7
Role of Ion Transporters and Intracellular Monovalent Cations

Cell physiology studies from several laboratories demonstrated a potent role of AII in monovalent ion handling. Thus, in VSMC, AII activates ubiquitous isoforms of Na^+-H^+ exchanger [NHE1 (Hatori et al. 1987; Berk et al. 1990)], Na^+,K^+,Cl^- cotransport [NKCC1 (Smith and Smith 1987; Orlov et al. 1992; Peiro et al. 1997; Akar et al. 1999)] and the Na^+,K^+ pump (Orlov et al. 1992). Activation of NHE1 and the Na^+,K^+ pump by AII was also revealed in secretory epithelial cells (Hou and Delamere 2002). In REC, AII activates the NHE3 isoform of the Na^+-H^+ exchanger (Tsuganezawa et al. 1998) and Na^+,HCO_3^- cotransport (Horita et al. 2002; Robey et al. 2002) and inhibits the Na^+,glucose cotransporter SGLT1 (Kawano et al. 2002). Activation of Na^+,HCO_3^- cotransport was also observed in AII-treated cardiomyocytes (Baetz et al. 2002). Recently, it was shown that similarly to blood cells (Avdonin et al. 1990), non-selective intracellular Ca^{2+} store-operated cation channels contribute to $[Na^+]_i$ elevation in AII-treated arterial myocytes (Arnon et al. 2000). In neuronal cells, AII activates delayed rectifier K^+ channels, thus leading to membrane hyperpolarization (Kang et al. 1994; Matsukawa and Ichikawa 1997).

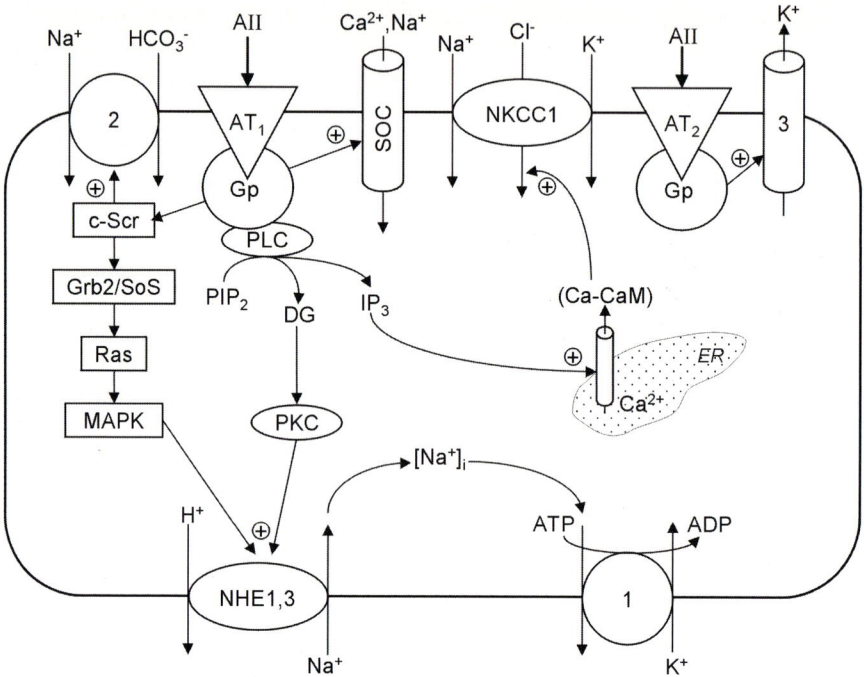

Fig. 4 Mechanism of the regulation by AII of intracellular monovalent ion handling. *Gp*, GTP-binding proteins; *1*, Na^+,K^+-ATPase; *2*, Na^+,HCO_3^- cotransporter; *3*, delayed rectifier K^+ channels. For more details, see text

Our current knowledge on the regulation of intracellular monovalent ion handling by AII is schematically presented in Fig. 4. In VSMC and the secretory epithelium, NHE1 activation is mediated by AT_1 receptors (Ye et al. 1996; Hou and Delamere 2002), protein kinase C (PKC) (Hatori et al. 1987; Grinstein et al. 1989), and ERK1/2 mitogen-activated protein kinase (MAPK) (Kusuhara et al. 1998). In VSMC, transient elevation of $[Ca^{2+}]_i$ contributes to enhanced NKCC1 (Smith and Smith 1987; Orlov et al. 1992) whereas in REC, NHE3, and Na^+, HCO_3^- cotransport is activated by a c-Src-dependent pathway (Tsuganezawa et al. 1998; Robey et al. 2002). In the majority of cells studied so far, activation of the Na^+,K^+ pump is a secondary event caused by the NHE1-mediated rise of $[Na^+]_i$ (Orlov et al. 1992; Hou and Delamere 2002; Rangel et al. 2002). Indeed, in VSMC and REC, AII led up to threefold elevation of $[Na^+]_i$ that was completely abolished by potent NHE1 blockers (Ye et al. 1996; Touyz and Schiffrin 1999; Hou and Delamere 2002). In contrast to the above-listed ion carriers, AII activates K^+ channels via an unknown signaling cascade triggered by AT_2 receptors (Kang et al. 1994; Matsukawa and Ichikawa 1997).

It is well-documented that AII signaling efficiency is increased in SHR compared to normotensive rats. Importantly, these data were obtained by analysis of the $[Ca^{2+}]_i$ response, inositol-4,5-triphosphate production, and MAPK activity

(Touyz et al. 1994; Lucchesi et al. 1996; Phan et al. 1997) as well as $[Na^+]_i$ and pH_i modulation triggered by AII in cultured VSMC (Touyz and Schiffrin 1997, 1999) and REC (Thomas et al. 1990; Cheng et al. 1998) from SHR. It should be stressed that apart from increased sensitivity to AII, augmented increment of $[Na^+]_i$ might be caused by enhanced baseline activity of NHE1 and other ion carriers documented in several cell types from SHR, the Milan hypertensive strain, and patients with essential hypertension (Rosskopf et al. 1993; Siffert and Dusing 1995; Orlov et al. 1999a). In addition, intracellular monovalent cation content is subjected to regulation by endogenous Na^+,K^+ pump inhibitors (ESPI) identified as immunoreactive ouabain-like (Ludens et al. 1991; Tymiak et al. 1993) and marinobufagenin-like (Bagrov et al. 1995; Hilton et al. 1996; Sich et al. 1996) steroids.

Several research teams have proposed that abnormal intracellular monovalent ion handling is involved in long-term maintenance of elevated BP via enhanced contractility of VSMC and exaggerated salt and water reabsorption in REC (Blaustein 1984; Postnov and Orlov 1985; Blaustein 1996; Cheng et al. 1998; Hall et al. 1999; Orlov et al. 1999a; Ortiz and Garvin 2001). The latter hypothesis was partially proved by identification of genes whose mutation leads to the development of monogenic (Mendelian) hypertension forms (Lifton et al. 2001). More recently, several research teams have reported modulatory effects of monovalent ion transport inhibitors on cell proliferation and death. Thus, it has been shown that ouabain at low concentrations increases the proliferation of cultured SMC (Golomb et al. 1994; Aydemir-Koksoy et al. 2001; Chueh et al. 2001). In essential hypertension, the level of ESPI correlated positively with left ventricular mass (Manunta et al. 1999). These results suggest that AII-induced remodeling in hypertension can be caused by the altered activity of ion transporters.

Based on data showing that tissue hypertrophy/hyperplasia in hypertension is limited to a few organs, such as the heart, blood vessels, and kidneys, we proposed that abnormalities of cell cycle progression and apoptosis, i.e., the two major counterparts of organ remodeling seen in hypertension, are evoked by mutations within genes encoding their cell type-specific regulators rather than ubiquitous intermediates, such as cyclins, cyclin-dependent kinase, p53, proteins of the Bcl-2 family, etc. (Orlov et al. 1999b, 2002). We focused our initial interest on the study of such regulators in VSMC apoptosis.

Wild-type VSMC are resistant to apoptosis, so that even after 24 h of serum deprivation, the amount of morphologically defined apoptotic cells does not exceed 2%-4% (Bennett et al. 1994, 1995). Therefore, to elaborate the mechanism of VSMC apoptosis, we adopted the chromatin cleavage assay which allows us to measure intracellular chromatin fragments in amounts less than 1% of total DNA content (Hamet et al. 1995; Orlov et al. 1996) and we used VSMC transfected with the functional c-myc analog, E1A-adenoviral protein (VSMC-E1A), which are highly sensitive to apoptotic stimuli (Bennett et al. 1995; Champagne et al. 1999). In these cells, apoptosis can be prevented by growth factors, such as insulin-like growth factor, platelet-derived growth factor, and epidermal growth factor (Orlov et al. 1999b), whose receptors are coupled to phospholipase C

(PLC-γ), polyphosphoinositide-3-kinase (PI3 K) and MAPK. Our team (Orlov et al. 1999b) and Bai with co-workers (1999) reported that PI3 K inhibition with wortmannin potentiated apoptosis in serum-supplied VSMC. We failed to observe the involvement of any other intermediates of intracellular signaling, such as PLC, $[Ca^{2+}]_i$, and MAPK, in the protection of serum-deprived VSMC against apoptosis (Orlov et al. 1999b).

In addition to the above-listed intermediates of intracellular signaling, activation of AT_2 receptors in REC leads to NO production and activation of cGMP signaling. Recently, Dulin with co-workers reported that vasoconstrictors, including AII, transiently activated PKA (Dulin et al. 2001; Davis et al. 2003). To study the role of this enzyme, we treated VSMC with activators of cyclic adenosine monophosphate signaling and observed that these compounds blocked the development of apoptosis during the initial 10-12 h of serum-deprivation at a step upstream of caspase-3 (Orlov et al. 1999c). In contrast, we did not detect any effect of activators of soluble (NO donors) and membrane-bound atrial natriuretic peptide guanylate cyclase on apoptosis in VSMC-E1A (Orlov et al. 1999c).

It was shown that inhibitors of NHE1 abolished the proliferative action of AII (Sachinidis et al. 1996) and AII-induced hypertrophy of cultured VSMC (Peiro et al. 1997). The role of intracellular monovalent cations in organ remodeling is also supported by data demonstrating the suppression of myocardial hypertrophy with NHE1 inhibitors (Cingolani 1999; Morris 2002) and parallel normalization of cardiac hypertrophy and NHE activity in cardiomyocytes from SHR subjected to antihypertensive therapy (Alvarez et al. 2002). It should be underlined, however, that long-term application of NHE blockers leads to numerous side effects, such as inhibition of PKA, PKG, and PKC (Orlov et al. 1989). Considering this, we used ouabain, i.e., a potent and selective Na^+,K^+-ATPase inhibitor, to study the role of intracellular monovalent ions in apoptosis.

We observed that preincubation of VSMC from the rat aorta with ouabain and in K^+-free medium blocks the development of the apoptotic machinery at a step upstream of caspase-3 (Orlov et al. 1999d). Importantly, Na^+,K^+ pump inhibition protects VSMC against distinct apoptotic stimuli, such as serum-deprivation, inhibitors of serine-threonine protein kinases (staurosporine), phosphoprotein phosphatases (okadaic acid) (Orlov et al. 1999d) and DNA damage caused by its extensive [^{3}H] labeling (Orlov et al. 2003). Soon after our publication, it was reported that ouabain prevents apoptosis in rat cerebellar granule cells (Isaev et al. 2000) and REC (Zhou et al. 2001). In contrast to these cell types, Na^+/K^+ pump inhibition potentiated the development of apoptosis in lymphocytes (Olej et al. 1998), Jurkat cells (Orlov et al. 1999d; Bortner et al. 2001), and several tumor cell lines (McConkey et al. 2000; Penning et al. 2000; Yeh et al. 2001). In rat VSMC (Orlov et al. 2001), human lymphocytes (Falciola et al. 1994), and HEK-293 cells (unpublished data), long-term inhibition with ouabain did not exhibit any cytotoxic action, whereas in REC (Sato and Ozawa 1977; Ledbetter et al. 1986; Bolivar et al. 1987), ischemic neuronal cells (Urenjak and Obrenovitch 1996), and the mesothelioma cell line (Marklund et al. 1999), it elicited massive cell death. The limited number of markers employed in these studies complicates a definitive conclusion on the relative contribution of necrosis and apoptosis to overall cell death. However, the prevalence of necrosis can be proposed based on the lack of protective effect of bcl-2 overexpression (Gilbert and Knox 1997) and insensitivity to caspase inhibitors (Pchejetski et al. 2003).

Two approaches have been employed to identify genes whose expression is triggered by elevated $[Na^+]_i/[K^+]_i$ ratio. The candidate gene approach suggests the expression of early response genes (ERG), i.e., transcriptional factors leading to massive RNA synthesis seen in ouabain-treated cells (Piechaczyk and Blanchard 1994; Tamai et al. 1997; Ahmad et al. 1998; Sylvester et al. 1998; Rivard and Andres 2000; Orlov et al. 2001). Indeed, it was shown that inhibition of the VSMC Na^+,K^+ pump with ouabain or in K^+-depleted medium leads to rapid and massive expression of c-Fos followed by the appearance of c-Jun protein (Taurin et al. 2002b). It might be suggested that modulation of $[Ca^{2+}]_i$ rather than monovalent cations is an upstream signal of ERG expression. However, we demonstrated that c-Fos expression in ouabain-treated cells was insensitive to the presence of extracellular Ca^{2+} and Ca^{2+} chelators, and the kinetics of c-Fos mRNA expression correlates with $[Na^+]_i$ elevation rather than with $[K^+]_i$ loss (Taurin et al. 2002b). Using luciferase *trans-* and *cis-* reporter assays, it was shown that the activity of Elk-1, serum response element, cAMP-response element binding protein, and activator protein-1 transcription factors interacting with major transcriptional control elements of the c-fos promoter was potentiated by serum but was insensitive to the presence of ouabain (Taurin et al. 2002b). These data suggest the presence of a novel $[Na^+]_i$-mediated, $[Ca^{2+}]_i$-independent mechanism of excitation-transcription coupling (Taurin et al. 2002b).

To search for $[Na^+]_i$-sensitive genes involved in the inhibition of apoptosis, the proteomes of control and ouabain-treated VSMC have been compared. This study led to the identification of more than 20 soluble proteins whose expression was increased by more than tenfold after a 3-h incubation with ouabain, including MOT-2 (Taurin et al. 2002a). We observed that similarly to ouabain, transient transfection with MOT-2 causes transient inhibition of apoptosis in VSMC, as discussed above.

The data presented in this section show that exposure of VSMC and REC to AII leads to transient $[Na^+]_i$ elevation. Both the increased activity of AII-sensitive ion carriers, such as NHE1 and NKCC1, and enhanced ESPI production contribute to the augmented AII-induced $[Na^+]_i$ response seen in hypertension. In VSMC, $[Na^+]_i$ elevation triggered by ouabain evokes the expression of MOT-2 and probably of other genes involved in the suppression of apoptosis (Fig. 5). In contrast to VSMC, ouabain leads to massive necrotic type of REC death. However, using Madin-Darby canine kidney cells, it was shown that this effect is mediated by conformational transition of Na^+,K^+-ATPase rather than by ion fluxes provided by this enzyme (Pchejetski et al. 2003). Which protein(s) interact(s) in REC with the Na^+,K^+-ATPase α-subunit in the presence of ouabain to induce the necrotic type of cell death? Which intermediate(s) of this signaling cascade are

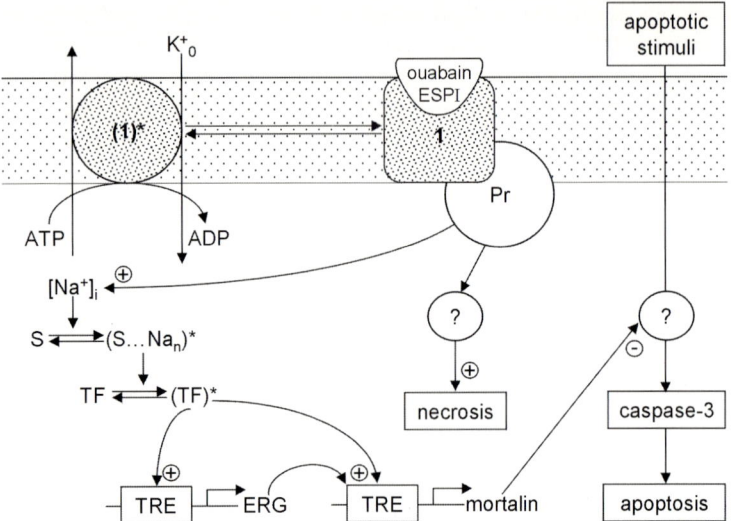

Fig. 5 Mechanism of the involvement of Na$^+$,K$^+$-ATPase and intracellular Na$^+$ in the apoptotic and necrotic modes of cell death. *1*, Na$^+$,K$^+$-ATPase; *Pr*, unknown protein interacting with inactivated Na$^+$,K$^+$-ATPase and triggering the death signal; *S*, [Na$^+$]$_i$ sensor; *TF*, transcription factor; *TRE*, transcription response element; *()**, activated state of the molecules; *?*, unknown intermediates. For more details, see text

missing/suppressed in VSMC and other cells resistant to the toxic action of ouabain? Does the enhanced production of ESPI seen in hypertension (Dmitrieva and Doris 2002) contribute to the premature death of REC and protection of VSMC against apoptosis? These questions should be addressed in forthcoming studies.

8
Antihypertensive Therapy, Cell Growth, and Apoptosis

As mentioned above, features of cell senescence include increased oxidative stress and a greater incidence of nuclear polyploidy. Both of these features are stimulated in the cardiovascular system not only with aging but also with hypertension. Moreover, AII stimulation of AT$_1$ receptors results in the potent activation and gene expression of oxidative stress-generating enzymes, such as nicotinamide adenine dinucleotide (phosphate), reduced [NAD(P)H] oxidase (Touyz et al. 2002). There is a growing body of pharmacological evidence suggesting that cardiovascular cell fate is modulated by antihypertensive therapies. The increased turnover of SMC in hypertension is prevented by these agents, including enalapril (Thorin-Trescases et al. 2001) (Table 4). This effect in SHR is achieved mainly because of decreased cell growth with no change in cumulative cell death rates. ACE inhibitors reduce cell turnover in the neonatal rat heart (Choi et al. 2002). Long-term treatment with the AT$_1$ receptor antagonist losartan in SHR

Table 4 Effect of chronic antihypertensive treatment on aortic weight, relative aortic weight corrected for body weight (BW), and aortic DNA content. Treatment with hydralazine (40 mg/kg/day), nifedipine (30 mg/kg/day), or enalapril (30 mg/kg/day) was initiated in 5-week-old SHR and maintained for 15 weeks. Data represent values for the different parameters in 20-week-old SHR compared with age-matched WKY. Data are mean±SEM

Rat (n)	Aorta weight (mg)	Aorta weight/BW (mg/g)	DNA content (μg)
WKY (9)	29.6±2.5	0.12±0.01	36.3±2.0
SHR (6)	42.0±3.5[a]	0.17±0.01[a]	51.9±6.2[a]
Hydralazine (8)	37.3±3.6	0.14±0.01	31.9±3.7[b]
	p=0.289	p=0.111	p=0.001
Nifedipine (7)	31.9±2.2[b]	0.12±0.09[b]	33.6±1.9[b]
	p=0.032	p=0.015	p=0.001
Enalapril (8)	31.8±2.7[b]	0.13±0.02[b]	31.5±1.5[b]
	p=0.026	p=0.034	p=0.001

SHR, spontaneously hypertensive rats; WKY, Wistar-Kyoto rats.
[a] $p < 0.05$ SHR vs WKY.
[b] $p < 0.05$ treated SHR vs untreated SHR.
Reproduced from Thorin-Trescases et al. (2001) with permission from *Journal of Cardiovascular Pharmacology*.

alters the phenotype and suppresses the proliferative behavior of SMC cultured from the aorta (Bravo et al. 2001). In the heart, cardiomyocyte polyploidy is reduced by RAS inhibitors (Panizo-Santos et al. 1995). We recently obtained evidence that cardiomyocyte polyploid nuclei may be removed by apoptosis during treatment with RAS inhibitors (Der Sarkissian et al. 2003).

Our studies have shown that, irrespective of the class of antihypertensive drug administered to SHR, those that are able to induce the regression of cardiac or aortic hypertrophy also evoke a transient increase of apoptosis in these organs, in large part via pressure-independent mechanisms (deBlois et al. 1997; Tea et al. 1999). Drugs effective in these regards include antagonists of AT_1 receptors, such as valsartan and losartan, and ACE inhibitors, such as enalapril. The induction of cardiovascular apoptosis in SHR is transient and significant only within the first weeks after the initiation of drug treatment, even when it is maintained for several weeks. Moreover, the induction of SMC apoptosis is temporally dissociated from inhibition of cell growth since the time window of apoptosis occurs before the sustained inhibition of SMC DNA synthesis and organ mass reduction. Polyploidy is a common feature of senescent cells (Ly et al. 2000). AII-dependent hypertension is associated with increased SMC polyploidy (Owens 1989, 1995). The age-dependent accumulation of cells arrested in G_2 (i.e., polyploid cells) occurs typically faster in hypertensive individuals (Barrett et al. 1983; Lee et al. 1992; Panizo-Santos et al. 1995), a feature that is reversed (Dominiczak et al. 1996) by the same classes of antihypertensive drugs that induce apoptosis and regression of cardiovascular hyperplasia (deBlois et al. 1997; Tea et al. 1999). Administration of quinapril decreases the percentage of tetraploid cardiomyocytes in SHR (Pochampally et al. 1999), an effect that reflects either nuclear division or polyploid cell deletion, suggesting that the accumulation of aberrant cells with aging and hypertension may be reversed by programmed cell death during drug therapy. While the AT_1 receptor has been implicated in the inhibition of SMC apoptosis (Pollman et al. 1996), recent evidence indicates that a subpopulation of SMC responds to AT_1 receptor stimulation with sustained transmembrane calcium influx and apoptosis induction rather than survival or growth (Bascands et al. 2001). This is reminiscent of the recent report that AII infusion for 7 days, a condition known to be associated with SMC growth (deBlois et al. 1996), also activates the Bax/caspase-3 pathway (Diep et al. 1999). Together, these observations provide further evidence that under certain conditions AII may be stimulating SMC turnover in vivo. SMC apoptosis induction, growth inhibition, and vascular regression evoked by AT_1 antagonists in the SHR aorta are completely dependent on AT_2 receptors for AII (Tea et al. 2000).

In the heart, AT_1 antagonists and ACE inhibitors increase apoptosis selectively in fibroblasts, an effect independent of AT_2 receptors and consistent with a possible role of AT_1 receptors in cell survival (Der Sarkissian et al. 2003). Cardiac apoptosis induction resulted in caspase-3 activation and selective deletion of cardiac fibroblasts without affecting cardiomyocyte number or capillary density (Der Sarkissian et al. 2003). We propose that the stimulation of apoptosis in fibroblasts is an important determinant of cardiac fibrosis suppression. In contrast, long-term prophylactic antihypertensive treatment (initiated before the onset of hypertension) inhibits vascular hyperplasia in adult rats without affecting the rate of cell death (Thorin-Trescases et al. 2001).

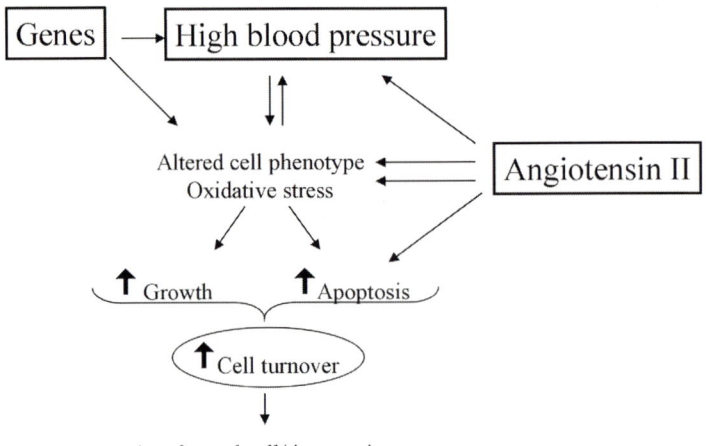

Fig. 6 Angiotensin II and senescence: a hypothesis. Key features of senescent cells are enhanced in hypertensive individuals, including cardiovascular cell polyploidy and oxidative stress. These features are in part genetically determined but can be modulated by pharmacotherapy. Recently, angiotensin II has emerged as a potent pro-oxidant factor and a key regulator of cell phenotype. Depending on cardiovascular cell phenotype, angiotensin II can stimulate growth (via AT_1 receptors) or apoptosis (via AT_1 and AT_2 receptors). Enhanced cell growth and apoptosis in tissues will result in enhanced cellular turnover and accelerated aging causing target organ damage

In conclusion, overall, complex epigenetic and genetic factors contribute to aging and age-related diseases, including hypertension. Part of this process involves the defective regulation of cell proliferation and death in cardiovascular target organs via a putative specific senescence pathway which may involve the RAS (Fig. 6).

Acknowledgements. This work was supported by grants from the Canadian Institutes of Health Research (MT-10803, MOP-43859), Valorisation-Recherche Québec (2200-015), the Heart and Stroke Foundation of Canada and the Czech Republic (GA CR K015/98). The authors thank Carole Long and Gilles Corbeil for their expert technical assistance, Ovid Da Silva for editing this manuscript, and Ginette Dignard for her secretarial work.

References

Ahmad M, Theofanidis P, Medford RM et al (1998) Role of activating protein-1 in the regulation of the vascular cell adhesion molecule-1 gene expression by tumor necrosis factor-alpha. J Biol Chem 273:4616-4621

Aisner DL, Wright DE, Shay JW et al (2002) Telomerase regulation: not just flipping the switch. Curr Opin Genet Dev 12:80-85

Akar F, Skinner E, Klein JD et al (1999) Vasoconstrictors and nitrovasodilators reciprocally regulate the Na+-K+-2Cl− cotransporter in rat aorta. Am J Physiol 276:C1383-C1390

Alvarez BV, Ennis IL, De Hurtado MC et al (2002) Effects of antihypertensive therapy on cardiac sodium/hydrogen ion exchanger activity and hypertrophy in spontaneously hypertensive rats. Can J Cardiol 18:667-672

Anversa P, Olivetti G, Leri A et al (1997) Myocyte cell death and ventricular remodeling. Curr Opin Nephrol Hypertens 6:169-176

Ariyoshi H, Okahara K, Sakon M et al (1998) Possible involvement of m-calpain in vascular smooth muscle cell proliferation. Arterioscler Thromb Vasc Biol 18:493-498

Arnon A, Hamlyn JM, Blaustein MP et al (2000) Na(+) entry via store-operated channels modulates Ca(2+) signaling in arterial myocytes. Am J Physiol Cell Physiol 278:C163-C173

Avdonin PV, Cheglakov IB, Tkachnuk VA (1990) Ionic permeability and regulation of receptor-operated channels in plasma membrane of human platelets. Biol Membr (Moscow) 7:12-92

Aydemir-Koksoy A, Abramowizt J, Allen JC (2001) Ouabain-induced signaling and vascular smooth muscle cell proliferation. J Biol Chem 276:46605-46611

Baetz D, Haworth RS, Avkiran M et al (2002) The ERK pathway regulates Na(+)-HCO(3)(−) cotransport activity in adult rat cardiomyocytes. Am J Physiol Heart Circ Physiol 283:H2102-H2109

Bagrov AY, Fedorova OV, Austin-Lane JL et al (1995) Endogenous marinobufagenin-like immunoreactive factor and Na$^+$,K$^+$ATPase inhibition during voluntary hypoventilation. Hypertension 26:781-788

Bai H, Pollman MJ, Inishi Y et al (1999) Regulation of vascular smooth muscle cell apoptosis. Modulation of bad by a phosphatidylinositol 3-kinase-dependent pathway. Circ Res 85:229-237

Barrett TB, Sampson P, Owens GK et al (1983) Polyploid nuclei in human artery wall smooth muscle cells. Proc Natl Acad Sci USA 80:882-885

Bascands JL, Girolami JP, Troly M et al (2001) Angiotensin II induces phenotype-dependent apoptosis in vascular smooth muscle cells. Hypertension 38:1294-1299

Bennett MR, Evan GI, Newby AC et al (1994) Deregulated expression of the *c-myc* oncogene abolishes inhibition of proliferation of rat vascular smooth muscle cells by serum reduction, interferon-a, heparin, and cyclic nucleotide analogues and induces apoptosis. Circ Res 74:525-536

Bennett MR, Evan GI, Schwartz SM et al (1995) Apoptosis of rat vascular smooth muscle cells is regulated by p53 dependent and independent pathways. Circ Res 77:266-273

Bennett MR, Macdonald K, Chan SW et al (1998) Cooperative interactions between RB and p53 regulate cell proliferation, cell senescence, and apoptosis in human vascular smooth muscle cells from atherosclerotic plaques. Circ Res 82:704-712

Berk BC, Elder E, Mitsuka M et al (1990) Hypertrophy and hyperplasia cause differing effects on vascular smooth muscle cell Na+/H+ exchange and intracellular pH. J Biol Chem 265:19632-19637

Blaustein MP (1984) Sodium transport and hypertension. Where are we going? Hypertension 6:445-453

Blaustein MP (1996) Endogenous ouabain: role in the pathogenesis of hypertension. Kidney Int 49:1748-1753

Bolivar JJ, Lazaro A, Fernandez S et al (1987) Rescue of a wild-type MDCK cell by a ouabain-resistant mutant. Am J Physiol 253:C151-C161

Bonnet F, Cao Z, Cooper ME et al (2001) Apoptosis and angiotensin II: yet another renal regulatory system? Exp Nephrol 9:295-300

Booz GW, Day JN, Baker KM et al (2002) Interplay between the cardiac renin angiotensin system and JAK-STAT signaling: role in cardiac hypertrophy, ischemia/reperfusion dysfunction, and heart failure. J Mol Cell Cardiol 34:1443-1453

Bortner CD, Gomez-Angelats M, Cidlowski JA (2001) Plasma membrane depolarization without repolarization is an early molecular event in anti-Fas-induced apoptosis. J Biol Chem 276:4304-4314

Bravo R, Somoza B, Ruiz-Gayo M et al (2001) Differential effect of chronic antihypertensive treatment on vascular smooth muscle cell phenotype in spontaneously hypertensive rats. Hypertension 37:E4-E10

Bueno OF, Molkentin JD (2002) Involvement of extracellular signal-regulated kinases 1/2 in cardiac hypertrophy and cell death. Circ Res 91:776-781

Bukoski RD (1990) Intracellular Ca^{2+} metabolism of isolated resistance arteries and cultured vascular myocytes of spontaneously hypertensive and Wistar-Kyoto normotensive rats. J Hypertens 8:37-43

Cao Y, Li H, Mu FT et al (2002) Telomerase activation causes vascular smooth muscle cell proliferation in genetic hypertension. FASEB J 16:96-98

Champagne MJ, Dumas P, Orlov SN et al (1999) Protection against necrosis but not apoptosis by HSPs in vascular smooth muscle cells: evidence for distinct modes of cell death. Hypertension 33:906-913

Cheng HF, Wang JL, Vinson GP et al (1998) Young SHR express increased type 1 angiotensin II receptors in renal proximal tubule. Am J Physiol 274:F10-F17

Choi JH, Yoo KH, Cheon HW et al (2002) Angiotensin converting enzyme inhibition decreases cell turnover in the neonatal rat heart. Pediatr Res 52:325-332

Choy JC, Granville DJ, Hunt DW et al (2001) Endothelial cell apoptosis: biochemical characteristics and potential implications for atherosclerosis. J Mol Cell Cardiol 33:1673-1690

Chueh SC, Guh JH, Chen J et al (2001) Dual effects of ouabain on the regulation of proliferation and apoptosis in human prostatic smooth muscle cells. J Urol 166:347-353

Cingolani HE (1999) Na+/H+ exchange hyperactivity and myocardial hypertrophy: are they linked phenomena? Cardiovasc Res 44:462-467

Clegg KB, Sambhi MP (1989) Inhibition of epidermal growth factor-mediated DNA synthesis by a specific tyrosine kinase inhibitor in vascular smooth muscle cells of the spontaneously hypertensive rat. J Hypertens 7(Suppl 6):S144-S145

Cowley Jr AW, Stoll M, Greene AS et al (2000) Genetically defined risk of salt sensitivity in an intercross of Brown Norway and Dahl S rats. Physiol Genomics 2:107-115

Davis A, Hogarth K, Fernandes D et al (2003) Functional significance of protein kinase A (PKA) activation by endothelin-1 and ATP: negative regulation of SRF-dependent gene expression by PKA. Cell Signal 15:597-604

deBlois D, Viswanathan M, Su JE et al (1996) Smooth muscle DNA replication in response to angiotensin II is regulated differently in the neointima and media at different times after balloon injury in the rat carotid artery. Role of AT1 receptor expression. Arterioscler Thromb Vasc Biol 16:1130-1137

deBlois D, Tea BS, Than VD et al (1997) Smooth muscle cell apoptosis during vascular regression in spontaneously hypertensive rats. Hypertension 29:340-349

Deng AY, Dene H, Rapp JP et al (1997) Congenic strains for the blood pressure quantitative trait locus on rat chromosome 2. Hypertension 30[part 1]:199-202

Der Sarkissian S, Marchand EL, Duguay D et al (2003) Reversal of interstitial fibroblast hyperplasia via apoptosis in hypertensive rat heart with valsartan or enalapril. Cardiovasc Res 57:775-783

Devlin AM, Gordon JM, Davidson AO et al (1995) The effects of perindopril on vascular smooth muscle polyploidy in stroke-prone spontaneously hypertensive rats. J Hypertens 13:211-218

Devlin AM, Clark JS, Reid JL et al (2000) DNA synthesis and apoptosis in smooth muscle cells from a model of genetic hypertension. Hypertension 36:110-115

Devlin AM, Solban N, Tremblay S et al (2003) HCaRG is a novel regulator of renal epithelial cell growth and differentiation causing G_2M arrest. Am J Physiol Renal Physiol 284:F753-F762

Diep QN, Li JS, Schiffrin EL (1999) In vivo study of AT(1) and AT(2) angiotensin receptors in apoptosis in rat blood vessels. Hypertension 34:617-624

Diez J, Panizo A, Hernandez M et al (1997) Cardiomyocyte apoptosis and cardiac angiotensin-converting enzyme in spontaneously hypertensive rats. Hypertension 30:1029-1034

Dimri GP, Lee X, Basile G et al (1995) A biomarker that identifies senescent human cells in culture and in aging skin in vivo. Proc Natl Acad Sci USA 92:9363-9367

Dmitrieva RI, Doris PA (2002) Cardiotonic steroids: potential endogenous sodium pump ligands with diverse function. Exp Biol Med 227:561-569

Domanico SZ, DeNagel DC, Dahlseid JN et al (1993) Cloning of the gene encoding peptide-binding protein 74 shows that it is a new member of the heat shock protein 70 family. Mol Cell Biol 13:3598-3610

Dominiczak AF, Devlin AM, Lee WK et al (1996) Vascular smooth muscle polyploidy and cardiac hypertrophy in genetic hypertension. Hypertension 27:752-759

Dulin NO, Niu J, Browning DD et al (2001) Cyclic AMP-independent activation of protein kinase A by vasoactive peptides. J Biol Chem 276:20827-20830

Dumas P, Pausova Z, Kren V et al (2000a) Contribution of autosomal loci and the Y chromosome to the stress response in rat. Hypertension 35:568-573

Dumas P, Sun Y, Corbeil G et al (2000b) Mapping of quantitative trait loci (QTL) of differential stress gene expression in rat recombinant inbred strains. J Hypertens 18:545-551

Eisen MB, Spellman PT, Brown PO et al (1998) Cluster analysis and display of genome-wide expression patterns. Proc Natl Acad Sci USA 95:14863-14868

Falciola J, Volet B, Anner RM et al (1994) Role of cell membrane Na,K-ATPase for survival of human lymphocytes in vitro. Biosci Rep 14:189-204

Flatt PM, Polyak K, Tang LJ et al (2000) p53-dependent expression of PIG3 during proliferation, genotoxic stress, and reversible growth arrest. Cancer Lett 156:63-72

Garrett MR, Dene H, Walder R et al (1998) Genome scan and congenic strains for blood pressure QTL using Dahl salt-sensitive rats. Genome Res 8:711-723

Gilbert M, Knox S (1997) Influence of Bcl-2 overexpression on Na+/K(+)-ATPase pump activity: correlation with radiation-induced programmed cell death. J Cell Physiol 171:299-304

Golomb E, Hill MR, Brown RG et al (1994) Ouabain enhances the mitogenic effect of serum in vascular smooth muscle cells. Am J Hypertens 7:69-74

Grinstein S, Smith JD, Benedict SH et al (1989) Activation of sodium-hydrogen exchange by mitogens. In: Hoffmann JF, Schulz SG, Glebisch G (eds) Cellular and molecular biology of sodium transport. Curr Top Membr Transp Series, vol 24. Academic Press, pp 331–343

Gu L, Dene H, Deng AY et al (1996) Genetic mapping of two blood pressure quantitative trait loci on rat chromosome 1. J Clin Invest 97:777-788

Guicheney P, Wauquier I, Paquet JL et al (1991) Enhanced response to growth factors and to angiotensin II of spontaneously hypertensive rat skin fibroblasts in culture. J Hypertens 9:23-27

Hadrava V, Tremblay J, Hamet P et al (1989) Abnormalities in growth characteristics of aortic smooth muscle cells in spontaneously hypertensive rats. Hypertension 13:589-597

Hadrava V, Tremblay J, Sekaly RP et al (1992) Accelerated entry of aortic smooth muscle cells from spontaneously hypertensive rats into the S phase of the cell cycle. Biochem Cell Biol 70:599-604

Hall JE, Brands MW, Henegar JR (1999) Angiotensin II and long-term arterial pressure regulation: the overriding dominance of the kidney. J Am Soc Nephrol 10 (Suppl 12):S258-S265

Hamada M, Nishio I, Baba A et al (1990) Enhanced DNA synthesis of cultured vascular smooth muscle cells from spontaneously hypertensive rats—difference of response to growth factor, intracellular free calcium concentration and DNA synthesizing cell cycle. Atherosclerosis 81:191-198

Hamet P (1995) Proliferation and apoptosis in hypertension. Curr Opin Nephrol Hypertens 4:1-7

Hamet P (1997) Cancer and hypertension: a potential for crosstalk? J Hypertens 15:1573-1577

Hamet P, Richard L, Dam TV et al (1995) Apoptosis in target organs of hypertension. Hypertension 26:642-648

Hamet P, Moreau P, Dam TV et al (1996a) The time window of apoptosis: a new component in the therapeutic strategy for cardiovascular remodeling. J Hypertens 14 (Suppl 5):S65-S70

Hamet P, Kaiser MA, Sun Y et al (1996b) HSP27 locus cosegregates with left ventricular mass independently of blood pressure. Hypertension 28:1112-1117

Hamet P, Pausova Z, Dumas P et al (1998) Newborn and adult recombinant inbred strains: a tool for the search of genetic determinants of target organ damage in hypertension. Kidney Int 53:1488-1492

Hamet P, Thorin-Trescases N, Moreau P et al (2001) Excess growth and apoptosis. Is hypertension a case of accelerated aging of cardiovascular cells? Hypertension 37:760-766

Harris EL, Grigor MR, Millar JA (1990) Differences in mitogenic responses to angiotensin II, calf serum and phorbol ester in vascular smooth muscle cells from two strains of genetically hypertensive rat. Biochem Biophys Res Commun 170:1249-1255

Harris EL, Phelan EL, Thompson CM et al (1995) Heart mass and blood pressure have separate genetic determinants in the New Zealand genetically hypertensive (GH) rat. J Hypertens 13:397-404

Hatori N, Fine BP, Nakamura A et al (1987) Angiotensin II effect on cytosolic pH in cultured rat vascular smooth muscle cells. J Biol Chem 262:5073-5078

Heessen S, Leonchiks A, Issaeva N et al (2002) Functional p53 chimeras containing the Epstein-Barr virus Gl-Ala repeat are protected from Mdm2- and HPV-E6-induced proteolysis. Proc Natl Acad Sci USA 99:1532-1537

Hilton PJ, White RW, Lord GA et al (1996) An inhibitor of the sodium pump obtained from human placenta. Lancet 348:303-305

Horita S, Zheng Y, Hara C et al (2002) Biphasic regulation of Na^+-HCO_3-cotransporter by angiotensin II type 1A receptor. Hypertension 40:707-712

Hou Y, Delamere NA (2002) Influence of ANG II on cytoplasmic sodium in cultured rabbit nonpigmented ciliary epithelium. Am J Physiol Cell Physiol 283:C552-C559

Intengan HD, Schiffrin EL (2000) Structure and mechanical properties of resistance arteries in hypertension: role of adhesion molecules and extracellular matrix determinants. Hypertension 36:312-318

Isaev NK, Stelmashook EV, Halle A et al (2000) Inhibition of $Na(+),K(+)$-ATPase activity in cultured rat cerebellar granule cells prevents the onset of apoptosis induced by low potassium. Neurosci Lett 283:41-44

Ito S, Nara Y, Yamori Y (1995) Distinction of endothelial cell growth and fibrinolytic activity between WKY/Izm and SHRSP/Izm in vitro. Clin Exp Pharmacol Physiol 22 (Suppl 1):S273-S274

Jacob H, Lindpaintner K, Lincoln SE et al (1991) Genetic mapping of a gene causing hypertension in the stroke-prone spontaneously hypertensive rat. Cell 67:213-224

Jeanclos E, Schork NJ, Kyvik KO et al (2000) Telomere length inversely correlates with pulse pressure and is highly familial. Hypertension 36:195-200

Johnson TE, Umbenhauer DR, Hill R et al (1992) Karyotypic and phenotypic changes during in vitro aging of human endothelial cells. J Cell Physiol 150:17-27

Kang J, Posner P, Sumners C (1994) Angiotensin II type 2 receptor stimulation of neuronal K+ currents involves an inhibitory GTP binding protein. Am J Physiol 267:C1389-C1397

Kaul SC, Reddel RR, Sugihara T et al (2000) Inactivation of p53 and life span extension of human diploid fibroblasts by mot-2. FEBS Lett 474:159-164

Kawano K, Ikari A, Nakano M et al (2002) Phosphatidylinositol 3-kinase mediates inhibitory effect of angiotensin II on sodium/glucose cotransporter in renal epithelial cells. Life Sci 71:1-13

King FW, Wawrzynow A, Hohfeld J et al (2001) Co-chaperones Bag-1, Hop and Hsp40 regulate Hsc70 and Hsp90 interactions with wild-type or mutant p53. EMBO J 20:6297-6305

Korniszewski L, Nowak R, Okninska-Hoffmann E et al (2001) Wiedemann-Rautenstrauch (neonatal progeroid) syndrome: new case with normal telomere length in skin fibroblasts. Am J Med Genet 103:144-148

Kren V, Pravenec M, Lu S et al (1997) Genetic isolation of a region of chromosome 8 that exerts major effects on blood pressure and cardiac mass in the spontaneously hypertensive rat. J Clin Invest 99:577-581

Krushkal J, Xiong M, Ferrell R et al (1998) Linkage and association of adrenergic and dopamine receptor genes in the distal portion of the long arm of chromosome 5 with systolic blood pressure variation. Hum Mol Genet 7:1379-1383

Kubbutat MH, Vousden KH (1997) Proteolytic cleavage of human p53 by calpain: a potential regulator of protein stability. Mol Cell Biol 17:460-468

Kusuhara M, Takahashi E, Peterson TE et al (1998) p38 Kinase is a negative regulator of angiotensin II signal transduction in vascular smooth muscle cells: effects on Na+/H+ exchange and ERK1/2. Circ Res 83:824-831

LaPointe MS, Ye M, Moe OW et al (1995) Na^+/H^+ antiporter (NHE-1 isoform) in cultured vascular smooth muscle from the spontaneously hypertensive rat. Kidney Int 47:78-87

Ledbetter ML, Young GJ, Wright ER (1986) Cooperation between epithelial cells demonstrated by potassium transfer. Am J Physiol 250:C306-C313

Lee RM, Conyers RB, Kwan CY (1992) Incidence of multinucleated and polyploid aortic smooth muscle cells cultured from different age groups of spontaneously hypertensive rats. Can J Physiol Pharmacol 70:1496-1501

Lifton RP, Gharavi AG, Geller DS (2001) Molecular mechanisms of human hypertension. Cell 104:545-556

Lucchesi PA, Bell JM, Willis LS et al (1996) Ca(2+)-dependent mitogen-activated protein kinase activation in spontaneously hypertensive rat vascular smooth muscle defines a hypertensive signal transduction phenotype. Circ Res 78:962-970

Ludens JH, Clark MA, DuCharme DW et al (1991) Purification of an endogenous digitalis-like factor from human plasma for structural analysis. Hypertension 17:923-929

Ly DH, Lockhart DJ, Lerner RA et al (2000) Mitotic misregulation and human aging. Science 287:2486-2492

Manunta P, Stella P, Rivera R et al (1999) Left ventricular mass, stroke volume, and ouabain-like factor in essential hypertension. Hypertension 34:450-456

Marklund L, Henriksson R, Grankvist K (1999) K^+-efflux modulation of cisplatin-induced apoptosis and cytotoxicity to cultured mesothelioma cells. Invest Ophthalmol Visual Sci 38:S109 (Abstract)

Matsukawa T, Ichikawa I (1997) Biological functions of angiotensin and its receptors. Ann Rev Physiol 59:395-412

McConkey DJ, Lin Y, Nutt LK et al (2000) Cardiac glycosides stimulate Ca2+ increases and apoptosis in androgen-independent, metastatic human prostate adenocarcinoma cells. Cancer Res 60:3807-3812

Moreau P, Tea BS, Dam TV et al (1997) Altered balance between cell replication and apoptosis in hearts and kidneys of newborn SHR. Hypertension 30[Part 2]:720-724

Morris K (2002) Targeting the myocardial sodium-hydrogen exchange for treatment of heart failure. Expert Opin Ther Targets 6:291-298

Moudgil R, Musat-Marcu S, Xu Y et al (2002) Increased AT(2)R protein expression but not increased apoptosis during cardioprotection induced by AT(1)R blockade. Can J Cardiol 18:1107-1116

Mueller C, Baudler S, Welzel H et al (2002) Identification of a novel redox-sensitive gene, Id3, which mediates angiotensin II-induced cell growth. Circulation 105:2423-2428

Nakayama KI, Hatakeyama S, Nakayama K (2001) Regulation of the cell cycle at the G1-S transition by proteolysis of cyclin E and p27Kip1. Biochem Biophys Res Commun 282:853-860

Nakayama M, Fukuda N, Watanabe Y et al (1999) Low dose of eicosapentaenoic acid inhibits the exaggerated growth of vascular smooth muscle cells from spontaneously hypertensive rats through suppression of transforming growth factor-beta. J Hypertens 17:1421-1430

Nouet S, Nahmias C (2000) Signal transduction from the angiotensin II AT2 receptor. Trends Endocrinol Metab 11:1-6

Ogihara T, Hata T, Tanaka K et al (1986) Hutchinson-Gilford progeria syndrome in a 45-year-old man. Am J Med 81:135-138

Okorokov AL, Ponchel F, Milner J (1997) Induced N- and C-terminal cleavage of p53: a core fragment of p53, generated by interaction with damaged DNA, promotes cleavage of the N-terminus of full-length p53, whereas ssDNA induces C-terminal cleavage of p53. EMBO J 16:6008-6017

Olej B, dos Santos NF, Leal L et al (1998) Ouabain induces apoptosis on PHA-activated lymphocytes. Biosci Rep 18:1-7

Orlov SN, Pokudin NI, Kotelevtsev YV et al (1989) Volume-dependent regulation of ion transport and membrane phosphorylation in human and rat erythrocytes. J Membr Biol 107:105-107

Orlov SN, Resink TJ, Bernhardt J et al (1992) Na^+-K^+ pump and Na^+-K^+ co-transport in cultured vascular smooth muscle cells from spontaneously hypertensive and normotensive rats: baseline activity and regulation. J Hypertens 10:733-740

Orlov SN, Dam TV, Tremblay J et al (1996) Apoptosis in vascular smooth muscle cells: role of cell shrinkage. Biochem Biophys Res Commun 221:708-715

Orlov SN, Adragna NC, Adarichev VA et al (1999a) Genetic and biochemical determinants of abnormal monovalent ion transport in primary hypertension. Am J Physiol 276:C511-C536

Orlov SN, deBlois D, Tremblay J et al (1999b) Apoptosis in hypertension: mechanisms and implications in vascular remodeling. Cardiovasc Risk Factors 9:67-79

Orlov SN, Thorin-Trescases N, Dulin NO et al (1999c) Activation of cAMP signaling transiently inhibits apoptosis in vascular smooth muscle cells in a site upstream of caspase-3. Cell Death Differ 6:661-672

Orlov SN, Thorin-Trescases N, Kotelevtsev SV et al (1999d) Inversion of the intracellular Na+/K+ ratio blocks apoptosis in vascular smooth muscle at a site upstream of caspase-3. J Biol Chem 274:16545-16552

Orlov SN, Taurin S, Tremblay J et al (2001) Inhibition of Na+,K+ pump affects nucleic acid synthesis and smooth muscle cell proliferation via elevation of the [Na+]i/[K+]i ratio: possible implication in vascular remodelling. J Hypertens 19:1559-1565

Orlov SN, Tremblay J, deBlois D et al (2002) Genetics of programmed cell death and proliferation. Semin Nephrol 22:161-171

Orlov SN, Pchejetski DV, Sarkissian SD et al (2003) [^{3}H]Thymidine labelling of DNA triggers apoptosis potentiated by E1A-adenoviral protein. Apoptosis 8:199-208

Ortiz PA, Garvin JL (2001) Intrarenal transport and vasoactive substances in hypertension. Hypertension 38:621-624

Owens GK (1989) Control of hypertrophic versus hyperplastic growth of vascular smooth muscle cells. Am J Physiol 257:H1755-H1765

Owens GK (1995) Regulation of differentiation of vascular smooth muscle cells. Physiol Rev 75:487-517

Pang SC, Long C, Poirier M et al (1986) Cardiac and renal hyperplasia in newborn genetically hypertensive rats. J Hypertens 4 (Suppl 3):S119-S122

Panizo-Santos A, Sola JJ, Pardo-Mindan FJ et al (1995) Angiotensin converting enzyme inhibition prevents polyploidization of cardiomyocytes in spontaneously hypertensive rats with left ventricular hypertrophy. J Pathol 177:431-437

Paquet JL, Baudouin-Legros M, Marche P et al (1989) Enhanced proliferating activity of cultured smooth muscle cells from SHR. Am J Hypertens 2:108-110

Park WY, Hwang CI, Kang MJ et al (2001) Gene profile of replicative senescence is different from progeria or elderly donor. Biochem Biophys Res Commun 282:934-939

Pchejetski D, Taurin S, Der Sarkissian S et al (2003) Inhibition of Na$^+$,K$^+$-ATPase by ouabain triggers epithelial cell death independently of inversion of the [Na$^+$]$_i$/[K$^+$]$_i$ ratio. Biochem Biophys Res Commun 301:735-744

Peiro C, Angulo J, Llergo JL et al (1997) Angiotensin II mediates cell hypertrophy in vascular smooth muscle cultures from hypertensive Ren-2 transgenic rats by an amiloride- and furosemide-sensitive mechanism. Biochem Biophys Res Commun 240:367-371

Peng Y, Chen L, Li C et al (2001) Inhibition of MDM 2 by hsp90 contributes to mutant p53 stabilization. J Biol Chem 276:40583-40590

Penning LC, Denecker G, Vercammen D et al (2000) A role for potassium in TNFinduced apoptosis and gene-induction in human and rodent tumour cell lines. Cytokine 12:747-750

Phan VN, Kusuhara M, Lucchesi PA et al (1997) A 90-kD Na(+)-H+ exchanger kinase has increased activity in spontaneously hypertensive rat vascular smooth muscle cells. Hypertension 29:1265-1272

Piechaczyk M, Blanchard JM (1994) c-fos proto-oncogene regulation and function. Crit Rev Oncol Hematol 17:93-131

Pochampally R, Fodera B, Chen L et al (1999) Activation of an MDM 2-specific caspase by p53 in the absence of apoptosis. J Biol Chem 274:15271-15277

Pollman MJ, Yamada T, Horiuchi M et al (1996) Vasoactive substances regulate vascular smooth muscle cell apoptosis. Countervailing influences of nitric oxide and angiotensin II. Circ Res 79:748-756

Postnov YV, Orlov SN (1985) Ion transport across plasma membrane in primary hypertension. Physiol Rev 65(4):904-945

Pravenec M, Klir P, Kren V et al (1989) An analysis of spontaneous hypertension in spontaneously hypertensive rats by means of new recombinant inbred strains. J Hypertens 7:217-222

Pravenec M, Gauguier D, Schott JJ et al (1995) Mapping of quantitative trait loci for blood pressure and cardiac mass in the rat by genome scanning of recombinant inbred strains. J Clin Invest 96:1973-1978

Rangel LB, Caruso-Neves C, Lara LS et al (2002) Angiotensin II stimulates renal proximal tubule Na(+)-ATPase activity through the activation of protein kinase C. Biochim Biophys Acta 1564:310-316

Rivard A, Andres V (2000) Vascular smooth muscle cell proliferation in the pathogenesis of atherosclerotic cardiovascular diseases. Histol Histopathol 15:557-571

Robey RB, Ruiz OS, Espiritu DJ et al (2002) Angiotensin II stimulation of renal epithelial cell Na/HCO3 cotransport activity: a central role for Src family kinase/classic MAPK pathway coupling. J Membr Biol 187:135-145

Rosman NP, Anselm I, Bhadelia RA (2001) Progressive intracranial vascular disease with strokes and seizures in a boy with progeria. J Child Neurol 16:212-215

Rosskopf D, Dusing R, Siffert W (1993) Membrane sodium-proton exchange and primary hypertension. Hypertension 21:607-617

Ruiz-Ortega M, Lorenzo O, Ruperez M et al (2000) Angiotensin II activates nuclear transcription factor kb through AT_1 and AT_2 in vascular smooth muscle cells. Circ Res 86:1266-1272

Ruiz-Ortega M, Lorenzo O, Ruperez M et al (2001) Role of the renin-angiotensin system in vascular diseases: expanding the field. Hypertension 38:1382-1387

Sachinidis A, Seul C, Ko Y et al (1996) Effect of the Na^+/H^+ antiport inhibitor Hoe 694 on the angiotensin II-induced vascular smooth muscle cell growth. Br J Pharmacol 119:787-796

Saltis J, Bobik A (1992) Vascular smooth muscle growth in genetic hypertension: evidence for multiple abnormalities in growth regulatory pathways. J Hypertens 60:635-644

Sarkar PK, Shinton RA (2001) Hutchinson-Guilford progeria syndrome. Postgrad Med J 77:312-317

Sato A, Ozawa K (1977) Effect of ouabain on the hyperosmolarity tolerant cells from rat kidney. Jpn J Pharmacol 27:168-170

Scott-Burden T, Resink TJ, Baur U et al (1989) Epidermal growth factor responsiveness in smooth muscle cells from hypertensive and normotensive rats. Hypertension 13:295-304

Sich B, Kirch U, Tepel M et al (1996) Pulse pressure correlates in humans with a proscillaridin A immunoreactive compound. Hypertension 27:1073-1078

Siffert W, Dusing R (1995) Sodium-proton exchange and primary hypertension. An update. Hypertension 26:649-655

Slagboom PE, Heijmans BT, Beekman M et al (2000) Genetics of human aging. The search for genes contributing to human longevity and diseases of the old. Ann NY Acad Sci 908:50-63

Smith JB, Smith L (1987) $Na^+/K^+/Cl^-$ cotransport in cultured vascular smooth muscle cells: stimulation by angiotensin II and calcium ionophores, inhibition by cyclic AMP and calmodulin antagonists. J Membr Biol 99:51-63

Solban N, Jia HP, Richard S et al (2000) HCaRG, a novel calcium-regulated gene coding for a nuclear protein, is potentially involved in the regulation of cell proliferation. J Biol Chem 275:32234-32243

Stehbens WE, Delahunt B, Shozawa T et al (2001) Smooth muscle cell depletion and collagen types in progeric arteries. Cardiovasc Pathol 10:133-136

Stoll M, Unger T (2001) Angiotensin and its AT2 receptor: new insights into an old system. Regul Pept 99:175-182

Stoll M, Steckelings UM, Paul M et al (1995) The angiotensin AT2-receptor mediates inhibition of cell proliferation in coronary endothelial cells. J Clin Invest 95:651-657

Stoll M, Cowley AW Jr, Tonellato PJ et al (2001) A genomic-systems biology map for cardiovascular function. Science 294:1723-1726

Sylvester AM, Chen D, Krasinski K et al (1998) Role of c-fos and E2F in the induction of cyclin A transcription and vascular smooth muscle cell proliferation. J Clin Invest 101:940-948

Tamai KT, Monaco L, Nantel F et al (1997) Coupling signalling pathways to transcriptional control: nuclear factors responsive to cAMP. Recent Prog Horm Res 52:121-139

Tanase H, Yamori Y, Hansen CT et al (1982) Heart size in inbred strains of rats. II: Genetic determination of the development of cardiovascular enlargement in rats. Hypertension 4:864-872

Taurin S, Seyrantepe V, Orlov SN et al (2002a) Proteome analysis and functional expression identify mortalin as an antiapoptotic gene induced by elevation of $[Na+]i/[K+]i$ ratio in cultured vascular smooth muscle cells. Circ Res 91:915-922

Taurin S, Dulin NO, Pchejetski D et al (2002b) c-Fos expression in ouabain-treated vascular smooth muscle cells from rat aorta: evidence for an intracellular-sodium-mediated, calcium-independent mechanism. J Physiol 543:835-847

Tea BS, Dam TV, Moreau P et al (1999) Apoptosis during regression of cardiac hypertrophy in spontaneously hypertensive rats. Temporal regulation and spatial heterogeneity. Hypertension 34:229-235

Tea BS, Der Sarkissian S, Touyz RM et al (2000) Pro-apoptotic and growth inhibitory role of angiotensin II type 2 receptor in vascular smooth muscle cells of spontaneously hypertensive rats *in vivo*. Hypertension 35:1069-1073

Thomas D, Harris PJ, Morgan TO (1990) Altered responsiveness of proximal tubule fluid reabsorption of peritubular angiotensin II in spontaneously hypertensive rats. J Hypertens 8:407-410

Thorin-Trescases N, deBlois D, Hamet P (2001) Evidence of an altered in vivo vascular cell turnover in spontaneously hypertensive rats and its modulation by long-term antihypertensive treatment. J Cardiovasc Pharmacol 38:764-774

Touyz RM, Schiffrin EL (1997) Angiotensin II regulates vascular smooth muscle cell pH, contraction, and growth via tyrosine kinase-dependent signaling pathways. Hypertension 30:222-229

Touyz RM, Schiffrin EL (1999) Activation of the Na(+)-H+ exchanger modulates angiotensin II-stimulated Na(+)-dependent Mg2+ transport in vascular smooth muscle cells in genetic hypertension. Hypertension 34:442-449

Touyz RM, Tolloczko B, Schiffrin EL (1994) Mesenteric vascular smooth muscle cells from spontaneously hypertensive rats display increased calcium responses to angiotensin II but not to endothelin-1. J Hypertens 12:663-673

Touyz RM, Chen X, Tabet F et al (2002) Expression of a functionally active gp91phox-containing neutrophil-type NAD(P)H oxidase in smooth muscle cells from human resistance arteries: regulation by angiotensin II. Circ Res 90:1205-1213

Tremblay J, Hadrava V, Kruppa U et al (1992) Enhanced growth-dependent expression of TGFb1 and *hsp70* genes in aortic smooth muscle cells from spontaneously hypertensive rats. Can J Physiol Pharmacol 70:565-572

Tsuganezawa H, Preisig PA, Alpern RJ (1998) Dominant negative c-Src inhibits angiotensin II induced activation of NHE3 in OKP cells. Kidney Int 54:394-398

Tsuruoka S, Nishiki K, Sugimoto K et al (2001) Chronotherapy with active vitamin D3 in aged stroke-prone spontaneously hypertensive rats, a model of osteoporosis. Eur J Pharmacol 428:287-293

Tymiak AA, Norman JA, Bolgar M et al (1993) Physicochemical characterization of a ouabain isomer isolated from bovine hypothalamus. Proc Natl Acad Sci USA 90:8189-8193

Tyner SD, Venkatachalam S, Choi J et al (2002) p53 mutant mice that display early ageing-associated phenotypes. Nature 415:45-53

Uehara Y, Numabe A, Kawabata Y et al (1991) Rapid smooth muscle cell growth and endogenous prostaglandin system in spontaneously hypertensive rats. Am J Hypertens 4:806-814

Urenjak J, Obrenovitch TP (1996) Pharmacological modulation of voltage-gated Na^+ channels: a rational and effective strategy against ischemic brain damage. Pharmacol Rev 48:22-67 (Abstract)

Venance SL, Watson MH, Wigle DA et al (1993) Differential expression and activity of p34cdc2 in cultured aortic adventitial fibroblasts derived from spontaneously hypertensive and Wistar-Kyoto rats. J Hypertens 11:483-489

Wadhwa R, Kaul SC, Ikawa Y et al (1993a) Identification of a novel member of mouse hsp70 family. Its association with cellular mortal phenotype. J Biol Chem 268:6615-6621

Wadhwa R, Kaul SC, Sugimoto Y et al (1993b) Induction of cellular senescence by transfection of cytosolic mortalin cDNA in NIH 3T3 cells. J Biol Chem 268:22239-22242

Walsh RA, Dorn II GW (1998) Growth and hypertrophy of the heart and blood vessels. McGraw-Hill, New York

Walter SV, Hamet P (1986) Enhanced DNA synthesis in heart and kidney of newborn spontaneously hypertensive rats. Hypertension 8:520-525

Watanabe T, Hashimoto M, Okuyama S et al (1999) Effects of targeted disruption of the mouse angiotensin II type 2 receptor gene on stress-induced hyperthermia. J Physiol 515 (Part 3):881-885

Wax SD, Tsao L, Lieb ME et al (1996) SM-20 is a novel 40-kd protein whose expression in the arterial wall is restricted to smooth muscle. Lab Invest 74:797-808

Wolf G, Harendza S, Schroeder R et al (2002) Angiotensin II's antiproliferative effects mediated through AT2-receptors depend on down-regulation of SM-20. Lab Invest 82:1305-1317

Yamagata K, Nara Y, Tagami M et al (1995) Demonstration of hereditarily accelerated proliferation in astrocytes derived from spontaneously hypertensive rats. Clin Exp Pharmacol Physiol 22:605-609

Yamori Y, Igawa T, Kanbe T et al (1981) Mechanisms of structural vascular changes in genetic hypertension: analyses on cultured vascular smooth muscle cells from spontaneously hypertensive rats. Clin Sci 61:121s-123s

Ye M, Flores G, Batlle D (1996) Angiotensin II and angiotensin-(1-7) effects on free cytosolic sodium, intracellular pH, and the Na(+)-H+ antiporter in vascular smooth muscle. Hypertension 27:72-78

Yeh JY, Huang WJ, Kan SF et al (2001) Inhibitory effects of digitalis on the proliferation of androgen dependent and independent prostate cancer cells. J Urol 166:1937-1942

Zhou X, Jiang G, Zhao A et al (2001) Inhibition of Na,K-ATPase activates PI3 kinase and inhibits apoptosis in LLC-PK1 cells. Biochem Biophys Res Commun 285:46-51

Zhu DL, Herembert T, Marche P (1991) Increased proliferation of adventitial fibroblasts from spontaneously hypertensive rat aorta. J Hypertens 9:1161-1168

Regulation of Angiogenesis by Angiotensin II

J. Haendeler · S. Dimmeler

Molecular Cardiology, Department of Internal Medicine IV,
University of Frankfurt, Theodor-Stern-Kai 7, Frankfurt, Germany
e-mail: Dimmeler@em.uni-frankfurt.de

1	Introduction	100
2	Role of the Renin–Angiotensin System in Angiogenesis	100
3	Pro-angiogenic Effects of Angiotensin II	102
3.1	Stimulation of VEGF, bFGF, and Nitric Oxide Synthase by Angiotensin II	102
3.2	Transactivation of Growth Factor Receptors by Angiotensin II	103
3.3	Regulation of Pro-angiogenic Transcription Factors by Angiotensin II	105
4	Antiangiogenic Effects of Angiotensin II Via the AT2R: Important Role for Apoptosis	106
5	Conclusions	107
References		107

Abstract The renin/angiotensin system can exert a double-edged role in angiogenesis signaling. The pro-angiogenic activity of angiotensin II is predominantly mediated by stimulation of the angiotensin II receptor 1 (AT1R), whereas the angiotensin receptor 2 (AT2R) inhibits angiogenesis. The pro-angiogenic activity of angiotensin II is believed to be predominantly mediated by stimulation of VEGF expression. Moreover, a transactivation of the growth promoting epidermal growth factor receptor (EGFR) and the enhanced expression of pro-angiogenic transcription factors such as Cyr61 and Krüppel-like factor may contribute to the pro-angiogenic effects of angiotensin II. In contrast, stimulation of the AT2R inhibits angiogenesis by preventing endothelial cell proliferation and inducing apoptosis of endothelial cells, which counteracts angiogenesis. The apoptosis induction is mediated by an activation of phosphatases and subsequent down-regulation of the anti-apoptotic protein Bcl-2.

Keywords Angiogenesis · Apoptosis · Cyr61 · Double-edged role · Transactivation · VEGF

1
Introduction

Adequate oxygen delivery is essential to maintain normal tissue function. Understanding of the signaling processes that regulate the formation of new blood vessels has become a major challenging objective of the last decade. Postnatal neovascularization is essential after tissue ischemia in order to adapt blood supply. Moreover, neovascularization plays an important role in wound healing. However, in a number of pathophysiological situations, neovascularization may accelerate disease progression (Folkman 1995). Thus, neovascularization of tumor tissue allows the tumor to grow beyond a diameter of 2 mm (Folkman 1995). Neovascularization has also been implicated in atherosclerotic plaque growth, although the contribution of neovascularization for plaque rupture and the initiation of myocardial infarction is controversial (Isner 2001). Taken together, strategies to either improve neovascularization in order to improve oxygen supply, e.g., after myocardial infarction, or concepts to inhibit neovascularization in order to limit tumor growth are important therapeutic options (Isner and Asahara 1999).

During the last few years, the understanding of the processes leading to neovascularization has dramatically changed. In the past, the vascularization of ischemic tissue was believed to be caused by the migration and proliferation of mature endothelial cells—a process termed "angiogenesis" (Isner and Asahara 1999; Carmeliet 2000). Meanwhile, increasing evidence suggests that circulating progenitor cells home to sites of ischemia and contribute to the formation of new blood vessels. In analogy to the embryonic development of blood vessels out of primitive endothelial progenitors (angioblasts) and the subsequent organization into a primary capillary plexus, this process is called "vasculogenesis" (Isner and Asahara 1999; Carmeliet 2000). Since these novel insights into the processes underlying post-natal vessel growth have become evident during the past 2 years, most of the cited studies do not discriminate between angiogenesis and vasculogenesis. Therefore, the use of the term angiogenesis in this review article does not necessarily exclude effects on vasculogenesis.

2
Role of the Renin–Angiotensin System in Angiogenesis

Pharmacological approaches revealed the existence of two subtypes of angiotensin receptors, namely angiotensin II types 1 and 2 receptors (AT1R and AT2R) (Unger et al. 1996). In mice, the AT1R is further subdivided into AT1aR and AT1bR (Martin et al. 1995). The AT1R is ubiquitously and abundantly distributed (Murphy et al. 1991; Sasaki et al. 1991). Most of the well-known functions in the cardiovascular system are mediated through the AT1R and especially through the AT1aR in rodents (Martin et al. 1995). In contrast, little is known regarding the physiological roles of the AT2R and its signal–transduction pathways. Several lines of evidence, however, suggest that the AT2R might mediate opposite effects to those related to the AT1R activation. It is very well established that angiotensin II-induced smooth muscle cell proliferation is mediated by the AT1R (Rao et al. 1994; Berk 1999). In contrast, studies on the AT2R demonstrate that stimulation of the AT2R induce apoptosis in vascular smooth muscle cells in vitro and in spontaneously hypertensive rats in vivo (Rao et al. 1994; Berk 1999), suggesting the opposing effects of the two receptors.

Another interesting field where angiotensin II has now been implicated is angiogenesis, the development of new vessels from preexisting vessels. The first reports on the effects of the renin–angiotensin system on angiogenesis were obtained from studies using angiotensin-converting enzyme (ACE) inhibitors regardless of the AT receptor subtypes. Volpert et al. showed that captopril suppresses tumor angiogenesis and growth (Volpert et al. 1996). In contrast, Fabre et al. reported that quinapril increases angiogenesis in a rabbit model of hindlimb ischemia (Fabre et al. 1999). Taken together these studies give no precise picture of the renin–angiotensin system in angiogenesis. Different reasons can be found to explain such differences. One rationale to explain the discrepancy of the two mentioned studies might be the use of different ACE inhibitors. Thus, quinapril was shown to more potently inhibit tissue ACE activity in comparison to captopril, and moreover, captopril contains a free SH-group which may inhibit Zn^{2+}-dependent metalloproteinase activity that is required to respond to an angiogenic stimulus (Volpert et al. 1996; Fabre et al. 1999).

Generally, ACE inhibition not only prevents angiotensin II synthesis, but also increases bradykinin production, which leads to stimulation of the endothelial nitric oxide synthase. Activation of the nitric oxide synthase has been reported to promote angiogenesis (Murohara et al. 1998). Therefore, the promoting effects of ACE inhibition could be rather a side effect than a direct function of angiotensin II blockade. Moreover, angiotensin II is not only produced by the angiotensin-converting enzyme in vivo, but also by other enzymes, such as chymase (Muramatsu et al. 2000).

Therefore, recent studies used the advantage of knockout mice either from the AT1aR or from the AT2R to understand the direct role of angiotensin II and the role of the two receptor subtypes in angiogenic processes. Sasaki and coworkers examined the role of the AT1R receptor in a mouse model of ischemia-induced angiogenesis (Sasaki et al. 2002). Well-developed collateral vessels were observed in wild-type mice in response to hindlimb ischemia, whereas ischemia-induced angiogenesis was significantly reduced in AT1aR knockout mice (Sasaki et al. 2002). Moreover, infiltration of mononuclear cells was suppressed in the ischemic tissues of AT1aR knockout mice. Interestingly, the phenotype of reduced angiogenesis observed in AT1aR knockout mice could be rescued by muscular transplantation of mononuclear cells from wild type mice (Sasaki et al. 2002). These data suggest that AT1aR knockout mice show an impairment of circulating progenitor cells or monocytic cells, which are known to contribute to vasculogenesis and arteriogenesis, respectively.

In contrast, AT2R knockout mice showed the opposite results as obtained in the AT1aR knockout mice. Angiogenesis, again analyzed in the model of hindlimb ischemia, was increased in AT2R knockout mice compared to their wild-type littermates (Silvestre et al. 2002). Of note is that Silvestre et al. did not detect a further enhancement of the angiogenic score in AT2R knockout mice by angiotensin II infusion (Silvestre et al. 2002). Likewise, no improvement of ischemia-induced angiogenesis was achieved in wild-type mice by infusion of the well-established AT2R blocker, PD123319 (Tamarat et al. 2002). These findings are in contrast to pharmacological blockade of the AT2R in the rat subcutaneous sponge granuloma, where angiotensin II-induced angiogenesis was reduced by PD1233319 (Walsh et al. 1997). But nevertheless, all these studies further contribute to the concept that the AT1R and the AT2R exert opposite effects on angiogenesis.

Therefore, the following sections will separately discuss the mechanisms underlying the pro-angiogenic effects of the AT1R and the inhibitory effect of AT2R.

3
Pro-angiogenic Effects of Angiotensin II

3.1
Stimulation of VEGF, bFGF, and Nitric Oxide Synthase by Angiotensin II

Over the past few years several angiogenic factors have been identified. Among them, vascular endothelial growth factor (VEGF), is one of the most important players in angiogenesis. This was most impressively demonstrated by the findings that targeted deletion of VEGF or the VEGF-receptors 1 and 2 showed lacks or abnormalities in vessel formation (Carmeliet et al. 1996). Recent studies suggest that the pro-angiogenic effects of angiotensin II depend on the stimulation of VEGF expression in vitro and in vivo (Tamarat et al. 2002). In cultivated bovine aortic endothelial cells, angiotensin II induced expression of the VEGF receptor and thereby enhanced VEGF-induced proliferation and tube formation of the cells. These effects of angiotensin II were blocked by the AT1R antagonist, losartan, demonstrating that these processes are regulated by the AT1R in endothelial cells (Otani et al. 1998). Moreover, several studies were undertaken to show the effect of ACE and/or AT1R blockade on VEGF expression. Microvascular angiogenesis in skeletal muscle induced by electrical stimulation and short-term exercise was inhibited by ACE inhibitors and AT1R blockade (Amaral et al. 2001). Another scenario of more pathological angiogenesis, namely tumor growth, resulted in differential effects of ACE inhibitors and AT1R blockade. Yoshiji et al. showed that the ACE inhibitor, perindopril, but not AT1R blockade significantly inhibited the progression of tumor growth with a concomitant suppression of VEGF expression (Yoshiji et al. 2001). Overall, these studies demonstrate that angiotensin II stimulates VEGF production; however, the up-stream signaling cascade involved is not clearly defined but may include the transactivation of growth factor receptors and stimulation of several transcription factors (see Sects. 6 and 7).

Another intermediate which could link the renin–angiotensin system and VEGF is thrombin. Thrombin is a tightly regulated serine proteinase that plays a central role in hemostasis. Besides the well-known functions of thrombin to convert fibrinogen into fibrin and to stimulate platelet aggregation and secretion, thrombin is a key factor in tissue repair and wound healing. Thrombin has been shown to increase VEGF expression via the thrombin receptor PAR-1 (Williams et al. 1995; Richard et al. 2000) and has been implicated as another angiogenic factor. Browder et al. suggested that angiotensin II-induced VEGF expression led to generation of the tissue factor which will increase intracellular thrombin production and thereby may act as a feedback loop to further enhance VEGF production by angiotensin II through thrombin (Browder et al. 2000).

Besides the stimulation of the expression of VEGF and its receptors, angiotensin II also increased the expression of the endothelial nitric oxide synthase (eNOS) protein content within the ischemic leg in the hindlimb ischemia model (Silvestre et al. 2002) and thereby exerts proangiogenic effects. Indeed, angiotensin II administration did not improve the angiographic score and blood perfusion within the ischemic leg of eNOS knockout mice (Tamarat et al. 2002), suggesting that angiotensin II-mediated VEGF release requires the nitric oxide system for promoting angiogenesis signaling. This is in accordance with various studies that demonstrate that endothelial-derived NO is necessary for VEGF-induced angiogenesis and endothelial function (Murohara et al. 1998; Dimmeler et al. 2000).

Basic fibroblast growth factor and one of its receptors, named endoglin, represent another target of angiotensin II action (Lastres et al. 1996). Recently, it has been demonstrated that angiotensin II stimulates the expression of endoglin (Li et al. 2001). Since endoglin knockout mice showed defects in angiogenesis (Li et al. 1999) and inhibition of endoglin expression in endothelial cells attenuates angiogenesis (Li et al. 2000), one may speculate about a potential contribution of endoglin up-regulation in the pro-angiogenic effects of angiotensin II. The mechanism, however, by which angiotensin II increases the expression of endoglin requires further investigation.

3.2
Transactivation of Growth Factor Receptors by Angiotensin II

It is well established that G protein-coupled receptors stimulate growth factor receptor transactivation, like EGFR and platelet-derived growth factor receptor (PDGFR) (Prenzel et al. 1999; Heeneman et al. 2000). Therefore, transactivation of growth factor receptors is one important signaling mechanism for angiotensin II-induced angiogenesis. Growth factor receptors belong to the family of receptor tyrosine kinases, which are tyrosine phosphorylated upon stimulation. Angiotensin II stimulated classical signaling pathways via transactivation of growth receptors (Fig. 1). Upon stimulation with angiotensin II, pro heparin-binding EGF is cleaved by metalloproteinases and HB-EGF is released, which results in activation of the EGFR (Prenzel et al. 1999). The EGFR is rapidly au-

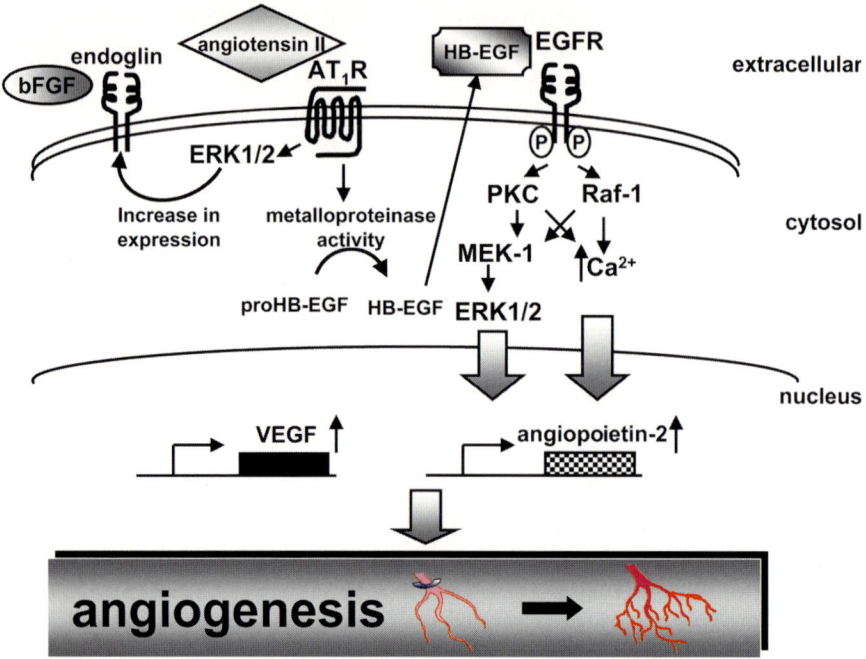

Fig. 1 Transactivation of growth factor receptors by angiotensin II leading to angiogenesis. Model illustrating the different signaling pathways involved. *bFGF*, basic fibroblast growth factor; *EGFR*, epidermal growth factor receptor; *ERK1/2*, extracellular signal-regulated kinase

tophosphorylated leading to protein kinase C activation, release of intracellular Ca^{2+}, and concomitant extracellular signal-regulated kinase (ERK)1/2 activation after 30 min. Activation of these kinases and the intracellular Ca^{2+} release results then in transcriptional activation of angiogenic factors like VEGF and angiopoietin-2. This finally leads to induction of angiogenic processes (Fujiyama et al. 2001). The AT1R as a G protein-coupled receptor requires heterotrimeric, small G proteins for activation. Heterotrimeric G proteins consist of an α, β, and γ subunit and are defined by the identity of their α subunit. Sixteen α subunit genes have been cloned, and on the basis of sequence similarities they are divided into four groups: $G\alpha_{I/o}$, $G\alpha_q$, $G\alpha_s$, $G\alpha_{12/13}$ (Simon et al. 1991). The angiotensin II/AT1R-activated increase in intracellular Ca^{2+} was selectively inhibited by injection of antisense oligonucleotides directed against mRNAs coding for the $G\alpha_{13}$, $G\beta_1$ and $G\gamma_3$ subunits (Macrez-Lepretre et al. 1997), suggesting an important role for these subunits in angiotensin II-induced angiogenesis. This has been most notably underscored by the disruption of the gene encoding for $G\alpha_{13}$. $G\alpha_{13}$ knockout mice indicate a recessive lethal phenotype. Homozygous embryos could only be recovered until day 9.5, showing an impaired ability of endothelial cells to develop into an organized vascular system and demonstrating an important role for $G\alpha_{13}$ in developmental angiogenesis (Offermanns et al.

1997). However, a direct involvement of $G\alpha_{13}$ in transactivation of growth receptors by the AT1R has not been described yet and requires further investigation.

3.3
Regulation of Pro-angiogenic Transcription Factors by Angiotensin II

Recent findings demonstrate that angiotensin II induces expression of the Krüppel-like zinc-finger transcription factor Klf5 (Shindo et al. 2002) (Fig. 2). Initially described as a molecular marker of phenotypically modulated smooth muscle cells, Klf5 also exerts diverse functions during cell differentiation and embryonic development (Aikawa et al. 1995; Bieker 2001). Importantly, heterozygous Klf5 mice showed impaired angiogenic activity in the hindlimb ischemia model, and moreover, angiogenic responses to implanted tumors were markedly attenuated (Shindo et al. 2002). Therefore, one may speculate that Klf5 might be a new candidate to mediate the pro-angiogenic effects of angiotensin II.

Recently, an another transcription factor termed Cyr61 has been considered as a downstream target of angiotensin II (Hilfiker et al. 2002) (Fig. 2). Cyr61 is a heparin-binding, secreted cysteine-rich protein that integrates into the extracellular matrix and binds directly to integrins (Perbal 2001). Cyr61 gene transfer potently stimulated limb revascularization, thereby promoting even greater improvement than achieved with VEGF (Fataccioli et al. 2002). Hilfiker et al. have demonstrated that Cyr61 is induced by angiotensin II in vascular cells and tissue and that the expression of Cyr61 colocalized with angiotensin II in small vessels of human arteriosclerotic lesions (Hilfiker et al. 2002).

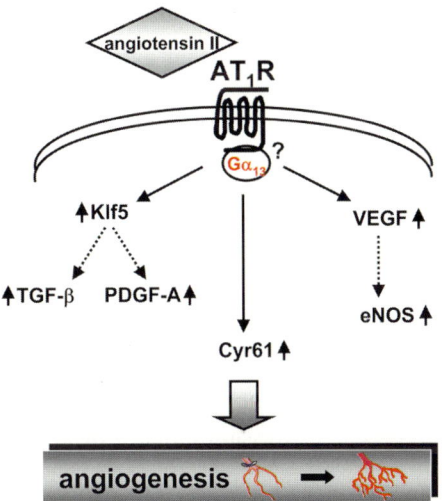

Fig. 2 Activation of transcription factors leading to angiogenesis. *eNOS*, endothelial NO-synthase; *Klf5*, Krüppel-like factor 5; *PDGF-A*, platelet-derived growth factor-A; *TGF-β*, transforming growth factor-β

Although the causal involvement of Klf5 and Cyr61 in angiotensin II-induced angiogenesis has to be proved, these factors might play a role in the complex mechanisms regulating angiogenesis signaling in response to angiotensin II.

4
Antiangiogenic Effects of Angiotensin II Via the AT2R: Important Role for Apoptosis

Little information is available regarding the physiological roles of the AT2R and its signaling pathways. In cultured cells the AT2R seems to be difficult to study, since its expression is unstable. However, initial studies in 1995 demonstrated that stimulation of the AT2R in endothelial cells inhibits proliferation (Stoll et al. 1995). Moreover, angiotensin II or pharmacological stimulation of the AT2 receptor were shown to induce apoptosis of cultivated endothelial cells (Dimmeler et al. 1997). In addition, similar pro-apoptotic effects are described in AT2R-expressing cells (Yamada et al. 1996). The inhibition of endothelial cell proliferation and the induction of endothelial cell death obviously can counteract angiogenesis (Dimmeler and Zeiher 2000). Indeed, recent in vivo studies support an anti-angiogenic activity mediated by the AT2R (Silvestre et al. 2002). AT2R knockout mice or pharmacological AT2R blockade by PD123319 revealed a pronounced angiogenic response (Walsh et al. 1997; Silvestre et al. 2002). In addition, endothelial cell apoptosis was reduced in mice lacking the AT2R (Silvestre et al. 2002). The pro-apoptotic activity of AT2R stimulation was mainly attributed to the activation of mitogen-activated protein (MAP) kinase phosphatases (MKP), which subsequently reduce the activity of the survival promoting ERK1/2 kinase and diminish Bcl-2 expression (Horiuchi et al. 1997). In endothelial cells mRNA expression of MKP-3 is increased by angiotensin II stimulation (Rössig et al. 2002), whereas other studies reported that angiotensin II induced the expression of another isoform, MKP-1, in smooth muscle cells (Horiuchi et al. 1997). In both cell types the expression of the anti-apoptotic protein Bcl-2 was reduced by AT2R stimulation. This has further been confirmed in AT2R knockout mice, which revealed significantly higher levels of Bcl-2 as compared to wild-type mice (Silvestre et al. 2002).

Beside the induction of apoptosis, AT2R stimulation also prevents angiotensin II-induced up-regulation of VEGF and eNOS. Thus, administration of PD123319 in mice inhibited the increase in VEGF protein expression and prevented transactivation of EGFR by angiotensin II (Fujiyama et al. 2001). The signaling seems to involve activation of protein tyrosine phosphatases, since tyrosine phosphorylation of the EGFR was inhibited by PD123319 (Fujiyama et al. 2001). Moreover, the increase in VEGF and eNOS expression induced by angiotensin II was abolished in AT2R knockout mice (Silvestre et al. 2002).

5
Conclusions

The renin/angiotensin system can exert a double-edged role in angiogenesis signaling. The pro-angiogenic activity is predominantly mediated by stimulation of the ATR1. In contrast, stimulation of the AT2R induces apoptosis of endothelial cells and prevents angiogenesis. The balance of receptor expression may thus determine the angiogenic response to angiotensin II.

Acknowledgements. We apologize for the failure to cite many of the important and relevant papers in this field due to space limitations.

References

Aikawa M, Kim HS, et al (1995) Phenotypic modulation of smooth muscle cells during progression of human atherosclerosis as determined by altered expression of myosin heavy chain isoforms. Ann NY Acad Sci 748:578–585

Amaral SL, Linderman JR, Morse MM, Greene AS (2001) Angiogenesis induced by electrical stimulation is mediated by angiotensin II and VEGF. Microcirculation 8:57–67

Berk BC (1999) Angiotensin II signal transduction in vascular smooth muscle: pathways activated by specific tyrosine kinases. J Am Soc Nephrol 10:S62–S68

Bieker JJ (2001) Kruppel-like factors: three fingers in many pies. J Biol Chem 276:34355–34358

Browder T, Folkman J, Pirie-Shepherd S (2000) The hemostatic system as a regulator of angiogenesis. J Biol Chem 275:1521–1524

Carmeliet P (2000) Mechanisms of angiogenesis and arteriogenesis. Nat Med 6:389–395

Carmeliet P, Ferreira V, et al (1996) Abnormal blood vessel development and lethality in embryos lacking a single VEGF allele. Nature 380:435–439

Dimmeler S, Dernbach E, Zeiher AM (2000) Phosphorylation of the endothelial nitric oxide synthase at Ser 1177 is required for VEGF-induced endothelial cell migration. FEBS Lett 477:258–262

Dimmeler S, Rippmann V, Weiland U, Haendeler J, Zeiher AM (1997) Angiotensin II induces apoptosis of human endothelial cells. Protective effect of nitric oxide. Circ Res 81:970–976

Dimmeler S, Zeiher AM (2000) Endothelial cell apoptosis in angiogenesis and vessel regression. Circ Res 87:434–439

Fabre JE, Rivard A, Magner M, Silver M, Isner JM (1999) Tissue inhibition of angiotensin-converting enzyme activity stimulates angiogenesis in vivo. Circulation 99:3043–3049

Fataccioli V, Abergel V, et al (2002) Stimulation of angiogenesis by cyr61 gene: a new therapeutic candidate. Hum Gene Ther 13:1461–1470

Folkman J (1995) Angiogenesis in cancer, vascular, rheumatoid and other disease. Nat Med 1:27–31

Fujiyama S, Matsubara H, et al (2001) Angiotensin AT(1) and AT(2) receptors differentially regulate angiopoietin-2 and vascular endothelial growth factor expression and angiogenesis by modulating heparin binding-epidermal growth factor (EGF)-mediated EGF receptor transactivation. Circ Res 88:22–29

Heeneman S, Haendeler J, Saito Y, Ishida M, Berk BC (2000) Angiotensin II induces transactivation of two different populations of the platelet-derived growth factor beta receptor. Key role for the p66 adaptor protein Shc. J Biol Chem 275:15926–15932

Hilfiker A, Hilfiker-Kleiner D, et al (2002) Expression of CYR61, an angiogenic immediate early gene, in arteriosclerosis and its regulation by angiotensin II. Circulation 106:254–260

Horiuchi M, Hayashida W, Kambe T, Yamada T, Dzau VJ (1997) Angiotensin type 2 receptor dephosphorylates Bcl-2 by activating mitogen-activated protein kinase phophatase-1 and induces apoptosis. J Biol Chem 272:19022–19026

Isner JM (2001) Still more debate over VEGF. Nat Med 7:639–641

Isner JM, Asahara T (1999) Angiogenesis and vasculogenesis as therapeutic strategies for postnatal neovascularization. J Clin Invest 103:1231–1236

Lastres P, Letamendia A, et al (1996) Endoglin modulates cellular responses to TGF-beta 1. J Cell Biol 133:1109–1121

Li C, Hampson IN, et al (2000) CD105 antagonizes the inhibitory signaling of transforming growth factor beta1 on human vascular endothelial cells. FASEB J 14:55–64

Li D, Chen H, Mehta JL (2001) Angiotensin II via activation of type 1 receptor upregulates expression of endoglin in human coronary artery endothelial cells. Hypertension 38:1062–1067

Li DY, Sorensen LK, et al (1999) Defective angiogenesis in mice lacking endoglin. Science 284:1534–1537

Macrez-Lepretre N, Kalkbrenner F, Morel JL, Schultz G, Mironneau J (1997) G protein heterotrimer Galpha13beta1gamma3 couples the angiotensin AT1A receptor to increases in cytoplasmic Ca2+ in rat portal vein myocytes. J Biol Chem 272:10095–10102

Martin MM, White CR, Li H, Miller PJ, Elton TS (1995) A functional comparison of the rat type-1 angiotensin II receptors (AT1AR and AT1BR). Regul Pept 60:135–147

Muramatsu M, Katada J, Hayashi I, Majima M (2000) Chymase as a proangiogenic factor. A possible involvement of chymase-angiotensin-dependent pathway in the hamster sponge angiogenesis model. J Biol Chem 275:5545–5552

Murohara T, Asahara T, et al (1998) Nitric oxide synthase modulates angiogenesis in response to tissue ischemia. J Clin Invest 101:2567–2578

Murphy TJ, Alexander RW, Griendling KK, Runge MS, Bernstein KE (1991) Isolation of a cDNA encoding the vascular type-1 angiotensin II receptor. Nature 351:233–236

Offermanns S, Mancino V, Revel JP, Simon MI (1997) Vascular system defects and impaired cell chemokinesis as a result of Galpha13 deficiency. Science 275:533–536

Otani A, Takagi H, Suzuma K, Honda Y (1998) Angiotensin II potentiates vascular endothelial growth factor-induced angiogenic activity in retinal microcapillary endothelial cells. Circ Res 82:619–628

Perbal B (2001) NOV (nephroblastoma overexpressed) and the CCN family of genes: structural and functional issues. Mol Pathol 54:57–79

Prenzel N, Zwick E, et al (1999) EGF receptor transactivation by G-protein-coupled receptors requires metalloproteinase cleavage of proHB-EGF. Nature 402:884–888

Rao GN, Lassegue B, Alexander RW, Griendling KK (1994) Angiotensin II stimulates phosphorylation of high-molecular-mass cytosolic phospholipase A2 in vascular smooth-muscle cells. Biochem J 299:197–201

Richard DE, Berra E, Pouyssegur J (2000) Nonhypoxic pathway mediates the induction of hypoxia-inducible factor 1alpha in vascular smooth muscle cells. J Biol Chem 275:26765–26771

Rössig L, Hermann C, et al (2002) Angiotensin II-induced upregulation of MAP kinase phosphatase-3 mRNA levels mediates endothelial cell apoptosis. Basic Res Cardiol 97:1–8

Sasaki K, Murohara T, et al (2002) Evidence for the importance of angiotensin II type 1 receptor in ischemia-induced angiogenesis. J Clin Invest 109:603–611

Sasaki K, Yamano Y, et al (1991) Cloning and expression of a complementary DNA encoding a bovine adrenal angiotensin II type-1 receptor. Nature 351:230–233

Shindo T, Manabe I, et al (2002) Kruppel-like zinc-finger transcription factor KLF5/BTEB2 is a target for angiotensin II signaling and an essential regulator of cardiovascular remodeling. Nat Med 8:856–863

Silvestre JS, Tamarat R, et al (2002) Antiangiogenic effect of angiotensin II type 2 receptor in ischemia-induced angiogenesis in mice hindlimb. Circ Res 90:1072–1079

Simon MI, Strathmann MP, Gautam N (1991) Diversity of G proteins in signal transduction. Science 252:802–808

Stoll M, Steckelings UM, et al (1995) The angiotensin AT2-receptor mediates inhibition of cell proliferation in coronary endothelial cells. J Clin Invest 95:651–657

Tamarat R, Silvestre JS, Durie M, Levy BI (2002) Angiotensin II angiogenic effect in vivo involves vascular endothelial growth factor- and inflammation-related pathways. Lab Invest 82:747–756

Tamarat R, Silvestre JS, et al (2002) Endothelial nitric oxide synthase lies downstream from angiotensin II-induced angiogenesis in ischemic hindlimb. Hypertension 39:830–835

Unger T, Chung O, et al (1996) Angiotensin receptors. J Hypertens Suppl 14:S95–S103

Volpert OV, Ward WF, et al (1996) Captopril inhibits angiogenesis and slows the growth of experimental tumors in rats. J Clin Invest 98:671–679

Walsh DA, Hu DE, et al (1997) Sequential development of angiotensin receptors and angiotensin I converting enzyme during angiogenesis in the rat subcutaneous sponge granuloma. Br J Pharmacol 120:1302–1311

Williams B, Baker AQ, Gallacher B, Lodwick D (1995) Angiotensin II increases vascular permeability factor gene expression by human vascular smooth muscle cells. Hypertension 25:913–917

Yamada T, Horiuchi M, Dzau VJ (1996) Angiotensin II type 2 receptor mediates programmed cell death. Proc Natl Acad Sci 93:156–160

Yoshiji H, Kuriyama S, et al (2001) The angiotensin-I-converting enzyme inhibitor perindopril suppresses tumor growth and angiogenesis: possible role of the vascular endothelial growth factor. Clin Cancer Res 7:1073–1078

Immunohistochemistry of Angiotensin

H. Imboden · D. Felix

Institute of Cell Biology, University of Bern, Baltzerstrasse 6, 3012 Bern, Switzerland
e-mail: hans.imboden@izb.unibe.ch

1	Introduction	112
2	The Specificity of Immunocytochemical Techniques.	113
2.1	Affinity Purification of Angiotensin Antiserum	113
2.2	Anti-idiotypic Angiotensin Antibody	115
2.3	The Use of Different Anti-peptide Antibodies	115
2.4	Intracellular Staining of Angiotensin Receptors	116
3	Combination of Immunocytochemical Methods with Other Techniques	118
4	Localization of the Components of the Renin–Angiotensin System	119
4.1	Angiotensinogen in Brain	119
4.2	Angiotensin Receptors	120
5	Colocalization of Central Angiotensinergic and Vasopressinergic Systems	121
6	A New Angiotensinergic System in the Central Nervous System	122
7	Concluding Remarks	123
References		124

Abstract Immunocytochemical approaches have achieved an outstanding level for characterizing angiotensinergic systems in the brain and periphery. The development of affinity-purified antibodies for angiotensin II has resulted in distinct loss of background activity and produced discrete staining of neurons and small angiotensinergic fibres. The improvement of immunocytochemical techniques has enhanced the ability to develop maps of angiotensinergic pathways. Furthermore, the production of specific monoclonal antibodies to angiotensin receptor subtypes, such as AT_1 and AT_2, has led to the characterization and localization of angiotensin hormonal and neurotransmitter systems in brain and peripheral tissues. Such affinity-purified antibodies have revealed that angiotensin and vasopressin are colocalized in hypothalamo-neurohypophysial systems, implying that both peptides are released together in the median eminence and the neurohypophysis. Recent studies have shown that in spite of using the same antigen molecule for immunization as well as for affinity purification, different staining pattern are obtained, suggesting that the different carrier proteins forced the production of different antibodies to different epitopes. Furthermore, the finding supports the hypothesis that angiotensin II is not the only effector peptide in the renin–angiotensin system but a member of a family of biologically active angiotensinergic peptides. The presented studies support once more that immunocytochemical approaches remain important tools for characterizing angiotensinergic systems in the central nervous and peripheral tissues.

Keywords Renin–angiotensin system · Immunocytochemistry · Affinity purification · Anti-idiotypic antibody · Monoclonal antibody · Colocalization · Angiotensinogen · Angiotensin receptor subtypes · AT_1 receptor · AT_2 receptor

1
Introduction

The brain renin–angiotensin system (RAS), which plays an important role in controlling cardiovascular function and maintenance of body fluid homeostasis, has been the subject of many reviews (Bunnemann et al. 1993; Bottari et al. 1993; Wright and Harding 1994; Mosimann et al. 1996; De Gasparo et al. 2000). The action of angiotensin II is mediated by the activation of its surface receptors, which are located in the central nervous system and in peripheral tissues.

In the rat brain, two angiotensin II receptor subtypes, AT_1 and AT_2 have been established using autoradiography and selective ligands (Tsutsumi and Saavedra 1991; Rowe et al. 1992; Song et al. 1992). Most of the known central actions of angiotensin II have been shown to be mediated through the AT_1 receptor. This receptor is coupled to G proteins and stimulates classical second messenger systems such as inhibition of adenylcyclase or activation of phospholipase C (Timmermans et al. 1992). Genetically different AT_1 receptor isoforms, AT_{1A} and AT_{1B} have been localized by in situ hybridization using specific riboprobes (Jöhren et al. 1995). AT_1 receptor expression has been demonstrated in human breast tissues, suggesting that the RAS may be involved in normal and tumour tissue function (Inwang et al. 1997). Recent results demonstrated the presence of AT_1 receptor in malignant glial cells and a favourable therapeutic response in glioma cells by a selective blockade of this receptor subtype (Rivera et al. 2001). Furthermore, blockade of the AT_{1A} receptor becomes an effective novel strategy for tumour chemoprevention, since such blockage reduced tumour growth, angiogenesis and metastasis (Fujita et al. 2002).

In contrast, the function of the AT_2 receptor is far from being understood. AT_2 is found to exert growth inhibitory and proapoptotic effects (Ishida and Fukamizu 1999). This receptor seems to exert opposite effects in terms of cell growth and cardiovascular haemodynamics (Xoriuchi et al. 1999). An upregulation of AT_2 expression under certain pathological conditions exists, such as vascular injury, myocardial infarction and heart failure (Nakamura et al. 1999; Malendowicz et al. 2000). Of high interest are results which suggest that the AT_2 receptor promotes vascular differentiation, contributes to vasculogenesis (Yamada et al. 1999) and cell migration (Cote et al. 1999).

Most published results about the RAS are obtained from radioreceptor assays of tissue homogenates or from autoradiography with specific antagonists in tissue sections. However, such methods are not suitable for studying exact cellular location of binding. In this respect, immunocytochemical approaches have achieved an outstanding level from both a methodological and scientific point of view. Especially, the specific affinity purification of polyclonal antibodies and the development of monoclonal antibodies offer a powerful tool for localization as well as colocalization studies at the cellular level. In addition, these antibodies can be used for Western blot analysis from tissue under distinct conditions. The combined immunocytochemical and biochemical approaches are certainly suitable for understanding principles of different angiotensin functions. Furthermore, with the aid of immunocytochemical techniques, possible interrelationship of angiotensin with other peptides may be clarified. For such purposes, the production and use of different monoclonal antibodies will be a prerequisite for the characterization and localization of the hormonal and neurotransmitter systems in the brain and in peripheral tissues.

2
The Specificity of Immunocytochemical Techniques

2.1
Affinity Purification of Angiotensin Antiserum

The sensitivity, specificity and also simplicity of immunocytochemical techniques have made them useful tools in the field of neurobiology over the past few years. The brain RAS involves selected populations of neurons within the brain. It is primarily the domain of immunocytochemical studies that determine the exact location of selected populations of neurons and their projection fields necessary for a complete understanding of the distribution of centrally produced angiotensin II (Imboden et al. 1987a,b). Implicit in the use of immunocytochemistry for such purposes is the belief that the staining produced is due to authentic angiotensins. The validity of this assumption deserves special scrutiny since only a small number of angiotensin antisera proved suitable for angiotensin immunocytochemistry. The use of crude angiotensin antiserum resulted in the staining of large varicosities and cell bodies. Fibres, however, were all but invisible owing to extensive background staining. Biochemical and histochemical analysis of crude angiotensin antiserum indicated that the background staining seen with the crude antiserum was neither angiotensin II nor several angiotensin derived fragments (Imboden et al. 1987a).

The enhancement in immunocytochemical resolution was drastically improved when Imboden and co-workers (1987a,b) prepared affinity-purified antisera. Antibodies were purified by chromatographing crude antiserum (rabbit DE) with an angiotensin II affinity column prepared with CH-sepharose 4B (Pharmacia). Such affinity purification yielded an antibody preparation that produced staining which confirms the presence of angiotensin II. This staining sufficiently improved the level of resolution to visualize extensive fine fibres as well as the more precise location of angiotensin immunoreactive neurons. Furthermore, the purified antibodies yielded only discrete staining without background.

A major advantage of the purified antibodies over the crude antiserum is the ability of the former to stain clearly fine fibres and their accompanying varicosities. These studies have been done without colchicine pretreatment of the animals. Such improved capability to visualize fibres allows one to directly map angiotensinergic pathways. The potential of this mapping capability is exemplified in Fig. 1 in which a part of the paraventricular-hypophysial pathway is shown (Imboden et al. 1987b).

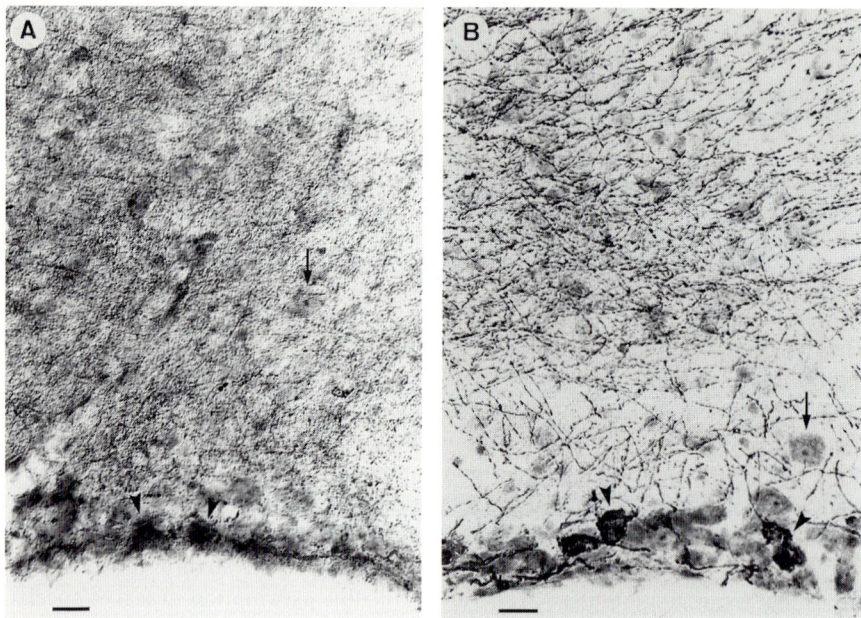

Fig. 1A, B Comparison of angiotensin II-like immunoreactive product in fibre tracts of the dorsal and ventral supraoptic commissure and in cells of the ventral base of the hypothalamus. *Bars*=20 μm. **A** Staining with the crude antiserum. Only a few cells at the hypothalamic border show faint and diffuse reaction (*arrowheads*). No clear fibre pathways were observed in the commissure. **B** Staining with the affinity-purified antibody. Angiotensin II-like immunoreactive fibre pathways are clearly visible. Immunoreactive cells (*arrowheads*) can be distinguished from toluidine blue-counterstained, nonreactive cells (*arrows*). (Reproduction by permission from Imboden et al. 1987b, by courtesy of Marcel Dekker, Inc., NY)

2.2
Anti-idiotypic Angiotensin Antibody

Another approach to the crucial angiotensin receptors has been the use of anti-idiotypic antibodies, which is based on the network theory of the immune system (Jerne 1974). One anti-idiotypic antibody reacting with angiotensin II receptors (Couraud 1987), whose capacity served as a reliable tool for mapping immunohistochemical angiotensin II receptor distribution, has been reported by Pfister et al. (1993). Specificity of these antibodies toward angiotensin II receptor has been demonstrated with a number of biochemical experiments. The conclusion of this immunohistochemical study shows it to be a reliable tool for mapping angiotensin II receptor distribution. The established experimental approaches to angiotensin II receptors are thus supplemented with the possibility of immunocytochemical investigations. Using anti-idiotypic antibodies, punctate immunoreactive granules in or on neurons of different hypothalamic nuclei were demonstrated (Pfister et al. 1993).

2.3
The Use of Different Anti-peptide Antibodies

As mentioned above, angiotensin II mediates its biological actions via two subtypes of receptors, termed AT_1 and AT_2. In rat and mouse, two further subtypes of AT_1, namely AT_{1A} and AT_{1B} exist (Iwai and Inagami 1992). Therefore, the primary goal for immunohistochemists is to raise monoclonal anti-peptide antibodies specific to the different angiotensin receptor subtypes and to angiotensin II itself and using them to determine the intraorganic and/or the intracellular localization. Until now, only scarce immunohistochemical investigations have been performed in which polyclonal anti-idiotypic antibodies were used; these studies were unable to distinguish between the different receptor subtypes (Pfister et al. 1993). Comparable results have been published concerning the use of an antibody to a portion of the third cytoplasmic loop of the AT_1 receptor (Phillips et al. 1993). Further polyclonal antibodies (Zelezna et al. 1992; Rakugi et al. 1997; Benarroch and Schmeichel 1998; Giles et al. 1999) and one monoclonal antibody (Barker et al. 1993; Harrison-Bernard et al. 1997) to the AT_1 receptor exist. Regarding the AT_2 receptor, only two polyclonal and no monoclonal antibody have been published (Yiu et al. 1997; Nora et al. 1998). So far, only the monoclonal antibody against the AT_1 receptor has been proved useful for immunohistochemical investigations (Harrison-Bernard et al. 1997). In a recent study, Frei et al. (2001) reported the production and immunocytochemical characterization of three novel murine monoclonal antibodies directed against angiotensin II receptor subtypes AT_1 and AT_2, and against angiotensin II. Immunocytochemical characterization were carried out in the rat adrenal, in which AT_1-reactive cells were found in the zona glomerulosa and the medulla, whereas the AT_2 antibody bound specifically to medullar cells. Cells in the medulla and the zona glomerulosa reacted both with the angiotensin II monoclonal antibody.

In Western blots, both AT_1 and AT_2 antibodies detected a major 73-kDa protein band from adrenal homogenates but did not recognize a protein of 41 or 44 kDa as reported for COS-7 cells transfected with AT_{1A} receptor cDNA (Barker et al. 1993; Rakugi et al. 1997). These observations suggest that the previously identified 75-kDa protein (Sen and Rajasekaran 1991) obtained from cross-linking experiments of [125 I]-labelled angiotensin II was the biological relevant glycosylated form of the AT_1 receptor. This finding is of importance, since the new monoclonal antibodies specifically recognize angiotensin II and AT_1 and AT_2 receptors, representing powerful tools for further subcellular and effector pathway investigations (Frei et al. 2001).

2.4
Intracellular Staining of Angiotensin Receptors

Despite the knowledge of the existence of angiotensin II itself as well as its receptors in specialized brain areas, the coexistence of components of angiotensin and angiotensin receptors in individual cells have been reported only recently. For such investigations, the lack of pretreatment of the animals with colchicine and the purification of the specific antibodies were important steps. With this procedure it was possible to stain not only neurons in hypothalamic regions but also to visualize angiotensinergic fibres in the hypothalamic paraventricular-hypophysial pathway. Spengler et al. (1996) used 5-μm thick sections, allowing a chance to get at least three adjacent sections out of a single neuron. With these serial sections they were able to carry out immunoreaction studies in one single cell using an anti-idiotypic angiotensin antibody, an affinity-purified angiotensin II antibody ("BODE") and an antibody against the third intracellular loop of the AT_1 receptor ("ELAINE"). As an example, Fig. 2 illustrates the coexistence of angiotensin and its receptor subtype AT_1 in individual neurons.

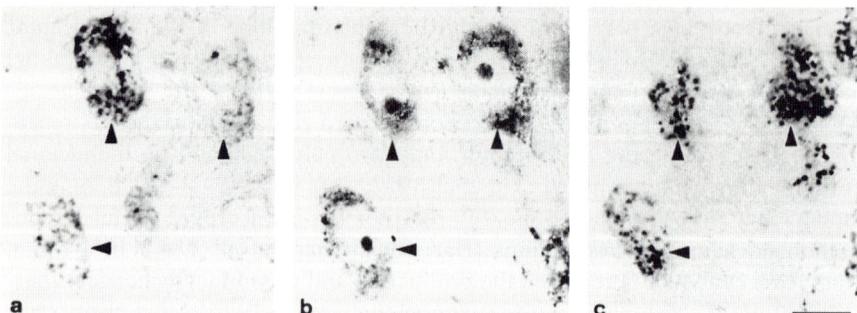

Fig. 2a–c Coexistence of angiotensin and angiotensin receptors in neurons of the supraoptic nucleus. **a** Immunoreactive staining with the anti-idiotypic antibody for angiotensin II receptors. **b** Immunoreactive staining with the affinity-purified antibody to angiotensin II. **c** Immunoreactive staining with the antibody to the angiotensin AT_1 receptor oligo-sequence. *Scale bar*=10 μm. (Reproduction by permission from Spengler et al. 1996, by courtesy of Birkhäuser Verlag Basel/Schweiz)

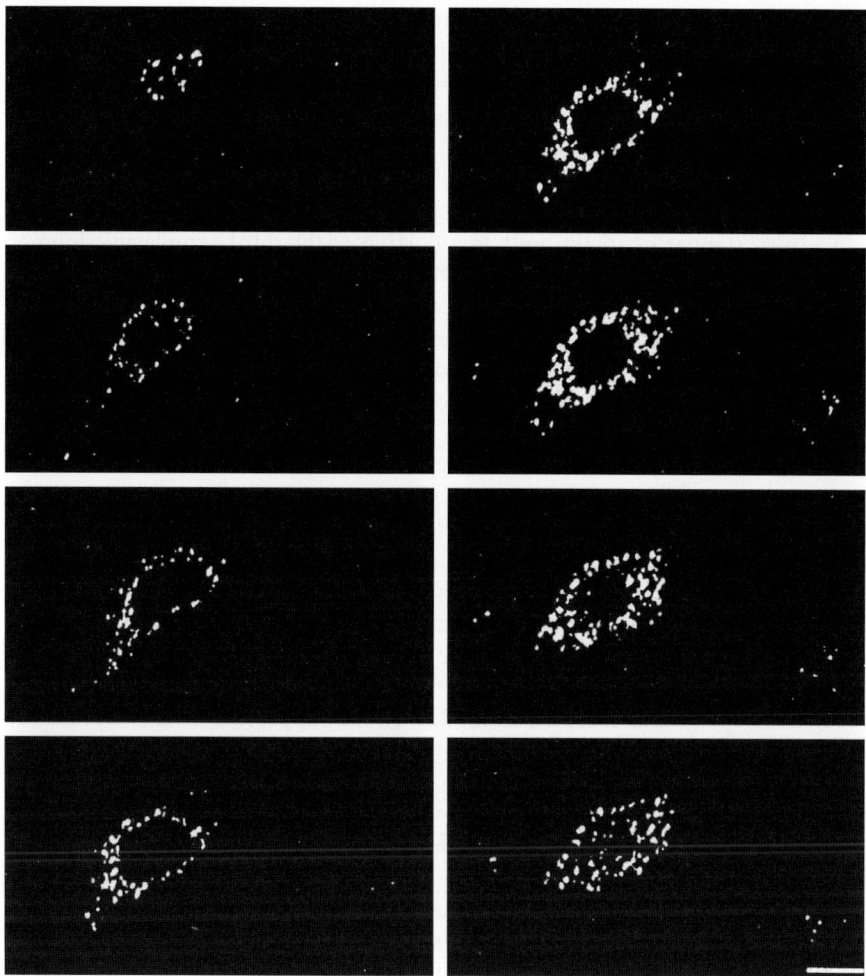

Fig. 3 Confocal laser-scan microscopic analysis of a neuron in the paraventricular nucleus (PVN). The sections show cytoplasmic localization of angiotensin II-receptor immunoreactivity. The series (*from left to right, top to bottom*) show optical sections in the z-direction of a neuron that was incubated with an anti-AT_1 receptor antibody. *Scale bar*=10 μm. (Reproduction by permission from Pfister et al. 1997, by courtesy of Elsevier Science)

Although several individual studies using immunocytochemical (Phillips et al. 1993) or autoradiographic (Tsutsumi and Saavedra 1991) techniques reported the existence of angiotensin or angiotensin receptors, this was the first report of coexistence of the peptide and its receptors within the same neuron. The fact that angiotensin-containing neurons possess receptor sites as well, raises the possibility of an angiotensinergic input on the same neuron. The advantage of using thin sections allowed testing the different antibodies in individual slices of the same neuron, thereby preventing any possible interaction with each other.

To summarize, the chosen approach offers a good tool for any further studies on colocalization of the components of hypothalamic peptidergic systems.

The immunocytochemical reaction product from intracellular staining of angiotensin receptors showed characteristic focal appearance. Using anti-idiotypic antibodies, such focal immunoreactive granules were described in or on neurons of the supraoptic nucleus and of the paraventricular nucleus in the hypothalamus of rats (Pfister et al. 1993). Comparable results have been published concerning the use of an antibody to a portion of the third cytoplasmic loop of the AT_1 receptor (Phillips et al. 1993). To clarify whether the mentioned immunoreactivity indeed resulted from intracellular staining, Pfister et al. (1997) tried three different approaches: (1) use of the fluorescein isothiocyanate (FITC)-conjugated anti-idiotypic antibody with confocal laser-scan microscopic analysis; (2) use of the anti-AT_1 receptor antibody ("ELAINE") in combination with a FITC-conjugated second antibody and subsequent confocal laser-scan microscopic analysis; (3) use of the anti-idiotypic antibody with the PAP/DAB-method and subsequent light-microscopic analysis of semi-thin sections. The use of the different approaches has shown punctate immunoreactive material homogeneously distributed in the cytoplasm shown in Fig. 3.

In contrast, the nuclei of marked neurons were devoid of staining. With the pre-embedding semi-thin sections, it was possible to localize punctate staining at the border of the cell membrane. The intracellular distribution pattern of angiotensin II receptor immunoreactivity could indicate that the epitope being recognized by the anti-receptor antibodies already appears during receptor biosynthesis and/or during degradation.

3
Combination of Immunocytochemical Methods with Other Techniques

The current knowledge on brain RAS has been derived mostly from a synthesis of many studies with different methods of investigations. The disadvantage of all these studies, however, is the lack of information to the same neuron. Such information can only be gained with a combination of different methods focused on a single structure and performed in one experimental approach. The advantage is to develop a relation between the results obtained. For instance with a combination of electrophysiology, morphology and immunocytochemistry, a neuron could be investigated with regard to its sensitivity to specific neurotransmitters, the content of different peptides, enzymes, receptors, and its morphology.

Egli et al. (2000) used different methods to characterize single neurons in the hypothalamic paraventricular area of the brain. Starting with intracellular recordings and a pharmacological identification, not only neurotransmitters but also involved receptors were defined with the application of receptor agonists and antagonists. Finally, the injection of Lucifer yellow into the electrophysiologically investigated cell using confocal laser-scan microscopically generated optical sections allowed a 3D-model to be created. The use of different antibodies on 1-μm semi-thin sections from the investigated cell showed a possible presence of diverse substances. The electrophysiological method was used to find angiotensin II-sensitive neurons in this brain area. An application of the angiotensin II-specific AT_1 receptor antagonist Losartan (Dup 753) during the angiotensin II application confirmed its evoked reaction, followed by a dye injection into the cell, defining its morphology. Finally, an immunocytochemical method was used to look for vasopressin and components of the angiotensin system inside or outside the neuron (Egli et al. 2000). Such techniques offer new insight into the function of neurotransmitter systems in general, and the results enables us to gain improved information on interactions of the angiotensinergic system within the RAS or other systems in specialized areas of the brain or in peripheral tissues.

4
Localization of the Components of the Renin–Angiotensin System

4.1
Angiotensinogen in Brain

Each of the components of the brain RAS including angiotensinogen (Lewicki et al. 1978; Campbell et al. 1984), renin (Ganten and Speck 1978; Ganten et al. 1984), converting enzyme (Ganten et al. 1984) and angiotensin II itself (Hermann et al. 1984) have been identified in the brain. Although the presence of all of the components in the brain seems indisputable, the exact cellular locations of several key components, such as angiotensinogen (Deschepper et al. 1986; Sernia 1995) have been disputed. While angiotensin II appears to be exclusively located in neurons (Lind et al. 1985; Imboden et al. 1987a), angiotensinogen has been reported to reside only in astrocytes and ependymal cells. In the hypoglossal nucleus of the rat, angiotensinogen was found in astrocytes, with small amounts in neuronal dendrites (Tham et al. 2001). The cellular localization of angiotensinogen mRNAs in the subfornical organ is predominantly expressed in astroglial cells (Lippoldt et al. 1993).

Imboden et al. (1987c) confirmed, with affinity-purified polyclonal antibodies against the tetradecapeptide, that a large portion of the total angiotensinogen immunoreactivity was present in astrocytes. However, it appeared that only a subpopulation of astrocytes contained angiotensinogen, having the highest concentration in the arcuate nucleus. Of potential importance was the tendency of angiotensinogen-containing astrocytes to contact brain microvessels. Therefore, astrocytic angiotensinogen could represent the precursor pool for a second brain-angiotensin system that is exclusively associated with blood vessels and involved with the regulation of cerebral blood flow. In contrast to earlier reports, Imboden et al. (1987c) could show that clear angiotensinogen staining of neurons was evident in paraventricular, supraoptic and accessory magnocellular hypothalamic nuclei. Interestingly, these same nuclei always costained for angiotensin II, suggesting a unique relationship between angiotensinogen-containing astrocytes and angiotensin-producing neurons. Such a system would require direct contact between these elements so that angiotensinogen could be taken up by the neurons. The final location of angiotensinogen is the choroid plexus, indicating that angiotensinogen is being actively secreted into the cerebrospinal fluid (Imboden et al. 1987c).

The localization of angiotensinogen in three cell types of the brain, using immunohistochemical methods, are consistent with multiple functions for brain angiotensinogen as a precursor for neuronal angiotensin II and as a potential source for angiotensin II that is locally produced in the brain (Schelling et al. 1983). Regional distribution of angiotensinogen in fetal and neonatal rat brain showed that most of the immunopositive cells in the hypothalamus and brainstem were astrocytes, while those in the cortex were almost exclusively neurons (Mungall et al. 1995).

4.2
Angiotensin Receptors

Although angiotensin II receptors were among the first peptide receptors to be characterized, the discovery of receptor subtypes and the elucidation of the signal transduction mechanisms linking receptors to effectors have revealed new possibilities concerning the role of an angiotensin system in the brain (Bottari et al. 1993). Most of the known central actions of angiotensin II, such as dipsogenic response, increase in blood pressure and vasopressin release, have been shown to be mediated through the AT_1 receptor. In the past, many immunohistochemical studies have focused on the distribution pattern of this receptor subtype. The precise localization and the physiological role of angiotensins in circumscribed areas of the brain especially in the third ventricle area (subfornical organ, organum vasculosum laminae terminalis, anterior third ventricle area), the hypothalamic areas and the hypothalamo-neurohypophysial axis have shed light on the understanding of the role of the brain RAS (Imboden and Felix 1991; Paxton et al. 1993; Phillips et al. 1993; Mosimann et al. 1996; Benarroch and Schmeichel 1998). Besides the third ventricular and hypothalamic area, direct evidence exists of neuronal localization of the AT_1 receptor in distinct areas of the brain stem (Yang et al. 1997; Benarroch and Schmeichel 1998) and the limbic system (von Bohlen et al. 1998).

As already mentioned, the function of the AT_2 receptor is far from being understood. High AT_2 receptor mRNA expression was localized at the cellular level in the lateral septum, in several thalamic nuclei, in the subthalamic nucleus, in the locus coeruleus and in the inferior olive. In all regions examined, AT_2 receptor immunoreactivity was associated with the cytoplasm and cell membrane and was not localized within the nucleus (Reagan et al. 1994). Schelman et al. (1997) reported an AT_2 receptor-mediated inhibition of NMDA (N-methyl-D-aspartate) receptor signalling in neuronal cells. The results suggest that angiotensin II inhibits NMDA-mediated nitric oxide and cyclic guanosine monophosphate (cGMP) production through a mechanism involving AT_2 receptor subtype.

In the adrenal gland both angiotensin receptor subtypes have been characterized. Immunocytochemical studies revealed AT_1 reactive cells in the zona glomerulosa and the medulla, whereas the AT_2 antibody bound specifically to medullar cells (Bird et al. 1996; Frei et al. 2001). Since the AT_2 is expressed in the adrenal medulla, it is tempting to speculate whether this receptor subtype might play a role in adrenal medullar apoptosis (Shenoy et al. 1999).

5
Colocalization of Central Angiotensinergic and Vasopressinergic Systems

Since the discovery of angiotensin receptor subtypes and the development of specific angiotensin receptor antagonists, new aspects concerning the functional role of these receptors in the central nervous system have been elucidated. There is a close interrelationship between angiotensin receptors and other neurotransmitter systems. Using microdialysis, Stadler et al. (1992) showed that stimulation of periventricular AT_1 receptors leads to a release of noradrenaline in hypothalamic nuclei. Noradrenaline, on the other hand, acts on postsynaptic adrenoceptors to release vasopressin from the neurohypophysis.

The existence of the two peptides angiotensin and vasopressin in individual magnocellular cell groups of the hypothalamus has been demonstrated by using immunocytochemical methods (Imboden and Felix 1991). These neurosecretory magnocellular groups consist of the paraventricular nucleus and the supraoptic nucleus, as well as different accessory cell groups (Riva et al. 1999). From a functional point of view, these peptide systems are closely related and have dramatic vasoconstrictive effects when given peripherally and both exhibit pressor activity when applied centrally (Matsuguchi et al. 1982; Phillips 1987; Langhans 1988; Janiak et al. 1989). Their axons project into the pituitary, hindbrain structure, and other regions of the central nervous system (Healy and Printz 1984; Lind et al. 1985; Imboden et al. 1989). An immunocytochemical comparison of the angiotensin and vasopressin hypothalamo-neurohypophisial systems in rats by Imboden and Felix (1991) confirmed the colocalization of the two peptides in neurons of the paraventricular nucleus and the supraoptic nucleus (Kilcoyne et al. 1980; Hoffman et al. 1982; Silverman and Zimmerman 1983; Lind et al. 1984).

Furthermore, with affinity-purified antibodies it has been shown that the angiotensin and vasopressin hypothalamo-neurohypophysial systems are located in the same fibre system (Imboden and Felix 1991) as shown in Fig. 4.

Arching fibre tracts were formed mainly by projections emanating from cell bodies in the paraventricular nucleus and supraoptic nucleus, extending as far as the median eminence and the neurohypophysis, where major terminal fields exist. The implication of this finding is that angiotensin and vasopressin may be released together from the median eminence and the neurohypophysis. An important question remains to be answered as to how the release of such colocalized peptides is controlled. Functional studies have indicated that the AT_2 receptor subtype may participate in the regulation of vasopressin release by angioten-

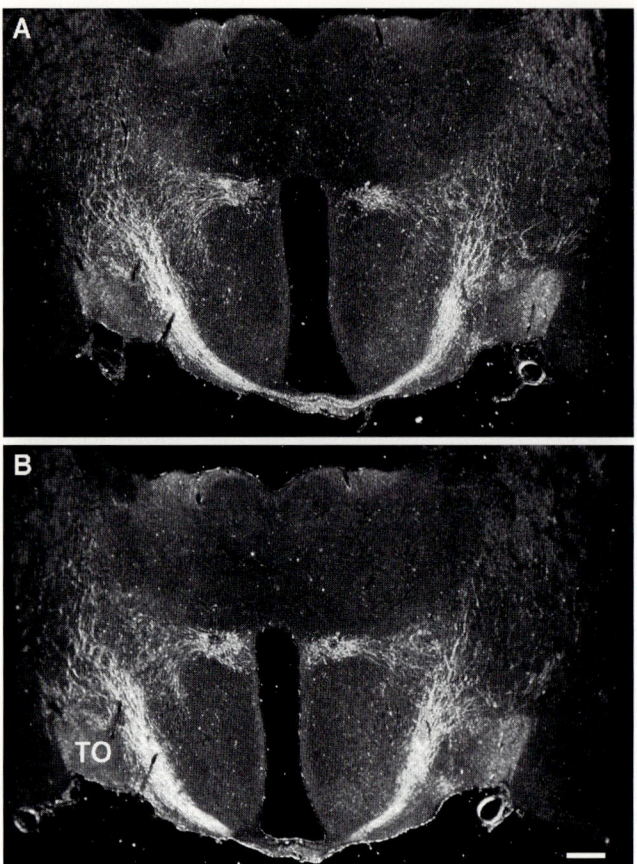

Fig. 4A, B The angiotensinergic (**A**) and vasopressinergic (**B**) fibre systems in coronal adjacent sections in the rat hypothalamus. *TO*, tractus opticus; *III*, third ventricle. *Scale bar*=500 μm

sin II. Furthermore, subtype AT_2 receptor-directed antiserum has been immunohistochemically detected AT_2 receptors in identified vasopressinergic neurons within the paraventricular and the supraoptic nuclei (Shelat et al. 1998).

6
A New Angiotensinergic System in the Central Nervous System

Through immunocytochemical techniques, the localization of angiotensin II-like immunoreactivity has mainly been demonstrated in the hypothalamo-neurohypophysial pathway, neurosecretory paraventricular nucleus, supraoptic nucleus, suprachiasmatic nucleus, as well as in the median eminence (Brownfield et al. 1982; Lind et al. 1985). Similar results were obtained by the use of a polyclonal, affinity-purified, monospecific antibody to angiotensin II called "BODE", produced by Imboden et al. (1987a,b). Recently, a new polyclonal rabbit antibody to angiotensin II, called "BODE 1" was developed by the same group of researchers (Burkhalter et al. 2001). Nitrocellulose tests revealed for both antibodies a high degree of crossreactivity with different angiotensin peptides. They did not, however, interact with other neuropeptides like vasopressin, indicating that both antibodies are uniquely specific for angiotensins. Interestingly, the use of the two antibodies revealed dissimilar distribution of angiotensin II immunoreactivity within the rat central nervous system. The angiotensin II-like material detected using BODE was concentrated in the neurosecretory hypothalamic nuclei, in the inner layer of the median eminence and in the posterior lobe of the pituitary. In contrast, the BODE 1 antibody did not stain the hypothalamo-neurohypophysial angiotensinergic system; the staining pattern was much more broadly distributed throughout the central nervous system. This distribution of angiotensin-like material is new and indicates the existence of a new angiotensinergic system. BODE 1 is the first antibody that can be used to verify the locations of endogenous angiotensin and their receptor sites in the central nervous system (Burkhalter et al. 2001).

The results, however, do not reveal which exact types of angiotensins are detected. It is interesting to note that in spite of using the same antigen molecule for immunization as well as the same antigen molecule for affinity purification, these two polyclonal antibodies showed a dissimilar staining pattern. The findings suggest that the different carrier proteins forced the production of different antibodies to different epitopes. Therefore, it has to be taken into account, that, depending on how the different antibodies are able to bind to the exposed epitopes in the paraformaldehyde-fixed sections, different staining pattern can be obtained.

7
Concluding Remarks

Early immunohistochemical studies that described the detection of angiotensin II-immunoreactive material were forced to rely on a small number of antisera because most angiotensin antibodies had proved unsuitable for immunohistochemistry. Although useful tools, these antisera have suffered from high background staining. With the development of affinity-purified antibodies, the resolution has been greatly enhanced. The purification results in a distinct loss of background activity and produces a discrete staining of neurons and even small angiotensin fibres in animals not treated with colchicine. Such low background and discrete staining obtained through the improvement of immunocytochemical techniques has enhanced our ability to develop accurate maps of angiotensinergic pathways.

Another important aspect of the immunocytochemical approach is the production of different specific monoclonal antibodies to angiotensin AT_1 and AT_2 receptor subtypes and to angiotensin II as well as to components of the RAS. The development and use of such angiotensin antibodies will be a prerequisite for the characterization and localization of angiotensin hormonal and/or neurotransmitter systems in the brain and in peripheral tissues. It has to be taken into account that in spite of using the same antigen molecule for immunization as well as the same antigen molecule for affinity purification, different staining patterns can be obtained. The finding that in the central nervous system at least two different angiotensin systems exist supports the hypothesis that angiotensin II is not the only effector peptide in the RAS, but a member of a family of biologically active angiotensin peptides. Nevertheless, immunohistochemical techniques remain important methodological tools for characterizing and localizing angiotensinergic systems in the central nervous system and in peripheral tissues.

Whereas most of the immunohistochemical studies concerning the RAS have dealt in the past with the central nervous system and the adrenal, there is an increasing need for immunocytochemical approaches in normal and pathological peripheral tissues, e.g. cancer. With few exceptions, such studies are missing. It will be of importance that clinical trials using angiotensin receptor antagonists for treatment, e.g. glioblastoma multiforme and other malignant tumours, take place. Therefore, the immunocytochemical approach will become an important tool to place such trials on a firm scientific basis.

Acknowledgements. This work was supported by the "Stiftung zur Förderung der wissenschaftlichen Forschung an der Universität Bern". We especially thank Susanne Gygax and Robert Mosimann for technical assistance.

References

Barker S, Marchant W, Ho MM, Puddefoot JR, Hinson JP, Clark AJ, and Vinson GP (1993) A monoclonal antibody to a conserved sequence in the extracellular domain recognizes the angiotensin II AT1 receptor in mammalian target tissues. J Mol Endocrinol 11:241–245

Benarroch EE, Schmeichel AM (1998) Immunohistochemical localization of the angiotensin II type 1 receptor in human hypothalamus and brainstem. Brain Res 812:292–296

Bird IM, Zheng J, Corbin CJ, Magness RR, Conley AJ (1996) Immunohistochemical analysis of AT1 receptor versus P450c17 and 3 beta HSD expression in ovine adrenals. Endocr Res 22:349–353

Bottari SP, de Gasparo M, Steckelings UM, Levens NR (1993) Angiotensin II receptor subtypes: characterization, signalling mechanisms, and possible physiological implications. Front Neuroendocrinol 14:123–171

Brownfield MS, Reid IA, Ganten D, Ganong WF (1982) Differential distribution of immunoreactive angiotensin and angiotensin- converting enzyme in rat brain. Neuroscience 7:1759–1769

Bunnemann B, Fuxe K, Ganten D (1993) The renin–angiotensin system in the brain: an update 1993. Regul Pept 46:487–509

Burkhalter J, Felix D, Imboden H (2001) A new angiotensinergic system in the CNS of the rat. Regul Pept 99:93–101

Campbell DJ, Bouhnik J, Menard J, Corvol P (1984) Identity of angiotensinogen precursors of rat brain and liver. Nature 308:206–208

Cote F, Do TH, Laflamme L, Gallo JM, Gallo-Payet N (1999) Activation of the AT(2) receptor of angiotensin II induces neurite outgrowth and cell migration in microexplant cultures of the cerebellum. J Biol Chem 274:31686–31692

Couraud PO (1987) Anti-angiotensin II anti-idiotypic antibodies bind to angiotensin II receptor. J Immunol 138:1164–1168

de Gasparo M, Catt KJ, Inagami T, Wright JW, Unger T (2000) International union of pharmacology. XXIII. The angiotensin II receptors. Pharmacol Rev 52:415–472

Deschepper CF, Bouhnik J, Ganong WF (1986) Colocalization of angiotensinogen and glial fibrillary acidic protein in astrocytes in rat brain. Brain Res 374:195–198

Egli M, Laurent JP, Mosimann R, Felix D, Imboden H (2000) Morphological and immunocytochemical characterization of electrophysiologically investigated neurons in the PVN of the rat. J Neurosci Methods 95:145–150

Frei N, Weissenberger J, Beck-Sickinger AG, Hofliger M, Weis J, Imboden H (2001) Immunocytochemical localization of angiotensin II receptor subtypes and angiotensin II with monoclonal antibodies in the rat adrenal gland. Regul Pept 101:149–155

Fujita M, Hayashi I, Yamashina S, Itoman M, Majima M (2002) Blockade of angiotensin AT1a receptor signaling reduces tumor growth, angiogenesis, and metastasis. Biochem Biophys Res Commun 294:441–447

Ganong WF (1995) Reproduction and the renin–angiotensin system. Neurosci Biobehav Rev 19:241–250

Ganten D, Speck G (1978) The brain renin–angiotensin system: a model for the synthesis of peptides in the brain. Biochem Pharmacol 27:2379–2389

Ganten D, Lang RE, Lehmann E, Unger T (1984) Brain angiotensin: on the way to becoming a well-studied neuropeptide system. Biochem Pharmacol 33:3523–3528

Giles ME, Fernley RT, Nakamura Y, Moeller I, Aldred GP, Ferraro T, Penschow JD, McKinley MJ, Oldfield BJ (1999) Characterization of a specific antibody to the rat angiotensin II AT1 receptor. J Histochem Cytochem 47:507–516

Harrison-Bernard LM, Navar LG, Ho MM, Vinson GP, el Dahr SS (1997) Immunohistochemical localization of ANG II AT1 receptor in adult rat kidney using a monoclonal antibody. Am J Physiol 273:F170-F177

Healy DP, Printz MP (1984) Distribution of immunoreactive angiotensin II, angiotensin I, angiotensinogen and renin in the central nervous system of intact and nephrectomized rats. Hypertension 6:I130-I136

Hermann K, McDonald W, Unger T, Lang RE, Ganten D (1984) Angiotensin biosynthesis and concentrations in brain of normotensive and hypertensive rats. J Physiol (Paris) 79:471–480

Hoffman DL, Krupp L, Schrag D, Nilaver G, Valiquette G, Kilcoyne MM, Zimmerman EA (1982) Angiotensin immunoreactivity in vasopressin cells in rat hypothalamus and its relative deficiency in homozygous Brattleboro rats. Ann N Y Acad Sci 394:135–141

Imboden H, Harding JW, Abhold RH, Ganten D, Felix D (1987a) Improved immunohistochemical staining of angiotensin II in rat brain using affinity purified antibodies. Brain Res 426:225–234

Imboden H, Harding JW, Ganten D, Felix D (1987b) Comparison of angiotensin II staining in rat brain using affinity purified and crude antisera. Clin Exp Hypertens A 9:1133–1139

Imboden H, Harding JW, Hilgenfeldt U, Celio MR, Felix D (1987c) Localization of angiotensinogen in multiple cell types of rat brain. Brain Res 410:74–77

Imboden H, Harding JW, Felix D (1989) Hypothalamic angiotensinergic fibre systems terminate in the neurohypophysis. Neurosci Lett 96:42–46

Imboden H, Felix D (1991) An immunocytochemical comparison of the angiotensin and vasopressin hypothalamo-neurohypophysial systems in normotensive rats. Regul Pept 36:197–218

Inwang ER, Puddefoot JR, Brown CL, Goode AW, Marsigliante S, Ho MM, Payne JG, Vinson GP (1997) Angiotensin II type 1 receptor expression in human breast tissues. Br J Cancer 75:1279–1283

Ishida J, Fukamizu A (1999) Angiotensin II and apoptosis. Nippon Rinsho 57:1117–1123

Iwai N, Inagami T (1992) Identification of two subtypes in the rat type I angiotensin II receptor. FEBS Lett 298:257-260

Janiak P, Kasson BG, Brody MJ (1989) Central vasopressin raises arterial pressure by sympathetic activation and vasopressin release. Hypertension 13:935-940

Jerne NK (1974) Towards a network theory of the immune system. Ann Immunol (Paris) 125C:373-389

Jöhren O, Inagami T, Saavedra JM (1995) AT1A, AT1B, and AT2 angiotensin II receptor subtype gene expression in rat brain. Neuroreport 6:2549-2552

Kilcoyne MM, Hoffman DL, Zimmerman EA (1980) Immunocytochemical localization of angiotensin II and vasopressin in rat hypothalamus: evidence for production in the same neuron. Clin Sci (Lond) 59 Suppl 6:57s-60 s

Langhans W (1988) The antidiuretic hormone: new aspects of an "old" peptide. Zentralbl Veterinarmed A 35:641-654

Lewicki JA, Fallon JH, Printz MP (1978) Regional distribution of angiotensinogen in rat brain. Brain Res 158:359-371

Lind RW, Swanson LW, Ganten D (1984) Angiotensin II immunoreactivity in the neural afferents and efferents of the subfornical organ of the rat. Brain Res 321:209-215

Lind RW, Swanson LW, Bruhn TO, Ganten D (1985) The distribution of angiotensin II-immunoreactive cells and fibers in the paraventriculo-hypophysial system of the rat. Brain Res 338:81-89

Lippoldt A, Bunnemann B, Iwai N, Metzger R, Inagami T, Fuxe K, Ganten D (1993) Cellular localization of angiotensin type 1 receptor and angiotensinogen mRNAs in the subfornical organ of the rat brain. Neurosci Lett 150:153-158

Malendowicz SL, Ennezat PV, Testa M, Murray L, Sonnenblick EH, Evans T, LeJemtel TH (2000) Angiotensin II receptor subtypes in the skeletal muscle vasculature of patients with severe congestive heart failure. Circulation 102:2210-2213

Matsuguchi H, Sharabi FM, Gordon FJ, Johnson AK, Schmid PG (1982) Blood pressure and heart rate responses to microinjection of vasopressin into the nucleus tractus solitarius region of the rat. Neuropharmacology 21:687-693

Mosimann R, Imboden H, Felix D (1996) The neuronal role of angiotensin II in thirst, sodium appetite, cognition and memory. Biol Rev Camb Philos Soc 71:545-559

Mungall BA, Shinkel TA, Sernia C (1995) Immunocytochemical localization of angiotensinogen in the fetal and neonatal rat brain. Neuroscience 67:505-524

Nakamura Y, Makino H, Morishita R (1999) [Distribution and function of angiotensin receptor subtypes in cardiovascular system]. Nippon Rinsho 57:1032-1035

Nora EH, Munzenmaier DH, Hansen-Smith FM, Lombard JH, Greene AS (1998) Localization of the ANG II type 2 receptor in the microcirculation of skeletal muscle. Am J Physiol 275:H1395-H1403

Paxton WG, Runge M, Horaist C, Cohen C, Alexander RW, Bernstein KE (1993) Immunohistochemical localization of rat angiotensin II AT1 receptor. Am J Physiol 264:F989-F995

Pfister J, Felix D, Imboden H (1993) Immunohistochemical demonstration of angiotensin II receptors in rat brain by use of an anti-idiotypic antibody. Regul Pept 44:109-117

Pfister J, Spengler C, Grouzmann E, Raizada MK, Felix D, Imboden H (1997) Intracellular staining of angiotensin receptors in the PVN and SON of the rat. Brain Res 754:307-310

Phillips MI (1987) Functions of angiotensin in the central nervous system. Annu Rev Physiol 49:413-435

Phillips MI, Shen L, Richards EM, Raizada MK (1993) Immunohistochemical mapping of angiotensin AT1 receptors in the brain. Regul Pept 44:95-107

Rakugi H, Okamura A, Kamide K, Ohishi M, Sasamura H, Morishita R, Higaki J, Ogihara T (1997) Recognition of tissue- and subtype-specific modulation of angiotensin II receptors using antibodies against AT1 and AT2 receptors. Hypertens Res 20:51-55

Reagan LP, Flanagan-Cato LM, Yee DK, Ma LY, Sakai RR, Fluharty SJ (1994) Immunohistochemical mapping of angiotensin type 2 (AT2) receptors in rat brain. Brain Res 662:45–59

Riva C, Eggli P, Felix D, Mosimann R, Imboden H (1999) Hypothalamic accessory nuclei and their relation to the angiotensinergic and vasopressinergic systems. Regul Pept 83:129–133

Rivera E, Arrieta O, Guevara P, Duarte-Rojo A, Sotelo J (2001) AT1 receptor is present in glioma cells; its blockage reduces the growth of rat glioma. Br J Cancer 85:1396–1399

Rowe BP, Saylor DL, Speth RC (1992) Analysis of angiotensin II receptor subtypes in individual rat brain nuclei. Neuroendocrinology 55:563–573

Saavedra JM (1992) Brain and pituitary angiotensin. Endocr Rev 13:329–380

Schelling P, Clauser E, Felix D (1983) Regulation of angiotensinogen in the central nervous system. Clin Exp Hypertens A 5:1047–1061

Schelman WR, Kurth JL, Berdeaux RL, Norby SW, Weyhenmeyer JA (1997) Angiotensin II type-2 (AT2) receptor-mediated inhibition of NMDA receptor signalling in neuronal cells. Brain Res Mol Brain Res 48:197–205

Sen I, Rajasekaran AK (1991) Angiotensin II-binding protein in adult and neonatal rat heart. J Mol Cell Cardiol 23:563–572

Sernia C (1995) Location and secretion of brain angiotensinogen. Regul Pept 57:1–18

Shelat SG, Reagan LP, King JL, Fluharty SJ, Flanagan-Cato LM (1998) Analysis of angiotensin type 2 receptors in vasopressinergic neurons and pituitary in the rat. Regul Pept 73:103–112

Shenoy UV, Richards EM, Huang XC, Sumners C (1999) Angiotensin II type 2 receptor-mediated apoptosis of cultured neurons from newborn rat brain. Endocrinology 140:500–509

Silverman AJ, Zimmerman EA (1983) Magnocellular neurosecretory system. Annu Rev Neurosci 6:357–380

Song K, Allen AM, Paxinos G, Mendelsohn FA (1992) Mapping of angiotensin II receptor subtype heterogeneity in rat brain. J Comp Neurol 316:467–484

Spengler C, Pfister J, Raizada M, Felix D, Imboden H (1996) Coexistence of angiotensin receptors and angiotensin in hypothalamic neurons of the rat. In: Krisch B and Mentlein R (eds) The peptidergic neuron. Birkhäuser, Basel, pp 151–155

Stadler T, Veltmar A, Qadri F, Unger T (1992) Angiotensin II evokes noradrenaline release from the paraventricular nucleus in conscious rats. Brain Res 569:117–122

Tham M, Sim MK, Tang FR (2001) Location of renin-angiotensin system components in the hypoglossal nucleus of the rat. Regul Pept 101:51–57

Timmermans PB, Chiu AT, Herblin WF, Wong PC, Smith RD (1992) Angiotensin II receptor subtypes. Am J Hypertens 5:406–410

Tsutsumi K, Saavedra JM (1991) Quantitative autoradiography reveals different angiotensin II receptor subtypes in selected rat brain nuclei. J Neurochem 56:348–351

von Bohlen und Halbach O, Albrecht D (1998) Mapping of angiotensin AT1 receptors in the rat limbic system. Regul Pept 78:51–56

Wright JW, Harding JW (1994) Brain angiotensin receptor subtypes in the control of physiological and behavioral responses. Neurosci Biobehav Rev 18:21–53

Xoriuchi M, Hamai M, Cui TX, Iwai M, Minokoshi Y (1999) Cross talk between angiotensin II type 1 and type 2 receptors: cellular mechanism of angiotensin type 2 receptor-mediated cell growth inhibition. Hypertens Res 22:67–74

Yamada H, Akishita M, Ito M, Tamura K, Daviet L, Lehtonen JY, Dzau VJ, Horiuchi M (1999) AT2 receptor and vascular smooth muscle cell differentiation in vascular development. Hypertension 33:1414–1419

Yang SN, Lippoldt A, Jansson A, Phillips MI, Ganten D, Fuxe K (1997) Localization of angiotensin II AT1 receptor-like immunoreactivity in catecholaminergic neurons of the rat medulla oblongata. Neuroscience 81:503–515

Yiu AK, Wong PF, Yeung SY, Lam SM, Luk SK, Cheung WT (1997) Immunohistochemical localization of type-II (AT2) angiotensin receptors with a polyclonal antibody against a peptide from the C-terminal tail. Regul Pept 70:15–21

Zelezna B, Richards EM, Tang W, Lu D, Sumners C, Raizada MK (1992) Characterization of a polyclonal anti-peptide antibody to the angiotensin II type-1 (AT1) receptor. Biochem Biophys Res Commun 183:781–788

Renin–Angiotensin System: Plasma Versus Tissues

A. H. J. Danser

Department of Pharmacology, Room EE1418b, Erasmus MC, Dr. Molewaterplein 50, 3015 GE, Rotterdam, The Netherlands
e-mail: a.danser@erasmusmc.nl

1	Introduction	130
2	Release of Angiotensins from Tissue Sites into the Circulation?	131
3	Origin of Tissue Angiotensin I and II	133
4	Origin of Renin in Tissues: (Pro)Renin Receptor(s)?	135
5	The Site of Tissue Angiotensin Generation: Intra- or Extracellular?	137
6	Plasma and Tissue Angiotensin Levels During Renin–Angiotensin System Blockade: How Do ACE Inhibitors and AT_1 Receptor Antagonists Work?	139
7	Conclusions	141
References		142

Abstract The classical view of the renin–angiotensin system (RAS) as an endocrine system, designed to deliver its active end-product, angiotensin II, to tissues via circulating blood, has been challenged in the past two decades by studies demonstrating high tissue levels of angiotensin II, as well as release of angiotensin I from tissue sites. The high tissue levels of angiotensin II could not be explained on the basis of diffusion and/or receptor-mediated uptake of circulating angiotensin II, thereby supporting the concept that tissue angiotensin II is largely, if not completely, synthesized locally. Remarkably, in many organs (in particular the heart and vessel wall) this local synthesis depends on the uptake of circulating (i.e. kidney-derived) renin and/or prorenin, either via diffusion into the interstitial space or through binding to (pro)renin receptors. Renin and/or prorenin binding to (pro)renin receptors located on the cell membrane, combined with the fact that angiotensin-converting enzyme (ACE) also is a membrane-associated enzyme, favours angiotensin generation on the cell surface. Although renin has also been demonstrated intracellularly, intracellular angiotensin generation is unlikely to occur, due to the lack of angiotensinogen internalization. Tissue angiotensin generation is further supported by clinical studies demonstrating that RAS blockade, with either ACE inhibitors or angiotensin receptor antagonists, results in beneficial cardiac, vascular and renal effects that are independent, at least in part, of the blood pressure-lowering effects of these drugs. Taken together therefore, angiotensin II is generated at tissue sites, where it acts as a paracrine and/or autocrine hormone, and RAS blockers exert their effects mainly through interference with tissue angiotensin II.

Keywords Angiotensin · Renin · Prorenin · AT_1 receptor · AT_2 receptor · Mannose 6-phosphate/insulin-like growth factor II receptor · Chymase · Intracellular · ACE inhibitor

1
Introduction

Classically, the renin–angiotensin system (RAS) has been viewed as an endocrine system, designed to deliver its effector peptide, angiotensin (Ang) II, via the circulation to target organs, where it binds to Ang II type 1 (AT_1) and type 2 (AT_2) receptors in order to induce effects, the most important of which are vasoconstriction, sodium and water retention, and growth stimulation. According to this concept, liver-derived angiotensinogen is cleaved in the circulating blood by kidney-derived renin, to form Ang I. Ang I is subsequently converted to Ang II by angiotensin-converting enzyme (ACE) located on the luminal side of the vascular endothelium. ACE also occurs as a soluble form in blood plasma, but the contribution of soluble ACE to Ang II generation is negligible (Ng and Vane 1968; Admiraal et al. 1993).

Following the introduction of the ACE inhibitors in the early 1980s, it soon became clear that these drugs exerted effects in the face of unaltered or even increased levels of Ang II in circulating blood (Nussberger et al. 1985, 1986; Mooser et al. 1990; van den Meiracker et al. 1992). Moreover, beneficial effects were observed that occurred independently of their blood pressure-lowering effects (Linz et al. 1992; Yusuf et al. 2000). Although the former may have been due, at least in part, to inaccurate measurements of Ang II (Nussberger et al. 1985, 1986), these observations have led to the concept that angiotensins are also synthesized outside the circulation, i.e. at tissue sites (Dzau 1988). Recent studies with other RAS blockers (renin inhibitors and AT_1 receptor antagonists) further support this view (Fischli et al. 1994; Linz et al. 2000).

Remarkably, despite an overwhelming number of studies on "tissue Ang II generation" in the past 20 years, the precise mechanism underlying this phenomenon is still not completely understood. Originally, it was thought that all components required to synthesize Ang II locally, i.e. renin, angiotensinogen and ACE, were also synthesized in tissues, thus allowing organs like the heart, vessel wall, brain and adrenal to synthesize Ang II independently of angiotensin generation in the circulation. Protein and mRNA measurements have been performed to support the existence of local RASs in organs. Protein measurements, however, cannot be taken as evidence that the protein is synthesized locally. Moreover, such measurements are often hampered by a lack of specificity. For example, the presence of "renin" activity in tissue extracts may be caused by enzymes other than renin that are capable of generating Ang I (Katz et al. 2001).

Furthermore, detection of mRNA, particularly when the concentrations are low, does not necessarily mean that the corresponding protein is synthesized locally in physiologically relevant amounts (von Lutterotti et al. 1994). These uncertainties, as well as observations that renin could not be detected in cardiac and vascular tissue following bilateral nephrectomy (Loudon et al. 1983; Danser et al. 1994), have led to the idea that tissue Ang II generation might depend on one or more RAS components taken up from the circulation. Consequently, the term local (or tissue) RAS has been replaced by "local Ang II-generating system".

2
Release of Angiotensins from Tissue Sites into the Circulation?

Local production of angiotensins has been studied extensively, both through in vitro and in vivo experiments. In studies of isolated perfused rat Langendorff hearts, cardiac production of Ang I and II could only be demonstrated after the addition of renin to the perfusion buffer (Lindpaintner et al. 1990; de Lannoy et al. 1997, 1998). This suggests that angiotensinogen and ACE, but not renin, are present in the isolated rat heart. Results from isolated perfused rat hindquarter experiments parallel these findings (Fig. 1), in that the spontaneous release of angiotensins from these preparations decreased to low or undetectable levels when the animals had been nephrectomized prior to hindquarter perfusion (Hilgers et al. 1993; Kato et al. 1993).

Attempts have been made to study tissue angiotensin production in the intact animal by measuring arteriovenous differences of Ang I and II across the various vascular beds. However, such differences are small (Admiraal et al. 1990, 1993; Danser et al. 1992a; Neri Serneri et al. 1996), and it should be taken into account that a large proportion of circulatory Ang I and II is metabolized during

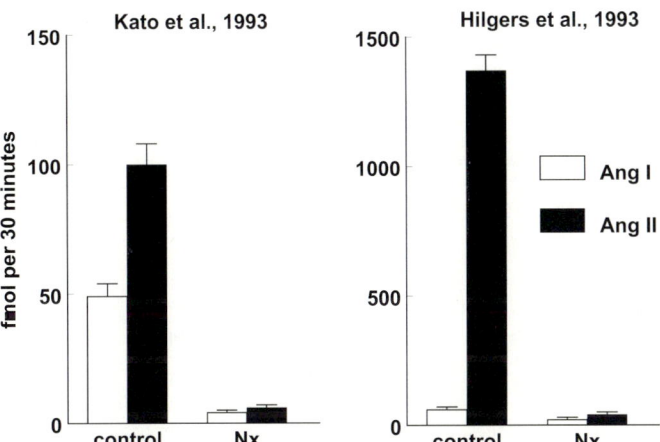

Fig. 1 Angiotensin (*Ang*) release from isolated perfused rat hindlimbs before and after bilateral nephrectomy (*Nx*). (Hilgers et al. 1993; Kato et al. 1993)

a single passage through an organ. In early studies in sheep (Fei et al. 1981), Ang II clearance across the pulmonary and combined systemic vascular beds was calculated from arteriovenous differences measured across these beds during infusion of high doses of Ang II into the pulmonary artery. These studies already indicated that Ang II is rapidly degraded and that plasma renin activity (PRA) is probably insufficient to generate the endogenous levels of circulating Ang II.

More recently, the regional clearance and metabolism of Ang I and II was quantified, both in humans and pigs, by constant infusions of radiolabelled ^{125}I-Ang I or II (Admiraal et al. 1990, 1993; Danser et al. 1992b, 1998; Neri Serneri et al. 1996). Blood was sampled from various arterial and venous sites for measurements of radiolabelled Ang I and II. Additional measurements of the levels of endogenous Ang I and II, and of PRA at physiological pH made it possible to estimate how much of venous Ang I could be attributed to arterial delivery, vs de novo synthesis, and what proportion of de novo synthesized Ang I depends on the action of circulating renin on circulating angiotensinogen (PRA). It was also possible to calculate how much of venous Ang II originated from arterial delivery, and how much from the conversion of arterially delivered Ang I.

It was found that in virtually all vascular beds studied (heart, lung, kidney, head, arm, leg, liver), a major proportion of venous Ang I originated from de novo production and that PRA contributed little to this de novo production

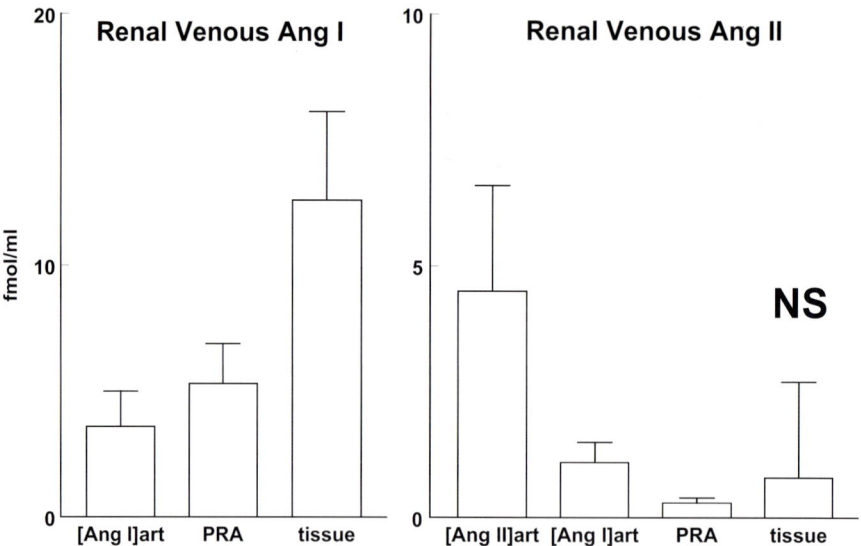

Fig. 2 Origin of renal venous angiotensin (Ang) I and II in humans (Danser et al. 1998). *[Ang I]art* and *[Ang II]art* represent arterial Ang I and Ang II, respectively. *PRA*, plasma renin activity. The majority of renal venous Ang I originates from tissue sites, whereas the release of Ang II from tissue sites is not significant (*NS*) from zero. The latter implies that de novo synthesized renal venous Ang II is derived only from arterial Ang I and Ang I generated by PRA during passage of the renal vascular bed

(Fig. 2). One can therefore conclude that Ang I is produced at tissue sites and that part of it is released into the circulation. While most of venous Ang I appears not to be generated by the action of circulating renin on circulating angiotensinogen, the level of venous Ang I produced at (and released from) tissue sites correlated strongly with the level of PRA (Danser et al. 1992b). This, together with findings in pigs and rats that tissue angiotensin levels are greatly reduced after nephrectomy (Campbell et al. 1993; Danser et al. 1994; Nussberger 2000), strongly suggests that it is kidney-derived renin that is a major determinant of tissue Ang I production.

For Ang II, we were unable to demonstrate release from tissue sites (Admiraal et al. 1993; Danser et al. 1998). Most, if not all, venous Ang II appeared to originate from delivered by the artery and from Ang II generated by conversion of arterially delivered Ang I. Thus, Ang I produced at tissue sites and released into the circulation may have escaped conversion to Ang II. It is possible that Ang I produced in the tissue enters the blood at a level distal to the site where arterially delivered Ang I is converted to Ang II by the vascular endothelium, so that this conversion site is bypassed. For instance, Ang I formed at tissue sites may enter the circulation at the level of the capillaries or venules, whereas Ang I to II conversion may occur only at the level of the arterioles. Alternatively, Ang II is produced in the tissue and remains there.

3
Origin of Tissue Angiotensin I and II

The studies described above are limited to measurements of angiotensins in the circulation. Although they support the concept of Ang I release from tissue sites into the circulation, further proof for Ang I (and II) generation at tissue sites can only be obtained by measuring tissue levels of Ang I and II. Such measurements are technically difficult because of ex vivo metabolism and generation of angiotensins, and because the antibodies used to quantify Ang I and II usually cross-react with other angiotensin metabolites and non-related substances. Thus, great care must be taken when measuring the "true" in vivo levels of Ang I and II. Furthermore, a correction should be made for (active) uptake of angiotensins from circulating blood, for instance through AT_1 receptor binding, as net tissue angiotensin levels are the sum of uptake from blood and local production. Uptake from the circulation can be quantified by measuring steady-state tissue and plasma levels of ^{125}I-labelled Ang I and II during ^{125}I-Ang I and II infusion.

The results obtained from pigs (van Kats et al. 1997, 1998, 2001) showed that ^{125}I-Ang II, but not ^{125}I-Ang I, accumulated in tissues (Fig. 3). The absence of significant tissue ^{125}I-Ang I accumulation already indicated that the presence of Ang I in tissues cannot be attributed to uptake from blood plasma. Steady-state levels of ^{125}I-Ang II at tissue sites were up to 20 times higher than those in blood plasma. This accumulation was largely prevented by pretreatment with an AT_1 receptor antagonist (Fig. 3), suggesting that it is mediated via AT_1 receptor-de-

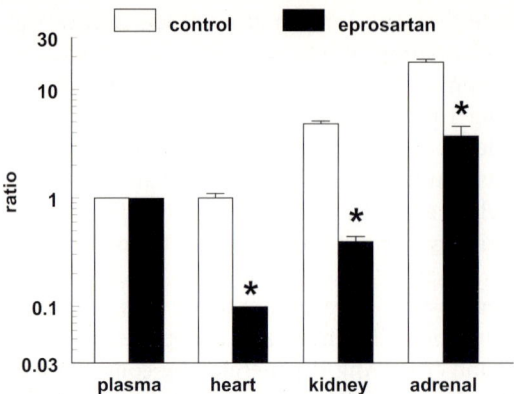

Fig. 3 ^{125}I-Angiotensin II levels (expressed as a ratio versus the steady-state plasma ^{125}I-angiotensin II level) during systemic infusion of ^{125}I-angiotensin II in pigs, both under control conditions and following AT$_1$ receptor blockade with eprosartan (van Kats et al. 1997, 2000, 2001). Eprosartan reduces the tissue/plasma concentrations of ^{125}I-angiotensin II by more than 80% without affecting plasma ^{125}I-angiotensin II. *, $p<0.05$ vs control pigs

pendent endocytosis. A role for AT$_2$ receptors in this process is unlikely because these receptors do not internalize Ang II (Matsubara 1998). Similar conclusions were drawn from Ang II infusion studies in rats (Zou et al. 1996a,b). Comparison of the ^{125}I-labelled and endogenous angiotensin levels (Fig. 4) revealed that, despite the significant uptake of ^{125}I-Ang II in various tissues, in all organs the majority (>90%) of tissue Ang II is not derived from circulation, i.e. is locally synthesized from locally generated Ang I.

Thus, both Ang I and II are generated at tissue sites, and of these locally generated angiotensins, only Ang I is released into the circulation, and Ang II remains in the tissue. The reason for this discrepancy may be that Ang II,

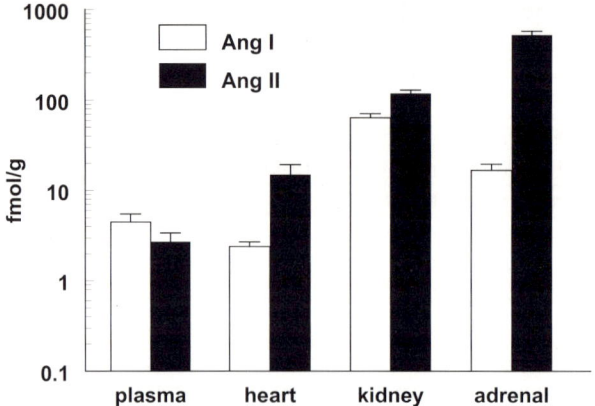

Fig. 4 Plasma and tissue levels of angiotensin I and II in normal pigs. (van Kats et al. 1997, 2000, 2001)

like ^{125}I-Ang II, rapidly binds to AT_1 receptors following its local generation, thereby preventing it from "leaking" into the circulation. Two lines of evidence support this view. First, the tissue Ang II content (adrenal>kidney>heart; Fig. 4), like the tissue ^{125}I-Ang II content (Fig. 3), parallels AT_1 receptor density in these organs (Whitebread et al. 1989; Zhuo et al. 1994; Regitz-Zagrosek et al. 1995). In this regard, it is important to note that the tissue Ang II levels, even in the adrenal (where the highest tissue Ang II levels are found), are too low to occupy all receptors, and thus that a large reserve of unoccupied AT_1 receptors exists. Second, the subcellular distribution of Ang II mimics that of ^{125}I-Ang II (van Kats et al. 2001), and as discussed above, the latter peptide accumulates in tissues via AT_1 receptor-mediated endocytosis only.

The half-life of AT_1 receptor-bound Ang II is roughly 15 min; i.e., 20–30 times longer than the half-life of Ang II in extracellular fluid (Schuijt et al. 1999; van Kats et al. 2001). A large number of free AT_1 receptors, combined with a significant increase of Ang II half-life following its binding to AT_1 receptors, would allow tissue Ang II to reach (much) higher levels than blood plasma Ang II, even when the level of Ang II generation is similar in both compartments.

4
Origin of Renin in Tissues: (Pro)Renin Receptor(s)?

Renin mRNA has been demonstrated in many tissues (Ekker et al. 1989; Iwai and Inagami 1992), in confirmation with the aforementioned studies providing direct evidence for tissue angiotensin generation. However, few attempts have been made to either combine renin mRNA measurement with tissue renin protein measurements (using renin-specific antibodies and inhibitors), to demonstrate renin release from tissues in vivo and/or from cultured cells, or to measure tissue renin following a bilateral nephrectomy. Tissues that convincingly show local renin synthesis are the ovary, testis, adrenal and eye (Itskovitz et al. 1987, 1992; Sealey et al. 1988; Danser et al. 1989; Peters et al. 1996; Wagner et al. 1996). Remarkably, in most of these tissues, prorenin (the inactive precursor of renin) rather than renin was detected. In addition, recent studies have reported the existence of a renin transcript that lacks the coding sequence for the prosegment; i.e. the transcript cannot be targeted to secretory pathways and thus remains intracellular (Clausmeyer et al. 1999, 2000; Lee-Kirsch et al. 1999). The functions of prorenin, without evidence for its local conversion to renin, and intracellular renin [which appears to be located in mitochondria (Clausmeyer et al. 1999)] remain unclear.

Renin and angiotensin disappear from cardiac and vascular tissues following bilateral nephrectomy, under both normal and pathological conditions (Hilgers et al. 1993; Danser et al. 1994; Katz et al. 1997). Moreover, 48 h after nephrectomy the angiotensin levels in muscle, liver, lung, and adrenal (but not in the brain) also decrease by 60%–95%, suggesting that they too, depend on renal renin (Campbell et al. 1993; Nussberger 2000). The fact that these levels, particularly in the adrenal, did not decrease to zero within 48 h, may either suggest that

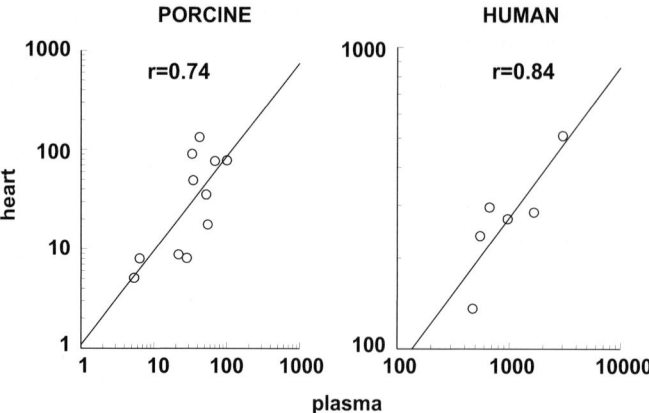

Fig. 5 Relationship between the plasma and cardiac tissue levels of renin in normal pigs (*left panel*) and in humans with dilated cardiomyopathy (*right panel*). (Danser et al. 1994, 1997)

renal renin, once present in tissues, has a long half-life, or that a percentage of angiotensin in these organs has indeed been synthesized by locally generated renin.

The large contribution of renal renin to angiotensin production in many tissues, in particular the heart and vascular wall, implies that tissues are capable of taking up circulating (i.e. kidney-derived) renin and/or prorenin, and possibly of locally converting prorenin to renin. Levels of renin in the heart, under both normal and pathological circumstances (Danser et al. 1994, 1997; Heller et al. 1998), correlate directly with plasma levels of renin (Fig. 5). "Uptake" might mean diffusion into interstitial space and/or binding to (pro)renin receptors. Evidence for renin diffusion/binding was obtained 20 years ago, when it was observed that vascular renin disappeared more slowly than circulating renin following a bilateral nephrectomy (Loudon et al. 1983). Subsequent studies confirmed that tissue renin is partly membrane-associated and that renin-binding proteins/receptors exist (Campbell and Valentijn 1994; Danser et al. 1994). Three such proteins have now been identified.

The first is an intracellular renin-binding protein (RnBP), that reduces the Ang I-generating activity of renin by more than 80% (Takahashi et al. 1992). This RnBP was found to be identical to the enzyme *N*-acyl-D-glucosamine 2-epimerase (Maru et al. 1996). It does not co-localize with renin, and mice lacking the RnBP display normal blood pressure and renin activity (Leckie et al. 2000; Schmitz et al. 2000). Thus, it is most likely unrelated to renin. The second is the mannose 6-phosphate insulin-like growth factor II (M6P/IGFII) receptor. This receptor not only binds and internalizes phosphomannosylated (i.e. M6P-containing) proteins like renin and prorenin with high affinity (van Kesteren et al. 1997; Admiraal et al. 1999; Saris et al. 2001; van den Eijnden et al. 2001), but it also activates prohormones (including native human prorenin) intracellularly (Helseth and Veis 1984). Its expression is developmentally regulated (Kornfeld

1992). The third is a high-affinity (pro)renin receptor that was originally discovered in human mesangial cells (Nguyen et al. 1996). It probably also occurs in membrane preparations obtained from rat tissues (Sealey et al. 1996). The gene encoding this binding site has recently been cloned (Nguyen et al. 2002). It encodes for a 350-amino acid protein showing no homology with known proteins. The binding site is predominantly present in vascular smooth muscle cells and glomerular mesangial cells. Interestingly, when bound to this receptor, renin displays increased catalytic activity as compared to soluble renin. Finally, there is evidence for a fourth receptor/binding site in adult rat cardiomyocytes that results in internalization of non-glycosylated prorenin (Peters et al. 2002). No attempts have been made so far to identify this protein.

The functional importance of these (pro)renin receptors should now be addressed, and the percentage of renin and prorenin containing the M6P signal should be determined under normal and pathological conditions. Studies in transgenic animals with liver-specific overexpression of renin and/or prorenin have already confirmed the uptake of circulating (pro)renin in cardiac tissue, as well as the subsequent generation of Ang I in the heart (Prescott et al. 2000, 2002). Unravelling the precise mechanism(s) underlying this phenomenon might result in the development of drugs that prevent generation of tissue Ang II by blocking (pro)renin uptake/binding.

5
The Site of Tissue Angiotensin Generation: Intra- or Extracellular?

Despite numerous reports suggesting that Ang II is an intracrine hormone, i.e. a hormone that is synthesized and acts intracellularly, convincing evidence for this concept is currently lacking. Despite the effects exerted by intracellularly applied Ang II (de Mello 1994; Filipeanu et al. 2001), and the reported presence of Ang II in and its release from cells (Dostal et al. 1992; Sadoshima et al. 1993; Leri et al. 1998), cellular Ang II synthesis remains unproven. Firstly Ang II accumulates in cells via AT_1 receptor-mediated internalization (van Kats et al. 1997), and at least part of its effects depend on such internalization (Griendling et al. 1987), and secondly, it cannot be excluded that cellularly released Ang II (e.g. after stretch) had been previously internalized (van Kesteren et al. 1999).

Intracellular angiotensin generation requires the presence of renin and angiotensinogen in the same subcellular compartment. As mentioned above, there is evidence for renin occurring intracellularly. An alternative renin transcript gives rise to a renin variant that lacks the coding sequence for the prosegment (and thus cannot be secreted); and M6P/IGFII receptor mediated internalization of renin and prorenin, followed by intracellular activation of prorenin, also results in renin accumulation in cells (Admiraal et al. 1999; Clausmeyer et al. 1999; Saris et al. 2001). Yet, the addition of angiotensinogen to the medium of cultured, renin-containing cardiomyocytes and endothelial cells did not result in intracellular Ang II generation (van den Eijnden et al. 2001; Saris et al. 2002). The main reason for this appeared to be a complete lack of angiotensinogen in-

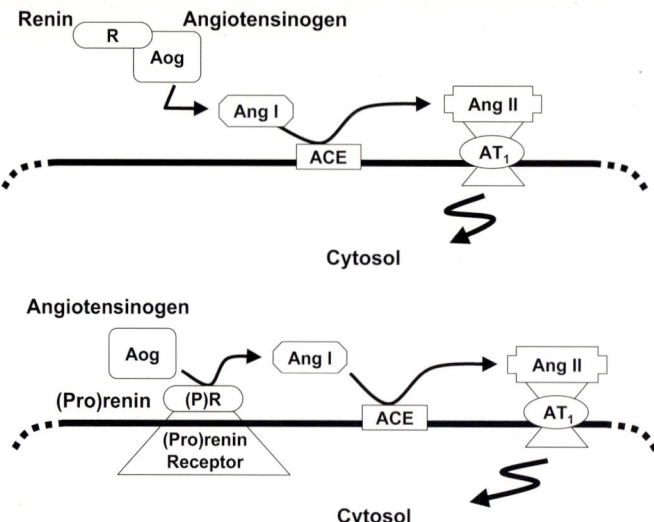

Fig. 6 Model of cell surface angiotensin (*Ang*) II generation and subsequent AT_1 receptor activation. Ang I synthesis from angiotensinogen (*Aog*) by renin (*R*) may either occur in the interstitium (*top panel*) or on the cell surface (*bottom panel*). The latter would require the existence of a (pro)renin [*(P)R*] binding protein/receptor. Prorenin, when bound to this receptor, may display enzymatic activity. (Nguyen et al. 2002)

ternalization (van den Eijnden et al. 2001; Saris et al. 2002). In view of this finding, and also taking into consideration that angiotensinogen-synthesizing cells normally release angiotensinogen into the extracellular fluid (without storing it intracellularly, due to the presence of a signal peptide that leads to secretion) (Klett et al. 1993), the question arises whether intracellular angiotensin generation is possible at all. In fact, a recent study in rat hepatoma cells demonstrated that such intracellular angiotensin generation could only occur following transfection of the cells with a mutated angiotensinogen cDNA that produces a non-secreted form of angiotensinogen (Cook et al. 2001). Taken together, intracellular Ang II generation is highly questionable under physiological conditions.

This conclusion leaves the possibility that tissue angiotensin generation occurs extracellularly, either in the interstitial fluid or on the cell surface (Fig. 6). Circulating renin diffuses into the interstitial fluid compartment, reaching steady-state levels that are equal to those in plasma (de Lannoy et al. 1997). The efficient cleavage of angiotensinogen by membrane-bound renin particularly supports cell surface angiotensin generation (Nguyen et al. 2002). In tissues where local angiotensin generation depends entirely on the sequestration of circulating renin, such efficient cleavage could result in interstitial angiotensin levels that are somewhat higher than the plasma angiotensin levels (de Lannoy et al. 1998; Hilgers et al. 1998; Schuijt et al. 2002). Larger differences, corresponding with several orders of magnitude (Siragy et al. 1995; Dell'Italia et al. 1997; Nishiyama et al. 2002), are unlikely to occur following sequestration of circulating renin only, and could therefore be suggestive for an important contribution of locally synthesized renin.

Extracellular angiotensin generation implies that Ang II is an autocrine and/or paracrine hormone. It does not argue against Ang II exerting intracellular effects following internalization of the AT_1 receptor-Ang II complex. If not leading to intracellular angiotensin generation, M6P/IGFII receptor-mediated (pro)renin internalization may represent (pro)renin clearance, thereby indirectly determining extracellular Ang I-generating activity. The function of mitochondrial renin remains unknown.

6
Plasma and Tissue Angiotensin Levels During Renin–Angiotensin System Blockade: How Do ACE Inhibitors and AT_1 Receptor Antagonists Work?

RAS blockade with ACE inhibitors or AT_1 receptor antagonists immediately results in renin release from the kidney, because it no longer allows Ang II to suppress renin release via AT_1 receptor activation. As a consequence, Ang I generation will increase (Mooser et al. 1990). This will lead to a rapid rise in the levels of Ang II during AT_1 receptor blockade (Fig. 7), and depending on the degree of ACE inhibition, a rise in Ang II levels during treatment with an ACE inhibitor (Campbell et al. 1993, 1995; van Kats et al. 2000). Other explanations for the rise in plasma and tissue Ang II during prolonged ACE inhibition (Fig. 7) are ACE upregulation and/or a role for alternative converting enzymes such as chymase (Urata et al. 1990; Balcells et al. 1997). ACE upregulation occurs both as a consequence of chronic ACE inhibitor therapy and during the progression of cardiovascular diseases (Farquharson and Struthers 2002). In support of the former,

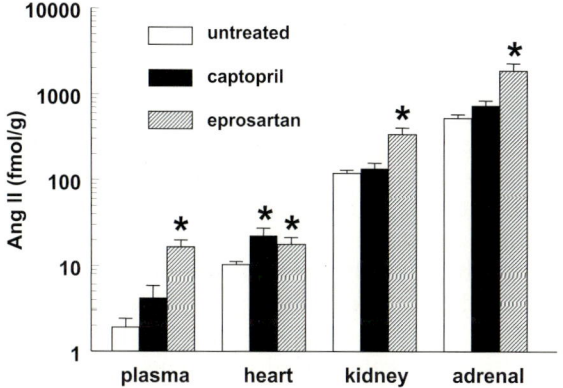

Fig. 7 Plasma and tissue levels of angiotensin II in untreated pigs and in pigs that were treated for 3 weeks with either the ACE inhibitor captopril or the AT_1 receptor antagonist eprosartan. (van Kats et al. 1997, 2000, 2001). *, $p<0.05$ vs untreated pigs

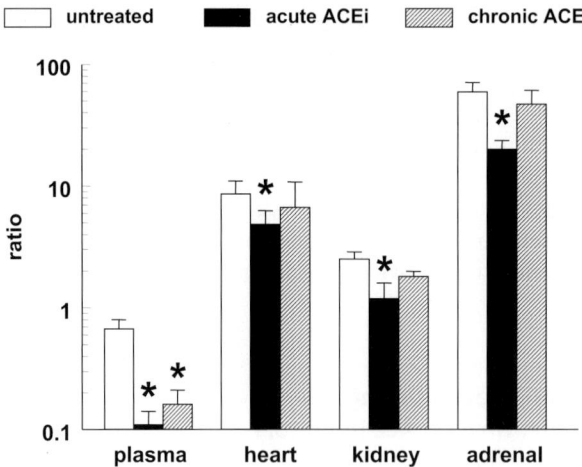

Fig. 8 Angiotensin II/angiotensin I ratios in untreated pigs and in pigs that were treated for either 3 days (acute ACEi) or 3 weeks (chronic ACEi) with the ACE inhibitor captopril. (van Kats et al. 1998, 2000, 2001). The ratio is taken as a measure for ACE activity. Note that the 45%–85% decrease in ACE activity following acute ACE inhibition disappears at tissue sites, but not in the circulation, upon prolonged treatment. *, $p<0.05$ vs untreated pigs

the tissue Ang II/I ratio in captopril-treated pigs is acutely decreased, but increases to normal values following prolonged ACE inhibitor treatment (Fig. 8) (van Kats et al. 1998, 2000, 2001). This phenomenon may be counteracted by increasing ACE inhibitor dosage (Jorde et al. 2000). As far as chymase is concerned, it is important to realize that chymase is largely located intracellularly (Urata et al. 1993). Studies demonstrating its significance, invariably measured Ang II generation in homogenized tissues, and evidence for chymase-mediated conversion of endogenous Ang I in humans is currently not available. In fact, when infusing large amounts of Ang I into the forearm of healthy volunteers (raising plasma Ang I by nearly four orders of magnitude), we were unable to demonstrate significant Ang II generation or Ang II-mediated vasoconstriction in the presence of the ACE inhibitor enalaprilat (Saris et al. 2000).

Furthermore, the phenotypes of ACE knockout mice and angiotensinogen knockout mice are similar (Tanimoto et al. 1994; Krege et al. 1995). This suggests that, also in mice, ACE is the main contributor to Ang II generation.

How do RAS blockers exert their effects if they result in increased or unaltered, rather than decreased plasma and tissue Ang II levels? One explanation might be that these increased Ang II levels result predominantly in AT_2 receptor activation. AT_2 receptors counteract the effects mediated by AT_1 receptors [they may in fact function as natural antagonists of AT_1 receptors (AbdAlla et al. 2001)], and their number increases under pathological conditions (Matsubara 1998; Siragy and Carey 1999; Schuijt et al. 2001) (Fig. 9). During AT_1 receptor blockade, it is obvious that increased Ang II levels result in AT_2 receptor stimulation. During ACE inhibition, increased Ang II levels will result in stimulation

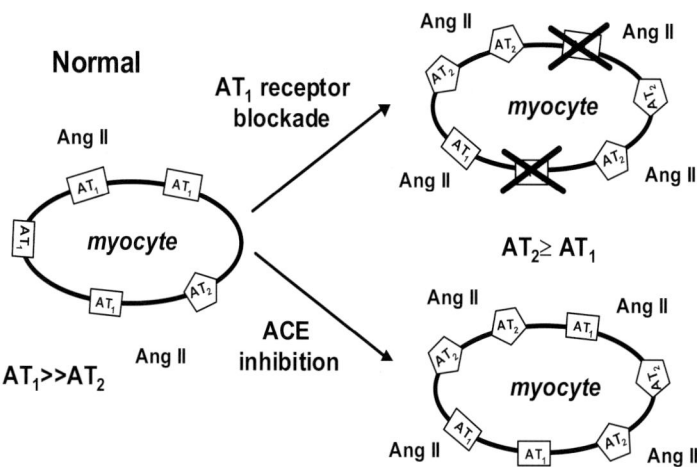

Fig. 9 Model explaining the cardiovascular effectiveness of ACE inhibitors and AT_1 receptor antagonists despite their ability to increase the levels of angiotensin (*Ang*) II at tissue sites. The rise in AT_2 receptor density under pathological conditions results in a predominance of AT_2 receptor-mediated effects, even when AT_1 receptors are not completely blocked

of both AT_1 and AT_2 receptors. However, the increase in AT_2 receptor density alters the net effect of Ang II. This concept is supported by studies showing that AT_2 receptor antagonism prevents the beneficial effects of AT_1 receptor blockade (Liu et al. 1997; Xu et al. 2002).

7
Conclusions

In conclusion, there is convincing evidence to support the production of both Ang I and Ang II at tissue sites. Intriguingly, in many tissues this production depends on circulating (i.e. kidney-derived) renin and/or prorenin. Circulating renin and prorenin diffuse into the interstitial space and bind to specific (pro)renin receptors, leading to generation of extracellular Ang I, either in the interstitium or on the cell surface. Although renin and prorenin are also internalized by various cells, the lack of intracellular angiotensinogen accumulation argues against significant intracellular angiotensin generation. Following its generation, tissue Ang II rapidly binds to AT_1 receptors and thereby increases its half-life from less than 30 s in the extracellular space to more than 10–15 min. This phenomenon may explain why tissue Ang II levels, particularly in organs with high AT_1 receptor densities (e.g. the adrenal), are several orders of magnitude higher than the levels of Ang II in blood, even when the level of Ang II generation is similar in both compartments. In addition, a large number of "free" AT_1 receptors, combined with low (i.e. comparable to blood plasma) interstitial Ang II levels, offers an explanation for the high sensitivity of organs like the kidney and adrenal, to modest changes in arterial Ang II. Finally, because of the AT_1—AT_2 receptor shift that occurs under pathological conditions, the increase in tissue Ang II generation that is sometimes observed during treatment with RAS blockers (including ACE inhibitors) may result in net AT_2 receptor stimulation, thereby counteracting AT_1 receptor-mediated vasoconstriction and growth stimulation.

References

AbdAlla S, Lother H, Abdel-tawab AM, Quitterer U (2001) The angiotensin II AT_2 receptor is an AT_1 receptor antagonist. J Biol Chem 276:39721–39726

Admiraal PJJ, Danser AHJ, Jong MS, Pieterman H, Derkx FHM, Schalekamp MADH (1993) Regional angiotensin II production in essential hypertension and renal artery stenosis. Hypertension 21:173–184

Admiraal PJJ, Derkx FHM, Danser AHJ, Pieterman H, Schalekamp MADH (1990) Intrarenal de novo production of angiotensin I in subjects with renal artery stenosis. Hypertension 16:555–563

Admiraal PJJ, Derkx FHM, Danser AHJ, Pieterman H, Schalekamp MADH (1990) Metabolism and production of angiotensin I in different vascular beds in subjects with hypertension. Hypertension 15:44–55

Admiraal PJJ, van Kesteren CAM, Danser AHJ, Derkx FHM, Sluiter W, Schalekamp MADH (1999) Uptake and proteolytic activation of prorenin by cultured human endothelial cells. J Hypertens 17:621–629

Balcells E, Meng QC, Johnson WH, Oparil S, Dell'Italia LJ (1997) Angiotensin II formation from ACE and chymase in human and animal hearts: methods and species considerations. Am J Physiol 273:H1769–H1774

Campbell DJ, Kladis A, Duncan AM (1993) Nephrectomy, converting enzyme inhibition, and angiotensin peptides. Hypertension 22:513–522

Campbell DJ, Kladis A, Valentijn AJ (1995) Effects of losartan on angiotensin and bradykinin peptides and angiotensin-converting enzyme. J Cardiovasc Pharmacol 26:233–240

Campbell DJ, Valentijn AJ (1994) Identification of vascular renin-binding proteins by chemical cross-linking: inhibition of binding of renin by renin inhibitors. J Hypertens 12:879–890

Clausmeyer S, Reinecke A, Farrenkopf R, Unger T, Peters J (2000) Tissue-specific expression of a rat renin transcript lacking the coding sequence for the prefragment and its stimulation by myocardial infarction. Endocrinology 141:2963–2970

Clausmeyer S, Sturzebecher R, Peters J (1999) An alternative transcript of the rat renin gene can result in a truncated prorenin that is transported into adrenal mitochondria. Circ Res 84:337–344

Cook JL, Zhang Z, Re RN (2001) In vitro evidence for an intracellular site of angiotensin action. Circ Res 89:1138–1146

Danser AHJ, Admiraal PJJ, Derkx FHM, Schalekamp MADH (1998) Angiotensin I-to-II conversion in the human renal vascular bed. J Hypertens 16:2051–2056

Danser AHJ, Koning MMG, Admiraal PJJ, Derkx FHM, Verdouw PD, Schalekamp MADH (1992a) Metabolism of angiotensin I by different tissues in the intact animal. Am J Physiol 263:H418–H428

Danser AHJ, Koning MMG, Admiraal PJJ, Sassen LMA, Derkx FHM, Verdouw PD, Schalekamp MADH (1992b) Production of angiotensins I and II at tissue sites in intact pigs. Am J Physiol 263:H429–H437

Danser AHJ, van den Dorpel MA, Deinum J, Derkx FHM, Franken AAM, Peperkamp E, de Jong PTVM, Schalekamp MADH (1989) Renin, prorenin, and immunoreactive renin in vitreous fluid from eyes with and without diabetic retinopathy. J Clin Endocrinol Metab 68:160–167

Danser AHJ, van Kats JP, Admiraal PJJ, Derkx FHM, Lamers JMJ, Verdouw PD, Saxena PR, Schalekamp MADH (1994) Cardiac renin and angiotensins. Uptake from plasma versus in situ synthesis. Hypertension 24:37–48

Danser AHJ, van Kesteren CAM, Bax WA, Tavenier M, Derkx FHM, Saxena PR, Schalekamp MADH (1997) Prorenin, renin, angiotensinogen, and angiotensin-converting enzyme in normal and failing human hearts. Evidence for renin binding. Circulation 96:220–226

de Lannoy LM, Danser AHJ, Bouhuizen AMB, Saxena PR, Schalekamp MADH (1998) Localization and production of angiotensin II in the isolated perfused rat heart. Hypertension 31:1111–1117

de Lannoy LM, Danser AHJ, van Kats JP, Schoemaker RG, Saxena PR, Schalekamp MADH (1997) Renin–angiotensin system components in the interstitial fluid of the isolated perfused rat heart. Local production of angiotensin I. Hypertension 29:1240–1251

de Mello WC (1994) Is an intracellular renin–angiotensin system involved in control of cell communication in heart? J Cardiovasc Pharmacol 23:640–646

Dell'Italia LJ, Meng QC, Balcells E, Wei CC, Palmer R, Hageman GR, Durand J, Hankes GH, Oparil S (1997) Compartmentalization of angiotensin II generation in the dog heart. Evidence for independent mechanisms in intravascular and interstitial spaces. J Clin Invest 100:253–258

Dostal DE, Rothblum KN, Conrad KM, Cooper GR, Baker KM (1992) Detection of angiotensin I and II in cultured rat cardiac myocytes and fibroblasts. Am J Physiol 263:C851–C863

Dzau VJ (1988) Circulating versus local renin–angiotensin system in cardiovascular homeostasis. Circulation 77:I4–I13

Ekker M, Tronik D, Rougeon F (1989) Extra-renal transcription of the renin genes in multiple tissues of mice and rats. Proc Natl Acad Sci USA 86:5155–5158

Farquharson CA, Struthers AD (2002) Gradual reactivation over time of vascular tissue angiotensin I to angiotensin II conversion during chronic lisinopril therapy in chronic heart failure. J Am Coll Cardiol 39:767–775

Fei DT, Scoggins BA, Tregear GW, Coghlan JP (1981) Angiotensin I, II, and III in sheep. A model of angiotensin production and metabolism. Hypertension 3:730–737

Filipeanu CM, Henning RH, de Zeeuw D, Nelemans A (2001) Intracellular angiotensin II and cell growth of vascular smooth muscle cells. Br J Pharmacol 132:1590–1596

Fischli W, Clozel JP, Breu V, Buchmann S, Mathews S, Stadler H, Vieira E, Wostl W (1994) Ciprokiren (Ro 44–9375). A renin inhibitor with increasing effects on chronic treatment. Hypertension 24:163–169

Griendling KK, Delafontaine P, Rittenhouse SE, Gimbrone MA, Jr., Alexander RW (1987) Correlation of receptor sequestration with sustained diacylglycerol accumulation in angiotensin II-stimulated cultured vascular smooth muscle cells. J Biol Chem 262:14555–14562

Heller LJ, Opsahl JA, Wernsing SE, Saxena R, Katz SA (1998) Myocardial and plasma renin–angiotensinogen dynamics during pressure-induced cardiac hypertrophy. Am J Physiol 274:R849–R856

Helseth DL, Jr., Veis A (1984) Cathepsin D-mediated processing of procollagen: lysosomal enzyme involvement in secretory processing of procollagen. Proc Natl Acad Sci USA 81:3302–3306

Hilgers KF, Bingener E, Stumpf C, Müller DN, Schmieder RE, Veelken R (1998) Angiotensinases restrict locally generated angiotensin II to the blood vessel wall. Hypertension 31:368–372

Hilgers KF, Hilgenfeldt U, Veelken R, Muley T, Ganten D, Luft FC, Mann JF (1993) Angiotensinogen is cleaved to angiotensin in isolated rat blood vessels. Hypertension 21:1030–1034

Itskovitz J, Rubattu S, Levron J, Sealey JE (1992) Highest concentrations of prorenin and human chorionic gonadotropin in gestational sacs during early human pregnancy. J Clin Endocrinol Metab 75:906–910

Itskovitz J, Sealey JE, Glorioso N, Rosenwaks Z (1987) Plasma prorenin response to human chorionic gonadotropin in ovarian- hyperstimulated women: correlation with the number of ovarian follicles and steroid hormone concentrations. Proc Natl Acad Sci USA 84:7285–7289

Iwai N, Inagami T (1992) Quantitative analysis of renin gene expression in extrarenal tissues by polymerase chain reaction method. J Hypertens 10:717–724

Jorde UP, Ennezat PV, Lisker J, Suryadevara V, Infeld J, Cukon S, Hammer A, Sonnenblick EH, Le Jemtel TH (2000) Maximally recommended doses of angiotensin-converting enzyme (ACE) inhibitors do not completely prevent ACE-mediated formation of angiotensin II in chronic heart failure. Circulation 101:844–846

Kato H, Iwai N, Inui H, Kimoto K, Uchiyama Y, Inagami T (1993) Regulation of vascular angiotensin release. Hypertension 21:446–454

Katz SA, Opsahl JA, Forbis LM (2001) Myocardial enzymatic activity of renin and cathepsin D before and after bilateral nephrectomy. Basic Res Cardiol 96:659–668

Katz SA, Opsahl JA, Lunzer MM, Forbis LM, Hirsch AT (1997) Effect of bilateral nephrectomy on active renin, angiotensinogen, and renin glycoforms in plasma and myocardium. Hypertension 30:259–266

Klett C, Nobiling R, Gierschik P, Hackenthal E (1993) Angiotensin II stimulates the synthesis of angiotensinogen in hepatocytes by inhibiting adenylylcyclase activity and stabilizing angiotensinogen mRNA. J Biol Chem 268:25095–25107

Kornfeld S (1992) Structure and function of the mannose 6-phosphate/insulinlike growth factor II receptors. Annu Rev Biochem 61:307–330

Krege JH, John SW, Langenbach LL, Hodgin JB, Hagaman JR, Bachman ES, Jennette JC, O'Brien DA, Smithies O (1995) Male-female differences in fertility and blood pressure in ACE-deficient mice. Nature 375:146–148

Leckie BJ, Lacy PS, Lidder S (2000) The expression of renin-binding protein and renin in the kidneys of rats with two-kidney one-clip hypertension. J Hypertens 18:935–943

Lee-Kirsch MA, Gaudet F, Cardoso MC, Lindpaintner K (1999) Distinct renin isoforms generated by tissue-specific transcription initiation and alternative splicing. Circ Res 84:240–246

Leri A, Claudio PP, Li Q, Wang X, Reiss K, Wang S, Malhotra A, Kajstura J, Anversa P (1998) Stretch-mediated release of angiotensin II induces myocyte apoptosis by activating p53 that enhances the local renin–angiotensin system and decreases the Bcl-2-to-Bax protein ratio in the cell. J Clin Invest 101:1326–1342

Lindpaintner K, Jin MW, Niedermaier N, Wilhelm MJ, Ganten D (1990) Cardiac angiotensinogen and its local activation in the isolated perfused beating heart. Circ Res 67:564–573

Linz W, Heitsch H, Schölkens BA, Wiemer G (2000) Long-term angiotensin II type 1 receptor blockade with fonsartan doubles lifespan of hypertensive rats. Hypertension 35:908–913

Linz W, Schaper J, Wiemer G, Albus U, Schölkens BA (1992) Ramipril prevents left ventricular hypertrophy with myocardial fibrosis without blood pressure reduction: a one year study in rats. Br J Pharmacol 107:970–975

Liu YH, Yang XP, Sharov VG, Nass O, Sabbah HN, Peterson E, Carretero OA (1997) Effects of angiotensin-converting enzyme inhibitors and angiotensin II type 1 receptor antagonists in rats with heart failure. Role of kinins and angiotensin II type 2 receptors. J Clin Invest 99:1926–1935

Loudon M, Bing RF, Thurston H, Swales JD (1983) Arterial wall uptake of renal renin and blood pressure control. Hypertension 5:629–634

Maru I, Ohta Y, Murata K, Tsukada Y (1996) Molecular cloning and identification of N-acyl-D-glucosamine 2-epimerase from porcine kidney as a renin-binding protein. J Biol Chem 271:16294–16299

Matsubara H (1998) Pathophysiological role of angiotensin II type 2 receptor in cardiovascular and renal diseases. Circ Res 83:1182–1191

Mooser V, Nussberger J, Juillerat L, Burnier M, Waeber B, Bidiville J, Pauly N, Brunner HR (1990) Reactive hyperreninemia is a major determinant of plasma angiotensin II during ACE inhibition. J Cardiovasc Pharmacol 15:276–282

Neri Serneri GG, Boddi M, Coppo M, Chechi T, Zarone N, Moira M, Poggesi L, Margheri M, Simonetti I (1996) Evidence for the existence of a functional cardiac renin–angiotensin system in humans. Circulation 94:1886–1893

Ng KKF, Vane JR (1968) Fate of angiotensin I in the circulation. Nature 218:144–150

Nguyen G, Delarue F, Berrou J, Rondeau E, Sraer JD (1996) Specific receptor binding of renin on human mesangial cells in culture increases plasminogen activator inhibitor-1 antigen. Kidney Int 50:1897–1903

Nguyen G, Delarue F, Burcklé C, Bouzhir L, Giller T, Sraer J-D (2002) Pivotal role of the renin/prorenin receptor in angiotensin II production and cellular responses to renin. J Clin Invest 109:1417–1427

Nishiyama A, Seth DM, Navar LG (2002) Renal interstitial fluid angiotensin I and angiotensin II concentrations during local angiotensin-converting enzyme inhibition. J Am Soc Nephrol 13:2207–2212

Nussberger J (2000) Circulating versus tissue angiotensin II. In: Epstein M, Brunner HR (eds) Angiotensin II Receptor Antagonists. Hanley & Belfus, Philadelphia, pp 69–78

Nussberger J, Brunner DB, Waeber B, Brunner HR (1986) Specific measurement of angiotensin metabolites and in vitro generated angiotensin II in plasma. Hypertension 8:476–482

Nussberger J, Brunner DB, Waeber B, Brunner HR (1985) True versus immunoreactive angiotensin II in human plasma. Hypertension 7:I1–I7

Peters J, Farrenkopf R, Clausmeyer S, Zimmer J, Kantachuvesiri S, Sharp MG, Mullins JJ (2002) Functional significance of prorenin internalization in the rat heart. Circ Res 90:1135–1141

Peters J, Kranzlin B, Schaeffer S, Zimmer J, Resch S, Bachmann S, Gretz N, Hackenthal E (1996) Presence of renin within intramitochondrial dense bodies of the rat adrenal cortex. Am J Physiol 271:E439–E450

Prescott G, Silversides DW, Chiu SM, Reudelhuber TL (2000) Contribution of circulating renin to local synthesis of angiotensin peptides in the heart. Physiol Genomics 4:67–73

Prescott G, Silversides DW, Reudelhuber TL (2002) Tissue activity of circulating prorenin. Am J Hypertens 15:280–285

Regitz-Zagrosek V, Auch-Schwelk W, Hess B, Klein U, Duske E, Steffen C, Hildebrandt AG, Fleck E (1995) Tissue- and subtype-specific modulation of angiotensin II receptors by chronic treatment with cyclosporin A, angiotensin-converting enzyme inhibitors and AT1 antagonists. J Cardiovasc Pharmacol 26:66–72

Sadoshima J, Xu Y, Slayter HS, Izumo S (1993) Autocrine release of angiotensin II mediates stretch-induced hypertrophy of cardiac myocytes in vitro. Cell 75:977–984

Saris JJ, Derkx FHM, de Bruin RJA, Dekkers DHW, Lamers JMJ, Saxena PR, Schalekamp MADH, Danser AHJ (2001) High-affinity prorenin binding to cardiac man-6-P/IGF-II receptors precedes proteolytic activation to renin. Am J Physiol 280:H1706-H1715

Saris JJ, van den Eijnden MMED, Lamers JMJ, Saxena PR, Schalekamp MADH, Danser AHJ (2002) Prorenin-induced myocyte proliferation: no role for intracellular angiotensin II. Hypertension 39:573–577

Saris JJ, van Dijk MA, Kroon I, Schalekamp MADH, Danser AHJ (2000) Functional importance of angiotensin-converting enzyme-dependent in situ angiotensin II generation in the human forearm. Hypertension 35:764–768

Schmitz C, Gotthardt M, Hinderlich S, Leheste JR, Gross V, Vorum H, Christensen EI, Luft FC, Takahashi S, Willnow TE (2000) Normal blood pressure and plasma renin activity in mice lacking the renin-binding protein, a cellular renin inhibitor. J Biol Chem 275:15357–15362

Schuijt MP, Basdew M, van Veghel R, de Vries R, Saxena PR, Schoemaker RG, Danser AHJ (2001) AT2 receptor-mediated vasodilation in the heart: effect of myocardial infarction. Am J Physiol 281:H2590–H2596

Schuijt MP, de Vries R, Saxena PR, Schalekamp MADH, Danser AHJ (2002) Vasoconstriction is determined by interstitial rather than circulating angiotensin II. Br J Pharmacol 135:275–283

Schuijt MP, van Kats JP, de Zeeuw S, Duncker DJ, Verdouw PD, Schalekamp MADH, Danser AHJ (1999) Cardiac interstitial fluid levels of angiotensin I and II in the pig. J Hypertens 17:1885–1891

Sealey JE, Catanzaro DF, Lavin TN, Gahnem F, Pitarresi T, Hu LF, Laragh JH (1996) Specific prorenin/renin binding (ProBP). Identification and characterization of a novel membrane site. Am J Hypertens 9:491–502

Sealey JE, Goldstein M, Pitarresi T, Kudlak TT, Glorioso N, Fiamengo SA, Laragh JH (1988) Prorenin secretion from human testis: no evidence for secretion of active renin or angiotensinogen. J Clin Endocrinol Metab 66:974–978

Siragy HM, Carey RM (1999) Protective role of the angiotensin AT2 receptor in a renal wrap hypertension model. Hypertension 33:1237–1242

Siragy HM, Howell NL, Ragsdale NV, Carey RM (1995) Renal interstitial fluid angiotensin. Modulation by anesthesia, epinephrine, sodium depletion, and renin inhibition. Hypertension 25:1021–1024

Takahashi S, Inoue H, Miyake Y (1992) The human gene for renin-binding protein. J Biol Chem 267:13007–13013

Tanimoto K, Sugiyama F, Goto Y, Ishida J, Takimoto E, Yagami K, Fukamizu A, Murakami K (1994) Angiotensinogen-deficient mice with hypotension. J Biol Chem 269:31334–31337

Urata H, Boehm KD, Philip A, Kinoshita A, Gabrovsek J, Bumpus FM, Husain A (1993) Cellular localization and regional distribution of an angiotensin II-forming chymase in the heart. J Clin Invest 91:1269–1281

Urata H, Kinoshita A, Misono KS, Bumpus FM, Husain A (1990) Identification of a highly specific chymase as the major angiotensin II-forming enzyme in the human heart. J Biol Chem 265:22348–22357

van den Eijnden MMED, Saris JJ, de Bruin RJA, de Wit E, Sluiter W, Reudelhuber TL, Schalekamp MADH, Derkx FHM, Danser AHJ (2001) Prorenin accumulation and activation in human endothelial cells. Importance of mannose 6-phosphate receptors. Arterioscler Thromb Vasc Biol 21:911–916

van den Meiracker AH, Man in 't Veld AJ, Admiraal PJJ, Ritsema van Eck HJ, Boomsma F, Derkx FHM, Schalekamp MADH (1992) Partial escape of angiotensin converting enzyme (ACE) inhibition during prolonged ACE inhibitor treatment: does it exist and does it affect the antihypertensive response? J Hypertens 10:803–812

van Kats JP, Danser AHJ, van Meegen JR, Sassen LM, Verdouw PD, Schalekamp MADH (1998) Angiotensin production by the heart: a quantitative study in pigs with the use of radiolabeled angiotensin infusions. Circulation 98:73–81

van Kats JP, de Lannoy LM, Danser AHJ, van Meegen JR, Verdouw PD, Schalekamp MADH (1997) Angiotensin II type 1 (AT1) receptor-mediated accumulation of angiotensin II in tissues and its intracellular half-life in vivo. Hypertension 30:42–49

van Kats JP, Duncker DJ, Haitsma DB, Schuijt MP, Niebuur R, Stubenitsky R, Boomsma F, Schalekamp MADH, Verdouw PD, Danser AHJ (2000) Angiotensin-converting enzyme inhibition and angiotensin II type 1 receptor blockade prevent cardiac remodeling in pigs after myocardial infarction: role of tissue angiotensin II. Circulation 102:1556–1563

van Kats JP, Schalekamp MADH, Verdouw PD, Duncker DJ, Danser AHJ (2001) Intrarenal angiotensin II: interstitial and cellular levels and site of production. Kidney Int 60:2311–2317

van Kats JP, van Meegen JR, Verdouw PD, Duncker DJ, Schalekamp MADH, Danser AHJ (2001) Subcellular localization of angiotensin II in kidney and adrenal. J Hypertens 19:583–589

van Kesteren CAM, Danser AHJ, Derkx FHM, Dekkers DHW, Lamers JMJ, Saxena PR, Schalekamp MADH (1997) Mannose 6-phosphate receptor-mediated internalization and activation of prorenin by cardiac cells. Hypertension 30:1389–1396

van Kesteren CAM, Saris JJ, Dekkers DHW, Lamers JMJ, Saxena PR, Schalekamp MADH, Danser AHJ (1999) Cultured neonatal rat cardiac myocytes and fibroblasts do not synthesize renin or angiotensinogen: evidence for stretch-induced cardiomyocyte hypertrophy independent of angiotensin II. Cardiovasc Res 43:148–156

von Lutterotti N, Catanzaro DF, Sealey JE, Laragh JH (1994) Renin is not synthesized by cardiac and extrarenal vascular tissues. A review of experimental evidence. Circulation 89:458–470

Wagner J, Danser AHJ, Derkx FHM, de Jong PTVM, Paul M, Mullins JJ, Schalekamp MADH, Ganten D (1996) Demonstration of renin mRNA, angiotensinogen mRNA, and angiotensin converting enzyme mRNA expression in the human eye: evidence for an intraocular renin–angiotensin system. Br J Ophthalmol 80:159–163

Whitebread S, Mele M, Kamber B, de Gasparo M (1989) Preliminary biochemical characterization of two angiotensin II receptor subtypes. Biochem Biophys Res Commun 163:284–291

Xu J, Carretero OA, Liu YH, Shesely EG, Yang F, Kapke A, Yang XP (2002) Role of AT(2) Receptors in the Cardioprotective Effect of AT(1) Antagonists in Mice. Hypertension 40:244–250

Yusuf S, Sleight P, Pogue J, Bosch J, Davies R, Dagenais G (2000) Effects of an angiotensin-converting-enzyme inhibitor, ramipril, on cardiovascular events in high-risk patients. The Heart Outcomes Prevention Evaluation Study Investigators. N Engl J Med 342:145–153

Zhuo J, Song K, Abdelrahman A, Mendelsohn FA (1994) Blockade by intravenous losartan of AT1 angiotensin II receptors in rat brain, kidney and adrenals demonstrated by in vitro autoradiography. Clin Exp Pharmacol Physiol 21:557–567

Zou LX, Hymel A, Imig JD, Navar LG (1996a) Renal accumulation of circulating angiotensin II in angiotensin II-infused rats. Hypertension 27:658–662

Zou LX, Imig JD, von Thun AM, Hymel A, Ono H, Navar LG (1996b) Receptor-mediated intrarenal angiotensin II augmentation in angiotensin II-infused rats. Hypertension 28:669–677

Angiotensin–Endothelin Interactions

P. Moreau[1] · M.-A. Laplante[2] · P. Beaucage[1] · J. de Champlain[2]

[1] Faculty of Pharmacy, Université de Montréal, PO Box 6128, Station Centre-ville, Montréal, QC, H3C 3J7, Canada

[2] Department of Physiologie, Faculty of Medicine, Université de Montréal, PO Box 6128, Station Centre-ville, Montréal, QC, H3C 3J7, Canada
e-mail: Jacques.de.champlain@umontreal.ca

1	Introduction	150
1.1	Biology of Endothelin	150
1.2	RAAS Modulation of ET Production	151
1.3	ET Modulation of the RAAS	152
2	Hemodynamic Effects	153
2.1	Angiotensin II-Induced Hypertension	153
2.2	Vascular Reactivity	155
3	Cardiovascular Remodeling	155
3.1	Small Arteries	156
3.2	Large Arteries	157
3.3	Cardiac Remodeling	158
4	Renal Interactions	159
5	Oxidative Stress	159
6	Combination of ETRA with Inhibitors of the RAAS	161
7	Conclusion	163
	References	164

Abstract Angiotensin II (Ang II) has been shown in several experimental conditions to stimulate the production of endothelin (ET), an endothelium-derived constricting factor. It is therefore not surprising that some of the physiological or pathological effects of angiotensin II are in fact mediated by endothelin through interaction with its specific receptors. For instance, ET receptor antagonists block part of the pressor effect of Ang II, especially when the treatment is initiated early. Accordingly, Ang II-induced vasoconstriction appears to be dependent on ET release, especially in smaller arteries. Moreover, blockade of ET receptors has a profound effect on Ang II-induced hypertrophic remodeling of small arteries. However, the contribution of ET appears to be limited to the initial phase of hypertrophy, as ET receptor antagonists do not regress already established vascular hypertrophy, but completely prevent it. Interestingly, ET also appears to mediate part of Ang II-induced vascular superoxide production in vivo. Thus, blocking ET receptors may provide similar cardiovascular benefits as inhibiting the renin–angiotensin–aldosterone system, and the combination of inhibitors of both systems therefore appears counter-intuitive. However, experimental data strongly support added benefits of a combined treatment. Therefore, the interaction between the renin–angiotensin system and ET may become clinically significant if clinical trials confirm the already available experimental data.

Keywords Endothelin · Vascular remodeling · Hypertension · Oxidative stress

1
Introduction

The biology of the endothelin (ET) system has been reviewed in details elsewhere (Schiffrin and Touyz 1998; Miyauchi and Masaki 1999), but highlights will be introduced to provide an appropriate background to discuss the interactions between this system and the renin–angiotensin–aldosterone system (RAAS). In addition, the effect of ET and its antagonists (ETRA for ET receptor antagonists) will be discussed in pathological conditions relevant to the RAAS. Thus, this chapter will not discuss all the effects of ET and ETRA, but only those involved in interactions with local or circulating elements of the RAAS.

1.1
Biology of Endothelin

Endothelin-1 (ET-1), a 21 amino acid peptide, is the most studied member of a family of three peptides that includes ET-2 and ET-3 (Yanagisawa et al. 1988); all three peptides are encoded by different genes (Inoue et al. 1989a). It is mainly, but not exclusively, produced by endothelial cells (Miyauchi and Masaki 1999). Indeed, other cell types, including vascular smooth muscle cells (VSMC) and cardiomyocytes have also been shown to produce the peptide (Hahn et al. 1990; Miyauchi and Masaki 1999). ET-1 is generally considered a local factor that has both autocrine and paracrine functions. Most of its release, in fact, occurs abluminally rather than in the circulation (Wagner et al. 1992).

The promoter region of the gene contains consensus sequences for several transcription factors, such as nuclear factor-1 and activator protein (AP)-1 (Inoue et al. 1989a,b). Induction of prepro-ET mRNA transcription results in a rapid onset, and appears to be the rate-limiting step (Yanagisawa et al. 1988). ET-1 is generated from its precursor prepro-ET, in two steps: the first intermediate generated is a 38 amino acid peptide called Big-ET (Kido et al. 1997). The second cleavage is made by an ET converting enzyme (ECE), and seems to occur in vesicles that may act as temporary storage (Harrison et al. 1995; Russell et al. 1998). Mechanical factors such as shear stress and pulsatile stretch are known to modulate ET transcription. Several endogenous substances stimulate ET production: thrombin, transforming growth factor β, tumor necrosis factor α, immunoglobulin-1, insulin, angiotensin (Ang) II, vasopressin, bradykinin and norepinephrine (Miyauchi and Masaki 1999). Interestingly, the signaling pathways required to produce ET are similar to those used by ET (and other vasoactive substances such as Ang II) to elicit its cellular responses (Schiffrin and Touyz 1998; Miyauchi and Masaki 1999). Indeed, increased intracellular calcium concentration, and phospholipase C (PLC), protein kinase C (PKC) and extracellular signal-regulated kinase (ERK) activation are linked to ET production and to its biological actions. In contrast, nitric oxide, atrial natriuretic peptide, prostacyclin, and heparin are known inhibitors of ET production (Schiffrin and Touyz 1998; Miyauchi and Masaki 1999).

Two ET receptors have been cloned in most mammalian species. Both belong to the superfamily of G protein-coupled receptors with seven-transmembrane domains. Both ET_A and ET_B receptors mediate contraction of VSMC, although the former seems the more important, efficaciously dominant (Seo et al. 1994; Moreau et al. 1997a). ET_B receptors on endothelial cells are also responsible for the release of endothelium-derived relaxing factors, such as nitric oxide and prostacyclin, which leads to vasodilatation (Moreau et al. 1997a). However, when injected systemically, ET elicits a short interval of hypotension followed by a long-lasting increase in blood pressure (Yanagisawa et al. 1988), suggesting that the vasoconstriction it induces is far more important than vasodilatation. In addition to its pressor effects, ET regulates cellular growth (hypertrophy) and proliferation, and could also influence cellularity by inhibiting apoptosis (Sharifi and Schiffrin 1997; Shichiri et al. 1997; Dao et al. 2001). In that respect, it is not clear if ET is a trophic factor or a mitogen in vivo, as in vitro data conflict (Hirata et al. 1989; Grainger et al. 1994; Miyauchi and Masaki 1999). Interestingly, a similar debate on Ang II has been going on for years. ET is also involved in wound healing, a process that may be of importance in cardiovascular remodeling (Guidry 1992). The physiological effects of ET will be discussed in more details in relation with the RAAS in the next sections.

1.2
RAAS Modulation of ET Production

Cultured endothelial cells stimulated with Ang II show an increased expression of prepro-ET mRNA, (Emori et al. 1991). This was later confirmed by in vivo administration of Ang II (Dohi et al. 1992). In addition, we have observed that in vivo (aorta, kidney, and mesenteric resistance arteries) ET content is increased when rats are chronically treated with Ang II (Barton et al. 1997, Moreau et al. 1997b). Furthermore, Larivière et al. have shown that tissue ET levels correlate better with local Ang II concentrations than with circulating Ang II levels (Larivière et al. 1998). Interestingly, cultured VSMC, fibroblasts, and cardiomyocytes also express ET-1 when stimulated with Ang II, extending the interaction to non-endothelial cells (Hahn et al. 1990; Resink et al. 1990; Ito et al. 1993; Sung et al. 1994). Moreover, in transgenic mice expressing both human renin and angiotensinogen genes, cardiac concentrations of ET are elevated (Maki et al. 1998). ET may amplify its own production by an autocrine mechanism operating through ET_A receptors, possibly by increasing ECE activity (Alberts et al. 1994; Fujisaki et al. 1995; Barton et al. 1997).

The opposing paradigm (i.e., inhibition of the RAAS) is associated with reductions of vascular and renal ET levels, demonstrating that the interaction is physiologically relevant and operates endogenously (Larivière et al. 1998; Dumont et al. 2001). Furthermore, ACE inhibitors have been found to reduce plasma ET-1 levels in patients with hypertension (Horky et al. 1993), congestive heart failure (Galatius-Jensen et al. 1996), and myocardial infarction (Dipasquale et al. 1997). Thus, in vitro and in vivo data suggest that the RAAS can stimulate endogenous ET production.

The question that arises from this positive interaction is whether endogenous ET mediates some of the effects attributed to Ang II. The best approach for determining the contribution of endogenous substances on physiological functions is to use their receptor antagonists. Fortunately, and because of the potential role of ET in several cardiovascular pathologies, its receptor antagonists (ETRA) were developed rather quickly after the discovery of the peptide in 1988. Orally active, non-peptidic ETRA are now available and one of them, bosentan, has recently been approved for treatment of pulmonary hypertension. Since Ang II stimulates ET production, it was logical to evaluate the contribution of ET in Ang II-induced cardiovascular effects.

1.3
ET Modulation of the RAAS

Contrary to the mechanisms discussed in the previous section, several lines of evidence also support the concept of ET-induced modulation of RAAS activity. Indeed, numerous studies have reported an inhibition of renin release from juxtaglomerular cells by ET, suggesting that ET can decrease the activity of circulating RAAS (see Rossi et al. 1999 for a review). Several mechanisms, including ET_B-mediated release of NO or prostacyclin, may contribute to this inhibition. However, the physiological relevance of this modulation remains unclear. It must be emphasized that ET should not be considered as a circulating hormone, but rather as a local regulator. As such, its effects on the RAAS activity should be considered only when local synthesis of the peptide is evident. Thus, in vitro results should be interpreted with caution. Accordingly, under control conditions or after renal clipping, the administration of an ETRA did not modulate renin expression or plasma renin activity (PRA) in rats (Schricker et al. 1995). In dogs, however, ETRA treatment increased PRA twofold and enhanced the sensitivity of pressure-dependent renin release at lower perfusion pressures (<80 mmHg), suggesting a physiological role for that interaction (Berthold et al. 1999).

In contrast, ET has been reported to stimulate the production of aldosterone equipotently to Ang II in the adrenal cortex (Rossi et al. 1997, 1999). Accordingly, zona glomerulosa cells express both subtypes of ET receptors (Imai et al. 1992). It also appears that, in the rat, ETB receptors are involved in the production of aldosterone (Belloni et al. 1996), while both subtypes are involved in man (Rossi et al. 1997). Thus, the ET system could amplify the release of aldosterone, which is now becoming an important facilitator of end-organ damage in cardiovascular diseases.

2
Hemodynamic Effects

2.1
Angiotensin II-Induced Hypertension

Elevation of arterial pressure is a global hemodynamic index that reflects more than just the balance between vasoconstriction and vasodilatation; it is also a clinically meaningful parameter that is partially under the control of the RAAS. Several mechanisms are thought to contribute to this hemodynamic effect of Ang II and aldosterone; notably, vascular constriction and regulation of renal sodium. Inhibition of ET receptors was reported to blunt Ang II-evoked elevations in blood pressure, in normotensive and hypertensive rats (Balakrishnan et al. 1996). Most studies on chronic administration of Ang II reported some hypotensive effects with ETRA treatment (Moreau et al. 1997b; Rajagopalan et al. 1997; Herizi et al. 1998; see Dao and Moreau 2001 for a review). As the blood pressure reduction was only partial (approximately 50%), it is likely that some hypertensive actions of Ang II are not ET-mediated. Interestingly, the hypotensive effect of ETRA is mainly observed when administered during the development of hypertension (Fig. 1). When treatment is undertaken after establishment of hypertension (10–14 days of Ang II administration), ETRA often fails to lower arterial pressure (Dao et al. 2001; Ficai et al. 2001). In the 2 K/1C model, where the early phase of hypertension development depends on the RAAS, only one in four chronic studies reported a partial reduction of hypertension (Li et al. 1996; Ehmke et al. 1999; Hocher et al. 1999; Bianciotti and de Bold 2001). In the positive study, the ETRA administration occurred early, i.e., concurrently with the clipping of the renal artery. In a 2-day study on the same model, a blunting of pressure elevation was observed following ETRA treatment started at the time of the clipping (Schricker et al. 1995).

In the transgenic (mREN2)27 model, which was obtained by inserting the mouse Ren-2 gene into the genome of normotensive Sprague-Dawley rats, acute ETRA treatments were effective in lowering arterial pressure (Gardiner et al. 2000). However, when started late (4 weeks of age), the chronic ETRA treatment did not induce hypotension in this model (Rossi et al. 2000). In another model in which both human renin and angiotensinogen genes were inserted, rats treated from week 4–7 with an ETRA had a reduction of 40% of their systolic pressure (Muller et al. 2000), while treatment from weeks 6–10 was ineffective

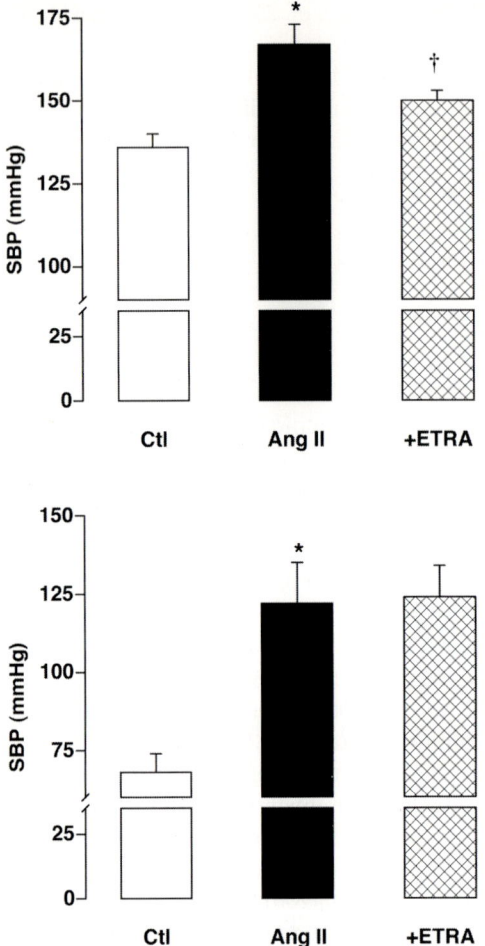

Fig. 1 *Upper panel*: systolic blood pressure (SBP) measured by tail cuff plethysmography in conscious rats treated for 2 weeks with angiotensin II alone (*Ang II*, 200 ng/kg/min) or in combination with an endothelin receptor antagonist (*+ETRA*, darusentan 30 mg/kg/day). (Adapted from Moreau et al. 1997b). The *lower panel* represents the same hemodynamic parameter measure in anesthetized rats treated for 4 weeks with Ang II alone or with an ETRA only for the last 2 weeks (weeks 3 and 4). (Adapted from Dao et al. 2001). *, $p<0.05$ as compared to control values; †, $p<0.05$ as compared to Ang II values

(Bohlender et al. 2000). This, again, suggests that early rather than late treatments with ETRA are more effective to lower arterial pressure.

Very recent observations also suggest an involvement of ET in the pressor effect of aldosterone (Park and Schiffrin 2001a). Indeed, simultaneous administration of an ETRA completely prevented the pressure elevation, suggesting a greater involvement of the peptide in aldosterone-induced hypertension than in the Ang II-induced form of the disease.

2.2
Vascular Reactivity

One mechanism whereby ET can participate in the hypertensive effect of Ang II is direct vasoconstriction. ET released from the endothelium was found to mediate part of the vasoconstriction induced by Ang II (Chen et al. 1995). In a very elegant study, Chen et al. showed that although the involvement of ET in Ang II-induced vasoconstriction is absent in the aorta, its contribution progressively increases as arteries decrease in size (Chen et al. 1995). This is of importance, considering that the control of arterial pressure lies downstream in the vascular tree. Moreover, in the human skin microcirculation, a selective ET_A-receptor antagonist blunted the Ang II vasoconstriction, suggesting that this interaction is also relevant in man (Wenzel et al. 2001).

Very low concentrations of ET, although not inducing vasoconstriction, have been shown to amplify the vascular response to other vasoconstrictors (Yang et al. 1990). Such an effect is physiologically important because in pathological conditions the peptide is present in low basal concentrations and its fluctuations are limited. Dohi et al. have reported that the ET produced by Ang II is sufficient to amplify norepinephrine (NE)-evoked contractions (Dohi et al. 1992).

Exogenous Ang II-induced hypertension is associated with blunted acetylcholine-induced relaxation (d'Uscio et al. 1997), and as ACE inhibitors generally improve endothelium-dependent vasorelaxation (Kahonen et al. 1995), endogenous Ang II was also postulated to influence this vasodilatatory homeostatic mechanism. Ang II-induced alteration of endothelium-dependent response was normalized by an ETRA selective for ET_A receptors (d'Uscio et al. 1997). The unopposed endothelial ET_B-receptor stimulation, and subsequent release of NO and prostacyclin, represents a probable mechanism for improvement of endothelium-dependent relaxations, considering the enhanced local ET levels produced by Ang II. Blocking ET receptors during Ang II administration could therefore lower arterial pressure by reducing ET-mediated vasoconstriction, preventing its amplification of other vasoconstrictors, and restoring endothelial vasodilatatory functions.

3
Cardiovascular Remodeling

Both Ang II and ET are trophic factors capable of influencing cardiovascular structure. Cardiovascular remodeling is believed to be initially beneficial, but has deleterious consequences for system integrity under chronic conditions. Since these changes are structural, they are not as easily reversible as vasoreactivity changes, and their prevention or regression still presents therapeutic challenge.

3.1
Small Arteries

Changes in the structure of small arteries induced by hemodynamic changes or vascular remodeling, can also contribute to the maintenance of hypertension (Heagerty et al. 1993; Mulvany et al. 1996). Ang II stimulates hypertrophic remodeling of resistance arteries (Griffin et al. 1991; Simon and Altman 1992; Moreau et al. 1997b; Simon et al. 1998), defined as increased media thickness, media/lumen ratio, and cross-sectional area (Heagerty et al. 1993). Since Ang II stimulates local ET production, we postulate the involvement of endogenous ET

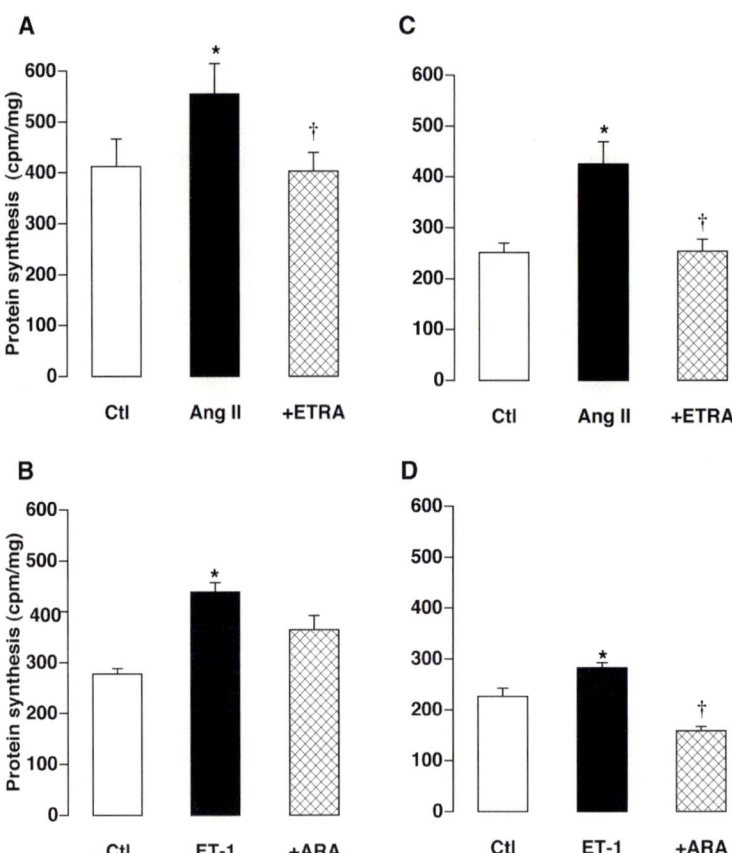

Fig. 2A–D In vivo protein synthesis measurements in small mesenteric arteries (**A**) and in the aorta (**C**) expressed as cpm of incorporated [^{3}H]-leucine in vivo in control rats (*Ctl*) and rats treated for 24 h with exogenous angiotensin II alone (*Ang II*, 400 ng/kg/min) or in combination with an endothelin receptor antagonist (*+ETRA*, darusentan of 30 mg/kg/day). The same measurements in rats treated with exogenous ET-1 alone (5 pg/kg/min) or combined with an angiotensin receptor antagonist (irbesartan 10 mg/kg/day) in mesenteric arteries (**B**) and in the aorta (**D**). (See Martens et al. 2002 for methodological details.) *, $p<0.05$ as compared to control values; †, $p<0.05$ as compared to Ang II or ET-1 values

in Ang II-induced vascular hypertrophy. The administration of an ETRA, which blocks ET receptors specifically, totally prevented Ang II-induced hypertrophic remodeling of small arteries (Moreau et al. 1997b). Similar results were obtained, with an ET_A-selective ETRA for intramyocardial arteries, in the 2 K/1C model (Hocher et al. 1999). The most likely explanation for this phenomenon is that Ang II stimulates local production of ET that, in turn, mediates vascular hypertrophy. Blockade of ET_A-receptors on VSMC then inhibits the trophic response. In a model developed in our laboratory to acutely measure in vivo vascular protein synthesis, a good predictor of hypertrophy, (Martens et al. 2002; Voisin et al. 2002), concomitant administration of an ETRA prevented Ang II-induced protein synthesis in small mesenteric arteries during the first 24 h of administration (Fig. 2A). In contrast, an AT_1 receptor antagonist failed to influence exogenous ET-induced protein synthesis, suggesting that the interaction is not reciprocal (Fig. 2B). Treatment with an ETRA also prevented vascular remodeling of small arteries in a model of aldosterone-induced hypertension (Park and Schiffrin 2001a).

Our group recently found that when hypertrophy of small arteries is allowed to proceed, the inhibition of ET receptors did not regress the process (Dao et al. 2001). Thus, it can be concluded that ET is transiently involved in development of Ang II-induced hypertrophy, but not in the maintenance phase. Similar conclusions can also be drawn from norepinephrine- and pulse pressure-induced hypertrophy (Dao et al. 2001, 2002). This concept may help explain some discrepancy in the literature: where early treatments with ETRA were more effective in limiting hypertrophic remodeling than those started after remodeling has occurred. Similarly, the hypotensive effects of ETRA in Ang II-dependent models of hypertension also appear to be optimal when treatments are started early (see above).

3.2
Large Arteries

One of the most striking observations concerning the interaction between the two peptides is that even in cell culture conditions, an ETRA can prevent Ang II-stimulated cell growth (Ito et al. 1993; Fujisaki et al. 1995). Thus when Ang II is used to stimulate hypertrophy in cultured VSMC, some signaling events are involved in the production of ET, while others mediate the trophic actions of ET. Consequently, using the more direct effector (ET in this example) as the stimulus would be more appropriate when studying mechanisms related to vascular hypertrophy. If this paradigm were transposed in vivo, VSMC should also use ET as an autocrine growth regulator. ETRAs have been shown to prevent the effect of Ang II-induced large artery remodeling (Herizi et al. 1998; Ficai et al. 2001). In the aorta, Ang II-induced in vivo protein synthesis was blunted by an ETRA (Fig. 2C). As in small arteries, the effect of the ETRA was restricted to early treatments, since giving the treatment after the establishment of hypertension with Ang II did not regress carotid artery hypertrophy (Ficai et al. 2001).

In contrast to the situation in small arteries, blocking AT_1 receptors also reduced ET-1-induced aortic protein synthesis (Fig. 2D). Thus, the trophic interaction between the two peptides may be mutual. However, to the best of our knowledge these preliminary observations, with ET as the agonist, were never tested in vivo. Exogenous ET amplifies neointimal formation after balloon injury in the rat (Barolet et al. 2001). Although treatment with an AT_1 receptor antagonist (ARA) could prevent or regress neointimal hyperplasia (Lemay et al. 2000; Lemay and Deblois 2002), we are not aware of any studies where ARA was administered in the presence of exogenous ET administration. Protein synthesis data following 24-h stimulation suggest that ET could be involved in both local Ang II production and action in the large arteries. Similar conclusions can be drawn for aldosterone, as ET has been shown to modulate aldosterone production (see above), and the effects of aldosterone on the cardiovascular system, such as fibrogenesis and hypertrophy, can be prevented by treatment with an ETRA (Park and Schiffrin 2001a,b).

3.3
Cardiac Remodeling

As previously reported, Ang II stimulates ET-1 production in cardiomyocytes and fibroblasts (Ito et al. 1993; Gray et al. 1998). Accordingly, ET receptor blockade abolished the hypertrophic effects of Ang II in these cell lines (Ito et al. 1993). This action may be due to either an autocrine or a paracrine mechanism, thus implicating an involvement of fibroblasts (Gray et al. 1998). It is logical to postulate that ET may, at least partially, mediate the cardiac trophic effects of Ang II in pathological conditions that implicate the RAAS. Exogenous Ang II administration does not increase the in vivo myocardial content of ET-1 (Barton et al. 1998). Furthermore, from in vivo studies it is not clear whether the reduction of left ventricular hypertrophy (LVH) with ETRA is only secondary to a reduction in blood pressure (see Dao and Moreau 1999 for a review). In a recent study, hydralazine was co-administered with Ang II to prevent blood pressure elevation, but an ETRA was not effective in preventing ventricular hypertrophy (Moser et al. 2002). In rats overexpressing the Ren2 gene, cardiac function was enhanced without amelioration of cardiac remodeling and fibrosis (Rothermund et al. 2000). At variance with these findings, two studies showed a pressure-independent effect of an ETRA, the first in the prevention of ventricular hypertrophy in 2K/1C rats (Ehmke et al. 1999) and the second in its regression in Ang II-induced hypertension (Ficai et al. 2001). Overall, the situation is complex and several factors must be considered when interpreting these results, including hemodynamics and the model of LVH used.

4
Renal Interactions

Renal protection by ETRA, both in context of hypertension and renal failure, is quite remarkable and clearly independent of effects on arterial pressure (Rabelink et al. 1998; Moreau and Rabelink 1999). Endothelin seems to be involved in key events leading to chronic renal failure, such as mesangial cell proliferation, extracellular matrix production, and renal fibrosis. When transgenic rats expressing the human renin and human angiotensinogen genes were treated with ETRA, expression of markers of renal inflammation were blunted (Muller et al. 2000). These results are in line with demonstrations that Ang II stimulates renal ET production (Barton et al. 1997, 1998; Alexander et al. 2001), and that ARA decrease glomerular ET levels in a model of renal insufficiency (Dumont et al. 2001). Blockade of ET receptors was found to improve renal function in models of Ang II hypertension (Herizi et al. 1998; Alexander et al. 2001; Riggleman et al. 2001). In addition, the increase in sodium sensitivity associated with chronic exogenous Ang II administration was reported to be mediated by ET_A receptors (Ballew and Fink 2001).

5
Oxidative Stress

Oxidative stress is a consequence of the oxidant properties of free radicals and molecules that are produced mainly by mitochondria, or enzymes such as NAD(P)H [nicotinamide adenine dinucleotide (phosphate), reduced] oxidase, xanthine oxidase or NO synthase (Wallace and Melov 1998). Reactive oxygen species (ROS), apart from their capacity to induce damages to cells and tissues, can impair endothelial function (Heitzer et al. 2001) and are associated with increased blood pressure (Wu et al. 2001). ROS also promote cellular growth, and are suspected of playing important roles in vascular remodeling (Mezzetti et al. 1999). A number of studies have clearly established that Ang II can stimulate superoxide anion production by activating NAD(P)H oxidase (Griendling et al. 1994; Rajagopalan et al. 1996; Touyz and Schiffrin 2001). Thus, it may be postulated that ROS production is likely to play an important role in cardiovascular disorders involving Ang II, such as cardiac and vessel wall hypertrophy or atherosclerosis.

Taking into consideration the aforementioned interactions between Ang II and ET, and the similarities between ET- and ROS-induced vasoconstriction and cell proliferation, it is likely that ROS are closely related either to the production or signaling of ET. Treatment of human aortic or coronary smooth muscle cells with hydrogen peroxide or xanthine/hypoxanthine oxidase (a superoxide anion producer) increased ET synthesis (Kahler et al. 2001; Ruef et al. 2001). This upregulation of ET production by oxidative stress was also observed in endothelial cells (Saito et al. 2001). Moreover, it was noted that treatment with tempol and vitamin E, two antioxidants, can block the increase of plasma ET levels following chronic Ang II administration (Ortiz et al. 2001). ROS thus appear to be important mediators of Ang II-induced ET production.

In various cell types, such as intestinal mucosae (Oktar et al. 2000), VSMC (Wedgwood et al. 2001a,b), and monocytes (Huribal et al. 1994), ET appears to participate in a positive feedback loop where it increases ROS production, and it is increased by oxidative stress. The mechanisms underlying superoxide anion production are still unclear, but several studies suggest that NAD(P)H oxidase can be directly activated by ET (Duerrschmidt et al. 2000; Fei et al. 2000). These data show that in addition to being stimulated by ROS, the ET system can also

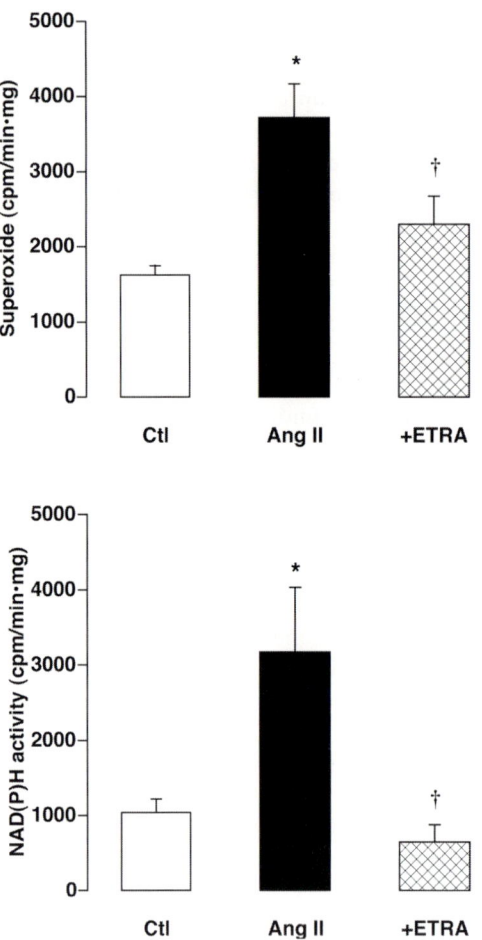

Fig. 3 *Upper panel*: measurement of superoxide production by lucigenin (5 mM)-enhanced chemiluminescence in the aorta of control (*Ctl*), angiotensin II (Ang II, 200 ng/kg/min for 12 days), or Ang II in combination with an endothelin receptor antagonist (*+ETRA*, LU302872 30 mg/kg/day). *Lower panel*: measurement of aortic NAD(P)H oxidase activity in the same groups, estimated by the superoxide anion formation upon stimulation with NADH (10^{-4} M) after subtraction of basal production. *, $p<0.05$ as compared to control values; †, $p<0.05$ as compared to Ang II or ET-1 values

contribute to ROS production by activating similar signaling pathways as Ang II; this suggests a positive interaction (and potential mutual amplification) between the two systems in promotion of oxidative stress.

Unfortunately, most studies investigating the interaction between the RAAS and ET system were performed in vitro. Superoxide production by NAD(P)H oxidase in aortic tissues taken from rats continually treated with Ang II (200 ng/kg day) for 12 days, was studied in our laboratory. Superoxide production was found to be greatly enhanced following chronic Ang II treatment (Fig. 3, upper panel). Concomitant in vivo treatment with LU302872, a non-selective ET_A and ET_B receptor antagonist, markedly blunted Ang II-induced superoxide anion production, supporting the hypothesis that this process is dependent on local production of ET. In addition, this Ang II-induced activation of NAD(P)H oxidase was prevented by blockade of ET receptors (Fig. 3, lower panel). Although, little is known about the functional interactions between ET, oxidative stress, and Ang II, our latest results suggest a strong and complex association between these three factors in the control of vascular functions.

6
Combination of ETRA with Inhibitors of the RAAS

Considering the participation of ET in some of Ang II in vivo action, the combination of an ETRA and ARA does not appear logical. Both antagonists would work on the same pathway and could have additive effects, at best. Accordingly, ACE inhibitors have been shown to decrease circulating ET levels in different pathological conditions, suggesting that blocking the RAAS functionally affects the ET system (Horky et al. 1993; Uemasu et al. 1994; Galatius-Jensen et al. 1996; Dipasquale et al. 1997).

However, the results of several preclinical and clinical studies suggest otherwise. In a detailed hemodynamic study performed in normotensive and transgenic (mRen-2)27 hypertensive rats, Gardiner et al. showed that the acute fall in blood pressure, over 8 h in hypertensive rats, was cumulative when combining treatments with ARA and ETRA: these treated hypertensive rats obtained a 60-mmHg reduction in blood pressure (Gardiner et al. 1995). In SHR and SHR-SP (stroke-prone), the combination of an ARA with an ETRA also showed an additive hypotensive effect (Ikeda et al. 2000). In Dahl salt-sensitive rats, the ARA, which had no effects on its own, amplified the hypotensive effects of the ETRA (Ikeda et al. 2000). In a study that explored the mechanisms whereby ETRA treatment can amplify the response to inhibitors of the RAAS (and vice versa), the authors reported that blockade of ET receptors increased the sensitivity of pressure-dependent renin release at lower perfusion pressures (Berthold et al. 1999). Thus, it appears that RAAS inhibition is more effective in the presence of ETRA than in its absence. In addition to their hypotensive effect, the combined inhibition of the two systems could also limit end-organ damage either indirectly, by favoring their hemodynamic effects, or directly, by blocking the effects of peptides on protein synthesis and/or ROS generation. Perhaps the most striking demonstration of the benefits of combined therapy comes from the study of Bohlender et al. (2000). They showed, in the hRen/hAgt transgenic rats, that individual treatments (low-dose ARA or therapeutic-dose ETRA) had little effect on pressure and the high mortality rate at 10 weeks (42%), but the combined therapy normalized blood pressure and totally prevented mortality in those rats.

In a model of hypertension-induced LVH in transition to heart failure, the combination of an inhibitor of the RAAS with an ETRA was effective in improving survival, possibly because each peptide plays variable pathophysiological roles during the transition (Iwanaga et al. 2001). In the acute phase following an experimental myocardial infarction (MI), the combined treatment was also superior to individual treatments, in the prevention of fibrosis (Tzanidis et al. 2001). Moreover, long-term combined ETRA and ACE inhibition was shown to more effectively improve the progression of cardiac failure following extensive MI, than monotherapy (Fraccarollo et al. 2002). A favorable cardiac hemodynamic effect was also observed in another study of post MI-induced heart failure rats, in which ETRA was administered acutely on top of chronic ACE inhibition (Qiu et al. 2001). In a model of congestive heart failure induced by rapid atrial pacing in pigs, the combination of an ARA and an ETRA provided the best recovery of LV function (New et al. 2000). In patients with congestive heart failure receiving an ACE inhibitor as standard therapy, the addition of an ETRA improved the hemodynamic profile of patients, suggesting that this combined therapy may provide an interesting alternate treatment (Kiowski et al. 1995; Love et al. 1996; Sutsch et al. 1998).

Several studies focusing on the kidneys have also revealed unexpected benefits for combining the inhibitors of the RAAS and ET system. Benigni et al. reported that an ACE inhibitor and an ETRA have beneficial additive effects, in a model of nephropathy with proteinuria (Benigni et al. 1998). Moreover, in a dog model of renovascular hypertension, the combined therapy of an ARA with ETRA produced additional hypotension, as compared to the treatment with ARA alone (Massart et al. 1998). More recently, the effectiveness of combined therapy to attenuate glomerulosclerosis was confirmed in a low-renin model of partial kidney resection, independent of pressure reduction (Amann et al. 2001). However, albuminuria was not improved by the combined treatment. Along the same lines, this combined therapy was effective in reducing blood pressure, but did not ameliorate renal damage in a low renin model of hypertension (DOCA-salt model) (Pollock et al. 2000). In partially nephrectomized rats, a model of progressive renal injury, the inhibition of the RAAS (by ACE inhibitor or ARA) improved glomerular filtration rate, proteinuria, glomerulosclerosis, and tubular injury; but the additional inhibition of the endothelin system did not produce any additional effect (Cao et al. 2000).

Overall, most of the preclinical data suggest that a combined therapy of RAAS and ET inhibitors can amplify the beneficial effect of each individual treatment; one possible exception is the treatment of certain renal diseases. Although the RAAS requires the ET system for several of its cardiovascular effects, inhibiting both systems appears favorable. Available observations indicate either that the interaction between the two systems is more complex than anticipated, or that the individual antagonist does not produce full inhibition of their respective system. In that respect, it is noteworthy that only one dose of each antagonist is generally used in most combination studies, and in some circumstances, the dosage was even selected to provide a minimal effect in order to favor an improvement. Although it remains to be established whether combined therapy would be better if one of the two systems was fully inhibited, the current therapeutic trend to decrease the dosage of single drugs in favor of low-dose combinations would support the clinical relevance of this combined therapy.

7
Conclusion

Interactions between the RAAS and ET system can have clinical significance in the evolution of cardiovascular diseases and their respective treatments. The RAAS has profound effects on the cardiovascular system, as exemplified by the numerous clinical applications of ACE inhibitors, ARA and aldosterone antagonists. However, at the vascular and renal levels, endothelial ET, acting mainly through its specific receptors located on VSMC and mesangial cells, seems to mediate many effects of Ang II, such as vasoconstriction, hypertrophy, fibrogenesis, and oxidative stress (Fig. 4). Thus, ETRA may represent a new therapy for

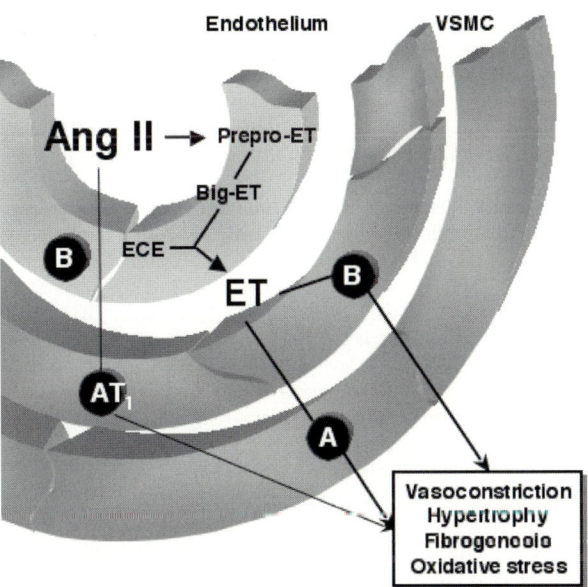

Fig. 4 Schematic representation of angiotensin II (*Ang II*) and endothelin (*ET*) interaction in the vessel wall. Although Ang II produces some direct effects, several of its actions are mediated, at least in part, by local ET and can be effectively blocked by endothelin receptor antagonists. *A*, endothelin A receptor; AT_1, angiotensin II type 1 receptor; *B*, endothelin B receptor

fine-tuning the RAAS at the level of the arteries and kidneys. Considering the rather low antihypertensive efficacy of ETRA and their superior ability to limit end-organ damage, both ETRA and inhibitors of the RAAS could be used most profitably in combination to maximize therapeutic benefits.

References

Alberts GF, Peifley KA, Johns A, Kleha JF, Winkles JA (1994) Constitutive endothelin-1 overexpression promotes smooth muscle cell proliferation via an external autocrine loop. J Biol Chem 269:10112–10118

Alexander BT, Cockrell KL, Rinewalt AN, Herrington JN, Granger JP (2001) Enhanced renal expression of preproendothelin mRNA during chronic angiotensin II hypertension. Am J Physiol Regul Integr Comp Physiol 280:R1388–R1392

Amann K, Simonaviciene A, Medwedewa T, Koch A, Orth S, Gross ML, Haas C, Kuhlmann A, Linz W, Scholkens B, Ritz E (2001) Blood pressure-independent additive effects of pharmacologic blockade of the renin–angiotensin and endothelin systems on progression in a low-renin model of renal damage. J Am Soc Nephrol 12:2572–2584

Balakrishnan SM, Wang HD, Gopalakrishnan V, Wilson TW, McNeill JR (1996) Effect of an endothelin antagonist on hemodynamic responses to angiotensin II. Hypertension 28:806–809

Ballew JR, Fink GD (2001) Role of endothelin ETB receptor activation in angiotensin II-induced hypertension: effects of salt intake. Am J Physiol Heart Circ Physiol 281:H2218–H2225

Barolet AW, Babaei S, Robinson R, Picard P, Tsui W, Nili N, Mohamed F, Ornatsky O, Sparkes JD, Stewart DJ, Strauss BH (2001) Administration of exogenous endothelin-1 following vascular balloon injury: early and late effects on intimal hyperplasia. Cardiovasc Res 52:468–476

Barton M, Shaw S, d'Uscio LV, Moreau P, Luscher TF (1998) Differential modulation of the renal and myocardial endothelin system by angiotensin II in Vivo. Effects of chronic selective ETA receptor blockade. J Cardiovasc Pharmacol 31 Suppl 1:S265–S268

Barton M, Shaw S, d'Uscio LV, Moreau P, Lüscher TF (1997) Angiotensin II increases vascular and renal endothelin-1 and functional endothelin converting enzyme activity in vivo: role of ET_A-receptors for endothelin regulation. Biochem Biophys Res Commun 238:861–865

Belloni AS, Rossi GP, Andreis PG, Neri G, Albertin G, Pessina AC, Nussdorfer GG (1996) Endothelin adrenocortical secretagogue effect is mediated by the B receptor in rats. Hypertension 27:1153–1159

Benigni A, Corna D, Maffi R, Benedetti G, Zoja C, Remuzzi G (1998) Renoprotective effect of contemporary blocking of angiotensin II and endothelin-1 in rats with membranous nephropathy. Kidney Int 54:353–359

Berthold H, Munter K, Just A, Kirchheim HR, Ehmke H (1999) Stimulation of the renin-angiotensin system by endothelin subtype A receptor blockade in conscious dogs. Hypertension 33:1420–1424

Bianciotti LG, de Bold AJ (2001) Modulation of cardiac natriuretic peptide gene expression following endothelin type A receptor blockade in renovascular hypertension. Cardiovasc Res 49:808–816

Bohlender J, Gerbaulet S, Kramer J, Gross M, Kirchengast M, Dietz R (2000) Synergistic effects of AT(1) and ET(A) receptor blockade in a transgenic, angiotensin II-dependent, rat model. Hypertension 35:992–997

Cao Z, Cooper ME, Wu LL, Cox AJ, Jandeleit-Dahm K, Kelly DJ, Gilbert RE (2000) Blockade of the renin–angiotensin and endothelin systems on progressive renal injury. Hypertension 36:561–568

Chen L, McNeill JR, Wilson TW, Gopalakrishnan V (1995) Heterogeneity in vascular smooth muscle responsiveness to angiotensin II: role of endothelin. Hypertension 26:83–88

d'Uscio LV, Moreau P, Shaw S, Takase H, Barton M, Lüscher TF (1997) Effects of chronic ETA-receptor blockade in angiotensin II-induced hypertension. Hypertension 29:435–441

Dao HH, Essalihi R, Graillon JF, Lariviere R, de Champlain J, Moreau P (2002) Pharmacological prevention and regression of arterial remodeling in a rat model of isolated systolic hypertension. J Hypertens 20:1597–1606

Dao HH, Lemay J, de Champlain J, deBlois D, Moreau P (2001) Norepineprhine-induced aortic hyperplasia and extracellular matrix deposition is endothelin-dependent. J Hypertens 19:1965–1973

Dao HH, Martens FMAC, Larivière R, Yamaguchi N, Cernacek P, de Champlain J, Moreau P (2001) Transient involvement of endothelin in hypertrophic remodeling of small arteries. J Hypertens 19:1801–1812

Dao HH, Moreau P (1999) Endothelin receptor antagonists: novel agents for the treatment of hypertension. Exp Opin Invest Drugs 8:1807–1821

Dao HH, Moreau P (2001) An update on the status of endothelin receptor antagonists for hypertension. Exp Opin Invest Drugs 10:1937–1946

Dipasquale P, Valdes L, Albano V, Bucca V, Scalzo S, Pieri D, Maringhini G, Paterna S (1997) Early captopril treatment reduces plasma endothelin concentrations in the acute and subacute phases of myocardial infarction: a pilot study. J Cardiovasc Pharmacol 29:202–208

Dohi Y, Hahn AWA, Boulanger CM, Bühler FR, Lüscher TF (1992) Endothelin stimulated by angiotensin II augments contractility of spontaneously hypertensive rat resistance arteries. Hypertension 19:131–137

Duerrschmidt N, Wippich N, Goettsch W, Broemme HJ, Morawietz H (2000) Endothelin-1 induces NAD(P)H oxidase in human endothelial cells. Biochem Biophys Res Commun 269:713–717

Dumont Y, D'Amours M, Lebel M, Lariviere R (2001) Blood pressure-independent effect of angiotensin AT1 receptor blockade on renal endothelin-1 production in hypertensive uremic rats. J Hypertens 19:1479–87

Ehmke H, Faulhaer J, Munter K, Kirchengast M, Wiesner RJ (1999) Chronic ETA receptor blockade attenuates cardiac hypertrophy independently of blood pressure effects in renovascular hypertensive rats. Hypertension 33:954–960

Emori T, Hirata Y, Ohta K, Kanno K, Eguchi S, Imai T, Shichiri M, Marumo F (1991) Cellular mechanism of endothelin-1 release by angiotensin and vasopressin. Hypertension 18:165–170

Fei J, Viedt C, Soto U, Elsing C, Jahn L, Kreuzer J (2000) Endothelin-1 and smooth muscle cells: induction of jun amino-terminal kinase through an oxygen radical-sensitive mechanism. Arterioscler Thromb Vasc Biol 20:1244–1249

Ficai S, Herizi A, Mimran A, Jover B (2001) Endothelin blockade in Angiotensin II hypertension: prevention and treatment studies in the rat. Clin Exp Pharmacol Physiol 28:1100–1103

Fraccarollo D, Bauersachs J, Kellner M, Galuppo P, Ertl G (2002) Cardioprotection by long-term ET(A) receptor blockade and ACE inhibition in rats with congestive heart failure: mono- versus combination therapy. Cardiovasc Res 54:85–94

Fujisaki H, Ito H, Hirata Y, Tanaka M, Hata M, Lin M, Adachi S, Akimoto H, Marumo F, Hiroe M (1995) Natriuretic peptides inhibit angiotensin II-induced proliferation of rat cardiac fibroblasts by blocking endothelin-1 gene expression. J Clin Invest 96:1059–65

Galatius-Jensen S, Wroblewski H, Emmeluth C, Bie P, Haunso S, Kastrup J (1996) Plasma endothelin in congestive heart failure: effect of the ACE inhibitor, fosinopril. Cardiovasc Res 32:1148–1154

Gardiner SM, March JE, Kemp PA, Bennett T (2000) Cardiovascular effects of endothelin-1 and endothelin antagonists in conscious, hypertensive ((mRen-2)27) rats. Br J Pharmacol 131:1732–1738

Gardiner SM, March JE, Kemp PA, Mullins JJ, Bennett T (1995) Haemodynamic effects of losartan and the endothelin antagonist, SB 209670, in conscious, transgenic ((mRen-2)27), hypertensive rats. Br J Pharmacol 116:2237–2244

Grainger DJ, Witchell CM, Weissberg PL, Metcalfe JC (1994) Mitogens for adult rat aortic vascular smooth muscle cells in serum-free primary culture. Cardiovasc Res 28:1238–1242

Gray MO, Long CS, Kalinyak JE, Li HT, Karliner JS (1998) Angiotensin II stimulates cardiac myocyte hypertrophy via paracrine release of TGF-beta 1 and endothelin-1 from fibroblasts. Cardiovasc Res 40:352–363

Griendling KK, Minieri CA, Ollerenshaw JD, Alexander RW (1994) Angiotensin II stimulates NADH and NADPH oxidase activity in cultured vascular smooth muscle cells. Circ Res 74:1141–1148

Griffin SA, Brown WCB, Macpherson F, McGrawth JC, Wilson VG, Korsgaard N, Mulvany MJ, Lever AF (1991) Angiotensin II causes vascular hypertrophy in part by a non-pressor mechanism. Hypertension 17:626–635

Guidry C (1992) Extracellular matrix contraction by fibroblasts: peptide promoters and second messengers. Cancer Metastasis Rev 11:45–54

Hahn AWA, Resnik TJ, Scott-Burden T, Powell J, Roni Y, Bühler FR (1990) Stimulation of endothelin mRNA and secretion in rat vascular smooth muscle cells: a novel autocrine function. Cell Regulation 1:649–659

Harrison VJ, Barnes K, Turner AJ, Wood E, Corder R, Vane JR (1995) Identification of endothelin-1 and big endothelin-1 in secretory vesicles isolated from bovine aortic endothelial cells. Proc Natl Acad Sci USA 92:6344–6348

Heagerty AM, Aalkjaer C, Bund SJ, Korsgaard N, Mulvany MJ (1993) Small artery structure in hypertension: dual process of remodelling and growth. Hypertension 21:391–397

Heitzer T, Schlinzig T, Krohn K, Meinertz T, Munzel T (2001) Endothelial dysfunction, oxidative stress, and risk of cardiovascular events in patients with coronary artery disease. Circulation 104:2673–2678

Herizi A, Jover B, Bouriquet N, Mimran A (1998) Prevention of the cardiovascular and renal effects of angiotensin II by endothelin blockade. Hypertension 31:10–14

Hirata Y, Takagi Y, Fukuda Y, Marumo F (1989) Endothelin is a potent mitogen for rat vascular smooth muscle cells. Atherosclerosis 78:225–228

Hocher B, George I, Rebstock J, Bauch A, Schwarz A, Neumayer HH, Bauer C (1999) Endothelin system-dependent cardiac remodeling in renovascular hypertension. Hypertension 33:816–822

Horky K, Jindra A, Peleska J, Jachymova M, Umnerova V, Savlikova J, Jarolim M (1993) Plasma concentrations of some cardiovascular humoral factors in essential hypertension and their changes during the treatment with converting enzyme inhibitor lisinopril. Sb Lek 94:155–161

Huribal M, Kumar R, Cunningham ME, Sumpio BE, McMillen MA (1994) Endothelin-stimulated monocyte supernatants enhance neutrophil superoxide production. Shock 1:184–187

Ikeda T, Ohta H, Okada M, Kawai N, Nakao R, Siegl PK, Kobayashi T, Miyauchi T, Nishikibe M (2000) Antihypertensive effects of a mixed endothelin-A- and -B-receptor antagonist, J-104132, were augmented in the presence of an AT1-receptor antagonist, MK-954. J Cardiovasc Pharmacol 36:S337–S341

Imai T, Hirata Y, Eguchi S, Kanno K, Ohta K, Emori T, Sakamoto A, Yanagisawa M, Masaki T, Marumo F (1992) Concomitant expression of receptor subtype and isopeptide of endothelin by human adrenal gland. Biochem Biophys Res Commun 182:1115–1121

Inoue A, Yanagisawa M, Kimura S, Kasuya Y, Miyauchi T, Goto K, Masaki T (1989a) The human endothelin family: three structurally and pharmacologically distinct isopeptides predicted by three separate genes. Proc Natl Acad Sci USA 86:2863–2867

Inoue A, Yanagisawa M, Takuwa Y, Mitsui Y, Kobayashi M, Masaki T (1989b) The human preproendothelin-1 gene. Complete nucleotide sequence and regulation of expression. J Biol Chem 264:14954–14959

Ito H, Hirata Y, Adachi S, Tanaka M, Tsujimo M, Koike A, Nogami A, Marumo F, Hiroe M (1993) Endothelin is an autocrine/paracrine factor in the mechanism of angiotensin II-induced hypertrophy in cultured rat cardiomyocytes. J Clin Invest 92:398–403

Iwanaga Y, Kihara Y, Inagaki K, Onozawa Y, Yoneda T, Kataoka K, Sasayama S (2001) Differential effects of angiotensin II versus endothelin-1 inhibitions in hypertrophic left ventricular myocardium during transition to heart failure. Circulation 104:606–612

Kahler J, Ewert A, Weckmuller J, Stobbe S, Mittmann C, Koster R, Paul M, Meinertz T, Munzel T (2001) Oxidative stress increases endothelin-1 synthesis in human coronary artery smooth muscle cells. J Cardiovasc Pharmacol 38:49–57

Kahonen M, Makynen H, Wu XM, Arvola P, Porsti I (1995) Endothelial function in spontaneously hypertensive rats. Influence of quinapril treatment. Br J Pharmacol 115:859–867

Kido T, Sawamura T, Hoshikawa H, D'Orleans-Juste P, Denault JB, Leduc R, Kimura J, Masaki T (1997) Processing of proendothelin-1 at the C-terminus of big endothelin-1 is essential for proteolysis by endothelin-converting enzyme-1 in vivo. Eur J Biochem 244:520–526

Kiowski W, Sutsch G, Hunziker P, Muller P, Kim J, Oechslin E, Schmitt R, Jones R, Bertel O (1995) Evidence for endothelin-1-mediated vasoconstriction in severe chronic heart failure. Lancet 346:732–736

Larivière R, Lebel M, Kingma I, Grose JH (1998) Increased tissue angiotensin II induces endothelin-1 production in blood vessels and glomeruli of hypertensive uremic rats. J Hypertens 16 (Suppl. 2):S96 (abstract)

Larivière R, Lebel M, Kingma I, Grose JH, Boucher D (1998) Effects of losartan and captopril on endothelin-1 production in blood vessels and glomeruli of rats with reduced renal mass. Am J Hypertens 11:989–997

Lemay J, Hamet P, deBlois D (2000) Losartan-induced apoptosis as a novel mechanism for the prevention of vascular lesion formation after injury. J Renin Angiotensin Aldosterone Syst 1:46–50

Li JS, Knafo L, Turgeon A, Garcia R, Schiffrin EL (1996) Effect of endothelin antagonism on blood pressure and vascular structure in renovascular hypertensive rats. Am J Physiol 40:H88–H93

Love MP, Haynes WG, Gray GA, Webb DJ, McMurray JJ (1996) Vasodilator effects of endothelin-converting enzyme inhibition and endothelin ETA receptor blockade in chronic heart failure patients treated with ACE inhibitors. Circulation 94:2131–2137

Maki S, Miyauchi T, Sakai S, Kobayashi T, Maeda S, Takata Y, Sugiyama F, Fukamizu A, Murakami K, Goto K, Sugishita Y (1998) Endothelin-1 expression in hearts of transgenic hypertensive mice overexpressing angiotensin II. J Cardiovasc Pharmacol 31 Suppl 1:S412–S416

Martens FMAC, Demeilliers B, Girardot D, Daigle C, Dao HH, deBlois D, Moreau P (2002) Vessel specific stimulation of protein synthesis by nitric oxide synthase inhibition: role of extracellular regulated kinases 1/2. Hypertension 39:16–21

Massart PE, Hodeige DG, Van Mechelen H, Charlier AA, Ketelslegers JM, Heyndrickx GR, Donckier JE (1998) Angiotensin II and endothelin-1 receptor antagonists have cumulative hypotensive effects in canine Page hypertension. J Hypertens 16:835–41

Mezzetti A, Guglielmi MD, Pierdomenico SD, Costantini F, Cipollone F, De Cesare D, Bucciarelli T, Ucchino S, Chiarelli F, Cuccurullo F, Romano F (1999) Increased systemic oxidative stress after elective endarterectomy: relation to vascular healing and remodeling. Arterioscler Thromb Vasc Biol 19:2659–2665

Miyauchi T, Masaki T (1999) Pathophysiology of endothelin in the cardiovascular system. Ann Rev Physiol 61:391–415

Moreau P, Nava E, Takase H, Lüscher TF (1997a) Local regulation of vascular function: focus on endothelium-dependent mechanisms in normotension, hypertension and atherosclerosis. In: Zanchetti A, Mancia G (eds) Pathophysiology of hypertension. Elsevier, Amsterdam, pp 975–1006

Moreau P, d'Uscio LV, Takase H, Shaw S, Barton M, Lüscher TF (1997b) Angiotensin II increases tissue endothelin and induced vascular hypertrophy in vivo: reversal by ET_A-receptor antagonist. Circulation 96:1593–1597

Moreau P, Rabelink TJ (1999) Endothelin and its antagonists in hypertension: can we foresee the future? Curr Hypertens Reports 1:79–78

Moser L, Faulhaber J, Wiesner RJ, Ehmke H (2002) Predominant activation of endothelin-dependent cardiac hypertrophy by norepinephrine in rat left ventricle. Am J Physiol Regul Integr Comp Physiol 282:R1389–R1394

Muller DN, Mervaala EM, Schmidt F, Park JK, Dechend R, Genersch E, Breu V, Loffler BM, Ganten D, Schneider W, Haller H, Luft FC (2000) Effect of bosentan on NF-kappaB, inflammation, and tissue factor in angiotensin II-induced end-organ damage. Hypertension 36:282–290

Mulvany MJ, Baumbach GL, Aalkjaer C, Heagerty AM, Korsgaard N, Schiffrin EL, Heistad DD (1996) Vascular remodeling (letter). Hypertension 28:505–506

New RB, Sampson AC, King MK, Hendrick JW, Clair MJ, McElmurray JH, 3rd, Mandel J, Mukherjee R, de Gasparo M, Spinale FG (2000) Effects of combined angiotensin II and endothelin receptor blockade with developing heart failure: effects on left ventricular performance. Circulation 102:1447–53

Oktar BK, Coskun T, Bozkurt A, Yegen BC, Yuksel M, Haklar G, Bilsel S, Aksungar FB, Cetinel U, Granger DN, Kurtel H (2000) Endothelin-1-induced PMN infiltration and mucosal dysfunction in the rat small intestine. Am J Physiol Gastrointest Liver Physiol 279:G483–G491

Ortiz MC, Manriquez MC, Romero JC, Juncos LA (2001) Antioxidants block angiotensin II-induced increases in blood pressure and endothelin. Hypertension 38:655–659

Park JB, Schiffrin EL (2001a) ET(A) receptor antagonist prevents blood pressure elevation and vascular remodeling in aldosterone-infused rats. Hypertension 37:1444–1449

Park JB, Schiffrin EL (2001b) Small artery remodeling is the most prevalent (earliest?) form of target organ damage in mild essential hypertension. J Hypertens 19:921–930

Pollock DM, Derebail VK, Yamamoto T, Pollock JS (2000) Combined effects of AT(1) and ET(A) receptor antagonists, candesartan, and A-127722 in DOCA-salt hypertensive rats. Gen Pharmac 34:337–342

Qiu CB, Qiu CS, Hess P, Clozel JP, Clozel M (2001) Additional effects of endothelin receptor blockade and angiotensin converting enzyme inhibition in rats with chronic heart failure. Acta Pharmacol Sin 22:541–548

Rabelink TJ, Stroes ESG, Banten KP, Morrison P (1998) Endothelin blockers and renal protection: a new strategy to prevent end-organ damage in CV diseases? Cardiovasc Res 39:543–549

Rajagopalan S, Kurz S, Munzel T, Tarpey M, Freeman BA, Griendling KK, Harrison DG (1996) Angiotensin II-mediated hypertension in the rat increases vascular superoxide production via membrane NADH/NADPH oxidase activation. Contribution to alterations of vasomotor tone. J Clin Invest 97:1916–1923

Rajagopalan S, Laursen JB, Borthayre A, Kurz S, Keiser J, Haleen S, Giaid A, Harrison DG (1997) Role for endothelin-1 in angiotensin II-mediated hypertension. Hypertension 30:29–34

Resink TJ, Hahn AW, Scott-Burden T, Powell J, Weber E, Buhler FR (1990) Inducible endothelin mRNA expression and peptide secretion in cultured human vascular smooth muscle cells. Biochem Biophys Res Commun 168:1303–1310

Riggleman A, Harvey J, Baylis C (2001) Endothelin mediates some of the renal actions of acutely administered angiotensin II. Hypertension 38:105–109

Rossi GP, Albertin G, Bova S, Belloni AS, Fallo F, Pagotto U, Trevisi L, Palu G, Pessina AC, Nussdorfer GG (1997) Autocrine-paracrine role of endothelin-1 in the regulation of aldosterone synthase expression and intracellular Ca2+ in human adrenocortical carcinoma NCI-H295 cells. Endocrinology 138:4421–4426

Rossi GP, Albertin G, Neri G, Andreis PG, Hofmann S, Pessina AC, Nussdorfer GG (1997) Endothelin-1 stimulates steroid secretion of human adrenocortical cells ex vivo via both ETA and ETB receptor subtypes. J Clin Endocrinol Metab 82:3445–3449

Rossi GP, Sacchetto A, Cesari M, Pessina AC (1999) Interactions between endothelin-1 and the renin–angiotensin-aldosterone system. Cardiovasc Res 43:300–307

Rossi GP, Sacchetto A, Rizzoni D, Bova S, Porteri E, Mazzocchi G, Belloni AS, Bahcelioglu M, Nussdorfer GG, Pessina AC (2000) Blockade of angiotensin II type 1 receptor and not of endothelin receptor prevents hypertension and cardiovascular disease in transgenic (mREN2)27 rats via adrenocortical steroid-independent mechanisms. Arterioscler Thromb Vasc Biol 20:949–956

Rothermund L, Pinto YM, Hocher B, Vetter R, Leggewie S, Kobetamehl P, Orzechowski HD, Kreutz R, Paul M (2000) Cardiac endothelin system impairs left ventricular function in renin- dependent hypertension via decreased sarcoplasmic reticulum Ca(2+) uptake. Circulation 102:1582–1588

Ruef J, Moser M, Kubler W, Bode C (2001) Induction of endothelin-1 expression by oxidative stress in vascular smooth muscle cells. Cardiovasc Pathol 10:311–315

Russell FD, Skepper JN, Davenport AP (1998) Human endothelial cell storage granules: a novel intracellular site for isoforms of the endothelin-converting enzyme. Circ Res 83:314–321

Saito T, Itoh H, Chun TH, Fukunaga Y, Yamashita J, Doi K, Tanaka T, Inoue M, Masatsugu K, Sawada N, Sakaguchi S, Arai H, Mukoyama M, Tojo K, Hosoya T, Nakao K (2001) Coordinate regulation of endothelin and adrenomedullin secretion by oxidative stress in endothelial cells. Am J Physiol Heart Circ Physiol 281:H1364–H1371

Schiffrin EL, Touyz RM (1998) Vascular biology of endothelin. J. Cardiovasc Pharmacol 32:S2–S13

Schricker K, Scholz H, Hamann M, Clozel M, Kramer BK, Kurtz A (1995) Role of endogenous endothelins in the renin system of normal and two-kidney, one clip rats. Hypertension 25:1025–1029

Seo B, Oemar BS, Siebenmann R, von Ludwig S, Lüscher TF (1994) Both ET_A and ET_B receptors mediate contraction to endothelin-1 in human blood vessels. Circulation 89:1203–1208

Sharifi AM, Schiffrin EL (1997) Apoptosis in aorta of deoxycorticosterone acetate-salt hypertensive rats: effect of endothelin receptor antagonism. J Hypertens 15:1441–1448

Shichiri M, Kato H, Marumo F, Hirata Y (1997) Endothelin-1 as an autocrine/paracrine apoptosis survival factor for endothelial cells. Hypertension 30:1198–1203

Simon G, Altman S (1992) Subpressor angiotensin II is a bifunctional growth factor of vascular muscle in rats. J Hypertens 10:1165–1171

Simon G, Illyes G, Csiky B (1998) Structural vascular changes in hypertension: role of angiotensin II, dietary sodium supplementation, blood pressure and time. Hypertension 32:654–660

Sung C-P, Arleth AJ, Storer BL, Ohlstein EH (1994) Angiotensin type 1 receptors mediate smooth muscle proliferation and endothelin biosynthesis in rat vascular smooth muscle. J Pharmacol Exp Ther 271:429–437

Sutsch G, Kiowski W, Yan XW, Hunziker P, Christen S, Strobel W, Kim JH, Rickenbacher P, Bertel O (1998) Short-term oral endothelin-receptor antagonist therapy in conventionally treated patients with symptomatic severe chronic heart failure. Circulation 98:2262–2268

Touyz RM, Schiffrin EL (2001) Increased generation of superoxide by angiotensin II in smooth muscle cells from resistance arteries of hypertensive patients: role of phospholipase D-dependent NAD(P)H oxidase-sensitive pathways. J Hypertens 19:1245–1254

Tzanidis A, Lim S, Hannan RD, See F, Ugoni AM, Krum H (2001) Combined angiotensin and endothelin receptor blockade attenuates adverse cardiac remodeling post-myocardial infarction in the rat: possible role of transforming growth factor beta(1). J Mol Cell Cardiol 33:969–981

Uemasu J, Munemura C, Fujihara M, Kawasaki H (1994) Inhibition of plasma endothelin-1 concentration by captopril in patients with essential hypertension. Clin Nephrol 41:150–152

Voisin L, Foisy S, Giasson E, Lambert C, Moreau P, Meloche S (2002) EGF receptor transactivation is obligatory for protein synthesis stimulation by G protein-coupled receptors. Am J Physiol Cell Physiol 283:C446–C455

Wagner O, Christ G, Wojta J, Vierhapper H, Parzer S, Nowotny P, Schneider B, Waldhäusl W, Bimder BR (1992) Polar secretion of endothelin-1 by cultured endothelial cells. J Biol Chem 267:16066–16068

Wallace DC, Melov S (1998) Radicals r'aging. Nat Genet 19:105–106

Wedgwood S, Dettman RW, Black SM (2001a) ET-1 stimulates pulmonary arterial smooth muscle cell proliferation via induction of reactive oxygen species. Am J Physiol Lung Cell Mol Physiol 281:L1058–L1067

Wedgwood S, McMullan DM, Bekker JM, Fineman JR, Black SM (2001b) Role for endothelin-1-induced superoxide and peroxynitrite production in rebound pulmonary hypertension associated with inhaled nitric oxide therapy. Circ Res 89:357–364

Wenzel RR, Ruthemann J, Bruck H, Schafers RF, Michel MC, Philipp T (2001) Endothelin-A receptor antagonist inhibits angiotensin II and noradrenaline in man. Br J Clin Pharmacol 52:151–157

Wu R, Millette E, Wu L, de Champlain J (2001) Enhanced superoxide anion formation in vascular tissues from spontaneously hypertensive and desoxycorticosterone acetate-salt hypertensive rats. J Hypertens 19:741–748

Yanagisawa M, Kurihara H, Kimura S, Tomobe Y, Kobayashi M, Mitsui Y, Yazaki Y, Goto K, Masaki T (1988) A novel potent vasoconstrictor peptide produced by vascular endothelial cells. Nature 332:411–415

Yang Z, Richard V, von Segesser L, Bauer E, Stulz P, Turina M, Lüscher TF (1990) Threshold concentrations of endothelin-1 potentiate contractions to norepinephrine and serotonin in human arteries: a new mechanism of vasospasm? Circulation 82:188–195

Part 2
Genetics

Genetics of the Human Renin–Angiotensin System

X. Jeunemaitre

INSERM U 36, Collège de France, 11 place Marcelin Berthelot, 75005 Paris, France
e-mail: xavier.jeunemaitre@college-de-france.fr

1	Introduction	174
2	The Angiotensinogen Gene	174
2.1	Gene Structure, Mutations and Polymorphisms	174
2.2	Relation with Plasma Angiotensinogen	176
2.3	Relation with Essential Hypertension	177
2.4	Relation with Other Phenotypes	178
3	The Renin Gene	179
3.1	Gene Structure, Mutations and Polymorphisms	179
3.2	Relation with Plasma Renin	180
3.3	Relation with Essential Hypertension	181
4	The Angiotensin I-Converting Enzyme	181
4.1	Gene Structure, Mutations and Polymorphisms	182
4.2	Relation with Plasma ACE Levels	182
4.3	Relation with Essential Hypertension	183
4.4	Relation with End-Organ Damage	184
5	The Angiotensin II Type 1 Receptor	185
5.1	Gene Structure, Mutations and Polymorphisms	186
5.2	Relation with Essential Hypertension	187
5.3	Relation with End-Organ Damage	188
6	The Angiotensin II Type 2 Receptor	189
6.1	Gene Localisation, Structure and Polymorphisms	189
6.2	Relation with Urogenital Abnormalities	190
6.3	Relation with Cardiovascular Diseases	190
6.4	Relation with X-Mental Retardation	191
7	Future Directions	192
7.1	Test Gene–Gene and Gene–Environment Interactions	192
7.1.1	Gene–Gene Interactions	192
7.1.2	Gene–Environment Interactions	193
7.2	Analysis of Other Components of the Renin–Angiotensin System	194
7.2.1	The Prorenin Receptor	194
7.2.2	The ACE2 Gene	194
7.2.3	The Aminopeptidases A and N	195
7.3	Pharmacogenetics	195
	References	196

Abstract Genes of the human renin–angiotensin system have been extensively studied within the last 10 years. A large number of polymorphisms have been described and tested in linkage and association studies in regard to cardiovascular traits such as essential hypertension, coronary disease, and diabetic nephropathy. Gene structure, polymorphisms, and associations are summarised for each gene. Association between polymorphisms of the angiotensinogen and angiotensin I-converting enzyme genes and the corresponding plasma concentrations have been well demonstrated. However, possible relationships with hypertension or end-organ damage are debated. Other surprising findings have been observed such as mutations of the angiotensin type 2 receptor in X-mental retardation. Future studies should use more integrated approaches, based on the genotyping of haplotypes, the measurement of sophisticated phenotypes, the evaluation of gene–gene and gene–environment interactions. As well, genetic investigation of new components of the renin–angiotensin system such as the prorenin receptor, the angiotensin-converting enzyme type 2, and the aminopeptidase A and N genes are necessary. Prediction of drug efficacy and/or of adverse drug reaction on a pharmacogenetic profile will need large, well-designed trials in which patients will be standardised on the drug dosage and metabolism.

Keywords Renin · Angiotensin · Polymorphism · Gene · Hypertension

1
Introduction

The genes of the renin–angiotensin system are probably those that have attracted the greatest attention of researchers working on the molecular pathophysiology of hypertension and cardiovascular diseases. During the last 10 years, considerable progress has been achieved, since the genes encoding all the proteins of the renin–angiotensin system have been cloned in humans and rodents, informative markers and polymorphisms identified, and numerous linkage and association studies performed using different phenotypes. We will not review here the important results that have been generated by experimental transgenic models (reviewed in other chapters of this book) and will focus, instead, on the results obtained on the human genes.

2
The Angiotensinogen Gene

2.1
Gene Structure, Mutations and Polymorphisms

The human angiotensinogen (AGT) gene belongs to the superfamily of serpins and is localised to chromosome 1q42.3. The human angiotensinogen gene contains 5 exons with an organisation that is similar to other serine protease inhibitors (Kageyama et al. 1984; Gaillard et al. 1989).

A large number of single nucleotide polymorphisms (SNPs) have been described in the 5′ flanking region, exons, introns and 3′ part of the AGT gene (www.ncbi.nih.nlm.gov/SNP). Among them, the coding SNP (M235T) in exon 2 and the G-6A nucleotide substitution at position −6 upstream from the initial transcription start, have been particularly studied. These two polymorphisms occur almost at the same frequency and are in complete linkage disequilibrium (Inoue et al. 1996; Jeunemaitre et al. 1997). Thus, in a given case-control study, the analysis of the M235T polymorphism will be equivalent to that of the G-6A, and vice-versa. Using several constructs and luciferase assays, Lalouel and colleagues were able to show that the G-6A substitution is associated in vitro with an increased expression of the AGT gene, a possible explanation to the association of the M235T polymorphism with increased plasma angiotensinogen (Inoue et al. 1997). Recently, Nakajima et al. (2002) showed a specific interaction between the G-6A substitutions and the nuclear factor YB1. In cotransfection experiments, YB1 reduced basal AGT promoter activity in a dose-dependent manner. Although these observations suggested a possible role for YB1 in modulating AGT expression, this function was thought likely to occur in the context of complex interactions involving other nuclear factors. However, the true in vivo biological effect of this polymorphism may be more complex, since other polymorphisms, C-532T, A-217G, C-18T, A-20C, T+31C, also in linkage disequilibrium with G-6A and M235T (Fig. 1), might play a role in the variation of transcription of the gene (Sato et al. 1997; Jeunemaitre et al. 1999; Paillard et al. 1999; Ishikawa et al. 2001).

The C/A polymorphic site at position −20 is located between the TATA box and the transcription initiation site, within a sequence that can bind the estrogen receptor (Zhao et al. 1999). The orphan receptor Arp-1 also binds to this sequence and reduces in vitro the oestrogen receptor-induced promoter activity (Narayanan et al. 1999). The A/G polymorphic site at position −217 has recently been studied in more details (Jain et al. 2002). The authors showed that the surrounding sequence corresponds to a consensus C/EBP binding site, and that reporter constructs containing the human AGT gene promoter with nucleoside A

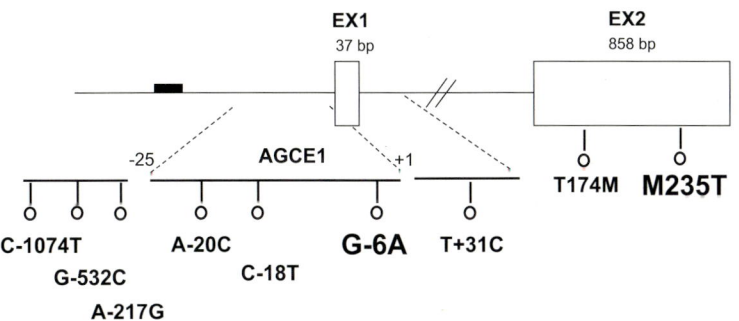

Fig. 1 Main polymorphisms of the human angiotensinogen gene

at −217 have increased basal promoter activity in Hep G2 cells. In addition, they found an increased frequency of the −217 A allele in 186 African-American hypertensive subjects compared to 156 normotensive (0.29 vs 0.19, $p=0.002$).

The ethnic origin has a strong effect on the allele frequency of most of the polymorphisms found at the AGT locus. For example, the 235T allele frequency varies from 40% in Caucasians to 70% in the Asian and 80%–90% in the African-American population and even 93% in Nigerians (Rotimi et al. 1994).

2.2
Relation with Plasma Angiotensinogen

AGT, the renin substrate, is mainly synthesised by the liver and is the unique substrate for renin. In humans, plasma AGT levels are around the K_m of renin, and therefore it is logical to suspect that a chronic state of increased plasma AGT might increase angiotensin I (Ang I) and facilitate hypertension and/or cardiovascular diseases. Its role in human hypertension was suspected in an epidemiological study where a strong correlation was found between plasma AGT concentration and blood pressure (BP) (Walker et al. 1979) and in another study where offspring of hypertensive patients had elevated plasma AGT levels (Fasola et al. 1968). Heritability of plasma AGT has been suggested. Segregation analysis in 685 members of 186 families recruited from a rural community in southwest Nigeria suggested that about 13% of the variance in plasma AGT concentration was due to the recessive gene segregation (Guo et al. 1999).

Among the 15 polymorphisms that we initially identified, two of them leading to amino acid changes, 174 M and M235T, were found to be associated with hypertension and plasma AGT concentration (Jeunemaitre et al. 1992c). This association between plasma AGT level and the M235T genotype was further confirmed in white children (Bloem et al. 1995). In African-American young individuals, Pratt and colleagues showed an association between a haplotype containing the T235 allele and a polymorphism located in the promoter region of the gene (Bloem et al. 1997). In a large sample of the Monica Augsburg cohort, a mild co-dominant and significant increase of plasma AGT concentration was also observed according to the M235T polymorphism (Schunkert et al. 1997). In the Danish general population, homozygosity for both the 235T and T174 allele was associated with a 10% increase in plasma AGT (Sethi et al. 2001). Because of the intra and interassay variability of the plasma AGT measurement and the mild association with the M235T polymorphism, a large number of individuals is required to detect this relation.

To analyse the influence of the M235T polymorphism on the ethinylestradiol-induced increase in plasma AGT concentration and on the resulting generation of Ang I and Ang II in plasma, we compared changes in the circulating renin–angiotensin system after short-term (2 days) and repeated (7 days) administration of 50 μg ethinylestradiol in homozygous normotensive men (TT and MM) (Azizi et al. 2000). In the 7-day study, TT subjects had higher peak plasma AGT concentrations than did MM subjects. In the short term, however, complete readjustment of the circulating renin–angiotensin system occurred through a decrease in renin release, which blunted the effects of the increase in AGT concentration.

2.3
Relation with Essential Hypertension

With Lalouel's group, we reported the first molecular arguments for a role of the AGT gene in human essential hypertension (Jeunemaitre et al. 1992c). An extensive study was performed in two large series of hypertensive sibships yielding a total of 379 sib pairs (Salt Lake City, Utah, USA and Paris, France) and using a highly polymorphic microsatellite marker at the AGT locus. An excess of AGT allele sharing was found mainly in severely hypertensive sib pairs and in men, suggesting a positive although mild relationship.

Since then, several linkage studies have led to controversial results. Caulfield et al. showed a strong linkage and an association of the AGT gene locus to essential hypertension in a set of British families (Caulfield et al. 1994, 1996). At the other extreme, no evidence for linkage was found in a large European study involving 630 affected sibling pairs, either in the whole panel or in family subsets selected for severity or early onset of disease (Brand et al. 1998). Linkage of the AGT gene to essential hypertension was also found in 63 affected African Caribbean sibling pairs (Caulfield et al. 1995). Similarly, positive albeit modest significant excess of AGT allele sharing was found in 46 extended Mexican American (Atwood et al. 1997). No linkage was found between the AGT locus and hypertension in 310 hypertensive Chinese sibling pairs (Niu et al. 1998).

The association between the M235T polymorphism and essential hypertension has been tested in a large number of case-control studies that have been reviewed recently (Jeunemaitre et al. 1999; Lalouel et al. 2001). A meta-analysis of case-control studies representing 5,493 Caucasian patients showed that the 235T allele was significantly but mildly associated with hypertension [odds ratio (OR): 1.20; 95% CI (confidence interval): 1.11–1.29; $p<.0001$], association which increased in studies with positive family history (OR: 1.42; 95% CI: 1.25 to 1.61, $p<.0001$) (Kunz et al. 1997). Another meta-analysis that included 10,720 white subjects showed a 32% increase in the risk of elevated BP associated with the 235T allele (Staessen et al. 1999). More recently, Sethi et al. studied 9,100 men and women from the Danish general population (Sethi et al. 2001). On multifactorial logistic regression analysis, women homozygous for the 235T allele versus non-carriers had an odds ratio for elevated BP of 1.29 which increased to 1.50 if they were also homozygous for the T174 allele. No significant association was found in men. The analysis of 4,322 subjects of the National Heart, Lung and Blood Institute (NHLBI) Family Blood Pressure Program (FBPP), showed only a weak association between the G-6A polymorphism (Province et al. 2000). In a Japanese population, the analysis of nine polymorphisms at the AGT gene confirmed that the G-6A, T+31C and M235T polymorphisms are in absolute linkage disequilibrium and that the haplotype containing −6A, +31C and 235T was associated with hypertension but not with plasma AGT in this particular study (Sato et al. 2000). More recently, the analysis of 1,425 subjects of urban populations of Vitoria, Brazil, showed a linear relation between 235T allele number and BP (Pereira et al. 2003). The magnitude of BP variations was about 3–4 mmHg per copy of the 235T allele.

The high prevalence of the 235T allele may explain why no relation between 235T or 174 M allele frequencies and hypertension was observed in a study of African-Americans (Rotimi et al. 1994). In the Japanese population, a significant association between this allele and high BP was found in several separate case control studies (Hata et al. 1994; Iwai et al. 1994; Kamitani et al. 1994; Morise et al. 1995). These differences in allele frequencies might also facilitate false-positive results in case of admixture in case-control studies. Altogether, these results highlight the modest effect of the AGT locus and the difficulty of identifying susceptibility genes by linkage analysis in complex diseases.

2.4
Relation with Other Phenotypes

Other studies have tested the influence of the AGT gene on various pathological conditions, such as coronary heart disease and diabetic nephropathy, for which the reader will find recent reviews (Jeunemaitre et al. 1999; Smithies et al. 2000; Wang and Staessen 2000; Katsuya and Ogihara 2001; Lalouel and Rohrwasser 2001). Several phenotypes probably merit a particular attention.

The first is the possible relationship between AGT gene polymorphisms and body weight. AGT is indeed abundantly expressed in human adipose tissue (Saye et al. 1990) and plasma AGT concentrations have been correlated to BP and body mass index (Eggena et al. 1991; Bloem et al. 1995; Cooper et al. 2000). The analysis of a group of young hypertensive adults on a 3-year follow-up period showed a positive interaction between body weight changes and the G-6A polymorphism (Chaves et al. 2002). Some studies have also suggested that the AGT M235T polymorphism effect could be highly sensitive to environmental context for physical activity (McCole et al. 2002; Rauramaa et al. 2002).

The second is the possible relationship with pregnancy-induced hypertension. Ward et al. (1993) found a significant association between the AGT 235T variant and pre-eclampsia in both Caucasian and Japanese samples. Using another strategy, analysis of the allelic inheritance of the GT repeat in 52 sibling pairs of preeclamptic sisters, Arngrimsson et al. (1993) showed a significant linkage between the AGT locus and preeclampsia in Icelandic and Scottish families. However, other studies found no indication of association or linkage between preeclampsia and the AGT gene, and recent genome-scan studies have reported possible loci on chromosomes 4q (Harrison et al. 1997), 2p (Arngrimsson et al. 1999; Moses et al. 2000) and 9p (Laivuori et al. 2003). Two molecular mechanisms could explain the relationship between AGT polymorphisms and pregnancy-induced hypertension. The first is its association with AGT expression in decidual spiral arteries (Morgan et al. 1997), reduced plasma volume during the follicular phase of the menstrual cycle (Bernstein et al. 1998), and intra-uterine growth (Zhang et al. 2003). The second is its possible influence on the formation of high-molecular-weight AGT (Gimenez-Roqueplo et al. 1998). Both mechanisms acting together, an increased AGT expression due to the G-6A variant, and an increased proportion of active monomeric form of AGT, could increase the local formation of Ang II, facilitate an hyperplasia of the spiral arteries and thus lead to a reduction of the uteroplacental blood flow.

3
The Renin Gene

3.1
Gene Structure, Mutations and Polymorphisms

The human renin gene is located on the short arm of chromosome 1 (1q32-1q42) (Cohen-Haguenauer et al. 1983). The primary structure of renin precursor was deduced from its cDNA sequence: it consists of 406 amino acids with a pre and a pro segment carrying 20 and 46 amino acids, respectively (Imai et al. 1983; Hobart et al. 1984). The human renin gene spans 12 kb of DNA and contains 8 introns (Miyazaki et al. 1984). The structure of the renin gene is similar to that of pepsinogen, a closely related aspartyl protease.

We identified an informative C/A repeat polymorphism in the 3' part of the gene (Lifton and Jeunemaitre 1993). From pairwise logarithm of differences (lod) score calculations, the sex-averaged recombination fraction between the renin and AGT locus was estimated at 0.26. Several restriction fragment-length polymorphisms (RFLPs) have been located throughout the renin gene: *Taq*I and *Bgl*I polymorphisms in the 5' region, *Hind*III in the 3' region, and *Hinf*I in the first intron (Jeunemaitre et al. 1992b). On the NCBI site, seven SNPs are present at the human renin gene—respectively at exons 1, 2, 3, 9 and 10—but have not all been validated (www.ncbi.nih.nlm.gov/SNP). A systematic screening of the coding sequences of the gene by single-strand conformation polymorphism analysis (SSCP) and direct sequencing allowed us the identification of a synonymous SNP at exon 2 [Thr(ACA)→Thr(ACC) at position 2 of the mature renin], and a polymorphism in intron 4 (T+17int4G). A trinucleotide repeat (CTG)8 is also present in intron 7, displaying 4 alleles with an average heterozygosity of 0.49 (Fig. 2). In the Japanese population, two novel SNPs have been detected, the one in intron 4 (T+17int4G), and one in exon 9 (G1051A) leading to an amino acid change (Val351Ile) (Hasimu et al. 2003). Our group identified two SNPs close to a strong enhancer element located more than 5 kb upstream of the promoter of the human gene (Germain et al. 1998; Fuchs et al. 2002). These two SNPs are located at positions −5434 and −5312 and seem to influence in vitro the levels of transcription, when the corresponding constructs are transfected in choriodecidual (Fuchs et al. 2002).

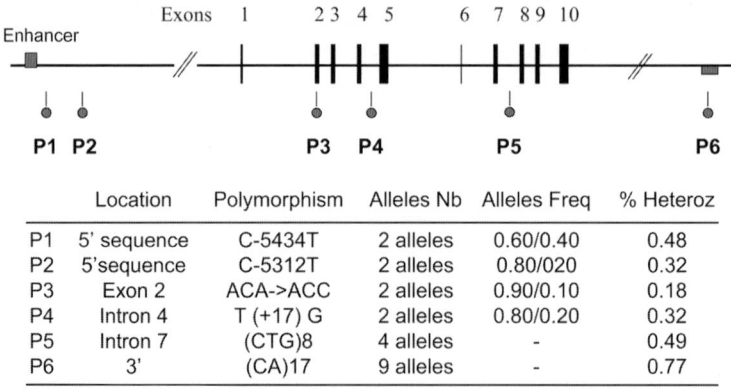

Fig. 2 Main polymorphisms of the human renin gene

3.2
Relation with Plasma Renin

Activation of the renin system depends on the renin–AGT reaction which is the first and rate-limiting step leading to Ang II production. Hypertensive individuals are usually classified according to their renin status. Interestingly, renin levels seem at least partly heritable, as observed by Grim and colleagues (1980) in twins submitted to well-standardised conditions of posture and sodium diet and more recently by Rossi et al. (1999). Our analysis of 175 sibling pairs and trios who participated to a multicentre trial showed very significant correlations of plasma renin levels either on a high- or a low-salt diet. Further analysis of this cohort showed the familial aggregation of low-renin hypertension with twice more low-renin families than expected (10.9% vs 5.5%), in contrast to the normal-renin state, where the observed and expected number of families was similar (61.0 vs 58.6%) (Fisher et al. 2002).

Testing 212 patients with EH and 209 age-matched normotensive subjects, Hasimu et al. (2003) found no difference in three renin polymorphisms either separately, or combining them into haplotypes. However, plasma renin activity (PRA) levels in patients with essential hypertension (EH) with the GG genotype at position 1051 were significantly higher than in subjects with GA and AA genotypes, thus suggesting that the amino acid change (Val351Ile) may affect the enzymatic function of renin or may be in linkage disequilibrium with polymorphisms affecting renin transcription.

A mutation responsible for a syndrome with hyperproreninaemia has been found in a unique Dutch family (Villard et al. 1994). All family members were normotensive and had normal PRA. A point mutation in the last exon of the gene at codon 387 of the preprorenin cDNA introduced a premature stop codon (TGA) in the renin gene sequence. This mutation directs the synthesis of a truncated form of renin, with 20 amino acids deleted from the carboxyl terminus. We found also a mutation on the signal peptide-encoding region of the preprofragment of renin (Pro 8 Ala) in a unique family (X. Jeunemaitre, unpublished). No particular biological phenotype could be associated with this mutation.

3.3
Relation with Essential Hypertension

Numerous studies have involved renin to some degree in experimental forms of hypertension. In human, one of the first case-control study was that performed by Soubrier et al. (1990). Renin gene allele and haplotype frequencies of 3 RFLPs were similar in 120 normotensive and 102 hypertensive subjects. To explore further the potential role of the renin gene as a genetic determinant of hypertension, we used the hypertensive sib pairs approach (Jeunemaitre et al. 1992b). Using the same clinical criteria and the same renin gene haplotypes as in the association study, no linkage was found between the renin gene and hypertension, suggesting again that the renin gene does not have a frequent and/or important role in the pathogenesis of essential hypertension.

In a single large Utah human pedigree with high prevalence of coronary disease and hypertension, there was no significant association between the renin RFLPs and BP or plasma renin activity (Naftilan et al. 1989). Interpretation was, however, limited by the very low number of patients studied. In another preliminary report, Morris and Griffiths (1988) compared the renin RFLPs of 29 subjects under antihypertensive treatment with those of 202 adult patients. No association was found between hypertension and the renin gene allele but, again, no definite conclusion could be drawn since only a few hypertensives were studied, clinical data were not available and the renin gene polymorphism was defined by a single and weakly informative RFLP. The most recent study corresponds to that performed in a Japanese population by Hasimu et al. (2003). No statistical difference was observed in the genotype distribution of three polymorphisms between the 212 patients and the 209 age-matched normotensive subjects.

4
The Angiotensin I-Converting Enzyme

Angiotensin I-converting enzyme (ACE) is a zinc metalloprotease whose main functions are to convert Ang I into Ang II, and to inactivate bradykinin. It is assumed that this step of the renin–angiotensin system is not limiting in plasma, and indeed there is no indication that plasma ACE levels are directly related to BP levels. However, the local generation of Ang I and the degradation of a bradykinin might depend on the level of ACE expressed in tissues.

4.1
Gene Structure, Mutations and Polymorphisms

The enzyme consists of two highly homologous and functionally active domains resulting from a gene duplication (Soubrier et al. 1988). There are two ACE promoters, a somatic promoter localised on the 5' side of the first exon of the gene and a germinal, intragenic, promoter located on the 5' side of the specific testicular ACE mRNA (Hubert et al. 1991). The two alternate promoters of the ACE gene exhibit highly contrasting cell specificities, as the somatic promoter is active in endothelial, epithelial and neuronal cell types, whereas the germinal promoter is only active in a stage-specific manner in male germinal cells (Howard et al. 1990). It is thought that plasma ACE concentration reflects the level of the synthesis of the somatic enzyme.

The most studied is an insertion/deletion (I/D) of a 287-base pair DNA fragment in the intron 16 of the gene, corresponding to an Alu sequence. Soubrier's group also identified eight new polymorphisms in 95 healthy nuclear families, most of them being in strong linkage disequilibrium with the I/D polymorphism (Villard et al. 1996). More recently, Rieder and colleagues performed a complete genomic scan of the ACE gene (24 kb) from 11 individuals (Rieder et al. 1999). They identified 78 varying sites in 22 chromosomes that resolved into 13 distinct haplotypes. Among these polymorphisms, 17 are in absolute linkage disequilibrium with the ACE I/D polymorphism, producing two distinct and distantly related clades.

4.2
Relation with Plasma ACE Levels

From a geneticist point of view, plasma ACE concentration is an interesting marker as it varies markedly between individuals (from 1 to 8 at the extremes of the distribution) but remains remarkably constant when measured repeatedly in a given subject (Alhenc-Gelas et al. 1991) This important variability is due, in large part, to a major genetic effect, as shown by Cambien et al. (1988). In a family study where there was an intrafamilial resemblance between plasma ACE levels, they estimated that this effect accounting for approximately 30% and 75% of the ACE variance in parents and in offspring, respectively.

The role of the ACE gene in the genetic control of plasma ACE has been clearly established. In their seminal observation of 80 healthy subjects, Rigat and colleagues showed that the serum ACE concentration of DD subjects was almost twice as high as that observed in patients homozygous for the I allele, whereas heterozygous patients were intermediate (Rigat et al. 1990). Like for serum ACE, T lymphocyte ACE levels are significantly higher in patients homozygous for the D allele than in the other subjects (Costerousse et al. 1993). Another study combining segregation and linkage analysis in 98 healthy nuclear families showed that the ACE I/D polymorphism is only a neutral marker in strong linkage disequilibrium with the putative functional variant (Tiret et al. 1992). After adjustment for the I/D polymorphism, all polymorphisms of the 5' group remained significantly associated with ACE levels, which suggested the existence of two quantitative trait loci (QTL) acting additively on ACE levels accounting for 38% and 49% of the ACE variance in parents and offspring, respectively. The authors suggested that the causal variant should be located within the 3' part of the gene. Despite these efforts, the causative variant responsible for the increase in ACE expression has yet to be found, which might foreshadow the difficulty of identifying causal molecular variants in complex traits.

A rare missense mutation (Pro1199Leu) in the stalk region of the mature ACE protein has been identified in eight Dutch families, in which plasma ACE values exceeded fourfold the upper limit of normal (Kramers et al. 2001). All affected individuals were heterozygous for the mutation which is supposed to lead to an alteration in the juxtamembrane region of ACE, thus in a more efficient cleavage of the protein. The physiology of the renin–angiotensin system and BP were not altered in affected individuals, indicating that in the presence of similar amounts of membrane-bound ACE, the higher extracellular concentration of ACE is of no clinical significance (Kramers et al. 2001).

4.3
Relation with Essential Hypertension

The observation that plasma ACE levels are under the direct control of an ACE gene variant rapidly made the ACE I/D polymorphism one of the most popular markers tested in cardiovascular diseases. An interrogation of the PubMed database with "angiotensin I-converting enzyme insertion/deletion polymorphism" retrieves 652 publications. We will only summarise here some of the main findings using the ACE gene as a candidate gene for human essential hypertension and cardiovascular disorder.

One association study comparing a normotensive and a hypertensive Australian population with two hypertensive parents, showed an association of hypertension with ACE gene polymorphism (Zee et al. 1992). This finding was interpreted as due to an over risk of cardiovascular events in hypertensive patients carrying the D allele (Morris et al. 1994). Harrap et al. (1993) investigated the distribution of the ACE I/D gene polymorphism in young adults with contrasting genetic predisposition to high BP (Watt et al. 1992) ("four-corners approach"): young adults with high BP and two parents with high BP did not show any significant difference in the I/D allele frequencies when compared with adults of the same age but with low BP and no genetic predisposition to high BP. Other association studies were also negative (Higashimori et al. 1993; Schmidt et al. 1993). In a large series of hypertensive sib pairs from Utah, we found no evidence of linkage between hypertension and a growth hormone gene polymorphic marker in complete linkage disequilibrium with the ACE gene (Jeunemaitre et al. 1992a). Taken together, these results suggest that the ACE gene does not play a major role on BP variance in these populations.

Some positive results suggest, however, that the ACE locus might influence BP variability in a sex-specific manner. In a logistic regression analysis of 3,095 participants in the Framingham Heart Study (O'Donnell et al. 1998), the adjusted odds ratios for hypertension among men for the DD and DI versus II genotypes were 1.59 and 1.18, respectively, whereas no effect was observed in women. Positive results were also reported by Fornage et al. in the analysis of a large population-based sample of 1,488 siblings having a mean age of 15 years and belonging to the youngest generation of 583 randomly ascertained three-generation pedigrees from Rochester, MN (Fornage et al. 1998). In sex-specific analyses, genetic variation in the region of the ACE gene explained as much as 35% of the interindividual BP variation, again in males but not in females. Finally, Julier et al. (1997) conducted an affected sib-pair analysis in French and U.K. families and explored the region of chromosome 17q 23–32, based on the location of the ACE locus and on the QTL observed in rat on the homologous region of chromosome 10. Significant evidence of linkage was found near two closely linked microsatellite markers, D17S183 and D17S934, that reside 18 cM proximal to the ACE locus in the homology region.

4.4
Relation with End-Organ Damage

A very large number of reported and unreported studies have been performed and have been reviewed (Cambien and Soubrier 1995; Butler 2000; Danser and Schunkert 2000; Rieder and Nickerson 2000; Wang and Staessen 2000). The ACE I/D polymorphism has been associated with coronary heart disease, including myocardial infarction, stable and unstable angina pectoris, restenosis following percutaneous coronary angioplasty and stenting, left ventricular hypertrophy, cardiac insufficiency, peripheral artery disease, stroke and Alzheimer disease, proteinuria, diabetic nephropathy and retinopathy, human performance to exercise, aggravation of the course of IgA nephropathy and of polycytic kidney disease, and a variety of other diseases.

Of course, when such an attractive polymorphism is available, it is quite simple for researchers and clinicians to set up a case-control study and perform a statistical test to see whether a difference exists between each group, even if the a priori hypothesis is not substantiated by a strong pathophysiological background. Thus, among all the studies performed (>700), it is likely that a large number of false positive studies exist, due to publication bias and to selection criteria that do not fulfil those suggested for "high-quality" association studies (Bogardus 1999; Sharma and Jeunemaitre 2000). If well-performed, a meta-analysis can also help in the estimation of the strength of the impact of a given polymorphism on a given trait. In that regard, it is remarkable to observe that meta-analyses of the effect of the ACE I/D polymorphism on myocardial infarction found no or very mild association when analysed on several thousand patients, with mainly a publication bias for the small and positive studies (Fig. 3) (Agerholm-Larsen et al. 2000; Keavney et al. 2000). A similar observation has

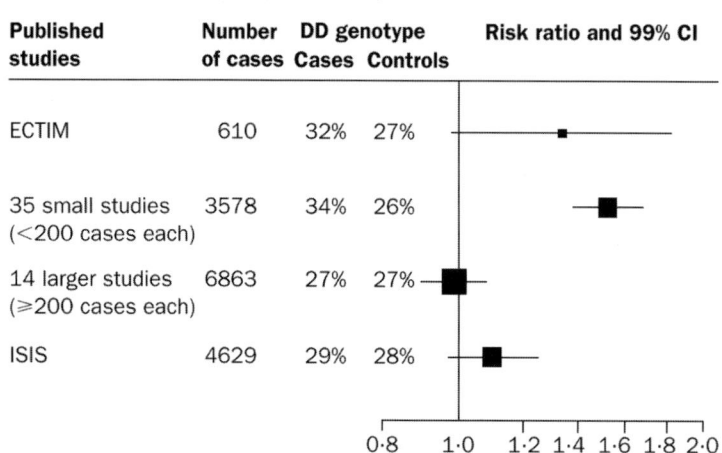

Fig. 3 Association between the human ACE gene and myocardial infarct: a meta-analysis. *CI*, confidence interval; *ECTIM*, Etude Cas-Temoin de l'Infarctus du Myocarde; *ISIS*, International Studies of Infarct Survival. From Kearney et al. (2000)

been made by meta-analyses of its association with coronary restenosis (Agema et al. 2002; Bonnici et al. 2002). A greater impact of the polymorphism might occur on ischemic stroke (Sharma 1998) and on diabetic nephropathy (Fujisawa et al. 1998). On the whole, it is also important to consider that associations might vary across gender or ethnic groups, or different socio-ecological settings, and that very few studies consider the potential gene–gene and gene–environment interactions.

Probably one of the most innovating findings has been the association between the I allele of the ACE I/D polymorphism with enhanced endurance performance in elite distance runners, rowers and mountaineers (Montgomery et al. 1997, 1999). The nature of the gene–environment interaction between ACE I/D polymorphisms and physical training, an overview of recent findings and a discussion of possible underlying mechanisms, have been reviewed by these authors (Woods et al. 2000).

5
The Angiotensin II Type 1 Receptor

Ang II receptors which mediate all the biological and physiological effects of the renin–angiotensin system are also candidate genes for essential hypertension. Most of the peripheral and central actions of Ang II, including vasoconstriction, facilitation of sympathetic transmission, modification of vascular and cardiac structure, renal salt and water retention, are mediated by the angiotensin type 1 (AT_1) receptor (AT1R). The AT1R belongs to the family of seven-transmembrane domain, G protein-coupled receptors. It is inserted into the plasma membrane of Ang II target, vascular smooth muscle, renal vasculature and mesangial, adrenal and brain cells.

5.1
Gene Structure, Mutations and Polymorphisms

The human gene has been cloned and is located on chromosome 3q21-3q25 (Furuta et al. 1992). It consists of at least five exons and spans more than 55 kb of genomic DNA (Guo et al. 1994). All the coding sequence is encoded by the last exon, the first four encoding 5' untranslated sequences, with multiple transcription initiation sites. There are at least four distinct alternatively spliced transcripts, whose relative abundance varies from one tissue to another one. Interestingly, a long AT1R isoform has been recently described, that has diminished activity for Ang II (Martin et al. 2001). Modifications of the relative abundance of the long and short isoforms of the AT1R might allow for a fine tuning of Ang II responsiveness in given circumstances or in a given tissue.

Several polymorphisms have been detected at the AT1R locus. An informative CA repeat polymorphism maps within 15 kb downstream the 3' end of the coding sequence (Davies et al. 1994). The interrogation of the human SNP database (www.ncbi.nlm.nih.gov.SNP) shows the presence of more than 20 SNPs, not all being validated nor their frequency determined. No polymorphism has been found to change the coding sequence of the receptor. The two SNPs at the AT1R gene that have been extensively used in the literature are those initially identified by Bonnardeaux et al. (1994). They consist of a non-synonymous change within the coding sequence (T573C) and a nucleotidic change in the 3' untranslated region (+1166A/C) whose frequency is 0.50–0.55 and 0.30–0.35 in Caucasian subjects. Other diallelic polymorphisms have been identified in the 3' untranslated region of the gene (+575T/C, +1062A/G, +1517G/T) in strong linkage disequilibrium (Bonnardeaux et al. 1994; Rolfs et al. 1994). Erdmann et al. performed a systematic screening of 2.5 kb of the 5' flanking region of the gene in a sample of 150 healthy subjects (Erdmann et al. 1999). They found eight polymorphic sites, six of them being in strong linkage disequilibrium but not with the +1166A/C polymorphism. Poirier et al. (1998) screened the first four exons and 2.2 kb in the 5' flanking region of the AT1R gene by SSCP and sequencing. Seven polymorphisms were detected in the 5' region at positions −1424, −810, −713, −521, −214, −213 and −153 upstream from the transcription start (http://genecanvas.idf.inserm.fr). All were in complete (or nearly complete) linkage disequilibrium. More recently, Antonellis et al. (2002) developed a technique for rapid identification of polymorphisms in long stretches of genomic DNA and analysed the AT1R gene. They genotyped 18 polymorphisms spanning the 60.5-kb AGTR1 locus, with an average spacing of 3.2 kb and an average minor allele frequency of 24%.

5.2
Relation with Essential Hypertension

The first study to investigate the relationship between the AT1R gene and essential hypertension was that performed by Bonnardeaux et al, with both an affected sibling pair approach and a case-control study (Bonnardeaux et al. 1994). There was no evidence for linkage between the microsatellite marker CA at the gene locus and essential hypertension in 267 hypertensive sib pairs from 138 pedigrees. The case-control study performed on a panel of 206 hypertensive subjects and 298 normotensive individuals was mainly negative, except for the 1166C allele, which was more frequent in hypertensive (0.36) than in normotensive subjects.

Other negative results were obtained by other groups. Castellano et al. (1996) examined a sample of 212 subjects randomly selected from a general population in Northern Italy. No statistically significant differences among AT1R A/C1166 genotypes were observed for ambulatory BP, left ventricular mass or carotid artery wall thickness. Several polymorphisms at the promoter region of the AT1R gene were tested by Zhang et al. in a large Caucasian population-based sample (Zhang et al. 2000). None of these polymorphisms showed evidence for association with hypertension. Negative results were obtained by two groups in Japan (Kikuya et al. 2003; Ono et al. 2003). The analysis of the A1166C polymorphism in 3,918 subjects recruited from the Suita study showed similar frequencies of the C allele (0.08), this frequency being markedly lower than in Caucasian subjects (Ono et al. 2003). The analysis of 802 subjects of the Ohasama study found no statistical difference in ambulatory BP according to the genotype, as soon as it was adjusted for confounding variables (age, sex, BMI) (Kikuya et al. 2003).

However, some studies found a positive association between essential hypertension and the A1166C polymorphism. Wang et al. (1997) analysed a well-characterised group of 108 Caucasian hypertensive subjects with a strong family history and 84 controls, in Australia. In this limited series, the 1166C allele was more frequent in hypertensives (0.40) than in controls (0.26). In the Finnish population, the sib pair analysis of 329 hypertensive individuals of 142 families gave a positive multipoint lod score of 2.9 (Kainulainen et al. 1999). Interestingly, a genome-wide scan performed in 47 families with two affected siblings and additional family members, showed the AT1R locus as the most significant locus in the Finnish population (Fig. 4) (Perola et al. 2000). However, no particular mutation or polymorphism was identified.

In a recent analysis of 218 Caucasian hypertensive patients and sib pairs, we found no direct relation with basal BP levels but a positive relationship between the acute BP response to Ang II and the AT1R locus (Vuagnat et al. 2001). Significant familial resemblances in the Ang II-induced systolic and diastolic BP response were observed, and the AT1R gene could participate to its genetic determination. However, no significant association was found between the A1166C polymorphism and the BP and aldosterone responses to infused Ang II in 116 male normal volunteers (Hilgers et al. 1999). Spiering et al. studied the reactivity

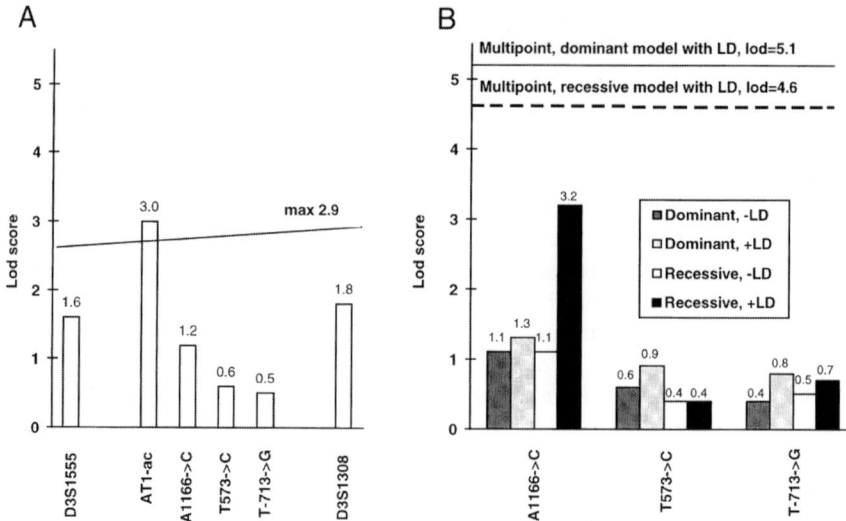

Fig. 4A, B Linkage and association between the AT$_1$ receptor gene and hypertension. In **A**, the line represents maximal logarithm of differences (lod) score obtained in the multipoint sib-pair linkage analyses of the six markers. In **B**, the lod scores have been obtained using different models in a subgroup of nonobese sib pairs. *Bars* represent two-point lod scores between hypertension and each marker. (From the Finnish study in Perola et al. 2000)

to infused Ang II of 42 subjects with essential hypertension, after 7 days of high-sodium diet (Spiering et al. 2000). They found that renal plasma flow, glomerular filtration rate and renal vascular resistance were more pronounced in patients with the CC genotype compared to those with the AA genotype. However, the changes of plasma renin, aldosterone and atrial natriuretic peptide (ANP) were not dependent on the genotype. It is interesting to observe that the effect of the genotype on the renal response to infused Ang II is probably dependent on the gender with a significant effect in males but not in females (Vuagnat et al. 2001; Reich et al. 2003).

5.3
Relation with End-Organ Damage

Taking into account the multiple actions of Ang II on multiple organs such as heart, vessels and brain, a large number of association studies have been performed to test the possible association of the AT1R gene polymorphisms and end-organ damage. It is difficult to summarise them in a short paragraph. On the whole, the polymorphisms that have been analysed are often unique in a given study, and/or tested on a too limited sample to be really contributive.

The A1166C polymorphism has been associated to aortic stiffness (Benetos et al. 1996; Lajemi et al. 2001), left ventricular mass (Osterop et al. 1998), coronary vasoconstriction (Amant et al. 1997), increased response to Ang II in isolated human arteries (Henrion et al. 1998; van Geel et al. 2000) and myocardial infarction in interaction with the ACE I/D polymorphism (Tiret et al. 1994). A positive association between the T573C polymorphisms and microalbuminuria was also found in a group of young adults with essential hypertension, the TT genotype being suggested as an independent protective factor for microalbuminuria (Chaves et al. 2001). An increased frequency of the 1166CC genotype was observed in a Polish group of 430 patients on dialysis compared to 260 healthy controls, with a significantly shorter time from the onset of renal disease to end-stage renal disease (ESRD) in patients with C allele compared to those with AA genotype (Buraczynska et al. 2002).

In a recent attempt to test the association between this polymorphism and cardiovascular events, Hindorff et al. (2002) analysed 800 African Americans and 1,372 random participants of the Cardiovascular Health Study. Although a marginal interaction was observed in the white treated participants with increased risk of congestive heart failure and stroke, on the whole the A1166C polymorphism was not associated with BP control or cardiovascular events on the entire populations. Finally, no relation was found between AT1R density in platelets—a possible biological intermediate phenotype—and the A1166C genotype (Paillard et al. 1999).

6
The Angiotensin II Type 2 Receptor

Since its discovery and molecular characterisation, the type 2 Ang II (AT_2) receptor (AT2R) has been enigmatic with respect to signalling pathways and function. These aspects are developed in another chapter of this book. Evidence has now emerged that Ang II exerts actions through the AT2Rs, which are directly opposed to those mediated by the AT1R. The AT2R appears to act as a modulator of complex biological programmes involved in embryonic development, cell differentiation, tissue protection and regeneration, as well as in programmed cell death (Gallinat et al. 2000). A number of studies suggest a role of AT(2) receptors in brain, renal, and cardiovascular functions and in the processes of apoptosis and tissue regeneration (Unger 1999).

6.1
Gene Localisation, Structure and Polymorphisms

The human AT2R gene was first isolated from a human genomic DNA library of leukocytes (Tsuzuki et al. 1994). The gene is organised in three exons but the entire coding region is contained in the third exon (Martin and Elton 1995) and encodes a 363 amino acid protein which is highly homologous to the sequences of the rat and mouse protein. Although they exhibit a similar binding affinity for Ang II, the AT1R and AT2R share only a 34% identity at the amino acid level. The AT2R is located on chromosome X in both human and rat species, the human AT2R gene being assigned to Xq22 (Koike et al. 1994).

6.2
Relation with Urogenital Abnormalities

Nishimura et al. (1999) reported that mice inactivated for the AT2R gene have phenotypes that remarkably resemble human congenital anomalies of the kidney and urinary tract. In the same study, the authors suggested that, in humans, a polymorphism could be a risk factor for the development of anomalies of the kidney and urinary tract (CAKUT). A substantial fraction of the human population (about 30%–50%) carry a A-to-G transition at position -1332 in intron 1 (A-1332G) in the lariat branchpoint motif of this intron, that might perturb the mRNA splicing efficiency. Nishimura et al. (1999) found a strong association between this polymorphism and the incidence of CAKUT: 10 of 13 Caucasian patients (77%) and 17 of 23 (74%) Caucasian German males with ureteropelvic junction stenosis or atresia, being hemizygous for this A-to-G transition, compared to an allelic frequency of 42% in controls.

However this association has been disputed. Hiraoka et al. (2001) compared the frequency of the intron 1 polymorphism in 66 Japanese boys with CAKUT and controls. The frequency of the A–G transition was not different between the control population (31 of 102, 30%) and the patients with CAKUT [23 of 66 (35%)]. Similar negative results were obtained by Yoneda et al. (2002) who found no evidence for the implication of the AT2R gene in the pathogenesis of primary familial vesicoureteral reflux. These authors evaluated the incidence of A-1332G transition in 82 male and 110 female patients and unaffected family members. The incidence of A-1332G transition was similar in male and female patients with primary familial vesicoureteral reflux and controls. Moreover, the transmission/disequilibrium test revealed no significant distortion of genotype transmission from mother to children.

6.3
Relation with Cardiovascular Diseases

Since the AT2R might oppose the vasoconstrictor and antinatriuretic effects of Ang II mediated through the AT1R, several genetic studies have investigated whether variations within the AT2R might play a counterregulatory protective role in the regulation of BP and in cardiovascular disorders such as neointima formation after vascular injury, cardiac hypertrophy and myocardial infarction.

In humans, only few association studies have been performed. Our laboratory was not able to find significant difference in a set of hypertensive and controls (M. Giacché, X. Jeunemaitre, unpublished results). Schmieder et al. analysed BP, left ventricular structure, plasma Ang II and aldosterone concentrations in 120 young male subjects (Schmieder et al. 2001). Presence of the A allele of the intronic +1675 G/A polymorphism was associated with an increase left ventricular mass index in hypertensive subjects (Schmieder et al. 2001). More recently, Herrmann et al. analysed the same polymorphism in two large cohorts of 1,968 individuals (the Glasgow Heart Scan, GLAECO and Glasgow Heart Scan Old, GLAOLD studies) with echocardiographically and electrocardiographically assessed left ventricular measurements (Herrmann et al. 2002). In both studies, the genotype frequencies were similar in hypertensive and non-hypertensive individuals (0.50). In the GLAOLD study and in females only, the AT2R +1675A allele was more common in females with episodes of coronary ischaemia and myocardial infarction. In the same cohort, the +1675A allele was associated with left ventricular hypertrophy in males only. However, these results could not be replicated in the GLAECO cohort, giving weak credibility for these genotype/phenotype interactions.

6.4
Relation with X-Mental Retardation

Whereas AT1Rs are abundant in a number of adult tissues, AT2Rs are mainly expressed during embryonic life and their expression declines rapidly after birth and is then restricted to few organs and mainly in some brain areas. The presence of high levels of AT2Rs within the neonatal brain supports the suggestion that this receptor has a role in development and possibly in cognitive function via its involvement in axonal regeneration (Lucius et al. 1998). This possible role was also suggested by the attenuated exploratory behaviour and anxiety-like behaviour of AT2R-deficient mice (Hein et al. 1995; Ichiki et al. 1995; Okuyama et al. 1999).

A recent human study further supports a role for the AT2R in brain development and cognitive function (Vervoort et al. 2002). These authors analysed a de novo balanced translocation 46,X,t(X;7)(q24;q22) in a female patient with moderate mental retardation. They mapped the X-chromosome breakpoint and demonstrated by RT-PCR that the AT2R gene was silenced in this patient, possibly through a position effect. The authors then screened a series of 33 males with X-linked mental retardation (MR) and a large cohort of 552 unrelated male patients with MR of unknown cause. Eight of the 590 unrelated male patients with MR were found to have sequence changes in the AGTR2 gene, including 1 frameshift and 3 missense mutations (Fig. 5). Although these results have not

- Balance X; 7 translocation in a female patient with mental retardation (MR)

- 8 / 590 patients with MR have AT2R mutations

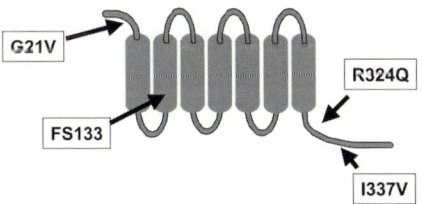

Fig. 5 Mutations of the AT$_2$ receptor gene and X-linked mental retardation

been replicated yet, they indicate that the AT2R might contribute to regulate central nervous system functions, including behaviour.

7
Future Directions

With the notable exception of few monogenic forms of cardiovascular diseases for which the molecular basis has been often already elucidated, there is no indication of the number of genetic loci involved in these complex traits, the frequency of deleterious alleles, their mode of transmission, and the quantitative effect of any single allele. Because of their aetiologic heterogeneity, it is likely that no single biochemical or genetic marker will help the clinician in the management of most of the patients. How can we progress in the comprehension of the involved molecular pathways, and in the possible application of genetic markers? Some of the avenues that could be explored are indicated below.

7.1
Test Gene–Gene and Gene–Environment Interactions

7.1.1
Gene–Gene Interactions

It is important to note that most of the studies performed so far have been conducted in a very simple design, by testing one by one the role of each gene (sometimes of only one polymorphism) of the renin–angiotensin system. However, it is recognised that the situation is much more complex, involving a multitude of genes controlling the susceptibility to cardiovascular diseases, acting in an additive or interactive manner together with environmental factors. A more integrated approach needs to be designed in which several genes and some environmental factors could be tested with or without a priori specific hypothesis.

In that regard, the first attempts that have been made by several authors to test the interaction between renin–angiotensin system genetic polymorphisms are of interest. For example, interactions between the A1166C polymorphism of the AT1R gene with the ACE I/D polymorphism has been found to explain a genetic susceptibility to increased diastolic BP (Henskens et al. 2003) and myocardial infarction in interaction (Tiret et al. 1994). A study performed by Staessen and colleagues (2001) showed the combined effect of the ACE I/D, the alpha-adducin Gly460Trp and the aldosterone synthase C-344T polymorphisms on BP level. Lovati and colleagues (2001) found an interaction between the AGT M235T polymorphism and the ACE I/D polymorphism for the occurrence of end-stage renal disease in diabetic patients. Such an interaction for the progression of renal insufficiency in patients with IgA nephropathy has been also discussed (Pei et al. 1997; Frimat et al. 2000). Although these studies face the problem of stratification and thus of increased probability of false positive results, their replication in other well-performed and powerful studies might help to better estimate the role of the RAS polymorphisms in these complex traits.

7.1.2
Gene–Environment Interactions

As mentioned above, functional polymorphisms exist at the renin–angiotensin system. Nevertheless, their effect on hypertension and cardiovascular diseases has been highly disputed. One of the reasons for this is probably that the phenotype itself has not been determined with enough precision nor the polymorphisms tested in non-standardised environmental conditions (especially posture, salt diet, therapy). If we admit, as it is probably true, that these polymorphisms exert their effect only in certain conditions, a "stress-the-genotype" approach might help us to unravel their true pathophysiological influence. One of the best example of this approach involves the studies performed by Montgomery and colleagues to evaluate the effect of the ACE I/D polymorphism on left ventricular mass in response to exercise (Montgomery et al. 1999, 2002). The strategy was to select subjects homozygous for the I or D allele from a large number of military recruits and precisely phenotype left ventricular mass before and after a 10-week physical training period. In the several studies performed, left ventricular mass increased with training, but with DD men showing roughly threefold greater growth than II men. When indexed to lean body mass, left ventricular growth in II subjects was essentially negligible.

Some studies have also suggested that the AGT genotype might influence the BP response to non-pharmacological therapy, especially be a marker of salt-sensitivity. In the Trials of Hypertension Prevention Phase II (TOHP-II), both sodium reduction and weight loss were tested in randomised trials as BP-lowering strategies. Participants were typed for the G-6A polymorphism. In the usual care group, the AA genotype was associated with a higher 3-year incidence of hypertension compared to GA or GG genotypes. In the salt reduction and weight loss intervention groups, this AA genotype was also associated with a larger reduction in BP compared to GA and GG genotypes (Hunt et al. 1998). The same polymorphism G-6A was tested with the BP response to the Dietary Approaches to Stop Hypertension (DASH) diet (Svetkey et al. 2001). After 8 weeks, net systolic and diastolic BP response to the DASH diet was significantly greater in individuals with the AA genotype compared to those with the GG genotype. Undoubtedly, the use of these "stress-the-genotype" approaches to explore gene-environment interactions might be the key to understanding the complex effects of the renin–angiotensin system gene polymorphisms.

7.2
Analysis of Other Components of the Renin–Angiotensin System

7.2.1
The Prorenin Receptor

Up to now, Ang II has been believed to be the main, if not exclusive, effector of the renin–angiotensin system through binding to its receptors AT_1 and AT_2. The recent demonstration and the cloning of a functional receptor of prorenin might change this view (Nguyen et al. 2002). The highest level of expression of this receptor is observed in heart, brain, placenta, kidney and eye. Immuno-histological studies of heart and kidney show that it is localised in vascular structures, associated to smooth-muscle cells in coronary and renal artery (Nguyen et al. 2002). This receptor might have cellular effects per se on vascular smooth-muscle cells, independent of Ang II. It might represent an essential mean to capture renin and prorenin from the circulation and to concentrate renin and prorenin in tissues, especially at the interface with endothelial cells. In this context, the search of functional genetic polymorphisms will be of importance regarding their potential impact on the modulation of smooth muscle cell tone and of coronary artery function (Nguyen et al. 2003).

7.2.2
The ACE2 Gene

The identification of an enzyme similar to ACE, called ACE2, adds further complexity to the system. The ACE2 gene was initially cloned from a human cardiac cDNA library (Donoghue et al. 2000). ACE has a single catalytic domain, whereas somatic ACE has two catalytic domains. The phylogenetic analysis clearly shows that they evolve by duplication of a single-domain ACE-like protein, before the divergence of insect and mammalian lineages (Cornell et al. 1995). ACE2 transcripts are found mainly in vascular endothelial cells of the heart and kidney and in testis.

The ACE2 gene encodes for a carboxypeptidase that is able to convert Ang I to angiotensin 1-9. Since angiotensin 1-9 has no known effect on blood vessel and can be converted by ACE in angiotensin 1–7, which is a blood vessel dilator, it has been suggested that ACE2 might prevent the formation of the vasopressor Ang II. The recent work, published in *Nature* (Crackower et al. 2002) tends to demonstrate the pathophysiological importance of this enzyme. The authors were able to show that, in the salt-sensitive Sabra hypertensive rat strain, there is a markedly decrease expression of the ACE2 gene, although no mutation was found in the coding sequence of the gene. In addition, targeted disruption of the ACE2 gene in mice resulted in a cardiac contractility defect, suggesting its role as an essential regulator of heart function in vivo. If this role is confirmed, the molecular analysis of the human ACE gene in a variety of cardiovascular diseases will be important.

7.2.3
The Aminopeptidases A and N

Among the main bioactive peptides of the brain renin–angiotensin system, angiotensin Ang II and Ang III exhibit the same affinity for type 1 and type 2 Ang II receptors. In the murine brain renin–angiotensin system, Ang II is converted by aminopeptidase A (APA) into Ang III, which is itself degraded by aminopeptidase N (APN). Both peptides, injected intracerebroventricularly, cause similar increases in vasopressin release and BP (Reaux et al. 1999). Recent pharmacological studies using specific APN and APA inhibitors have demonstrated the predominant role of brain Ang III in the control of vasopressin release and BP (Reaux et al. 2001). These genes are therefore attractive candidates for human hypertension.

7.3
Pharmacogenetics

The increasing knowledge of polymorphisms in genes encoding for proteins that are drugs targets, the uncertainty of the individual efficacy and tolerance of these drugs, has fuelled a new molecular discipline, pharmacogenetics. It is expected that pharmacogenetics will help clinicians gauge drugs efficacy and tolerance and will help to give the "right medicine for the right patient" (Roses 2000).

In addition to polymorphisms in drug metabolising enzymes, there are cases where the response to drug treatment might be correlated with gene variants involved in the drug's mode of action. The renin–angiotensin system is a good example of such possibility, since functional polymorphisms exist as well as drugs blocking or stimulating the system. Although promising, only few therapeutic studies have been correctly conducted in patients according to renin–angiotensin genotypes. We will just take the example of the ACE I/D polymorphism. Its relationship with cough, the major side-effect of ACE inhibitors, is uncertain (O'Toole et al. 1998). Some studies found an increased frequency of the I allele in coughers (Lee and Tsai 2001), others an increased frequency of the D allele, still others no difference in genotype frequencies (Zee et al. 1998). It is therefore likely that this polymorphism does not explain this side-effect, even though it may be part of a particular combination of susceptibility variants belonging to several genes related to the renin–angiotensin system and kallikrein–bradykinin system.

The analysis of the efficacy of ACE inhibitors led also to controversial results. A pharmacokinetic interaction was observed in at least two studies. Todd et al. (1995) showed that the ACE I/D polymorphism might predict residual ACE activity after administration of a single dose of enalapril. Similarly, Ueda et al. found that compared to the II subjects, subjects with the DD genotype had a greater and longer effect of a unique dose of enalaprilat (Ueda et al. 1998). Despite this interaction, reports of BP decrease after chronic administration of an ACE inhibitor have been controversial (O'Byrne and Caulfield 1998).

Interaction with BP responses to hydrochlorothiazide according to the ACE I/D genotype have also been analysed. In a study of 376 patients with essential hypertension, Schwartz and colleagues (2002) found a significant interaction between the effects of the ACE genotype and gender on the responses of both systolic and diastolic BP to hydrochlorothiazide (HCTZ) 25 mg daily. However, the effects of the genotypes were opposed in men and women. Sciarrone et al. (2003) evaluated the BP response to HCTZ in 87 never-treated individuals with mild essential hypertension, according to ACE gene I/D and alpha-adducin Gly460Trp polymorphisms. They found that patients carrying at least one I allele of ACE and one 460Trp allele of alpha-adducin had the largest mean BP (MBP) decrease with treatment, the effect of the combination of genotypes being additive but not epistatic.

Obviously, prediction of drug efficacy and/or of adverse drug reaction with a pharmacogenetic profile will need larger, well-designed trials in which patients will be standardised on the drug dosage and metabolism (Rioux 2000).

Acknowledgements. This work was supported by Grants form INSERM, Collège de France, Bristol-Myers Squibb, Association Claude Bernard and Association Naturalia and Biologia.

References

Agema WR, Jukema JW, Zwinderman AH, van der Wall EE (2002) A meta-analysis of the angiotensin-converting enzyme gene polymorphism and restenosis after percutaneous transluminal coronary revascularization: evidence for publication bias. Am Heart J 144:760–768

Agerholm-Larsen B, Nordestgaard BG, Tybjaerg-Hansen A (2000) ACE gene polymorphism in cardiovascular disease: meta-analyses of small and large studies in whites. Arterioscler Thromb Vasc Biol 20:484–492

Alhenc-Gelas F, Richard J, Courbon D, Warnet JM, Corvol P (1991) Distribution of plasma angiotensin I-converting enzyme levels in healthy men: relationship to environmental and hormonal parameters. J Lab Clin Med 117:33–39

Amant C, Hamon M, Bauters C, Richard F, Helbecque N, McFadden EP, Escudero X, Lablanche JM, Amoyel P, Bertrand ME (1997) The angiotensin II type 1 receptor gene polymorphism is associated with coronary artery vasoconstriction. J Am Coll Cardiol 29:486–490

Antonellis A, Rogus JJ, Canani LH, Makita Y, Pezzolesi MG, Nam M, Ng D, Moczulski D, Warram JH, Krolewski AS (2002) A method for developing high-density SNP maps and its application at the type 1 angiotensin II receptor (AGTR1) locus. Genomics 79:326–332

Arngrimsson R, Purandare S, Connor M, Walker JJ, Bjornsson S, Soubrier F, Kotelevtsev YV, Geirsson RT, Bjornsson H (1993) Angiotensinogen: a candidate gene involved in preeclampsia? Nat Genet 4:114–115

Arngrimsson R, Sigurard ttir S, Frigge ML, Bjarnad ttir RI, Jonsson T, Stefansson H, Baldursdottir A, Einarsdottir AS, Palsson B, Snorradottir S, Lachmeijer AM, Nicolae D, Kong A, Bragason BT, Gulcher JR, Geirsson RT, Stefansson K (1999) A genome-wide scan reveals a maternal susceptibility locus for pre- eclampsia on chromosome 2p13. Hum Mol Genet 8:1799–1805

Atwood LD, Kammerer CM, Samollow PB, Hixson JE, Shade RE, MacCluer JW (1997) Linkage of essential hypertension to the angiotensinogen locus in Mexican Americans. Hypertension 30:326–330

Azizi M, Hallouin MC, Jeunemaitre X, Guyene TT, Menard J (2000) Influence of the M235T polymorphism of human angiotensinogen (AGT) on plasma AGT and renin concentrations after ethinylestradiol administration. J Clin Endocrinol Metab 85:4331–4337

Benetos A, Ricard S, Topouchian J, Asmar R, Poirier O, Larosa E, Guize L, Safar M, Soubrier F, Cambien F (1996) Influence of angiotensin-converting enzyme and angiotensin II type 1 receptor gene polymorphisms on aortic stiffness in normotensive and hypertensive patients. Circulation 94:698–703

Bernstein IM, Ziegler W, Stirewalt WS, Brumsted J, Ward K (1998) Angiotensinogen genotype and plasma volume in nulligravid women. Obstet Gynecol 92:171–173

Bloem LJ, Foroud TM, Ambrosius WT, Hanna MP, Tewksbury DA, Pratt JH (1997) Association of the angiotensinogen gene to serum angiotensinogen in blacks and whites. Hypertension 29:1078–1082

Bloem LJ, Manatunga AK, Tewksbury DA, Pratt JH (1995) The serum angiotensinogen concentration and variants of the angiotensinogen gene in white and black children. J Clin Invest 95:948–953

Bogardus ST Jr, Concato J, Feinstein AR (1999) Clinical epidemiological quality in molecular genetic research: the need for methodological standards. JAMA 281:1919–1926

Bonnardeaux A, Davies E, Jeunemaitre X, Fery I, Charru A, Clauser E, Tiret L, Cambien F, Corvol P, Soubrier F (1994) Angiotensin II type 1 receptor gene polymorphisms in human essential hypertension. Hypertension 24:63–69

Bonnici F, Keavney B, Collins R, Danesh J (2002) Angiotensin converting enzyme insertion or deletion polymorphism and coronary restenosis: meta-analysis of 16 studies. BMJ 325:517–520

Brand E, Chatelain N, Keavney B, Caulfield M, Citterio L, Connell J, Grobbee D, Schmidt S, Schunkert H, Schuster H, Sharma AM, Soubrier F (1998) Evaluation of the angiotensinogen locus in human essential hypertension: a European study. Hypertension 31:725–729

Buraczynska M, Ksiazek P, Zaluska W, Spasiewicz D, Nowicka T, Ksiazek A (2002) Angiotensin II type 1 receptor gene polymorphism in end-stage renal disease. Nephron 92:51–55

Butler R (2000) The DD-ACE genotype and cardiovascular disease. Pharmacogenomics 1:153–167

Cambien F, Alhenc-Gelas F, Herbeth B, Andre JL, Rakotovao R, Gonzales MF, Allegrini J, Bloch C (1988) Familial resemblance of plasma angiotensin-converting enzyme level: the Nancy Study. Am J Hum Genet 43:774–780

Cambien F, Soubrier F (1995) The angiotensin-converting enzyme: molecular biology and implication of the gene polymorphism in cardiovascular diseases. In: Laragh JH (ed) Hypertension: pathophysiology, prognosis and treatment, 2nd edn. Raven Press, New York, pp 1667–1682

Castellano M, Muiesan ML, Beschi M, Rizzoni D, Cinelli A, Salvetti M, Pasini G, Porteri E, Bettoni G, Zulli R, Agabiti-Rosei E (1996) Angiotensin II type 1 receptor A/C1166 polymorphism. Relationships with blood pressure and cardiovascular structure. Hypertension 28:1076–1080

Caulfield M, Lavender P, Farrall M, Munroe P, Lawson M, Turner P, Clark AJ (1994) Linkage of the angiotensinogen gene to essential hypertension. N Engl J Med 330:1629–1633

Caulfield M, Lavender P, Newell-Price J, Farrall M, Kamdar S, Daniel H, Lawson M, De Freitas P, Fogarty P, Clark AJ (1995) Linkage of the angiotensinogen gene locus to human essential hypertension in African Caribbeans. J Clin Invest 96:687–692

Caulfield M, Lavender P, Newell-Price J, Kamdar S, Farrall M, Clark AJ (1996) Angiotensinogen in human essential hypertension. Hypertension 28:1123–1125

Chaves FJ, Giner V, Corella D, Pascual J, Marin P, Armengod ME, Redon J (2002) Body weight changes and the A-6G polymorphism of the angiotensinogen gene. Int J Obes Relat Metab Disord 26:1173–1178

Chaves FJ, Pascual JM, Rovira E, Armengod ME, Redon J (2001) Angiotensin II AT1 receptor gene polymorphism and microalbuminuria in essential hypertension. Am J Hypertens 14:364–170

Cohen-Haguenauer O, Soubrier F, N'Guyene VC, Serero S, Turleau C, Jegou C, Gross MS, Corvol P, Frezal J (1983) Regional mapping of the human renin gene to $1q^{32}$ by in situ hybridization. Ann Genet (Paris) 32:16–20

Cooper RS, Guo X, Rotimi CN, Luke A, Ward R, Adeyemo A, Danilov SM (2000) Heritability of angiotensin-converting enzyme and angiotensinogen: a comparison of US blacks and Nigerians. Hypertension 35:1141–1147

Cornell MJ, Williams TA, Lamango NS, Coates D, Corvol P, Soubrier F, Hoheisel J, Lehrach H, Isaac RE (1995) Cloning and expression of an evolutionary conserved single-domain angiotensin converting enzyme from Drosophila melanogaster. J Biol Chem 270:13613–13619

Costerousse O, Allegrini J, Lopez M, Alhenc-Gelas F (1993) Angiotensin I-converting enzyme in human circulating mononuclear cells: genetic polymorphism of expression in T-lymphocytes. Biochem J 290:33–40

Crackower MA, Sarao R, Oudit GY, Yagil C, Kozieradzki I, Scanga SE, Oliveira-dos-Santos AJ, da Costa J, Zhang L, Pei Y, Scholey J, Ferrario CM, Manoukian AS, Chappell MC, Backx PH, Yagil Y, Penninger JM (2002) Angiotensin-converting enzyme 2 is an essential regulator of heart function. Nature 417:822–828

Danser AH, Schunkert H (2000) Renin–angiotensin system gene polymorphisms: potential mechanisms for their association with cardiovascular diseases. Eur J Pharmacol 410:303–316

Davies E, Bonnardeaux A, Lathrop GM, Corvol P, Clauser E, Soubrier F (1994) Angiotensin II (type-1) receptor locus: CA repeat polymorphism and genetic mapping. Hum Mol Genet 3:838

Donoghue M, Hsieh F, Baronas E, Godbout K, Gosselin M, Stagliano N, Donovan M, Woolf B, Robison K, Jeyaseelan R, Breitbart RE, Acton S (2000) A novel angiotensin-converting enzyme-related carboxypeptidase (ACE2) converts angiotensin I to angiotensin 1–9. Circ Res 87:E1–E9

Eggena P, Sowers JR, Maxwell MH, Barrett JD, Golub MS (1991) Hormonal correlates of weight loss associated with blood pressure reduction. Clin Exp Hypertens A 13:1447–1456

Erdmann J, Riedel K, Rohde K, Folgmann I, Wienker T, Fleck E, Regitz-Zagrosek V (1999) Characterization of polymorphisms in the promoter of the human angiotensin II subtype 1 (AT1) receptor gene. Ann Hum Genet 63:369–374.

Fasola AF, Martz BL, Helmer OM (1968) Plasma renin activity during supine exercice in offsprings of hypertensive parents. J Appl Physiol 25:410–415

Fisher N, Hurwitz S, Jeunemaitre X, Hopkins PN, Hunt SC, Hollenberg NH, Williams GH (2002) Familial aggregation of low-renin hypertension. Hypertension 39:914–918

Fornage M, Amos CI, Kardia S, Sing CF, Turner ST, Boerwinkle E (1998) Variation in the region of the angiotensin-converting enzyme gene influences interindividual differences in blood pressure levels in young white males. Circulation 97:1773–1779

Frimat L, Philippe C, Maghakian MN, Jonveaux P, Hurault de Ligny B, Guillemin F, Kessler M (2000) Polymorphism of angiotensin converting enzyme, angiotensinogen, and angiotensin II type 1 receptor genes and end-stage renal failure in IgA nephropathy: IGARAS–a study of 274 Men. J Am Soc Nephrol 11:2062–2067

Fuchs S, Philippe J, Germain S, Mathieu F, Jeunemaitre X, Corvol P, Pinet F (2002) Functionality of two new polymorphisms in the human renin gene enhancer region. J Hypertens 20:2391–2398

Fujisawa T, Ikegami H, Kawaguchi Y, Hamada Y, Ueda H, Shintani M, Fukuda M, Ogihara T (1998) Meta-analysis of association of insertion/deletion polymorphism of angiotensin I-converting enzyme gene with diabetic nephropathy and retinopathy. Diabetologia 41:47–53

Furuta H, Guo DF, Inagami T (1992) Molecular cloning and sequencing of the gene encoding human angiotensin II type 1 receptor. Biochem Biophys Res Commun 183:8-13

Gaillard I, Clauser E, Corvol P (1989) Structure of human angiotensinogen gene. DNA 8:87–99

Gallinat S, Busche S, Raizada MK, Sumners C (2000) The angiotensin II type 2 receptor: an enigma with multiple variations. Am J Physiol Endocrinol Metab 278:E357–E374

Germain S, Bonnet F, Philippe J, Fuchs S, Corvol P, Pinet F (1998) A novel distal enhancer confers chorionic expression on the human renin gene. J Biol Chem 273:25292–25300

Gimenez-Roqueplo AP, Celerier J, Schmid G, Corvol P, Jeunemaitre X (1998) Role of cysteine residues in human angiotensinogen. Cys232 is required for angiotensinogen-pro major basic protein complex formation. J Biol Chem 273:34480–34487

Grim CE, Luft FC, Miller JZ, Rose RJ, Christian JC, Weinberger MH (1980) An approach to the evaluation of genetic influences on factors that regulate arterial blood pressure in man. Hypertension 2[suppl 2]:I-34–I-42

Guo DF, Furuta H, Mizukoshi M, Inagami T (1994) The genomic organization of human angiotensin II type 1 receptor. Biochem Biophys Res Commun 200:313–319

Guo X, Rotimi C, Cooper R, Luke A, Elston RC, Ogunbiyi O, Ward R (1999) Evidence of a major gene effect for angiotensinogen among Nigerians. Ann Hum Genet 63:293–300

Harrap SB, Davidson HR, Connor JM, Soubrier F, Corvol P, Fraser R, Foy CJ, Watt GC (1993) The angiotensin I converting enzyme gene and predisposition to high blood pressure. Hypertension 21:455–460

Harrison GA, Humphrey KE, Jones N, Badenhop R, Guo G, Elakis G, Kaye JA, Turner RJ, Grehan M, Wilton AN, Brennecke SP, Cooper DW (1997) A genomewide linkage study of preeclampsia/eclampsia reveals evidence for a candidate region on 4q. Am J Hum Genet 60:1158–1167

Hasimu B, Nakayama T, Mizutani Y, Izumi Y, Asai S, Soma M, Kokubun S, Ozawa Y (2003) Haplotype analysis of the human Renin gene and essential hypertension. Hypertension 41:308–312

Hata A, Namikawa C, Sasaki M, Sato K, Nakamura T, Tamura K, Lalouel JM (1994) Angiotensinogen as a risk factor for essential hypertension in Japan. J Clin Invest 93:1285–1287

Hein L, Barsh GS, Pratt RE, Dzau VJ, Kobilka BK (1995) Behavioural and cardiovascular effects of disrupting the angiotensin II type-2 receptor in mice. Nature 377:744–747

Henrion D, Amant C, Benessiano J, Philip I, Plantefeve G, Chatel D, Hwas U, Desmont JM, Durand G, Amouyel P, Levy BI (1998) Angiotensin II type 1 receptor gene polymorphism is associated with an increased vascular reactivity in the human mammary artery in vitro. J Vasc Res 35:356–362

Henskens LH, Spiering W, Stoffers HE, Soomers FL, Vlietinck RF, de Leeuw PW, Kroon AA (2003) Effects of ACE I/D and AT1R-A1166C polymorphisms on blood pressure in a healthy normotensive primary care population: first results of the Hippocates study. J Hypertens 21:81–86

Herrmann SM, Nicaud V, Schmidt-Petersen K, Pfeifer J, Erdmann J, McDonagh T, Dargie HJ, Paul M, Regitz-Zagrosek V (2002) Angiotensin II type 2 receptor gene polymorphism and cardiovascular phenotypes: the GLAECO and GLAOLD studies. Eur J Heart Fail 4:707–712

Higashimori K, Zhao Y, Higaki J, Kamitani A, Katsuya T, Nakura J, Miki T, Mikami H, Ogihara T (1993) Association analysis of a polymorphism of the angiotensin converting enzyme gene with essential hypertension in the Japanese population. Biochem Biophys Res Commun 191:399–404

Hilgers KF, Langenfeld MR, Schlaich M, Veelken R, Schmieder RE (1999) 1166 A/C polymorphism of the angiotensin II type 1 receptor gene and the response to short-term infusion of angiotensin II. Circulation 100:1394–1399

Hindorff LA, Heckbert SR, Tracy R, Tang Z, Psaty BM, Edwards KL, Siscovick DS, Kronmal RA, Nazar-Stewart V (2002) Angiotensin II type 1 receptor polymorphisms in the cardiovascular health study: relation to blood pressure, ethnicity, and cardiovascular events. Am J Hypertens 15:1050–1056

Hiraoka M, Taniguchi T, Nakai H, Kino M, Okada Y, Tanizawa A, Tsukahara H, Ohshima Y, Muramatsu I, Mayumi M (2001) No evidence for AT2R gene derangement in human urinary tract anomalies. Kidney Int 59:1244–1249

Hobart PM, Fogliano M, O'Connor BA, Schaefer IM, Chirgwin JM (1984) Human renin gene: structure and sequence analysis. Proc Natl Acad Sci USA 81:5026–5030

Howard TE, Shai SY, Langford KG, Martin BM, Bernstein KE (1990) Transcription of testicular angiotensin-converting enzyme (ACE) is initiated within the 12th intron of the somatic ACE gene. Mol Cell Biol 10:4294–4302

Hubert C, Houot AM, Corvol P, Soubrier F (1991) Structure of the angiotensin I-converting enzyme gene. Two alternate promoters correspond to evolutionary steps of a duplicated gene. J Biol Chem 266:15377–15383

Hunt SC, Cook NR, Oberman A, Cutler JA, Hennekens CH, Allender PS, Walker WG, Whelton PK, Williams RR (1998) Angiotensinogen genotype, sodium reduction, weight loss, and prevention of hypertension: trials of hypertension prevention, phase II. Hypertension 32:393–401

Ichiki T, Labosky PA, Shiota C, Okuyama S, Imagawa Y, Fogo A, Niimura F, Ichikawa I, Hogan BL, Inagami T (1995) Effects on blood pressure and exploratory behaviour of mice lacking angiotensin II type-2 receptor. Nature 377:748–750

Imai T, Miyazaki H, Hirose S, Hori H, Hayashi T, Kageyama R, Ohkubo H, Nakanishi S, Murakami K (1983) Cloning and sequence analysis of cDNA for human renin precursor. Proc Natl Acad Sci USA 80:7405–7409

Inoue I, Nakajima T, Williams CS, Quackenbush J, Puryear R, Powers M, Cheng T, Ludwig EH, Sharma AM, Hata A, Jeunemaitre X, Lalouel JM (1997) A nucleotide substitution in the promoter of human angiotensinogen is associated with essential hypertension and affects basal transcription in vitro. J Clin Invest 99:1786–1797

Inoue K, Osaka H, Sugiyama N, Kawanishi C, Onishi H, Nezu A, Kimura K, Yamada Y, Kosaka K (1996) A duplicated PLP gene causing Pelizaeus-Merzbacher disease detected by comparative multiplex PCR. Am J Hum Genet 59:32–39

Ishikawa K, Baba S, Katsuya T, Iwai N, Asai T, Fukuda M, Takiuchi S, Fu Y, Mannami T, Ogata J, Higaki J, Ogihara T (2001) T+31C polymorphism of angiotensinogen gene and essential hypertension. Hypertension 37:281–285

Iwai N, Ohmichi N, Nakamura Y, Mitsunnami X, Kinoshita M (1994) Molecular variants of the angiotensinogen gene and hypertension in a Japanese population. Hypertens Res 17:117–121

Jain S, Tang X, Narayanan CS, Agarwal Y, Peterson SM, Brown CD, Ott J, Kumar A (2002) Angiotensinogen gene polymorphism at −217 affects basal promoter activity and is associated with hypertension in African-Americans. J Biol Chem 277:36889–36896

Jeunemaitre X, Gimenez-Roqueplo AP, Celerier J, Corvol P (1999) Angiotensinogen variants and human hypertension. Curr Hypertens Rep 1:31–41

Jeunemaitre X, Inoue I, Williams C, Charru A, Tichet J, Powers M, Sharma AM, Gimenez-Roqueplo AP, Hata A, Corvol P, Lalouel JM (1997) Haplotypes of angiotensinogen in essential hypertension. Am J Hum Genet 60:1448–1460

Jeunemaitre X, Lifton RP, Hunt SC, Williams RR, Lalouel JM (1992a) Absence of linkage between the angiotensin converting enzyme locus and human essential hypertension. Nat Genet 1:72–75

Jeunemaitre X, Rigat B, Charru A, Houot AM, Soubrier F, Corvol P (1992b) Sib pair linkage analysis of renin gene haplotypes in human essential hypertension. Hum Genet 88:301–306

Jeunemaitre X, Soubrier F, Kotelevtsev YV, Lifton RP, Williams CS, Charru A, Hunt SC, Hopkins PN, Williams RR, Lalouel JM, et al (1992c) Molecular basis of human hypertension: role of angiotensinogen. Cell 71:169–180

Julier C, Delepine M, Keavney B, Terwilliger J, Davis S, Weeks DE, Bui T, Jeunemaitre X, Velho G, Froguel P, Ratcliffe P, Corvol P, Soubrier F, Lathrop GM (1997) Genetic susceptibility for human familial essential hypertension in a region of homology with blood pressure linkage on rat chromosome 10. Hum Mol Genet 6:2077–2085

Kageyama R, Ohkubo H, Nakanishi S (1984) Primary structure of human preangiotensinogen deduced from the cloned cDNA sequence. Biochemistry 23:3603–3609

Kainulainen K, Perola M, Terwilliger J, Kaprio J, Koskenvuo M, Syvanen AC, Vartiainen E, Peltonen L, Kontula K (1999) Evidence for involvement of the type 1 angiotensin II receptor locus in essential hypertension. Hypertension 33:844–849

Kamitani A, Rakugi H, Higaki J, Yi Z, Mikami H, Miki T, Ogihara T (1994) Association analysis of a polymorphism of the angiotensinogen gene with essential hypertension in Japanese. J Hum Hypertens 8:521–524

Katsuya T, Ogihara T (2001) Current genetics of essential hypertension. Curr Hypertens Rep 3:1–2

Keavney B, McKenzie C, Parish S, Palmer A, Clark S, Youngman L, Delepine M, Lathrop M, Peto R, Collins R (2000) Large-scale test of hypothesised associations between the angiotensin- converting-enzyme insertion/deletion polymorphism and myocardial infarction in about 5000 cases and 6000 controls. International Studies of Infarct Survival (ISIS) Collaborators. Lancet 355:434–442

Kikuya M, Sugimoto K, Katsuya T, Suzuki M, Sato T, Funahashi J, Katoh R, Kazama I, Michimata M, Araki T, Hozawa A, Tsuji I, Ogihara T, Yanagisawa T, Imai Y, Matsubara M (2003) A/C1166 gene polymorphism of the angiotensin II type 1 receptor (AT1) and ambulatory blood pressure: the Ohasama Study. Hypertens Res 26:141–145

Koike G, Horiuchi M, Yamada T, Szpirer C, Jacob HJ, Dzau VJ (1994) Human type 2 angiotensin II receptor gene: cloned, mapped to the X chromosome, and its mRNA is expressed in the human lung. Biochem Biophys Res Commun 203:1842–1850

Kramers C, Danilov SM, Deinum J, Balyasnikova IV, Scharenborg N, Looman M, Boomsma F, de Keijzer MH, van Duijn C, Martin S, Soubrier F, Adema GJ (2001) Point mutation in the stalk of angiotensin-converting enzyme causes a dramatic increase in serum angiotensin-converting enzyme but no cardiovascular disease. Circulation 104:1236–1240

Kunz R, Kreutz R, Beige J, Distler A, Sharma AM (1997) Association between the angiotensinogen 235T-variant and essential hypertension in whites: a systematic review and methodological appraisal. Hypertension 30:1331–1337

Laivuori H, Lahermo P, Ollikainen V, Widen E, Haiva-Mallinen L, Sundstrom H, Laitinen T, Kaaja R, Ylikorkala O, Kere J (2003) Susceptibility loci for preeclampsia on chromosomes 2p25 and 9p13 in Finnish families. Am J Hum Genet 72:168–177

Lajemi M, Labat C, Gautier S, Lacolley P, Safar M, Asmar R, Cambien F, Benetos A (2001) Angiotensin II type 1 receptor-153A/G and 1166A/C gene polymorphisms and increase in aortic stiffness with age in hypertensive subjects. J Hypertens 19:407–413

Lalouel JM, Rohrwasser A (2001) Development of genetic hypotheses in essential hypertension. J Hum Genet 46:299–306

Lalouel JM, Rohrwasser A, Terreros D, Morgan T, Ward K (2001) Angiotensinogen in essential hypertension: from genetics to nephrology. J Am Soc Nephrol 12:606–615

Lee YJ, Tsai JC (2001) Angiotensin-converting enzyme gene insertion/deletion, not bradykinin B2 receptor −58T/C gene polymorphism, associated with angiotensin-converting enzyme inhibitor-related cough in Chinese female patients with non-insulin-dependent diabetes mellitus. Metabolism 50:1346–1350

Lifton RP, Jeunemaitre X (1993) Finding genes that cause human hypertension. J Hypertens 11:231–236

Lovati E, Richard A, Frey BM, Frey FJ, Ferrari P (2001) Genetic polymorphisms of the renin–angiotensin-aldosterone system in end-stage renal disease. Kidney Int 60:46–54

Lucius R, Gallinat S, Rosenstiel P, Herdegen T, Sievers J, Unger T (1998) The angiotensin II type 2 (AT2) receptor promotes axonal regeneration in the optic nerve of adult rats. J Exp Med 188:661–670

Martin MM, Elton TS (1995) The sequence and genomic organization of the human type 2 angiotensin II receptor. Biochem Biophys Res Commun 209:554–562

Martin MM, Willardson BM, Burton GF, White CR, McLaughlin JN, Bray SM, Ogilvie JW, Jr., Elton TS (2001) Human angiotensin II type 1 receptor isoforms encoded by messenger RNA splice variants are functionally distinct. Mol Endocrinol 15:281–293

McCole SD, Brown MD, Moore GE, Ferrell RE, Wilund KR, Huberty A, Douglass LW, Hagberg JM (2002) Angiotensinogen M235T polymorphism associates with exercise hemodynamics in postmenopausal women. Physiol Genomics 10:63–69

Miyazaki H, Fukamizu A, Hirose S, Hayashi T, Hori H, Ohkubo H, Nakanishi S, Murakami K (1984) Structure of the human renin gene. Proc Natl Acad Sci USA 81:5999–6003

Montgomery H, Brull D, Humphries SE (2002) Analysis of gene-environment interactions by "stressing-the-genotype" studies: the angiotensin converting enzyme and exercise-induced left ventricular hypertrophy as an example. Ital Heart J 3:10–14

Montgomery H, Clarkson P, Barnard M, Bell J, Brynes A, Dollery C, Hajnal J, Hemingway H, Mercer D, Jarman P, Marshall R, Prasad K, Rayson M, Saeed N, Talmud P, Thomas L, Jubb M, World M, Humphries S (1999) Angiotensin-converting-enzyme gene insertion/deletion polymorphism and response to physical training. Lancet 353:541–545

Montgomery HE, Clarkson P, Dollery CM, Prasad K, Losi MA, Hemingway H, Statters D, Jubb M, Girvain M, Varnava A, World M, Deanfield J, Talmud P, McEwan JR, McKenna WJ, Humphries S (1997) Association of angiotensin-converting enzyme gene I/D polymorphism with change in left ventricular mass in response to physical training. Circulation 96:741–747

Morgan T, Craven C, Nelson L, Lalouel JM, Ward K (1997) Angiotensinogen T235 expression is elevated in decidual spiral arteries. J Clin Invest 100:1406–1415

Morise T, Takeuchi Y, Takeda R (1995) Rapid detection and prevalence of the variants of the angiotensinogen gene in patients with essential hypertension. J Intern Med 237:175–180

Morris BJ, Griffiths LR (1988) Frequency in hypertensives of alleles for a RFLP associated with the renin gene. Biochem Biophys Res Commun 150:219–224

Morris BJ, Zee RY, Schrader AP (1994) Different frequencies of angiotensin-converting enzyme genotypes in older hypertensive individuals. J Clin Invest 94:1085–1089

Moses EK, Lade JA, Guo G, Wilton AN, Grehan M, Freed K, Borg A, Terwilliger JD, North R, Cooper DW, Brennecke SP (2000) A genome scan in families from Australia and New Zealand confirms the presence of a maternal susceptibility locus for pre-eclampsia, on chromosome 2. Am J Hum Genet 67:1581–1585

Naftilan AJ, Williams R, Burt D, Paul M, Pratt RE, Hobart P, Chirgwin J, Dzau VJ (1989) A lack of genetic linkage of renin gene restriction fragment length polymorphisms with human hypertension. Hypertension 14:614–618

Nakajima T, Inoue I, Cheng T, Lalouel JM (2002) Molecular cloning and functional analysis of a factor that binds to the proximal promoter of human angiotensinogen. J Hum Genet 47:7-13

Narayanan CS, Cui Y, Zhao YY, Zhou J, Kumar A (1999) Orphan receptor Arp-1 binds to the nucleotide sequence located between TATA box and transcriptional initiation site of the human angiotensinogen gene and reduces estrogen induced promoter activity. Mol Cell Endocrinol 148:79-86

Nguyen G, Burckle C, Sraer JD (2003) The renin receptor: the facts, the promise and the hope. Curr Opin Nephrol Hypertens 12:51-55

Nguyen G, Delarue F, Burckle C, Bouzhir L, Giller T, Sraer JD (2002) Pivotal role of the renin/prorenin receptor in angiotensin II production and cellular responses to renin. J Clin Invest 109:1417-1427

Nishimura H, Yerkes E, Hohenfellner K, Miyazaki Y, Ma J, Hunley TE, Yoshida H, Ichiki T, Threadgill D, Phillips JA, 3rd, Hogan BM, Fogo A, Brock JW, 3rd, Inagami T, Ichikawa I (1999) Role of the angiotensin type 2 receptor gene in congenital anomalies of the kidney and urinary tract, CAKUT, of mice and men. Mol Cell 3:1-10

Niu T, Xu X, Rogus J, Zhou Y, Chen C, Yang J, Fang Z, Schmitz C, Zhao J, Rao VS, Lindpaintner K (1998) Angiotensinogen gene and hypertension in Chinese. J Clin Invest 101:188-194

O'Byrne S, Caulfield M (1998) Genetics of hypertension. Therapeutic implications. Drugs 56:203-214

O'Donnell CJ, Lindpaintner K, Larson MG, Rao VS, Ordovas JM, Schaefer EJ, Myers RH, Levy D (1998) Evidence for association and genetic linkage of the angiotensin-converting enzyme locus with hypertension and blood pressure in men but not women in the Framingham Heart Study. Circulation 97:1766-1772

Okuyama S, Sakagawa T, Chaki S, Imagawa Y, Ichiki T, Inagami T (1999) Anxiety-like behavior in mice lacking the angiotensin II type-2 receptor. Brain Res 821:150-159

Ono K, Mannami T, Baba S, Yasui N, Ogihara T, Iwai N (2003) Lack of association between angiotensin II type 1 receptor gene polymorphism and hypertension in Japanese. Hypertens Res 26:131-134

Osterop AP, Kofflard MJ, Sandkuijl LA, Ten Cate FJ, Krams R, Schalekamp MA, Danser AH (1998) AT1 receptor A/C1166 polymorphism contributes to cardiac hypertrophy in subjects with hypertrophic cardiomyopathy. Hypertension 32:825-830

O'Toole L, Stewart M, Padfield P, Channer K (1998) Effect of the insertion/deletion polymorphism of the angiotensin- converting enzyme gene on response to angiotensin-converting enzyme inhibitors in patients with heart failure. J Cardiovasc Pharmacol 32:988-994

Paillard F, Chansel D, Brand E, Benetos A, Thomas F, Czekalski S, Ardaillou R, Soubrier F (1999) Genotype-phenotype relationships for the renin–angiotensin-aldosterone system in a normal population. Hypertension 34:423-429

Pei Y, Scholey J, Thai K, Suzuki M, Cattran D (1997) Association of angiotensinogen gene T235 variant with progression of immunoglobin A nephropathy in Caucasian patients. J Clin Invest 100:814-820

Pereira AC, Mota GF, Cunha RS, Herbenhoff FL, Mill JG, Krieger JE (2003) Angiotensinogen 235T allele "dosage" is associated with blood pressure phenotypes. Hypertension 41:25-30

Perola M, Kainulainen K, Pajukanta P, Terwilliger JD, Hiekkalinna T, Ellonen P, Kaprio J, Koskenvuo M, Kontula K, Peltonen L (2000) Genome-wide scan of predisposing loci for increased diastolic blood pressure in Finnish siblings. J Hypertens 18:1579-1585

Poirier O, Georges JL, Ricard S, Arveiler D, Ruidavets JB, Luc G, Evans A, Cambien F, Tiret L (1998) New polymorphisms of the angiotensin II type 1 receptor gene and their associations with myocardial infarction and blood pressure: the ECTIM study. Etude Cas-Temoin de l'Infarctus du Myocarde. J Hypertens 16:1443-1447

Province MA, Boerwinkle E, Chakravarti A, Cooper R, Fornage M, Leppert M, Risch N, Ranade K (2000) Lack of association of the angiotensinogen-6 polymorphism with blood pressure levels in the comprehensive NHLBI Family Blood Pressure Program. National Heart, Lung and Blood Institute. J Hypertens 18:867–876

Rauramaa R, Kuhanen R, Lakka TA, Vaisanen SB, Halonen P, Alen M, Rankinen T, Bouchard C (2002) Physical exercise and blood pressure with reference to the angiotensinogen M235T polymorphism. Physiol Genomics 10:71–77

Reaux A, Fournie-Zaluski MC, David C, Zini S, Roques BP, Corvol P, Llorens-Cortes C (1999) Aminopeptidase A inhibitors as potential central antihypertensive agents. Proc Natl Acad Sci USA 96:13415–13420

Reaux A, Fournie-Zaluski MC, Llorens-Cortes C (2001) Angiotensin III: a central regulator of vasopressin release and blood pressure. Trends Endocrinol Metab 12:157–162

Reich H, Duncan JA, Weinstein J, Cattran DC, Scholey JW, Miller JA (2003) Interactions between gender and the angiotensin type 1 receptor gene polymorphism. Kidney Int 63:1443–1449

Rieder MJ, Nickerson DA (2000) Hypertension and single nucleotide polymorphisms. Curr Hypertens Rep 2:44–49

Rieder MJ, Taylor SL, Clark AG, Nickerson DA (1999) Sequence variation in the human angiotensin converting enzyme. Nat Genet 22:59–62

Rigat B, Hubert C, Alhenc-Gelas F, Cambien F, Corvol P, Soubrier F (1990) An insertion/deletion polymorphism in the angiotensin I-converting enzyme gene accounting for half the variance of serum enzyme levels. J Clin Invest 86:1343–1346

Rioux PP (2000) Clinical trials in pharmacogenetics and pharmacogenomics: methods and applications. Am J Health Syst Pharm 57:887–98; quiz 899–901

Rolfs A, Weber-Rolfs I, Regitz-Zagrosek V, Kallisch H, Riedel K, Fleck E (1994) Genetic polymorphisms of the angiotensin II type 1 (AT1) receptor gene. Eur Heart J 15 Suppl D:108–112

Roses AD (2000) Pharmacogenetics and future drug development and delivery. Lancet 355:1358–1361

Rossi GP, Narkiewicz K, Cesari M, Winnicki M, Bigda J, Chrostowska M, Szczech R, Pawlowski R, Pessina AC (1999) Genetic determinants of plasma ACE and renin activity in young normotensive twins. J Hypertens 17:647–655

Rotimi C, Morrison L, Cooper R, Oyejide C, Effiong E, Ladipo M, Osotemihen B, Ward R (1994) Angiotensinogen gene in human hypertension. Lack of an association of the 235T allele among African Americans. Hypertension 24:591–594

Sato N, Katsuya T, Nakagawa T, Ishikawa K, Fu Y, Asai T, Fukuda M, Suzuki F, Nakamura Y, Higaki J, Ogihara T (2000) Nine polymorphisms of angiotensinogen gene in the susceptibility to essential hypertension. Life Sci 68:259–272

Sato N, Katsuya T, Rakugi H, Takami S, Nakata Y, Miki T, Higaki J, Ogihara T (1997) Association of variants in critical core promoter element of angiotensinogen gene with increased risk of essential hypertension in Japanese. Hypertension 30:321–325

Saye J, Lynch KR, Peach MJ (1990) Changes in angiotensinogen messenger RNA in differentiating 3T3-F442A adipocytes. Hypertension 15:867–871

Schmidt S, van Hooft IM, Grobbee DE, Ganten D, Ritz E (1993) Polymorphism of the angiotensin I converting enzyme gene is apparently not related to high blood pressure: Dutch Hypertension and Offspring Study. J Hypertens 11:345–348

Schmieder RE, Erdmann J, Delles C, Jacobi J, Fleck E, Hilgers K, Regitz-Zagrosek V (2001) Effect of the angiotensin II type 2-receptor gene (+1675 G/A) on left ventricular structure in humans. J Am Coll Cardiol 37:175–182

Schunkert H, Hense HW, Gimenez-Roqueplo AP, Stieber J, Keil U, Riegger GA, Jeunemaitre X (1997) The angiotensinogen T235 variant and the use of antihypertensive drugs in a population-based cohort. Hypertension 29:628–633

Schwartz GL, Turner ST, Chapman AB, Boerwinkle E (2002) Interacting effects of gender and genotype on blood pressure response to hydrochlorothiazide. Kidney Int 62:1718–1723

Sciarrone MT, Stella P, Barlassina C, Manunta P, Lanzani C, Bianchi G, Cusi D (2003) ACE and alpha-adducin polymorphism as markers of individual response to diuretic therapy. Hypertension 41:398–403

Sethi AA, Tybjaerg-Hansen A, Gronholdt ML, Steffensen R, Schnohr P, Nordestgaard BG (2001) Angiotensinogen mutations and risk for ischemic heart disease, myocardial infarction, and ischemic cerebrovascular disease. Six case- control studies from the Copenhagen City Heart Study. Ann Intern Med 134:941–954

Sharma AM, Jeunemaitre X (2000) The future of genetic association studies in hypertension: improving the signal-to-noise ratio. J Hypertens 18:811–814

Sharma P (1998) Meta-analysis of the ACE gene in ischaemic stroke. J Neurol Neurosurg Psychiatry 64:227–230

Smithies O, Kim HS, Takahashi N, Edgell MH (2000) Importance of quantitative genetic variations in the etiology of hypertension. Kidney Int 58:2265–2280

Soubrier F, Alhenc-Gelas F, Hubert C, Allegrini J, John M, Tregear G, Corvol P (1988) Two putative active centers in human angiotensin I-converting enzyme revealed by molecular cloning. Proc Natl Acad Sci USA 85:9386–9390

Soubrier F, Jeunemaitre X, Rigat B, Houot AM, Cambien F, Corvol P (1990) Similar frequencies of renin gene restriction fragment length polymorphisms in hypertensive and normotensive subjects. Hypertension 16:712–717

Spiering W, Kroon AA, Fuss-Lejeune MM, Daemen MJ, de Leeuw PW (2000) Angiotensin II sensitivity is associated with the angiotensin II type 1 receptor A(1166)C polymorphism in essential hypertensives on a high sodium diet. Hypertension 36:411–416

Staessen JA, Kuznetsova T, Wang JG, Emelianov D, Vlietinck R, Fagard R (1999) M235T angiotensinogen gene polymorphism and cardiovascular renal risk. J Hypertens 17:9–17

Staessen JA, Wang JG, Brand E, Barlassina C, Birkenhäger WH, Herrmann SM, Fagard R, Tizzoni L, Bianchi G (2001) Effects of three candidate genes on prevalence and incidence of hypertension in a Caucasian population. J Hypertens 19:1349–1358

Svetkey LP, Moore TJ, Simons-Morton DG, Appel LJ, Bray GA, Sacks FM, Ard JD, Mortensen RM, Mitchell SR, Conlin PR, Kesari M (2001) Angiotensinogen genotype and blood pressure response in the Dietary Approaches to Stop Hypertension (DASH) study. J Hypertens 19:1949–1956

Tiret L, Bonnardeaux A, Poirier O, Ricard S, Marques-Vidal P, Evans A, Arveiler D, Luc G, Kee F, Ducimetiere P, et al (1994) Synergistic effects of angiotensin-converting enzyme and angiotensin-II type 1 receptor gene polymorphisms on risk of myocardial infarction. Lancet 344:910–913

Tiret L, Rigat B, Visvikis S, Breda C, Corvol P, Cambien F, Soubrier F (1992) Evidence, from combined segregation and linkage analysis, that a variant of the angiotensin I-converting enzyme (ACE) gene controls plasma ACE levels. Am J Hum Genet 51:197–205

Todd GP, Chadwick IG, Higgins KS, Yeo WW, Jackson PR, Ramsay LE (1995) Relation between changes in blood pressure and serum ACE activity after a single dose of enalapril and ACE genotype in healthy subjects. Br J Clin Pharmacol 39:131–134

Tsuzuki S, Ichiki T, Nakakubo H, Kitami Y, Guo DF, Shirai H, Inagami T (1994) Molecular cloning and expression of the gene encoding human angiotensin II type 2 receptor. Biochem Biophys Res Commun 200:1449–1454

Ueda S, Meredith PA, Morton JJ, Connell JM, Elliott HL (1998) ACE (I/D) genotype as a predictor of the magnitude and duration of the response to an ACE inhibitor drug (enalaprilat) in humans. Circulation 98:2148–2153

Unger T (1999) The angiotensin type 2 receptor: variations on an enigmatic theme. J Hypertens 17:1775–1786 van Geel PP, Pinto YM, Voors AA, Buikema H, Oosterga M, Crijns HJ, van Gilst WH (2000) Angiotensin II type 1 receptor A1166C gene polymorphism is associated with an increased response to angiotensin II in human arteries. Hypertension 35:717–721

Vervoort VS, Beachem MA, Edwards PS, Ladd S, Miller KE, de Mollerat X, Clarkson K, DuPont B, Schwartz CE, Stevenson RE, Boyd E, Srivastava AK (2002) AGTR2 mutations in X-linked mental retardation. Science 296:2401–2403

Villard E, Lalau JD, van Hooft IS, Derkx FH, Houot AM, Pinet F, Corvol P, Soubrier F (1994) A mutant renin gene in familial elevation of prorenin. J Biol Chem 269:30307–30312

Villard E, Tiret L, Visvikis S, Rakotovao R, Cambien F, Soubrier F (1996) Identification of new polymorphisms of the angiotensin I-converting enzyme (ACE) gene, and study of their relationship to plasma ACE levels by two-QTL segregation-linkage analysis. Am J Hum Genet 58:1268–1278

Vuagnat A, Giacche M, Hopkins PN, Hunt SC, Azizi M, Fisher NDL, Williams GH, Corvol P, Jeunemaitre X (2001) Plasma LDL cholesterol is a strong predictor of the blood pressure increase following an acute infusion of angiotensin II. J Mol Med 79:175–183

Walker WG, Whelton PK, Saito H, Russell RP, Hermann J (1979) Relation between blood pressure and renin, renin substrate, angiotensin II, aldosterone and urinary sodium and potassium in 574 ambulatory subjects. Hypertension 1:287–291

Wang JG, Staessen JA (2000) Genetic polymorphisms in the renin–angiotensin system: relevance for susceptibility to cardiovascular disease. Eur J Pharmacol 410:289–302

Wang WY, Zee RY, Morris BJ (1997) Association of angiotensin II type 1 receptor gene polymorphism with essential hypertension. Clin Genet 51:31–34

Ward K, Hata A, Jeunemaitre X, Helin C, Nelson L, Namikawa C, Farrington PF, Ogasawara M, Suzumori K, Tomoda S, et al (1993) A molecular variant of angiotensinogen associated with preeclampsia. Nat Genet 4:59–61

Watt GC, Harrap SB, Foy CJ, Holton DW, Edwards HV, Davidson HR, Connor JM, Lever AF, Fraser R (1992) Abnormalities of glucocorticoid metabolism and the renin–angiotensin system: a four-corners approach to the identification of genetic determinants of blood pressure. J Hypertens 10:473–482

Woods DR, Humphries SE, Montgomery HE (2000) The ACE I/D polymorphism and human physical performance. Trends Endocrinol Metab 11:416–420

Yoneda A, Cascio S, Green A, Barton D, Puri P (2002) Angiotensin II type 2 receptor gene is not responsible for familial vesicoureteral reflux. J Urol 168:1138–1141

Zee RY, Lou YK, Griffiths LR, Morris BJ (1992) Association of a polymorphism of the angiotensin I-converting enzyme gene with essential hypertension. Biochem Biophys Res Commun 184:9–15

Zee RY, Rao VS, Paster RZ, Sweet CS, Lindpaintner K (1998) Three candidate genes and angiotensin-converting enzyme inhibitor- related cough: a pharmacogenetic analysis. Hypertension 31:925–928

Zhang X, Erdmann J, Regitz-Zagrosek V, Kurzinger S, Hense HW, Schunkert H (2000) Evaluation of three polymorphisms in the promoter region of the angiotensin II type I receptor gene. J Hypertens 18:267–272

Zhang XQ, Varner M, Dizon-Townson D, Song F, Ward K (2003) A molecular variant of angiotensinogen is associated with idiopathic intrauterine growth restriction. Obstet Gynecol 101:237–242

Zhao YY, Zhou J, Narayanan CS, Cui Y, Kumar A (1999) Role of C/A polymorphism at −20 on the expression of human angiotensinogen gene. Hypertension 33:108–115

Knockout Models of the Renin–Angiotensin System

M. Brede · L. Hein

Institut für Pharmakologie und Toxikologie, Universität Würzburg,
Versbacher Strasse 9, 97078 Würzburg, Germany
e-mail: hein@toxi.uni-wuerzburg.de

1	Introduction	208
2	Angiotensinogen	212
3	Renin	213
4	Angiotensin-Converting Enzymes (ACE, ACE-2)	214
5	Angiotensin II AT_1 Receptors	216
5.1	AT_1 Receptors and Cardiac Hypertrophy	216
5.2	Renal AT_1 Receptor Function	217
5.3	Brain AT_1 Receptors	218
6	Angiotensin II AT_2 Receptors	218
6.1	Cardiac AT_2 Receptors	220
6.2	AT_2 Receptors and Brain Function	221
7	Conclusions	222
	References	223

Abstract Angiotensin II, generated from its precursor angiotensinogen by renin and angiotensin-converting enzyme (ACE), mediates its biological effects via two different classes of G protein-coupled receptors, termed angiotensin II AT_1 receptors and AT_2 receptors. Transgenic mouse models have greatly advanced our understanding of the specific functions of the individual parts of the renin angiotensin aldosterone system (RAAS). Recently, all components of the RAAS have been deleted by homologous recombination in the mouse genome. This review summarizes the in vivo significance of the available knockout mouse lines of the RAAS. While most of the classical functions of the RAAS are mediated via AT_1 receptor, recent in vivo evidence suggests that AT_2 receptors may antagonize the cardiovascular effects of AT_1 receptor activation. Most importantly, AT_2 receptors are required to inhibit the growth-promoting effects of AT_1 receptors in vascular smooth muscle cells and in cardiac myocytes. Targeted deletions of angiotensinogen and ACE genes have extended our knowledge of the embryonic functions of the renin angiotensin system. Most recently, a novel angiotensin-converting-enzyme (ACE2) has been cloned and deleted by homologous recombination in mice. Thus, the experimental findings from transgenic animal models offer new insights into the physiological regulation of the RAAS and help in the development of novel therapeutic strategies for treating human diseases.

Keywords Transgenic mice · Gene targeting · Angiotensin receptors · Angiotensin · Converting enzyme · Renin

1
Introduction

Angiotensin II (Ang II) is generated by serial cleavage of angiotensinogen in two enzymatic steps (Fig. 1): First renin cleaves angiotensin I (Ang I) from angiotensinogen, then an angiotensin converting enzyme (ACE) removes two amino acids to generate the active octapeptide angiotensin II. Ang II exerts its effects via two classes of G protein-coupled receptors, termed AT_1 and AT_2 with most of its classical actions mediated via the AT_1 receptor. Two subtypes of AT_1 receptors, AT_{1A} and AT_{1B} have been identified in rodents, but humans have only a single AT_1 receptor gene.

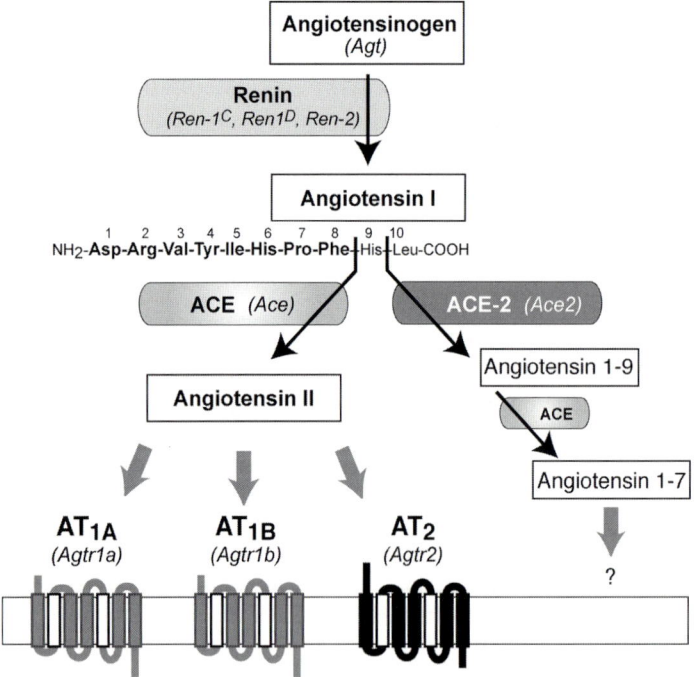

Fig. 1 Components of the renin–angiotensin system in the mouse. Angiotensin II is generated by serial cleavage of angiotensinogen by renin and angiotensin-converting enzyme (ACE). In the mouse, angiotensin II activates three different G protein-coupled receptors, AT_{1A}, AT_{1B}, and AT_2. A homolog of ACE, termed ACE-2, was recently identified to convert angiotensin I to angiotensin 1-9, which is further cleaved to angiotensin 1-7 by ACE. The receptor for angiotensin 1-7 has not been identified yet. Murine gene symbols are given as *italicized abbreviations*

In this chapter, we summarize recent progress in the molecular physiology and pharmacology of the renin–angiotensin system which was derived from mouse models carrying targeted gene deletions ("knockout mice"). Until recently, all components of the renin–angiotensin system which had been identified were targeted for deletion in mice. Mostly, these models are constitutive "knockouts" of the entire gene leading to the complete absence of the respective protein. However, a few modifications of the standard gene-targeting technique were used to generate gene duplications (e.g., 1–4 copies of the angiotensinogen gene; Kim et al. 1995), or tissue- or endothelial-specific deletion of ACE expression (Esther et al. 1996; Cole et al. 2002). The major phenotypes of mice with targeted deletions of the renin–angiotensin system are summarized in Fig. 2 and in Table 1.

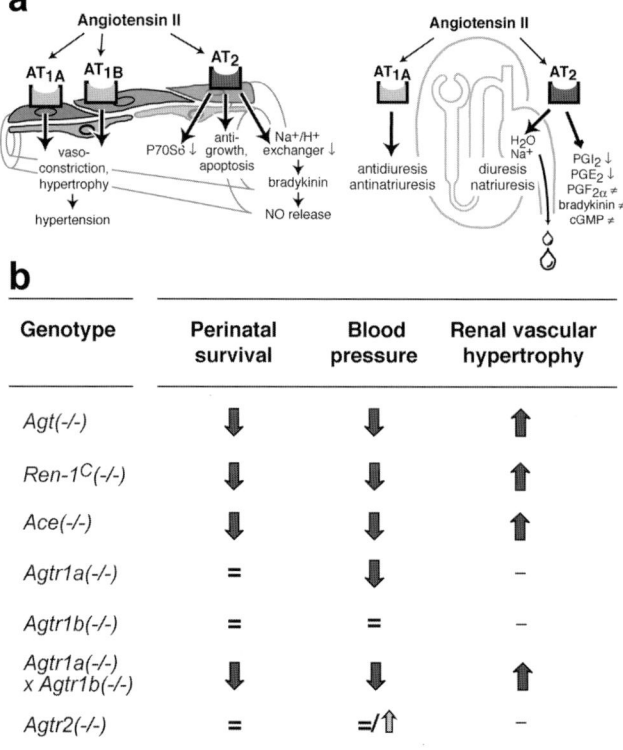

Fig. 2a, b Effects of gene deletion of components of the renin–angiotensin system on survival, blood pressure, and renal development. **a** Vascular and renal functions of AT_1 and AT_2 receptor subtypes derived from studies in gene-targeted mice. **b** Overview of renal, vascular, and mortality phenotypes in mice with deletions in the renin–angiotensin system. While deletion of the genes for angiotensinogen, ACE, and renin-1^C show largely overlapping phenotypes in these systems, the function of angiotensin receptor subtypes are mostly complementary (=, similar to wild-type; –, not detectable). References for the observed phenotypes are given in the text and in Table 1

Table 1 Phenotypes of mice with targeted deletions of the genes of the renin–angiotensin system

Protein	Genetic model	Phenotype	Reference(s)
Angio-tensinogen	Agt(−/−)	Hypotension; kidney: vascular dilatation and hypertrophy, tubular atrophy and interstitial fibrosis; postnatal lethality (before weaning)	Tanimoto et al. 1994; Kim et al. 1995; Niimura et al. 1995
		Diuresis	Kihara et al. 1998
		Ang-independent increase in aldosterone after Na^+ restriction	Okubo et al. 1997
		Delayed repair of injured blood–brain barrier	Kakinuma et al. 1998
		Smaller infarct size after ligation of middle cerebral artery	Maeda et al. 1999
		Decreased metabolic efficiency in Agt(−/−) mice, decreased weight gain after high fat diet	Massiera et al. 2001
	1-2-3-4 Agt copies	8 mmHg blood pressure increase per gene copy	Kim et al. 1995
Renin	Ren-1^D(−/−)	Defect in granule synthesis in juxtaglomerular cells	Clark et al. 1997
	Ren-1^C KO	Hypotension, increased urine and drinking volume, renal defects	Yanai et al. 2000
	Ren-2 KO	Normal development	
ACE	ACE(−/−)	Hypotension, no blood pressure response to Ang I, enhanced hypotension after bradykinin	Krege et al. 1995; Esther et al. 1996
		Developmental defects of the kidney (atrophic tubules, dilated and thickened arteries)	Esther et al. 1996; Hilgers et al. 1997
		Decreased male fertility	Esther et al. 1996
	ACE.2(−/−)	Deletion of membrane anchor of ACE ("tissue ACE KO") →hypotension, renal defects	Esther et al. 1997
		Normocytic anemia	Cole et al. 2000
	ACE.3(−/−)	Switch of transcriptional control to albumin promoter ("endothelial ACE KO")→87-fold increase in hepatic ACE expression, normal blood pressure, normal plasma Ang II	Cole et al. 2002
ACE-2	Ace2(−/−)	Decreased cardiac contractility	Crackower et al. 2002
	Ace(−/−) xAce2(−/−)	Normal cardiac function	Crackower et al. 2002
AT_{1A} receptor	Agtr1a(−/−)	Hypotension	Ito et al. 1995; Sugaya et al. 1995
		Decreased pressure response to Ang II	Oliverio et al. 1997
		Decreased blood pressure response to central Ang II	Davisson et al. 2000
		Reduced effective plasma and extracellular fluid volume	Cervenka et al. 1999
		No cardiac hypertrophy after Ang II infusion	Harada et al. 1998b

Table 1 (continued)

Protein	Genetic model	Phenotype	Reference(s)
	Agtr1a(−/−)	Normal cardiac hypertrophy after aortic constriction	Hamawaki et al. 1998; Harada et al. 1998b
		Normal infarct size after coronary artery ligation, lower incidence of arrhythmias, improved survival, decreased remodeling	Harada et al. 1998a,b; Nakamura et al. 2002
		Normal degree of neointima formation	Harada et al. 1999a
		Attenuated antiglomerular basement membrane nephritis	Hisada et al. 2001
		Blunted renal damage after proteinuria	Suzuki et al. 2001
		Increased activity and drinking periods with intact circadian rhythm	Mistlberger et al. 2001
AT_{1B} receptor	Agtr1b(−/−)	Normal development, normal blood pressure	Chen et al. 1997
		Blunted dipsogenic response to central Ang II	Davisson et al. 2000
	Agtr1a(−/−) xAgtr1b(−/−)	Hypotension, renal defects	Oliverio et al. 1998a; Tsuchida et al. 1998
AT_2 receptor	Agtr2(−/−)	Normal development	Hein et al. 1995; Ichiki et al. 1995
		Low penetrance defect in urinary tract	Nishimura et al. 1999
		Normal blood pressure	Hein et al. 1995
		Moderately increased blood pressure	Ichiki et al. 1995
		Rightward shift in pressure diuresis and natriuresis curves	Gross et al. 2000b
		Hypertension after deoxycorticosterone and salt-loading	Gross et al. 2000a
		Altered baroreflex sensitivity	Gross et al. 2002
		Increased vasopressor effects, vascular hypertrophy, increased vascular P70S6 kinase phosphorylation	Hein et al. 1995; Ichiki et al. 1995; Akishita et al. 1999; Brede et al. 2001
		Normal cardiac function at rest	Brede et al. 2001; Gross et al. 2001
		Normal cardiac hypertrophy after aortic banding, increased coronary artery hypertrophy and perivascular fibrosis	Akishita et al. 2000b
		No cardiac hypertrophy after Ang II infusion or aortic banding	Senbonmatsu et al. 2000; Ichihara et al. 2001
		Normal infarct size and post-infarct remodeling after coronary artery ligation	Xu et al. 2002
		Increased neointima formation	Akishita et al. 2000a; Wu et al. 2001
		Attenuated susceptibility to azoxymethane-induced colon tumorigenesis	Takagi et al. 2002
		Impaired drinking behavior in response to water deprivation, reduced locomotor activity	Hein et al. 1995; Ichiki et al. 1995
		Anxiety-like behavior, increased sensitivity to pain	Okuyama et al. 1999; Sakagawa et al. 2000

2
Angiotensinogen

The first gene of the renin–angiotensin system which was deleted in embryonic stem cells and in mice was the angiotensinogen gene (*Agt*) (Tanimoto et al. 1994; Kim et al. 1995; Niimura et al. 1995). As angiotensinogen is the only (known) substrate for the generation of Ang II, *Agt*(−/−), mice were indeed completely deficient in Ang II. This resulted in a marked hypotension and blood pressure was reduced from 100 mmHg in wild-type mice to 67 mmHg in *Agt*(−/−) mice (Tanimoto et al. 1994). In studies with mice carrying 1–4 copies of the *Agt* gene, blood pressure was increased by approximately 8 mmHg per gene copy with plasma angiotensinogen levels ranging between 35% (single *Agt* gene copy) and 145% (4 copies) of the values detected in normal mice with two *Agt* copies (Kim et al. 1995).

One of the most important initial findings in angiotensinogen-deficient mice was the fact that animals without a functional renin–angiotensin system were born alive and did not show any major developmental defects at birth (Tanimoto et al. 1994; Kim et al. 1995; Niimura et al. 1995). However, within 3 weeks after birth, most of the *Agt*(−/−) mice died and only very few animals survived until adulthood. These adult *Agt*(−/−) mice developed severe vascular abnormalities in the kidney. Maturation of the glomeruli was delayed, interlobular and afferent arterioles had thickened walls, juxtaglomerular cells were hyperplastic, and interstitial fibrosis and inflammation developed together with cortical thinning and tubular atrophy. These structural defects were associated with functional renal impairment and a marked inability of *Agt*(−/−) mice to concentrate urine (Kihara et al. 1998). Histologically, the vascular lesions in kidneys from *Agt*(−/−) mice resembled those found in hypertensive nephrosclerosis, where increased expression of transforming growth factor (TGF)-β was implicated in the pathophysiology. Indeed, the endothelial cells of hypertrophic renal arteries from *Agt*(−/−) mice showed elevated expression of TGF-β, indicating that altered growth factor expression may be responsible for the vascular defects in these mice (Niimura et al. 1995).

However, the precise mechanism of renal vascular hypertrophy in *Agt*(−/−) remains to be elucidated. Normally, vascular hypertrophy is associated with hypertension and results from pressure- or Ang II-mediated proliferation of vascular smooth muscle cells. In contrast, *Agt*(−/−) mice were hypotensive and were completely deficient in Ang II. While medial thickening with concentric hypertrophy is characteristic for hypertensive arteries, *Agt*(−/−) vessels showed non-concentric medial thickening with luminal dilatation, suggesting that renal vascular hypertrophy differs from hypertension-induced hypertrophy (Nagata et al. 1996). It is unclear at present whether the vascular defect is due to a deficiency in Ang II signal transduction or due to decreased blood pressure and lowered renal perfusion. In order to address this question, Matsusaka and colleagues generated chimeric mice which were made up by cells with intact and disrupted genes for the AT$_{1A}$ receptor (Matsusaka et al. 1996). These chimeric mice had normal blood pressure and did not show any signs of hypertrophy in juxtaglomerular cells or afferent arterioles, indicating that the lowered blood pressure may indeed be important for the vascular defects in $Agt(-/-)$ mice.

As angiotensin II is an important stimulus for the release of aldosterone from the adrenal cortex, the concept of an integrated "renin–angiotensin–aldosterone system" has been suggested to indicate that aldosterone mediates the long-term effects of the renin–angiotensin system on blood pressure by retaining Na^+ and water. Studies in $Agt(-/-)$ mice showed that a powerful non-angiotensin mechanism exists (at least in mice) to increase circulating aldosterone levels after Na^+ restriction even in the absence of angiotensin II (Okubo et al. 1997). During Na^+ restriction in mice, suppression of K^+, but not angiotensin, led to a marked attenuation of hyperaldosteronism, suggesting that an angiotensin-independent mechanism exists to increase aldosterone during low salt diet (Okubo et al. 1997).

Several studies were performed in $Agt(-/-)$ mice which extended the previously known functions of the renin–angiotensin system. In the brain, angiotensinogen was required for normal repair of injury to the blood–brain barrier (Kakinuma et al. 1998). In contrast, $Agt(-/-)$ mice had a smaller ischemic core and larger penumbra after surgical induction of a brain infarct (Maeda et al. 1999), indicating that inhibition of angiotensin formation may prolong the therapeutic window for treatment of cerebral infarcts. Furthermore, angiotensinogen seems to be involved in the control of body fat mass by a combination of effects on lipogenesis and locomotor activity (Massiera et al. 2001).

3
Renin

Renin is the rate-limiting enzyme in the biosynthesis of Ang II—the majority of circulating renin is secreted from the juxtaglomerular cells of the kidney. Mice differ from other mammals as some mouse strains have two renin genes, $Ren-1^D$ and $Ren-2$ (129, DBA mice) while other mice have only one renin gene, $Ren-1^C$ (C57BL/6, BALB/c). The second renin gene resulted from a gene duplication on chromosome 1, which presumably occurred 3–10 million years ago, after the speciation of the mouse (Dickinson et al. 1984). The murine renin genes are highly homologous, but they give rise to distinct proteins with different sites for post-translational glycosylation. Glycosylation may be physiologically essential for renin storage in juxtaglomerular cells, as $Ren-1^D(-/-)$ mice did not have electron-dense secretory granules in juxtaglomerular cells despite the synthesis of renin-2 in these cells (Clark et al. 1997).

Mice deficient for renin-1^C (Yanai et al. 2000) displayed the same perinatal mortality, renal abnormalities, and hypotension as those mice lacking angiotensinogen (Tsuchida et al. 1998) or ACE (Krege et al. 1995; Carpenter et al. 1996) (see below). $Ren-1^D(-/-)$ mice showed hypotension in females, whereas blood pressure was normal in male knockout (KO) mice. In contrast, genetic deletion of the $Ren-2$ gene did not result in any obvious phenotype: $Ren-2(-/-)$ mice were viable and healthy and they had normal blood pressure despite an increase in active plasma renin (Sharp et al. 1996). At present, only few data from gene-targeted mice suggest that Ang I may be generated by a non-renin pathway: $Ren\text{-}1^C(-/-)$ mice did not show the impaired blood–brain barrier function which was observed in $Agt(-/-)$ mice, suggesting that renin may be dispensable for the blood–brain barrier function of the renin–angiotensin system (Yanai et al. 2000).

4
Angiotensin-Converting Enzymes (ACE, ACE-2)

Mice deficient in the angiotensin-converting enzyme (ACE) showed the expected hypotension as they failed to generate the vasopressor angiotensin II from endogenous or exogenous Ang I (Krege et al. 1995; Esther et al. 1996). The second function of ACE, i.e., breakdown of bradykinin, was also diminished, as $ACE(-/-)$ mice exhibited an exaggerated hypotensive response to infusion of bradykinin (Krege et al. 1995).

$ACE(-/-)$ mice showed developmental defects of the kidney which were similar to the reported effects of perinatal pharmacologic inhibition of the renin–angiotensin system (Guron and Friberg 2000). In analogy to the $Agt(-/-)$ mice, $ACE(-/-)$ mice had atrophic and dilated kidney tubules, some kidneys were hydronephrotic and all intrarenal arteries including afferent arterioles were dilated and thickened (Hilgers et al. 1997). The defect of $ACE(-/-)$ mice to concentrate urine may not only be related to structural defects of the kidney, but may also be caused by altered expression of renal transporter proteins: in mice lacking ACE, decreased expression of the aquaporin APQ1 water channel, the 117-kDa urea transporter UT-A1, and the $Na^+\text{-}K^+\text{-}2Cl^-$ cotransporter could be detected (Klein et al. 2002).

Surprisingly, male $ACE(-/-)$ mice had reduced fertility (Carpenter et al. 1996; Esther et al. 1996) which is in contrast to the observation in angiotensinogen-deficient mice, which showed normal reproductive behavior. Male mice deficient in somatic ACE retain testis ACE and are fertile, thus demonstrating that somatic ACE is not essential for normal male fertility (Hagaman et al. 1998). These results suggest that testicular ACE may process a non-angiotensin substrate which is necessary for reproduction.

ACE is a membrane-bound enzyme, which may be released into the interstitial fluid or into the circulation after proteolytic cleavage. In order to investigate the physiological significance of tissue-bound versus circulating ACE, mice carrying a selective deletion in its membrane anchoring domain have been generated to eliminate all tissue-bound ACE. These mice were termed "$ACE.2(-/-)$" mice (Esther et al. 1997). Surprisingly, these mice developed the same defects (reduced male fertility, hypotension, inability to concentrate urine) as mice with a complete deletion of ACE (Esther et al. 1997). In these "tissue ACE KO" mice, circulating ACE levels were 34% of normal. Thus, tissue-ACE is essential for maintenance of blood pressure and for renal structure and development. In addition, *ACE.2(−/−)* mice had normocytic anemia with elevated plasma erythropoietin levels, indicating that Ang II stimulates erythropoiesis (Cole et al. 2000). These data may explain clinical observations that ACE inhibitors may reduce hematocrit values in some patients.

Another mouse strain, termed "*ACE.3(−/−)*", was generated to substitute transcriptional control of somatic ACE transcription from the ACE promoter to the albumin promoter (Cole et al. 2002). Switching the promoter in *ACE.3(−/−)* mice resulted in complete absence of ACE in the lung, aorta, or any vasculature ("endothelial ACE KO"). However, the albumin promoter caused a 87-fold increase in hepatic ACE expression. Surprisingly, blood pressure, plasma Ang II levels and renal structure and function were normal in these mice. Usually one would assume that the lung contains enough ACE to entirely convert blood Ang I to Ang II during a single transit (Ng and Vane 1967), but the data from *ACE.3(−/−)* mice suggest that endothelial ACE is not essential to maintain normal blood pressure. Heterologous expression of ACE in the liver is sufficient to substitute for endothelial ACE. Subjecting this mouse strain to models of cardiovascular injury should be an important step to delineate the role of endothelial ACE.

Among the greatest surprises of recent research in the renin–angiotensin system was the identification of a novel homolog of the angiotensin-converting enzyme, ACE-2 (Donoghue et al. 2000; Tipnis et al. 2000). This homolog of ACE was first cloned from human heart failure ventricle and from lymphoma cDNA libraries. ACE-2 cleaves the C-terminal Leu from Ang I to generate Ang 1-9, which is further converted to Ang 1-7 by "classical" ACE. As outlined in Fig. 1, ACE-2 differs from ACE as it does not generate Ang II or degrade bradykinin. ACE-2 is not inhibited by classical ACE-inhibitors, including captopril and lisinopril. ACE-2 shares 40% amino acid identity and 61% similarity with human ACE (Tipnis et al. 2000), thus similar to renin-2, ACE-2 may be the result of a gene duplication.

When compared with the widespread expression pattern of ACE, ACE-2 is much more restricted in its tissue distribution with primary sites of expression in endothelial cells of coronary and intrarenal vessels, in renal tubules, and in the testis. Until very recently, the function of ACE-2 was completely unknown. In an elegant series of experiments, Crackower and colleagues demonstrated that ACE-2 is essential for the development of cardiac function (Crackower et al. 2002).

The pathophysiological relevance of ACE-2 is highlighted by the fact that the rat ACE-2 gene was mapped to a region of the X chromosome that has previously been implicated in several rat models of high blood pressure (Crackower et al. 2002). Mice with a targeted deletion in their *Ace2* gene had normal blood pressure, but cardiac contractility was severely depressed and circulating Ang II levels were elevated. The precise mechanism of this phenotype remains unclear at present. However, crossing of *Ace2(−/−)* mice with mice lacking "classical" ACE rescued the cardiac phenotype, indicating that altered Ang II generation may be responsible for the pathomechanism in *Ace2(−/−)* mice.

5
Angiotensin II AT_1 Receptors

At present, two types of G protein-coupled receptors have been cloned which mediate the biological effects of angiotensin II, type 1 receptors (AT_1) and type 2 receptors (AT_2). Whereas most species, including humans, have only one AT_1 receptor gene, two AT_1 genes which arose by gene duplication (*Agtr1a, Agtr1b*) have been identified in mice and rats, termed AT_{1A} and AT_{1B} receptors. All three angiotensin receptor genes were deleted in mice, revealing separate and distinct functions for these receptors.

Most of the classical functions of Ang II, vasoconstriction, hypertension, aldosterone release, and cardiovascular hypertrophy, are mediated by AT_1 receptors in mice. These functions are predominantly controlled by the AT_{1A} receptor. Mice deficient in AT_{1A} had significantly lower resting blood pressure and a blunted pressure response to infusion of Ang II (Ito et al. 1995; Sugaya et al. 1995). However, AT_{1B} receptors contribute to blood pressure control as Ang II infusion could still elicit moderate hypertensive responses in *Agtr1a(−/−)* mice (Oliverio et al. 1997). In contrast, mice lacking AT_{1B} receptors had normal blood pressure at rest and a normal response to Ang II, indicating that the AT_{1A} receptor is the major receptor involved in blood pressure regulation (Chen et al. 1997).

AT_{1A} and AT_{1B} receptors have similar signal transduction pathways as Ang II induced transient increases in intracellular Ca^{2+} concentration in vascular smooth muscle cells cultured from AT_{1A}-deficient mice (Zhu et al. 1998). However, both AT_1 receptor subtypes may have distinct physiological functions, as the blood pressure response to centrally administered Ang II was completely mediated by AT_{1A} receptors, whereas the dipsogenic response to central Ang II was mediated via the AT_{1B} receptor (Davisson et al. 2000). Mice lacking both AT_{1A} and AT_{1B} receptors showed greater lowering of blood pressure than single receptor-deficient mice (Oliverio et al. 1998a; Tsuchida et al. 1998). The mechanism of hypotension in AT_1 receptor-deficient mice may be related to a reduction in effective plasma and extracellular fluid volume rather than a decrease in peripheral vascular resistance, as arterial hypotension could be rapidly restored in *Agtr1a(−/−)* mice by infusion of isotonic saline (Cervenka et al. 1999).

5.1
AT_1 Receptors and Cardiac Hypertrophy

Cardiac AT_1 receptors play an important role in the regulation of left ventricular contractility and hypertrophic remodeling. In the murine heart, AT_{1A} receptors predominate over the AT_{1B} subtype. In *Agtr1a(−/−)* mice, only very little AT_1 receptor mRNA and Ang II binding sites remained, probably reflecting a very low density of AT_{1B} receptors (Harada et al. 1998b). Infusion of Ang II did not induce *c-fos* gene expression (Harada et al. 1998c) or cardiac hypertrophy in *Agtr1a(−/−)* mice (Harada et al. 1998b). In contrast, *Agtr1a(−/−)* mice still developed cardiac hypertrophy after left ventricular pressure overload, indicating that AT_{1A}-mediated signaling is not essential for the development of pressure overload-induced hypertrophy (Hamawaki et al. 1998; Harada et al. 1998b). A similar response was found in isolated cardiac myocytes from AT_{1A}-deficient mice in vitro, which showed a normal degree of hypertrophy to mechanical stretch (Kudoh et al. 1998). However, it remains to be determined whether cardiac AT_{1B} receptors may compensate for the loss of AT_{1A} receptors to induce left ventricular hypertrophy.

In contrast to the situation after cardiac pressure overload, AT_{1A} receptors played a role in early and late adaptation to experimental myocardial infarction in mice. While the infarct size did not differ between $Agtr1a(-/-)$ and control mice after occlusion of the left coronary artery, the incidence of premature arrhythmic beats was significantly reduced during the reperfusion phase in $Agtr1a(-/-)$ mice (Harada et al. 1998a) and survival was also greater in $Agtr1a(-/-)$, mice which was attributed to improved systolic and diastolic function and decreased left ventricular remodeling (Harada et al. 1999b; Nakamura et al. 2002). Similar to the situation in the heart, the AT_{1A} receptor by itself was not essential for neointima formation after vascular injury of the common carotid artery (Harada et al. 1999a).

5.2
Renal AT_1 Receptor Function

$Agtr1a(-/-)$ mice exhibit hydronephrosis due to defective development of the urinary peristaltic machinery during the perinatal period (Miyazaki and Ichikawa 2001). Mice which were lacking AT_{1A} and AT_{1B} receptors showed impaired growth and marked renal abnormalities (Oliverio et al. 1998a; Tsuchida et al. 1998) which were similar to the alterations observed in $Ace(-/-)$ and $Agt(-/-)$ mice (see above). Despite the hypotension of $Agtr1a(-/-)$ mice, renal blood flow and perfusion was normal in these mice and residual renal vasoconstriction could be induced by infusion of Ang II (Ruan et al. 1999). The kidneys of $Agtr1a(-/-)$ mice developed normally and showed only mild signs of mesangial expansion and juxtaglomerular cell hypertrophy (Oliverio et al. 1998b). Micropuncture experiments revealed that the tubuloglomerular feedback response was completely absent in $Agtr1a(-/-)$ mice (Schnermann et al. 1997). AT_{1A} receptors may also play an important role in the control of expression of renal tubular transporters and channels (Brooks et al. 2002). In experimental models of renal injury, deletion of the AT_{1A} receptor attenuated the development of antiglomerular basement membrane nephritis (Hisada et al. 2001) and blunted the tubulointerstitial damage after chronic proteinuria (Suzuki et al. 2001).

5.3
Brain AT_1 Receptors

Several functions of Ang II in the central nervous system were investigated in receptor-deficient mice. Deletion of the AT_{1A} receptor attenuated the increase in plasma vasopressin levels after dehydration (Morris et al. 2001), indicating that AT_{1A} receptors are involved in the central regulation of fluid and electrolyte balance.

In the mouse, AT_1 receptors are normally expressed in the suprachiasmatic nucleus, the essential regulator of circadian rhythmicity. Mean daily activity and drinking periods were increased in $Agtr1a(-/-)$ mice, but their overall circadian rhythm remained intact (Mistlberger et al. 2001).

Occlusion of the middle cerebral artery was used to induce focal cerebral ischemia. In this model, $Agtr1a(-/-)$ mice showed a smaller lesion area of energy failure and a significantly larger penumbra (Walther et al. 2002). Cultured $Agtr1a(-/-)$ neurons were more susceptible to cell damage in vitro than wild-type neurons (Walther et al. 2002).

6
Angiotensin II AT_2 Receptors

Despite high levels of AT_2 receptor expression in the developing embryo, genetic deletion of the AT_2 receptor gene was not associated with gross morphological defects or decreased survival (Hein et al. 1995; Ichiki et al. 1995). However, the development of the kidney and urinary tract showed structural defects with very low penetrance in AT_2 KO mice, resembling a human congenital anomaly of the kidney and urinary tract (CAKUT syndrome) (Nishimura et al. 1999).

Some dispute remains with respect to the significance of AT_2 receptors for blood pressure regulation and cardiovascular hypertrophy under pathological situations. While initial reports showed that one strain of $Agtr2(-/-)$ mice had normal blood pressure (Hein et al. 1995) the other strain had slightly elevated blood pressure at rest (Ichiki et al. 1995). This discrepancy has not been resolved yet, but it may be related to the fact that the two $Agtr2(-/-)$ strains were generated on different genetic backgrounds [FVB/N (Hein et al. 1995) and C57BL/6 (Ichiki et al. 1995)]. While in the initial publications variable genetic contributions of the 129 strain—from the embryonic stem cells used to generate the $Agtr2$ gene deletion—were present, later studies employed mice which were crossed back onto FVB/N or C57BL/6 backgrounds (Ichiki et al. 1995; Siragy et al. 1999a; Gross et al. 2000a; Brede et al. 2001; Ichihara et al. 2001). In addition to genetic backgrounds, environmental conditions may affect the blood pressure response in AT_2-deficient mice. When blood pressure was recorded by telemetry, Gross and coworkers did not detect differences in initial blood pressure between wild-type and $Agtr2(-/-)$ mice (Gross et al. 2002). However, after several cycles of low- and high-salt diet followed by administration of deoxycorticosterone (DOCA), blood pressure remained higher in $Agtr2(-/-)$ mice after returning to a normal Na^+ chow. The importance of renal AT_2 receptors was demonstrated by the fact that pressure diuresis and natriuresis curves were shifted rightward in $Agtr2(-/-)$ mice. Thus at similar renal perfusion pressures, normal mice excreted three times more Na^+ and water than mice lacking AT_2 receptors (Gross et al. 2000b). In addition, $Agtr2(-/-)$ mice were more susceptible to develop hypertension after DOCA-salt loading than normal mice (Gross et al. 2000a). In addition to the renal effects of AT_2, altered baroreceptor reflex sensitivity may be important for the cardiovascular function of $Agtr2(-/-)$ mice (Gross et al. 2002a).

One remarkable finding in $Agtr2(-/-)$ mice was the increased sensitivity to the vasopressor effects of Ang II (Hein et al. 1995; Ichiki et al. 1995). Isolated aortic ring segments as well as muscular arteries from $Agtr2(-/-)$ mice showed increased vasoconstrictor responses to Ang II (Fig. 3), norepinephrine and K^+ depolarization (Akishita et al. 1999; Brede et al. 2001). Some reports have sug-

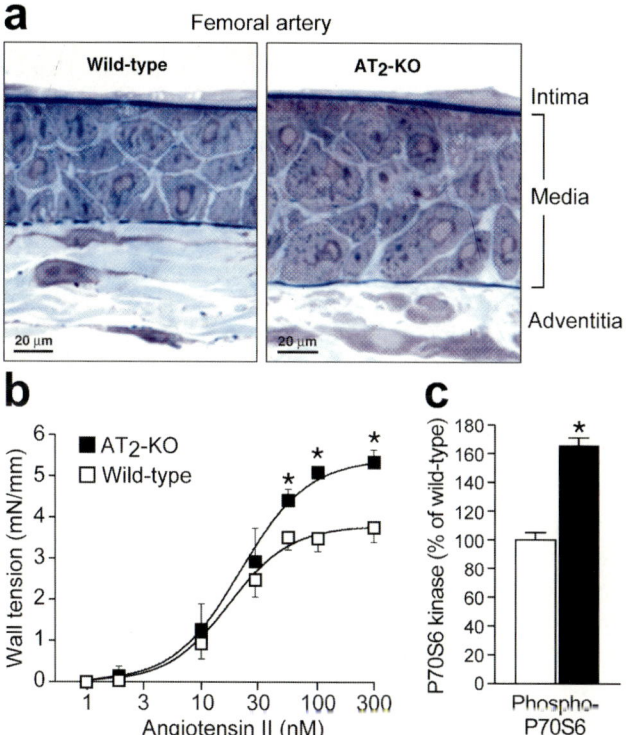

Fig. 3a–c Vascular hypertrophy and increased P70S6 kinase in AT_2 receptor-deficient mice. **a** Longitudinal sections through femoral arteries show enlarged vascular smooth muscle cells and thickened media in specimens from AT_2-deficient mice as compared with wild-type control. **b, c** Maximal vasoconstriction induced by angiotensin II (**b**) and abundance of the phosphorylated (i.e., active) form of the ribosomal P70S6 kinase (**c**) were significantly greater in isolated artery segments from AT_2-deficient mice then from wild-type animals. (Reproduced with permission from Brede at al. 2001)

gested that upregulation of vascular AT_1 receptors (Tanaka et al. 1999) or increased levels of circulating ACE (Hunley et al. 2000) may be responsible for the enhanced vascular reactivity of *Agtr2(−/−)* mice. Histological analyses of different blood vessel types demonstrated that *Agtr2(−/−)* mice developed media smooth muscle cell hypertrophy which was associated with increased phosphorylation of the ribosomal P70S6 kinase (Brede et al. 2001). These findings are in line with the observation that AT_2 activation promotes vascular differentiation during development and contributes to vasculogenesis (Yamada et al. 1999). Additional studies suggest that AT_2 may counteract the hypertrophic effects of AT_1 receptor activation in the vasculature in vivo: After aortic constriction, *Agtr2(−/−)* mice developed greater coronary arterial thickening and perivascular fibrosis than wild-type control mice (Akishita et al. 2000b). After vascular injury, the neointimal lesion and smooth muscle cell proliferation were greater in *Agtr2(−/−)* mice compared with normal mice (Akishita et al. 2000a; Wu et al. 2001). Treatment with an AT_1 receptor antagonist decreased neointima formation to a smaller extent in *Agtr2(−/−)* than in control mice, indicating that AT_2 receptor activation after pharmacological AT_1 receptor blockade is important for the beneficial effect of the AT_1 antagonist on neointima growth (Wu et al. 2001).

In blood vessels, murine AT_2 receptors may negatively regulate ischemia-induced angiogenesis through activation of apoptotic processes (Silvestre et al. 2002). In a mouse model of surgically induced hindlimb ischemia (femoral artery ligation) to study angiogenesis, microangiographic vessel density and laser Doppler perfusion showed significant improvement in *Agtr2(−/−)* mice when compared with wild-type mice. AT_2 receptor activation also induced apoptosis in other cell/tissue types, including embryonic fibroblasts (Li et al. 1998) and vascular smooth muscle cells after intima injury of the femoral artery (Suzuki et al. 2002).

6.1
Cardiac AT_2 Receptors

Cardiac function under normal resting conditions was not affected by deletion of the AT_2 receptor gene (Brede et al. 2001; Gross et al. 2001). The contribution of AT_2 receptor signaling to cardiac hypertrophy remains a matter of debate, as some conflicting data have been reported recently. Whereas pressure overload or chronic infusion of Ang II failed to induce left ventricular hypertrophy in one *Agtr2(−/−)* mouse strain (Senbonmatsu et al. 2000; Ichihara et al. 2001), another report did not find any effect of *Agtr2* gene deletion on cardiac hypertrophy after aortic constriction (Akishita et al. 2000b).

After experimental myocardial infarction by ligation of the left coronary artery, infarct size and postinfarct remodeling did not differ between control and AT_2-deficient mice (Xu et al. 2002). In contrast to wild-type mice, *Agtr2(−/−)* mice did not benefit from treatment with an AT_1 receptor antagonist after myocardial infarction. These data demonstrate that—at least in mice—AT_2 receptors play an important role in the therapeutic effect of AT_1 receptor antagonists (Xu et al. 2002). Compensatory changes may occur in the absence of AT_2 receptors, e.g., upregulation of AT_1 receptor gene expression has been reported in Agtr2(−/−) hearts (Gross et al. 2001), in the adrenal medulla and zona glomerulosa (Saavedra et al. 2001a), in kidney glomeruli (Saavedra et al. 2001b), and in arteries (Tanaka et al. 1999).

In Agtr2(−/−) mice, infusion of angiotensin II for 7 days produced a marked and sustained increase in systolic blood pressure and a significant reduction in urinary sodium excretion compared with control mice (Siragy et al. 1999a). Agtr2(−/−) mice had low basal levels of renal interstitial fluid bradykinin and cyclic guanosine monophosphate (cGMP), an index of nitric oxide production, compared with wild-type mice (Siragy et al. 1999a). In addition, Agtr2(−/−) mice had increased renal vasodilator prostanoids, including higher renal interstitial fluid 6-keto-prostaglandin $(PG)F_{1\alpha}$ (a stable hydrolysis product of prostacyclin) and PGE_2 levels than did wild-type mice (Siragy et al. 1999b). Treatment of mice with indomethacin increased blood pressure to hypertensive levels in Agtr2(−/−) mice but was without effect in control mice (Siragy et al. 1999b). Absence of the AT_2 receptor led to vascular and renal hypersensitivity to angiotensin II, including sustained antinatriuresis and hypertension. These results strongly suggest that the AT_2 receptor plays a counterregulatory protective role mediated via bradykinin and nitric oxide against the antinatriuretic and pressor actions of angiotensin II.

Agtr2(−/−) mice were also used to investigate tumor growth and metastasis. Surprisingly, mice lacking AT_2 receptors are less susceptible to develop colon tumors after exposure to azoxymethane (Takagi et al. 2002). Azoxymethane is activated to a toxic metabolite in the liver by the cytochrome P450 isoenzyme CYP2E1. While basal CYP2E1 levels were similar in wild-type and in Agtr2(−/−) mice, azoxymethane caused more robust downregulation in wild-type mice than in AT_2-deficient animals.

6.2
AT_2 Receptors and Brain Function

Both angiotensin receptor subtypes may also have distinct roles in structural development of the central nervous system. While Agtr2(−/−) mice had a higher cell number in certain brain areas (cortex, hippocampus, amygdala, thalamus), Agtr1a(−/−) animals had a lower cell number in other brain regions (lateral geniculate and medial amygdaloid nucleus) (von Bohlen und Halbach et al. 2001). The functional relevance of these findings is unclear at present. However, subtle differences in several behavioral parameters were observed between Agtr2(−/−) and wild-type mice: Agtr2(−/−) animals had impaired drinking behavior in response to water deprivation (Hein et al. 1995), they showed reduced exploratory behavior and locomotor activity (Hein et al. 1995; Ichiki et al. 1995), they exhibited anxiety-like behavior (Okuyama et al. 1999), and were more sensitive to pain stimuli (Sakagawa et al. 2000).

7
Conclusions

Targeted deletion of the genes of the renin–angiotensin system has provided substantial novel information about the physiological function and pathophysiological significance of the individual components of this system. While some targeting experiments have confirmed previous data and sometimes long-established concepts (e.g., hypotension in ACE- or angiotensinogen-deficient mice), other results were quite unexpected [e.g., increased pressure sensitivity of AT_2-deficient mice, cardiac dysfunction of $Ace2(-/-)$ mice]. Experimental evidence from gene-targeted mice has helped to identify that the two principal angiotensin receptor types, AT_1 and AT_2, have opposing roles in the control of cardiovascular growth and hypertrophy. Furthermore, data from mice have even provided evidence that AT_2 receptors mediated part of the beneficial therapeutic effect of pharmacological AT_1 receptor antagonists.

The power of mouse molecular genetics was also highlighted after the unexpected cloning of the homolog ACE-2 of the angiotensin-converting enzyme. As classical ACE inhibitors did not influence enzymatic activity of the newly identified ACE-2 enzyme, targeted deletion of the mouse gene encoding for ACE-2 was the most straightforward approach to rapidly identify the physiological relevance of this enzyme.

To date, only mouse models with constitutive gene deletions of the renin–angiotensin system have been reported. As the genes are not only inactive at the time of the experiments but throughout embryonic and postnatal development until the experimental time point, mechanisms of genetic and functional compensation have to be considered. Indeed, several reports have demonstrated that AT_1 receptor expression is upregulated in AT_2-deficient mice. In this situation, careful experiments are required to distinguish whether the observed phenotypes are due to AT_2 receptor deletion or whether they are mediated via enhanced AT_1 receptor signaling.

Despite the fact that genetic mouse models have unraveled novel concepts about the renin angiotensin system, many question remain unanswered at present. The pathophysiological role of AT_2 receptors and of ACE-2 warrants further study, the receptor(s) required to mediate the biological, angiotensin-dependent functions of ACE-2 are still unknown, and phenotypes of gene-targeted mice in other organ systems (kidney, brain, skin, etc.) may guide the development of novel therapeutic strategies.

Acknowledgements. The authors own work was supported by the Deutsche Forschungsgemeinschaft (SFBs 355 and 487).

References

Akishita M, Horiuchi M, Yamada H, Zhang L, Shirakami G, Tamura K, Ouchi Y, Dzau VJ (2000a) Inflammation influences vascular remodeling through AT2 receptor expression and signaling. Physiol Genomics 2:13–20

Akishita M, Iwai M, Wu L, Zhang L, Ouchi Y, Dzau VJ, Horiuchi M (2000b) Inhibitory effect of angiotensin II type 2 receptor on coronary arterial remodeling after aortic banding in mice. Circulation 102:1684–1689

Akishita M, Yamada H, Dzau VJ, Horiuchi M (1999) Increased vasoconstrictor response of the mouse lacking angiotensin II type 2 receptor. Biochem Biophys Res Commun 261:345–349

Brede M, Hadamek K, Meinel L, Wiesmann F, Peters J, Engelhardt S, Simm A, Haase A, Lohse MJ, Hein L (2001) Vascular hypertrophy and increased P70S6 kinase in mice lacking the angiotensin II AT(2) receptor. Circulation 104:2602–2607

Brooks HL, Allred AJ, Beutler KT, Coffman TM, Knepper MA (2002) Targeted proteomic profiling of renal Na(+) transporter and channel abundances in angiotensin II type 1a receptor knockout mice. Hypertension 39:470–473

Carpenter C, Honkanen AA, Mashimo H, Goss KA, Huang P, Fishman MC, Asaad M, Dorso CR, Cheung H (1996) Renal abnormalities in mutant mice. Nature 380:292

Cervenka L, Mitchell KD, Oliverio MI, Coffman TM, Navar LG (1999) Renal function in the AT1A receptor knockout mouse during normal and volume-expanded conditions. Kidney Int 56:1855–1862

Chen X, Li W, Yoshida H, Tsuchida S, Nishimura H, Takemoto F, Okubo S, Fogo A, Matsusaka T, Ichikawa I (1997) Targeting deletion of angiotensin type 1B receptor gene in the mouse. Am J Physiol 272:F299–304

Clark AF, Sharp MG, Morley SD, Fleming S, Peters J, Mullins JJ (1997) Renin-1 is essential for normal renal juxtaglomerular cell granulation and macula densa morphology. J Biol Chem 272:18185–18190

Cole J, Ertoy D, Lin H, Sutliff RL, Ezan E, Guyene TT, Capecchi M, Corvol P, Bernstein KE (2000) Lack of angiotensin II-facilitated erythropoiesis causes anemia in angiotensin-converting enzyme-deficient mice. J Clin Invest 106:1391–1398

Cole J, Quach du L, Sundaram K, Corvol P, Capecchi MR, Bernstein KE (2002) Mice lacking endothelial angiotensin-converting enzyme have a normal blood pressure. Circ Res 90:87–92

Crackower MA, et al (2002) Angiotensin-converting enzyme 2 is an essential regulator of heart function. Nature 417:822–828

Davisson RL, Oliverio MI, Coffman TM, Sigmund CD (2000) Divergent functions of angiotensin II receptor isoforms in the brain. J Clin Invest 106:103–106

Dickinson DP, Gross KW, Piccini N, Wilson CM (1984) Evolution and variation of renin genes in mice. Genetics 108:651–667

Donoghue M, Hsieh F, Baronas E, Godboit K, Gosselin M, Stagliano N, Donovan M, Woolf B, Robison K, Jeyaseelan R, Breitbart RE, Acton S (2000) A novel angiotensin-converting enzyme-related carboxypeptidase (ACE2) converts angiotensin I to angiotensin 1–9. Circ Res 87:e1-e9

Esther CR, Jr., Howard TE, Marino EM, Goddard JM, Capecchi MR, Bernstein KE (1996) Mice lacking angiotensin-converting enzyme have low blood pressure, renal pathology, and reduced male fertility. Lab Invest 74:953–965

Esther CR, Marino EM, Howard TE, Machaud A, Corvol P, Capecchi MR, Bernstein KE (1997) The critical role of tissue angiotensin-converting enzyme as revealed by gene targeting in mice. J Clin Invest 99:2375–2385

Gross V, Milia AF, Plehm R, Inagami T, Luft FC (2000a) Long-term blood pressure telemetry in AT2 receptor-disrupted mice. J Hypertens 18:955–961

Gross V, Plehm R, Tank J, Jordan J, Diedrich A, Obst M, Luft FC (2002) Heart rate variability and baroreflex function in AT2 receptor- disrupted mice. Hypertension 40:207–213

Gross V, Schunck WH, Honeck H, Milia AF, Kargel E, Walther T, Bader M, Inagami T, Schneider W, Luft FC (2000b) Inhibition of pressure natriuresis in mice lacking the AT2 receptor. Kidney Int 57:191–202

Gross V, Walther T, Milia AF, Walter K, Schneider W, Luft FC (2001) Left ventricular function in mice lacking the AT2 receptor. J Hypertens 19:967–976

Guron G, Friberg P (2000) An intact renin–angiotensin system is a prerequisite for normal renal development. J Hypertens 18:123–137

Hagaman JR, Moyer JS, Bachman ES, Sibony M, Magyar PL, Welch JE, Smithies O, Krege JH, O'Brien DA (1998) Angiotensin-converting enzyme and male fertility. Proc Natl Acad Sci USA 95:2552–2557

Hamawaki M, Coffman TM, Lashus A, Koide M, Zile MR, Oliverio MI, DeFreyte G, Cooper Gt, Carabello BA (1998) Pressure-overload hypertrophy is unabated in mice devoid of AT1A receptors. Am J Physiol 274:H868–873

Harada K, Komuro I, Hayashi D, Sugaya T, Murakami K, Yazaki Y (1998a) Angiotensin II type 1a receptor is involved in the occurrence of reperfusion arrhythmias. Circulation 97:315–317

Harada K, Komuro I, Shiojima I, Hayashi D, Kudoh S, Mizuno T, Kijima K, Matsubara H, Sugaya T, Murakami K, Yazaki Y (1998b) Pressure overload induces cardiac hypertrophy in angiotensin II type 1A receptor knockout mice. Circulation 97:1952–1959

Harada K, Komuro I, Sugaya T, Murakami K, Yazaki Y (1999a) Vascular injury causes neointimal formation in angiotensin II type 1a receptor knockout mice. Circ Res 84:179–185

Harada K, Komuro I, Zou Y, Kudoh S, Kijima K, Matsubara H, Sugaya T, Murakami K, Yazaki Y (1998c) Acute pressure overload could induce hypertrophic responses in the heart of angiotensin II type 1a knockout mice. Circ Res 82:779–785

Harada K, Sugaya T, Murakami K, Yazaki Y, Komuro I (1999b) Angiotensin II type 1A receptor knockout mice display less left ventricular remodeling and improved survival after myocardial infarction. Circulation 100:2093–2099

Hein L, Barsh GS, Pratt RE, Dzau VJ, Kobilka BK (1995) Behavioural and cardiovascular effects of disrupting the angiotensin II type-2 receptor in mice. Nature 377:744–747

Hilgers KF, Reddi V, Krege JH, Smithies O, Gomez RA (1997) Aberrant renal vascular morphology and renin expression in mutant mice lacking angiotensin-converting enzyme. Hypertension 29:216–221

Hisada Y, Sugaya T, Tanaka S, Suzuki Y, Ra C, Kimura K, Fukamizu A (2001) An essential role of angiotensin II receptor type 1a in recipient kidney, not in transplanted peripheral blood leukocytes, in progressive immune-mediated renal injury. Lab Invest 81:1243–1251

Hunley TE, Tamura M, Stoneking BJ, Nishimura H, Ichiki T, Inagami T, Kon V (2000) The angiotensin type II receptor tonically inhibits angiotensin- converting enzyme in AT2 null mutant mice. Kidney Int 57:570–577

Ichihara S, Senbonmatsu T, Price E, Jr., Ichiki T, Gaffney FA, Inagami T (2001) Angiotensin II type 2 receptor is essential for left ventricular hypertrophy and cardiac fibrosis in chronic angiotensin II-induced hypertension. Circulation 104:346–351

Ichiki T, Labosky PA, Shiota C, Okuyama S, Imagawa Y, Fogo A, Niimura F, Ichikawa I, Hogan BL, Inagami T (1995) Effects on blood pressure and exploratory behaviour of mice lacking angiotensin II type-2 receptor. Nature 377:748–750

Ito M, Oliverio MI, Mannon PJ, Best CF, Maeda N, Smithies O, Coffman TM (1995) Regulation of blood pressure by the type 1A angiotensin II receptor gene. Proc Natl Acad Sci USA 92:3521–3525

Kakinuma Y, Hama H, Sugiyama F, Yagami K, Goto K, Murakami K, Fukamizu A (1998) Impaired blood-brain barrier function in angiotensinogen-deficient mice. Nat Med 4:1078–1080

Kihara M, Umemura S, Sumida Y, Yokoyama N, Yabana M, Nyui N, Tamura K, Murakami K, Fukamizu A, Ishii M (1998) Genetic deficiency of angiotensinogen produces an impaired urine concentrating ability in mice. Kidney Int 53:548–555

Kim HS, Krege JH, Kluckman KD, Hagaman JR, Hodgin JB, Best CF, Jennette JC, Coffman TM, Maeda N, Smithies O (1995) Genetic control of blood pressure and the angiotensinogen locus. Proc Natl Acad Sci USA 92:2735–2739

Klein JD, Le Quach D, Cole JM, Disher K, Mongiu AK, Wang X, Bernstein KE, Sands JM (2002) Impaired urine concentration and absence of tissue ACE: involvement of medullary transport proteins. Am J Physiol Renal Physiol 283:F517–524

Krege JH, John SW, Langenbach LL, Hodgin JB, Hagaman JR, Bachman ES, Jennette JC, O'Brien DA, Smithies O (1995) Male-female differences in fertility and blood pressure in ACE- deficient mice. Nature 375:146–148

Kudoh S, Komuro I, Hiroi Y, Zou Y, Harada K, Sugaya T, Takekoshi N, Murakami K, Kadowaki T, Yazaki Y (1998) Mechanical stretch induces hypertrophic responses in cardiac myocytes of angiotensin II type 1a receptor knockout mice. J Biol Chem 273:24037–24043

Li W, Ye Y, Fu B, Wang J, Yu L, Ichiki T, Inagami T, Ichikawa I, Chen X (1998) Genetic deletion of AT2 receptor antagonizes angiotensin II-induced apoptosis in fibroblasts of the mouse embryo. Biochem Biophys Res Commun 250:72–76

Maeda K, Hata R, Bader M, Walther T, Hossmann KA (1999) Larger anastomoses in angiotensinogen-knockout mice attenuate early metabolic disturbances after middle cerebral artery occlusion. J Cereb Blood Flow Metab 19:1092–1098

Massiera F, Seydoux J, Geloen A, Quignard-Boulange A, Turban S, Saint-Marc P, Fukamizu A, Negrel R, Ailhaud G, Teboul M (2001) Angiotensinogen-deficient mice exhibit impairment of diet-induced weight gain with alteration in adipose tissue development and increased locomotor activity. Endocrinology 142:5220–5225

Matsusaka T, Nishimura H, Utsunomiya H, Kakuchi J, Niimura F, Inagami T, Fogo A, Ichikawa I (1996) Chimeric mice carrying 'regional' targeted deletion of the angiotensin type 1A receptor gene. Evidence against the role for local angiotensin in the in vivo feedback regulation of renin synthesis in juxtaglomerular cells. J Clin Invest 98:1867–1877

Mistlberger RE, Antle MC, Oliverio MI, Coffman TM, Morris M (2001) Circadian rhythms of activity and drinking in mice lacking angiotensin II 1A receptors. Physiol Behav 74:457–464

Miyazaki Y, Ichikawa I (2001) Role of the angiotensin receptor in the development of the mammalian kidney and urinary tract. Comp Biochem Physiol A Mol Integr Physiol 128:89–97

Morris M, Means S, Oliverio MI, Coffman TM (2001) Enhanced central response to dehydration in mice lacking angiotensin AT(1a) receptors. Am J Physiol Regul Integr Comp Physiol 280:R1177–1184

Nagata M, Tanimoto K, Fukamizu A, Kon Y, Sugiyama F, Yagami K, Murakami K, Watanabe T (1996) Nephrogenesis and renovascular development in angiotensinogen-deficient mice. Lab Invest 75:745–753

Nakamura Y, Yoshiyama M, Omura T, Yoshida K, Kim S, Takeuchi K, Iwao H, Yoshikawa J (2002) Transmitral inflow pattern assessed by Doppler echocardiography in angiotensin II type 1A receptor knockout mice with myocardial infarction. Circ J 66:192–196

Ng KK, Vane JR (1967) Conversion of angiotensin I to angiotensin II. Nature 216:762–766

Niimura F, Labosky PA, Kakuchi J, Okubo S, Yoshida H, Oikawa T, Ichiki T, Naftilan AJ, Fogo A, Inagami T, et al (1995) Gene targeting in mice reveals a requirement for angiotensin in the development and maintenance of kidney morphology and growth factor regulation. J Clin Invest 96:2947–2954

Nishimura H, et al (1999) Role of the angiotensin type 2 receptor gene in congenital anomalies of the kidney and urinary tract, CAKUT, of mice and men. Mol Cell 3:1–10

Okubo S, Niimura F, Nishimura H, Takemoto F, Fogo A, Matsusaka T, Ichikawa I (1997) Angiotensin-independent mechanism for aldosterone synthesis during chronic extracellular fluid volume depletion. J Clin Invest 99:855–860

Okuyama S, Sakagawa T, Chaki S, Imagawa Y, Ichiki T, Inagami T (1999) Anxiety-like behavior in mice lacking the angiotensin II type-2 receptor. Brain Res 821:150–159

Oliverio MI, Best CF, Kim HS, Arendshorst WJ, Smithies O, Coffman TM (1997) Angiotensin II responses in AT1A receptor-deficient mice: a role for AT1B receptors in blood pressure regulation. Am J Physiol 272:F515–520

Oliverio MI, Kim HS, Ito M, Le T, Audoly L, Best CF, Hiller S, Kluckman K, Maeda N, Smithies O, Coffman TM (1998a) Reduced growth, abnormal kidney structure, and type 2 (AT2) angiotensin receptor-mediated blood pressure regulation in mice lacking both AT1A and AT1B receptors for angiotensin II. Proc Natl Acad Sci USA 95:15496–15501

Oliverio MI, Madsen K, Best CF, Ito M, Maeda N, Smithies O, Coffman TM (1998b) Renal growth and development in mice lacking AT1A receptors for angiotensin II. Am J Physiol 274:F43–50

Ruan X, Oliverio MI, Coffman TM, Arendshorst WJ (1999) Renal vascular reactivity in mice: AngII-induced vasoconstriction in AT1A receptor null mice. J Am Soc Nephrol 10:2620–2630

Saavedra JM, Armando I, Terron JA, Falcon-Neri A, Johren O, Hauser W, Inagami T (2001a) Increased AT(1) receptors in adrenal gland of AT(2) receptor gene- disrupted mice. Regul Pept 102:41–47

Saavedra JM, Hauser W, Ciuffo G, Egidy G, Hoe KL, Johren O, Sembonmatsu T, Inagami T, Armando I (2001b) Increased AT(1) receptor expression and mRNA in kidney glomeruli of AT(2) receptor gene-disrupted mice. Am J Physiol Renal Physiol 280:F71–78

Sakagawa T, Okuyama S, Kawashima N, Hozumi S, Nakagawasai O, Tadano T, Kisara K, Ichiki T, Inagami T (2000) Pain threshold, learning and formation of brain edema in mice lacking the angiotensin II type 2 receptor. Life Sci 67:2577–2585

Schnermann JB, Traynor T, Yang T, Huang YG, Oliverio MI, Coffman T, Briggs JP (1997) Absence of tubuloglomerular feedback responses in AT1A receptor- deficient mice. Am J Physiol 273:F315–320

Senbonmatsu T, Ichihara S, Price E, Jr., Gaffney FA, Inagami T (2000) Evidence for angiotensin II type 2 receptor-mediated cardiac myocyte enlargement during in vivo pressure overload. J Clin Invest 106:R25–29

Sharp MG, Fettes D, Brooker G, Clark AF, Peters J, Fleming S, Mullins JJ (1996) Targeted inactivation of the Ren-2 gene in mice. Hypertension 28:1126–1131

Silvestre JS, Tamarat R, Senbonmatsu T, Icchiki T, Ebrahimian T, Iglarz M, Besnard S, Duriez M, Inagami T, Levy BI (2002) Antiangiogenic effect of angiotensin II type 2 receptor in ischemia- induced angiogenesis in mice hindlimb. Circ Res 90:1072–1079

Siragy HM, Inagami T, Ichiki T, Carey RM (1999a) Sustained hypersensitivity to angiotensin II and its mechanism in mice lacking the subtype-2 (AT2) angiotensin receptor. Proc Natl Acad Sci USA 96:6506–6510

Siragy HM, Senbonmatsu T, Ichiki T, Inagami T, Carey RM (1999b) Increased renal vasodilator prostanoids prevent hypertension in mice lacking the angiotensin subtype-2 receptor. J Clin Invest 104:181–188

Sugaya T, et al (1995) Angiotensin II type 1a receptor-deficient mice with hypotension and hyperreninemia. J Biol Chem 270:18719–18722

Suzuki J, Iwai M, Nakagami H, Wu L, Chen R, Sugaya T, Hamada M, Hiwada K, Horiuchi M (2002) Role of angiotensin II-regulated apoptosis through distinct AT1 and AT2 receptors in neointimal formation. Circulation 106:847–853

Suzuki Y, Lopez-Franco O, Gomez-Garre D, Tejera N, Gomez-Guerrero C, Sugaya T, Bernal R, Blanco J, Ortega L, Egido J (2001) Renal tubulointerstitial damage caused by persistent proteinuria is attenuated in AT1-deficient mice: role of endothelin-1. Am J Pathol 159:1895–1904

Takagi T, Nakano Y, Takekoshi S, Inagami T, Tamura M (2002) Hemizygous mice for the angiotensin II type 2 receptor gene have attenuated susceptibility to azoxymethane-induced colon tumorigenesis. Carcinogenesis 23:1235–1241

Tanaka M, Tsuchida S, Imai T, Fujii N, Miyazaki H, Ichiki T, Naruse M, Inagami T (1999) Vascular response to angiotensin II is exaggerated through an upregulation of AT1 receptor in AT2 knockout mice. Biochem Biophys Res Commun 258:194–198

Tanimoto K, Sugiyama F, Goto Y, Ishida J, Takimoto E, Yagami K, Fukamizu A, Murakami K (1994) Angiotensinogen-deficient mice with hypotension. J Biol Chem 269:31334–31337

Tipnis SR, Hooper NM, Hyde R, Karran E, Christie G, Turner AJ (2000) A human homolog of angiotensin converting enzyme. Cloning and functional expression as a captopril-insensitive carboxypeptidase. J Biol Chem 275:33238–33243

Tsuchida S, Matsusaka T, Chen X, Okubo S, Niimura F, Nishimura H, Fogo A, Utsunomiya H, Inagami T, Ichikawa I (1998) Murine double nullizygotes of the angiotensin type 1A and 1B receptor genes duplicate severe abnormal phenotypes of angiotensinogen nullizygotes. J Clin Invest 101:755–760

von Bohlen und Halbach O, Walther T, Bader M, Albrecht D (2001) Genetic deletion of angiotensin AT2 receptor leads to increased cell numbers in different brain structures of mice. Regul Pept 99:209–216

Walther T, Olah L, Harms C, Maul B, Bader M, Hortnagl H, Schultheiss HP, Mies G (2002) Ischemic injury in experimental stroke depends on angiotensin II. Faseb J 16:169–176

Wu L, Iwai M, Nakagami H, Li Z, Chen R, Suzuki J, Akishita M, de Gasparo M, Horiuchi M (2001) Roles of angiotensin II type 2 receptor stimulation associated with selective angiotensin II type 1 receptor blockade with valsartan in the improvement of inflammation-induced vascular injury. Circulation 104:2716–2721

Xu J, Carretero OA, Liu YH, Shesely EG, Yang F, Kapke A, Yang XP (2002) Role of AT2 receptors in the cardioprotective effect of AT1 antagonists in mice. Hypertension 40:244–250

Yamada H, Akishita M, Ito M, Tamura K, Daviet L, Lehtonen JY, Dzau VJ, Horiuchi M (1999) AT2 receptor and vascular smooth muscle cell differentiation in vascular development. Hypertension 33:1414–1419

Yanai K, Saito T, Kakinuma Y, Kon Y, Hirota K, Taniguchi-Yanai K, Nishijo N, Shigematsu Y, Horiguchi H, Kasuya Y, Sugiyama F, Yagami K, Murakami K, Fukamizu A (2000) Renin-dependent cardiovascular functions and renin-independent blood- brain barrier functions revealed by renin-deficient mice. J Biol Chem 275:5–8

Zhu Z, Zhang SH, Wagner C, Kurtz A, Maeda N, Coffman T, Arendshorst WJ (1998) Angiotensin AT1B receptor mediates calcium signaling in vascular smooth muscle cells of AT1A receptor-deficient mice. Hypertension 31:1171–1177

Transgenics of the RAS

M. Bader · D. Ganten

Max-Delbrück-Center for Molecular Medicine (MDC), Robert-Rössle-Strasse 10,
13092 Berlin-Buch, Germany
e-mail: mbader@mdc-berlin.de

1	Transgenic Technology	230
2	Ubiquitous Overexpression	232
2.1	Rodent Angiotensinogen	232
2.2	Human Angiotensinogen and Human Renin	233
2.3	Mouse Renin	234
3	Brain-Specific Models	236
3.1	Human Angiotensinogen and Human Renin	236
3.2	AT_{1A} Receptor	237
3.3	Angiotensin II	237
3.4	Antisense RNA Against Angiotensinogen	237
4	Kidney-Specific Models	238
5	Heart- and Vessel-Specific Models	239
5.1	Angiotensinogen and Angiotensin II	239
5.2	Angiotensin-Converting Enzyme	239
5.3	AT_1 Receptor	239
5.4	AT_2 Receptor	240
5.5	Chymase	240
6	Fat-Specific Models	240
7	Testis-Specific Models	240
8	Promoter Studies	241
8.1	Renin	241
8.2	Angiotensin-Converting Enzyme	242
9	Conclusions	242
	References	243

Abstract The introduction of transgenes into the genome by microinjection of DNA constructs in zygotes is a powerful technology to characterize the function of genes in rodents. It allows overexpression and—by the use of antisense constructs—also attenuation of gene expression in rats and mice. Transgenic technology has been extensively used to study the functions of the renin–angiotensin system (RAS). Most transgenic rat or mouse models overexpressing renin or angiotensinogen became hypertensive. When the human RAS components were used for the generation of transgenic animals, both renin and angiotensinogen had to be present for angiotensin synthesis and hypertension development due to the species-specificity of the RAS. Transgenic rats with reduced angiotensinogen in the brain are hypotensive and accordingly mice with enhanced angiotensin-II generation in the central nervous system exhibited increased blood pressure, supporting an important role of the central renin–angiotensin system in cardiovascular control. Enhanced angiotensin II generation exclusively in the kidney of transgenic mice also led to hypertension. Transgenic animals with cardiac overexpression of RAS components were prone to cardiac hypertrophy and fibrosis. When the angiotensin-II generating enzyme chymase was expressed as transgene in the vascular wall, hypertension and smooth muscle cell proliferation was elicited. Hypertension was also observed in mice overexpressing angiotensinogen only in fat tissue. Furthermore, transgenic technology was employed to define DNA elements in the promoters of the renin and the angiotensin-converting enzyme (ACE) genes, which are responsible for correct tissue specificity and regulation of their expression. Thus, transgenic technology has been extremely useful to analyze the regulation and the physiological functions of the circulating and particularly the tissue-bound RAS. Novel methods of gene suppression such as RNA interference will render this technique even more utilitarian for this purpose.

Keywords Transgenic rat · Transgenic mouse · Tissue-specific promoter · Microinjection · Renin · Angiotensinogen · Angiotensin

1
Transgenic Technology

Transgenic technology was established for the mouse more than 20 years ago. Most of the experiments since then have also been performed in mice (reviewed in Palmiter and Brinster 1986; Hanahan 1989; Mockrin et al. 1991; Franz et al. 1997). However, the technique has also been extended to other mammals, such as the rat (Mullins et al. 1990; Hammer et al. 1990; Ganten et al. 1992; Matsumoto et al. 1995). The most common method to produce transgenic mammals is the microinjection of DNA constructs into the paternal pronucleus of a fertilized oocyte (Fig. 1). The injected zygotes are implanted into the oviduct of foster mothers and brought to term. For mice and rats, 5%–20% of the offspring have integrated the transgene into their genome and pass it to their offspring, thereby establishing a transgenic line. Mostly, several copies of the foreign DNA are integrated at one site in a chromosome. The expression of the transgene, however, does not only depend on the copy number but also on the chromosomal environment of the integration site and is therefore not absolutely predictable.

In most of the cases it is intended to overexpress the gene of interest. Ubiquitous overexpression of genes is achieved by linking cDNAs to strong viral promoters with little tissue-specificity or by the use of large genomic constructs containing most or all regulatory elements of the gene, which leads to an in-

1. Microinjection of DNA into zygotes

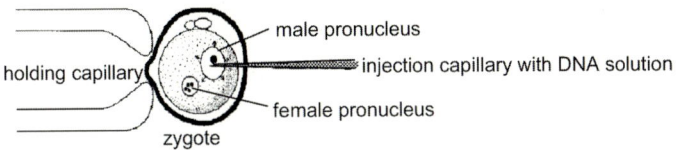

2. Transfer of injected zygotes into the oviduct of a foster mother

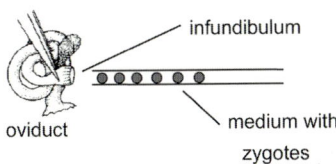

3. DNA-analysis of offspring

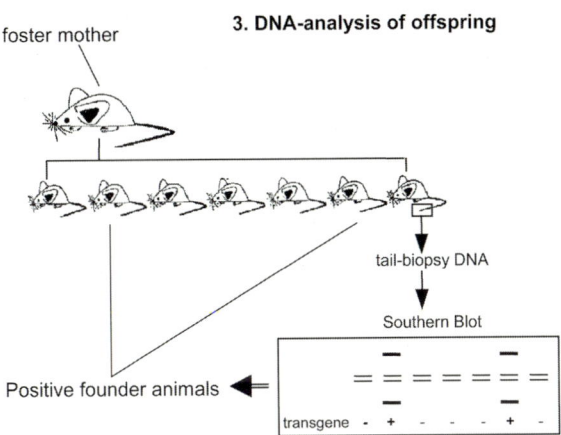

Fig. 1 Scheme for the generation of transgenic animals. About 100 copies of the DNA of interest are injected into a pronucleus of a fertilized oocyte (*1*). The injected oocytes are transferred into the oviduct of a pseudopregnant foster mother (*2*). The genomic DNA of the offspring is analyzed, e.g., by Southern blot, and transgene-carrying founder animals are identified (*3*) and bred to generate the transgenic lines

creased copy number of the gene in the genetically modified animal and to a higher expression with conserved tissue specificity. In order to overexpress a gene in a specific organ or cell type, the respective cDNA is fused to a tissue-specific promoter, and this construct is used for the production of transgenic animals. When oncogenes like the SV40 T antigen are expressed in a specific cell type of transgenic animals, the cells are transformed and can eventually be taken in permanent cell culture. Furthermore, inducible promoter systems are available, which allow for the control of transgene expression by the application of substances to the genetically modified animals, e.g., ecdysone (No et al. 1996), tetracyclines (Kistner et al. 1996), or indole-3 carbinol (Kantachuvesiri et al. 2001).

The transgenic expression of genes for proteins like luciferase, β-galactosidase (lacZ), green fluorescent protein (GFP), or chloramphenicol acetyltransferase (CAT), which can easily be visualized and quantitatively assessed, allows for the analysis of promoter regions important for tissue-specificity and inducibility of transcription of a gene.

The microinjection technique has also been applied to downregulate gene expression in mice and rats. The method used is the expression of antisense RNAs in transgenic animals by the microinjection of constructs in which a part of the cDNA of interest has been fused in opposite orientation to a strong promoter. This results in the synthesis of an RNA complementary to the transcript from the gene of interest. The two RNAs will form a duplex in cells, which blocks translation of the mRNA. Despite the fact that the ablation of gene expression by antisense RNA is never complete and sometimes even fails (Jaquet et al. 1996), animal models with physiological alterations have been successfully developed using this method (Katsuki et al. 1988; Munir et al. 1990; Pepin et al. 1992; Matsumoto et al. 1995; Schinke et al. 1999; Beggah et al. 2002).

Because of the complexity of the cardiovascular system and its regulation, the functional analysis of genes involved in its regulation has been grossly limited to whole organism models. Therefore, transgenic technology has also been of highest importance for characterizing the role of the renin–angiotensin system (RAS) in cardiovascular regulation and in the etiology of diseases. The expression of all RAS components has been altered by genetic manipulation in rats and mice. In particular, the physiological functions of angiotensin II, locally generated in tissues such as brain, heart, and kidney, have been the focus of transgenic research.

2
Ubiquitous Overexpression

2.1
Rodent Angiotensinogen

The first transgenic model with alteration in angiotensinogen expression was a mouse overexpressing the rat gene under the control of the mouse metallothionein promoter (Ohkubo et al. 1990). Although exhibiting high levels of circulating angiotensin II, these mice were normotensive. Our group has generated a transgenic mouse with the rat angiotensinogen gene under the control of its own regulatory sequences (Kimura et al. 1992). These animals [TGM(rAOGEN)123] exhibit marked transgene expression in brain and liver and high levels of circulating angiotensinogen and angiotensin II, that cause high blood pressure (Kimura et al. 1992) and the typical signs of end-organ damage also

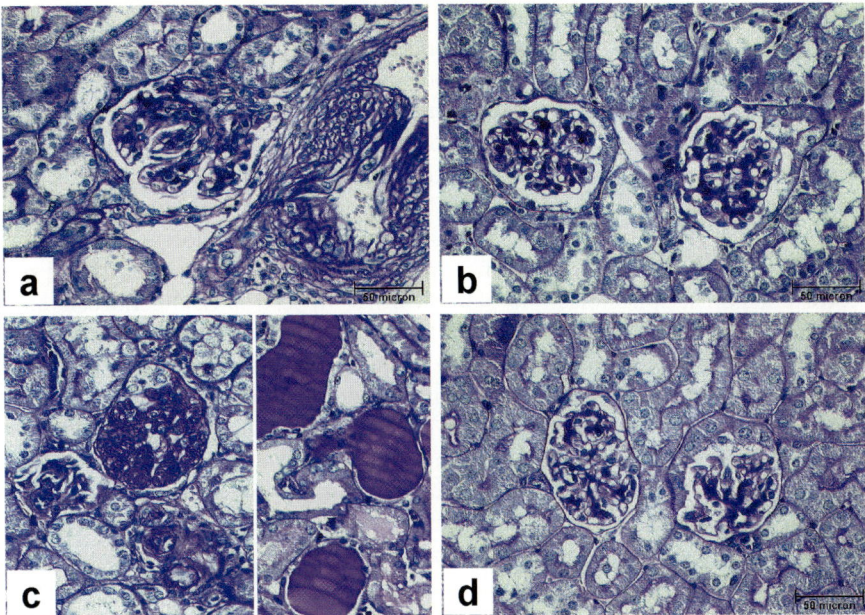

Fig. 2a–d Kidney damage in mice with modified angiotensinogen expression. Periodic acid–Schiff staining of kidney sections from angiotensinogen-knockout (**a**), wild-type (**b**), TGM(rAOGEN)123 (**c**), and TGM(rAOGEN)123 mice on a knockout background (**d**). TGM(rAOGEN)123 (**c**) show the typical signs of tubular and glomerular damage also observed in hypertensive patients, e.g., hyalin depositions. When kidney expression of angiotensinogen is blunted by ablation of the endogenous gene (**d**) the damage is hardly detectable. Angiotensinogen-deficient mice (**a**) show thickening of the elastic vascular lamina and of the glomerular basement membrane as described previously. (Niimura et al. 1995)

observed in human hypertensives, such as cardiac hypertrophy and renal fibrosis (Fig. 2) (Kang et al. 2002).

2.2
Human Angiotensinogen and Human Renin

All rodents transgenic for human angiotensinogen remain normotensive even though some of them exhibit very high levels of the human protein in plasma (Takahashi et al. 1991; Ganten et al. 1992; Yang et al. 1994). The same is true for all animals carrying the human renin gene as transgene (Fukamizu et al. 1989; Ganten et al. 1992; Sigmund et al. 1992; Thompson et al. 1996; Sinn et al. 1999a). These findings corroborate previous biochemical studies showing that human renin and angiotensinogen do not interact with their rodent counterparts (Oliver and Gross 1966). Because of this species specificity of the enzyme-substrate reaction, only double transgenic mice and rats carrying both the human renin and angiotensinogen genes as transgenes become hypertensive (Fukamizu et al. 1993; Merrill et al. 1996; Thompson et al. 1996; Bohlender et al. 1997; Sinn et al.

1999a). These "humanized" rodent models independently developed by several groups are useful to test human-renin inhibitors, which because of the species specificity cannot be tested in normal rodents (Ganten et al. 1992).

Transgenic rats expressing the two human RAS genes develop a fulminant form of malignant hypertension and die at about 7 weeks of age. Because of that, they have become an important model to study the pathophysiology of end-organ damage caused by hypertension and high levels of angiotensin II in tissues (Luft et al. 1999). The work of Luft et al. has shown that angiotensin II acts in concert with mineralocorticoids, since treatment with the mineralocorticoid receptor antagonist spironolactone blunts cardiac hypertrophy and fibrosis and rescues the animals (Fiebeler et al. 2001). Furthermore, the double transgenic rats have furthered the concept that the immune system is intimately involved in the pathogenesis of hypertensive renal damage probably via the activation of nuclear factor (NF)-κB by angiotensin II (Mervaala et al. 1999). Inhibitors of NF-κB such as PDTC and immunosuppressing agents such as dexamethasone, cyclosporine A, and aspirin abolish kidney damage without much effect on blood pressure (Mervaala et al. 2000; Müller et al. 2000, 2001).

Interestingly, when female transgenic mice and rats with the human angiotensinogen gene are bred with male animals carrying the human renin gene, they develop hypertension during pregnancy, which is completely cured after delivery (Takimoto et al. 1996; Bohlender et al. 2000). Thus, such animals may also be relevant models for specific forms of pregnancy-induced hypertension.

2.3
Mouse Renin

The most frequently used transgenic model for the functional analysis of the RAS has been a rat carrying the murine renin gene, *Ren-2* [TGR(mREN2)27] (Mullins et al. 1990). The same transgene containing all exons and introns as well as 5.3-kb upstream and 9.5-kb downstream sequences had been tested before in transgenic mice (Mullins et al. 1989) and in both species it is highly expressed in a variety of tissues including kidney, adrenal gland, and brain. Despite relatively low levels of circulating angiotensin II, TGR(mREN2)27 rats develop severe hypertension already starting at early age (5 weeks after birth) reaching 240 mmHg at 12 weeks of age (Lee et al. 1991, 1996; Kreutz et al. 1998). In some genetic backgrounds or after duplication by homozygosity breeding, the transgene even causes malignant forms of hypertension with a high mortality rate at the same age (Whitworth et al. 1994; Lee et al. 1996; Kantachuvesiri et al. 1999). Also in the surviving heterozygous TGR(mREN2)27, hypertension is accompanied by end-organ damage as evidenced by a marked cardiac hypertrophy and typical signs of hypertensive renal damage such as tubulointerstitial fibrosis, glomerulosclerosis, and albuminuria (Bachmann et al. 1992; Böhm et al. 1995; Böhm et al. 1996). The animals have been very useful for elucidating the role of angiotensin II in the etiology of end-organ damage. Even very low levels of angiotensin-converting enzyme (ACE) inhibitors and angiotensin II receptor

AT$_1$ antagonists—which do not lower blood pressure significantly—blunt the development of cardiac hypertrophy and renal damage in this model (Böhm et al. 1995, 1996). These results explain the extreme effectiveness of these drugs for organ protection in hypertensive and diabetic patients (Brown and Vaughan 1998). Obviously, angiotensin II exerts blood pressure-independent effects on target organs. Concordantly, alterations in Ca^{2+} handling and adrenoceptor signaling have been observed in cardiac cells of TGR(mREN2)27 similar to the ones diagnosed in heart failure patients (Zolk et al. 1998).

The adrenal gland is the organ with the highest transgene expression in TGR(mREN2)27 (Mullins et al. 1990; Zhao et al. 1993). The high renin activity in this organ disturbs steroidogenesis and leads to increased concentrations in particular of mineralocorticoids in the plasma of the animals (Sander et al. 1992; Peters et al. 1993; Sander et al. 1994). This effect may participate in the etiology of the hypertension observed in the transgenic rats, since suppression of adrenal steroidogenesis by dexamethasone normalizes blood pressure (Djavidani et al. 1995).

The hypertension developed by TGR(mREN2)27 may also partly be of neurogenic origin. These animals generate up to tenfold more angiotensin II in the central nervous system than control rats (Senanayake et al. 1994). When they are anesthetized by chloralose-urethane, blood pressure drops to normal, arguing in favor of this central angiotensin generation as cause for their hypertension (Diz et al. 1998). Furthermore, when the central angiotensin generation in TGR(mREN2)27 is blunted by crossbreeding with transgenic rats exhibiting a brain-specific deficiency in angiotensinogen [TGR(ASrAOGEN), see Sect. 3.4], a significant reduction in blood pressure is observed (Schinke et al. 1999).

Interestingly, TGR(mREN2)27 exhibit significant perturbances of the circadian rhythm of blood pressure. While in normal rats blood pressure is highest in the active phase, which is at night, the transgenic rats show highest blood pressure levels at noon (Lemmer et al. 1993). The same shift in blood pressure acrophase is observed after chronic angiotensin II infusion into normal rats (Baltatu et al. 2001). Thus, it may be caused by enhanced angiotensin II levels in peripheral tissues. However, the central RAS also is involved in this shift, since TGR(ASrAOGEN) are resistant to the influence of chronic angiotensin II infusion [(Baltatu et al. 2001), see Sect. 3.4].

Furthermore, TGR(mREN2)27 rats contain drastically enhanced prorenin levels in plasma, which may also cause the cardiovascular phenotype of this animal model. This protein is probably mainly released from the adrenal gland, since adrenalectomy reduces its levels to normal (Bachmann et al. 1992; Tokita et al. 1994). In a different transgenic rat model, Veniant et al. (1996) have shown that prorenin, when expressed in the liver of transgenic rats by using the α1-antitrypsin promoter, reaches similarly high circulating levels as in TGR(mREN2)27. As a consequence, these rats show a comparable degree of cardiovascular hypertrophy as TGR(mREN2)27, probably because prorenin may be activated in peripheral tissues. The recent discovery of a renin (and prorenin) receptor with signaling function may help to explain the physiological actions of elevated circulating prorenin (Nguyen et al. 2002).

Again using the *Ren-2* gene, a transgenic rat model with inducible hypertension has been developed. To this purpose, the promoter of the cytochrome P450 enzyme CYP1A1 was employed which can be induced mainly in the liver by treatment of the animals with a xenobiotic drug, indole-3 carbinol (Kantachuvesiri et al. 2001). After application of the drug, the upregulated prorenin and renin levels lead to an increase in blood pressure and vascular injury.

Recently, a novel method to express a transgene in a defined tissue was applied to generate a comparable model (Caron et al. 2002). A chimeric mRNA of mouse *Ren-1^C* and *Ren-2* was integrated by homologous recombination into a site between the apolipoprotein C3 and A1 gene of the mouse, a region which is strongly influenced by a liver-specific enhancer. As expected, the gene is exclusively expressed in the liver leading to high levels of circulating renin and prorenin that is not controlled by the physiological renin regulators. The animals exhibit significantly increased blood pressure, water intake, and urine output and develop all signs of end-organ damage in kidney and heart similar to the above-mentioned models.

3
Brain-Specific Models

3.1
Human Angiotensinogen and Human Renin

Transgenic mice with overexpression of the human RAS only in the brain become hypertensive (Morimoto et al. 2001, 2002a). Interestingly, it is not important from which cell type the two proteins, human renin and angiotensinogen, are produced. Combined synthesis of both RAS components in neurons by the use of the synapsin-I promoter or in astrocytes employing the promoter for the glial fibrillary acidic protein (GFAP) equally causes hypertension and increases water and salt intake (Morimoto et al. 2002b). In these models, the high blood pressure can be reduced by intracerebroventricular injection of the angiotensin II receptor AT_1 antagonist, losartan, suggesting that the brain RAS is a major determinant of hypertension. Part of the effect seems to be mediated by the release of vasopressin, since intravenous injection of a V1 receptor antagonist attenuates the hypertensive phenotype (Davisson et al. 1998; Morimoto et al. 2002b). In some of the models but not in all, ganglion blockade also affects blood pressure, indicating that enhanced sympathetic outflow may also be involved in the central angiotensin action (Davisson et al. 1998; Morimoto et al. 2002b).

3.2
AT$_{1A}$ Receptor

Mice overexpressing the AT$_{1A}$ receptor in most neurons, by using the neuron-specific enolase promoter, remain normotensive but are very sensitive to intracerebroventricular injections of angiotensin II and losartan, indicating that blood pressure in this model is balanced by the baroreflex (Lazartigues et al. 2002).

3.3
Angiotensin II

Recently, a transgenic mouse was presented with eight times more angiotensin II in the brain but normal levels in the circulation (Lochard et al. 2003). The peptide is liberated during secretion from an artificial chimeric protein (Methot et al. 1997) expressed under the control of the GFAP promoter. Also these animals are hypertensive.

3.4
Antisense RNA Against Angiotensinogen

The already-mentioned transgenic rat model, TGR(ASrAOGEN), has provided numerous insights in the functionality of the brain RAS. These rats carry a transgene expressing an antisense RNA against angiotensinogen specifically in astrocytes of the brain under the control of the GFAP promoter (Schinke et al. 1999). This causes a reduction of local angiotensinogen levels by 90% and reduced central angiotensin generation (Huang et al. 2001) without affecting the circulating RAS. The rats are slightly hypotensive and excrete increased amounts of diluted urine (Fig. 3). This mild diabetes insipidus is probably caused by reduced vasopressin levels in the circulation, again supporting a central involvement of brain angiotensin II in vasopressin secretion.

Probably due to decreased central angiotensin levels, TGR(ASrAOGEN) exhibit an increased expression of AT$_1$ receptors in most brain areas beyond the

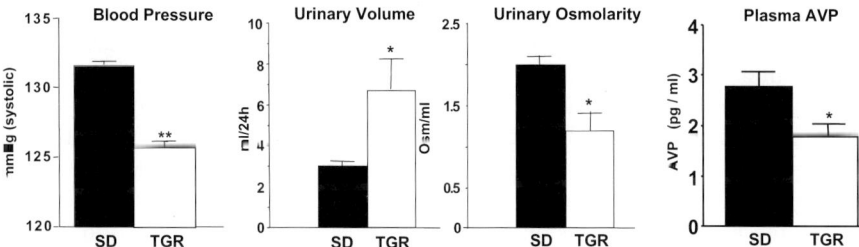

Fig. 3 Phenotype of TGR(ASrAOGEN). TGR(ASrAOGEN) (*TGR*) exhibit reduced blood pressure and excrete a high amount of urine with low osmolarity, probably due to lowered plasma levels of vasopressin (*AVP*) compared to Spraque-Dawley control rats (*SD*)

blood–brain barrier, which causes an augmented drinking response to intracerebroventricularly injected angiotensin II (Monti et al. 2001).

Furthermore, TGR(ASrAOGEN) show alterations in short-term and long-term blood pressure variability and an increased baroreflex sensitivity due to an imbalance of the parasympathetic and sympathetic nervous system (Baltatu et al. 2001). Together with the findings that TGR(mREN2)27 (Borgonio et al. 2001) and mice expressing the human RAS (Davisson et al. 1998) exhibit a decreased baroreflex sensitivity, these data characterize central angiotensin as a relevant moderator of the baroreflex.

As mentioned above, central angiotensin is also significantly involved in cardiovascular rhythm control. Renin overexpression in TGR(mREN2)27 (Lemmer et al. 1993), as well as low-dose peripheral infusions of angiotensin II in normal rats (Baltatu et al. 2001) cause an inversion of the circadian blood pressure rhythm. This effect of increased peripheral angiotensin II is absent in TGR (ASrAOGEN) (Baltatu et al. 2001). Thus, peripheral angiotensin II requires central angiotensin II as a mediator of the rhythm shift. Since only the rhythm of blood pressure but not of heart rate is altered by angiotensin II, the peptide does not appear to affect the main oscillator in the suprachiasmatic nucleus, but seems to affect its output pathways or its synchronization with peripheral oscillators in cardiovascular organs.

Since the pineal gland is considered to regulate circadian rhythms, we studied the content of melatonin and its indole precursors in pineals of TGR(ASrAOGEN) rats (Baltatu et al. 2002). The levels of these indoles were significantly decreased in TGR(ASrAOGEN). Moreover, the mRNA for the first enzyme in melatonin synthesis, tryptophan hydroxylase, was significantly reduced in TGR (ASrAOGEN) rats in contrast to the transcript of the *N*-acetyltransferase gene which was unaffected. These results demonstrate that a local pineal RAS exerts a tonic modulation of indole synthesis by influencing expression and activity of tryptophan hydroxylase.

4
Kidney-Specific Models

In addition to the brain, locally generated angiotensin in the kidney is also relevant for blood pressure regulation. Sigmund et al. (Davisson et al. 1999; Ding et al. 2001) recently showed that overexpression of human angiotensinogen in the kidney, in the presence of ubiquitously expressed human renin without spillover of angiotensin II into the circulation, leads to hypertension. The importance of local angiotensin generation in end-organs has been confirmed in a hybrid mouse model carrying a rat angiotensinogen transgene on an angiotensinogen knockout background (Kang et al. 2002). These animals became hypertensive due to the exclusive expression of rat angiotensinogen in the liver and brain. However, due to the absence of local angiotensinogen synthesis in the kidney and heart, end-organ damage was attenuated (Fig. 2). Thus, the local synthesis of angiotensin II is pivotal for a full development of end-organ damage.

5
Heart- and Vessel-Specific Models

5.1
Angiotensinogen and Angiotensin II

Mice overexpressing angiotensinogen only in the heart remained normotensive but nevertheless developed cardiac hypertrophy, indicating that the local formation of angiotensin II induces cardiac damage independent of blood pressure elevation (Mazzolai et al. 1998, 2000). By contrast, transgenic mice that generate angiotensin II from the aforementioned chimeric protein (Methot et al. 1997) exclusively in the heart do not develop hypertrophy unless spillover of the peptide into the circulation raises blood pressure (van Kats et al. 2001). Cardiac fibrosis is, however, detected in all angiotensinogen-transgenic mice, independent of hypertension, suggesting a direct effect of cardiac angiotensin II on this parameter.

5.2
Angiotensin-Converting Enzyme

Transgenic rats overexpressing ACE predominantly in the heart have been produced (Tian et al. 1996). In spite of very high levels of ACE activity in the heart, there are no morphological alterations of this organ unless it is pressure-overloaded by aortic banding. This treatment results in a significantly higher hypertrophic response in the ACE-transgenic rats than in control animals, supporting the important role of angiotensin II in this process.

5.3
AT_1 Receptor

The α-myosin heavy-chain promoter has been employed to overexpress AT_1 receptors in the heart of transgenic animal models (Hein et al. 1997; Paradis et al. 2000; Hoffmann et al. 2001). However, the phenotypes of the transgenic animals generated were markedly different. Mouse models exhibit a drastic cardiac hypertrophy and die within several days (Hein et al. 1997) or months (Paradis et al. 2000) of age. Despite high levels of AT_1 receptors in the heart (Rosenkranz et al. 1997; Hoffmann et al. 2001), the transgenic rats appear absolutely normal unless the heart is pressure- or volume-overloaded by aortic banding or aortocaval shunt, respectively, which, as for the ACE-transgenic rats, elicit a more pronounced hypertrophy than in control animals (Hoffmann et al. 2001). The difference between the models might be related to a differential sensitivity of mouse and rat hearts for angiotensin II effects that may be linked to a high level of uncoupled AT_1 receptors in the transgenic rats (Rosenkranz et al. 1997).

5.4
AT$_2$ Receptor

When the AT$_2$ receptor is overexpressed in the heart, the resulting transgenic mice show no obvious morphological alterations but they are less sensitive to angiotensin II-induced blood-pressure elevation and less susceptible to remodeling of the left ventricle after myocardial infarction, indicating that the AT$_2$ receptor counteracts the AT$_1$ receptor (Masaki et al. 1998; Sugino et al. 2001; Yang et al. 2002). The possible mechanism of AT$_2$ action involves activation of kinins, as has been shown in another transgenic mouse model overexpressing the AT$_2$ receptor in vascular smooth muscle cells (Tsutsumi et al. 1999). These animals did not increase blood pressure after angiotensin II infusion due to a counter-regulatory hypotensive action of the transgenic AT$_2$ receptor via bradykinin and NO.

5.5
Chymase

Vascular smooth muscle overexpression of rat vascular chymase, an enzyme that like ACE metabolizes angiotensin I to II, caused hypertension and smooth muscle cell proliferation in transgenic mice (Ju et al. 2001). The authors used a tetracycline-regulated transgene (Kistner et al. 1996), allowing them to switch on and off transgene expression by doxycycline application. These animals underscore the possible importance of chymases for pathophysiological angiotensin II generation.

6
Fat-Specific Models

Adipose tissue is a major site of angiotensinogen expression. In order to analyze the role of this protein in fat tissue, Massiera et al. (2001) generated a transgenic mouse overexpressing angiotensinogen in adipocytes using the aP2 promoter. The resulting animals showed adipocyte hypertrophy, enlarged total fat mass, and hypertension, the latter probably being caused by a spillover of angiotensinogen into the circulation. Thus, angiotensinogen generation in fat may be physiological relevant for adipocyte development and regulation but may also represent a significant source for circulating renin substrate.

7
Testis-Specific Models

The only RAS component studied by transgenic technology in testis was ACE. Besides the promoter responsible for endothelial expression of the enzyme, a second promoter in intron 13 of the gene drives expression of the testis isoform of ACE. Male knockout mice for ACE are sterile, indicating an essential role of this isoform (Krege et al. 1995). Sen et al. have studied this role by rescuing testicular ACE expression in knockout mice by introducing a transgene with a sperm-specific promoter linked to the cDNA coding for the testis isoform of ACE (Ramaraj et al. 1998). When the endothelial isoform is introduced in the same way, no restoration of male fertility is observed (Kessler et al. 2000). However, the exact function of the testis isoform of ACE in male fertility is still unresolved.

8
Promoter Studies

8.1
Renin

Transgenic experiments have shown that DNA sequences in the promoter as well as in transcribed regions of the *Ren-2* gene are necessary for correct tissue specificity of expression. Transgenic mice with 5.3 kb (Mullins et al. 1989) or 2.5 kb (Tronik et al. 1987; Tronik and Rougeon 1988) of the promoter region and all exons and introns, or with a 4.6-kb promoter fragment and SV40 T antigen as a reporter gene, express the transgene correctly in the kidneys, reproductive organs, and submandibular and adrenal glands; and transformed renin-producing cells have been developed from these mice (Sigmund et al. 1990a,b,c, 1991; Jacob et al. 1991). Animals with only 2.5 kb of the promoter and the same reporter gene, however, show ectopic expression and do not develop renin-producing tumors (Sola et al. 1989). Thus, there appears to be some redundancy of tissue-specific elements in the promoter and in the transcribed region, since 2.5 kb of promoter are only sufficient for correct expression in concert with all exons and introns, while longer 5′-flanking regions are independent of other parts of the *Ren-2* gene. Comparable results have also been published for the *Ren-1^D* gene, since 5 kb of the promoter leads to correct expression in transgenic mice, predominantly in the kidney in the presence of the entire transcribed region but not when fused to a CAT reporter gene (Miller et al. 1989).

Furthermore, the human renin gene is expressed in a partially ectopic pattern in transgenic animals. Gene regulation is inadequate in transgenic mice (Fukamizu et al. 1989; Sigmund et al. 1992; Thompson et al. 1996; Yan et al. 1998a; Sinn et al. 1999b) and rats (Ganten et al. 1992; Bohlender et al. 1997) if only up to 3 kb of upstream sequences are included in the construct. While the kidney is the major tissue of expression, other organs such as the spleen and testis, contain considerable amounts of transgenic mRNA. However, when constructs are used with more than 25 kb of 5′-flanking sequences, transgene expression is largely confined to the kidney and to a lesser extent to the lung and other organs which also express renin at low levels in humans (Yan et al. 1998a,b; Catanzaro et al. 1999; Sinn et al. 1999a). Furthermore, bilateral nephrectomy depletes circulating human renin and prorenin in these mice, as expected, while animals with shorter transgene constructs paradoxically show increased plasma prorenin levels after the operation (Yan et al. 1998a). These differential sources of circulating renin may also be responsible for observed differences in blood pressure after coexpression of human angiotensinogen in human renin transgenic animals. Animals with short transgene constructs become hypertensive (Fukamizu et al. 1993; Merrill et al. 1996; Thompson et al. 1996; Bohlender et al. 1997) and mice with long promoter regions develop either only slightly elevated blood pressure or even remain normotensive (Catanzaro et al. 1999; Sinn et al. 1999a). This indicates that the normal downregulation of the human renin transgene by elevated blood pressure occurs only in animals harboring long promoter regions.

Recently, Fuchs et al. (2002) showed that 12 kb of the upstream region, but neither 2.8 kb nor 5.8 kb, are sufficient to drive correct tissue-specific expression of the human renin gene. In their transgenic mice, the lacZ reporter gene is expressed in juxtaglomerular cells and in a striped pattern in renal arteries, but surprisingly they also detected stripes of transgenic lacZ activity and endogenous mouse renin immunoreactivity in arteries outside the kidney, e.g., in the aorta. Furthermore, classical stimulators of renin synthesis such as two-kidney, one-clip hypertension and salt depletion increase the expression of the lacZ transgene even more pronounced than the endogenous renin gene, indicating that all DNA elements necessary for tissue specificity and upregulation of human renin gene expression reside within the 12-kb fragment employed in the lacZ construct.

Accordingly, angiotensin II infusion in human renin transgenic mice with only a 900-bp upstream region results in a paradoxical upregulation of human renin expression, while mouse renin is expectedly suppressed (Keen and Sigmund 2001).

8.2
Angiotensin-Converting Enzyme

Using lacZ as a reporter gene in transgenic mice, the group of Bernstein et al. has restricted the testis-specific promoter of ACE to 91 bp in intron 13 of the gene (Langford et al. 1991; Howard et al. 1993). However, the testis-specific elements could not yet be determined (Zhou et al. 1996; Esther et al. 1997).

9
Conclusions

Transgenic technology in concert with gene targeting (see the previous chapter, by Brede and Hein, this volume) has been extremely useful to determine the physiological functions of the RAS. In particular, the analysis of tissue RAS relies mainly on these techniques since pharmacological approaches cannot be easily targeted to specific organs. When novel methods of gene suppression such as RNA interference (Tuschl 2002) will be available for use in transgenic models, transgenic technology will become even more utilitarian for this purpose.

References

Bachmann S, Peters J, Engler E, Ganten D, Mullins J (1992) Transgenic rats carrying the mouse renin gene—morphological characterization of a low renin hypertension model. Kidney Int 41:24–36

Baltatu O, Afeche SC, Santos SHJ, Campos LA, Barbosa R, Michelini LC, Bader M, Cipolla-Neto J (2002).Locally synthesized angiotensin modulates pineal melatonin generation. J Neurochem 80:328–334

Baltatu O, Janssen BJ, Bricca G, Plehm R, Monti J, Ganten D, Bader M (2001) Alterations in blood pressure and heart rate variability in transgenic rats with low brain angiotensinogen. Hypertension 37:408–413

Beggah AT, Escoubet B, Puttini S, Cailmail S, Delage V, Ouvrard-Pascaud A, Bocchi B, Peuchmaur M, Delcayre C, Farman N, Jaisser F (2002) Reversible cardiac fibrosis and heart failure induced by conditional expression of an antisense mRNA of the mineralocorticoid receptor in cardiomyocytes. Proc Natl Acad Sci USA 99:7160–7165

Bohlender J, Fukamizu A, Lippoldt A, Nomura T, Dietz R, Menard J, Murakami K, Luft FC, Ganten D (1997) High human renin hypertension in transgenic rats. Hypertension 29:428–434

Bohlender J, Ganten D, Luft FC (2000) Rats transgenic for human renin and human angiotensinogen as a model for gestational hypertension. J Am Soc Nephrol 11:2056–2061

Borgonio A, Pummer S, Witte K, Lemmer B (2001) Reduced baroreflex sensitivity and blunted endogenous nitric oxide synthesis precede the development of hypertension in TGR(mREN2)27 rats. Chronobiol Int 18:215–226

Böhm M, Lee M, Kreutz R, Kim S, Schinke M, Djavidani B, Wagner J, Kaling M, Wienen W, Bader M, Ganten D (1995) Angiotensin II receptor blockade in TGR(mREN2)27: Effects on renin–angiotensin-system gene expression and cardiovascular functions. J Hypertens 13:891–899

Böhm M, Lippoldt A, Wienen W, Ganten D, Bader M (1996) Reduction of cardiac hypertrophy in TGR(mREN2)27 by angiotensin II receptor blockade. Mol Cell Biochem 163–164:217–221

Brown NJ, Vaughan DE (1998) Angiotensin-converting enzyme inhibitors. Circulation 97:1411–1420

Caron KM, James LR, Kim HS, Morham SG, Sequeira Lopez ML, Gomez RA, Reudelhuber TL, Smithies O (2002) A genetically clamped renin transgene for the induction of hypertension. Proc Natl Acad Sci USA 99:8248–8252

Catanzaro DF, Chen R, Yan Y, Hu LF, Sealey JE, Laragh JH (1999) Appropriate regulation of renin and blood pressure in 45-kb human renin/human angiotensinogen transgenic mice. Hypertension 33:318–322

Davisson RL, Ding Y, Stec DE, Catterall JF, Sigmund CD (1999) Novel mechanism of hypertension revealed by cell-specific targeting of human angiotensinogen in transgenic mice. Physiol Genomics 1:3–9

Davisson RL, Yang G, Beltz TG, Cassell MD, Johnson AK, Sigmund CD (1998) The brain renin–angiotensin system contributes to the hypertension in mice containing both the human renin and human angiotensinogen transgenes. Circ Res 83:1047–1058

Ding Y, Stec DE, Sigmund CD (2001) Genetic evidence that lethality in angiotensinogen-deficient mice is due to loss of systemic but not renal angiotensinogen. J Biol Chem 276:7431–7436

Diz DI, Westwood B, Bosch SM, Ganten D, Ferrario C (1998) NK1 receptor antagonist blocks angiotensin II responses in renin transgenic rat medulla oblongata. Hypertension 31:473–479

Djavidani B, Sander M, Kreutz R, Zeh K, Bader M, Mellon SH, Vecsei P, Peters J, Ganten D (1995) Chronic dexamethasone treatment suppresses hypertension development in the transgenic rat TGR(mREN2)27. J Hypertens 13:637–645

Esther CR, Jr., Semeniuk D, Marino EM, Zhou Y, Overbeek PA, Bernstein KE (1997) Expression of testis angiotensin-converting enzyme is mediated by a cyclic AMP responsive element. Lab Invest 77:483–488

Fiebeler A, Schmidt F, Muller DN, Park JK, Dechend R, Bieringer M, Shagdarsuren E, Breu V, Haller H, Luft FC (2001) Mineralocorticoid receptor affects AP-1 and nuclear factor-kappab activation in angiotensin II-induced cardiac injury. Hypertension 37:787–793

Franz WM, Mueller OJ, Hartong R, Frey N, Katus HA (1997) Transgenic animal models: new avenues in cardiovascular physiology. J Mol Med 75:115–129

Fuchs S, Germain S, Philippe J, Corvol P, Pinet F (2002) Expression of renin in large arteries outside the kidney revealed by human renin promoter/LacZ transgenic mouse. Am J Pathol 161:717–725

Fukamizu A, Seo MS, Hatae T, Yokoyama M, Nomura T, Katsuki M, Murakami K (1989) Tissue-specific expression of the human renin gene in transgenic mice. Biochem Biophys Res Commun 165:826–832

Fukamizu A, Sugimura K, Takimoto E, Sugiyama F, Seo MS, Takahashi S, Hatae T, Kajiwara N, Yagami K, Murakami K (1993) Chimeric renin angiotensin system demonstrates sustained increase in blood pressure of transgenic mice carrying both human renin and human angiotensinogen genes. J Biol Chem 268:11617–11621

Ganten D, Wagner J, Zeh K, Bader M, Michel J-B, Paul M, Zimmermann F, Ruf P, Hilgenfeldt U, Ganten U, Kaling M, Bachmann S, Fukamizu A, Mullins JJ, Murakami K (1992) Species specificity of renin kinetics in transgenic rats harboring the human renin and angiotensinogen genes. Proc Natl Acad Sci USA 89:7806–7810

Hammer RE, Maika SD, Richardson JA, Tang J, Taurog JD (1990) Spontaneous inflammatory disease in transgenic rats expressing HLA-B27 and human β2m: An animal model of HLA-B27-associated human disorders. Cell 63:1099–1112

Hanahan D (1989) Transgenic mice as probes into complex systems. Science 246:1265–1275

Hein L, Stevens ME, Barsh GS, Pratt RE, Kobilka BK, Dzau VJ (1997) Overexpression of angiotensin AT1 receptor transgene in the mouse myocardium produces a lethal phenotype associated with myocyte hyperplasia and heart block. Proc Natl Acad Sci USA 94:6391–6396

Hoffmann S, Krause T, van Geel PP, Willenbrock R, Pagel I, Pinto YM, Buikema H, Van Gilst WH, Lindschau C, Paul M, Inagami T, Ganten D, Urata H (2001) Overexpression of the human angiotensin II type 1 receptor in the rat heart augments load induced cardiac hypertrophy. J Mol Med 79:601–608

Howard T, Balogh R, Overbeek P, Bernstein KE (1993) Sperm-specific expression of angiotensin-converting enzyme (ACE) is mediated by a 91-base-pair promoter containing a CRE-like element. Mol Cell Biol 13:18–27

Huang BS, Ganten D, Leenen FH (2001) Responses to central Na(+) and ouabain are attenuated in transgenic rats deficient in brain angiotensinogen. Hypertension 37:683–686

Jacob HJ, Sigmund CD, Shockley TR, Gross KW, Dzau VJ (1991) Renin promoter SV40 T-antigen transgenic mouse. Hypertension 17:1167–1172

Jaquet V, Gow A, Tosic M, Suchanek G, Breitschopf H, Lassmann H, Lazzarini RA, Matthieu JM (1996) An antisense transgenic strategy to inhibit the myelin oligodendrocyte glycoprotein synthesis. Mol Brain Res 43:333–337

Ju H, Gros R, You X, Tsang S, Husain M, Rabinovitch M (2001) Conditional and targeted overexpression of vascular chymase causes hypertension in transgenic mice. Proc Natl Acad Sci USA 98:7469–7474

Kang N, Walther T, Tian XL, Bohlender J, Fukamizu A, Ganten D, Bader M (2002) Reduced hypertension-induced end-organ damage in mice lacking cardiac and renal angiotensinogen synthesis. J Mol Med 80:359–366

Kantachuvesiri S, Fleming S, Peters J, Peters B, Brooker G, Lammie AG, McGrath I, Kotelevtsev Y, Mullins JJ (2001) Controlled hypertension, a transgenic toggle switch reveals differential mechanisms underlying vascular disease. J Biol Chem 276:36727–36733

Kantachuvesiri S, Haley CS, Fleming S, Kurian K, Whitworth CE, Wenham P, Kotelevtsev Y, Mullins JJ (1999) Genetic mapping of modifier loci affecting malignant hypertension in TGRmRen2 rats. Kidney Int 56:414–420

Katsuki M, Sato M, Kimura M, Yokoyama M, Kobayashi K, Nomura T (1988) Conversion of normal behavior to shiverer by myelin basic protein antisense cDNA in transgenic mice. Science 241:593–595

Keen HL, Sigmund CD (2001) Paradoxical regulation of short promoter human renin transgene by angiotensin II. Hypertension 37:403–407

Kessler SP, Rowe TM, Gomos JB, Kessler PM, Sen GC (2000) Physiological non-equivalence of the two isoforms of angiotensin- converting enzyme. J Biol Chem 275:26259–26264

Kimura S, Mullins JJ, Bunnemann B, Metzger R, Hilgenfeldt U, Zimmermann F, Jacob H, Fuxe K, Ganten D, Kaling M (1992) High blood pressure in transgenic mice carrying the rat angiotensinogen gene. EMBO J 11:821–827

Kistner A, Gossen M, Zimmermann F, Jerecic J, Ullmer C, Lubbert H, Bujard H (1996) Doxycycline-mediated quantitative and tissue-specific control of gene expression in transgenic mice. Proc Natl Acad Sci USA 93:10933–10938

Krege JH, John SW, Langenbach LL, Hodgin JB, Hagaman JR, Bachman ES, Jennette JC, O'Brien DA, Smithies O (1995) Male-female differences in fertility and blood pressure in ACE- deficient mice. Nature 375:146–148

Kreutz R, Fernandez-Alfonso MS, Paul M, Peters J (1998) Differential development of early hypertension in heterozygous transgenic TGR(mREN2)27 rats. Clin Exp Hypertens 20:273–282

Langford KG, Shai S-Y, Howard TE, Kovac MJ, Overbeek PA, Bernstein KE (1991) Transgenic mice demonstrate a testis-specific promoter for angiotensin-converting enzyme. J Biol Chem 266:15559–15562

Lazartigues E, Dunlay SM, Loihl AK, Sinnayah P, Lang JA, Espelund JJ, Sigmund CD, Davisson RL (2002) Brain-selective overexpression of angiotensin (AT1) receptors causes enhanced cardiovascular sensitivity in transgenic mice. Circ Res 90:617–624

Lee M, Zhao Y, Peters J, Ganten D, Zimmermann F, Ganten U, Bachmann S, Bader M, Mullins JJ (1991) Preparation and analysis of transgenic rats expressing the mouse Ren-2 gene. J Vasc Med Biol 3:50–54

Lee MA, Böhm M, Paul M, Bader M, Ganten U, Ganten D (1996) Physiological characterization of the hypertensive transgenic rat TGR(mREN2)27. Am J Physiol 270: E919-E929

Lemmer B, Mattes A, Böhm M, Ganten D (1993) Circadian blood pressure variation in transgenic hypertensive rats. Hypertension 22:97–101

Lochard N, Silversides DW, van Kats JP, Mercure C, Reudelhuber TL (2003) Brain-specific restoration of angiotensin II corrects renal defects seen in angiotensinogen-deficient mice. J Biol Chem 278:2184–2189

Luft FC, Mervaala E, Müller DN, Gross V, Schmidt F, Park JK, Schmitz C, Lippoldt A, Breu V, Dechend R, Dragun D, Schneider W, Ganten D, Haller H (1999) Hypertension-induced end-organ damage: new transgenic approach to an old problem. Hypertension 33:212–218

Masaki H, Kurihara H, Yamaki A, Inomata N, Nozawa Y, Mori Y, Murasawa S, Kizima K, Maruyama K, Horiuchi M, Dzau VJ, Takahashi H, Iwasaka T, Inada M, Matsubara H (1998) Cardiac-specific overexpression of angiotensin II AT2 receptor causes attenuated response to AT1 receptor-mediated pressor and chronotropic effects. J Clin Invest 101:527–535

Massiera F, Bloch-Faure M, Ceiler D, Murakami K, Fukamizu A, Gasc JM, Quignard-Boulange A, Negrel R, Ailhaud G, Seydoux J, Meneton P, Teboul M (2001) Adipose angiotensinogen is involved in adipose tissue growth and blood pressure regulation. FASEB J 15:2727–2729

Matsumoto K, Kakidani H, Anzai M, Nakagata N, Takahashi A, Takahashi Y, Miyata K (1995) Evaluation of an antisense RNA transgene for inhibiting growth hormone gene expression in transgenic rats. Dev Genet 16:273–277

Mazzolai L, Nussberger J, Aubert JF, Brunner DB, Gabbiani G, Brunner HR, Pedrazzini T (1998) Blood pressure-independent cardiac hypertrophy induced by locally activated renin–angiotensin system. Hypertension 31:1324–1330

Mazzolai L, Pedrazzini T, Nicoud F, Gabbiani G, Brunner HR, Nussberger J (2000) Increased cardiac angiotensin II levels induce right and left ventricular hypertrophy in normotensive mice. Hypertension 35:985–991

Merrill DC, Thompson MW, Carney CL, Granwehr BP, Schlager G, Robillard JE, Sigmund CD (1996) Chronic hypertension and altered baroreflex responses in transgenic mice containing the human renin and human angiotensinogen genes. J Clin Invest 97:1047–1055

Mervaala E, Müller DN, Park JK, Dechend R, Schmidt F, Fiebeler A, Bieringer M, Breu V, Ganten D, Haller H, Luft FC (2000) Cyclosporin A protects against angiotensin II-induced end-organ damage in double transgenic rats harboring human renin and angiotensinogen genes. Hypertension 35:360–366

Mervaala EM, Müller DN, Park JK, Schmidt F, Lohn M, Breu V, Dragun D, Ganten D, Haller H, Luft FC (1999) Monocyte infiltration and adhesion molecules in a rat model of high human renin hypertension. Hypertension 33:389–395

Methot D, Lapointe MC, Touyz RM, Yang XP, Carretero OA, Deschepper CF, Schiffrin EL, Thibault G, Reudelhuber TL (1997) Tissue targeting of angiotensin peptides. J Biol Chem 272:12994–12999

Miller CCJ, Carter AT, Brooks JI, Lovell-Badge RH, Brammar WJ (1989) Differential extra-renal expression of the mouse renin genes. Nucl Acids Res 17:3117–3128

Mockrin SC, Dzau VJ, Gross KW, Horan MJ (1991) Transgenic animals: new approaches to hypertension research. Hypertension 17:394–399

Monti J, Schinke M, Böhm M, Ganten D, Bader M, Bricca G (2001) Glial angiotensinogen regulates brain angiotensin II receptors in transgenic rats TGR(ASrAOGEN) Am J Physiol 280: R233-R240

Morimoto S, Cassell MD, Beltz TG, Johnson AK, Davisson RL, Sigmund CD (2001) Elevated blood pressure in transgenic mice with brain-specific expression of human angiotensinogen driven by the glial fibrillary acidic protein promoter. Circ Res 89:365–372

Morimoto S, Cassell MD, Sigmund CD (2002b) Glia- and neuron-specific expression of the renin–angiotensin system in brain alters blood pressure, water Intake, and salt preference. J Biol Chem 277:33235–33241

Morimoto S, Cassell MD, Sigmund CD (2002a) The brain renin–angiotensin system in transgenic mice carrying a highly regulated human renin transgene. Circ Res 90:80–86

Mullins JJ, Peters J, Ganten D (1990) Fulminant hypertension in transgenic rats harbouring the mouse Ren-2 gene. Nature 344:541–544

Mullins JJ, Sigmund CD, Kane-Haas C, McGowan RA, Gross KW (1989) Expression of the DBA/2 J Ren-2 gene in the adrenal gland of transgenic mice. EMBO J 8:4065–4072

Munir MI, Rossiter BJF, Caskey CT (1990) Antisense RNA production in transgenic mice. Somat Cell Mol Genet 16:383–394

Müller DN, Dechend R, Mervaala EM, Park JK, Schmidt F, Fiebeler A, Theuer J, Breu V, Ganten D, Haller H, Luft FC (2000) NF-kappaB inhibition ameliorates angiotensin II-induced inflammatory damage in rats. Hypertension 35:193–201

Müller DN, Heissmeyer V, Dechend R, Hampich F, Park JK, Fiebeler A, Shagdarsuren E, Theuer J, Elger M, Pilz B, Breu V, Schroer K, Ganten D, Dietz R, Haller H, Scheidereit C, Luft FC (2001) Aspirin inhibits NF-kappaB and protects from angiotensin II-induced organ damage. FASEB J 15:1822–1824

Nguyen G, Delarue F, Burckle C, Bouzhir L, Giller T, Sraer JD (2002) Pivotal role of the renin/prorenin receptor in angiotensin II production and cellular responses to renin. J Clin Invest 109:1417–1427

Niimura F, Labosky PA, Kakuchi J, Okubo S, Yoshida H, Oikawa T, Ichiki T, Naftilan AJ, Fogo A, Inagami T, Hogan BLM, Ichikawa I (1995) Gene targeting in mice reveals a requirement for angiotensin in the development and maintenance of kidney morphology and growth factor regulation. J Clin Invest 96:2947–2954

No D, Yao TP, Evans RM (1996) Ecdysone-inducible gene expression in mammalian cells and transgenic mice. Proc Natl Acad Sci USA 93:3346–3351

Ohkubo H, Kawakami H, Kakehi Y, Takumi T, Arai H, Yokota Y, Iwai M, Tanabe Y, Masu M, Hata J, Iwao H, Okamoto H, Yokoyama M, Nomura T, Katsuki M, Nakanishi S (1990) Generation of transgenic mice with elevated blood pressure by introduction of the rat renin and angiotensinogen genes. Proc Natl Acad Sci USA 87:5153–5157

Oliver WJ, Gross F (1966) Unique specificity of mouse angiotensinogen to homologous renin. Proc Soc Exp Biol Med 122:923–926

Palmiter RD, Brinster RL (1986) Germ-line transformation of mice. Ann Rev Genet 20:465–499

Paradis P, Dali-Youcef N, Paradis FW, Thibault G, Nemer M (2000) Overexpression of angiotensin II type I receptor in cardiomyocytes induces cardiac hypertrophy and remodeling. Proc Natl Acad Sci USA 97:931–936

Pepin M-C, Pothier F, Barden N (1992) Impaired type II glucocorticoid-receptor function in mice bearing antisense RNA transgene. Nature 355:725–728

Peters J, Münter K, Bader M, Hackenthal E, Mullins JJ, Ganten D (1993) Increased adrenal renin in transgenic hypertensive rats, TGR(mREN2)27, and its regulation by cAMP, angiotensin II, and calcium. J Clin Invest 91:742–747

Ramaraj P, Kessler SP, Colmenares C, Sen GC (1998) Selective restoration of male fertility in mice lacking angiotensin- converting enzymes by sperm-specific expression of the testicular isozyme. J Clin Invest 102:371–378

Rosenkranz S, Nickenig G, Flesch M, Cremers B, Schnabel P, Lenz O, Krause T, Ganten D, Hoffmann S, Bohm M (1997) Cardiac angiotensin II receptors: studies on functional coupling in Sprague-Dawley rats and TGR(alphaMHC-hAT1) transgenic rats. Eur J Pharmacol 330:35–46

Sander M, Bader M, Djavidani B, Maser-Gluth C, Vecsei P, Mullins J, Ganten D, Peters J (1992) The role of the adrenal gland in the hypertensive transgenic rats TGR (mREN2)27. Endocrinology 131:807–814

Sander M, Ganten D, Mellon SH (1994) Role of adrenal renin in the regulation of adrenal steroidogenesis by corticotropin. Proc Natl Acad Sci USA 91:148–152

Schinke M, Baltatu O, Böhm M, Peters J, Rascher W, Bricca G, Lippoldt A, Ganten D, Bader M (1999) Blood pressure reduction and diabetes insipidus in transgenic rats deficient in brain angiotensinogen. Proc Natl Acad Sci USA 96:3975–3980

Senanayake P, Moriguchi A, Kumagai H, Ganten D, Ferrario CM, Brosnihan KB (1994) Increased expression of angiotensin peptides in the brain of transgenic hypertensive rats. Peptides 15:919–926

Sigmund CD, Jones CA, Fabian JR, Mullins JJ, Gross KW (1990a) Tissue and cell specific expression of a renin promoter-reporter gene construct in transgenic mice. Biochem Biophys Res Commun 170:344–350

Sigmund CD, Jones CA, Jacob HJ, Ingelfinger J, Kim U, Gamble D, Dzau VJ, Gross KW (1991) Pathophysiology of vascular smooth muscle in renin promoter-T-antigen transgenic mice. Am J Physiol 260: F249–F257

Sigmund CD, Jones CA, Kane CM, Wu C, Lang JA, Gross KW (1992) Regulated tissue- and cell-specific expression of the human renin gene in transgenic mice. Circ Res 70:1070–1079

Sigmund CD, Jones CA, Mullins JJ, Kim U, Gross KW (1990b) Expression of murine renin genes in subcutaneous connective tissue. Proc Natl Acad Sci USA 87:7993–7997

Sigmund CD, Okuyama K, Ingelfinger J, Jones CA, Mullins JJ, Kane C, Kim U, Wu C, Kenny L, Rustum Y, Dzau VJ, Gross KW (1990c) Isolation and characterization of renin-expression cell lines from transgenic mice containing a renin-promoter viral oncogene fusion construct. J Biol Chem 265:19916–19922

Sinn PL, Davis DR, Sigmund CD (1999a) Highly regulated cell type-restricted expression of human renin in mice containing 140- or 160-kilobase pair P1 phage artificial chromosome transgenes. J Biol Chem 274:35785–35793

Sinn PL, Zhang X, Sigmund CD (1999b) JG cell expression and partial regulation of a human renin genomic transgene driven by a minimal renin promoter. Am J Physiol 277: F634-F642

Sola C, Tronik D, Dreyfus M, Babinet C, Rougeon F (1989) Renin-promoter SV40 large T-antigen transgenes induce tumors irrespective of normal cellular expression of renin genes. Oncogene Res 5:149–153

Sugino H, Ozono R, Kurisu S, Matsuura H, Ishida M, Oshima T, Kambe M, Teranishi Y, Masaki H, Matsubara H (2001) Apoptosis is not increased in myocardium overexpressing type 2 angiotensin II receptor in transgenic mice. Hypertension 37:1394–1398

Takahashi S, Fukamizu A, Hasegawa T, Yokoyama M, Nomura T, Katsuki M, Murakami K (1991) Expression of the human angiotensinogen gene in transgenic mice and transfected cells. Biochem Biophys Res Commun 180:1103–1109

Takimoto E, Ishida J, Sugiyama F, Horiguchi H, Murakami K, Fukamizu A (1996) Hypertension induced in pregnant mice by placental renin and maternal angiotensinogen. Science 274:995–998

Thompson MW, Smith SB, Sigmund CD (1996) Regulation of human renin mRNA expression and protein release in transgenic mice. Hypertension 28:290–296

Tian X-L, Costerousse O, Urata H, Franz W-M, Paul M (1996) A new transgenic rat model overexpressing human angiotensin-converting enzyme in the heart. Hypertension 28:520

Tokita Y, Franco-Saenz R, Mulrow PJ, Ganten D (1994) Effects of nephrectomy and adrenalectomy on the renin–angiotensin system on transgenic rats TGR(mRen2)27. Endocrinology 134:253–257

Tronik D, Dreyfus M, Babinet C, Rougeon F (1987) Regulated expression of the Ren-2 gene in transgenic mice derived from parental strains carrying only the Ren-1 gene. EMBO J 6:983–987

Tronik D, Rougeon F (1988) Thyroxine and testosterone transcriptionally regulate renin gene expression in the submaxillary gland of normal and transgenic mice carrying extra copies of the Ren2 gene. FEBS Lett 234:336–340

Tsutsumi Y, Matsubara H, Masaki H, Kurihara H, Murasawa S, Takai S, Miyazaki M, Nozawa Y, Ozono R, Nagakawa K, Miwa T, Kawada N, Mori Y, Shibasaki Y, Tanaka Y, Fujiyama S, Koyama Y, Fujiyama A, Takahashi H, Iwasaka T (1999) Angiotensin II type 2 receptor overexpression activates the vascular kinin system and causes vasodilation. J Clin Invest 104:925–935

Tuschl T (2002) Expanding small RNA interference. Nat Biotechnol 20:446–448

van Kats JP, Methot D, Paradis P, Silversides DW, Reudelhuber TL (2001) Use of a biological peptide pump to study chronic peptide hormone action in transgenic mice. Direct and indirect effects of angiotensin II on the heart. J Biol Chem 276:44012–44017

Veniant M, Menard J, Bruneval P, Morley S, Gonzales MF, Mullins J (1996) Vascular damage without hypertension in transgenic rats expressing prorenin exclusively in the liver. J Clin Invest 98:1966–1970

Whitworth CE, Fleming S, Cumming AD, Morton JJ, Burns NJT, Williams BC, Mullins JJ (1994) Spontaneous development of malignant phase hypertension in transgenic Ren-2 rats. Kidney Int 46:1528–1532

Yan Y, Chen R, Pitarresi T, Sigmund CD, Gross KW, Sealey JE, Laragh JH, Catanzaro DF (1998a) Kidney is the only source of human plasma renin in 45-kb human renin transgenic mice. Circ Res 83:1279–1288

Yan Y, Hu L, Chen R, Sealey JE, Laragh JH, Catanzaro DF (1998b) Appropriate regulation of human renin gene expression and secretion in 45-kb human renin transgenic mice. Hypertension 32:205–214

Yang G, Merrill DC, Thompson MW, Robillard JE, Sigmund CD (1994) Functional expression of the human angiotensinogen gene in transgenic mice. J Biol Chem 269:32497–32502

Yang Z, Bove CM, French BA, Epstein FH, Berr SS, DiMaria JM, Gibson JJ, Carey RM, Kramer CM (2002) Angiotensin II type 2 receptor overexpression preserves left ventricular function after myocardial infarction. Circulation 106:106–111

Zhao Y, Bader M, Kreutz R, Fernandez-Alfonso M, Zimmermann F, Ganten U, Metzger R, Ganten D, Mullins JJ, Peters J (1993) Ontogenetic regulation of mouse Ren-2d renin gene in transgenic hypertensive rats, TGR(mREN2)27. Am J Physiol 265:E699–E707

Zhou Y, Overbeek PA, Bernstein KE (1996) Tissue specific expression of testis angiotensin converting enzyme is not determined by the −32 nonconsensus TATA motif. Biochem Biophys Res Commun 223:48–53

Zolk O, Flesch M, Nickenig G, Schnabel P, Bohm M (1998) Alteration of intracellular Ca2(+)-handling and receptor regulation in hypertensive cardiac hypertrophy: insights from Ren2-transgenic rats. Cardiovasc Res 39:242–256

Gene Therapy and the Renin–Angiotensin System

M. I. Phillips · B. Kimura

Department of Physiology and Functional Genomics,
University of Florida, College of Medicine, Box 100274, Gainesville, FL 32610, USA
e-mail: MIP@ufl.edu

1	Introduction	252
2	Antisense Oligonucleotides	255
3	Plasmid Vectors	255
4	Viral Vectors	255
5	RAS Gene Therapy	256
6	AT_1 Receptors	257
7	Angiotensinogen	258
8	Angiotensin-Converting Enzyme	258
9	Renin	259
10	AT_2 Receptors	259
11	Discussion	259
References		260

Abstract The success of gene therapy requires specific targets. For the potential treatment of cardiovascular disease by gene therapy, the renin–angiotensin system offers targets that are know to be effective based on pharmaceutical drugs that have targeted angiotensin-converting enzyme (ACE) angiotensin type I receptors (AT_1R). Other targets include renin and angiotensinogen. It is clear that lowering overactive amounts of angiotensin II has positive effects on reducing high blood pressure, decreasing left ventricular hypertrophy, and lowering the risk of heart attacks, stroke, and kidney failure. To inhibit the RAS system we have used antisense (AS) inhibitors which can be delivered either as oligonucleotides and used like small molecule drugs or in vectors with a sequence of DNA in the antisense direction. This review summarizes the effects of these two approaches on reducing high blood pressure. Gene therapy with antisense offers long-lasting, highly specific inhibition of gene products such as the components of RAS when their overactivity needs to be reduced in order to ameliorate disease.

Keywords Gene therapy · Angiotensin-converting enzyme (ACE) · Angiotensin type I · Renin · Angiotensinogen · Antisense (AS) · Oligonucleotides

1
Introduction

The renin–angiotensin system (RAS) is important in blood pressure regulation, volume regulation, and vascular tissue growth. Angiotensin II (Ang II), an octapeptide, is the active peptide of the system. It is formed from angiotensin I (Ang I) by angiotensin-converting enzyme (ACE). Ang I is formed from angiotensinogen (AGT) by renin. Ang II is the ligand for Ang II type 1 receptors (AT_1R) and Ang II type 2 receptors (AT_2R). In addition, angiotensin II metabolites, Ang III, Ang IV, and Ang 1-7, are active and may have independent receptors (Fig. 1). All the components of RAS are present in the brain and in the periphery; however, renin levels in the brain are very low. Both the brain and blood-borne RAS are important for blood pressure regulation. The brain RAS is also involved in drinking, salt intake, the baroreflex, and hormonal release from the paraventricular nucleus (PVN). The peripheral tissue RAS is involved in cardiac hypertrophy and hyperplasia. AT_1 receptors have been shown to mediate blood pressure and the growth effects of Ang II (Benetos et al. 1996; Chung and Unger 1999; Kurland et al. 2002; Unger 2002). The role of the AT_2 receptor is still uncertain, although it has been implicated in apoptosis and has effects opposing those of the AT_1 receptor (De Paepe et al. 2001, 2002; Schmieder et al. 2001; Unger 2002). However, mice with AT_1 receptors, but lacking AT_2

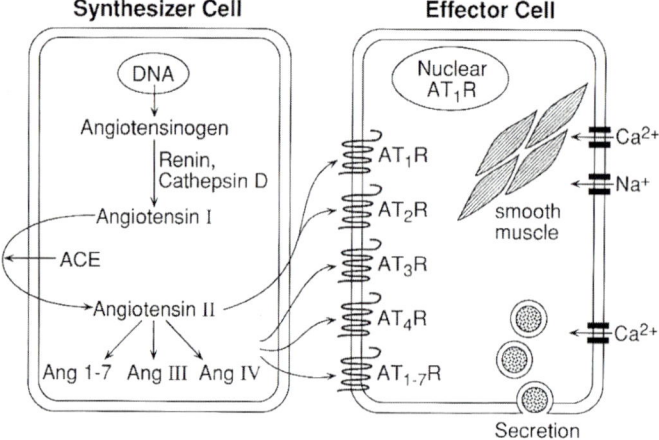

Fig. 1 The renin–angiotensin system (RAS) at the cellular level. In the brain and other tissues, all evidence points to the system being formed by a paracrine mechanism involving cells with genes for synthesis of angiotensin II (Ang II) and cells for uptake of metabolites through different, specific receptors. As angiotensin-converting enzyme (ACE) is an ectoenzyme, its action on the conversion of Ang I to Ang II is extracellular. Understanding of the specific functional roles of these receptors and metabolites is the key to specific gene targeting for cardiovascular gene therapies

receptors, did not develop hypertrophy in response to Ang II infusion (Ichihara et al. 2001), suggesting that the relationship between the two receptor types is more complex.

The rate-limiting step in the RAS cascade is the conversion of AGT to the decapeptide, Ang I. Increases in AGT production have been shown to effect blood pressure both in man and in experimental animals (Walker et al. 1979; Kim et al. 1995; Bloem et al. 1997). Transgenic mice that produce high levels of Ang II are hypertensive (Ohkubo et al. 1990; Fukamizu et al. 1993; Merrill et al. 1996; Davisson et al. 1999; Morimoto et al. 2001; Stec et al. 2002), and hypertensive rat models have increased levels of Ang II (Morton and Wallace 1983; Phillips and Kimura 1986, 1988; Morishita et al. 1992; Navar et al. 1995). Spontaneously hypertensive rats (SHR) also have an increased density of Ang II receptors (Gutkind et al. 1988; Brown et al. 1997).

Human genetic studies have shown that the AGT gene is linked to hypertension. In French and Utahan populations, the AGT 235T variant is more frequent in hypertensives than in controls (Atwood et al. 1997; Corvol and Jeunemaitre 1997; Niu et al. 1998; Jain et al. 2002). The ACE gene insertion/deletion variant has also been implicated, but the association of this gene with hypertension may depend on both ethnicity and gender (O'Donnell 1998; Agerholm-Larsen et al. 2000). There is also some evidence for the involvement of AT_1 receptor gene polymorphism involvement in human hypertension and arterial stiffness (Bonnardeaux et al. 1994; Benetos et al. 1996; Kurland et al. 2002).

The pressor effects of circulating Ang II have been known since the 1930s (Phillips and Schmidt-Ott 1999) and ACE inhibitors are a preferred class of drugs used to treat high blood pressure. Both ACE inhibitors and the newer AT_1 receptor antagonists decrease left ventricular hypertrophy (LVH) and hypertension (Chung and Unger 1999; Phillips 2001a,b; Unger 2002). However, these drugs have to be taken daily, and despite being effective in controlling hypertension, only 27% of patients with hypertension take the drugs consistently (Phillips 2000, 2001a,b). A shocking 73% of patient with hypertension do not comply with their drug treatment (Kaplan 1998). Cardiovascular disease is the leading cause of death in the United States and Europe, and WHO estimates that 17 million people worldwide die of cardiovascular diseases every year (www.americanheart.org). Hodgson and Cai reported that the cost of hypertension was $108.8 billion in the U.S. in 1998, and the American Heart Association estimated the direct and indirect costs of cardiovascular diseases to be $329 billion in the U.S. in 2002 (Hodgson and Cai 2001). Clearly, new treatments for hypertension are needed. We propose that a gene therapy approach would offer several advantages that may increase compliance: First, the treatment would be long-lasting (weekly, month or longer); and second, the high specificity of gene targeting would produce few side effects.

One approach is to target the mRNA of components involved in hypertension. Even though hypertension is a multifactorial disease, inhibition of RAS remains a promising strategy, as pharmacological depression of the system is known to decrease blood pressure. We propose antisense inhibition of well-documented drug targets such as ACE, AGT, and AT_1R. Preclinically, we have tested antisense oligonucleotides (AS-ODN), plasmids and viral vectors to decrease levels of components of the RAS (Tables 1 and 2) (Phillips et al. 1994; Kagiyama et al. 2001; Phillips 2001a,b).

Table 1 Gene therapy for hypertension: AS against brain RAS vasoconstrictor genes

Target gene	Construct	Route of delivery	Animal model	Max Δ BP (mmHg)	Duration of effect	Reference(s)
AT_1 receptor	AS-ODN	ICV	SHR	−45	7 days	Gyurko et al. 1993, 1997; Piegari et al. 2000
AT_1 receptor	AS-ODN	ICV	CIH	−35	4 days	Peng et al. 1998
AT_1 receptor	AS-ODN	ICV	2K1C chronic	−20	>5 days	Kagiyama et al. 2001
AT_1 receptor	AAV	ICV	SHR	−40	>9 weeks	Phillips et al. 1997
AGT	AS-ODN	ICV	SHR	−35	n.d.	Gyurko et al. 1993; Wielbo et al. 1995; Kagiyama et al. 1998
AGT	AS-ODN	ICV	CIH	−40	2 days	Peng et al. 1998
Renin	AS-ODN	ICV	SHR	−20	3 days	Kubo et al. 2001
AGE 2	Decoy ODN	ICV	SHR	−30	7 days	Nishii et al. 1999

Table 2 Gene therapy for hypertension: AS against peripheral RAS vasoconstrictor genes

Target gene	Construct	Route of delivery	Animal model	Max Δ BP (mmHg)	Duration of effect	Reference(s)
AT_1 receptor	AS-ODN	IC	CIH	−35	n.d	Peng et al. 1998
AT_1 receptor	AS-ODN	IV	2K1C acute	−30	>7 days	Galli and Phillips 2001
AT_1 receptor	LNSV	IC in 5-day-old rats	SHR	−45	>120 days	Iyer et al. 1996; Lu et al. 1997
AT_1 receptor	LNSV	IC in 5-day-old rats	60% fructose	−20	>2 weeks	Katovich et al. 2001
AGT	AS-ODN	Portal vein	SHR	−20	4 days	Tomita et al. 1995
AGT	AS-ODN	IV	SHR	−30	7 days	Wielbo et al. 1996; Makino et al. 1998, 1999; Sugano et al. 2000
AGT	P AS-AGT	IV	SHR	−20	8 days	Tang et al. 1999
AGT	AAV	IC in 5-day-old rats	SHR	−25	6 months	Kimura et al. 2001
AGE 2	Decoy ODN	Portal vein	SHR	−20	6 days	Morishita et al. 1996

n.d., Not determined.

2
Antisense Oligonucleotides

AS-ODNs consist of short DNA sequences of 12–20 bases that are complimentary to the mRNA producing the protein of interest. They bind to the mRNA and prevent translation of the specific protein encoded in the mRNA. To prevent degradation while in the circulation, the oligonucleotides are phosphorothioated, or otherwise modified, to increase stability. AS-ODNs can be solely administered, but delivery in liposomes, liposomes coupled to Sendai virus, or in carrier molecules increases uptake and prolongs the effect of AS-ODN (Morishita et al. 1993; Hughes et al. 1994; Dzau et al. 1996; Makino et al. 1999; Clare et al. 2000; Zhang et al. 2000, 2001; Fillion et al. 2001). In vitro experiments have shown that AS-ODNs enter the cell and the nucleus (Li et al. 1997).

3
Plasmid Vectors

Full-length antisense mRNA can be manufactured in plasmid vectors under the control of a promoter (Mohuczy and Phillips 2000). Theoretically, plasmids should be effective for a longer duration than AS-ODNs, but practically, the difference is negligible. Although uptake efficiency is an issue, recent studies using liposomes and receptor-mediated uptake have shown adequate effects of antisense mRNA (Tang et al. 1999; Merdan et al. 2002). Plasmid vectors have the potential to express antisense mRNA in specific cell types, if cell-specific promoters or cell-specific delivery systems are used (Zhang et al. 2002). Thus, they would be advantageous when transient expression is required.

4
Viral Vectors

Viral vectors containing cDNA in the antisense orientation, can potentially integrate into the genome and express antisense mRNA for components of RAS. Our results show long-term attenuation of hypertension and cardiac hypertrophy (Tables 1–3) (Phillips et al. 1997; Mohuczy and Phillips 2000; Kimura et al. 2001; Phillips 2001a,b). The adeno-associated virus (AAV) was used because it is safe, stable, long acting, and appropriate for gene therapy in adult models. Retroviruses preferentially infect dividing cells and integrate randomly into the genome (Mohuczy and Phillips 2000; Phillips 2001a,b). They have been used in infant models that study the development of hypertension (Lu et al. 1997; Wang et al. 2000; Metcalfe et al. 2002). However, they are not suitable for adult gene therapy. Lentiviruses (based on HIV, SIV, or FIV) can infect non-dividing cells and have a large carrying capacity; however, they integrate randomly and thus may disrupt other genes and cause mutagenesis (Hauswirth and McInnes 1998; Mohuczy and Phillips 2000; Sinnayah et al. 2002). Adenoviruses have very good uptake and do not integrate into the host genome. They show high levels of short-term expression. Their major disadvantage, as a long-term therapy, is the

Table 3 AS to RAS: effects on growth

Target gene	Construct	Route of delivery	Animal model	Effect studied	Magnitude	Reference
AT$_1$	LNSV	IC in 5-day-old rats	TGR mRen2	Hypertrophy	90% decrease	Pachori et al. 2002
AGT	AS-ODN	IV	SHR	Hypertrophy	60% decrease	Makino et al. 1999
AGT	AS-ODN	IV	SHR	Media of aorta	32% decrease	Sugano et al. 2000
AGT	AAV	IC in 5-day-old rats	SHR	Hypertrophy	About 25% decrease	Kimura et al. 2001
ACE	AS-ODN	Into injured artery	SD balloon catheter injury	Neointima formation	Injured control: 0.24 mm^2 AS treated: 0.1 mm^2	Morishita et al. 1992

immune and inflammatory responses they elicit (Hauswirth and McInnes 1998; Mohuczy and Phillips 2000; Phillips 2001a; Sinnayah et al. 2002). AAV is, therefore, the viral vector of choice for human use as it does not cause an immune response. It may integrate into the genome when it is modified and certainly has very long-lasting effects. The wild-type AAV does not cause any known diseases, and cannot proliferate without a helper virus. It is likely the safest of viral vectors. The primary disadvantage of the AAV vector is its small carrying capacity. AAV can only accommodate 4.4 kb, and thus the number of promoters, enhancers and length of AS-mRNA is limited. In addition, the deletion of the *rep* sequence removes site-specific integration of the vector (Hauswirth and McInnes 1998; Mohuczy and Phillips 2000; Phillips 2001a,b). Nevertheless, one of the advantages of antisense approach is that it is not necessary to have a full-length DNA, and therefore, shorter AS-DNA sequences can be used effectively in AAV

5
RAS Gene Therapy

We began targeting AT$_1$R and AGT with AS-ODNs in 1993, as gene therapy for hypertension (Gyurko et al. 1993; Phillips et al. 1994; Sinnayah et al. 2002). In the past 10 years, all the components of the RAS have been targeted for gene therapy and antisense inhibition. Some studies have aimed to understand the mechanisms of actions in RAS; other investigations have studied cardiovascular changes, hypertension, myocardial dysfunction, and the growth effects of the RAS; and still others have studied the hormones of RAS and behavior.

6
AT$_1$ Receptors

Using antisense oligonucleotides in an intact animal showed that AT$_1$R antisense injected into the brain lateral ventricle of SHR decreased blood pressure by about 25 mmHg within 24 h, and caused a 16%–40% decrease in AT$_1$ receptors in the PVN and organum vasculosum lamina terminalis (OVLT) (Ambuhl et al. 1995; Gyurko et al. 1997). These results have subsequently been corroborated by studies from our laboratory, as well as other researchers (Piegari et al. 2000; Sinnayah et al. 2002). AT$_1$R AS-ODN applied to the central RAS can also attenuate blood pressure in non-genetic models of hypertension. These include the surgical model, two-kidney-one-clip (2K1C), and the environmental model, cold-induced hypertension (CIH) (Peng et al. 1998; Kagiyama et al. 2001).The data are summarized in Table 1. In addition to the effect on blood pressure, spontaneous drinking, Ang II, and isoproterenol-induced drinking is decreased by the AT$_1$R antisense (Meng et al. 1994; Sakai et al. 1995; Peng et al. 1998). Saline-induced (i.e., osmotic) drinking is not affected (Sakai et al. 1995). In early studies, AS-ODN applied to the peripheral circulation failed to elicit a response. When mixed with liposomes, uptake of AT$_1$-AS-ODN into peripheral organs increased, blood pressure decreased by 25–35 mmHg, and AT$_1$Rs in kidney and arteries, in both CIH and 2K1C hypertensive rats, also decreased (Peng et al. 1998; Galli and Phillips 2001). The data are summarized in Table 2. AT$_1$-AS-ODN administered prior to ischemia–reperfusion also protected against myocardial dysfunction (Yang et al. 1998). The effects of AT$_1$-AS-ODN are transient, lasting for about a week, with the maximum effect seen after 2–3 days.

Viral vectors, on the other hand, enabled very long-term expression of antisense. We have used AAV to deliver AT$_1$AS mRNA both to the central and the peripheral system of SHR, and obtained attenuation of hypertension by approximately 25 mmHg (see Table 1). The reduction in blood pressure lasted at least 9 weeks (Phillips et al. 1997) and we recorded normalization of blood pressure in double transgenic mice for up to 6 months. These mice have a gene for human renin and another for human angiotensin, and therefore, they constantly overexpress Ang II and are hypertensive. AAV delivery of AS to AT$_1$R dramatically reduces blood pressure within a few days, and the effect persists for as long as the mice were tested. Our colleagues have used a lentiviral vector (LNSV) to deliver AT$_1$AS mRNA to 5 day-old SHR, and obtained similar results (see Table 2) (Iyer et al. 1996; Lu et al. 1997; Wang et al. 2000). The attenuation of hypertension and of hypertrophy persisted in the offspring of treated rats. This is probably the result of postnatal age hnumber –"Sec8"(Metcalfe et al. 2002; Pachori et al. 2002). We have found that AAV vectors enter the germ line only if the viral vectors are given at a very early age (0–5 days). At this time the blood–testis barrier is open. After postnatal day 5 or 6, rats injected with AAV do not transmit AAV to their offspring.

7
Angiotensinogen

The earliest studies targeting AGT in the brain showed a substantial, up to 40 mmHg, blood pressure decrease in SHR and a decrease of hypothalamic AGT (Wielbo et al. 1995; Gyurko et al. 1993). Subsequent research has confirmed that intracerebroventricular (ICV) injections of AGT-AS-ODN decrease blood pressure in SHR. Injections into the PVN did not affect blood pressure, although it did decrease vasopressin release (Kagiyama et al. 1998). The CIH and 2K1C hypertensive models also respond to AGT-AS-ODN ICV treatment with a decrease in blood pressure (Peng et al. 1998; Kagiyama et al. 2001). The data are summarized in Table 1. The drinking response in CIH animals is also attenuated (Peng et al. 1998). In a normotensive rat, the drinking response to renin and isoproterenol is attenuated by ICV AGT-AS-ODN injection, while the drinking responses to carbachol, Ang II and water depravation are unaffected (Sinnayah et al. 1997a,b). Early studies report that AS-ODN injected into the peripheral circulation had no effects, likely due to the failure to reach target organs in sufficient amounts. We compared the effects of naked AS-ODN and liposome encapsulated AS-ODN directed against AGT, on blood pressure, AGT and AngII concentration, and hepatic uptake. We found that naked AS-ODN was without effect, whereas liposome-encapsulated AS-ODN lowered blood pressure in SHR. This was accompanied by lower peripheral AGT and Ang II in the liver after injection (Wielbo et al. 1996). Similar results were obtained when AS-ODN was coupled to carrier molecules that targeted delivery to the liver (see Table 2) (Tomita et al. 1995; Makino et al. 1998; Sugano et al. 2000). In addition to effects on protein levels and blood pressure, AGT-AS-ODN also attenuated hypertrophy of the heart and aortic smooth muscles (see Table 3) (Makino et al. 1999; Sugano et al. 2000).

AGT-AS-ODN attenuated hypertension for 4–5 days (Tomita et al. 1995; Makino et al. 1998). When a full-length AGT cDNA was inserted into a plasmid under control of CMV promoter, and injected with liposomes into SHR, the blood pressure decrease lasted for 8 days (Tang et al. 1999). The same construct delivered by AAV to 5-day-old SHR caused a delay in hypertension development, attenuated hypertrophy, and reduced the degree of hypertension for at least 6 months (Kimura et al. 2001).

8
Angiotensin-Converting Enzyme

AS-ODN directed against ACE mRNA have been used to reduce the vascular ACE concentration and formation of neointima after balloon catheter injury (Table 3) (Morishita et al. 2000). ACE-AS-ODN have also been shown to improve cardiac performance after ischemia–reperfusion injury (Chen et al. 2001).

ACE is present at low to moderate levels in large areas of the brain, and in high levels in the NTS, and is likely to generate Ang II locally (Phillips and

Kimura 1999). Increasing ACE levels in the brain, by transfection with a plasmid containing the human ACE gene, caused an increase in blood pressure, heart rate, Ang II, and ACE levels that lasted for 2 weeks (Nakamura et al. 1999). Unpublished results from our lab testing three different sequences of ACE-AS-ODN showed a decrease of 15–25 mmHg in blood pressure in hypertensive SHR.

9
Renin

Renin-AS-ODN decreased blood pressure by about 20 mmHg for 2 days, when injected into the lateral ventricle of SHR. Expression of renin mRNA was also suppressed by this treatment (Kubo et al. 2001). So far, no systemic studies with renin-AS-ODN have been performed. A study with β_1-AS-ODN showed that renin release was inhibited and blood pressure decreased for up to 1 month in SHR after a single injection of the ODN delivered in liposome (Zhang et al. 2000).

10
AT$_2$ Receptors

AT$_2$-AS-ODN infused into the kidney of uninephrectomized normotensive rats, increased blood pressure by about 20 mmHg throughout the infusion period (Moore et al. 2001).

11
Discussion

The use of gene therapy to correct genetic abnormalities and to treat diseases is becoming more relevant, clinically. At least one AS-ODN has been approved by the FDA to treat cytomegalovirus retinitis (de Smet et al. 1999; Orr 2001), and AAV-RPE65 has been used to restore sight in a canine model of blindness (Acland et al. 2001). Adenovirus is being used in phase I and phase II clinical trials in cancer patients (Lamont et al. 2000; Reid et al. 2001; Teh et al. 2001; Freytag et al. 2002; Harvey et al. 2002). AS-ODNs and viral vectors have been introduced into both the periphery and central system of the brain. Peripheral administration is more likely to be the more clinically acceptable therapy. The aforementioned preclinical studies have demonstrated that AS to AGT, AT$_1$, and ACE can successfully lower blood pressure in hypertensive rats, and attenuate hypertrophy in adult animals when administered systemically. The decrease in blood pressure varies between 15 and 40 mmHg (Tomita et al. 1995; Wielbo et al. 1996; Peng et al. 1998; Makino et al. 1999; Tang et al. 1999; Kimura et al. 2001). While these effects are highly advantageous clinically, it is probably impossible to achieve a "cure" for hypertension with AS-ODN because the mechanism of antisense inhibition is a competition between copies of mRNA and the amount of ODN delivered to cells. This means that antisense treatment can be effective by reducing overactive receptors or hormones but not interfere with normal physiology. Clearly, there is a dose-dependent effect. We found a correlation between AGT-AS plasmid dose and blood pressure, and there was also a large decrease in blood pressure following AGT-AS-ODN and carrier protein targeting of the liver (Makino et al. 1999; Tang et al. 1999). Another factor favoring the dose-dependant mechanism is that pharmacological drugs, both ACE inhibitors and AT_1R antagonists, are able to normalize blood pressure alone, although some new approaches have tried combinations.

The next step, clinically, is toxicology tests and phase I, I,I and III trials. In the case of viral vectors, the next step is to ensure that they do not cause adverse effects over a very long time, such as immune responses or tumorigenesis. To this end, we are developing gene switches to turn vectors on and off as needed without resorting to drugs (Phillips et al. 2002). AS gene therapy has the potential of providing extended protection against hypertension, cardiovascular disease, and a multitude of other chronic diseases. Here we have reviewed its use on the RAS as a target system in hypertension and cardiovascular disease, but clearly any target with a known DNA sequence is a target for antisense inhibition.

References

Acland GM, Aguirre GD, Ray J et al (2001) Gene therapy restores vision in a canine model of childhood blindness. Nat Genet 28:92–95
Agerholm-Larsen B, Nordestgaard BG, Tybjarg-Hansen A (2000) ACE gene polymorphism in cardiovascular disease: meta-analyses of small and large studies in whites. Arterioscler Thromb Vasc Biol 20:484–492
Ambuhl P, Gyurko R, Phillips MI (1995) A decrease in angiotensin receptor binding in rat brain nuclei by antisense oligonucleotides to the angiotensin AT1 receptor. Regul Pept 59:171–182
Atwood LD, Kammerer CM, Samollow PB et al (1997) Linkage of essential hypertension to the angiotensinogen locus in Mexican Americans. Hypertension 30:326–330
Benetos A, Gautier S, Ricard S et al (1996) Influence of angiotensin-converting enzyme and angiotensin II type 1 receptor gene polymorphisms on aortic stiffness in normotensive and hypertensive patients. Circulation 94:698–703
Bloem LJ, Foroud TM, Ambrosius WT et al (1997) Association of the angiotensinogen gene to serum angiotensinogen in blacks and whites. Hypertension 29:1078–1082
Bonnardeaux A, Davies E, Jeunemaitre X et al (1994) Angiotensin II type 1 receptor gene polymorphisms in human essential hypertension. Hypertension 24:63–69
Brown L, Passmore M, Duce B et al (1997) Angiotensin receptors in cardiac and renal hypertrophy in rats. J Mol Cell Cardiol 29:2925–2929
Chen H, Mohuczy D, Li D et al (2001) Protection against ischemia/reperfusion injury and myocardial dysfunction by antisense-oligodeoxynucleotide directed at angiotensin-converting enzyme mRNA. Gene Ther 8:804–810
Chung O, Unger T (1999) Angiotensin II receptor blockade and end-organ protection. Am J Hypertens 12:S150–S156

Clare ZY, Kimura B, Shen L et al (2000) New beta-blocker: prolonged reduction in high blood pressure with beta(1) antisense oligodeoxynucleotides. Hypertension 35:219–224

Corvol P, Jeunemaitre X (1997) Molecular genetics of human hypertension: role of angiotensinogen. Endocr Rev 18:662–677

Davisson RL, Ding Yuem, Stec DE et al (1999) Novel mechanism of hypertension revealed by cell-specific targeting of human angiotensinogen in transgenic mice. Physiol Genomics 1:3–9

De Paepe B, Verstraeten VL, De Potter CR et al (2001) Growth stimulatory angiotensin II type-1 receptor is upregulated in breast hyperplasia and in situ carcinoma but not in invasive carcinoma. Histochem Cell Biol 116:247–254

De Paepe B, Verstraeten VM, De Potter CR et al (2002) Increased angiotensin II type-2 receptor density in hyperplasia, DCIS and invasive carcinoma of the breast is paralleled with increased iNOS expression. Histochem Cell Biol 117:13–19

de Smet MD, Meenken CJ, van den Horn GJ (1999) Fomivirsen—a phosphorothioate oligonucleotide for the treatment of CMV retinitis. Ocul Immunol Inflamm 7:189–198

Dzau VJ, Mann MJ, Morishita R et al (1996) Fusigenic viral liposome for gene therapy in cardiovascular diseases. Proc Natl Acad Sci USA 93:11421–11425

Fillion P, Desjardins A, Sayasith K et al (2001) Encapsulation of DNA in negatively charged liposomes and inhibition of bacterial gene expression with fluid liposome-encapsulated antisense oligonucleotides. Biochim Biophys Acta 1515:44–54

Freytag SO, Khil M, Stricker H et al (2002) Phase I study of replication-competent adenovirus-mediated double suicide gene therapy for the treatment of locally recurrent prostate cancer. Cancer Res 62:4968–4976

Fukamizu A, Sugimura K, Takimoto E et al (1993) Chimeric renin–angiotensin system demonstrates sustained increase in blood pressure of transgenic mice carrying both human renin and human angiotensinogen genes. J Biol Chem 268:11617–11621

Galli SM, Phillips MI (2001) Angiotensin II AT(1A) receptor antisense lowers blood pressure in acute 2-kidney, 1-clip hypertension. Hypertension 38:674–678

Gutkind JS, Kurihara M, Saavedra JM (1988) Increased angiotensin II receptors in brain nuclei of DOCA-salt hypertensive rats. Am J Physiol 255:H646–H650

Gyurko R, Tran D, Phillips MI (1997) Time course of inhibition of hypertension by antisense oligonucleotides targeted to AT1 angiotensin receptor mRNA in spontaneously hypertensive rats. Am J Hypertens 10:56S–62S

Gyurko R, Wielbo D, Phillips MI (1993) Antisense inhibition of AT1 receptor mRNA and angiotensinogen mRNA in the brain of spontaneously hypertensive rats reduces hypertension of neurogenic origin. Regul Pept 49:167–174

Harvey BG, Maroni J, O'Donoghue KA et al (2002) Safety of local delivery of low- and intermediate-dose adenovirus gene transfer vectors to individuals with a spectrum of morbid conditions. Hum Gene Ther 13:15–63

Hauswirth WW, McInnes RR (1998) Retinal gene therapy 1998: summary of a workshop. Mol Vis 4:11

Hodgson TA, Cai L (2001) Medical care expenditures for hypertension, its complications, and its comorbidities. Med Care 39:599–615

Hughes JA, Bennett CF, Cook PD et al (1994) Lipid membrane permeability of 2'-modified derivatives of phosphorothioate oligonucleotides. J Pharm Sci 83:597–600

Ichihara S, Senbonmatsu T, Price E Jr et al (2001) Angiotensin II type 2 receptor is essential for left ventricular hypertrophy and cardiac fibrosis in chronic angiotensin II-induced hypertension. Circulation 104:346–351

Iyer SN, Lu D, Katovich MJ et al (1996) Chronic control of high blood pressure in the spontaneously hypertensive rat by delivery of angiotensin type 1 receptor antisense. Proc Natl Acad Sci USA 93:9960–9965

Jain S, Tang X, Chittampalli S. N et al (2002) Angiotensinogen gene polymorphism at −217 affects basal promoter activity and is associated with hypertension in African-Americans. J Biol Chem M204732200

Kagiyama S, Kagiyama T, Phillips MI (2001) Antisense oligonucleotides strategy in the treatment of hypertension. Curr Opin Mol Ther 3:258–264

Kagiyama S, Tsuchihashi T, Abe I et al (1998) Antisense inhibition of angiotensinogen attenuates vasopressin release in the paraventricular hypothalamic nucleus of spontaneously hypertensive rats. Brain Res 829:120–124

Kagiyama S, Varela A, Phillips MI et al (2001) Antisense inhibition of brain renin–angiotensin system decreased blood pressure in chronic 2-kidney, 1 clip hypertensive rats. Hypertension 37:371–375

Kaplan NM (1998) Clinical hypertension. Williams and Williams, Baltimore

Katovich MJ, Reaves PY, Francis SC et al (2001) Gene therapy attenuates the elevated blood pressure and glucose intolerance in an insulin-resistant model of hypertension. J Hypertens 19:1553–1558

Kim HS, Krege JH, Kluckman KD et al (1995) Genetic control of blood pressure and the angiotensinogen locus. Proc Natl Acad Sci USA 92:2735–2739

Kimura B, Mohuczy D, Tang X et al (2001) Attenuation of hypertension and heart hypertrophy by adeno-associated virus delivering angiotensinogen antisense. Hypertension 37:376–380

Kubo T, Ikezawa A, Kambe T et al (2001) Renin antisense injected intraventricularly decreases blood pressure in spontaneously hypertensive rats. Brain Res Bull 56:23–28

Kurland L, Melhus H, Karlsson J et al (2002) Polymorphisms in the angiotensinogen and angiotensin II type 1 receptor gene are related to change in left ventricular mass during antihypertensive treatment: results from the Swedish Irbesartan Left Ventricular Hypertrophy Investigation versus Atenolol (SILVHIA) trial. J Hypertens 20:657–663

Lamont JP, Nemunaitis J, Kuhn JA et al (2000) A prospective phase II trial of ONYX-015 adenovirus and chemotherapy in recurrent squamous cell carcinoma of the head and neck (the Baylor experience). Ann Surg Oncol 7:588–592

Li B, Hughes JA, Phillips MI (1997) Uptake and efflux of intact antisense phosphorothioate deoxyoligonucleotide directed against angiotensin receptors in bovine adrenal cells. Neurochem Int 31:393–403

Lu D, Raizada MK, Iyer S et al (1997) Losartan versus gene therapy: chronic control of high blood pressure in spontaneously hypertensive rats. Hypertension 30:363–370

Makino N, Sugano M, Ohtsuka S et al (1998) Intravenous injection with antisense oligodeoxynucleotides against angiotensinogen decreases blood pressure in spontaneously hypertensive rats. Hypertension 31:1166–1170

Makino N, Sugano M, Ohtsuka S et al (1999) Chronic antisense therapy for angiotensinogen on cardiac hypertrophy in spontaneously hypertensive rats. Cardiovasc Res 44:543–548

Meng H, Wielbo D, Gyurko R et al (1994) Antisense oligonucleotide to AT1 receptor mRNA inhibits central angiotensin induced thirst and vasopressin. Regul Pept 54:543–551

Merdan T, Kopecek J, Kissel T (2002) Prospects for cationic polymers in gene and oligonucleotide therapy against cancer. Adv Drug Del Rev 54:715–758

Merrill DC, Thompson MW, Carney CL et al (1996) Chronic hypertension and altered baroreflex responses in transgenic mice containing the human renin and human angiotensinogen genes. J Clin Invest 97:1047–1055

Metcalfe BL, Raizada M, Katovich MJ (2002) Genetic targeting of the renin–angiotensin system for long-term control of hypertension. Curr Hypertens Rep 4:25–31

Mohuczy D, Phillips MI (2000) Designing antisense to inhibit the renin–angiotensin system. Mol Cell Biochem 212:145–153

Moore AF, Heiderstadt NT, Huang E et al (2001) Selective inhibition of the renal angiotensin type 2 receptor increases blood pressure in conscious rats. Hypertension 37:1285–1291

Morimoto S, Cassell MD, Beltz TG et al (2001) Elevated blood pressure in transgenic mice with brain-specific expression of human angiotensinogen driven by the glial fibrillary acidic protein promoter. Circ Res 89:365–372

Morishita R, Gibbons GH, Ellison KE et al (1993) Single intraluminal delivery of antisense cdc2 kinase and proliferating-cell nuclear antigen oligonucleotides results in chronic inhibition of neointimal hyperplasia. Proc Natl Acad Sci USA 90:8474–8478

Morishita R, Gibbons GH, Tomita N et al (2000) Antisense oligodeoxynucleotide inhibition of vascular angiotensin-converting enzyme expression attenuates neointimal formation: evidence for tissue angiotensin-converting enzyme function. Arterioscler Thromb Vasc Bio 20:915–922

Morishita R, Higaki J, Miyazaki M et al (1992) Possible role of the vascular renin–angiotensin system in hypertension and vascular hypertrophy. Hypertension 19:II62–II67

Morishita R, Higaki J, Tomita N et al (1996) Role of transcriptional cis-elements, angiotensinogen gene-activating elements, of angiotensinogen gene in blood pressure regulation. Hypertension 27:502–507

Morton JJ, Wallace EC (1983) The importance of the renin–angiotensin system in the development and maintenance of hypertension in the two-kidney one-clip hypertensive rat. Clin Sci (Lond) 64:359–370

Nakamura S, Moriguchi A, Morishita R et al (1999) Activation of the brain angiotensin system by in vivo human angiotensin-converting enzyme gene transfer in rats. Hypertension 34:302–308

Navar LG, Von Thun AM, Zou L et al (1995) Enhancement of intrarenal angiotensin II levels in 2 kidney 1 clip and angiotensin II induced hypertension. Blood Press Suppl 2:88–92

Nishii T, Moriguchi A, Morishita R et al (1999) Angiotensinogen gene-activating elements regulate blood pressure in the brain. Circ Res 85:257–263

Niu T, Xu X, Rogus J et al (1998) Angiotensinogen gene and hypertension in Chinese. J Clin Invest 101:188–194.

O'Donnell CJ, Lindpaintner K, Larson MG et al (1998) Evidence for association and genetic linkage of the angiotensin-converting enzyme locus with hypertension and blood pressure in men but not women in the Framingham Heart Study. Circulation 97:1766–1772

Ohkubo H, Kawakami H, Kakehi H et al (1990) Generation of transgenic mice with elevated blood pressure by introduction of the rat renin and angiotensinogen genes. PNAS 87:5153–5157

Orr RM (2001) Technology evaluation: fomivirsen, Isis Pharmaceuticals Inc/CIBA vision. Curr Opin Mol Ther 3:288–294

Pachori AS, Numan MT, Ferrario CM et al (2002) Blood pressure-independent attenuation of cardiac hypertrophy by AT(1)R-AS gene therapy. Hypertension 39:969–975

Peng JF, Kimura B, Fregly MJ et al (1998) Reduction of cold-induced hypertension by antisense oligodeoxynucleotides to angiotensinogen mRNA and AT1-receptor mRNA in brain and blood. Hypertension 31:1317–1323

Phillips MI (2000) Somatic gene therapy for hypertension. Braz J Med Biol Res 33:715–721

Phillips MI (2001a) Gene therapy for hypertension: sense and antisense strategies. Expert Opin Biol Ther 1:655–662

Phillips MI (2001b) Gene therapy for hypertension: the preclinical data. Hypertension 38:543–548

Phillips MI, Kimura B (1988) Brain angiotensin in the developing spontaneously hypertensive rat. J Hypertens 6:607–612

Phillips MI, Kimura B (1999) Central nervous system and angiotensin in the development of hypertension. In: McCarty R, Blizard DA, Chevalier RL (eds) Development of the hypertensive phenotype: basic and clinical studies. Elsevier Science BV 383–411

Phillips MI, Kimura BK (1986) Levels of brain angiotensin in the spontaneously hypertensive rat and treatment with ramiprilat. J Hypertens Suppl 4:S391–S394

Phillips MI, Mohuczy-Dominiak D, Coffey M et al (1997) Prolonged reduction of high blood pressure with an in vivo, nonpathogenic, adeno-associated viral vector delivery of AT1-R mRNA antisense. Hypertension 29:374–380

Phillips MI, Schmidt-Ott KM (1999) The discovery of renin 100 years ago. News Physiol Sci 14:271–274

Phillips MI, Wielbo D, Gyurko R (1994) Antisense inhibition of hypertension: a new strategy for renin–angiotensin candidate genes. Kidney Int 46:1554–1556

Piegari E, Galderisi U, Berrino L et al (2000) In vivo effects of partial phosphorothioated AT1 receptor antisense oligonucleotides in spontaneously hypertensive and normotensive rats. Life Sci 66:2091–2099

Reid T, Galanis E, Abbruzzese J et al (2001) Intra-arterial administration of a replication-selective adenovirus (dl1520) in patients with colorectal carcinoma metastatic to the liver: a phase I trial. Gene Ther 8:1618–1626

Sakai RR, Ma LY, He PF et al (1995) Intracerebroventricular administration of angiotensin type 1 (AT1) receptor antisense oligonucleotides attenuate thirst in the rat. Regulatory Peptides 59:183–192

Schmieder RE, Erdmann J, Delles C et al (2001) Effect of the angiotensin II type 2-receptor gene (+1675 G/A) on left ventricular structure in humans. J Am Coll Cardiol 37:175–182

Sinnayah P, Kachab E, Haralambidis J et al (1997a) Effects of angiotensinogen antisense oligonucleotides on fluid intake in response to different dipsogenic stimuli in the rat. Brain Res Mol Brain Res 50:43–50

Sinnayah P, Lindley TE, Staber PD et al (2002) Selective gene transfer to key cardiovascular regions of the brain: comparison of two viral vector systems. Hypertension 39:603–608

Sinnayah P, McKinley MJ, Coghlan JP (1997b) Angiotensinogen antisense oligonucleotides and fluid intake. Clin Exp Hypertens 19:993–1007

Stec DE, Keen HL, Sigmund CD (2002) Lower blood pressure in floxed angiotensinogen mice after adenoviral delivery of Cre-recombinase. Hypertension 39:629–633

Sugano M, Tsuchida K, Sawada S et al (2000) Reduction of plasma angiotensin II to normal levels by antisense oligodeoxynucleotides against liver angiotensinogen cannot completely attenuate vascular remodeling in spontaneously hypertensive rats. J Hypertens 18:725–731

Tang X, Mohuczy D, Zhang YC et al (1999) Intravenous angiotensinogen antisense in AAV-based vector decreases hypertension. Am J Physiol 277:H2392–H2399

Teh BS, Aguilar-Cordova E, Kernen K et al (2001) Phase I/II trial evaluating combined radiotherapy and in situ gene therapy with or without hormonal therapy in the treatment of prostate cancera preliminary report. Int J Radiat Oncol Biol Phys 51:605–613

Tomita N, Morishita R, Higaki J et al (1995) Transient decrease in high blood pressure by in vivo transfer of antisense oligodeoxynucleotides against rat angiotensinogen. Hypertension 26:131–136

Unger T (2002) The role of the renin–angiotensin system in the development of cardiovascular disease. Am J Cardiol 89:3A–9A

Walker WG, Whelton PK, Saito H et al (1979) Relation between blood pressure and renin, renin substrate, angiotensin II, aldosterone and urinary sodium and potassium in 574 ambulatory subjects. Hypertension 1:287–291

Wang H, Lu D, Reaves PY et al (2000) Retrovirally mediated delivery of angiotensin II type 1 receptor antisense in vitro and in vivo. Methods Enzymol 314:581–590

Wielbo D, Sernia C, Gyurko R et al (1995) Antisense inhibition of hypertension in the spontaneously hypertensive rat. Hypertension 25:314–319

Wielbo D, Simon A, Phillips MI et al (1996) Inhibition of hypertension by peripheral administration of antisense oligodeoxynucleotides. Hypertension 28:147–151

Yang B, Li D, Phillips MI et al (1998) Myocardial angiotensin II receptor expression and ischemia-reperfusion injury. Vasc Med 3:121–130

Zhang Y, Jeong LH, Boado RJ et al (2002) Receptor-mediated delivery of an antisense gene to human brain cancer cells. J Gene Med 4:183–194

Zhang YC, Bui JD, Shen L et al (2000) Antisense inhibition of beta(1)-adrenergic receptor mRNA in a single dose produces a profound and prolonged reduction in high blood pressure in spontaneously hypertensive rats. Circulation 101:682–688

Zhang YM, Rusckowski M, Liu N et al (2001) Cationic liposomes enhance cellular/nuclear localization of 99mTc-antisense oligonucleotides in target tumor cells. Cancer Biother Radiopharm 16:411–419

**Part 3
ANG Receptors**
AT_1

AT₁ Receptor Molecular Aspects

S. Conchon · E. Clauser

Département d'Endocrinologie, INSERM U567, CNRS UMR8104,
Faculté de Médecine Cochin, 24 rue du Fg St Jacques, 75014 Paris, France
e-mail: clauser@cochin.inserm.fr

1	Introduction	270
2	Structure of the AT$_1$ Receptor Gene, mRNA and Protein	270
2.1	Some History	270
2.2	Structure of the Gene	271
2.3	Structure of the mRNAs	272
2.4	Structure of the Protein	273
2.5	Phylogenic Aspects of the AT$_1$ Receptor	274
3	Structure–Function Relationships	275
3.1	The Ligand Binding Site	275
3.1.1	AngII Binding Site	276
3.1.2	Non-peptide Ligand Binding Site	277
3.2	Receptor Activation	279
3.3	Interaction with G Proteins	282
3.4	Interactions with Other Signalling Molecules	283
3.5	Regulation of AT$_1$ Receptor Activity	284
3.5.1	Phosphorylation	285
3.5.2	Phosphorylation and Regulation of AT$_1$ Receptor Activity	285
3.5.3	Phosphorylation and Activation of the AT$_1$ Receptor	286
3.5.4	AT$_1$ Phosphorylation and β-Arrestin Binding	286
3.6	Internalization and Trafficking	287
3.6.1	Molecular Determinants of Internalization	288
3.6.2	Internalization and Activation	288
3.6.3	Phosphorylation, β-Arrestin Binding and Receptor Trafficking	289
	References	290

Abstract In the past 12 years, cloning of the AngII AT$_1$ receptor cDNA and gene have elucidated the primary structure of the protein, but also the structure of the mRNA and gene in different species. Using this cDNA as a tool, the different motifs and sequences involved in functions of the receptor has been established. The binding of AngII implicates several sequences and amino acids of the extracellular loops, the N-terminus and upper segments of transmembrane domains (TM)4 to TM7. In contrast, the binding site of nonpeptide inverse agonist losartan is composed of polar residues of TM2 to TM7, and is deeply buried in the lipid bilayer. After agonist binding, the receptor is activated by a conformational change involving residues of the TMs. In the active state, sequences of intracellular loops two and three and of the C-terminus, which are adjacent to the TM, interact with the G protein to activate it. Sequences of the C-terminus are also involved in activation of other signalling pathways, such as the Jak-STAT pathway. In parallel, seryl residues of the C-terminus are phosphorylated and interact with β-arrestins, resulting in receptor internalization.

Keywords G protein-coupled receptor · AT_1 receptor · Angiotensin II · Site-directed mutagenesis · Ligand binding · Activation · Phosphorylation · Internalization

1
Introduction

The first cloning of the angiotensin II (AngII) AT_1 receptor in 1991 elucidated the primary sequence of the protein and was the starting point of extensive analysis of the structure function relationships of this G protein-coupled receptor (GPCR). The first part of this chapter will review the molecular aspects of the AT_1 receptors, including the structure of the gene(s), mRNAs and protein with their known species differences.

The receptor sequences and amino acids involved in ligand binding, activation and the G protein coupling, signalling, regulation and trafficking of the AT_1 receptor, have been extensively investigated using site-directed mutagenesis of its cDNA and expression of the mutated recombinant proteins. This investigation has produced a precise map of the structural molecular determinants of the different receptor functions. They are reviewed in the second part of this chapter.

Some conformational and mechanistic hypotheses derived from this mutagenesis work and computer modelling are also summarized, despite the absence of structural three-dimensional (NMR, crystallography) data concerning this receptor.

2
Structure of the AT_1 Receptor Gene, mRNA and Protein

2.1
Some History

The presence of AngII binding sites in target tissues such as adrenal cortex (Glossmann et al. 1974) and vascular smooth muscle cells (Gunther et al. 1980) was first demonstrated in the 1970s, when radiolabelled iodinated AngII was first synthesized. These experiments demonstrated the presence of membrane-bound receptors with some pharmacological differences from one tissue to another, but there was no definitive evidence at this time for different molecular species. Despite the preliminary biochemical characterization of the protein, attempts to purify the receptor were essentially unsuccessful due to loss of AngII-binding properties, after solubilization of the tissues. The first clues on the protein structure came from functional studies demonstrating, using guanosine nucleotide analogues, that the AngII receptor belongs to the protein superfamily of the GPCRs (Wright et al. 1982).

The development of new pharmacological tools in the late 1980s has permitted the identification of two AngII receptor types:

- The AngII AT_1 receptor, which is the standard functional receptor for AngII, binds with high-affinity imidazolic compounds such as Dup753 (losartan) and is sensitive to reducing agents such as dithiothreitol (DTT) (Chiu et al. 1989).
- The AngII AT_2 receptor, a newly identified membrane-bound receptor, binds L-spinacine derivatives (PD123319) (Chiu et al. 1989) and AngII pseudopeptidic analogues (CGP42112A) (Whitebread et al. 1989) with high affinity, is insensitive to DTT, and is expressed in adrenal medulla and myometrial cells and fetal mesenchymal tissues.

The next step in the elucidation of the AT_1 receptor structure was the cloning of the receptor cDNA. Concurrently, two American labs (Murphy et al. 1991; Sasaki et al. 1991) identified and sequenced the cDNA for, respectively, the bovine adrenal and the rat vascular AT_1 receptors. Their successful strategies used the screening, by AngII binding, of expression cDNA libraries divided into pools and expressed in COS cells. Later, it was demonstrated that there are two subtypes (called AT_{1A} and AT_{1B}) of the AT_1 receptor in rodents (Sandberg et al. 1992) but not in other species, including humans.

2.2
Structure of the Gene

Identification of the genomic sequences encoding the AT_1 receptor followed shortly after the cloning of their cDNAs. Among mammalian species, the gene structure has general conserved features. The sequence coding for the protein is contained in a single exon, which also contains part of the 5' and 3' untranslated sequences. However, the gene contains additional 5' or 3' untranslated exons, whose number and location vary from one species to another.

The human gene, which is almost 50 kb long and located on the chromosome 3q22, consists of four 5' untranslated exons followed by a fifth exon, which contains the receptor open reading frame (Guo et al. 1994; Curnow et al. 1995). Exons 2, 3 and 4 are short (from 58 to 157 bp) and can be alternatively spliced producing eight potential different mRNA species of very similar sizes (Fig. 1).

The two rat AT_1 receptor genes are located on different chromosomes: AT_{1A} on chromosome 17 and AT_{1B} on chromosome 2. The rat AngII AT_{1A} receptor gene consists of four exons and is more than 84 kb in size (Langford et al. 1992; Takeuchi et al. 1993). Exons 1 and 2 are 5' untranslated exons, exon 3 contains

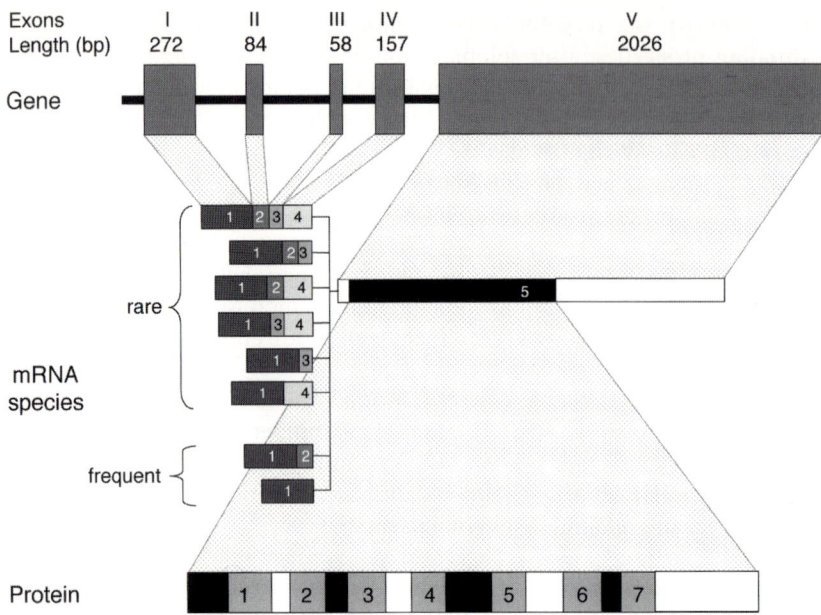

Fig. 1 Structure of the human AT$_1$ receptor gene, mRNAs and protein. The *upper panel* represents the structure of the gene with the exons (*grey boxes* numbered *I* to *V*) separated by the introns (*black lines*). Correspondence of each exon with the mRNA sequence or of mRNA with the protein are indicated by *light grey area*. The *middle panel* represents the different mRNA species. They all contain exon 1 (*dark grey*) and exon 5 (*white* for the non coding sequence and *black* for the coding sequence), but exon 2 (*grey*), 3 (*light grey*) and 4 (*very light grey*) are alternatively spliced. The open reading frame codes for the AT$_1$ protein (359 amino acids, *lower panel*) with its transmembrane domains (*grey*), extracellular (*black*) and intracellular (*white*) sequences

the coding sequence and exon 4 corresponds to 1 kbp of 3′ untranslated sequence. The rat AngII AT$_{1B}$ receptor gene has a similar structure, but only three exons have been described so far: the two 5′ untranslated exons and exon 3 coding for the protein (Guo and Inagami 1994). This gene is greater than 15 kb in length.

2.3
Structure of the mRNAs

The gene structure predicts eight potential AT$_1$ receptor mRNA species in humans, depending on the alternative splicing of exons 2, 3 and 4. These different mRNA species have been experimentally identified by RT-PCR (Curnow et al. 1995). However, Northern blot analysis identifies only one band for human AT$_1$ receptor mRNA, corresponding to a sequence 2.4 kb in length. This apparent discrepancy can be easily explained since exons 2–4 are very short, and therefore the different mRNA species are undistinguishable on Northern blot. The most abundant mRNA species are those containing exons 1 and 5 or 1, 2 and 5.

The functional and expressive differences of these alternatively spliced mRNAs are not known. Exon 1 presents a very complicated GC rich, double-stranded secondary structure, which could be involved in the regulation of translation. Exon 2 contains minicistrons, which may also modify the receptor translation rate. Finally, the mRNA species, in which exon 4 is spliced, may produce a receptor with an amino terminal extension of 32–35 residues, encoded by the end of exon 3 and the beginning of exon 5. The expression of this modified receptor has not been unambiguously demonstrated in vivo, and there is no evidence that this first methionine in exon 3 is used as an alternative translation start point of the protein. In addition, the differences in the functions of this potential alternate AT_1 receptor are completely unknown.

Similarly, exon 2 and 4 of the rat AT_{1A} receptor gene are also alternatively spliced. The alternative splicing of exon 4, which contains 1 kb of 3′ untranslated sequence, explains the presence of two mRNA species of 2.3 and 3.3 kb on Northern blot (Takeuchi et al. 1993).

The shorter mRNA is expressed in all target tissues of AngII, whereas the 3.3-kb species is abundantly expressed in smooth muscle cells, but not in neurons. It is not known if these various mRNA species have differences in their translation efficiency or in their stability, which would justify a regulation in the proportion of their expression.

2.4
Structure of the Protein

The cloning of the AT_1 receptor cDNA and its sequencing have shown that the receptor is a 359-amino-acid-long integral transmembrane glycoprotein. This has seven stretches of 20 to 26 mostly hydrophobic amino acids, which are assumed to form alpha helices integrated in the lipid bilayer of the cell membrane (Murphy et al. 1991; Sasaki et al. 1991). These seven transmembrane domains (TMs) delineate an extracellular NH_2 terminal segment and an intracellular COOH terminal segment and are connected by three extracellular and three intracellular loops (Fig. 1). This initial characterization was recently completed with the identification of a short eighth hydrophobic alpha helix (14 amino acids) in the COOH-terminal segment of the receptor, which is assumed to be in close contact with the inner face of the lipid bilayer. Like many other GPCRs, the AT_1 receptor is devoid of any N-terminal signal sequence, since its first TM (TM1) fulfils this function and allows proper membrane insertion of the protein.

Further analysis of the primary sequence of the AT_1 receptor identifies several specific features of this protein:

- Four extracellular cysteinyl residues in the primary sequence. Two of them are located in the first (Cys^{101}) and second (Cys^{180}) extracellular loops and are thought to form a disulfide bridge, which is well conserved in other GPCRs. The two additional cysteines are in the N-terminus segment (Cys^{18})

and the third (Cys^{274}) extracellular loop and may form an additional disulfide bridge, which may explain the specific structural sensitivity of this receptor to reducing agents such as dithiotreitol (DTT). There is one cysteine in each of TM2, TM3 and TM4 and two in TM7, but whether they form disulphide bridges or not is unknown. The intracellular segments do not contain cysteinyl residues, except the C-terminal segment, which contains a distal cysteine, which might be palmitoylated and therefore attached to the lipid bilayer, transforming the C-terminus segment in a fourth intracellular loop. Interestingly enough, this last cysteine is present in the rodent AT_{1A} subtype but not in the AT_{1B} subtype, suggesting a difference in the functions or regulation of the two AT_1 subtypes in rodent.
- After the alpha helix VIII, the COOH terminal segment has a sequence extremely rich in seryl and threonyl residues (12 out of 24 amino acids between residues 326 and 349). This cluster of serines and threonines is a potential site of phosphorylation, which regulates the receptor functions.
- The comparison of the AT_1 sequence with those of other GPCRs reveals the conservation of numerous residues in the different TMs, including several prolines (Pro^{162}, Pro^{207} and Pro^{255}) and tryptophanes (Trp^{153} and Trp^{253}). These residues may play a role in the general architecture of the receptor. In addition, a canonical sequence ^{125}Asp Arg Tyr^{127} (DRY) sequence is identified in the proximal part of the second intracellular loop. This sequence is the signature of the specific subclass 1 of the GPCR superfamily.

In addition to these structural features of the protein, analysis of the AT_1 receptor primary sequence indicates that it is a glycoprotein. Three potential asparagine-linked glycosylation sites (^{4}Asn-Ser-Ser, ^{176}Asn-Ile-Thr and ^{188}Asn-Ser-Thr) are located on the NH2-terminal segment (1 site) and second extracellular loop (2 sites). Site-directed mutagenesis of each of these sites reveals that they are all linked to oligosaccharides, the absence of which does not modify the pharmacological and signalling properties of the AT_1 receptor. However, the unglycosylated form of the AT_1 receptor, resulting from the combined mutation of the three sites, has a major defect in biosynthesis and membrane trafficking (Lanctot et al. 1999).

2.5
Phylogenic Aspects of the AT$_1$ Receptor

Since the first cloning of the rat AT_{1A} receptor cDNA, the amino acid sequences of AT_1 receptors have been identified in several mammalian species and also in birds and reptiles. The sequence identity between the AT_{1A} and AT_{1B} receptors is 95% in the rat and 94% in the mouse (Fig. 2). The sequence identities of the rat AT_{1A} receptor with other mammalian AT_1 receptors range between 99% (mouse AT_{1A}) and 92% (bovine AT_1) and these comparisons include rabbit, pig, dog and human sequences. Other AngII receptors have been identified in birds (turkey and chicken) and *Xenopus laevis*, but the absence of binding of specific

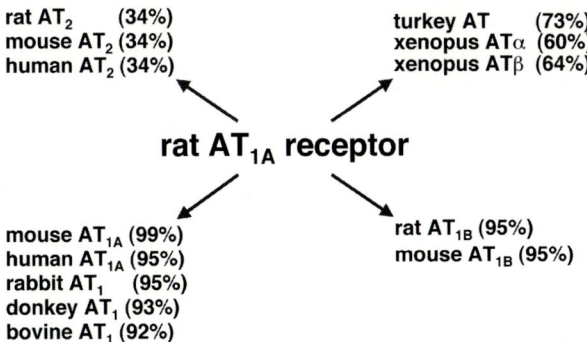

Fig. 2 Amino acid identities of the rat AT_{1A} receptor sequence with other angiotensin II receptors

AT_1 or AT_2 ligands to these receptors explains that they were called AT receptors, despite a canonical coupling to Gq protein and phospholipase C. The sequence identities of these receptors with the rat AT_{1A} receptor are between 73% and 60% at the amino acid level. These sequence homologies are rather high compared to that existing between rat AT_{1A} and AT_2 sequences, which is only 34%. However, this last homology is higher than those with other GPCRs for peptide ligands (25%–30%) or bioamine ligands (20%) or with the mas oncogene (23%), which was for a time considered as a potential angiotensin receptor.

3
Structure–Function Relationships

The cloning of AngII receptors cDNA from various species was the starting point for an extensive work of characterization of its structural features and their involvement in functional aspects of the AT_1 receptor. Site-directed mutagenesis of the AT_1 receptors allowed the identification of many amino acid sequences involved in the ligand binding, activation, G protein coupling, internalization and desensitization of the AT_1 receptors.

3.1
The Ligand Binding Site

The renin–angiotensin system, and the AT_1 receptors in particular, are key targets for drug discovery. Both peptide and non-peptide AT_1 agonists and antagonists are known and their binding sites have been well studied using site directed mutagenesis. Among the mutations altering the ligand-binding capacities of the receptor, it is difficult to discriminate those directly involved in the interaction from those participating in the architecture of the binding pocket and those required for the general architecture of the receptor. It is generally accepted that the docking of the agonist AngII stabilizes a conformational state of the receptor, which allows the initiation of intracellular signalling events. The interactions between the peptide and the exposed residues are more important in the stabilization of the hormone–receptor complex, whereas the interactions with buried amino acids are required for the first steps of the receptor activation switches to occur.

3.1.1
AngII Binding Site

The octapeptide AngII (Asp^1-Arg^2-Val^3-Tyr^4-Ile^5-His^6-Pro^7-Phe^8) has a high-affinity interaction with the AT_1 receptor ($K_d \approx 1$ nM). This relies on multiple contacts between the amino acid side chains of the peptide, and residues primarily located in the extracellular regions (amino terminus and the three extracellular loops; Fig. 3) (Hjorth et al. 1994). AngII docking also involves polar and charged residues located in the upper parts of the receptor TMs. For example, it is now accepted that C-terminal Phe^8 of AngII interacts with Lys^{199} in the upper part of TM5 (Underwood et al. 1994; Noda et al. 1995; Yamano et al. 1995). AngII seems to adopt a hairpin conformation for its docking to the receptor (Matsoukas et al. 1994; Nikiforovich et al. 1994), and there might be a salt-linked triad between Lys^{199} and the carboxyl group of Asp^1 and Phe^8 of AngII (Joseph et al. 1995a). In addition, Trp^{253} (TM6) stabilizes the ionic bridge formed between Lys^{199} and Phe^8 of AngII and Asp^1 of the peptide interacts with His^{183} in the second extracellular loop of AT_1 (Yamano et al. 1995). Asp^{281}, at the junction between the third extracellular loop and TM7, has been identified as a major docking point for AngII through its charge interaction with Arg^2 (Feng et al. 1995). Data and modelling studies suggest that Arg^{167} might be an important contact point with Tyr^4 (Yamano et al. 1995). This residue might also disrupt the hydrogen bonding between Asn^{111} in TM3 and Tyr^{292} in TM7, by competing with Tyr^{292}. This would permit the latter to interact with Asp^{74} in TM2 during receptor activation (Joseph et al. 1995a,b). Phe^{259} and Asp^{263} in TM6 might provide a docking site for His^6 of the peptide (Yamano et al. 1995). AngII binding also involves a residue in the outer part of TM3, Lys^{102} (Monnot et al. 1996) and others located in TM7 and more deeply buried in the plasma membrane, Asn^{295} (Schambye et al. 1994) and Phe^{301} (Hunyady et al. 1995b). However whether these residues are in direct contact with the peptide, or are involved in intramolecular interactions that define the intramembrane binding pocket for the ligand, is not known. The four cysteinyl residues that form the two extracellular disulphide bridges, Cys^{18}–Cys^{274} and Cys^{101}–Cys^{180} (Ohyama et al. 1995) are essential for the general architecture of the receptor and therefore for AngII binding, which is suppressed when these residues are mutated.

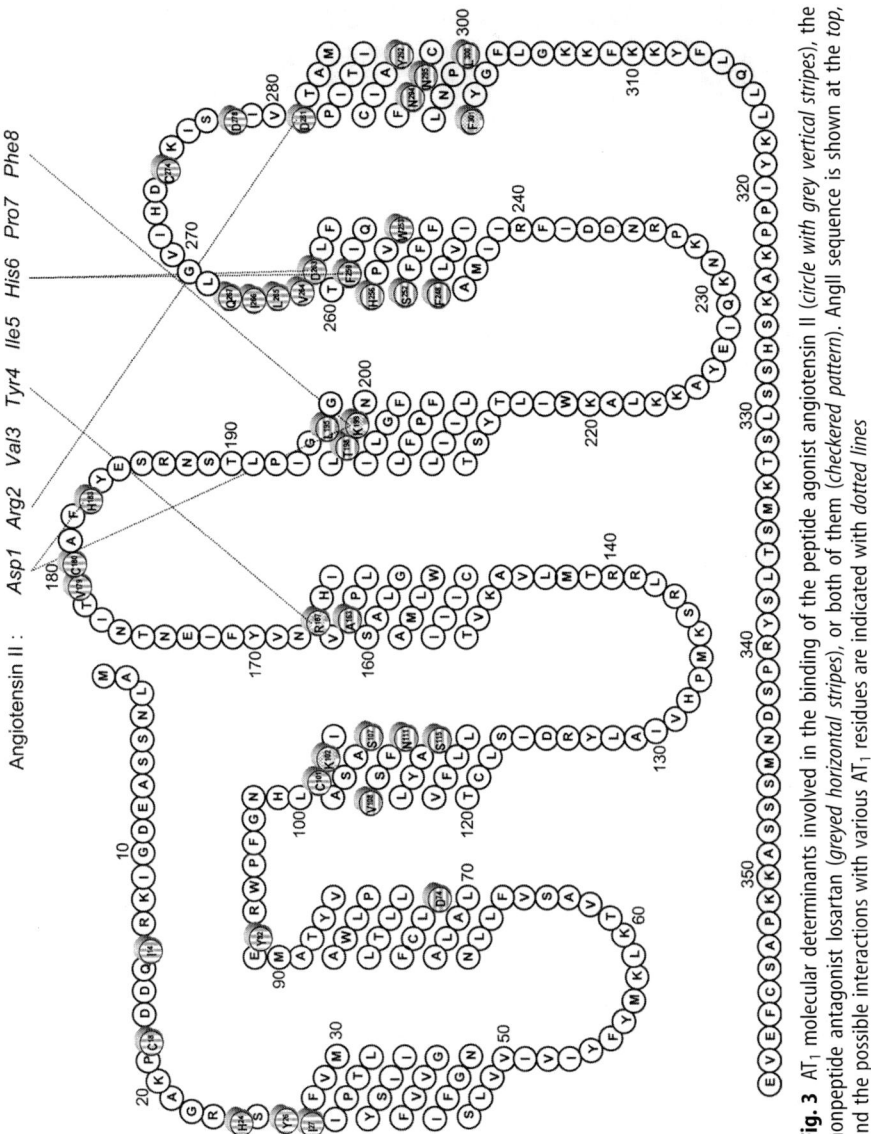

Fig. 3 AT₁ molecular determinants involved in the binding of the peptide agonist angiotensin II (*circle with grey vertical stripes*), the nonpeptide antagonist losartan (*greyed horizontal stripes*), or both of them (*checkered pattern*). AngII sequence is shown at the *top*, and the possible interactions with various AT₁ residues are indicated with *dotted lines*

3.1.2
Non-peptide Ligand Binding Site

Early attempts to develop therapeutic agents able to block the AngII receptor were complicated by the peptidic nature of antagonists, such as saralasin, which are devoid of oral activity. Based on imidazole derivatives first described in 1982 (Furukawa et al, US patent 4,340,598), nonpeptide AT₁ antagonists were synthesized (Timmermans et al. 1993; Goodfriend et al. 1996). The first of this series to reach the clinic, losartan, was followed by a large number of AT_1 antagonists that were orally active.

Identification of the losartan binding site was facilitated by the functional comparison of mammalian AT_1 and amphibian AngII receptors. Both present similar signal transduction mechanisms, but drastically differ in their binding of nonpeptide ligands such as losartan. The amino acids involved in losartan binding to the mammalian AT_1 were identified by analysing the binding properties of rat AT_1 mutant receptors in which the nonconserved residues had been substituted by their Xenopus counterparts (Ji et al. 1994). Most of these mutants preserved their affinity for AngII or for the peptide antagonist [Sar^1,Ile^8]AngII, suggesting that the general conformation of the receptor was unaltered by such replacements. These residues are mainly located in the TMs of the receptor (Val^{108} in TM3, Ala^{163} in TM4, Thr^{198} in TM5, Ser^{252} in TM6, Leu^{300} and Phe^{301} in TM7). Other amino acids have been identified in the various TMs, as shown in Fig. 3 (Bihoreau et al. 1993; Schambye et al. 1994; Noda et al. 1995; Monnot et al. 1996; Balmforth et al. 1997a; Hoe and Saavedra 2002). The residues known to be part of the losartan binding site in the mammalian receptor have been transferred into an amphibian receptor creating a "gain of function" mutant that binds the nonpeptide ligand with an affinity comparable to the rat AT_1 receptor (Ji et al. 1995).

These findings demonstrate that losartan binds to a site defined by amino acids located quite deeply in the membrane-spanning region of the receptor. These residues are mainly distinct from those involved in AngII binding, but all are located in a defined area lying between TM3, 5, 6 and 7 of the receptor. This binding site is similar to the intramembrane binding pocket of GPCRs for smaller ligands, such as catecholamines and acetylcholine.

By comparing the amino acids involved in their respective binding sites, it can be seen that AngII and nonpeptide analogues have common determinants on the receptor such as Lys^{102}, Ser^{105}, Arg^{167} and Lys^{199} (Groblewski et al. 1995; Noda et al. 1995; Yamano et al. 1995). The case of Asn^{295} appears to be more complex, since its substitution by a series of residues indicates its involvement as a direct contact point with the non peptide ligands of the antagonist type (losartan) but not of the agonist type (L-163,491, L-163,313) (Hunyady et al. 1998).

In summary, similarly to the AngII binding site, it is difficult to discriminate between residues directly involved in a molecular interaction with losartan and those involved in the general architecture of the binding site. The present picture of the ligand-binding sites shows for AngII the involvement of some transmembrane residues and mostly extracellular sequences, whereas smaller ligands such as losartan interact with a transmembrane binding pocket. The smaller ligands need less energy than peptides to achieve high affinity binding and the extracellular regions of the receptor provide additional contact points, stabilizing the binding of the peptide ligands. This fact is not limited to the AT_1 receptor, but applies for other peptide and protein hormones (Gershengorn and Osman 2001).

3.2
Receptor Activation

Several models based on allostery have been suggested to explain the process by which a ligand binding to a receptor leads to a conformational change, which results in intracellular signalling. According to these models, a receptor such as AT_1 exists in at least two interconverting conformations, one active and one inactive, and undergoes spontaneous isomerization between these two forms. In the absence of ligand (basal conditions), the proportion of receptors in the activated state is generally low and the basal conditions can be considered as inactive. This applies to the AT_1 receptor; however, for several individual GPCRs, such as muscarinic receptors, the basal conditions are associated with a high level of signalling activity and the receptor is considered to be constitutively active (Burstein et al. 1997). In the presence of AngII, the equilibrium is shifted towards this active state which allows efficient coupling to the intracellular signalling partners such as the Gq protein. This is due either to a change in the conformation of the receptor induced by AngII or, more probably, to a better affinity of AngII for the active conformation, which is then stabilized. This definition of the agonist, which has a greater affinity for the active state of the receptor, has been extended to other pharmacological compounds. Among the standard antagonists, it has been possible to distinguish the true antagonists, which have a similar affinity for the active and inactive states of a receptor and the inverse agonists, which have a greater affinity for the inactive states. These later compounds are able to shift the equilibrium towards the inactive state when the receptor is in a spontaneous active state, whereas a true antagonist is always neutral for this equilibrium but is a competitor for agonists.

Information from receptor mutants blocked in either the inactive or the active conformations are precious in understanding the activation process of the receptor. Inactivating mutations were the first to be reported. Mutations of polar residues of TM2 (Asp^{74}), TM3 (Ser^{115}), TM5 (Tyr^{215}) and TM7 (Tyr^{292}) (Bihoreau et al. 1993; Marie et al. 1994; Hunyady et al. 1995a; Monnot et al. 1996) have been shown to result in inactive receptors, unable to activate the G protein without greatly affecting their agonist-binding properties (Fig. 4). These residues appear to be involved in the activation process of the receptor. Some of these residues (Asp^{74} and Tyr^{215}) are conserved among the GPCRs, suggesting that they could be involved in the activating conformational switch leading to the intracellular coupling to the G proteins. Some modelling experiments suggest a possible interaction between Asn^{74} and Tyr^{292} that would stabilize the active conformation of the receptor (Joseph et al. 1995a).

A $NPX_{2-3}Y$ motif, located at the C terminal end of TM7 and very conserved among GPCRs, was identified as functionally important in the aminergic GPCRs. The tyrosyl residue is important for receptor internalization (Barak et al. 1994) and the whole sequence participates in the transmission of a signal from agonist-induced conformational changes in the ligand-binding region (Donnelly et al. 1994). The role of this sequence ^{298}Asn-Pro-Leu-Phe-Tyr^{302} in

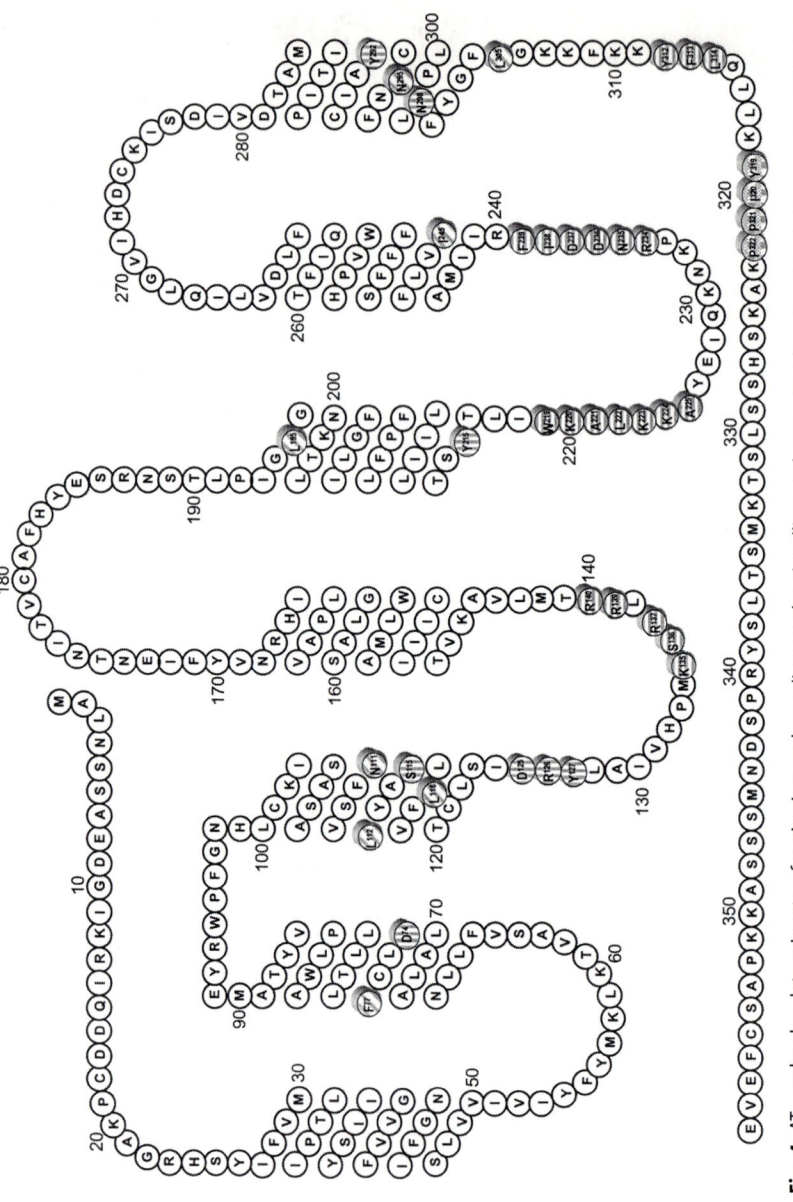

Fig. 4 AT$_1$ molecular determinants of activation and coupling to the signalling pathways. Residues involves in the receptor activation (*circle filled with grey diagonal stripes*)/inactivation (*grey vertical stripes*) process are located mainly in the transmembrane domains. Residues implicated in the coupling to the Gαq protein (*background horizontal stripes*) and to the Jak2/STAT1 and PLCγ pathways (*background checkered pattern*) are shown in the IC2, IC3 and C-terminal segment of the receptor

the AT$_1$ receptor was analysed similarly (Hunyady et al. 1995b), demonstrating that the mutation of Tyr^{302} does not impair ligand-induced internalization of the receptor. However, this mutation and the replacement of Pro^{299} reduce—and the mutation of Asn^{298} suppresses—the G protein-coupling of the receptor and the agonist-induced production of second messengers, demonstrating the role of this motif in receptor activation.

Beside loss-of-function mutations of the AT_1 receptor, gain-of-function mutants have also been identified and are informative for our understanding of the receptor activation mechanisms. First identified by site-directed mutagenesis of the adrenergic receptor (Cotecchia et al. 1990; Kjelsberg et al. 1992), such active mutations have also been found as natural mutations involved in the pathophysiology of several endocrine diseases such as thyroid toxic adenoma (Parma et al. 1993), precocious puberty and chondrodysplasia (Schipani et al. 1995).

Defined as mutations responsible for a permanent and agonist-independent activity of the receptor, higher than that of the wild-type receptor, these gain-of-function mutations are also called constitutively active mutations (CAM). These CAMs are potent pharmacological tools for discriminating true antagonists from inverse agonists, since only the latter are able to inactivate such mutants. In addition, apparent affinity and efficacy of (partial) agonists are better for these mutants blocked in their active state than for the wild-type receptors. The study of such receptors has enriched our understanding not only of GPCR activation mechanisms but also of GPCR pharmacology.

For the AT_1 receptor, in vitro studies on constitutively active mutants have implicated Asn^{111} in TM3 (Groblewski et al. 1997; Feng et al. 1998; Miura et al. 1999). Mutation of Asn^{295} into Ser^{295} in TM7 has been shown to increase the basal activity of the receptor and a direct interaction between Asn^{111} and Asn^{295} has been suggested (Balmforth et al. 1997a). However, mutation of this residue in alanine or aspartic acid does not lead to a constitutive activation of the receptor (Perlman et al. 1997; Hunyady et al. 1998).

More recently, the exhaustive cartography of the residues involved in AT_1 receptor activation has been established by identifying a repertoire of CAM after screening a randomly mutated AT_1 cDNA library using a functional test based on the higher sensitivity of these mutants to partial agonists (Parnot et al. 2000). Sixteen mutants with a higher sensitivity to the partial agonist CGP42112A were identified and among these seven showed a significantly higher basal inositol phosphate production ($Phe^{77}Tyr$ in TM2, $Asn^{111}Ser$, $Leu^{112}His$ and $Leu^{118}His$ in TM3, $Leu^{195}Pro$ in TM5, $Ile^{245}Thr$ in TM6 and $Leu^{305}Gln$). The mutants should be useful for molecular modelling of AT_1 activation and should help researchers to define the role of the mutated residues in the activation mechanism. The amino acids involved are not conserved among GPCRs, suggesting that if there are common mechanisms for activation of GPCRs, they do not imply highly conserved amino acid motives. As for many CAM receptors identified to date, these residues are mainly located in transmembrane regions, confirming that TM movements play a key role in the transition between the inactive and active states. A striking cluster of mutations was found in TM3, around Asn^{111}, indicating that this helix plays an important role in activation.

A proposed model of GPCR activation predicts that rigid body movement of TM3, 6 and 7 induces conformational changes in the cytoplasmic loops that permit G protein interaction with the agonist-activated receptor (Gether and Kobilka 1998). AT_1 mutagenesis results, where many residues involved in receptor activation are located in the TMs, seem to confirm this idea. Mutations of residues or motives located in the intracellular regions of the receptor have also been shown to impair the receptor coupling. However, for these amino acids it is difficult to determine if, like the transmembrane mutations, they participate in the general structural architecture of the receptor's active conformation or if they represent a direct contact point with the intracellular partner.

3.3
Interaction with G Proteins

The binding of AngII to the AT_1 receptor has been shown to trigger several intracellular signalling pathways. AT_1's main partner is the Gq/11 protein, which activates the phospholipase C-β (PLC-β). PLC-β generates the second messengers inositol (1,4,5) trisphosphate (IP) and diacylglycerol (DAG), which allow the release of calcium from intracellular stores and activate protein kinase C, respectively. The major interaction of the receptor is with the N- and C-terminal domains of the α-subunit of Gq. Several cytosolic molecular determinants of the AT_1 receptor have been shown to interact with the Gαq protein (Fig. 4). In the second intracellular loop (IC2), in addition to the extremely conserved ^{125}DRY^{127} motif at the junction with TM3, Ohyama et al. have shown that the mutation of five other residues (Lys^{135}, Ser^{136}, Arg^{137}, Arg^{139} and Arg^{140}) prevents the receptor from activating the G protein (Ohyama et al. 1992). The third intracellular loop (IC3) is considered as a major determinant in GPCR coupling selectivity and efficiency. The replacement of residues 219–225 and 234–237 of the AT_2 IC3 domain with the corresponding residues of the AT_1 receptor results in an efficient coupling to the Gq protein (Wang et al. 1995). Conversely, the replacement in AT_1 of residues 234–240 by their AT_2 homologues suppresses the Gq coupling of the receptor (Conchon et al. 1997). The coupling of the AT_1 receptor to the heterotrimeric G proteins is also dependent on the proximal region of its carboxyl-terminal cytoplasmic region. More specifically, site-directed mutations of individual amino acids of the hydrophobic cluster Tyr^{312}, Phe^{313}, Leu^{314} produce receptors that were insensitive to the guanosine triphosphate (GTP) analogue GTPγS and were unable to stimulate IP production after stimulation by AngII, demonstrating the absence of G protein coupling. Moreover, a purified peptide containing the wild-type sequence (residues 306–320) was able to directly stimulate the binding of GTPγS to purified G proteins, whereas similar peptides containing mutations of Tyr^{312}, Phe^{313} or Leu^{314} could not (Sano et al. 1997). It seems therefore that the Gq-protein binding site of the AT_1 receptor consists of a complex spatial arrangement of several determinants from the IC2, IC3 and C-terminus segment.

AT_1 receptors have also been reported to interact with other G proteins, such as G_i-proteins. Peptides derived from the sequence of the amino-terminal half of IC3 (residues 216–231) or from the C-terminus (residues 306–320) stimulate the binding of GTPγS to purified Gi1, Gi2 and Go proteins (Shirai et al. 1995). More recently, a mutation of TM4 abolished the Gq coupling to AT_1 (no IP production after AngII stimulation), but revealed the coupling to another, as-yet-unidentified G protein, since the AngII binding was still sensitive to GTP analogues (Feng and Karnik 1999).

3.4
Interactions with Other Signalling Molecules

Recently, the AT_1 receptor has been shown to activate cellular tyrosine kinases including Src, Fyn and Pyk2 (Sayeski et al. 1998). The binding of AngII to AT_1 also activates the intracellular tyrosine kinase Jak2, leading to tyrosine phosphorylation and nuclear translocation of the transcription factor signal transducer and activator of transcription (STAT1) (Marrero et al. 1995). Despite some data indicating that the Jak2/STAT1 pathway is activated via the coupling to Gq protein in vascular smooth muscle cells (Frank et al. 2002), it is generally accepted that Jak2 associates independently from G proteins with the AT_1 receptor and the interacting sequence ^{319}Tyr-Ile-Pro-Pro^{322} (YIPP) is located in the carboxyl-terminus segment of the receptor (Fig. 4) (Ali et al. 1997). This interaction is independent of AT_1 tyrosine phosphorylation since the mutation of all the intracellular tyrosine residues for phenylalanine does not change its co-immunoprecipitation with Jak2 and STAT1. The YIPP motif has also been implicated in the recruitment of phospholipase C (PLC)-γ1, via its Src-homology 2 (SH2) domains, in an AngII and tyrosine phosphorylation-dependant manner (Venema et al. 1998). However, the tyrosine phosphorylation of AT_1 is still controversial (Smith et al. 1998), which might cast a slight doubt on the final interpretation of these results.

The proximity of the ^{319}YIPP^{322} motif, which interacts with the Jak2-STAT1 complex and maybe the PLCγ and the ^{312}Tyr-Phe-Ile^{314} motif which is part of the G protein binding site, suggests either a competition between these signalling molecules or their interaction with different active conformations of the AT_1 receptor. The interaction of Jak2 and Gq with two different active conformations of the AT_1 receptor is suggested by two pieces of experimental data:

- A receptor with the Asp^{74} in TM3 mutated for a Gln, which can no longer activate the Gq protein, is still able to activate the Jak/STAT pathway (Doan et al. 2001).
- Conversely, the deletion of the ^{319}YIPP^{322} motif indicates that it is not necessary for Gq protein coupling (Thomas et al. 1995b).

In summary, the intracellular sequences of the AT_1 receptor are involved not only in the interactions with G proteins but also with other signalling proteins. The recent cloning of a novel protein (ATRAP), which interacts with the C-terminus of the AT_1 receptor and apparently inhibits the growth properties of AngII, is another example of these interactions with multiple signalling pathways (Daviet et al. 1999; Cui et al. 2000).

3.5
Regulation of AT₁ Receptor Activity

Receptor desensitization is a complex phenomenon leading to decoupling from G proteins, down-regulation and insensitivity to ligands. This phenomenon is associated with receptor phosphorylation, interactions with regulatory proteins and internalization. It has been extensively demonstrated in the model of the β_2-adrenergic receptor that the active conformations of the receptor allow not only G protein coupling but also phosphorylation by specific G protein receptor

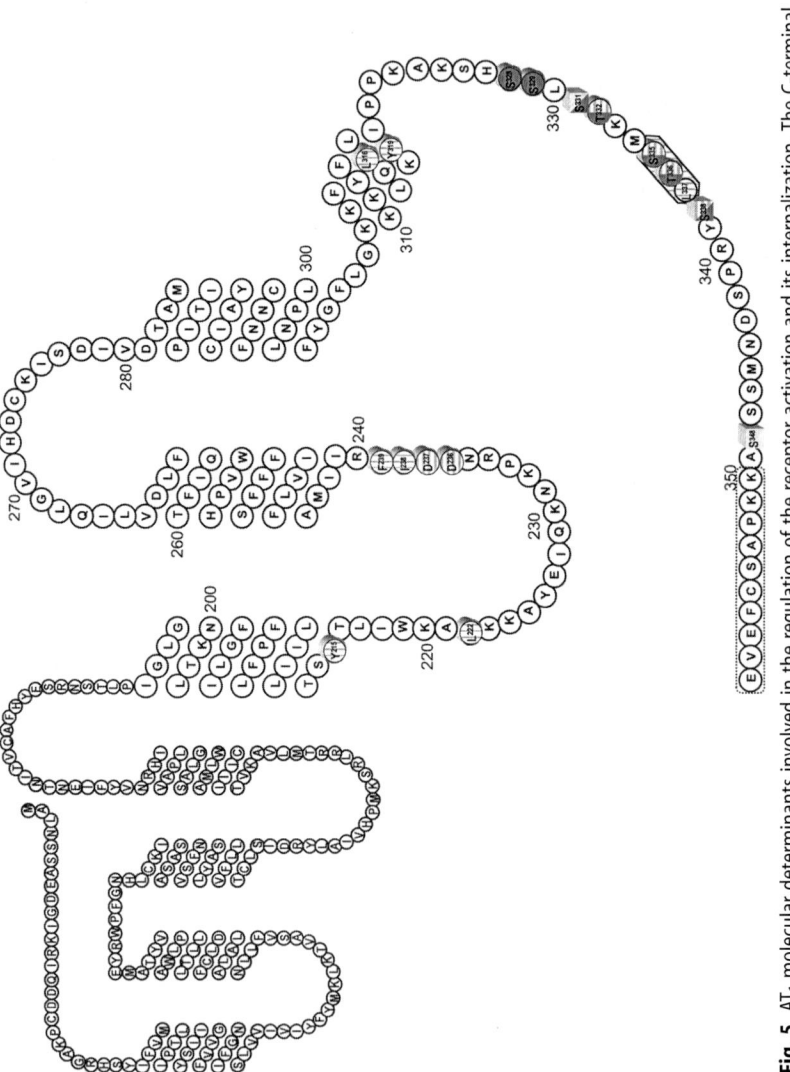

Fig. 5 AT₁ molecular determinants involved in the regulation of the receptor activation and its internalization. The C-terminal region of the receptor presents a high concentration of sequences and residues implicated in receptor phosphorylation [by PKC (*circle filled with horizontal stripes*) and GRK (*cube*)], interactions with partners of its trafficking [β-arrestin (*filled grey circle*) and rab5 (*area within dotted line*)] and more generally in its desensitization and internalization (*circle filled with grey vertical stripes*) processes

kinases (GRKs) and by second messenger activated kinases [protein kinases A and C (PKA, PKC)]. The phosphorylated receptor can then interact with cytoplasmic proteins, called arrestins, which prevent association between the receptor and the G protein, leading to desensitization. The binding of arrestin to the activated receptor might also serve to target the latter to clathrin-coated pits for internalization from the cell surface. Studies to identify determinants involved in the regulation of AT_1 receptor activity have focused on the carboxyl-terminal cytoplasmic segment which seems to concentrate a lot of information in less than 50 residues (Fig. 5).

3.5.1
Phosphorylation

The phosphorylation of AT_1 receptors has been shown to be agonist-, time- and dose-dependent in transfected cells and target tissues (Oppermann et al. 1996; Balmforth et al. 1997b; Smith et al. 1998) The AngII-stimulated AT_1 receptor phosphorylation occurs on serine and threonine residues by specific kinases (GRK) and by PKC. The involved GRKs differ from one tissue to another since GRK2, 3 and 5 have been shown to phosphorylate the receptor in HEK293 cells, (Oppermann et al. 1996), but not in Chinese hamster ovary (CHO) cells (Tang et al. 1998).

The phosphorylation sites of the AT_1 receptor for these different serine/threonine kinases have been investigated. Three potential PKC phosphorylation sites (Ser^{331}, Ser^{338}, Ser^{348}) are located in the carboxyl-terminal segment of the receptor and apparently all of them are utilized to some extent following homologous (AngII) and heterologous [phorbol myristoyl acetate (PMA)] stimulation (Qian et al. 1999). There is no consensus sequence for phosphorylation via GRK, but it is usual to observe acidic residues next to the seryl or threonyl residue they phosphorylate. The sequence ^{332}Thr-X-X-Ser-Thr-X-Ser^{338} in the middle of the AT_1 receptor C-terminus domain seems to be a major site of phosphorylation by GRK, since the mutation of the serines and threonines to alanines in this sequence results in a drastically decreased level of AngII-stimulated receptor phosphorylation (Thomas et al. 1998; Qian et al. 2001). However, this sequence is not flanked by acidic residues. Asp^{236} and Asp^{237} located in the third cytoplasmic loop may play this role, since their mutation largely impairs AngII-stimulated GRK-mediated phosphorylation of the receptor (Olivares-Reyes et al. 2001).

3.5.2
Phosphorylation and Regulation of AT_1 Receptor Activity

This major GRK phosphorylation site belongs to a larger sequence of the AT_1 receptor C-terminus extremely rich in seryl and threonyl residues (10 of these residues between amino acids 326 and 348). This sequence is involved not only in receptor phosphorylation but also in its desensitization (i.e. the amplitude of the signalling response to repeated agonist stimulations). Progressive C-terminal truncations of the receptor show that deletion after residue 328, but not after residue 336, produces a receptor with an amplification of AngII-induced intracellular signalling and an absence of heterologous desensitization (Conchon et al. 1998). Even if this amplified signal of the truncated receptor is not observed by all authors (Tang et al. 1998), all agree on the role of the sequence Ser^{328} to Ser^{348} in homologous and heterologous desensitization of the receptor.

3.5.3
Phosphorylation and Activation of the AT_1 Receptor

The increased phosphorylation of the AT_1 receptor after AngII binding suggests that, as for many other GPCRs, the active conformations of the receptor are the substrate of the kinases. However, this interpretation has not been confirmed for the AT_1 receptor by experimental data. Several lines of evidence seem to indicate that phosphorylation of AT_1 requires conformational states at least partly distinct from the active states responsible for signalling:

- Gain-of-function mutants of the AT_1 receptor ($Asn^{111}Ala$, $Asn^{111}Gly$), which activate constitutively the Gq protein signalling pathway, show a reduced basal and AngII-stimulated level of phosphorylation.
- AngII analogues, such as [Sar1,Ile8]AngII, do not induce G protein coupling of the wild-type receptor, but induce its phosphorylation. (Thomas et al. 1998, 2000).
- Partial agonists, such as [Sar1,Ile4,Ile8]AngII, do not induce wild-type receptor coupling to G protein but induce its phosphorylation. Conversely, they activate the G protein coupling but do not induce phosphorylation of constitutively active mutants of the AT_1 receptor.

This implies that the receptor does not have to be in an active state to be phosphorylated, and some active conformational states of the receptor are not subjected to phosphorylation, even after AngII stimulation. Altogether, these results indicate that the molecular switches required for phosphorylation and G protein-mediated signalling are partially distinct.

3.5.4
AT_1 Phosphorylation and β-Arrestin Binding

By interacting with the receptor, the β-arrestins play a major role in uncoupling the receptor from the G protein signalling pathways, in its internalization via the clathrin-coated pits and also in the scaffolding of new signalling complexes. Indirect evidence of an interaction of β-arrestins 1 and 2 with the AT_1 receptor has been given by confocal microscopy imaging. AngII is able to recruit enhanced green fluorescent protein (EGFP)-tagged versions of these proteins to the plasma membrane of cells expressing the AT_1 receptor. The β-arrestins are then internalized in endosomal vesicles together with the receptor (Oakley et al.

2000). Direct evidence of physical interaction between AT_1 and β-arrestin1 and 2 was obtained by co-immunoprecipitation experiments (Luttrell et al. 2001; Qian et al. 2001).

The structural requirements for β-arrestin interaction with AT_1 have been investigated. The crucial role of the AT_1 C-terminus in this interaction has been demonstrated through several experiments:

- An AT_1 mutant, truncated after Lys^{325} produces a receptor that can still reach the active state, but which is no longer phosphorylated and does not interact with β-arrestin1.
- After agonist-stimulation, β-arrestin can still be co-immunoprecipitated from endosomes with a chimeric receptor consisting of the β_2-adrenergic receptor with the carboxyl terminal tail of the AT_1 receptor but not with the wild-type $\beta2AR$ (Anborgh et al. 2000).

The serine/threonine cluster (^{328}Ser-Ser-X-Ser-Thr-X-X-Ser-Thr-X-Ser^{338}) in the central region of the C-terminus is a major molecular determinant of β-arrestin binding. Indeed, mutation of the four distal serines and threonines of this cluster into alanines results in a mutant receptor which is poorly phosphorylated and presents an AngII-induced interaction with β-arrestin reduced by 80%. Receptors mutated at the four proximal or three distal Ser/Thr residues can still translocate the β-arrestin to the plasma membrane, but in a less stable complex which dissociates rapidly (Oakley et al. 2001). A mutant in which the three PKC phosphorylation sites have been mutated (Ser331, 338 and 348) has a reduced phosphorylation but normal interaction with β-arrestin1. Altogether, these data indicate that it is not the active conformation, but the phosphorylation of the AT_1 receptor by GRKs, which is the crucial event for its interaction with β-arrestin (Qian et al. 2001).

3.6
Internalization and Trafficking

After activation, G protein coupling and phosphorylation, the AT_1 receptor becomes rapidly internalized ($t_{1/2}$~5 min) and then slowly recycled ($t_{1/2}$~1 h). This internalization is involved in the desensitization/resensitization process and also probably in the activation of some signalling pathways (Sorkin and von Zastrow 2002). In addition, several studies suggest that internalization of the ligand–receptor complex may participate in the sequestration of AngII in some target tissues from the circulation (van Kats et al. 1997). This intracellularly sequestered AngII could be released later for autocrine/paracrine functions, or might bind cytoplasmic or nuclear AngII receptors (Booz et al. 1992; Sugiura et al. 1992).

3.6.1
Molecular Determinants of Internalization

The receptor sequences involved in AngII-induced AT_1 internalization have been extensively studied by site-directed mutagenesis. Internalization motifs have been identified in the third intracellular loop (IC3) and in the C-terminal domain.

The role of the C-terminus in the internalization process has been clearly demonstrated by the deletion of this segment, which drastically impairs the receptor internalization (Hunyady et al. 1994; Balmforth et al. 1995; Thomas et al. 1995a). Further analysis of the carboxyl terminus has identified two short segments that are necessary for receptor internalization (Hunyady et al. 1994, 1995a,b, 1996; Balmforth et al. 1995; Thomas et al. 1995a,b; Conchon et al. 1998; Tang et al. 1998).

- The proximal sequence is the ^{316}Leu-Leu-Lys-Tyr^{319} motif, located closer to the plasma membrane. The substitution of Leu^{316} alone gives the strongest internalization impairment phenotype described for a single point mutant (Thomas et al. 1995a). This leucine is part of a dileucine motif, which has been involved in internalization of several membrane receptors, among which the β_2-adrenergic receptor.
- The distal sequence consists of the amino acid triplet ^{335}Ser-Thr-Leu^{337} in the middle of AT_1 carboxyl terminus, which has been identified as playing an important role in receptor endocytosis, maybe via phosphorylation and interaction with the β-arrestins.

The role of these different sequences, in either receptor interactions with the endocytic machinery or in the general conformation required for internalization, has not been clearly established.

Recently, AT_1 has been shown to interact with rab5, a well-known protein involved in intracellular trafficking. This interaction is not necessary for receptor internalization but for its further redistribution into larger vesicles. The last ten residues of the AT_1 C-terminal tail are involved in this interaction (Seachrist et al. 2002).

3.6.2
Internalization and Activation

Whether the internalization is strictly dependent on the active conformations of the receptor has been investigated and experimental data clearly demonstrate that activation and internalization of the receptor have different conformational requirements:

- A receptor mutant, Asp^{74}Asn, which does not couple to the Gq protein is normally internalized after AngII stimulation.

- The peptidic antagonist [Sar1Ile8]AngII elicits a strong and rapid internalization after binding to the receptor (Conchon et al. 1994).
- Some of the truncated mutants have impaired internalization but can still efficiently activate the Gq-protein.

3.6.3
Phosphorylation, β-Arrestin Binding and Receptor Trafficking

While the activation of the AT_1 receptor is not a prerequisite for its internalization, the phosphorylation of the receptor, followed by the binding of β-arrestin, seems to be necessary. Usually, GPCR binds β-arrestin, which acts as an adaptor protein linking the receptor to the coated pit machinery, involving clathrin- and dynamin-dependant processes. Several experimental observations indicate that the involvement of arrestin in AT_1 internalization is a controversial issue:

- Several dominant negative (DN) mutants of β-arrestin (β-arrV^{53}D) and dynamin (K^{44}A) have marginal effects on AT_1 receptor internalization (Zhang et al. 1996).
- The overexpression of GIT1 and Arf6, two factors involved in clathrin-coated vesicle budding from the plasma membrane, is without effect on AT_1 internalization, whereas it impairs endocytosis of the $β_2$-adrenergic receptor (Claing et al. 2000).
- Several mutants of the C-terminal region of AT_1 are still normally internalized but have a drastically reduced capacity to form a stable complex with β-arrestin after AngII binding (Oakley et al. 2001).
- As already mentioned, the binding of [Sar1,Ile4,Ile8]AngII, which induce receptor phosphorylation and its interaction with β-arrestin1, surprisingly does not induces AT_1 internalization (Qian et al. 2001).

In direct contrast, a series of recent studies argues in favour of a β-arrestin-, dynamin- and clathrin-dependent mechanism of AT_1 internalization. DN mutants of both arrestin (β-arr1V53D and β-arr1–349) and dynamin inhibit AngII-stimulated internalization (Werbonat et al. 2000; Gaborik et al. 2001). In addition, the phosphorylation mutants described above show a defect not only in β-arrestin interaction but also in internalization (Qian et al. 2001). Despite some contradictory results, the correlation between β-arrestin binding and internalization of the AT_1 receptor is now accepted by a majority of authors and agrees with the clear demonstration of a physical interaction between the two proteins.

Altogether, these data indicate that receptor phosphorylation and β-arrestin binding and internalization share some molecular determinants, but that some conformational states of the receptor can be specific to each of them. These observations lead to the conclusion that the receptor exists in a repertoire of active conformations, each one more or less specific of one or several functions: coupling to various signalling pathways, desensitization, phosphorylation, internalization, etc.

In conclusion, the elucidation of the structure of the gene and the primary sequence of the AT_1 receptor has opened the way to the understanding of the molecular mechanisms responsible for its different functions. A large number of receptor mutants has allowed the dissection of the different functions and interactions of the receptor with other proteins and the identification of the amino acids involved. However, the absence of specific information on the three-dimensional structure of the receptor does not allow a clear comprehension of the molecular mechanisms of its activation. In addition, there is no doubt that many other, yet-unidentified proteins are able to interact with the inactive or active conformations of the receptors. Discovering these proteins is one of the future challenges of AT_1 molecular research.

References

Ali MS, Sayeski PP, Dirksen LB, Hayzer DJ, Marrero MB, Bernstein KE (1997) Dependence on the motif YIPP for the physical association of Jak2 kinase with the intracellular carboxyl tail of the angiotensin II AT1 receptor. J Biol Chem 272:23382–23388

Anborgh PH, Seachrist JL, Dale LB, Ferguson SS (2000) Receptor/beta-arrestin complex formation and the differential trafficking and resensitization of beta2-adrenergic and angiotensin II type 1A receptors. Mol Endocrinol 14:2040–2053

Balmforth AJ, Lee AJ, Bajaj BP, Dickinson CJ, Warburton P, Ball SG (1995) Functional domains of the C-terminus of the rat angiotensin AT1A receptor. Eur J Pharmacol 291:135–141

Balmforth AJ, Lee AJ, Warburton P, Donnelly D, Ball SG (1997a) The conformational change responsible for AT1 receptor activation is dependent upon two juxtaposed asparagine residues on transmembrane helices III and VII. J Biol Chem 272:4245–4251

Balmforth AJ, Shepherd FH, Warburton P, Ball SG (1997b) Evidence of an important and direct role for protein kinase C in agonist-induced phosphorylation leading to desensitization of the angiotensin AT1A receptor. Br J Pharmacol 122:1469–1477

Barak LS, Tiberi M, Freedman NJ, Kwatra MM, Lefkowitz RJ, Caron MG (1994) A highly conserved tyrosine residue in G protein-coupled receptors is required for agonist-mediated beta 2-adrenergic receptor sequestration. J Biol Chem 269:2790–2795

Bihoreau C, Monnot C, Davies E, Teutsch B, Bernstein KE, Corvol P, Clauser E (1993) Mutation of Asp74 of the rat angiotensin II receptor confers changes in antagonist affinities and abolishes G-protein coupling. Proc Natl Acad Sci USA 90:5133–5137

Booz GW, Conrad KM, Hess AL, Singer HA, Baker KM (1992) Angiotensin-II-binding sites on hepatocyte nuclei. Endocrinology 130:3641–3649

Burstein ES, Spalding TA, Brann MR (1997) Pharmacology of muscarinic receptor subtypes constitutively activated by G proteins. Mol Pharmacol 51:312–319

Chiu AT, Herblin WF, McDall DE, Ardecky RJ, Carini DJ, Duncia JV, Pease LJ, Wong PC, Wexler RR, Johnson AL, Timmermans PBMWM (1989) Identification of angiotensin II receptor subtypes. Biochem Biophys Res Commun 165:196–203

Claing A, Perry SJ, Achiriloaie M, Walker JK, Albanesi JP, Lefkowitz RJ, Premont RT (2000) Multiple endocytic pathways of G protein-coupled receptors delineated by GIT1 sensitivity. Proc Natl Acad Sci USA 97:1119–1124

Conchon S, Barrault MB, Miserey S, Corvol P, Clauser E (1997) The C-terminal third intracellular loop of the rat AT_{1A} angiotensin II receptor plays a key role in G protein coupling specificity and transduction of the mitogenic signal. J Biol Chem 272:25566–25572

Conchon S, Monnot C, Teutsch B, Corvol P, Clauser E (1994) Internalization of the rat AT_{1a} and AT_{1b} receptors: pharmacological and functional requirements. FEBS Letter 349:365-370

Conchon S, Peltier N, Corvol P, Clauser E (1998) A non-internalized, non-desensitized truncated AT1A receptor transduces an amplified angiotensin II signal. Am J Physiol 274: E336-E345

Cotecchia S, Exum S, Caron MG, Lefkowitz RJ (1990) Regions of the α1-adrenergic receptor involved in coupling to phosphatidylinositol hydrolysis and enhanced sensitivity of biological function. Proc Natl Acad Sci USA 87:2896-2900

Cui T, Nakagami H, Iwai M, Takeda Y, Shiuchi T, Tamura K, Daviet L, Horiuchi M (2000) ATRAP, novel AT1 receptor associated protein, enhances internalization of AT1 receptor and inhibits vascular smooth muscle cell growth. Biochem Biophys Res Commun 279:938-941

Curnow KM, Pascoe L, Davies E, White PC, Corvol P, Clauser E (1995) Alternative splicing of the human type 1-angiotensin II receptor (AT1) gene leads to a novel receptor isoform and regulates the transcriptional efficiency of the mRNA. Mol Endocrinol 9:1250-1262

Daviet L, Lehtonen JY, Tamura K, Griese DP, Horiuchi M, Dzau VJ (1999) Cloning and characterization of ATRAP, a novel protein that interacts with the angiotensin II type 1 receptor. J Biol Chem 274:17058-17062

Doan TN, Ali MS, Bernstein KE (2001) Tyrosine kinase activation by the angiotensin II receptor in the absence of calcium signaling. J Biol Chem 276:20954-20958

Donnelly D, Findlay JB, Blundell TL (1994) The evolution and structure of aminergic G protein-coupled receptors. Receptors Channels 2:61-78

Feng YH, Karnik SS (1999) Role of transmembrane helix IV in G-protein specificity of the angiotensin II type 1 receptor. J Biol Chem 274:35546-35552

Feng YH, Miura S, Husain A, Karnik SS (1998) Mechanism of constitutive activation of the AT1 receptor: influence of the size of the agonist switch binding residue Asn(111). Biochemistry 37:15791-15798

Feng YH, Noda K, Saad Y, Liu XP, Husain A, Karnik SS (1995) The docking of Arg2 of angiotensin II with Asp281 of AT1 receptor is essential for full agonism. J Biol Chem 270:12846-12850

Frank GD, Saito S, Motley ED, Sasaki T, Ohba M, Kuroki T, Inagami T, Eguchi S (2002) Requirement of Ca(2+) and PKCdelta for Janus kinase 2 activation by angiotensin II: involvement of PYK2. Mol Endocrinol 16:367-377

Gaborik Z, Szaszak M, Szidonya L, Balla B, Paku S, Catt KJ, Clark AJ, Hunyady L (2001) Beta-arrestin- and dynamin-dependent endocytosis of the AT1 angiotensin receptor. Mol Pharmacol 59:239-247

Gershengorn MC, Osman R (2001) Minireview: insights into G protein-coupled receptor function using molecular models. Endocrinology 142:2-10

Gether U, Kobilka BK (1998) G protein-coupled receptors. II. Mechanism of agonist activation. J Biol Chem 273:17979-17982

Glossmann H, Baukal AJ, Catt KJ (1974) Properties of angiotensin II receptors in the bovine and rat adrenal cortex. J Biol Chem 249:825-834

Goodfriend TL, Elliott ME, Catt KJ (1996) Angiotensin receptors and their antagonists. N Engl J Med 334:1649-1654

Groblewski T, Maigret B, Larguier R, Lombard C, Bonnafous JC, Marie J (1997) Mutation of Asn111 in the third transmembrane domain of the AT1A angiotensin II receptor induces its constitutive activation. J Biol Chem 272:1822-1826

Groblewski T, Maigret B, Nouet S, Larguier R, Lombard C, Bonnafous JC, Marie J (1995) Amino acids of the third transmembrane domain of the AT1A angiotensin II receptor are involved in the differential recognition of peptide and nonpeptide ligands. Biochem Biophys Res Commun 209:153-160

Gunther S, Gimbrone MAJ, Alexander RW (1980) Identification and characterization of the high affinity vascular angiotensin II receptor in rat mesenteric artery. Circ Res 47:278–286

Guo DF, Furuta H, Mizukoshi M, Inagami T (1994) The genomic organization of human angiotensin II type 1 receptor. Biochem Biophys Res Commun 200:313–319

Guo DF, Inagami T (1994) The genomic organization of the rat angiotensin II receptor AT_{1B}. Biochim Biophys Acta 1218:91–94

Hjorth SA, Schambye HT, Greenlee WJ, Schwartz TW (1994) Identification of peptide binding residues in the extracellular domains of the AT_1 receptor. J Biol Chem 269:30953–30959

Hoe KL, Saavedra JM (2002) Site-directed mutagenesis of the gerbil and human angiotensin II AT(1) receptors identifies amino acid residues attributable to the binding affinity for the nonpeptidic antagonist losartan. Mol Pharmacol 61:1404–1415

Hunyady L, Bor M, Balla T, Catt KJ (1994) Identification of a cytoplasmic Ser-Thr-Leu motif that determines agonist-induced internalization of the AT1 angiotensin receptor. J Biol Chem 269:31378–31382

Hunyady L, Bor M, Balla T, Catt KJ (1995a) Critical role of a conserved intramembrane tyrosine residue in angiotensin II receptor activation. J Biol Chem 270:9702–9705

Hunyady L, Bor M, Baukal AJ, Balla T, Catt KJ (1995b) A conserved NPLFY sequence contributes to agonist binding and signal transduction but is not an internalization signal for the type 1 angiotensin II receptor. J Biol Chem 270:16602–16609

Hunyady L, Ji H, Jagadeesh G, Zhang M, Gaborik Z, Mihalik B, Catt KJ (1998) Dependence of AT1 angiotensin receptor function on adjacent asparagine residues in the seventh transmembrane helix. Mol Pharmacol 54:427–434

Hunyady L, Zhang M, Jagadeesh G, Bor M, Balla T, Catt KJ (1996) Dependence of agonist activation on a conserved apolar residue in the third intracellular loop of the AT1 angiotensin receptor. Proc Natl Acad Sci USA 93:10040–10045

Ji H, Leung M, Zhang Y, Catt KJ, Sandberg K (1994) Differential structural requirements for specific binding of nonpeptide and peptide antagonists to the AT1 angiotensin receptor. J Biol Chem 269:16533–16536

Ji H, Zheng W, Zhang Y, Catt KJ, Sandberg K (1995) Genetic transfer of a nonpeptide antagonist binding site to a previously unresponsive angiotensin receptor. Proc Natl Acad Sci USA 92:9240–9244

Joseph MP, Maigret B, Bonnafous JC, Marie J, Scheraga HA (1995a) A computer modeling postulated mechanism for angiotensin II receptor activation. J Protein Chem 14:381–398

Joseph MP, Maigret B, Scheraga HA (1995b) Proposals for the angiotensin II receptor-bound conformation by comparative computer modeling of AII and cyclic analogs. Int J Pept Protein Res 46:514–526

Kjelsberg MA, Cotecchia S, Ostrowski J, Caron MG, Lefkowitz RJ (1992) Constitutive activation of the α1b-adrenergic receptor by all amino acid substitutions at a single site. Evidence for a region which constrains receptor activation. J Biol Chem 267:1430–1433

Lanctot PM, Leclerc PC, Escher E, Leduc R, Guillemette G (1999) Role of N-glycosylation in the expression and functional properties of human AT1 receptor. Biochemistry 38:8621–8627

Langford K, Frenzel K, Martin BM, Bernstein KE (1992) The genomic organization of the rat AT1 angiotensin receptor. Biochem Biophys Res Commun 183:1025–1032

Luttrell LM, Roudabush FL, Choy EW, Miller WE, Field ME, Pierce KL, Lefkowitz RJ (2001) Activation and targeting of extracellular signal-regulated kinases by beta-arrestin scaffolds. Proc Natl Acad Sci U S A 98:2449–2454

Marie J, Maigret B, Joseph MP, Larguier R, Nouet S, Lombard C, Bonnafous JC (1994) Tyr^{292} in the seventh transmembrane domain of the AT1A angiotensin II receptor is essential for its coupling to phospholipase C. J Biol Chem 269:20815–20818

Marrero MB, Schieffer B, Paxton WG, Heerdt L, Berk BC, Delafontaine P, Bernstein KE (1995) Direct stimulation of Jak/STAT pathway by the angiotensin II AT1 receptor. Nature 375:247–250

Matsoukas JM, Hondrelis J, Keramida M, Mavromoustakos T, Makriyannis A, Yamdagni R, Wu Q, Moore GJ (1994) Role of the NH2-terminal domain of angiotensin II (ANG II) and [Sar1]angiotensin II on conformation and activity. NMR evidence for aromatic ring clustering and peptide backbone folding compared with [des-1,2,3]angiotensin II. J Biol Chem 269:5303–5312

Miura S, Feng YH, Husain A, Karnik SS (1999) Role of aromaticity of agonist switches of angiotensin II in the activation of the AT1 receptor. J Biol Chem 274:7103–7110

Monnot C, Bihoreau C, Conchon S, Corvol P, Clauser E (1996) Polar Residues in the transmembrane domains of the AT_{1A} angiotensin receptor are required for binding and coupling. J Biol Chem 271:1507–1513

Murphy TJ, Alexander RW, Griendling KK, Runge MS, Bernstein KE (1991) Isolation of a cDNA encoding the vascular type-1 angiotensin II receptor. Nature 351:233–236

Nikiforovich GV, Kao JL, Plucinska K, Zhang WJ, Marshall GR (1994) Conformational analysis of two cyclic analogs of angiotensin: implications for the biologically active conformation. Biochemistry 33:3591–3598

Noda K, Saad Y, Kinoshita A, Boyle TP, Graham RM, Husain A, Karnik SS (1995) Tetrazole and carboxylate groups of angiotensin receptor antagonists bind to the same subsite by different mechanisms. J Biol Chem 270:2284–2289

Oakley RH, Laporte SA, Holt JA, Barak LS, Caron MG (2001) Molecular determinants underlying the formation of stable intracellular G protein-coupled receptor-beta-arrestin complexes after receptor endocytosis. J Biol Chem 276:19452–19460

Oakley RH, Laporte SA, Holt JA, Caron MG, Barak LS (2000) Differential affinities of visual arrestin, beta arrestin1, and beta arrestin2 for G protein-coupled receptors delineate two major classes of receptors. J Biol Chem 275:17201–17210

Ohyama K, Yamano Y, Chaki S, Kondo T, Inagami T (1992) Domains for G protein coupling in angiotensin II receptor type I: studies by site-directed mutagenesis. Biochem Biophys Res Comm 189:677–683

Ohyama K, Yamano Y, Sano T, Nakagomi Y, Hamakubo T, Morishima I, Inagami T (1995) Disulfide bridges in extracellular domains of angiotensin II receptor type IA. Regul Pept 57:141–147

Olivares-Reyes JA, Smith RD, Hunyady L, Shah BH, Catt KJ (2001) Agonist-induced signaling, desensitization, and internalization of a phosphorylation-deficient AT1A angiotensin receptor. J Biol Chem 276:37761–37768

Oppermann M, Freedman NJ, Alexander RW, Lefkowitz RJ (1996) Phosphorylation of the type 1A angiotensin II receptor by G protein-coupled receptor kinases and protein kinase C. J Biol Chem 271:13266–13272

Parma J, Duprez L, van Sande J, Cochaux P, Gervy C, Mockel J, Dumont J, Vassart G (1993) Somatic mutations in the thyrotropin receptor gene cause hyperfunctioning thyroid adenomas. Nature 365:649–651

Parnot C, Bardin S, Miserey-Lenkei S, Guedin D, Corvol P, Clauser E (2000) Systematic identification of mutations that constitutively activate the angiotensin II type 1A receptor by screening a randomly mutated cDNA library with an original pharmacological bioassay. Proc Natl Acad Sci U S A 97:7615–7620

Perlman S, Costa-Neto CM, Miyakawa AA, Schambye HT, Hjorth SA, Paiva AC, Rivero RA, Greenlee WJ, Schwartz TW (1997) Dual agonistic and antagonistic property of nonpeptide angiotensin AT1 ligands: susceptibility to receptor mutations. Mol Pharmacol 51:301–311

Qian H, Pipolo L, Thomas WG (1999) Identification of protein kinase C phosphorylation sites in the angiotensin II (AT1A) receptor. Biochem J 343:637–644

Qian H, Pipolo L, Thomas WG (2001) Association of beta-Arrestin 1 with the type 1A angiotensin II receptor involves phosphorylation of the receptor carboxyl terminus and correlates with receptor internalization. Mol Endocrinol 15:1706–1719

Sandberg K, Ji H, Clark AJ, Shapira H, Catt KJ (1992) Cloning and expression of a novel angiotensin II receptor subtype. J Biol Chem 267:9455–9458

Sano T, Ohyama K, Yamano Y, Nakagomi Y, Nakazawa S, Kikyo M, Shirai H, Blank JS, Exton JH, Inagami T (1997) A domain for G protein coupling in carboxyl-terminal tail of rat angiotensin II receptor type 1A. J Biol Chem 272:23631–23636

Sasaki K, Yamano Y, Bardhan S, Iwai N, Murray JJ, Hasegawa M, Matsuda Y, Inagami T (1991) Cloning and expression of a complementary DNA encoding a bovine adrenal angiotensin II type-1 receptor. Nature 351:230–233

Sayeski PP, Ali MS, Semeniuk DJ, Doan TN, Bernstein KE (1998) Angiotensin II signal transduction pathways. Regul Pept 78:19–29

Schambye HT, Hjorth SA, Bergsma DJ, Sathe G, Schwartz TW (1994) Differentiation between binding sites for angiotensin II and nonpeptide antagonists on the angiotensin II type 1 receptors. Proc Natl Acad Sci USA 91:7046–7050

Schipani E, Kruse K, Jüppner H (1995) A constitutively active mutant PTH-PTHrP receptor in Jansen type metaphyseal chondrodysplasia. Science 98–100

Seachrist JL, Laporte SA, Dale LB, Babwah AV, Caron MG, Anborgh PH, Ferguson SS (2002) Rab5 association with the angiotensin II type 1A receptor promotes Rab5 GTP binding and vesicular fusion. J Biol Chem 277:679–685

Shirai H, Takahashi K, Katada T, Inagami T (1995) Mapping of G protein coupling sites of the angiotensin II type 1 receptor. Hypertension 25:726–730

Smith RD, Baukal AJ, Zolyomi A, Gaborik Z, Hunyady L, Sun L, Zhang M, Chen HC, Catt KJ (1998) Agonist-induced phosphorylation of the endogenous AT1 angiotensin receptor in bovine adrenal glomerulosa cells. Mol Endocrinol 12:634–644

Sorkin A, Von Zastrow M (2002) Signal transduction and endocytosis: close encounters of many kinds. Nat Rev Mol Cell Biol 3:600–614

Sugiura N, Hagiwara H, Hirose S (1992) Molecular cloning of porcine soluble angiotensin-binding protein. J Biol Chem 267:18067–18072

Takeuchi K, Alexander RW, Nakamura Y, Tsujino T, Murphy TJ (1993) Molecular structure and transcriptional function of the rat vascular AT1A angiotensin receptor gene. Circ Res 73:612–621

Tang H, Guo DF, Porter JP, Wanaka Y, Inagami T (1998) Role of cytoplasmic tail of the type 1A angiotensin II receptor in agonist- and phorbol ester-induced desensitization. Circ Res 82:523–531

Thomas WG, Baker KM, Motel TJ, Thekkumkara TJ (1995a) Angiotensin II receptor endocytosis involves two distinct regions of the cytoplasmic tail. A role for residues on the hydrophobic face of a putative amphipathic helix. J Biol Chem 270:22153–22159

Thomas WG, Motel TJ, Kule CE, Karoor V, Baker KM (1998) Phosphorylation of the angiotensin II (AT1A) receptor carboxyl terminus: a role in receptor endocytosis. Mol Endocrinol 12:1513–1524

Thomas WG, Qian H, Chang CS, Karnik S (2000) Agonist-induced phosphorylation of the angiotensin II (AT(1A)) receptor requires generation of a conformation that is distinct from the inositol phosphate-signaling state. J Biol Chem 275:2893–2900

Thomas WG, Thekkumkara TJ, Motel TJ, Baker KM (1995b) Stable expression of a truncated AT1A receptor in CHO.K1 cells. J Biol Chem 270:207–213

Timmermans PBMWM, Wong PC, Chiu AT, Herblin WF, Benfield P, Carini DJ, Lee RJ, Wexler RR, Saye JAM, Smith RD (1993) Angiotensin II receptors and angiotensin II receptor antagonists. Pharmacol Rev 45:205–251

Underwood DJ, Strader CD, Rivero R, Patchett AA, Greenlee W, Prendergast K (1994) Structural model of antagonist and agonist binding to the angiotensin II, AT1 subtype, G protein coupled receptor. Chem Biol 1:211–221 van Kats JP, de Lannoy LM, Jan Danser AH, van Meegen JR, Verdouw PD, Schalekamp MA (1997) Angiotensin II type 1 (AT1) receptor-mediated accumulation of angiotensin II in tissues and its intracellular half-life in vivo. Hypertension 30:42–49

Venema RC, Ju H, Venema VJ, Schieffer B, Harp JB, Ling BN, Eaton DC, Marrero MB (1998) Angiotensin II-induced association of phospholipase Cgamma1 with the G-protein-coupled AT1 receptor. J Biol Chem 273:7703–7708

Wang C, Jayadev S, Escobedo JA (1995) Identification of a domain of the angiotensin II type 1 receptor determining Gq coupling by the use of receptor chimeras. J Biol Chem 270:16677–16682

Werbonat Y, Kleutges N, Jakobs KH, van Koppen CJ (2000) Essential role of dynamin in internalization of M2 muscarinic acetylcholine and angiotensin AT1A receptors. J Biol Chem 275:21969–21974

Whitebread S, Mele M, Kamber B, De Gasparo M (1989) Preliminary biochemical characterization of two angiotensin II receptor subtypes. Biochem Biophys Res Commun 163:284–291

Wright GB, Alexander RW, Ekstein LS, Gimbrone MAJ (1982) Sodium, divalent cations, and guanine nucleotides regulate the affinity of the rat mesenteric artery angiotensin receptor. Circ Res 50:462–469

Yamano Y, Ohyama K, Kikyo M, Sano T, Nakagomi Y, Inoue Y, Nakamura N, Morishima I, Guo DF, Hamakubo T, et al (1995) Mutagenesis and the molecular modeling of the rat angiotensin II receptor (AT1). J Biol Chem 270:14024–14030

Zhang J, Ferguson SS, Barak LS, Menard L, Caron MG (1996) Dynamin and beta-arrestin reveal distinct mechanisms for G protein-coupled receptor internalization. J Biol Chem 271:18302–18305

AT₁ Receptor Interactions

G. Vauquelin · P. Vanderheyden

Department of Molecular and Biochemical Pharmacology,
Institute for Molecular Biology and Biotechnology, Free University of Brussels (VUB),
Gebouw E.5.10, Pleinlaan 2, 1050 Brussels, Belgium
e-mail: gvauquel@vub.ac.be

1	Introduction	298
2	Interaction with Naturally Occurring Angiotensin Fragments	299
3	Molecular Aspects of AT$_1$ Receptor Activation	300
4	Interaction with Nonpeptide Antagonists	303
5	Cellular and Molecular Aspects of Antagonist–AT$_1$ Receptor Binding	307
6	Inverse Agonism and Multiple AT$_1$ Receptor Conformations	311
References		312

Abstract Angiotensin II and antagonist–AT$_1$ receptor interactions have traditionally been studied on isolated tissues in organ bath experiments. Cells that express endogenous or transfected AT$_1$ receptors are now also increasingly used to this end. Angiotensin II and angiotensin III interact with the AT$_1$ receptor in a similar fashion. Other naturally occurring fragments like angiotensin IV or angiotensin-(1-7) display only low affinity for the receptor and the latter even acts as an antagonist. Combination of such structure–activity relationship and receptor mutation studies led to a more complete picture of the molecular events that take place during receptor activation. At least two steps take place: a pre-activation step, in which constraining intramolecular interactions within the receptor are broken by Arg^2 of angiotensin II, and a subsequent activation step in which the C-terminal side of the hormone plays an essential role. The outcome is a conformational change in the receptor that promotes its interaction with G proteins. Interaction with nonpeptide, biphenyltetrazole antagonists is also a multi-step process, and a "two-state, two-step" model is proposed in which the initial attraction to the receptor is fairly similar for all antagonists and in which a more stable, tight binding antagonist–receptor complex can be formed subsequently. This explains the often-mentioned distinction between "surmountable" and "insurmountable" antagonists in, e.g. vascular smooth muscle contraction experiments. This tight biding state is unrelated to receptor internalisation, but mutation studies reveal the implication of Lys^{199}. These recent observations comply with the increasing awareness that AT$_1$ receptors and other G protein-linked receptors may adopt a number of agonist- and antagonist-bound conformations and/or states, each with its own characteristic properties.

Keywords AT_1 receptor · Angiotensin fragments · Antagonists · Insurmountable · Interaction mechanisms · Cell lines · Receptor mutants

Abbreviations

ACE	Angiotensin-converting enzyme
Ang	Angiotensin
CHO-hAT$_1$ cells	Chinese hamster ovary cells expressing the human AT_1 receptor
GPCR	G protein-coupled receptor
L	Ligand (antagonist)
$N^{111}G$ mutant	AT_1 receptor mutant in which Asn^{111} is substituted by Gly
R, R′, R*, R$_I$	Receptor in stable inactive-, preactivated-, fully active and tight antagonist-binding states, respectively
TM	Transmembrane-spanning helix

1
Introduction

The octapeptide angiotensin (Ang) II is well known for its hypertensive effect and its ability to stimulate cardiac remodelling. Receptors of the AT_1 subtype play a major role in these processes (de Gasparo et al. 2000). Prevention of the hypertensive and trophic actions of Ang II has proved to be among the most successful strategies for the treatment of hypertension and congestive heart failure. To this end, Ang-converting enzyme (ACE) inhibitors have been introduced to decrease the plasma level of Ang II and, during the last decade, a fairly large number of nonpeptide antagonists have been developed to selectively block the AT_1 receptor (Timmermans 1999a,b; Unger 1999). This chapter deals with pharmacological and molecular aspects of the binding of agonists and antagonists to the AT_1 receptor. Traditionally, such interactions have been tested on isolated tissues in organ bath experiments. Cells that endogenously express the AT_1 receptor and cell lines that have been transfected with the gene encoding such receptor are also increasingly used to this end (Vanderheyden et al. 1999). Compared to isolated tissues, they allow the investigation of ligand–receptor interactions by radioligand binding and by measuring receptor-evoked responses under similar experimental conditions. This allows an objective comparison of both sets of data. Transfected cell systems offer additional advantages. They provide a homogenous receptor system, the untransfected parent cells can serve as a control for the detection of receptor-unrelated phenomena, and mutation studies allow the exploration of the role of specific amino acids in the binding and functional properties of the receptor. The physiological relevance of such intact cell experiments may be questioned, since the AT_1 receptors are often expressed at high density and in a foreign cellular environment (e.g. Chinese hamster ovary cells expressing the human AT_1 receptor, CHO-hAT_1 cells). Fortunately, however, the functional behaviour of the AT_1 receptors in such cell lines is remarkably similar to that in more complex in vitro experimental systems like vascular smooth muscle preparations.

2
Interaction with Naturally Occurring Angiotensin Fragments

Although Ang II has long been considered to represent the end product of the renin–angiotensin system, there is accumulating evidence that this system encloses additional effector peptides with diverse functions. In this respect, shorter Ang II fragments such as Ang III (deletion of the N-terminal amino acid), Ang IV [deletion of the two G protein-coupled receptor (GPCR) amino acids] and Ang-(1-7) (deletion of the C-terminal amino acid) have been found to accomplish central, cardiovascular and renal actions as well (Reaux et al. 2001) (Fig. 1). Similar to Ang II, Ang III is able to act both via AT_1 and AT_2 receptors. Both peptides elicit similar physiological effects, such as stimulation of aldosterone secretion, vasoconstriction and dipsogenic activity. They also bind with similar affinity to the AT_1 receptor and are equally potent agonists for this re-

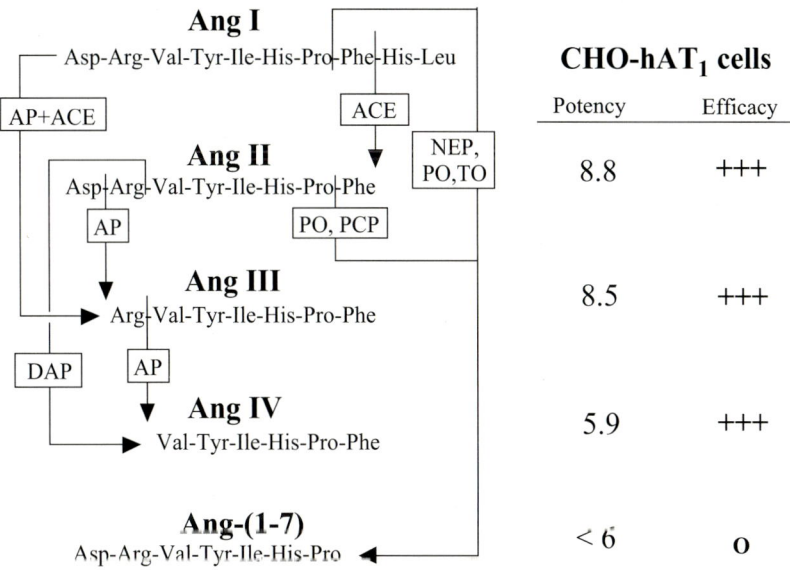

Fig. 1 *Left panel*: overview of the chemical structure and the enzymes involved in the synthesis of Ang II and its different fragments. Abbreviations: *ACE*, angiotensin-converting enzyme; *AP*, aminopeptidase; *DAP*, dipeptidyl aminopeptidases; *PO*, prolyl oligopeptidase; *PCP*, carboxypeptidase; *NEP*, neprilysin; *TO*, thimet oligopeptidase. *Right panel*: potency (in pEC_{50} for agonists and pIC_{50} for antagonists) and efficacy (+++, full agonism; *0*, antagonism) of Ang II and its different fragments for inducing inositol phosphate production in CHO cells expressing human AT_1 receptors. (Le et al. 2002, 2003)

ceptor (Le et al. 2002) (Fig. 1). Interestingly, Ang III was shown to be even more effective than Ang II in stimulating the firing rate of certain neurons (Harding and Felix 1987). There is now also good evidence that the modulation of the pressor response and vasopressin release by centrally administered Ang II in rats depends on its prior conversion to Ang III (Zini et al. 1996; Reaux et al. 2001). These findings support the hypothesis that Ang III might be the predominant effector peptide in the brain (Harding et al. 1986).

Ang IV is also able to stimulate the AT_1 receptor as a full agonist, albeit with much lower potency when compared to Ang II and Ang III (Pendleton et al. 1989a; Le et al. 2002) (Fig. 1). This interaction provides a likely explanation for some of the physiological effects elicited by high, micromolar concentrations of this angiotensin fragment and for the ability of AT_1 receptor-selective antagonists such as losartan to block such effects. Yet, most of the physiological effects of Ang IV are already observed at nanomolar concentrations and some of them, such as its facilitating role in memory acquisition and retrieval, are of potential therapeutic interest (Wright et al. 1999; Kramár et al. 2001). The involved cellular targets acquired the status of "AT_4 receptors" (de Gasparo et al. 2000; Mustafa et al. 2001). Very recently, compelling evidence has been presented that at least some of them may be related to a membrane-associated aminopeptidase formerly known as the insulin-regulated aminopeptidase/oxytocinase (Albiston et al. 2001).

Ang-(1-7) has been reported to elicit physiological effects in the vasculature, the CNS and the kidney (Ferrario and Iyer 1998). In this respect, it even emerged that ACE inhibitors may exert their cardiovascular effects not only by blocking Ang II formation but also by promoting the accumulation of Ang-(1-7) (Luque et al. 1996). The effects of this angiotensin fragment have often been investigated in systems which also contain AT_1 receptors and/or AT_2 receptors and, in this respect, it is of interest that some of them could be blocked by the AT_1 antagonist losartan (Yamada et al. 1998; Potts et al. 2000). However, AT_1 receptor interactions do not offer a plausible explanation for such effects. It has been consistently found that Ang-(1-7) displays very low affinity for the AT_1 receptor and even that it acts as an antagonist (Mahon et al. 1994; Bouley et al. 1998; Le et al. 2002) (Fig. 1). Hence, alternative pathways and/or cellular recognition sites need to be invoked to explain the effects of Ang-(1-7).

3
Molecular Aspects of AT_1 Receptor Activation

Molecular cloning studies established that the AT_1 receptor belongs to the GPCR superfamily and, as such, it contains seven transmembrane-spanning α-helices (Hunyady at al. 2001) (Fig. 2). It is primarily coupled through G proteins of the $G_{q/11}$ family to the activation of phospholipase C and calcium signalling. This second messenger system mediates smooth muscle contraction, aldosterone secretion, and the control of ion transport in renal tubule cells (de Gasparo et al. 2000). Whereas only one AT_1 receptor subtype is expressed in the human, two

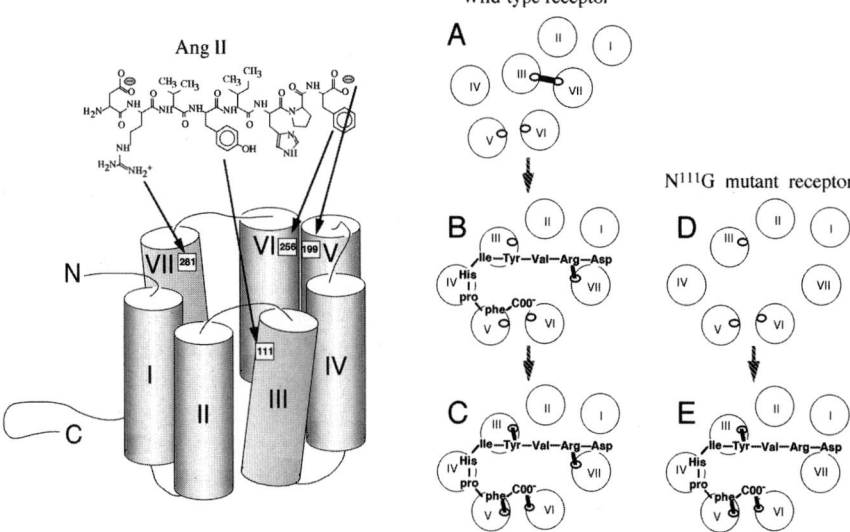

Fig. 2 *Left panel*: side view of the Ang II–AT$_1$ receptor interaction with reference to crucial parts of both molecules. (According to Hunyady et al. 2001 and Le et al. 2002). Both molecules are separated for the sake of clarity. *Roman and Arabic numbers* refer to transmembrane (TM) helices and amino acid residues of the receptor, respectively. *Right panel*: top view of the different steps involved in the Ang II–AT$_1$ receptor interaction. (According to Le et al. 2002). A–C: wild-type receptor. A: the basal state of the receptor is stabilised by the interaction of Asn^{111} with TM VII. B: Arg^2 of Ang II interacts with Asp^{281} of the receptor to relax the receptor into a pre-activated state (R′) and to provide optimal binding of carboxy-teminal portion of Ang II. C: full receptor activation due to additional interactions with Tyr^4, Phe^8 and the terminal carboxyl group of Ang II. D and E: N^{111}G receptor mutant. D: the basal state of the receptor is constitutively activated (and similar to the R′ state of the wild-type receptor) due to the absence of a Gly^{111}-TM VII interaction. E: full receptor activation by Ang II without the need of Arg^2

subtypes (i.e. AT$_{1A}$ and AT$_{1B}$ receptors) with similar pharmacological properties but with different anatomical distribution have been found in rodents like the rat (Kakar et al. 1992).

Structure–function relationship studies have led to the proposal that the side chains of Arg^2, His^6, and Pro^7 along with the charged carboxyl-terminus of Ang II are important for its binding to the AT$_1$ receptor. In addition, Tyr^4 and the aromatic group of Phe^8 are important for receptor activation (Noda et al. 1995,1996; Miura et al. 1999; Hunyady et al. 2001) (Fig. 2). This explains why Ang IV and Ang-(1-7) display low affinity for the receptor and why the latter is an antagonist. On the other hand, Asp^1 of Ang II does not appear to be involved in AT$_1$ receptor binding and activation since Ang II and Ang III display the same characteristics. Site-directed mutagenesis studies in which single amino acids of the AT$_1$ receptor are substituted by less reactive ones (e.g. glycine and alanine) suggest that extracellular regions of the receptor play an important role in the binding of Ang II (Hunyady et al. 1996) (Fig. 2, left). This is consistent with the general belief that peptide and glycoprotein hormones bind to the extracellular regions of their GPCRs, as opposed to the binding of small nonpeptide ligands within the central cleft of such receptors.

The outcome of the Ang II binding is a conformational change in the receptor molecule that promotes its interaction with G proteins. Whereas the classic theory of receptor activation assumes that receptors can only adopt a single inactive (R) and active (R*) state, there is accumulating evidence that GPCRs may adopt a range of additional conformations. These may even include states with intermediate activity (R') in addition to the inactive and fully active ones. AT_1 receptor mutation studies provided a major and pioneering contribution to this notion and it is now widely admitted that full activation of this receptor by Ang II proceeds in at least two steps (Noda et al. 1996).

According to these models, the inactive form(s) of the receptor is stabilised by intramolecular bonds between Asn^{111} of its third transmembrane domain (TM III) and amino acid residues present in TM VII (Hunyady at al. 2001) (Fig. 2A). A similar stabilising interaction has also been proposed to occur for rhodopsin, the only true GPCR for which the spatial structure has been elucidated by X-ray diffraction studies (Cohen et al. 1992). When Asn^{111} is substituted by Gly ($N^{111}G$ mutation) or by other amino acids with smaller side-chains, such interhelical bonds can no longer take place, and this allows the mutant AT_1 receptor to adopt a more "relaxed" conformation (Fig. 2D). These mutated receptors can still be fully activated by Ang II, but their basal activity is already well above that of the wild-type receptor (a phenomenon referred to as "constitutive activation") (Groblewski et al. 1997). These mutants are regarded to mimic an intermediate, not yet fully activated state (R') of the wild-type AT_1 receptor (Noda et al. 1996; Hunyady et al. 2001) (Fig. 2B). For the wild-type receptor, the disruption of the bonds between Asn^{111} and TM VII is thought to be the first consequence of Ang II binding. Initially, it was advanced that the interaction between Tyr^4 of Ang II and Asn^{111} of the receptor should play an important role in its pre-activation (Noda et al. 1996). Yet, more recent studies revealed that Arg^2 of Ang II and Asp^{281} of the receptor might be even more important actors (Le et al. 2002). In the model pictured in Fig. 2, the interaction between these two amino acids will induce a conformational change in TM VII of the receptor, which, similar to the $N^{111}G$ mutation, eliminates the constraining interaction between Asn^{111} and TM VII.

In line with the notion that the $N^{111}G$ mutation causes a conformational change that is associated with receptor pre-activation, it causes a slight increase of the potency of Ang II and Ang III when compared to the wild-type receptor. Interestingly, the $N^{111}G$ mutated AT_1 receptor can be fully activated by Ang IV with a potency that is similar to Ang II and Ang III (Le et al. 2002). This implies that, once the receptor has acquired the preactivated state R', Arg^2 will no longer play a major role in the binding and full activation of the receptor (Fig. 2E). Instead, the pre-activation is suggested to go along with a conformational change of the receptor to provide optimal binding of carboxy-terminal portion of Ang II (Le et al. 2002). It is this side of the hormone that will take over to produce full receptor activation (to yield R*, Fig. 2C,E). In this respect, the very low potency and efficacy of Ang-(5-8) to activate the wild-type AT_1 receptor as well as the $N^{111}G$ mutant highlights the importance of Tyr^4 in this part of the activation process (Noda et al. 1996; Miura et al. 1999). Asn^{111} is therefore considered to play a dual role; it stabilises R by interhelical interactions and, once liberated (in R′), it contributes as an "agonist switch" to the formation and/or stabilisation of R* by interacting with Tyr^4 of Ang II. In addition, evidence has been presented for the terminal carboxyl group of Ang II to form a salt bridge with Lys^{199} located in TM V of the receptor. This allows an appropriate positioning of the aromatic side chain of Phe^8 to undergo hydrogen bonding with His^{256} in TM VI of the receptor, an interaction which is considered to be of prime importance for the activation process (Noda et al. 1995,1996; Miura et al. 1999). Hence, the Phe^8 side chain represents the second "agonist switch" of Ang II.

Many synthetic Ang II analogues have been developed by substituting one or more of the original amino acids of Ang II either by other natural amino acids or by synthetic ones. Sarile ($[Sar^1,Ile^8]$-Ang II) constitutes a typical example and, in its radioiodinated form, it is widely used for the labelling of AT_1 receptors in radioligand binding studies. Although it has been considered by some to behave as an antagonist, it is able to produce modest activation of the wild-type AT_1 receptor in cell lines (Noda et al. 1996). Such partial agonistic activity also constitutes one of the reasons for the failure of the therapeutic use of saralasin ($[Sar^1,Val^5,Ala^8]$-Ang II) a close analogue of sarile (Pals et al. 1979). Interestingly, sarile has been found to fully activate the $N^{111}G$ receptor mutant (Noda et al. 1996). This example illustrates the use of constitutively activated AT_1 receptor mutants to distinguish between true antagonists and weak partial agonists.

4
Interaction with Nonpeptide Antagonists

In the past decade, nonpeptide AT_1 receptor antagonists have been routinely tested for their ability to affect Ang II dose-contractile response curves of vascular smooth muscle preparations. Such organ bath experiments have often been carried out with rabbit aortic rings/strips, a system with very small receptor reserve (Zhang et al. 1993; Robertson 1998). They constitute a major paradigm for comparing the behaviour of different antagonists. Losartan, the prototype of the biphenyl-tetrazole class of antagonists, produces parallel rightward shifts of the Ang II dose–response curve without depressing the maximal response. This behaviour is denoted as "surmountable" (Fig. 3, top). Yet, many structurally related antagonists are also capable of depressing the maximal response. This behaviour is denoted as "insurmountable" (Vauquelin et al. 2001a, 2002). The extent by which the maximal response is depressed is quite variable (Fig. 3, top). For example, it is only partial for irbesartan, valsartan and EXP3174 (the active metabolite of losartan) but almost complete for antagonists such as GR117289, KRH-594 and candesartan (Liu et al. 1992; Robertson et al. 1992; Cazaubon et al. 1993; Criscione et al. 1993; Noda et al. 1993; Mochizuki et al. 1995; Tamura et al. 1997). More recently, such experiments have been repeated with cell lines

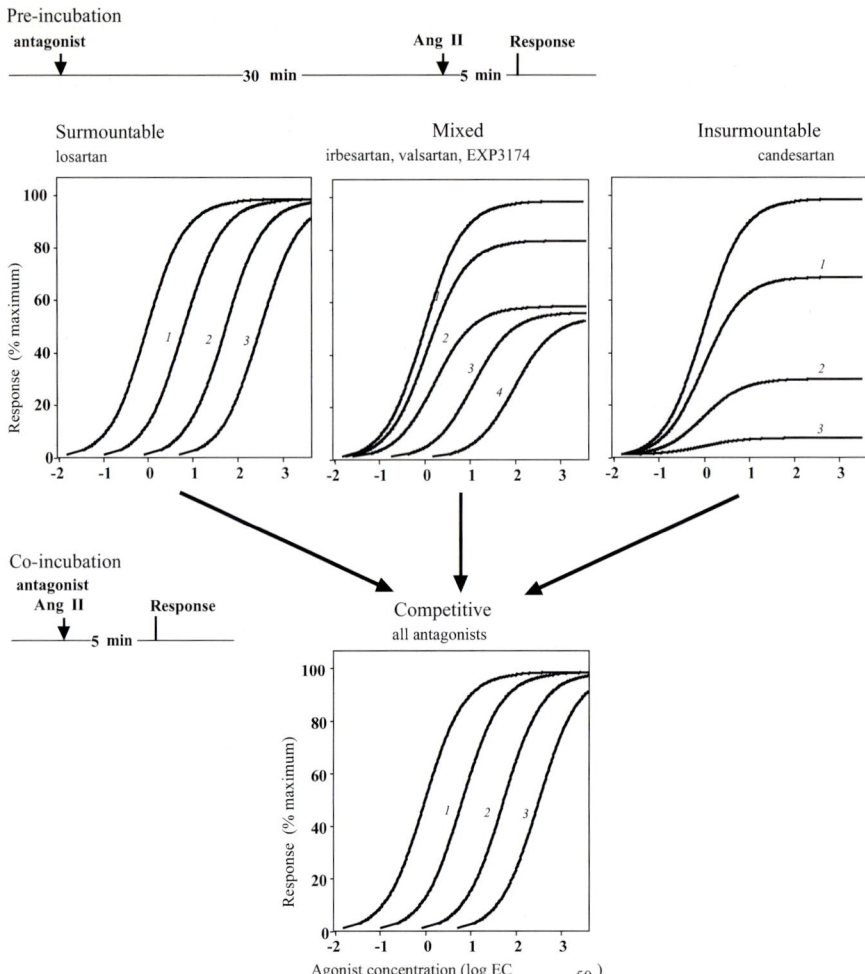

Fig. 3 *Top panel*: theoretical Ang II dose–response curves for antagonist-pretreated AT_1 receptor preparations. *Curve at the left*: no antagonist pretreatment. *Curves with italic numbers*: pretreatment with increasing antagonist concentrations of surmountable, partially insurmountable and nearly fully insurmountable antagonists. (Vanderheyden et al. 1999; Verheyen et al. 2000). *Lower panel*: theoretical Ang II dose–response curves with the same antagonists under co-incubation conditions. (Fierens et al. 1999a; Verheyen et al. 2000)

that endogenously express the AT_1 receptor or that have been transfected with the gene encoding such receptors as CHO-hAT_1 cells. Although these experiments rely on the measurement of receptor-evoked biochemical events within the cell, the provided information is consistent with that obtained by smooth muscle contraction studies, at least when the antagonist is administered ahead of Ang II (Fierens et al. 1999a; Vanderheyden et al. 1999; Verheyen et al. 2000).

Although most AT$_1$ receptor antagonists possess a biphenyltetrazole moiety (Fig. 5), several of them are structurally different. Such antagonists can also be classified as surmountable (e.g. eprosartan) and insurmountable (e.g. telmisartan and triacid 4-aminoimidazole derivatives) (Wienen et al. 1993; Timmermans et al. 1999b; Vanderheyden et al. 2000a; Hollenberg 2001). Additionally, peptide analogues of Ang II with partial agonistic activity, such as sarile, have also been shown to display insurmountable AT$_1$ receptor blockade (Pendleton et al. 1989b). Presently, insurmountable antagonism has only been investigated in detail for antagonists of the biphenyltetrazole class. The present discussion will therefore focus on such antagonists.

Insurmountable antagonism does not constitute the sole privilege of the AT$_1$ receptors either. Similar behaviour has been observed for many other GPCR antagonists, and several theories have been advanced in the past (Vauquelin et al. 2001a, 2002). One major explanation endorses that insurmountable antagonists are non-competitive. The other one takes into account that tissues are invariably pre-incubated with the antagonist before their challenge with the agonist. This second explanation allows competitive antagonists to be insurmountable provided that they bind to the receptor in an irreversible fashion (so that the receptor number is permanently reduced) or that they dissociate sufficiently slowly (so that only a limited number of receptors are liberated before the response is measured) (Kenakin 1987). Co-incubation experiments, in which the agonist and the antagonist are added simultaneously to the receptor, have the potential to provide a clear-cut discrimination between both explanations (Vauquelin et al. 2001a). Indeed, under those conditions, a decrease of the maximal agonist-evoked response is only to be expected in the case of non-competitive antagonism. Whereas such experiments are extremely laborious when dealing with contraction studies, they are commonplace for radioligand binding and functional studies with isolated cells or membranes thereof. In co-incubation experiments, surmountable and insurmountable antagonists were all found to produce parallel rightward shifts of the Ang II dose–response curves in bovine adrenal glomerulosa cells (Criscione et al. 1993) and in CHO-hAT$_1$ cells (Fierens et al. 1999a; Verheyen et al. 2000) (Fig. 3, lower panel). The same profile also emerges when comparing radioligand binding studies involving pre- and co-incubation experiments (Pendleton et al. 1989b; Hara et al. 1995; De Arriba et al. 1996). Such studies have now clearly established that nonpeptide AT$_1$ receptor antagonists are competitive with Ang II and even among each other. This also explains the ability of surmountable and insurmountable AT$_1$–receptor antagonists to counteract each other's effect in a dose-dependent fashion (Wong et al. 1991; Vauquelin et al. 2001b).

Direct measurements based on the binding of radiolabelled antagonists as well as indirect measurements based on the recovery of receptor functionality in wash-out experiments confirm that surmountable antagonists such as losartan dissociate very rapidly from the AT$_1$ receptor and that insurmountable antagonists dissociate slowly. For intact CHO-hAT$_1$ cells, the dissociation half-life of insurmountable antagonists at 37°C was estimated to be 7 min for irbesartan, 17 min for valsartan, 30 min for EXP3174 and 120 min for candesartan (Fierens et al. 1999b; Vanderheyden et al. 2000b,c; Verheyen et al. 2000). These determinations clearly establish that the blockade by insurmountable antagonists is sufficiently long lasting to impair the accessibility of a proportion of the AT_1 receptor molecules if the ensuing challenge with Ang II only takes place for relatively short time-span. Obviously, the occupancy of the receptors by such antagonists will readjust with time until, ultimately, both the Ang II–receptor and the antagonist–receptor interactions reach equilibrium. Therefore, the insurmountability of slow dissociating antagonists is only "apparent", and it can theoretically be avoided if the response is measured after a sufficient lag of time. However, such delays are precluded for practical considerations such as the necessity to avoid fading of the response.

Many AT_1 receptor antagonists display a rather complex pattern of insurmountability when examining systems as diverse as muscle strips and transfected cell lines. At low concentrations, they decrease the maximal response to an extent that is dependent on the antagonist in question. When the concentration is further increased, these antagonists still produce rightward shifts of the dose-response curve, but the maximal stimulation will not decline any further. Such mixed insurmountable/surmountable antagonism has also been documented for several other GPCRs (Vauquelin et al. 2002). The simplest explanation would be that the antagonists already liberated of part of the receptor sites during the ensuing agonist exposure. This permits a partial restoration of the response to take place. However, the fast liberation of only part of the antagonist-pretreated AT_1 receptor sites and the much slower liberation of the remainder in washout experiments (Fierens et al. 1999a) cannot be explained by this simple model. Instead, this phenomenon is readily explained when the classical theory of receptor activity with only one thermodynamically stable inactive form of the receptor, R, is extended to include a second inactive form, R_I, which cannot be recognised and stimulated by agonists (Fierens et al. 1999a). Several such "two-state" models have been elaborated over the years according to such a principle (Lucas et al. 1979; de Chaffoy de Courcelles et al. 1986; Liu et al. 1992; Robertson et al. 1996). In their most general formulation (Fig. 4), such models allow R to form fast reversible/surmountable complexes with all antagonists (L). In addition, these models agree on the implication of $L.R_I$ in insurmountable antagonism and allow an equilibrium between L.R and $L.R_I$ to take place (Vauquelin et al. 2002). Finally, the proportion between the surmountable and insurmountable antagonist–receptor complexes (i.e. the $[L.R]/[L.R_I]$ ratio) may differ from one antagonist to another. The major differences between the published two-state models reside in the explanations about the precise nature of L.R' and how it mediates insurmountable antagonism.

Insurmountable AT_1 receptor antagonism can obviously be explained by a slow dissociation of the $L.R_I$ complex. Alternatively, insurmountable antagonism might also be perceived when $L.R_I$ dissociates swiftly, provided that R_I is only slowly reconverted into R (Fierens et al. 1999a). According to this latter explanation, there is no strict obligation for the antagonist to remain bound to the receptor to produce a long-lasting effect. In the case of AT_1 receptor antagonists, a

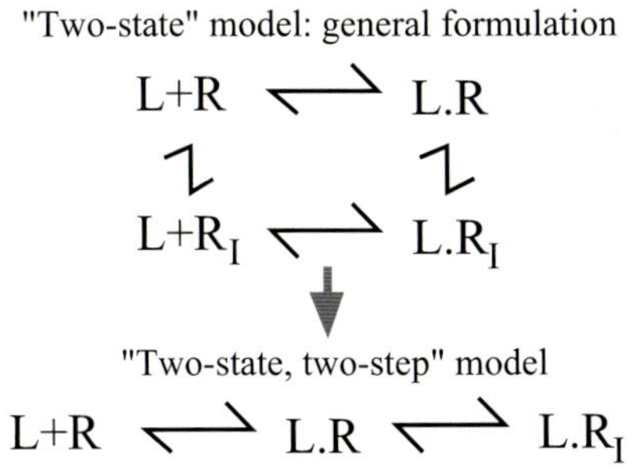

"Two-state" model: general formulation

$$L+R \rightleftarrows L.R$$
$$\downarrow \qquad \downarrow$$
$$L+R_I \rightleftarrows L.R_I$$
$$\downarrow$$

"Two-state, two-step" model

$$L+R \rightleftarrows L.R \rightleftarrows L.R_I$$

losartan	100 %	0 %
irbesartan	70 %	30 %
valsartan	50 %	50 %
EXP3174	30 %	70 %
candesartan	5 %	95 %

Fig. 4 Two-state models for insurmountable AT$_1$ receptor antagonism. The receptor can be stimulated by the agonist when present in the R state/conformation but not when present in the R$_I$ state. In the general formulation (Fierens et al. 1999a), insurmountable antagonism might take place with fast-dissociating antagonist-R$_I$ complexes as well as with fast-dissociating complexes, provided that R$_I$ is only slowly reconverted into R. In the two-step model (Vauquelin et al. 2001a,b), the R$_I$ state is slow-dissociating. Percentages of antagonist–receptor complexes in each state are estimated from inhibition data (Fierens et al. 1999a; Verheyen et al. 2000)

close match has been observed between the dissociation of [^{3}H]-candesartan and [^{3}H]-valsartan from CHO-hAT$_1$ cells and the recovery of functional receptors in these cells (Fierens et al. 1999b; Verheyen et al. 2000). In other words, the insurmountable effect of such antagonist is directly linked to their receptor occupancy. This supports the first explanation and constitutes a rationale for simplifying the general "two-state" model formulation into the "two-state, two-step" model (Vauquelin et al. 2001a,b) (Fig. 4). In this simplified model, all antagonists initially bind to R to form a fast-dissociating L.R complex. For insurmountable binding, this complex must further be converted into a tight-binding L.R$_I$ complex. Whereas insurmountable antagonists can induce this conversion with various degrees of effectiveness, surmountable antagonists like losartan are unable to do so. Hence, losartan–AT$_1$ receptor complexes are committed to reside in the fast-dissociating L.R state. Computer simulations were performed to match the "two-state, two-step" model with the experimental behaviour of the antagonists losartan, irbesartan, EXP3174 and candesartan (Vauquelin et al. 2001b). It was found that the initial "attraction" to the receptor (i.e. formation of L.R) is fairly similar for each of the investigated antagonists. The subsequent formation of L.R$_I$ also appears to proceed with the same "ease" for all the insurmountable antagonists but that the stability of this complex is different for each antagonist. Moreover, the model also provides an explanation for the observed proportionality between the experimental [L.R]/L.R$_I$] ratio (Fig. 4) and the dissociation rate of each involved antagonist (Vauquelin et al. 2001a).

5
Cellular and Molecular Aspects of Antagonist–AT$_1$ Receptor Binding

At present, it is not known why the antagonist–AT$_1$ receptor complexes may adopt distinct states. Differences between L.R and L.R$_I$ may reside at the level of the receptor conformation, its association with other proteins or even its subcellular localisation.

Differences in the subcellular localisation of the receptor, with L.R at the cell surface and $L.R_I$ inside the cell, have been invoked to explain insurmountable AT_1 receptor antagonism. Indeed, receptor internalisation is well known to occur with Ang II. When stimulated by this agonist, the AT_1 receptor has been shown to undergo rapid internalisation into the cell as part of their recycling process (Hein et al. 1997; Hunyady et al. 2001). Initial studies concluded that antagonist–receptor complexes are able to internalise as well (Crozat et al. 1986; Liu et al. 1992; Conchon et al. 1994), but more recently, it emerged that the presented criteria for such internalisation are not adequate in the case of nonpeptide AT_1 receptor antagonists. Instead, work with mutant receptors that had lost their ability to undergo internalisation (Fierens et al. 2001), as well as with green fluorescent protein-conjugated receptors (Hein et al. 1997; Hunyady et al. 2001), rather suggests that antagonist-bound AT_1 receptors always remain at the cell surface.

It is well known that AT_1 receptors and other GPCRs may exist as monomeric and dimeric forms with quite distinct functional properties and that such receptors also acquire a tight agonist-binding conformation when they undergo functional coupling to G proteins (De Lean et al. 1980). It is therefore also plausible that L.R and $L.R_I$ display differences in their association with other proteins which make part of the membrane or the cellular matrix, or even diffuse freely in the cytosol. At present, still very little is known about this issue, but it could be of great interest since the insurmountable antagonist $[^3H]$-candesartan has been found to dissociate much faster from isolated CHO-hAT_1 cell membranes as compared to the intact cells (Fierens et al. 2002). Thermodynamic characterisation also shows differences for the $[^3H]$-candesartan binding, with an almost completely enthalpy-driven interaction for intact cells as compared to a mixed contribution of both enthalpy and entropy for the membranes. As the most obvious explanations (proteolytic degradation and oxidative phenomena during the membrane preparation) were ruled out, these differences suggest a yet-to-be-disclosed impact of the structural organisation of living cells on the binding properties of AT_1 receptor antagonists (Fierens et al. 2002).

Additional information about the status of L.R and $L.R_I$ is provided by antagonist structure–binding activity relationship studies and receptor mutation studies. In this respect, it is noteworthy that the synthesis of nonpeptide AT_1 receptor antagonists has been sparked by the discovery that Ang II-mediated vasoconstriction could be antagonised by imidazole-5-acetic acid derivatives such as S-8307, albeit with low potency (Duncia et al. 1990; Timmermans et al. 1993). Starting from such lead compounds, more potent antagonists were designed with the major premises that the carboxyl group of lead and the terminal carboxyl group of Ang II should both point to the same positive charge of the receptor and that the lead requires an additional acidic functionality to mimic the carboxyl group in Asp^1 or the hydroxyl group in Tyr^4 of Ang II (Fig. 5, top). Further substitutions of S-8307, including the introduction of an anionic tetrazolium led to the development of losartan, the prototype of the biphenyltetrazole-containing sartans (Duncia et al. 1990; Timmermans et al. 1993). The carboxyl group of the lead compounds was replaced by an alcoholic function in

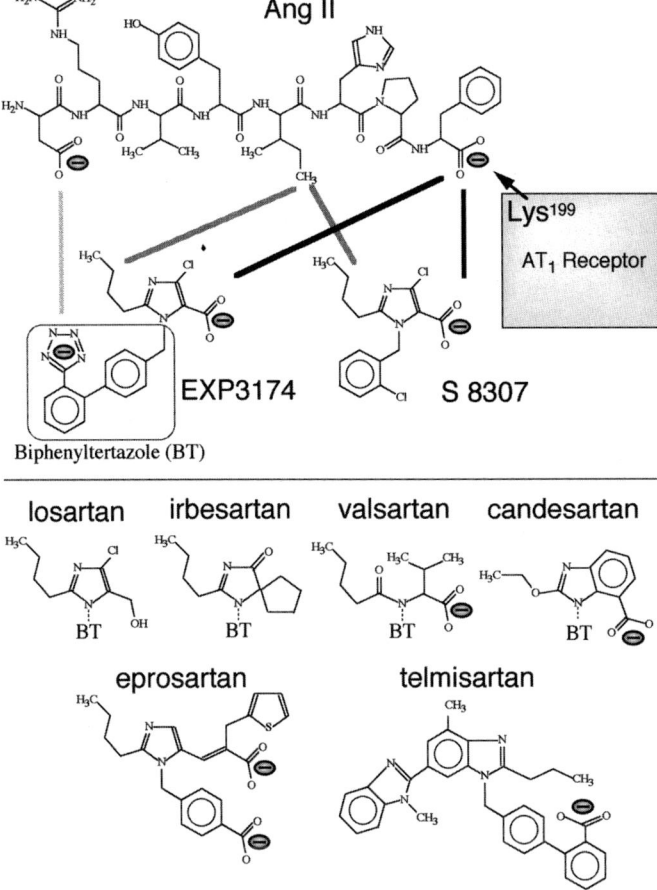

Fig. 5 *Top panel*: supposed structural homologies between Ang II and the nonpeptide AT_1 receptor antagonists S8307 and EXP3174. (According to Timmermans et al. 1993). *Lower panel*: structure of other nonpeptide AT_1 receptor antagonists of the "sartan" class. *BT*, biphenyltetrazole moiety

losartan but, after oxidative transformation in the liver, this group reappears in EXP3174, the active metabolite of losartan (Wong et al. 1990) (Fig. 5, top). Since losartan is fully surmountable, the carboxyl group of EXP3174 is clearly involved in its insurmountable character. The same distinction can be made between candesartan and its surmountable precursor molecule candesartan cilexetil (Noda et al. 1993; Mochizuki et al. 1995). In the same way, many other AT_1 receptor antagonists with a pronounced insurmountable character appear to possess both a carboxyl group and a tetrazole moiety (Vauquelin et al. 2000). However, the mere presence of a carboxyl group is not sufficient for biphenyltetrazole-containing AT_1 receptor antagonists to be insurmountable. Experiments with candesartan analogues clearly revealed that this group needs to be positioned correctly (Noda et al. 1993).

Taken together, aforementioned structure–binding activity relationship studies are compatible with the following model. Because of the similarity in the initial "attraction" of biphenyltetrazole-containing sartans to the receptor (Vauquelin et al. 2001b), this moiety is likely to play an important role in the formation of L.R. When present, the carboxyl group of such antagonists should not important for their initial binding, but it should play a major role in the stabilisation of the $L.R_I$ complex by interacting with basic amino acids that are present in a well-defined binding pocket of the receptor (Vauquelin et al. 2000). Obviously, additional interactions may also be of importance for the stabilisation of $L.R_I$ and, in this respect, it has been reported that the nature of the alkyl substituent next to the carboxyl group on analogues of UR-7280 may determine whether they are insurmountable or not (De Arriba et al. 1996).

The carboxyl group of insurmountable AT_1 receptor antagonists like EXP3174 is also supposed to mimic the terminal carboxyl group of Ang II (Timmermans et al. 1993) (Fig. 5, top). AT_1 receptor mutation studies support this conclusion. Indeed, Lys^{199} is known to be involved in the binding of the carboxy-terminus of Ang II. Its substitution by Gln (an amino acid with a neutral but still polar residue) causes a marked loss of binding affinity for Ang II, similar to that observed when its carboxyl group is replaced by a neutral amide group (Noda et al. 1995,1996). In contrast, this mutation has only little to moderate effect on the affinity of losartan (Schambye et al. 1994; Noda et al. 1995). In intact CHO-hAT_1 cells, this mutation produces affinity decreases of the nonpeptide antagonists that followed their degree of insurmountability: i.e. about 5-fold for losartan, 10-fold for irbesartan, 18-fold for EXP3174 and 45-fold for candesartan (Fierens et al. 2000). In addition, the dissociation of $[^3H]$-candesartan from the mutated receptor is also appreciably faster than from the native receptor (Fierens et al. 2000). These findings plead in favour of electrostatic bonding between Lys^{199} and the carboxyl group of insurmountable antagonists like candesartan and EXP3174 (Fig. 5, top). Yet, whereas such bonding will mediate receptor activation in the case of Ang II and related agonists, it will merely contribute to the stabilisation of a tight binding state of the receptor in the case of insurmountable nonpeptide antagonists.

Yamano et al. (1995) proposed that the anionic tetrazolium substituent of nonpeptide antagonists does not bind to Lys^{199}. Based on additional mutation studies, these investigators put forward that the tetrazolium moiety might selectively bind to the anionic side chain of Arg^{167}. However, as the substitution of Arg^{167} by Ala abolishes $[^3H]$-Ang II as well as $[^3H]$-candesartan binding to intact CHO-hAT_1 cells (Vauquelin et al. 2001c), the possibility also arises that this amino acid is only important for the structural integrity of the AT_1 receptor. Hence, there is no strict necessity for this amino acid to directly participate in any binding process itself. Based on similar mutation studies, several additional amino acids belonging to TM domains III, IV, V, VI and VII have been found to affect the binding of losartan (Ji et al. 1994). Most of them are positioned within a small distance from each other within a plane that is one or two helical turns below the membrane surface.

6
Inverse Agonism and Multiple AT_1 Receptor Conformations

Whereas Asn^{111}-substituted AT_1 receptor mutants exhibit increased affinity for Ang II and related agonists, they have been shown to display decreased affinity for losartan (Monnot et al. 1996; Groblewski et al. 1997). Such mutations are believed to improve the positioning of residues required for agonist binding and bring about a misalignment of the residues required for losartan binding (Hunyady et al. 2001). Recent studies on intact CHO-hAT_1 cells reveal that $N^{111}G$ and related mutant receptors display a similar decrease in affinity for losartan, irbesartan, EXP3174 and candesartan (Le et al. 2003). Provided that the wild-type AT_1 receptor can spontaneously adopt a related preactivated state (R'), these findings predict that nonpeptide antagonists have the potential to act as inverse agonists but that this effect should be independent of their degree of insurmountability. So far, there are no direct indications for endogenously expressed AT_1 receptors to undergo such spontaneous pre-activation, but other GPCRs have been reported to do so (Leurs et al. 2000). As Ang IV displays much higher potency for the $N^{111}G$ mutant compared to the wild-type receptor (Le et al. 2002), this peptide might constitute a novel powerful tool for the quest for preactivated AT_1 receptors under normal and pathophysiological conditions.

Opposed to this scenario, there is the opinion that agonists do not stabilise pre-existing active states of the AT_1 receptor but that their interaction with the receptor is required to induce receptor activation (Noda et al. 1996; Hunyady et al. 2001). This opens the possibility for different agonists, or even for the same agonist, to cause different conformational rearrangements of the receptor (Thomas et al. 2001). This provides a plausible explanation for the ability of [Sar^1,Ile^4,Ile^8]Ang II to promote AT_1 receptor phosphorylation without going through the conformation required for inositol phosphate signalling. It may also explain the fact that multiple forms of signal transduction can participate in the effects of Ang II on cell growth and on the remodelling of cardiac and vascular cells (de Gasparo et al. 2000; Dostal et al. 2000; Eguchi et al. 2000; Haendeler et al. 2000; Hunyady et al. 2001).

Taken together, the current concepts about agonist– and antagonist–AT_1 receptor interaction comply with the increasing awareness that the classical models for receptor activation, in which receptors isomerise between single inactive and active conformations, are over-simplistic (Clark et al. 1999; Onaran et al. 1999). Instead, a number of agonist-bound and antagonist-bound conformations and/or states of the receptor are suspected to occur, each with its own characteristic properties.

Acknowledgements. We are most obliged to the Queen Elisabeth Foundation Belgium, the Fonds voor Wetenschappelijk Onderzoek Vlaanderen, AstraZeneca and the Onderzoeksraad of the Vrije Universiteit Brussel for their kind support.

References

Albiston AL, McDowall SG, Matsacos D, Sim P, Clune E, Mustafa T et al (2001) Evidence that the Angiotensin IV (AT$_4$) receptor is the enzyme insulin-regulated aminopeptidase. J Biol Chem 276:48623–48626

Bouley R, Pérodin J, Plante H, Rihakova L, Bernier SG, Maletínská L, Guillemette G, Escher E (1998) N- and C-terminal structure-activity study of angiotensin II on the angiotensin AT$_2$ receptor. Eur J Pharmacol 343:23–31

Cazaubon C, Gougat J, Bousquet F, Guiraudou P, Gayraud R, Lacour C et al (1993) Pharmacological characterization of SR 47436, a new non-peptide AT$_1$ subtype angiotensin II receptor antagonist. J Pharmacol Exp Ther 265:826–834

Clark RB, Knoll BJ, Barber R (1999) Partial agonists and G protein-coupled receptor desensitization. Trends Pharmacol Sci 20:279–286

Cohen GB, Oprian DD, Robinson PR (1992) Mechanism of activation and inactivation of opsin: role of Glu113 and Lys296. Biochemistry 3:12592–12601

Conchon S, Monnot C, Teutsch B, Corvol P, Clauser E (1994) Internalization of the rat AT$_{1a}$ and AT$_{1b}$ receptors: pharmacological and functional requirements. FEBS Lett 349:365–370

Criscione L, de Gasparo M, Bühlmayer P, Whitebread S, Ramjoué HP, Wood J (1993) Pharmacological profile of valsartan: a potent, orally active, nonpeptide antagonist of the angiotensin II AT$_1$ receptor subtype. Br J Pharmacol 110:761–771

Crozat A, Penhoat A, Saez JM (1986) Processing of angiotensin II (A-II) and (sar^1,Ala^8) A-II by cultured bovine adrenocortical cells. Endocrinol 118:2312–2318

De Arriba AF, Gomez-Casajus LA, Cavalcanti F, Almansa C, Garcia-Rafanel J, Forn J (1996) In vitro pharmacological characterization of a new selective angiotensin AT$_1$ receptor antagonist, UR-7280. Eur J Pharmacol 318:341–347

de Chaffoy de Courcelles D, Leysen JE, Roevens P, Van Belle H (1986) The serotonin-S$_2$ receptor: a receptor, transducer coupling model to explain insurmountable antagonist effects. Drug Dev Res 8:173–178

de Gasparo M, Catt KJ, Inagami T, Wright JW, Unger T (2000) International Union of Pharmacology. XXIII. The angiotensin II receptors. Pharmacol Rev 52:415–472

De Lean A, Stadel JM, Lefkowitz RJ (1980) A ternary complex model explains the agonist-specific properties of the adenylate cyclase-coupled β adrenergic receptor. J Biol Chem 255:7108–7117

Duncia JV, Chiu AT, Carini DJ, Gregory GB, Johnson AL, Price WA et al (1990) The discovery of potent non-peptide angiotensin II receptors antagonists: a new class of potent antihypertensives. J Med Chem 33:1312–1329

Ferrario CM, Iyer SN (1998) Angiotensin-(1-7): a bioactive fragment of the renin-angiotensin system. Regul Pept 78:13–18

Fierens FLP, Vanderheyden PML, De Backer J-P, Vauquelin G (1999a) Insurmountable angiotensin II AT$_1$ receptor antagonists: the role of tight antagonist binding. Eur J Pharmacol 372:199–206

Fierens FLP, Vanderheyden PML, De Backer J-P, Vauquelin G (1999b) Binding of the antagonist [^{3}H]candesartan to angiotensin II AT$_1$ receptor-transfected Chinese hamster ovary cells. Eur J Pharmacol 367:413–422

Fierens FLP, Vanderheyden PML, Gaborik Z, Le Minh T, DeBacker J-P, Hunyady L et al (2000) Lys^{199} Mutation of the human angiotensin II AT$_1$ receptor differently affects the binding of surmountable and insurmountable non-peptide antagonists. J Renin Angiotensin Aldosterone System 1:283–288

Fierens FLP, Vanderheyden PML, De Backer J-P, Thekkumkara TJ, Vauquelin G (2001) Tight binding of the angiotensin AT$_1$ receptor antagonist [^{3}H]candesartan is independent of receptor internalization. Biochem Pharmacol 61:1227–1235

Fierens F, Vanderheyden PML, Roggeman C, Vande Gucht P, De Backer J-P, Vauquelin G (2002) Distinct binding properties of the AT_1 receptor antagonist [^{3}H]candesartan to intact cells and membrane preparations. Biochem Pharmacol 63:1273–1279

Groblewski T, Maigret B, Larguier R, Lombard C, Bonnafous JC, Marie J (1997) Mutation of Asn^{111} in the third transmembrane domain of the AT_{1A} angiotensin II receptor induces its constitutive activation. J BioI Chem 272:1822–1826

Hara M, Kiyama R, Nakajima S, Kawabata T, Kawakami M, Ohtani K et al (1995) Kinetic studies on the interaction of nonlabeled antagonists with the angiotensin II receptor. Eur J Pharmacol 289:267–273

Harding JW, Felix D (1987) Angiotensin-sensitive neurons in the rat paraventricular nucleus: relative potencies of angiotensin II and angiotensin III. Brain Res 410:130–134

Hein L, Meinel L, Pratt RE, Dzau VJ, Kobilka BK (1997) Intracellular trafficking of angiotensin II and its AT_1 and AT_2 receptors: evidence for selective sorting of receptor and ligand. Mol Endocrinol 11:1266–1277

Hollenberg NK (2001) Potential of the angiotensin II receptor 1 blocker eprosartan in the management of patients with hypertension or heart failure. Curr Hypertens Rep 3:S25–S28

Hunyady L, Balla T, Catt KJ (1996) The ligand binding site of the angiotensin AT_1 receptor. Trends Pharmacol Sci 17:135–140

Hunyady L, Gáborik Z, Vauquelin G, Catt KJ, Clark AJL (2001) Structural requirements for signaling and regulation of AT_1 angiotensin receptors. J Renin Angiotensin Aldosterone Syst 2:S16–S23

Ji H, Leung M, Zhang Y, Catt KJ, Sandberg K (1994) Differential structural requirements for specific binding of nonpeptide and peptide antagonists to the AT_1 angiotensin receptor. Identification of amino acid residues that determine binding of the antihypertensive drug losartan. J Biol Chem 269:16533–16536

Kakar SS, Sellers JC, Devor DC, Musgrove LC, Neill JD (1992) Angiotensin II type-1 receptor subtype cDNAs: differential tissue expression and hormonal regulation. Biochem Biophys Res Commun 183:1090–1096

Kenakin TP (1987) Analysis of dose-response data. In: Kenakin TP (ed) Pharmacologic analysis of drug receptor interaction. Raven Press, New York, pp 129–162

Kramar EA, Armstrong DL, Ikeda S, Wayner MJ, Harding JW, Wright JW (2001) The effects of angiotensin IV analogs on long-term potentiation within the CA1 region of the hippocampus in vitro. Brain Res 897:114–21

Le MT, Vanderheyden PML, Szaszák M, Hunyady L, Vauquelin G (2002) Distinct roles of the N- and C-terminal amino acid residues of angiotensin II in human AT_1 receptor activation. J Biol Chem 277:23107–23110

Le MT, Vanderheyden PML, Szaszák M, Hunyady L, Kersemans V, Vauquelin G (2003) Peptide and nonpeptide antagonist interaction with constitutively active human AT1 receptors. Biochem Pharmacol 65:1329–1338

Leurs R, Rodriguez Pena MS, Bakker RA, Alewijnse AE, Timmerman H (2000) Constitutive activity of G protein coupled receptors and drug action. Pharm Acta Helv 74:327–331

Liu YJ, Shankley NP, Welsh NJ, Black JW (1992) Evidence that the apparent complexity of receptor antagonism by angiotensin II analogues is due to a reversible and synoptic action. Br J Pharmacol 106:233–241

Lucas M, Homburger V, Dolphin A, Bockaert J (1979) In vitro and in vivo kinetic analysis of the interaction of a norbornyl derivative of propranolol with beta-adrenergic receptors of brain and C6 glioma cells; an irreversible or slowly reversible ligand. Mol Pharmacol 15:588–597

Luque M, Martin P, Martell N, Fernandez C, Brosnihan KB, Ferrario CM (1996) Effects of captopril related to increased levels of prostacyclin and angiotensin-(1-7) in essential hypertension. J Hypertens 14:799–805

Mahon JM, Carr RD, Nicol AK, Henderson IW (1994) Angiotensin(1-7) is an antagonist at the type 1 angiotensin II receptor. J Hypertens 12:1377–1381

Miura S, Feng YH, Husain A, Karnik SS (1999) Role of aromaticity of agonist switches of angiotensin II in the activation of the AT_1 receptor. J Biol Chem 274:7103–7110

Mochizuki S, Sato T, Furata K, Hase K, Ohkura Y, Fukai C et al (1995) Pharmacological properties of KT3-671, a novel nonpeptide angiotensin II receptor antagonist. J Cardiovasc Pharmacol 25:22–29

Monnot C, Bihoreau C, Conchon S, Curnow KM, Corvol P, Clauser E (1996) Polar residues in the transmembrane domains of the type 1 angiotensin II receptor are required for binding and coupling. Reconstitution of the binding site by co-expression of two deficient mutants. J Biol Chem 271:1507–1513

Mustafa T, Hyung Lee J, Yeen Chai S, Albiston AI, McDowall SG, Mendelsohn FAO (2001) Bioactive angiotensin peptides: focus on angiotensin IV. J Renin Angiotensin Aldosterone System 2:205–210

Noda M, Shibouta Y, Inada Y, Ojima M, Wada T, Sanada T et al (1993) Inhibition of rabbit aoritc angiotensin II (AII) receptor by CV-11974, a new neuropeptide AII antagonist. Biochem Pharmacol 46:311–318

Noda K, Saad Y, Karnik SS (1995) Interaction of Phe^8 of angiotensin II with Lys^{199} and His^{256} of AT_1 receptor in agonist activation. J Biol Chem 270:28511–28514

Noda K, FengYH, Liu XP, Saad Y, Husain A, Karnik SS (1996) The active state of the AT_1 angiotensin receptor is generated by angiotensin II induction. Biochemistry 35:16435–16442

Onaran HO, Gurdal H (1999) Ligand efficacy and affinity in an interacting 7TM receptor model. Trends Pharmacol Sci 20:274–278

Pals DT, Denning GS Jr, Keenan RE (1979) Historical development of saralasin. Kidney Int 15: S7-S10

Pendleton RG, Gessner G, Homer E (1989a) Studies defining minimal receptor domains for angiotensin II. J Pharmacol Exp Ther 250:31–36

Pendleton RG, Gessner G, Horner E (1989b) Studies on inhibition of angiotensin II receptors in rabbit adrenal and aorta. J Pharmacol Exp Ther 248:637–643

Potts PD, Horiuchi J, Coleman MJ, Dampney RA (2000) The cardiovascular effects of angiotensin-(1-7) in the rostral and caudal ventrolateral medulla of the rabbit. Brain Res 877:58–64

Reaux A, Fournie-Zaluski M-C, Llorens-Cortes C (2001) Angiotensin III: a central regulator of vasopressin release and blood pressure. Trends Endocrinol Metabol 12:157–162

Robertson MJ, Barnes JC, Drew GM, Clark KL, Marshall FH, Michel A et al (1992) Pharmacological profile of GR 117289 in vitro: a novel, potent and specific non-peptide angiotensin AT_1 receptor antagonist. Br J Pharmacol 107:1173–1180

Robertson MJ, Dougall IG, Harper D, Mckechnie KCW, Leff P (1994) Agonist-antagonist interactions at angiotensin receptors: application of a two-state receptor model. Trends Pharmacol Sci 15:364–369

Robertson, MJ (1998) Angiotensin antagonists. In: Leff P (ed) Receptor-based drug design. Marcel Dekker, New York, pp 207–223

Schambye HT, Hjorth SA, Bergsma DJ, Sathe G, Schwartz TW. (1994) Differentiation between binding sites for angiotensin II and nonpeptide antagonists on the angiotensin II type 1 receptors. Proc Natl Acad Sci USA 91:7046–7050

Tamura K, Okuhira M, Mikoshiba I, Hashimoto K (1997) In vitro pharmacological properties of KRH-594, a novel angiotensin II type 1 receptor antagonist. Biol Pharmacol Bull 20:850–855

Thomas WG, Qian H, Chang CS, Karnik S (2000) Agonist-induced phosphorylation of the angiotensin II (AT(1A)) receptor requires generation of a confomation that is distinct from the inositol phosphate-signaling state. J Biol Chem 275:2893–2900

Timmermans PBMWM, Wong PC, Chiu AT, Herblin WF, Benfield P, Carini DJ et al (1993) Angiotensin II receptors and angiotensin II receptor antagonists. Pharmacol Rev 45:205–251

Timmermans PBMWM (1999a) Angiotensin II receptor antagonists. An emerging new class of cardiovascular therapeutics. Hypertens Res 22:147–153

Timmermans PBMWM (1999b) Pharmacological properties of angiotensin II receptor antagonists. Can J Cardiol 15:26F–28F

Unger T (1999) Significance of angiotensin type 1 receptor blockade: why are angiotensin II receptor blockers different? Am J Cardiol 84:9S–15S

Vanderheyden P, Fierens FLP, De Backer J-P, Frayman N, Vauquelin G (1999) Distinction between surmountable and insurmountable selective AT_1 receptor antagonists by use of CHO-K1 cells expressing human angiotensin II AT_1 receptors. Brit J Pharmacol 126:1057–1065

Vanderheyden PML, Verheijen I, Fierens FLP, De Backer J-P, Vauquelin G (2000a) Inhibition of angiotensin II induced inositol phosphate production in CHO cells expressing human AT_1 receptors by triacid nonpeptide antagonists. Pharm Res 17:1482–1488

Vanderheyden PML, Fierens FLP, De Backer J-P, Vauquelin G (2000b) Reversible and syntopic interaction between angiotensin II AT_1 receptor antagonists and human AT_1 receptors expressed in CHO-K1 cells. Biochem Pharmacol 59:927–935

Vanderheyden PML, Fierens FLP, Verheijen I, De Backer J-P, Vauquelin G (2000c) Binding characteristics of [^{3}H]-irbesartan to human recombinant angiotensin II type 1 receptors. J Renin Angiotensin Aldosteron Syst 1:159–165

Vauquelin G, Fierens FLP, Verheijen I, Vanderheyden PML (2000) Mechanisms of Angiotensin II Antagonism. Competitive vs. Non-Competitive Inhibition. In: Epstein M, Brunner HR (eds) Angiotensin II Receptor Antagonists. Harley and Belfus, Philadelphia, pp 105–118

Vauquelin G, Fierens FLP, Verheijen I, Vanderheyden PML (2001a) Distinctions between non-peptide angiotensin II AT_1 receptor antagonists. J Renin Angiotensin Aldosterone Syst 2:S24–S31

Vauquelin G, Morsing P, Fierens FLP, De Backer J-P, Vanderheyden PML (2001b) A two-state receptor model for the interaction between angiotensin II AT_1 receptors and their non-peptide antagonists. Biochem Pharmacol 61:277–284

Vauquelin G, Fierens FLP, Gáborik Z, Le Minh T, De Backer J-P, Hunyady L, Vanderheyden PML (2001c) Role of basic amino acids of the human angiotensin type 1 receptor in the binding of the non-peptide antagonist candesartan. J Renin Angiotensin Aldosterone Syst 2:S32–S36

Vauquelin G, Van Liefde I, Vanderheyden PML (2003) Models and methods for studying insurmountable antagonism. Trends Pharmacol Res 23:514–518

Verheijen I, Fierens FLP, De Backer J-P, Vauquelin G, Vanderheyden PML (2000) Interaction between the partially insurmountable antagonist valsartan and human recombinant angiotensin II type 1 receptors. Fund Clin Pharmacol 14:577–585

Wienen W, Hauel N, Van Meel JC, Narr B, Ries U, Entzeroth M (1993) Pharmacological characterization of the novel nonpeptide angiotensin II receptor antagonist, BIBR 277. Br J Pharmacol 110:245–252

Wong PC, Timmermans PB (1991) Nonpeptide angiotensin II receptor antagonists: insurmountable angiotensin II antagonism of EXP3892 is reversed by the surmountable antagonist DuP 753. J Pharmacol Exp Ther 258:49–57

Wong PC, Price WA, Chiu AT, Thoolen MJMC, Duncia JV, Carini DJ et al (1990) Non-peptide angiotensin II receptor antagonists. IX. Pharmacology of EXP3174: an active metabolite of DuP 753, an orally active antihypertensive agent. J Pharmacol Exp Ther 255:211–217

Wright JW, Stubley L, Pedersen ES, Kramar EA, Hanesworth JM, Harding JW (1999) Contributions to the brain angiotensin IV-AT_4 receptor subtype system to spatial learning. J Neurosci 19:3952–3961

Yamada K, Iyer SN, Chappell MC, Ganten D, Ferrario CM (1998) Converting enzyme determines plasma clearance of angiotensin-(1-7). Hypertension 32:496–502

Yamano Y, Ohyama K, Kikyo M, Sano T, Nakagomi Y, Inoue Y et al (1995) Mutagenesis and the molecular modeling of the rat angiotensin II receptor (AT_1). J Biol Chem 270:14024–14030

Zhang JC, Van Meel A, Pfaffendorf M, Van Zwieten P (1993) Different types of angiotensin II receptor antagonism induced by BIBS 222 in the rat portal vein and rabbit aorta; the Influence of receptor reserve. J Pharmacol Exp Ther 269:509–514

Zini S, Fournie-Zaluski M-C, Chauvel E, Roques BP, Corvol P, Lorens-Cortes C (1996) Identification of metabolic pathways of brain angiotensin II and III using specific aminopeptidase inhibitors: Predominant role of angiotensin III in the control of vasopressin release. Proc Natl Acad Sci USA 93: 11968–11973

AT_1 Receptor Regulation

S. Wassmann · G. Nickenig

Medizinische Klinik und Poliklinik—Innere Medizin III,
Universitätskliniken des Saarlandes, Homburg/Saar, Germany
e-mail: nickenig@med-in.uni-sb.de

1	Introduction	318
2	Regulation of AT_1 Receptor Expression	318
3	Mechanisms of AT_1 Receptor Regulation	320
4	Posttranscriptional AT_1 Receptor Regulation	321
5	Mechanisms of Posttranscriptional AT_1 Receptor mRNA Regulation	322
6	Signal Transduction of AT_1 Receptor Regulation	326
7	Pathophysiological Relevance of Dysregulated AT_1 Receptor Expression	327
	References	329

Abstract Clinical and experimental studies have demonstrated that AT_1 receptor expression and regulation play an important role in the pathogenesis of cardiovascular diseases such as atherosclerosis and hypertension. The expression levels of the AT_1 receptor define the biological efficacy of angiotensin II. Many agonists, as for example angiotensin II, growth factors, low-density lipoprotein cholesterol, insulin, estrogen, progesterone, reactive oxygen species, cytokines, nitric oxide, and many others are known to regulate AT_1 receptor expression. Mechanisms of AT_1 receptor regulation include receptor internalization, desensitization, alternative splicing of receptor pre-mRNA, and most importantly modulation of its gene expression by transcriptional and posttranscriptional mechanisms. Posttranscriptional mechanisms predominate AT_1 receptor regulation. Binding of RNA-binding proteins to the 5′ and 3′ untranslated region of the AT_1 receptor mRNA has been shown to be involved in the regulation of mRNA stability. Recent studies revealed that a region just adjacent to the poly-A tail of the AT_1 receptor mRNA, which has a secondary structure that forms a stem loop, is important for the protein–mRNA interaction. The first identified AT_1 receptor binding protein is calreticulin, which, if phosphorylated, binds to this cognate sequence of the AT_1 receptor mRNA and leads to mRNA destabilization. Signal transduction molecules such as cAMP, reactive oxygen species, MAP kinases, nitric oxide, PI-3 kinase, and others have been shown to be involved in AT_1 receptor regulation by several agonists. The pathophysiological relevance of dysregulated AT_1 receptor expression has been demonstrated in many in vitro, in vivo and human studies. Hypercholesterolemia, estrogen deficiency, and hyperinsulinemia are associated with enhanced AT_1 receptor expression and increased oxidative stress, and AT_1 receptor antagonism inhibits pathological cellular processes and the development of endothelial dysfunction and atherosclerotic lesions. AT_1 receptor overexpression very likely represents, among other things, a potential molecular mechanism that links exogenous risk factors to cellular events in chronic vascular disease.

Keywords AT_1 receptor · Regulation · Transcriptional · Posttranscriptional · Agonists · mRNA binding proteins · Hypercholesterolemia

1
Introduction

A series of recent experimental and clinical studies have demonstrated that altered regulation and expression of the angiotensin (Ang) II type 1 (AT_1) receptor plays an important role in the pathogenesis of chronic cardiovascular diseases. The modulation of AT_1 receptor expression levels contributes on the one hand to the adaptation of the renin–angiotensin system to chronic agonist stimulation and serves on the other hand as a possible explanation for the association of various hormonal and metabolic disorders with hypertension and with the development and accelerated progression of atherosclerotic lesions (Peach 1977; Griendling et al. 1993).

2
Regulation of AT_1 Receptor Expression

Back in 1980, it was discovered that the vasoconstriction induced by AngII in resistance vessels is variable (Gunther et al. 1980). Further investigations revealed that AT_1 receptor expression is subject to a negative feedback regulation. Increased levels of AngII reduce AT_1 receptor expression, whereas decreased AngII concentrations enhance AT_1 receptor expression levels (Douglas and Brown 1982; Schiffrin et al. 1984; Griendling et al. 1987; Lassègue et al. 1995). More recently, it has been shown that various agonists other than AngII modulate AT_1 receptor expression. This phenomenon, referred to as heterologous AT_1 receptor regulation, is induced by various growth factors such as platelet-derived growth factor (PDGF), epidermal growth factor (EGF), or fibroblast growth factor (FGF), all of which downregulate AT_1 receptor expression (Nickenig and Murphy 1994). Numerous other factors, including glucocorticoids, aldosterone, forskolin, tumor necrosis factor-α, cytokines, nitric oxide, insulin, low-density lipoprotein (LDL) cholesterol, estrogen, progesterone, sodium chloride, reactive oxygen species, insulin-like growth factor-1, and isoprenaline are known to influence AT_1 receptor expression (Table 1) (Kitamura et al. 1986; Sumners et al. 1986; Douglas 1987; Linas et al. 1990; Kakar et al. 1992; Ullian et al. 1992; Sun and Weber 1993; Tufroet et al. 1993; Lu et al. 1994; Bird et al. 1995;

Table 1 Regulation of AT_1 receptor expression. Various agonists modulate AT_1 receptor expression in different tissues or cell types. Few studies have delineated the participating signal transduction pathways and characterized the mechanism of regulation

Cell/tissue	Agonist	Regulation ↑↓	Mechanism	Signal transduction
VSMC (rat)	IL1α	↑		
	TNFα+IFNγ	↓		
VSMC (rat)	LDL	↑	Posttranscriptional	
VSMC (rat)	AngII, α-thrombin, ATP, forskolin	↓	Posttranscriptional and transcriptional	Phenylarsine oxide
VSMC (rat)	Forskolin, isoproterenol	↓	Posttranscriptional	PKA
VSMC	SNAP (NO donor)	↓	Posttranscriptional	
VSMC	IFNγ	↓		MAP kinase, JAK2
VSMC (rat)	Insulin	↑	Posttranscriptional	Tyrosine kinase, calcium
VSMC (rat)	Estrogen	↓	Posttranscriptional	NO synthase
	Progesterone	↑	Posttranscriptional and transcriptional	PI-3 kinase
VSMC (rat)	Idoxifene (SERM)	↓	Posttranscriptional	NO synthase
VSMC (rat)	NaCl	↑		
VSMC (rat)	Free radicals	↓	Posttranscriptional	p38 MAP kinase
VSMC (rat)	*all-trans* retinoic acid (vitamin A)	↓	Transcriptional	
VSMC (rat)	Statins	↓	Posttranscriptional	Geranylgeranylation
VSMC (rat)	EGF	↓	Posttranscriptional and transcriptional	
	FGF	↓		
	PDGF	↓		
Heart (rat)	Dexamethasone	↑		
	Deoxycorticosterone	AT_{1A}↓		
Cardiac fibroblasts	TNF-α	↑		Inositol phosphat
	IL1β	↑		
	IL2, IL6	→		
Neuronal cells (WKY and SHR)	Phorbol ester	↑	Transcriptional	
	Forskolin	↑		
Brain (mouse)	Haloperidol	↑		
Astrocytes (rat)	GH	AT_{1A}↑	Transcriptional	
Pituitary, brain (rat)	Estrogen	↓		
Neuroblastoma cells (mouse)	AngII	↓		
Pituitary	Estrogen		Posttranscriptional	
Adrenal gland		↓		
Uterus		↑		
Bovine zona glomerulosa cells	ACTH	↓		
	cAMP	↓		
	IGF1	↑		
	KCl	↑		
	Cortisol	↓		
	Aldosterone	→		
Mesangial cells (rat)	Glucose	↓		PKC
Mesangial cells (rat)	Dexamethasone	AT_{1A}→, AT_{1B}↓		
Placenta, trophoblast cells (human)	Progesterone	↓		
	Estradiol	→		

Bruna et al. 1995; Nickenig et al. 1996, 1997a, 1998a,b,c, 2000a,b; Cheng et al. 1996; Kalenga et al. 1996; Sechi et al. 1996; Yang et al. 1996; Nishimura et al. 1997; Sasamura et al. 1997; Ullian et al. 1997; Wang et al. 1997; Ichiki et al. 1998; Amiri and Garcia 1999; Gurantz et al. 1999; Ikeda et al. 1999; Kisley et al. 1999; Krishnamurthi et al. 1999; Müller et al. 2000; Takeda et al. 2000; Bäumer et al. 2001; Wassmann et al. 2001a).

3
Mechanisms of AT$_1$ Receptor Regulation

Most of the AngII effects are mediated by stimulation of AT$_1$ receptors. Thus, the number of AT$_1$ receptors defines the biological efficacy of AngII. There are at least four different aspects involved in AT$_1$ receptor regulation. First, activation of AT$_1$ receptors with AngII may evoke internalization of the receptor protein and reduces receptor numbers on the cell surface (Kai et al. 1994). Second, chronic AngII stimulation reduces AngII signaling via protein kinase C-depen-

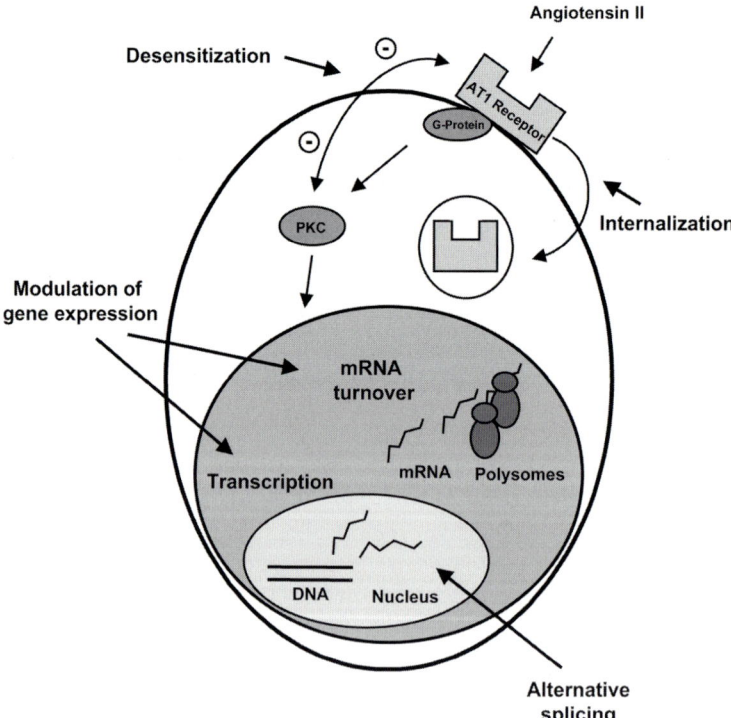

Fig. 1 Principal mechanisms involved in regulation of AT$_1$ receptor expression. AT$_1$ receptor expression may be modulated by AngII-induced internalization of the receptor protein or by desensitization of signal transduction pathways downstream of the AT$_1$ receptor. Furthermore, AT$_1$ receptor gene expression may be altered at both transcriptional and posttranscriptional levels. Finally, alternative splicing may modulate AT$_1$ receptor expression

dent pathways, which is referred to as desensitization (Curnow 1996). Third, alternative splicing of the AT_1 receptor pre-mRNA can alter AT_1 receptor protein translation. Fourth, and likely the most important mechanism regulating AT_1 receptors, is modulation of its gene expression (Lassègue et al. 1995; Nickenig and Murphy 1994). These principal mechanisms of AT_1 receptor regulation are summarized in Fig. 1.

Modulations of AT_1 receptor gene expression are generally obvious several hours after agonist stimulation and are sustained for variable periods of time thereafter. Recently, it has been shown that internalization of the AT_1 receptor seems to regulate expression of AT_1 receptor mRNA (Adams et al. 1999).

Gene expression is predominantly modulated by transcriptional and posttranscriptional mechanisms. Among other stimuli, growth factors and AngII are known to reduce the rate of AT_1 receptor mRNA transcription (Lassègue et al. 1995; Nickenig and Murphy 1994). Although numerous consensus sequences were discovered within the promoter region of the AT_1 receptor [e.g., AP-1, SP-1, estrogen-response element, cyclic adenosine monophosphate (cAMP)-response element], the exact mechanisms of transcriptional control of AT_1 receptor mRNA synthesis are poorly understood (Murphy et al. 1992).

4
Posttranscriptional AT_1 Receptor Regulation

In general, the abundance of a particular mRNA transcript and its resulting protein product is not only governed by its transcription rate but also by its half-life (also referred to as mRNA stability). Of relevance for AT_1 receptor regulation has been the finding that numerous agonists that modulate AT_1 receptor expression also affect posttranscriptional processing of its mRNA. Estrogens, AngII, and cAMP-stimulating agents decrease AT_1 receptor expression by stimulating degradation of the AT_1 receptor mRNA (Nickenig and Murphy 1996; Wang et al. 1997; Nickenig et al. 2000b). In contrast, progesterone, LDL, and insulin upregulate AT_1 receptor expression by decreasing its mRNA decay (Nickenig et al. 1997a, 1998a,b, 2000b). The data concerning growth factors such as PDGF are inconsistent. It is well established that these factors reduce AT_1 receptor mRNA transcription rate (Nickenig and Murphy 1994). Experiments following mRNA degradation after blockade of transcription with actinomycin D revealed that growth factors induce degradation of the AT_1 receptor mRNA (Nickenig and Murphy 1994). Other data with recombinant, retroviral approaches have suggested that PDGF may have no effect on AT_1 receptor mRNA stability (Wang and Murphy 1998; Xu and Murphy 2000). These discrepancies are likely due to difficulties in obtaining unambiguous measures of inducible mRNA turnover.

Despite these inconsistencies, there is strong evidence that posttranscriptional mechanisms predominate AT_1 receptor regulation. As is the case for many other mRNAs, this process involves binding of proteins to both the 5' and 3' untranslated region of the AT_1 receptor mRNA. Data from recombinant retroviral AT_1 receptor mRNA species and experiments performed in brain tissue suggest that proteins interacting with the 5′ untranslated region of the AT_1 receptor mRNA are involved in both cAMP- and estrogen-induced modulation of AT_1 receptor regulation (Krishnamurthi et al. 1998, 1999; Wang and Murphy 1998; Xu and Murphy 2000). To date, however, the identity of these proteins remains undefined, as do the precise regions of the AT_1 receptor mRNA involved in binding these proteins.

Several families of RNA binding proteins have been implicated in the regulation of steady-state RNA levels. Packaging, splicing, and nuclear translocation are arranged through actions of nuclear ribonucleoprotein particles (Dreyfuss 1986; Swanson and Dreyfuss 1988; Wilusz and Shenk 1990). Small nuclear RNA-binding proteins are thought to manage further splicing and processing of the 5′ cap and the 3′ poly-A-tail (Steitz et al. 1983; Konarska and Sharp 1987). In addition, cytosolic RNA-binding proteins have been identified that bind their corresponding mRNA in the 3′ untranslated region and that transfer (de)stabilization induced by the respective receptor agonist (Malter et al. 1989; Bohjanen et al. 1991; Brewer 1991; Vakalopoulou et al. 1991; Port et al. 1992). These mRNA binding proteins may bind to distinct mRNA consensus sequences, of which the AUUUA motif has attracted the most attention so far (Malter et al. 1989; Brewer 1991). It was demonstrated that the AT_1 receptor mRNA also binds to RNA-binding proteins and is destabilized in the polyribosomes of vascular smooth muscle cells after stimulation with AngII (Nickenig and Murphy 1996).

5
Mechanisms of Posttranscriptional AT_1 Receptor mRNA Regulation

It was recently shown that a family of proteins residing in the polysomal compartment bind to the 3′ untranslated region of the AT_1 receptor mRNA (Fig. 2). Detailed analysis revealed that a region between bases 2175–2195 within the AT_1 receptor mRNA, just adjacent to the poly-A-tail, is responsible for the protein–mRNA interaction. Transfection experiments in vascular smooth muscle cells and in vitro decay assays within the polysomal compartment derived from vas-

Fig. 2A, B Posttranscriptional regulation of AT_1 receptor expression. **A** UV crosslinking assay of polysomal proteins to the AT_1 receptor mRNA. Representative autoradiogram of the AT_1 receptor mRNA base 1864–2213 crosslinked to polysomal proteins isolated from VSMC. Bands indicate interactions between the labeled mRNA and polysomal proteins. As indicated, specific competition between portions of the AT_1 3′ untranslated region (UTR) and the full-length UTR can be observed. A non-relevant mRNA does not compete. **B** Scheme of posttranscriptional AT_1 receptor expression. Agonists such as AngII (*Ang II*) activate various signaling pathways which involve, for example, reactive oxygen species, nitric oxide, MAP kinases and PI-3-kinase. These events activate/induce proteins residing within the polysomal compartment which bind the AT_1 receptor mRNA in its very 3′UTR at bases 2175–2195. This cognate AT_1 receptor mRNA region forms a stem loop characteristic of mRNA sequences that interact with RNA-binding proteins. This protein–AT_1 receptor mRNA interaction increases or decreases AT_1 receptor mRNA decay, leading ultimately to either up- or downregulation of AT_1 receptor expression. *ORF*, open reading frame; *UTR*, untranslated region

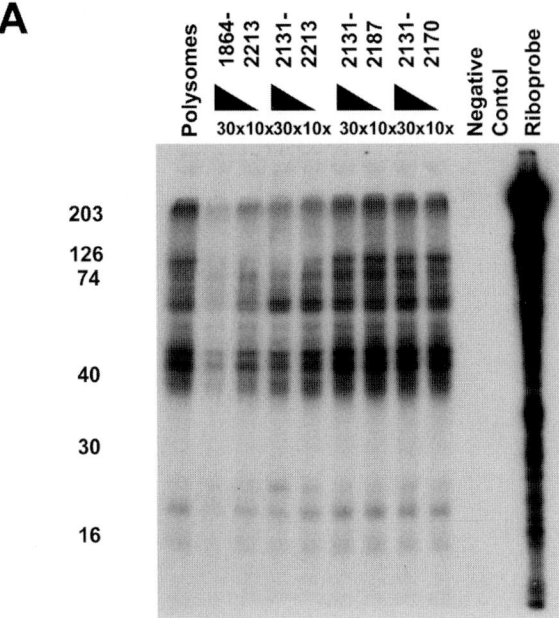

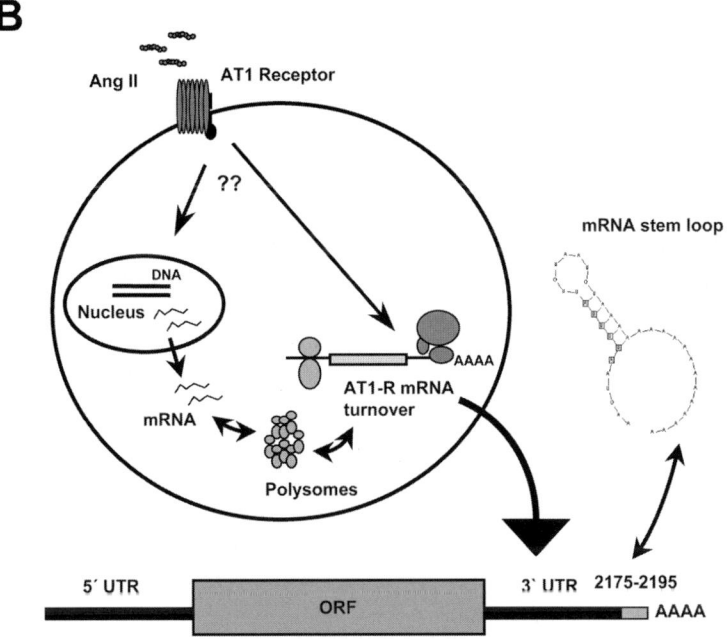

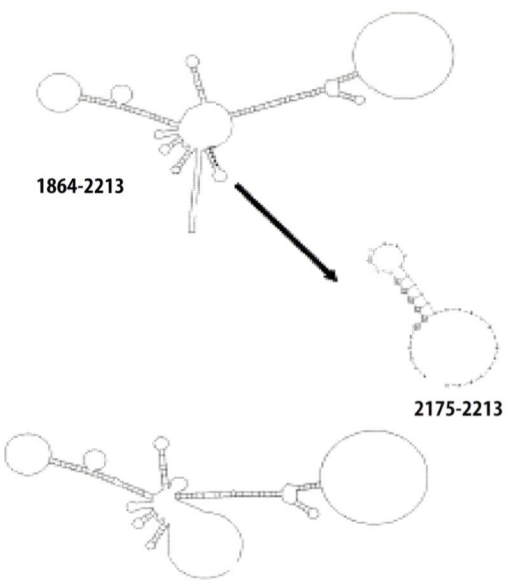

1864-2213 minus 2175-2195

Fig. 3 Computer modeling of AT_1 receptor mRNA. The secondary structures of the wild-type AT_1 receptor mRNA, the AT_1 receptor transcript bases 2175–2195, and an AT_1 receptor mRNA mutant lacking the motif bases 2175–2195 were imaged with the RNAdraw software under standard conditions. The 1864–2213 and the 2175–2213 fragments correspond to the wild-type AT_1 receptor mRNA displaying the stem loop structure. The fragment 1864–2213 minus 2175–2195 shows a mutant in which the stem loop formation is omitted

cular smooth muscle cells (VSMC) demonstrated that protein binding to the bases 2175–2195 of the AT_1 receptor mRNA mediates AT_1 receptor regulation (Fig. 2) (Nickenig et al. 2001). This cognate sequence contains an AUUUUA hexamer which shows intriguing similarities to the β_2-adrenergic receptor mRNA (Tholanikunnel and Malbon 1997), although the flanking region of this mRNA binding region differs slightly between the genes. The bases 2175–2195 of the AT_1 receptor mRNA are almost completely composed of A and U nucleotides (except for three Gs), a feature that has been shown for most mRNA binding sequences residing in the 3′ untranslated region.

Binding of mRNA binding proteins to their corresponding mRNA is profoundly influenced by secondary and tertiary structures of the mRNA. Hairpins

Fig. 4A–C The AT_1 receptor mRNA binding protein calreticulin. **A** Identification of the AT_1 receptor mRNA-binding protein calreticulin. Phosphorylated calreticulin was incubated and crosslinked with the AT_1 receptor riboprobe bases 1864–2213 in the presence of the AT_1 receptor mRNA competitors bases 1864–2213, 2175–2195, and the unspecific mRNA competitor GAPDH (30× and 10× excess over the riboprobe). The binding pattern reveals that calreticulin binds the AT_1 receptor mRNA at bases 2175–

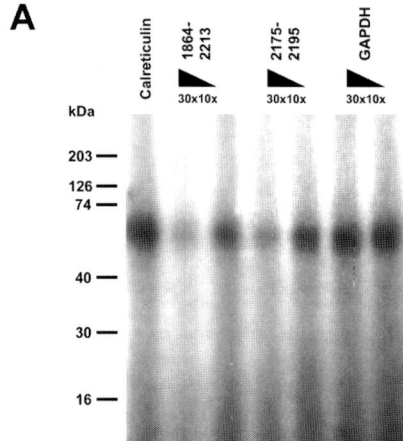

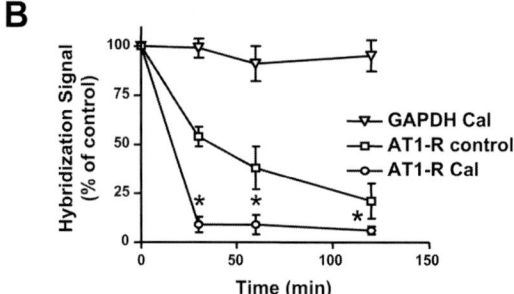

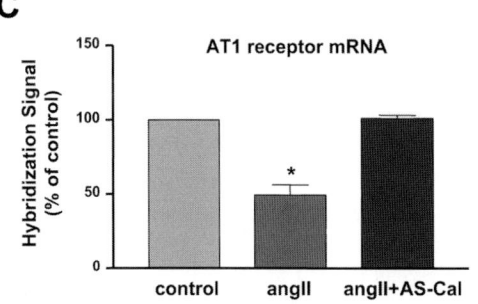

2195 and not the GAPDH mRNA. **B** Calreticulin destabilizes the AT$_1$ receptor mRNA. Either recombinant phosphorylated calreticulin (*AT1-R Cal*) or the kinase reaction without calreticulin (*AT1-R control*) were added to an in vitro decay assay including the in vitro transcribed AT$_1$ receptor mRNA. Degradation of AT$_1$ receptor mRNA was assessed by real-time PCR. As internal control, a GAPDH mRNA was included in the reaction and concomitantly quantified (*GAPDH-Cal*). **C** Cells were transfected with either an insertless pcDNA3 vector (*control* and *AngII*) or an antisense calreticulin construct (*AS-Cal*). The effect of antisense calreticulin and sense calreticulin construct transfection was monitored by Western blot. Twenty-four hours later, VSMC were incubated with vehicle (*control*) or 1 µmol/l AngII (*angII*) for 4 h. AT$_1$ receptor mRNA and GAPDH mRNA (not shown) were quantified by real-time PCR. Calreticulin antisense transfection abolishes the destabilizing effect of AngII on AT$_1$ receptor mRNA

or stem loops formed by the RNA region of interest may interact with neighboring proteins. In the case of the AT_1 receptor, computer modeling showed that the identified AT_1 receptor mRNA binding motif (bases 2175–2195) forms such a stem loop (Fig. 2). That holds true for the entire AT_1 receptor mRNA and also for the isolated 20-base transcript. Deletion of this motif abolishes the stem loop which is in concert with the finding that such a mutated mRNA binds no longer to polysomal proteins, suggesting the importance of secondary structure for protein–mRNA interaction (Fig. 3).

Recently, the first AT_1 receptor mRNA binding protein was identified (Nickenig et al. 2002). Calreticulin, which is known to be involved in intracellular calcium homeostasis and acts as chaperone-like molecule and as receptor for nuclear export, binds, if phosphorylated, to the cognate sequence bases 2175–2195 of the AT_1 receptor mRNA and leads to destabilization of the AT_1 receptor mRNA (Fig. 4). Stimulation with AngII, which causes destabilization of AT_1 receptor mRNA, leads to phosphorylation and thereby to activation of the mRNA binding properties of calreticulin (Nickenig et al. 2002). Phosphorylation causes either a conformational change of calreticulin or activates a preformed binding site of calreticulin which enables the interaction with the mRNA. Furthermore, calreticulin may also bind other protein factors before or while interacting with the target mRNA, it may itself activate RNases which realize the actual AT_1 receptor mRNA decay, or it may induce a change in the tertiary structure of the mRNA leading to a more or less pronounced interference with nucleases. This mechanism of regulation could represent a general mechanism involved in mRNA processing and could be applicable to other genes.

6
Signal Transduction of AT_1 Receptor Regulation

No uniform signal transduction pathway has been defined that inevitably results in modulation of AT_1 receptor expression. In vascular smooth muscle cells, cAMP participates in isoprenaline and possibly in AngII-induced AT_1 receptor downregulation (Lassègue et al. 1995; Wang et al. 1997). Furthermore, superoxide radicals and hydrogen peroxide are involved in AngII-induced AT_1 receptor regulation. The p38 mitogen-activated protein (MAP) kinase mediates free radical-induced AT_1 receptor regulation, whereas p42/44 MAP kinase is presumably involved in insulin-driven AT_1 receptor overexpression (Nickenig et al. 1998a, 2000a). In VSMC, superoxide as well as hydrogen peroxide downregulate AT_1 receptor mRNA expression mediated through posttranscriptional mechanisms (Nickenig et al. 2000a). Again, transcriptional regulation seems to be not of importance. Moreover, nitric oxide mediates estrogen-dependent AT_1 receptor downregulation, whereas phosphoinositide (PI)-3 kinase has been implicated in progesterone-caused AT_1 receptor upregulation (Table 1) (Nickenig et al. 2000b). Thus, various signal transduction pathways have been implicated; however, the detailed cascade between a cell surface impulse and the ultimate modulation of gene expression is only partially understood. Particularly, the steps immediately upstream of the described AT_1 receptor mRNA binding proteins are not known.

7
Pathophysiological Relevance of Dysregulated AT_1 Receptor Expression

Dysregulated expression of AT_1 receptors may profoundly participate in the development of vascular damage (Nickenig and Harrison 2002). Hypercholesterolemia and in particular elevated LDL cholesterol plasma concentrations play a fundamental role in the pathogenesis of atherosclerosis. Despite a large body of epidemiological evidence, the molecular events leading from hypercholesterolemia to hypertension and atherosclerosis are only partially understood. In recent studies, it has been shown that exposure of vascular smooth muscle cells to LDL markedly augments AT_1 receptor mRNA and protein expression (Nickenig et al. 1997a), and in rabbits with diet-induced atherosclerosis or heritable hyperlipidemia, aortic AT_1 receptor expression is increased twofold (Nickenig et al. 1997b; Warnholtz et al. 1999). Vascular superoxide production is also increased in hypercholesterolemia, which is associated with profound alteration of endothelium-dependent vasodilation. These abnormalities are normalized by blockade of the AT_1 receptor, despite the fact that blood pressure and lipoprotein levels are not changed by this treatment (Warnholtz et al. 1999). Most importantly, development of atherosclerotic lesions can be inhibited by AT_1 receptor antagonism (Warnholtz et al. 1999; Strawn et al. 2000). Subsequent work has provided evidence that hypercholesterolemia increases AT_1 receptor expression in humans (Nickenig et al. 1999). In hypercholesterolemic subjects, AngII infusion produced more than twice the increase in blood pressure as observed in normocholesterolemic subjects. In keeping with this finding, the expression of AT_1 receptors was increased two- to threefold in hypercholesterolemic subjects. Moreover, hypercholesterolemia-associated endothelial dysfunction is improved by AT_1 receptor antagonist treatment in humans (Wassmann et al. 2002a). These findings strongly support the concept that hypercholesterolemia increases AT_1 receptor expression and illustrate how this phenomenon may contribute to the development of atherosclerosis.

On the other hand, hypercholesterolemic patients are frequently treated with 3-hydroxy-3-methylglutaryl (HMG) coenzyme A (CoA) reductase inhibitors (statins). Treatment with statins for 4 weeks normalized the pressor response to AngII infusion and completely normalized AT_1 receptor expression (Nickenig et al. 1999). Further experiments in normocholesterolemic spontaneously hypertensive rats and cultured VSMC demonstrated that statins directly downregulate AT_1 receptor expression independent of cholesterol lowering (Wassmann et al. 2001a,b, 2002b). Thus, it may be speculated that the beneficial effects of statins relate not only to cholesterol lowering but also to so-called pleiotropic effects such as AT_1 receptor downregulation.

Many epidemiological studies indicate that premenopausal women have a low incidence of vascular disease but that the risk of cardiovascular events and the incidence of hypertension increases rapidly after the menopause (Colditz et al. 1987). Observational studies have shown that estrogen replacement therapy may exert beneficial effects on cardiovascular morbidity and mortality, suggesting an important role of estrogens in the pathogenesis of vascular diseases (Mendelsohn and Karas 1999). Estrogen has been shown to have beneficial effects on vascular cells (Mendelsohn and Karas 1999). Recently, it was demonstrated that in ovariectomized rats, AngII-induced vasoconstriction was significantly increased and AT_1 receptor density and mRNA levels were upregulated twofold. Estrogen replacement therapy normalized AT_1 receptor expression in these animals (Nickenig et al. 1998c). Overexpression of AT_1 receptors induced by estrogen deficiency was associated with endothelial dysfunction and increased vascular production of superoxide, and these abnormalities were normalized by either estrogen replacement or AT_1 receptor blockade (Wassmann et al. 2001c). Parallel studies in cultured vascular smooth muscle cells revealed that

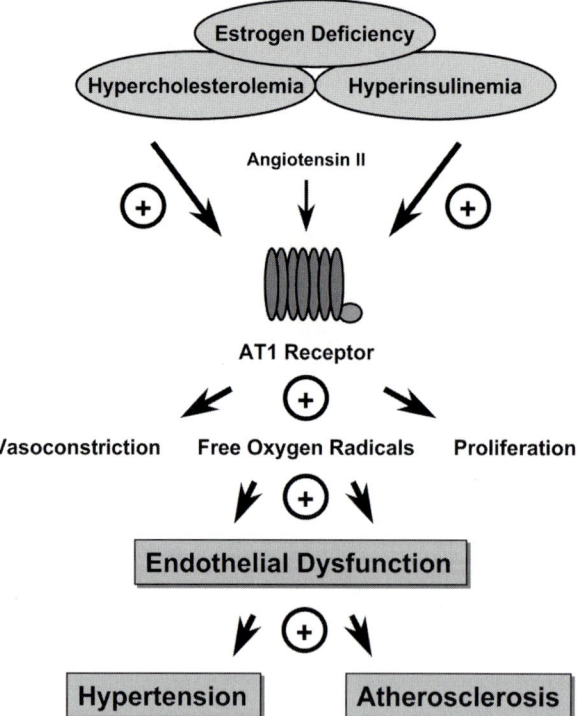

Fig. 5 Interaction of risk factors with AT_1 receptor expression and implications for the pathogenesis of atherosclerosis. Risk factors such as hypercholesterolemia, estrogen deficiency, or hyperinsulinemia may increase AT_1 receptor expression. Increased AT_1 receptor expression results in increased oxidative stress, accelerated growth of vascular smooth muscle cells, and enhanced vasoconstriction. Presumably, AT_1 receptor overexpression is accompanied by upregulation of vascular ACE and potentially other local components of the renin–angiotensin system. These events contribute among other conditions to elevated blood pressure, endothelial dysfunction, and progression of the atherosclerotic process

estradiol causes a downregulation of AT_1 receptor gene expression (Nickenig et al. 2000b). Estrogen-induced downregulation of AT_1 receptor expression could help to explain the association between estrogen-deficiency, hypertension, and atherosclerosis observed in many clinical studies.

Finally, hyperinsulinemia, which is also known to be a risk factor for atherosclerotic diseases, was also found to be associated with profound AT_1 receptor overexpression in VSMC (Nickenig et al. 1998a).

AT_1 receptor regulation very likely represents, among others, a molecular switch connecting traditional risk factors such as hypercholesterolemia, estrogen deficiency, and hyperinsulinemia with hypertension and atherosclerosis (Fig. 5). From this point of view, AT_1 receptor overexpression is one potential molecular mechanism that links a variety of exogenous risk factors to cellular events in vascular disease. Presently, it is not certain if increased activity of other local components of the renin–angiotensin system such as ACE and AngII is concomitantly required, or whether AT_1 receptor overexpression itself is sufficient to propagate vascular dysfunction and lesion formation.

References

Adams B, Obertone TS, Wang X, Murphy TJ (1999) Relationship between internalization and mRNA decay in down-regulation of recombinant type 1 angiotensin II receptor (AT1) expression in smooth muscle cells. Mol Pharmacol 55:1028–1036

Amiri F, Garcia R (1999) Regulation of angiotensin II receptors and PKC isoforms by glucose in rat mesangial cells. Am J Physiol 276:F691–F699

Bäumer AT, Wassmann S, Ahlbory K, Strehlow K, Müller C, Sauer H, Böhm M, Nickenig G (2001) Reduction of oxidative stress and AT1 receptor expression by the selective oestrogen receptor modulator idoxifene. Br J Pharmacol 134:579–584

Bird IM, Word A, Clyne C, Mason JI, Raincy WE (1995) Potassium negatively regulates angiotensin II type 1 receptor expression in human adrenocortical H295R cells. Hypertension 25:1129–1134

Bohjanen PR, Petryniak B, June CH et al (1991) An inducible cytoplasmic factor (AU-B) binds selectively to AUUUA multimers in the 3′ untranslated region of lymphokine mRNA. Mol Cell Biol 11:3288–3295

Brewer G (1991) An A + U-rich element RNA-binding factor regulates c-myc mRNA stability in vitro. Mol Cell Biol 11:2460–2466

Bruna RD, Ries S, Himmelstoss C, Kurtz A (1995) Expression of cardiac angiotensin II AT1 receptor genes in rat hearts is regulated by steroids but not by angiotensin II. J Hypertens 13:763–769

Cheng HF, Becker BN, Harris RC (1996) Dopamine decreases expression of type-1 angiotensin II receptors in renal proximal tubule. J Clin Invest 97:2745–2752

Colditz GA, Willett WC, Stampfer MJ, Rosner B, Speizer FE, Hennekens CH (1987) Menopause and the risk of coronary heart disease in women. N Eng J Med 316:1105–1110

Curnow KM (1996) Human type-1 angiotensin II (AT1) receptor gene structure and function. Clin Exp Pharmacol Physiol 3:S67–S73

Douglas JG (1987) Estrogen effects on angiotensin receptors are modulated by pituitary in female rats. Am J Physiol 252:E57–E62

Douglas JG, Brown GP (1982) Effect of prolonged low dose infusion of angiotensin II and aldosterone on rat smooth muscle and adrenal angiotensin II receptors. Endocrinology 111:988–992

Dreyfuss G (1986) Structure and function of nuclear and cytoplasmic ribonucleoprotein particles. Annu Rev Cell Biol 2:459–498

Griendling KK, Delafontaine P, Rittenhouse SE, Gimbrone MA Jr, Alexander RW (1987) Correlation of receptor sequestration with sustained diacylglycerol accumulation in angiotensin II-stimulated cultured vascular smooth muscle cells. J Biol Chem 262:14555–14562

Griendling KK, Murphy TJ, Alexander RW (1993) Molecular biology of the renin-angiotensin system. Circulation 87:1816–1828

Gunther S, Gimbrone MA Jr, Alexander RW (1980) Regulation by angiotensin II of its receptors in resistance blood vessels. Nature 287:230–232

Gurantz D, Cowling RT, Villarreal FJ, Greenberg BH (1999) Tumor necrosis factor-alpha upregulates angiotensin II type 1 receptors on cardiac fibroblasts. Circ Res 85:272–279

Ichiki T, Usui M, Kato M, Funakoshi Y, Ito K, Egashira K, Takeshita A (1998) Downregulation of angiotensin II type 1 receptor gene transcription by nitric oxide. Hypertension 31:342–8

Ikeda Y, Takeuchi K, Kato T, Taniyama Y, Sato K, Takahashi N, Sugawara A, Ito S (1999) Transcriptional suppression of rat angiotensin AT1a receptor gene expression by interferon-gamma in vascular smooth muscle cells. Biochem Biophys Res Commun 262:494–8

Kai H, Griendling KK, Lassègue B, Ollerenshaw JD, Runge MS, Alexander W (1994) Agonist-induced phosphorylation of vascular type 1 angiotensin II receptor. Hypertension 24:523–527

Kakar SS, Sellers JC, Devor DC, Musgrove LC, Neill JD (1992) angiotensin II type-1 receptor subtype cDNAs: differential tissue expression and hormonal regulation. Biochem Biophys Res Commun 183:1090–1096

Kalenga MK, de Gasparo M, Thomas K, de Hertogh R (1996) Down-regulation of angiotensin AT1 receptor by progesterone in human placenta. J Clin Endocrinol Metab 81:998–1002

Kisley LR, Sakai RR, Fluharty SJ (1999) Estrogen decreases hypothalamic angiotensin II AT1 receptor binding and mRNA in the female rat. Brain Res 844:34–42

Kitamura E, Kikkawa R, Fujiwara Y, Imai T, Shigeta Y (1986) Effect of angiotensin II infusion on glomerular angiotensin II receptor in rats. Bioch Biophys Acta 885:309–316

Konarska MM, Sharp PA (1987) Interactions between small nuclear ribonucleoprotein particles in formation of spliceosomes. Cell 49 763–774

Krishnamurthi K, Verbalis JG, Zheng W, Wu Z, Clerch LB, Sandberg K (1999) Estrogen regulates angiotensin AT1 receptor expression via cytosolic proteins that bind to the 5′ leader sequence of the receptor mRNA. Endocrinology 140:5435–8

Krishnamurthi K, Zheng W, Verbalis AD, Sandberg K (1998) Regulation of cytosolic proteins binding cis elements in the 5′ leader sequence of the angiotensin AT1 receptor mRNA. Biochem Biophys Res Commun 245:865–870

Lassègue B, Alexander RW, Nickenig G, Clark M, Murphy TJ, Griendling KK (1995) Angiotensin II down-regulates the vascular smooth muscle AT1 receptor by transcriptional and posttranscriptional mechanisms: evidence for homologous and heterologous regulation. Mol Pharmacol 48:601–609

Linas SL, Marzec-Calvert R, Ullian ME (1990) K depletion alters angiotensin II receptor expression in vascular smooth muscle cells. Am J Physiol 258:C849–C854

Lu D, Sumners C, Raizada MK (1994) Regulation of angiotensin II type 1 receptor mRNA in neuronal cultures of normotensive and spontaneously hypertensive rat brains by phorbol esters and forskolin. J Neurochem 62:2079–2084

Malter JS (1989) Identification of an AUUUA-specific messenger RNA binding protein. Science 246:664–666

Mendelsohn ME, Karas RH (1999) The protective effects of estrogen on the cardiovascular system. N Engl J Med 340:1801–1811

Müller C, Reddert A, Wassmann S, Strehlow K, Böhm M, Nickenig G (2000) Insulin-like growth factor induces up-regulation of AT1 receptor gene expression in vascular smooth muscle cells. J Renin Angiotensin Aldost Syst 1:273–277
Murphy TJ, Takeuchi K, Alexander RW (1992) Molecular cloning of AT1 angiotensin receptors. Am J Hypertens 5:236S–242S
Nickenig G, Bäumer AT, Grohé C, Kahlert S, Strehlow K, Rosenkranz S, Stäblein A, Beckers F, Smits JF, Daemen MJ, Vetter H, Böhm M (1998c) Estrogen modulates AT1 receptor gene expression in vitro and in vivo. Circulation 97:2197–201
Nickenig G, Bäumer AT, Temur Y, Kebben D, Jockenhövel F, Böhm M (1999) Statin-sensitive dysregulated AT1 receptor function and density in hypercholesterolemic men. Circulation 100:2131–2134
Nickenig G, Harrison DG (2002) The AT1-type angiotensin receptor in oxidative stress and atherogenesis. Part I: oxidative stress and atherogenesis. Circulation 105:393–396
Nickenig G, Jung O, Strehlow K, Zolk O, Linz W, Schölkens BA, Böhm M (1997b) Hypercholesterolemia is associated with enhanced angiotensin AT1-receptor expression. Am J Physiol 272:H2701–H2707
Nickenig G, Michaelsen F, Müller C, Berger A, Vogel T, Sachinidis A, Vetter H, Böhm M (2002) Destabilization of AT1 receptor mRNA by calreticulin. Circ Res 90:53–58
Nickenig G, Michaelsen F, Müller C, Vogel T, Strehlow K, Böhm M (2001) Post-transcriptional regulation of the AT1 receptor mRNA. Identification of the mRNA binding motif and functional characterization. FASEB J 10:1490–1492
Nickenig G, Murphy TJ (1994) Down-regulation by growth factors of vascular smooth muscle angiotensin receptor gene expression. Mol Pharmacol 46:653–9
Nickenig G, Murphy TJ (1996) Enhanced AT1 receptor mRNA degradation and induction of polyribosomal mRNA binding proteins by angiotensin II in vascular smooth muscle cells. Mol Pharmacol 50:743–751
Nickenig G, Röling J, Strehlow K, Schnabel P, Böhm M (1998a) Insulin induces upregulation of vascular AT1 receptor gene expression by posttranscriptional mechanisms. Circulation 98:2453–60
Nickenig G, Sachinidis A, Ko Y, Vetter H (1996) Regulation of angiotensin AT1 receptor gene expression during cell growth of vascular smooth muscle cells. Eur J Pharmacol 297:307–12
Nickenig G, Sachinidis A, Michaelsen F, Böhm M, Seewald S, Vetter H (1997a) Upregulation of vascular angiotensin II receptor gene expression by low-density lipoprotein in vascular smooth muscle cells. Circulation 95:473–478
Nickenig G, Strehlow K, Bäumer AT, Baudler S, Wassmann S, Sauer H, Böhm M (2000a) Negative feedback regulation of reactive oxygen species on AT1 receptor gene expression. Br J Pharmacol 131:795–803
Nickenig G, Strehlow K, Röling J, Zolk O, Knorr A, Böhm M (1998b) Salt induces vascular AT1 receptor overexpression in vitro and in vivo. Hypertension 31:1272–1277
Nickenig G, Strehlow K, Wassmann S, Bäumer AT, Ahlbory K, Sauer H, Böhm M (2000b) Differential effects of estrogen and progesterone on AT1-receptor gene expression in vascular smooth muscle cells. Circulation 102:1828–1833
Nishimura H, Matsusaka T, Fogo A, Kon V, Ichikawa I (1997) A novel in vivo mechanism for angiotensin type 1 receptor regulation. Kidney Internat 52:345–355
Peach MJ (1977) Renin-angiotensin system: biochemistry and mechanisms of action. Physiol Rev 57:313–370
Port JD, Huang LY, Malbon CC (1992) Beta-adrenergic agonists that down-regulate receptor mRNA up-regulate a M(r) 35,000 protein(s) that selectively binds to beta-adrenergic receptor mRNAs. J Biol Chem 267:24103–24108
Sasamura H, Nakazato Y, Hayashida T, Kitamura Y, Hayashi M, Saruta T (1997) Regulation of vascular type 1 angiotensin receptors by cytokines. Hypertension 30:35–41

Schiffrin EL, Gutkowska J, Genest J (1984) Effect of angiotensin II and deoxycorticosterone infusion on vascular angiotensin II receptors in rats. Am J Physiol 246:H608–H614

Sechi LA, Griffin CA, Giaccetti G, Valentin JP, Llorens-Cortes C, Corvol P, Schambelan M (1996) Tissue-specific regulation of type 1 angiotensin II receptor mRNA levels in the rat. Hypertension 28:403–408

Steitz JA, Wolin SL, Rinke J, Pettersson I, Mount SM, Lerner EA, Hinterberger M, Gottlieb E (1983) Small ribonucleoproteins from eukaryotes: structures and roles in RNA biogenesis. Cold Spring Harb Symp Quant Biol 47:893–900

Strawn WB, Chapell MC, Dean RH, Kivlighn S, Ferrario CM (2000) Inhibition of early atherogenesis by losartan in monkeys with diet-induced hypercholesterolemia Circulation 101:1586–1593

Sumners C, Watkins LL, Raizada MK (1986) α_1-Adrenergic receptor-mediated downregulation of angiotensin II receptors in neuronal cultures. J Neurochem 47:1117–1126

Sun Y, Weber KT (1993) Angiotensin II and aldosterone receptor binding in rat heart and kidney: response to chronic angiotensin II or aldosterone administration. J Lab Cun Med 122:404–411

Swanson MS, Dreyfuss G (1988) Classification and purification of proteins of heterogeneous nuclear ribonucleoprotein particles by RNA-binding specificities. Mol Cell Biol 8:2237–2241

Takeda K, Ichiki T, Funakoshi Y, Ito K, Takeshita A (2000) Downregulation of angiotensin II type 1 receptor by all-trans retinoic acid in vascular smooth muscle cells. Hypertension 35:297–302

Tholanikunnel BG, Malbon CC (1997) A 20-nucleotide (A + U)-rich element of beta2-adrenergic receptor (beta2AR) mRNA mediates binding to beta2AR-binding protein and is obligate for agonist-induced destabilization of receptor mRNA. J Biol Chem. 272:11471–11478

Tufro-McReddie A, Chevalier RL, Everett AD, Gomez RA (1993) Decreased perfusion pressure modulates renin and ang II type 1 receptor gene expression in the rat kidney. Am J Physiol 264:R696–R702

Ullian ME, Schelling JR, Linas SL (1992) Aldosterone enhances angiotensin II receptor binding and inositol phosphate responses. Hypertension 20:67–73

Ullian ME, Raymond JR, Willingham MC, Paul RV (1997) Regulation of vascular angiotensin II receptors by EGF. Am J Physiol 273:C1241–C1249

Vakalopoulou E, Schaack J, Shenk T (1991) A 32-kilodalton protein binds to AU-rich domains in the 3′ untranslated regions of rapidly degraded mRNAs. Mol Cell Biol 11:3355–3364

Wang FW, Nickenig G, Murphy TJ (1997) The vascular smooth muscle AT1 receptor mRNA is destabilized by cAMP-elevating agents. Mol Pharmacol 52:781–787

Wang X, Murphy TJ (1998) Inhibition of cyclic AMP-dependent kinase by expression of a protein kinase inhibitor/enhanced green fluorescent fusion protein attenuates angiotensin II-induced type 1 AT1 receptor mRNA down-regulation in vascular smooth muscle cells. Mol Pharmacol 54:514–524

Warnholtz A, Nickenig G, Schulz E, Macharzina R, Bräsen JH, Skatchkov M, Heitzer T, Stasch JT, Griendling KK, Harrison DG, Böhm M, Meinertz T, Münzel T (1999) Increased NADH-oxidase mediated superoxide production in the early stages of atherosclerosis: evidence for involvement of the renin angiotensin system. Circulation 99:2027–2033

Wassmann S, Bäumer AT, Strehlow K, van Eickels M, Grohé C, Ahlbory K, Rösen R, Böhm M, Nickenig G (2001c) Endothelial dysfunction and oxidative stress during estrogen deficiency in spontaneously hypertensive rats. Circulation 103:435–441

Wassmann S, Hilgers S, Laufs U, Böhm M, Nickenig G (2002a) Angiotensin II type 1 receptor antagonism improves hypercholesterolemia-associated endothelial dysfunction. Arterioscler Thromb Vasc Biol 22:1208–1212

Wassmann S, Laufs U, Bäumer AT, Müller K, Konkol C, Sauer H, Böhm M, Nickenig G (2001a) Inhibition of geranylgeranylation reduces angiotensin II-mediated free radical production in vascular smooth muscle cells: involvement of angiotensin AT1 receptor expression an rac1 GTPase. Mol Pharmacol 59:646–654

Wassmann S, Laufs U, Müller K, Bäumer AT, Ahlbory K, Linz W, Itter G, Rösen R, Böhm M, Nickenig G (2001b) HMG-CoA reductase inhibitors improve endothelial dysfunction in normocholesterolemic hypertension via reduced production of reactive oxygen species. Hypertension 37:1450–1457

Wassmann S, Laufs U, Müller K, Konkol C, Ahlbory K, Bäumer AT, Linz W, Böhm M, Nickenig G (2002b) Cellular antioxidant effects of atorvastatin in vitro and in vivo. Arterioscler Thromb Vasc Biol 22:300–305

Wilusz J, Shenk TA (1990) Uridylate tract mediates efficient heterogeneous nuclear ribonucleoprotein C protein-RNA cross-linking and functionally substitutes for the downstream element of the polyadenylation signal. Mol Cell Biol 10:6397–6407

Xu K, Murphy TJ (2000) Reconstitution of angiotensin receptor mRNA down-regulation in vascular smooth muscle. J Biol Chem 275:7604–7611

Yang H, Lu D, Raizada MK (1996) Lack of cross talk between α_1-adrenergic and angiotensin type 1 receptors in neurons of spontaneously hypertensive rat brain. Hypertension 27:1277–1283

Angiotensin AT$_1$ Receptor Signal Transduction

C. Maric · K. Sandberg

Department of Medicine, Georgetown University, Bldg D, 4000 Reservoir Road, NW, Washington, DC, 20057 USA
e-mail: sandberg@georgetown.edu

1	Introduction	336
2	**G Protein Coupling**	336
2.1	Multiple Signaling Pathways	336
2.1.1	Phospholipase C	336
2.1.2	Adenylate Cyclase	337
2.2	G Protein Types	337
2.3	Determinants of G Protein Coupling	338
2.4	Receptor Activation Models	339
3	**Protein Kinases**	340
3.1	Tyrosine Kinases	340
3.1.1	Receptor Tyrosine Kinases	341
3.1.2	Non-receptor Tyrosine Kinases	341
3.2	Mitogen-Activated Protein Kinases	344
4	**Small G Proteins**	344
5	**Conclusions**	345
	References	345

Abstract The octapeptide hormone, angiotensin II (Ang II) binds to type 1 angiotensin (AT$_1$) receptors in target tissues and through a number of signal transduction pathways, elicits its effects on blood pressure control and fluid and electrolyte homeostasis. AT$_1$ receptor antagonists are widely used to treat hypertension and associated cardiovascular and renal disease. In this review, we focus on the signal transduction mechanisms of AT$_1$ receptors. The list of known effectors through which the AT$_1$ receptor can signal has dramatically increased over the past 10 years and points to the complexity of receptor signaling to this effector. This complexity, however, may not be surprising given the various actions of Ang II at the cellular level.

Keywords Renin–angiotensin system · Angiotensin AT$_1$ receptor · Phospholipase C · Adenylate cyclase · G proteins · G protein coupling · Protein kinase · Tyrosine kinase · Mitogen-activated protein kinase (MAPK) · Small G proteins

1
Introduction

Angiotensin II (Ang II) mediates the biological effects of the renin angiotensin system by signaling through its two receptor subtypes (AT_1 and AT_2) in the cell membrane of target tissues. These two receptors play an important role in blood pressure control and fluid and electrolyte homeostasis through modulating the actions of Ang II on vasoconstriction (Griendling et al. 1997), aldosterone secretion (Vallotton 1987), thirst (Fitzsimons 1998), and vascular smooth muscle cell (VSMC) growth (Thomas et al. 1996; Marrero et al. 1997; Xi et al. 1999). Hypotension, hypertension, hyponatremia, and hemorrhage all alter the activity of the renin–angiotensin system. In this review, we focus on the various mechanisms by which AT_1 receptors signal.

2
G Protein Coupling

2.1
Multiple Signaling Pathways

2.1.1
Phospholipase C

The AT_1 receptor belongs to the family of calcium mobilizing G protein-coupled receptors (GPCRs). Upon binding Ang II, the AT_1 receptor stimulates phospho-

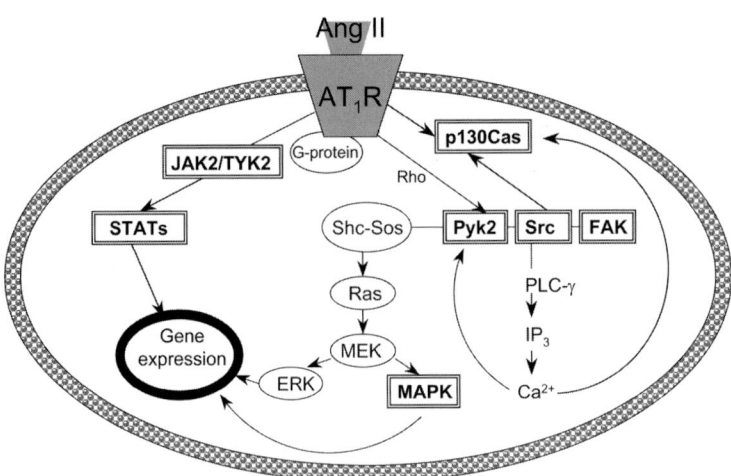

Fig. 1 Multiple AT_1 receptor signaling pathways. Binding of Ang II to the G protein-coupled AT_1 receptor activates PLC, which either leads to the generation of IP_3 with subsequent Ca^{2+} mobilization or generation of diacylglycerol (DAG) with subsequent protein kinase C (PKC) activation. Activation of the AT_1 receptor also leads to activation of the phospholipase A (PLA_2) pathway with subsequent release of arachidonic acid and activation of tyrosine kinase pathways

lipase C (PLC)-β; hydrolysis of phosphatidylinositol 4,5-bisphosphate by PLC then yields inositol 1,4,5-trisphosphate (IP$_3$). Subsequently, IP$_3$ binds to its receptor on the sarcoplasmic reticulum, which induces the release of intracellular calcium (Peach and Dostal 1990; Israel et al. 1995) (Fig. 1).

2.1.2
Adenylate Cyclase

The AT$_1$ receptor also inhibits adenylate cyclase (Catt et al. 1988; Douglas et al. 1990). AT$_1$ receptors can couple to more than one signal transduction pathway. When rat (r) AT$_{1a}$ receptors were stably transfected into Chinese hamster ovary cells, they not only coupled to the PLC pathway, but in addition, they coupled to inhibition of forskolin-evoked cyclic adenosine monophosphate (cAMP) accumulation (Ohnishi et al. 1992; Teutsch et al. 1992).

2.2
G Protein Types

The AT$_1$ receptor interacts with the pertussis-insensitive Gq$_{11}$ class of G proteins (Kai et al. 1996) as well as the Gq$_{12/13}$ family (Macrez-Lepretre et al. 1997). By turning on the α-subunit of the G protein, Ang II-activation of AT$_1$ receptors stimulates PLC-β_1 (Schelling et al. 1997). Studies in rat portal vein myocytes using antisense oligonucleotides directed against mRNAs encoding the G protein subunits α, β, and γ suggest that rAT$_{1a}$ receptors activate the PLC pathway by coupling to the G protein heterotrimer, G$\alpha_{13},\beta_1\gamma_3$ (Macrez-Lepretre et al. 1997). However, the rAT$_{1a}$ receptor couples to different G proteins in rat VSMCs. Electroporated antibodies directed against various G protein subunits indicate that the initial receptor–PLC coupling is mediated by G$\alpha_{q/11}\beta\gamma$ and G$\alpha_{12}\beta\gamma$ but not by G$\alpha_{13}\beta\gamma$ (Ushio-Fukai et al. 1998). Furthermore, these authors showed that the rAT$_{1a}$ receptor sequentially couples to PLC-β_1 followed by coupling to PLC-γ. Thus, these studies indicate that AT$_1$ receptors can couple to different G protein heterotrimers depending upon the cell type and growth stage and that Ang II signaling is temporally controlled.

The binding of guanosine triphosphate (GTP)γS is influenced by synthetic peptides that represent intracellular cytosolic regions of the AT$_1$ receptor. These representative synthetic peptides could also activate purified Gi$_1$, Gi$_2$, and Go proteins (Shirai et al. 1995). Furthermore, site-directed mutagenesis of the human (h) AT$_1$ receptor showed that a mutant deficient in coupling to Gi could still retain its ability to bind Gq (Shibata et al. 1996). Taken together, these data suggest the AT$_1$ receptor can couple to Gi, Go, and Gq.

Most GPCRs possess a number of receptor subtypes, which couple to distinct signaling pathways (Carman and Benovic 1998; Selbie and Hill 1998; Wess 1998). In contrast, many studies suggest that only two AT receptor classes (AT$_1$ and AT$_2$) exist. Thus, the finding that AT$_1$ receptors can signal in different ways depending on the coupling G protein explains how Ang II can activate multiple signal transduction pathways.

2.3
Determinants of G Protein Coupling

Intracellular loop (IC) 3 of the AT_1 receptor plays a key role in G protein coupling. Site-directed mutagenesis studies demonstrate that residues 219–225 in IC3 are essential for Gq coupling (Wang et al. 1995). Within this loop, Leu^{222} in the KALKK amino acid sequence is not only essential for G protein interaction, it also is crucial for signal transduction (Hunyady et al. 1996; Laporte et al. 1998). A leucine in this position (or sometimes another apolar residue) is commonly found in Gq-coupled receptors (Hunyady et al. 1996). Structure–function studies and homology comparisons of AT_1 receptors across species predict that an amphipathic α-helix is formed within this loop. On one side, all the residues are basic while the opposite side contains many uncharged residues. The IC3 of the M3 muscarinic GPCR is also predicted to form an amphipathic α-helix. Evidence that orientation of basic residues within the α-helix is critical for GTPase stimulation and subsequent activation of G proteins has come from mutagenesis structure–function studies and from studies of peptide sequences that correspond to IC3 in GPCRs and which stimulate GTPase activity (Okamoto et al. 1990; Okamoto and Nishimoto 1992; Oppi et al. 1992). This IC3 region is also critical for G protein recognition and coupling to Gi in the muscarinic M3 receptor (Bluml et al. 1994; Wess 1998). AT_1 receptor activation of Gq/Gq_{11}-dependent phosphoinositide hydrolysis is controlled by residues in the N-terminus region of IC3 and residues (Ile^{238} and Phe^{239}) in the C-terminal region of IC3 adjacent to the sixth transmembrane domain (TMVI) (Zhang et al. 2000).

IC2 also plays a crucial role in G protein coupling. The highly conserved DRYM amino acid sequence in IC2 is essential for G protein coupling in many GPCRs (Wess 1998) and is also shown to be critical for Gq and Gi coupling in the hAT_1 receptor (Shibata et al. 1996). Furthermore, evidence for the importance of IC2 arises from mapping studies with synthetic peptides. Peptides representing receptor sequences corresponding to the N-terminal region of the IC2 (125–137) inhibited receptor G protein coupling in the rAT_{1a} receptor (Shirai et al. 1995; Kai et al. 1996). Residues in the external third region of TMVI have also been shown to play a key role in signaling without interfering with ligand binding and include residues Val^{254}, His^{256}, and Phe^{259} in the rAT_{1a} receptor (Han et al. 1998).

G protein coupling in the AT_1 receptor is also influenced by residues in the C-terminal tail. GTP binding to Gi was either stimulated or inhibited depending upon the region of the carboxy terminal cytoplasmic tail that the peptide represented, suggesting Gi and the rAT_{1a} receptor directly interact within this region (Shirai et al. 1995; Kai et al. 1996). Receptor–G protein specificity is apparently controlled by specific sequences in the carboxy tail. Mutagenesis analysis showed that a truncated hAT_1 receptor, which lacked the carboxyl terminal 50 residues coupled to Gq but was unable to couple to Gi (Shibata et al. 1996). There are species differences in the residues that govern the specific determinants of AT_1 receptor G protein coupling. Unlike in the hAT_1 receptor, Tyr^{312}, Phe^{313}, and Leu^{324} in the rAT_{1a} cytoplasmic tail are essential for coupling to Gq (Sano et al. 1997). In conclusion, multiple contacts within IC2 and IC3, the cytoplasmic tail, and TMVI determine AT_1 receptor binding and activation of G proteins. Furthermore, homology comparisons among GPCRs indicate that residues in similar positions are important in G protein activation in general (Ren et al. 1993).

2.4
Receptor Activation Models

Thus far, there are no X-ray crystallographic structures of AT_1 receptors or any other peptide hormone GPCR. Thus, receptor activation models are commonly based on structure–function studies of mutant receptors. Many of these studies are based on loss-of-function analysis, which is problematic since loss of function could reflect nonspecific effects due to local losses in receptor structural integrity; however, gain-of-function mutagenesis studies and other approaches, such as biochemical mapping, can help in model development.

Mutagenesis studies suggest that Ang II signaling is initiated when Tyr^4 within the Ang II peptide interacts with Asn^{111} within TMIII of the receptor. This interaction induces a conformational change, which disrupts the interaction between Asn^{111} within TMIII and Tyr^{292} and Asn^{295} within TMVII (Marie et al. 1994; Balmforth et al. 1997). This model of receptor activation, implicating TMVII in the initiation site of agonist activation, is similar to models proposed for many GPCRs (Baldwin 1993; Roth et al. 1997). Numerous serine, threonine, and tyrosine residues within the intracellular carboxy tail are phosphorylated in an agonist-dependent manner, and evidence suggests they play a key role in receptor activation, internalization, and desensitization (see Thomas 1999 for an excellent review of the topic).

Commonly, the highly conserved residues within GPCRs play similar roles in G protein activation such as Asp^{74} in TMII (Bihoreau et al. 1993), the DRY amino acid sequence in TMIII (Ohyama et al. 1995) and Tyr^{215} in TMV (Hunyady et al. 1995). Residues in IC3 (Zhang et al. 2000), TMVI (Han et al. 1998), and TMVII (Hunyady et al. 1998) facilitate the rearrangement of helices that lead to receptor activation, while having no affect on ligand binding. While many similarities exist among GPCRs, not all highly conserved sequences regulate the same activities. In the intracellular domain of GPCRs, the $NPX_{ii}Y$ sequence is highly conserved. Receptor internalization and resensitization in the β_2-adrenergic receptor is prevented if the tyrosine in this sequence is substituted with another residue (Barak et al. 1994). In the AT_1 receptor, the corresponding residue in the NPLFY sequence is critical for PLC activation but is not necessary for agonist binding or receptor internalization (Laporte et al. 1996). The residues that determine G protein coupling are not identical to those residues that govern AT_1 receptor internalization. Ang II-induced receptor endocytosis has been shown to occur in non-signaling mutants of TMII, IC3, and the cytoplasmic tail (Hunyady et al. 1994a,b,c; Thomas 1999).

3
Protein Kinases

Less well understood pathways of AT_1 receptor signaling are through phosphorylation of receptor and non-receptor tyrosine kinases and activation of mitogen-activated protein kinases (MAPK). The fact that these signaling cascades are commonly associated with signaling by growth factors and cytokines, supports the accumulating data suggesting that Ang II plays a pivotal role in mediating long-term growth regulation and inflammatory processes under both physiological and pathophysiological conditions (Hunyady et al. 1994a,b,c; Thomas 1999). In addition to its important role in the control of systemic blood pressure, water, and salt homeostasis, Ang II is a potent growth factor in a number of tissues and cell types (Schnee and Hsueh 2000; Enseleit et al. 2001; Klahr 2001; De Gasparo 2002). To date, the best-defined Ang II-induced tyrosine phosphorylation and MAPK cascade is described in VSMCs. Ang II induces a multitude of actions in these cells that are typical of growth factors and cytokines, including adhesion, migration, synthesis of extracellular matrix, and regulation of long-term cell growth. Abnormal regulation of these pathways has been associated with cardiovascular disease and hypertension (Border and Noble 2001; Deblois et al. 2001; Schiffrin 2002; Unger 2002).

3.1
Tyrosine Kinases

The ability of Ang II to induce tyrosine phosphorylation is still a puzzle, since the AT_1 receptor is a member of the GPCR family and not a member of the tyrosine kinase family of receptors. Apparently, AT_1 receptors do not have intrinsic tyrosine kinase activity. Instead, accumulating experimental data suggest that its tyrosine kinase activity is achieved through activation of both receptor and non-receptor tyrosine kinases. Receptor tyrosine kinases include platelet-derived growth factor (PDGF) and epidermal growth factor (EGF) (Fujiyama et al. 2001; Saito and Berk 2001). Non-receptor tyrosine kinases include PLC-γ, Src family kinases, focal adhesion kinase (FAK), proline-rich tyrosine kinase (Pyk2), Janus family kinases (JAK), tyrosine kinase (TYK), signal transducers and activators of transcription (STATs), paxillin and p130Cas (Marrero et al. 1996; Berk and Corson 1997; Kim and Iwao 2000; Touyz and Berry 2002).

3.1.1
Receptor Tyrosine Kinases

Another mechanism by which Ang II exerts cell regulatory properties is by activating receptor tyrosine kinases. Accumulating evidence suggests that even though the AT_1 receptor does not have a direct physical association with receptor tyrosine kinases, Ang II-activated AT_1 receptors regulate receptor tyrosine kinase activity by transactivation. Signaling by AT_1 receptor transactivation has been shown for a number of receptor tyrosine kinases, including both Ca^{2+}-dependent (e.g., EGF) and independent (e.g., PDGF) receptors (Kalmes et al. 2001; Saito and Berk 2001).

The mechanisms of Ang II-induced receptor transactivation are still poorly understood. Studies suggest that Ang II-activated AT_1 receptors cause cleavage of pro-heparin-bound EGF via metalloproteinases (Haendeler and Berk 2000; Kalmes et al. 2001). The free, heparin-bound EGF then binds to its receptor and becomes autophosphorylated. This process is associated with activation of Pyk2 and Src as well as downstream activation of extracellular signal-related kinase (ERK)1/2 (Eguchi and Inagami 2000; Eguchi et al. 2001). The process of Ang II-induced transactivation of EGF has been reported to be important in VSMC hyperplasia and contraction (Eguchi et al. 2001).

3.1.2
Non-receptor Tyrosine Kinases

a. PLC-γ. Ang II activation of PLC-γ depends upon phosphorylation of the AT_1 receptor (Fig. 2). Ang II binding to the AT_1 receptor causes phosphorylation of Tyr^{319} in the YIPP motif within the C-terminal intracellular domain of the receptor. The C-terminal region of the phosphorylated receptor is then able to bind to two Src homology 2 domains in PLC-γ1 (Ali et al. 1997; Venema et al. 1998). Interestingly, evidence also suggests that the same YIPP motif is required for AT_1 receptor binding to the tyrosine phosphatase SHP-2 and to the JAK2 tyrosine kinase; however, the functional significance of these observations remains unknown. Activation of PLC-γ1 leads to IP_3 production and intracellular Ca^{2+} release (Harp et al. 1997; De Gasparo et al. 2000). The relative contribution of the various PLC isoforms including PLC-γ1 in AT_1 receptor signaling is not well understood. Species differences in the expression of PLC isoforms exists. In the rat, studies suggest that AT_1 receptors couple to PLC-β, then subsequently to PLC-γ (Marrero et al. 1994; Ushio-Fukai et al. 1998), leading to increased intracellular IP_3 and subsequent Ca^{2+} mobilization as well as tyrosine phosphorylation. Activation of non-receptor tyrosine kinases mediates a number of cellular functions in VSMCs, including cell growth and differentiation (Marrero et al. 1994; De Gasparo et al. 2000).

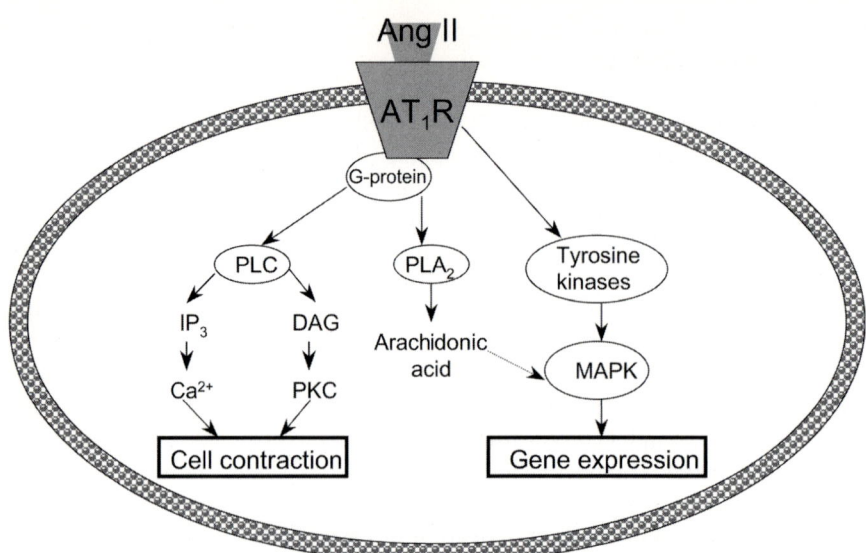

Fig. 2 AT$_1$ receptor activated tyrosine kinase pathways. Upon binding to the AT$_1$ receptor, Ang II triggers multiple tyrosine kinase signaling pathways. Ang II activates Src, which in turn regulates PLC-γ activity leading to IP$_3$ production and subsequent Ca^{2+} mobilization. Ang II activates the Pyk2 pathway, via Ca^{2+}-dependent mechanisms, which regulate ERK-dependent signaling pathways. Presumably via Rho GTPases, Ang II activates FAK. Ang II activation of the JAK2/TYK2-STAT pathway regulates gene transcription. Ang II also activates p130Cas via Ca^{2+}-dependent mechanisms

b. Src family kinases. To date, the Src kinase family consists of 14 kinase members. Src-mediated phosphorylation of PLC-γ leads to increased intracellular IP$_3$ and Ca^{2+} mobilization as well as phosphorylation of the cell adhesion signaling molecule p130Cas (Fig. 2). Evidence suggests Src also plays a role in Ang II-induced activation of FAK and Pyk2 as well as downstream signaling molecules, such as ERK, paxillin, JAK/STAT, and Shc (Erpel and Courtneidge 1995; Touyz and Berry 2002). An Ang II increase in Src-mediated signaling is involved in the regulation of focal adhesion, cytoskeletal reorganization, cell migration, and cell growth (Sayeski et al. 1998; Touyz and Berry 2002), and increased Src-mediated signaling has been implicated in the pathogenesis and development of hypertension (Yamazaki et al. 1999; Touyz and Berry 2002).

c. FAK and Pyk2. As its name implies, FAK is associated with focal adhesion points between the cell membrane and the extracellular matrix. Among its other roles, Pyk2 provides molecular scaffolding for the docking of signal transduction molecules. FAK is characterized by extracellular matrix-dependent tyrosine auto-phosphorylation, which triggers its interactions with other non-receptor tyrosine kinases, such as Src. These interactions with non-receptor tyrosine kinases lead to further phosphorylation of FAK and subsequent association with Grb2 and the guanosine diphosphate (GDP)–

GTP exchange proteins, Sos and Ras (Fig. 2). Although the direct link between FAK and the AT_1 receptor is still unclear, the potential role for RhoA or other Rho family GTPases has often been implicated. The Ang II-activated FAK cascade plays an important role in the control of cell migration, cell shape, and volume regulation, via phosphorylation of paxillin and talin (Leduc and Meloche 1995; Guan 1997), and overactivation of FAK has been linked to VSMC hypertrophy and atherosclerosis (Taylor et al. 2001; Abbi and Guan 2002). Pyk2 is related to FAK and is also known as Ca^{2+}-dependent tyrosine kinase. Studies suggest that Pyk2 associates with GPCRs including the AT_1 receptor and regulates tyrosine kinase-activated cell contraction, migration, and growth (Tang et al. 2000; Rocic and Lucchesi 2001).

d. JAK/STAT. Ang II-activation of AT_1 receptors leads to the coupling of two members of the JAK family, namely JAK2 and TYK2 and subsequent phosphorylation of several transcription factors in the STAT protein family, including STAT1α/β, STAT2, and STAT3 (Mascareno and Siddiqui 2000) (Fig. 2). The phosphorylated STAT proteins are then translocated to the nucleus, where they initiate transcription of a number of genes involved in physiological and pathophysiological cell proliferation (Marrero et al. 1998; Meloche et al. 2000). The activation and regulation of the JAK/STAT pathway has been extensively studied over the past few years because of its key role in regulating Ang II-induced cell growth in cardiovascular disease (see Mascareno and Siddiqui 2000; Meloche et al. 2000; Booz et al. 2002 for extensive reviews).

e. p130Cas. Previous reports suggested that Ang II-activated p130Cas tyrosine phosphorylation is Ca^{2+}-, Src- and PKC-dependent; however, emerging experimental data contradict these findings (Sayeski et al. 1998; Touyz and Berry 2002). The precise role of Ang II-activated p130Cas remains unclear. Some studies suggest that this proline-rich protein is involved in the regulation of integrin-mediated cell adhesion through activation of cytoskeletal elements, including FAK, paxillin and α-actin (Takahashi et al. 1998; Ishida et al. 1999) and subsequent involvement in cell proliferation (Ishida et al. 1999; Forte et al. 2002).

f. Paxillin. An integral component of focal adhesions, paxillin performs as an adapter protein by acting as an anchor in the formation of multiprotein complexes (Bouillier et al. 2001). Ang II-activated paxillin tyrosine phosphorylation, in concert with FAK and Src kinase, is associated with re-organization of the cytoskeletal network, such as formation of new focal adhesions and stress fibers. Some studies suggest that alterations in the pattern of paxillin expression is involved in the pathogenesis of genetic hypertension and cardiac hypertrophy (Sabri et al. 1998; Bouillier et al. 2001).

3.2
Mitogen-Activated Protein Kinases

MAPKs are a family of serine/threonine protein kinases associated with nuclear transduction of extracellular signals in a variety of cell types and under a number of stimuli, including Ang II. Ang II is known to activate three major members of the MAPK family, including ERK1/2, c-Jun N-terminal kinases (JNKs), and p38 MAPK (Touyz and Schiffrin 2000; Touyz and Schiffrin 2001) (Fig. 1). Activation of MAPK is highly complex and involved in mediating several aspects of cell growth and protein synthesis. This cascade begins with phosphorylation by MAPK kinase (MEK), which is under the control of Raf1, both of which regulate ERK1/2 activity. ERK1/2, in turn, translocates to the nucleus where it regulates gene expression by altering the activity of transcription factors through their phosphorylation. In addition to activating ERKs, MAPK is also known to directly regulate the activity of cyclo-oxygenase (COX)-2, Ca^{2+} channels, and the Na^+/H^+ exchanger (Robinson and Cobb 1997).

The interaction between Ang II and JNK begins with phosphorylation of JNK/stress-activated protein kinase (SAPK) via the Ca^{2+}- and PKC-dependent p21-activated kinase. Once phosphorylated, JNK1 and JNK2 translocate to the nucleus and activate several transcription factors, including c-Jun, ATF-2 and Elk-1 (Schmitz et al. 1998), which subsequently leads to regulation of cell death through apoptosis (Sadoshima et al. 2002). Although studies indicate that Ang II-induced apoptosis is mainly associated with the AT_2 receptor, the AT_1 receptor-mediated JNK/SAPK signaling pathway may also play an important role in promoting cell death.

A third mechanism by which Ang II activates the MAPK signaling cascade is via interaction with p38 MAPK. Similarly to the JNK/SAPK pathway, the Ang II-mediated MAPK signaling cascade is thought to be important in the regulation of apoptosis in the process of cardiac ischemia, hypertrophy, and remodeling (Kim and Iwao 2000; Sharov et al. 2003). The precise mechanisms by which Ang II activates p38 MAPK remain unknown; however, some data suggest this pathway plays an opposing role to Ang II-induced ERK1/2 activation.

4
Small G Proteins

In addition to signaling through activation of multiple heterotrimeric G proteins, tyrosine kinases, and MAPK, Ang II has now been shown to interact with small (21-kDa) guanine nucleotide-binding proteins (also termed small G proteins). So far, five subfamilies of small G proteins have been identified. These include Rho, Ras, Rab, Ran, and adenosine diphosphate (ADP) ribosylation factors. The Rho subfamily comprises RhoA, Rac1, and Cdc42 and is the best characterized with respect to AT_1 receptor signaling (Laufs et al. 2002). RhoA is a cytoplasmic protein coupled to GDP, which becomes activated upon its association with GTP. Following activation, RhoA/GTP interacts with Rho kinase, which has been implicated in regulating cell contraction via modulating myosin light chain phosphatase activity (Park et al. 2002; Cavarape et al. 2003) and in abnormal Ang II-induced growth regulatory mechanisms in atherosclerosis (Shimokawa 2002). In addition, studies suggest that Ang II regulates cytoskeletal organization, cell growth, and inflammation by activating Rac and consequent involvement with p21-activated kinase and JNK (Laufs et al. 2002; Park et al. 2002).

5
Conclusions

For a long time, it has puzzled many scientists how Ang II could elicit such a wide range of cellular functions. It was initially believed that Ang II-induced signaling pathways were discrete and mainly of short duration. Much evidence now indicates that cross-talk interactions among the myriad of AT_1 receptor signaling pathways mediate the various biological functions of Ang II. This complex interplay of signaling events is triggered by direct receptor activation and also by indirect mechanisms involving tyrosine phosphorylation and MAPK. These signaling pathways are tightly regulated events under normal physiological conditions, and their abnormal regulation has been associated with a number of pathophysiological conditions in almost every tissue and cell type. Much current research focuses on selectively targeting these discrete signaling pathways with the hope of developing new treatment strategies in prevention and treatment of hypertension and cardiovascular and renal disease.

References

Abbi S, Guan J (2002) Focal adhesion kinase: protein interactions and cellular functions. Histol Histopathol 17:1163–1171
Ali MS, Sayeski P, Dirksen L, Hayzer DJ, Marrero MB, Bernstein KE (1997) Dependence on the motif YIPP for the physical association of Jak2 kinase with the intracellular carboxyl tail of the angiotensin II AT_1 receptor. J Biol Chem 272:23382–23388
Baldwin JM (1993) The probable arrangement of the helices in G protein-coupled receptors. EMBO J 12:1693–1703
Balmforth AJ, Lee AJ, Warburton P, Donnelly D, Ball SG (1997) The conformational change responsible for AT_1 receptor activation is dependent upon two juxtaposed asparagine residues on transmembrane helices III and VII. J Biol Chem 272:4245–4251
Barak LS, Tiberi M, Freedman NJ, Kwatra MM, Lefkowitz RJ, Caron MG (1994) A highly conserved tyrosine residue in G protein-coupled receptors is required for agonist-mediated β_2-adrenergic receptor sequestration. J Biol Chem 269:2790–2795
Berk BC, Corson MA (1997) Angiotensin II signal transduction in vascular smooth muscle: role of tyrosine kinases. Circ Res 80:607–616
Bihoreau C, Monnot C, Davies E, Teutsch B, Bernstein KE, Corvol P, Clauser E (1993) Mutation of Asp[74] of the rat angiotensin II receptor confers changes in antagonist affinities and abolishes G-protein coupling. Proc Natl Acad Sci USA 90:5133–5137

Bluml K, Mutschler E, Wess J (1994) Insertion mutagenesis as a tool to predict the secondary structure of a muscarinic receptor domain determining specificity of G-protein coupling. Proc Natl Acad Sci USA 91:7980–7984

Booz GW, Day JN, Baker KM (2002) Interplay between the cardiac renin angiotensin system and JAK-STAT signaling: role in cardiac hypertrophy, ischemia/reperfusion dysfunction, and heart failure. J Mol Cell Cardiol 34:1443–1453

Border WA, Noble N (2001) Maximizing hemodynamic-independent effects of angiotensin II antagonists in fibrotic diseases. Semin Nephrol 21:563–572

Bouillier H, Samain E, Rucker-Martin C, Renaud JF, Safar M, Dagher G (2001) Effect of extracellular matrix elements on angiotensin II-induced calcium release in vascular smooth muscle cells from normotensive and hypertensive rats. Hypertension 37:1465–1472

Carman CV, Benovic JL (1998) G-protein-coupled receptors: turn-ons and turn-offs. Curr Opin Neurobiol 8:335–344

Catt KJ, Balla T, Baukal AJ, Hausdorff WP, Aguilera G (1988) Control of glomerulosa cell function by angiotensin II: transduction by G-proteins and inositol polyphosphates. Clin Exp Pharmacol Physiol 15:501–515

Cavarape A, Endlich N, Assaloni R, Bartoli E, Steinhausen M, Parekh N, Endlich K (2003) Rho-kinase inhibition blunts renal vasoconstriction induced by distinct signaling pathways in vivo. J Am Soc Nephrol 14:37–45

de Gasparo M (2002) Angiotensin II and nitric oxide interaction. Heart Fail Rev 7:347–358

de Gasparo M, Catt KJ, Inagami T, Wright JW, Unger T (2000) International union of pharmacology. XXIII. The angiotensin II receptors. Pharmacol Rev 52:415–472

deBlois D, Orlov SN, Hamet P (2001) Apoptosis in cardiovascular remodeling–effect of medication. Cardiovasc Drugs Ther 15:539–545

Douglas JG, Romero M, Hopfer U (1990) Signaling mechanisms coupled to the angiotensin receptor of proximal tubule endothelium. Kidney Int 30:S43–S47

Eguchi S, Inagami T (2000) Signal transduction of angiotensin II type 1 receptor through receptor tyrosine kinase. Regul Pept 91:13–20

Eguchi S, Dempsey PJ, Frank GD, Motley ED, Inagami T (2001) Activation of MAPKs by angiotensin II in vascular smooth muscle cells. Metalloprotease-dependent EGF receptor activation is required for activation of ERK and p38 MAPK but not for JNK. J Biol Chem 276:7957–7962

Enseleit F, Hurlimann D, Luscher TF (2001) Vascular protective effects of angiotensin converting enzyme inhibitors and their relation to clinical events. J Cardiovasc Pharmacol 37 Suppl 1:S21–S30

Erpel T, Courtneidge SA (1995) Src family protein tyrosine kinases and cellular signal transduction pathways. Curr Opin Cell Biol 7:176–182

Fitzsimons JT (1998) Angiotensin, thirst, and sodium appetite. Physiol Rev 78:583–686

Forte A, Di Micco G, Galderisi U, De Feo M, Esposito F, Esposito S, Renzulli A, Berrino L, Cipollaro M, Agozzino L, Cotrufo M, Rossi F, Cascino A (2002) Gene expression and morphological changes in surgically injured carotids of spontaneously hypertensive rats. J Vasc Res 39:114–121

Fujiyama S, Matsubara H, Nozawa Y, Maruyama K, Mori Y, Tsutsumi Y, Masaki H, Uchiyama Y, Koyama Y, Nose A, Iba O, Tateishi E, Ogata N, Jyo N, Higashiyama S, Iwasaka T (2001) Angiotensin AT(1) and AT(2) receptors differentially regulate angiopoietin-2 and vascular endothelial growth factor expression and angiogenesis by modulating heparin binding-epidermal growth factor (EGF)-mediated EGF receptor transactivation. Circ Res 88:22–29

Griendling KK, Ushio-Fukai M, Lassegue B, Alexander RW (1997) Angiotensin II signaling in vascular smooth muscle. New concepts. Hypertension 29:366–373

Guan JL (1997) Focal adhesion kinase in integrin signaling. Matrix Biol 16:195–200

Haendeler J, Berk BC (2000) Angiotensin II mediated signal transduction. Important role of tyrosine kinases. Regul Pept 95:1–7

Han HM, Shimuta SI, Kanashiro CA, Oliveira L, Han SW, Paiva AC (1998) Residues Val^{254}, His^{256}, and ^{Phe259} of the angiotensin II AT$_1$ receptor are not involved in ligand binding but participate in signal transduction. Mol Endocrinol 12:810–814

Harp JB, Sayeski PP, Scanlon M, Bernstein KE, Marrero MB (1997) Role of intracellular calcium in the angiotensin II-mediated tyrosine phosphorylation and dephosphorylation of PLC-γ1. Biochem Biophys Res Commun 232:540–544

Hunyady L, Bor M, Balla T, Catt KJ (1994a) Identification of a cytoplasmic Ser-Thr-Leu motif that determines agonist-induced internalization of the AT$_1$ angiotensin receptor. J Biol Chem 269:31378–31382

Hunyady L, Baukal AJ, Balla T, Catt KJ (1994b) Independence of type I angiotensin II receptor endocytosis from G protein coupling and signal transduction. J Biol Chem 269:24798–24804

Hunyady L, Bor M, Balla T, Catt KJ (1995) Critical role of a conserved intramembrane tyrosine residue in angiotensin II receptor activation. J Biol Chem 270:9702–9705

Hunyady L, Tian Y, Sandberg K, Balla T, Catt KJ (1994c) Divergent conformational requirements for angiotensin II receptor internalization and signaling. Kidney Int 46:1496–1498

Hunyady L, Zhang M, Jagadeesh G, Bor M, Balla T, Catt KJ (1996) Dependence of agonist activation on a conserved apolar residue in the third intracellular loop of the AT$_1$ angiotensin receptor. Proc Natl Acad Sci USA 93:10040–10045

Hunyady L, Ji H, Jagadeesh G, Zhang M, Gaborik Z, Mihalik B, Catt KJ (1998) Dependence of AT$_1$ angiotensin receptor function on adjacent asparagine residues in the seventh transmembrane helix. Mol Pharmacol 54:427–434

Ishida T, Ishida M, Suero J, Takahashi M, Berk BC (1999) Agonist-stimulated cytoskeletal reorganization and signal transduction at focal adhesions in vascular smooth muscle cells require c-Src. J Clin Invest 103:789–797

Israel A, Stromberg C, Tsutsumi K, del Rosario Garrido M, Torres M, Saavedra JM (1995) Angiotensin II receptor subtypes and phosphoinositide hydrolysis in rat adrenal medulla. Brain Res Bull 38:441–446

Kai H, Fukui T, Lassegue B, Shah A, Minieri CA, Griendling KK (1996) Prolonged exposure to agonist results in a reduction in the levels of the Gq/G11 alpha subunits in cultured vascular smooth muscle cells. Mol Pharmacol 49:96–104

Kalmes A, Daum G, Clowes AW (2001) EGFR transactivation in the regulation of SMC function. Ann NY Acad Sci 947:42–54; discussion 54–55

Kim S, Iwao H (2000) Molecular and cellular mechanisms of angiotensin II-mediated cardiovascular and renal diseases. Pharmacol Rev 52:11–34

Klahr S (2001) Progression of chronic renal disease. Heart Dis 3:205–209

Laporte SA, Roy SF, Escher E, Guillemette G, Leduc R (1998) Essential role of leucine^{222} in mediating signal transduction of the human angiotensin II type 1 receptor. Recept Channels 5:103–112

Laporte SA, Servant G, Richard DE, Escher E, Guillemette G, Leduc R (1996) The tyrosine within the NPXnY motif of the human angiotensin II type 1 receptor is involved in mediating signal transduction but is not essential for internalization. Mol Pharmacol 49:89–95

Laufs U, Kilter H, Konkol C, Wassmann S, Bohm M, Nickenig G (2002) Impact of HMG CoA reductase inhibition on small GTPases in the heart. Cardiovasc Res 53:911–920

Leduc I, Meloche S (1995) Angiotensin II stimulates tyrosine phosphorylation of the focal adhesion-associated protein paxillin in aortic smooth muscle cells. J Biol Chem 270:4401–4404

Macrez-Lepretre N, Kalkbrenner F, Morel JL, Schultz G, Mironneau J (1997) G protein heterotrimer G$\alpha_{13}\beta_1\gamma_3$ couples the angiotensin AT$_{1A}$ receptor to increases in cytoplasmic Ca^{2+} in rat portal vein myocytes. J Biol Chem 272:10095–10102

Marie J, Maigret B, Joseph M-P, Larguier R, Nouett S, Lombard C, Bonnafous JC (1994) Tyr^{292} in the seventh transmembrane domain of the AT$_{1A}$ angiotensin II receptor is essential for its coupling to phospholipase C. J Biol Chem 269:20815–20818

Marrero MB, Paxton WG, Duff JL, Berk BC, Bernstein KE (1994) Angiotensin II stimulates tyrosine phosphorylation of phospholipase C-γ1 in vascular smooth muscle cells. J Biol Chem 269:10935–10939

Marrero MB, Paxton WG, Schieffer B, Ling BN, Bernstein KE (1996) Angiotensin II signalling events mediated by tyrosine phosphorylation. Cell Signal 8:21–26

Marrero MB, Schieffer B, Li B, Sun J, Harp JB, Ling BN (1997) Role of Janus kinase/signal transducer and activator of transcription and mitogen-activated protein kinase cascades in angiotensin II- and platelet-derived growth factor-induced vascular smooth muscle cell proliferation. J Biol Chem 272:24684–24690

Marrero MB, Venema VJ, Ju H, Eaton DC, Venema RC (1998) Regulation of angiotensin II-induced JAK2 tyrosine phosphorylation: roles of SHP-1 and SHP-2. Am J Physiol 275:C1216–C1223

Mascareno E, Siddiqui MA (2000) The role of Jak/STAT signaling in heart tissue renin-angiotensin system. Mol Cell Biochem 212:171–175

Meloche S, Pelletier S, Servant MJ (2000) Functional cross-talk between the cyclic AMP and Jak/STAT signaling pathways in vascular smooth muscle cells. Mol Cell Biochem 212:99–109

Ohnishi J, Ishido M, Shibata T, Inagami T, Murakami K, Miyazaki H (1992) The rat angiotensin II AT$_{1A}$ receptor couples with three different signal transduction pathways. Biochem Biophys Res Commun 186:1094–1101

Ohyama K, Yamano Y, Sano T, Nakagomi Y, Hamakubo T, Morishima I, Inagami T (1995) Disulfide bridges in extracellular domains of angiotensin II receptor type IA. Regul Pept 57:141–147

Okamoto T, Nishimoto I (1992) Detection of G protein-activator regions in M4 subtype muscarinic, cholinergic, and α2-adrenergic receptors based upon characteristics in primary structure. J Biol Chem 267:8342–8346

Okamoto T, Katada T, Murayama Y, Ui M, Ogata E, Nishimoto I (1990) A simple structure encodes G protein-activating function of the IGF- II/mannose 6-phosphate receptor. Cell 62:709–717

Oppi C, Wagner T, Crisari A, Camerini B, Tocchini Valentini GP (1992) Attenuation of GTPase activity of recombinant G(o) alpha by peptides representing sequence permutations of mastoparan. Proc Natl Acad Sci USA 89:8268–8272

Park JK, Lee SO, Kim YG, Kim SH, Koh GY, Cho KW (2002) Role of rho-kinase activity in angiotensin II-induced contraction of rabbit clitoral cavernosum smooth muscle. Int J Impot Res 14:472–477

Peach MJ, Dostal DE (1990) The angiotensin II receptor and the actions of angiotensin II. J Cardiovasc Pharmacol 16:S25–S30

Ren Q, Kurose H, Lefkowitz RJ, Cotecchia S (1993) Constitutively active mutants of the α2-adrenergic receptor [published erratum appears in J Biol Chem 1994 Jan 14;269(2):1566]. J Biol Chem 268:16483–16487

Robinson MJ, Cobb MH (1997) Mitogen-activated protein kinase pathways. Curr Opin Cell Biol 9:180–186

Rocic P, Lucchesi PA (2001) Down-regulation by antisense oligonucleotides establishes a role for the proline-rich tyrosine kinase PYK2 in angiotensin II-induced signaling in vascular smooth muscle. J Biol Chem 276:21902–21906

Roth BL, Shoham M, Choudhary MS, Khan N (1997) Identification of conserved aromatic residues essential for agonist binding and second messenger production at 5-hydroxytryptamine$_{2A}$ receptors. Mol Pharmacol 52:259–266

Sabri A, Govindarajan G, Griffin TM, Byron KL, Samarel AM, Lucchesi PA (1998) Calcium- and protein kinase C-dependent activation of tyrosine kinase PYK2 by angiotensin II in vascular smooth muscle. Circ Res 83:841–851

Sadoshima J, Montagne O, Wang Q, Yang G, Warden J, Liu J, Takagi G, Karoor V, Hong C, Johnson GL, Vatner DE, Vatner SF (2002) The MEKK1-JNK pathway plays a protective role in pressure overload but does not mediate cardiac hypertrophy. J Clin Invest 110:271–279

Saito Y, Berk BC (2001) Angiotensin II-mediated signal transduction pathways. Curr Hypertens Rep 4:167–171

Sano T, Ohyama K, Yamano Y, Nakagomi Y, Nakazawa S, Kikyo M, Shirai H, Blank JS, Exton JH, Inagami T (1997) A domain for G protein coupling in carboxyl-terminal tail of rat angiotensin II receptor type 1A. J Biol Chem 272:23631–23636

Sayeski PP, Ali MS, Harp JB, Marrero MB, Bernstein KE (1998) Phosphorylation of p130Cas by angiotensin II is dependent on c-Src, intracellular Ca^{2+}, and protein kinase C. Circ Res 82:1279–1288

Schelling JR, Nkemere N, Konieczkowski M, Martin KA, Dubyak GR (1997) Angiotensin II activates the $\beta 1$ isoform of phospholipase C in vascular smooth muscle cells. Am J Physiol 272:C1558–1566

Schiffrin EL (2002) Beyond blood pressure: the endothelium and atherosclerosis progression. Am J Hypertens 15:115S–122S

Schmitz U, Ishida T, Ishida M, Surapisitchat J, Hasham MI, Pelech S, Berk BC (1998) Angiotensin II stimulates p21-activated kinase in vascular smooth muscle cells: role in activation of JNK. Circ Res 82:1272–1278

Schnee JM, Hsueh WA (2000) Angiotensin II, adhesion, and cardiac fibrosis. Cardiovasc Res 46:264–268

Selbie LA, Hill SJ (1998) G protein-coupled-receptor cross-talk: the fine-tuning of multiple receptor-signalling pathways. Trends Pharmacol Sci 19:87–93

Shibata T, Suzuki C, Ohnishi J, Murakami K, Miyazaki H (1996) Identification of regions in the human angiotensin II receptor type 1 responsible for Gi and Gq coupling by mutagenesis study. Biochem Biophys Res Commun 218:383–389

Sharov VG, Todor A, Suzuki G, Morita H, Tanhehco EJ, Sabbah HN (2003) Hypoxia, angiotensin-II, and norepinephrine mediated apoptosis is stimulus specific in canine failed cardiomyocytes: a role for p38 MAPK, Fas-L and cyclin D1. Eur J Heart Fail 5:121–129

Shimokawa H (2002) Rho-kinase as a novel therapeutic target in treatment of cardiovascular diseases. J Cardiovasc Pharmacol 39:319–327

Shirai H, Takahashi K, Katada T, Inagami T (1995) Mapping of G protein coupling sites of the angiotensin II type 1 receptor. Hypertension 25:726–730

Takahashi T, Kawahara Y, Taniguchi T, Yokoyama M (1998) Tyrosine phosphorylation and association of p130Cas and c-Crk II by ANG II in vascular smooth muscle cells. Am J Physiol 274:H1059–H1065

Tang H, Zhao ZJ, Landon EJ, Inagami T (2000) Regulation of calcium-sensitive tyrosine kinase Pyk2 by angiotensin II in endothelial cells. Roles of Yes tyrosine kinase and tyrosine phosphatase SHP-2. J Biol Chem 275:8389–8396

Taylor JM, Mack CP, Nolan K, Regan CP, Owens GK, Parsons JT (2001) Selective expression of an endogenous inhibitor of FAX regulates proliferation and migration of vascular smooth muscle cells. Mol Cell Biol 21:1565–1572

Teutsch B, Bihoreau C, Monnot C, Bernstein KE, Murphy TJ, Alexander RW, Corvol P, Clauser E (1992) A recombinant rat vascular AT_1 receptor confers growth properties to angiotensin II in Chinese hamster ovary cells. Biochem Biophys Res Commun 187:1381–1388

Thomas WG (1999) Regulation of angiotensin II type 1 (AT_1) receptor function. Regul Pept 79:9–23

Thomas WG, Thekkumkara TJ, Baker KM (1996) Cardiac effects of AII. AT_{1A} receptor signaling, desensitization, and internalization. Adv Exp Med Biol 396:59–69

Touyz RM, Schiffrin EL (2000) Signal transduction mechanisms mediating the physiological and pathophysiological actions of angiotensin II in vascular smooth muscle cells. Pharmacol Rev 52:639–672

Touyz RM, Schiffrin EL (2001) Increased generation of superoxide by angiotensin II in smooth muscle cells from resistance arteries of hypertensive patients: role of phospholipase D-dependent NAD(P)H oxidase-sensitive pathways. J Hypertens 19:1245–1254

Touyz RM, Berry C (2002) Recent advances in angiotensin II signaling. Braz J Med Biol Res 35:1001–10015

Unger T (2002) The role of the renin-angiotensin system in the development of cardiovascular disease. Am J Cardiol 89:3A–9A; discussion 10A

Ushio-Fukai M, Griendling KK, Akers M, Lyons PR, Alexander RW (1998) Temporal dispersion of activation of phospholipase C-$\beta 1$ and -γ isoforms by angiotensin II in vascular smooth muscle cells. Role of $\alpha q/11$, α_{12}, and $\beta \gamma$ G protein subunits. J Biol Chem 273:19772–119777

Vallotton MB (1987) The renin-angiotensin system. Trends Pharmacol Sci 8:69–74

Venema RC, Ju H, Venema VJ, Schieffer B, Harp JB, Ling BN, Eaton DC, Marrero MB (1998) Angiotensin II-induced association of phospholipase C$\gamma 1$ with the G-protein-coupled AT_1 receptor. J Biol Chem 273:7703–7708

Wang C, Jayadev S, Escobedo JA (1995) Identification of a domain in the angiotensin II type 1 receptor determining Gq coupling by the use of receptor chimeras. J Biol Chem 270:16677–16682

Wess J (1998) Molecular basis of receptor/G-protein-coupling selectivity. Pharmacol Ther 80:231–264

Xi X-P, Graf K, Goetze S, Fleck E, Hsueh WA, Law RE (1999) Central role of the MAPK pathway in Ang II-mediated DNA synthesis and migration in vascular smooth muscle cells. Arterioscler Thromb Vasc Biol 19:73–82

Yamazaki T, Komuro I, Shiojima I, Yazaki Y (1999) The molecular mechanism of cardiac hypertrophy and failure. Ann NY Acad Sci 874:38–48

Zhang M, Zhao X, Chen HC, Catt KJ, Hunyady L (2000) Activation of the AT_1 angiotensin receptor is dependent on adjacent apolar residues in the carboxyl terminus of the third cytoplasmic loop. J Biol Chem 275:15782–15788

Sympathetic Interactions of AT₁ Receptors

J. C. Balt · M. Pfaffendorf

Department Pharmacotherapy, Academic Medical Center,
University of Amsterdam, Meibergdreef 15, 1105 AZ, Amsterdam, The Netherlands
e-mail: jcbalt@hotmail.com

1	Interactions Between the Sympathetic Nervous System and the Renin–Angiotensin System	352
1.1	Influence of the Sympathetic Nervous System on the Renin–Angiotensin System	352
1.2	Influence of the Renin–Angiotensin System on the Sympathetic Nervous System	353
1.2.1	Central Nervous System	353
1.2.2	Adrenal Medulla	355
1.2.3	Sympathetic Ganglia	355
1.2.4	Sympathetic Nerve Terminals	355
2	Prejunctional and Postjuntional AT₁ Blockade	358
3	Interactions Between the Sympathetic Nervous System and the Renin–Angiotensin System in Hypertension	362
4	Interactions Between the Sympathetic Nervous System and the Renin–Angiotensin System in Congestive Heart Failure	364
5	Conclusions	366
	References	366

Abstract The present chapter deals with the interaction between the renin–angiotensin system (RAS) and the sympathetic nervous system (SNS). Numerous studies have shown that (locally produced) angiotensin II (Ang II), the main effector of the RAS, can enhance sympathetic nervous transmission. Ang II has been shown to enhance ganglionic transmission, to facilitate NA release from synaptic nerve terminals, block NA uptake, enhance NA synthesis, and enhance the postsynaptic effects of noradrenaline. Angiotensin-converting enzyme (ACE) inhibitors and AT₁ receptor blockers have been shown to attenuate the facilitating effect of Ang II on sympathetic neurotransmission. It is generally believed that Ang II exerts its effects on sympathetic neurotransmission through the activation of AT₁ receptors. Several new, non-peptidergic AT₁-selective receptor blockers are now available. The potency of these drugs with respect to the attenuation of the facilitation by Ang II of sympathetic neurotransmission has been the subject in several studies, discussed in the present chapter. In some studies, the order of potency regarding this sympatho-inhibition by AT_1 blockers differs from the order of potency regarding inhibition of the direct pressor effect of Ang II. These findings suggest differences in affinity of the AT_1 blockers for the pre- and postsynaptic AT_1 receptor. Sympatho-inhibition is a class effect of the AT_1 receptor antagonists. In conditions in which the sympathetic nervous system plays a pathophysiological role, such as hypertension and congestive heart failure, this property may well be of therapeutic relevance.

Keywords Angiotensin II · ATRBs · Prejunctional AT_1 receptor blockade · Sympathetic nervous system · Hypertension · Heart failure

1
Interactions Between the Sympathetic Nervous System and the Renin–Angiotensin System

For several decades it has been well known that the renin–angiotensin system (RAS) and the sympathetic nervous system (SNS) can interact at various levels (van Zwieten and de Jonge 1986; Story and Ziogas 1987; Reid 1992; Saxena 1992). The recognition of the deleterious effects of the activation of both the SNS and the RAS in hypertension and heart failure has generated renewed interest in the reciprocal positive feedback loop between these two regulatory systems.

1.1
Influence of the Sympathetic Nervous System on the Renin–Angiotensin System

It is well known that, via the activation of β_1-adrenoceptors in the juxtaglomerular apparatus, noradrenaline (NA) can provoke the release of renin from the kidney (Taher et al. 1976). Additionally, at the level of the local or tissue RAS in blood vessels, both renin and angiotensin II (Ang II) can be released through a β-adrenoceptor-mediated mechanism. In the rat isolated perfused mesenteric artery, the β-adrenoceptor agonist isoproterenol induced an enhancement of pressor responses to nerve stimulation (Fig. 1), as well as release of Ang II. These effects could be suppressed by propranolol, captopril, and the peptide AT receptor antagonist [Sar^1-Ile^8]Ang II (Nakamaru et al. 1986), indicating a β-adrenoceptor stimulation-sensitive local formation of Ang II, which in turn enhances the neuronal NA release and/or the postsynaptic signal transduction of this neurotransmitter. This facilatory effect of β-adrenoceptor stimulation on the activation of systemic or local RAS might play a role in the pathology of cardiovascular diseases, since in spontaneously hypertensive rats, β-adrenoceptor-mediated activation of the local vascular RAS is enhanced compared to normotensive WKY rats (Kawasaki et al. 1984). Besides animal data, there is sufficient evidence that the interaction takes place in the human situation as well, as shown in the human saphenous vein by Mölderings et al., who demonstrated that saralasin attenuated the facilitatory effect on stimulation-evoked NA release of the β_2-adrenoceptor agonist procaterol (Mölderings et al. 1988). Furthermore, in the human forearm

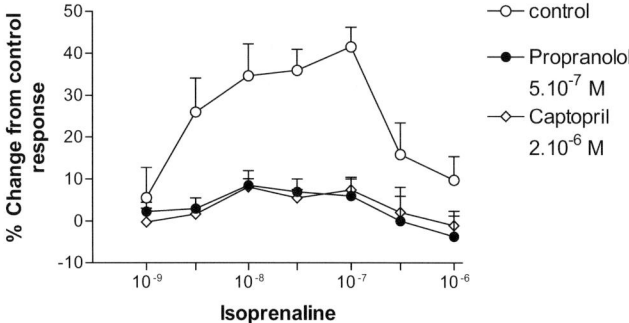

Fig. 1 Facilitation of stimulation-induced pressor responses by β-adrenoceptor stimulation in the rat perfused mesenteric artery. Both β-blockade by propanolol and ACE-inhibition by captopril inhibit the effects of isoprenaline. (From Nakamaru et al. 1986)

of hypertensive patients, isoproterenol-mediated NA overflow was significantly reduced by captopril infusion (Taddei et al. 1991). In conclusion, in a variety of experimental set-ups including human models, a facilitatory, β-adrenoceptor-dependent effect of the SNS on the systemic and the local RAS has been demonstrated which most likely contributes to the pathology of cardiovascular diseases.

1.2
Influence of the Renin–Angiotensin System on the Sympathetic Nervous System

On the other hand, Ang II exerts a marked influence on the sympathetic nervous system. This has been shown on the level of the central nervous system, the adrenal medulla, and the sympathetic ganglia. However, by far the most evidence for the RAS/SNS interaction has been gathered at the level of peripheral sympathetic nerve terminals, both, pre- and postsynaptically (Fig. 2). Functional tissue responses as well as the quantification of released catecholamines have been used to identify and characterize this important interplay and the components involved in its effectuation.

1.2.1
Central Nervous System

In a dog cross circulation study, it was shown that Ang II, infused in the carotid inflow of the cranial circulation, which was vascularly isolated but neuronally connected with its trunk, caused an increase in blood pressure (BP) in both the donor and in the recipient animal (Bickerton and Buckley 1961), pointing to an Ang II-sensitive, central activation of the sympathetic nervous system. It is now becoming clear that a functional RAS is intrinsic to the brain. Central administration of Ang II, in concentrations that are ineffective when administered peripherally, causes an increase in BP and sympathetic outflow as well as a suppression of baroreflex control (Reid 1992; Raizada et al. 1995). Catecholaminer-

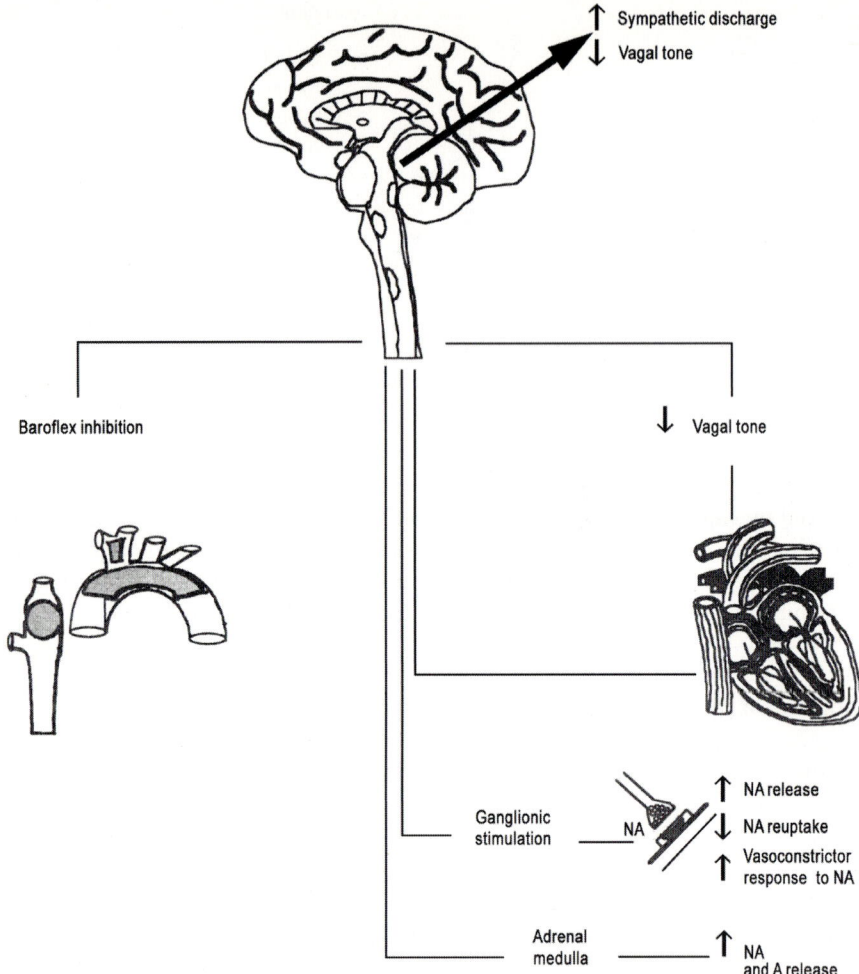

Fig. 2 Sites and mechanisms of interaction between the renin–angiotensin system and the sympathetic nervous system. Abbreviations: *NA*, noradrenaline; *A*, adrenaline. (From Grassi 2001)

gic neurons are involved in the central mechanism by which Ang II causes BP increase, since central effects of Ang II can be inhibited by the α-antagonist phentolamine (Jones 1984). Furthermore, in conscious rats, Ang II causes release of NA from the paraventricular nucleus (PVN), which is involved in BP control (Stadler et al. 1992). As with most of the other Ang II-induced effects, these central actions are mediated by the AT_1 receptor subtype. These receptors are also localized in other areas that are involved in BP control such as the nucleus tractus solitarii and the area postrema, whereas AT_1 receptor blockade can inhibit the effects of centrally applied Ang II (Raizada et al. 1995).

1.2.2
Adrenal Medulla

Angiotensin II stimulates the release of catecholamines from the adrenal glands (Peach 1971). A facilitatory effect of Ang II on stimulation-induced adrenaline release was demonstrated in anaesthetized dogs (Foucart et al. 1991). However, the doses needed are high and the physiological relevance of these observations remains uncertain. In pithed rats, adrenalectomy failed to alter the responses to Ang II (Knape 1986) and an infusion of Ang II did not alter plasma adrenaline levels in humans (Mendelsohn et al. 1980).

1.2.3
Sympathetic Ganglia

Injection of Ang II into the blood supply of the caudal cervical ganglia of dogs causes positive chronotropic and inotropic responses. The direct positive inotropic effect of Ang II needed doses that were 60 times higher (Farr and Grupp 1971). In pithed rats, this was shown to be an AT_1 receptor-mediated mechanism (Wong et al. 1990a). Binding sites for Ang II have been identified in rat sympathetic ganglia (Castren et al. 1987). Propranolol was demonstrated to inhibit the tachycardic response to Ang II, suggesting a role of the β-receptor (Finch and Leach 1969; Knape and van Zwieten 1988). Activation of post-ganglionic renal sympathetic nerves by Ang II, mediated by AT_1 receptors, was demonstrated in anaesthetized mice (Ma et al. 2001) (Fig. 3).

1.2.4
Sympathetic Nerve Terminals

In various models, the facilitating effect of Ang II on sympathetic neurotransmission at the peripheral level was demonstrated. This facilitation has been shown to be caused by presynaptic mechanisms, such as an increase of NA synthesis (Roth 1972), an increase of stimulation-induced release of NA (Starke et al. 1977), and the inhibition of its uptake (Panisset and Bourdois 1968; Raasch et al. 2001). The enhancing effect of Ang II on stimulation-induced NA release and vasoconstrictor responses could be antagonized by the peptidergic AT_1/AT_2 receptor antagonist saralasin (Endo et al. 1977; Collis and Keddie 1981) and by various selective AT_1 receptor antagonists such as losartan (Tofovic et al. 1991; Wong et al. 1992; Cox et al. 1996), irbesartan (Christophe et al. 1995), and eprosartan (Ohlstein et al. 1997), but not by the AT_2 receptor antagonists PD123177 or PD123319 (Tofovic et al. 1991; Wong et al. 1992; Brasch et al. 1993). From these experiments, it is generally concluded that Ang II exerts its effects on sympathetic neurotransmission through the activation of AT_1 receptors (Fig. 4).

In addition, an increased responsiveness of vascular smooth muscle to NA (postsynaptic facilitation) has been demonstrated (Purdy and Weber 1988).

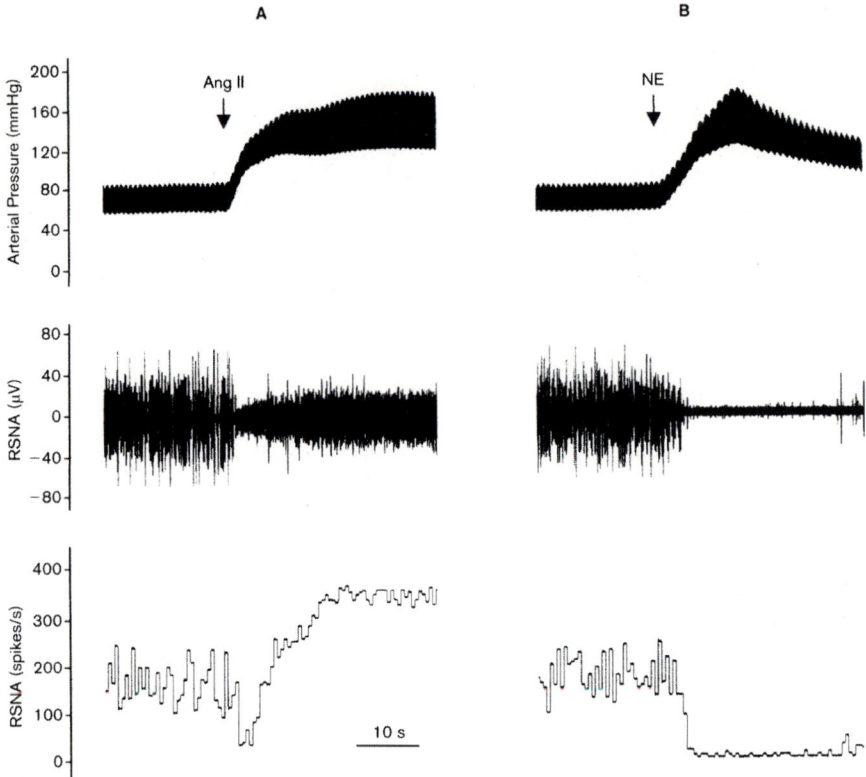

Fig. 3A, B Responses of arterial pressure and renal sympathetic nerve activity (*RSNA*) to a bolus injection of Ang II (4 ng/g, i.v.) in an anaesthetized mouse. **A** Responses to angiotensin II; **B** Responses to noradrenaline. After an initial inhibition of RNSA, Ang II persistently increases RSNA. An equipotent dose of noradrenaline abolished RSNA. AT_1 blockade by losartan inhibited the Ang II-induced increase in RSNA (not shown). From these data, it is concluded that the BP increase by noradrenaline causes a baroreflex-mediated inhibition of sympathetic outflow, while Ang II—through AT_1-receptors—increases RSNA. (From Ma et al. 2001)

Both the α_1- and the α_2-adrenoceptor have been shown to be involved (de Jonge et al. 1982; Richer et al. 1996). In some studies, however, such a postsynaptic facilitation could not be confirmed (Duckles 1981; Seidelin et al. 1991; Moreau et al. 1993a; Yokoyama et al. 1997). It is generally believed that the presynaptic facilitation is quantitatively more important than the postsynaptic enhancement.

Locally produced Ang II can enhance noradrenergic neurotransmission. This has been demonstrated in the guinea pig isolated atria and the rat caudal artery (Ziogas et al. 1984), isolated rat mesenteric arteries (Malik and Nasjletti 1976), kidneys (Boke and Malik 1983), and perfused hearts of rats, guinea pigs, and rabbits (Xiang et al. 1985). Such a local interaction was also demonstrated in the human forearm (Taddei et al. 1991). These studies indicate a significant involvement of the local RAS in the facilitation of sympathetic neurotransmission.

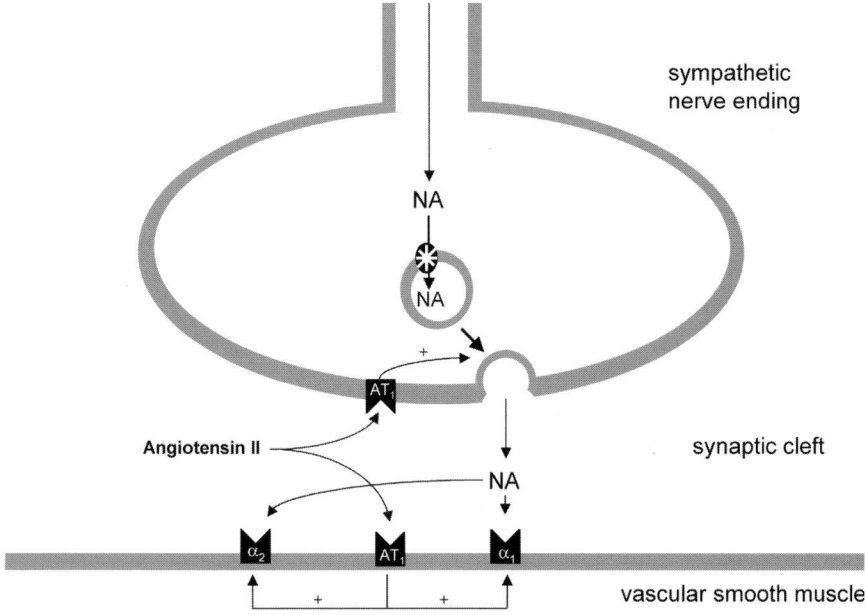

Fig. 4 Facilitatory actions of Ang II at the sympathetic nerve terminal and vascular smooth muscle cell. *NA*, noradrenaline

In the human forearm, Clemson et al. found that intra-arterial Ang II caused an increase of NA spillover (Clemson et al. 1994). Other authors, however, could not confirm these results (Chang et al. 1995; Goldsmith et al. 1998). In healthy volunteers, Ang II was shown to augment vasoconstriction as elicited by lower body negative pressure (LBNP) (Seidelin et al. 1991). Conversely, chronic angiotensin-converting enzyme (ACE) inhibition had already been demonstrated to diminish the decrements in forearm blood flow caused by LBNP (Morganti et al. 1989)

The mechanism by which Ang II facilitates sympathetic neurotransmission has been sparsely investigated. Ang II, via the AT_1 receptor, activates receptor-linked phospholipase C (PLC), resulting in the hydrolysis of phosphatidyl-inositol-4,5-biphosphate (PIP_2), producing inositol-1,4,5-triphosphate (IP_3) and diacylglycerol (DAG). IP_3 releases Ca^{2+} from the sarcoplasmic reticulum (SR), but this mechanism appears to have little effect on neurotransmission (Majewski and Musgrave 1995). DAG however, activates protein kinase C (PKC), which through a number of mechanisms plays an important part in transmitter release: (1) by prolonging Ca^{2+} influx, (2) by modulating NA synthesis (through phosphorylation of tyrosine hydroxylase), and (3) by modulation of exocytosis (through phosphorylation of exocytotic proteins) (Verhage et al. 1994).

At the postsynaptic level, the PKC pathway was shown to be involved in the facilitation by Ang II of α-adrenoceptor-mediated responses (Purdy and Weber 1988; Laher et al. 1990).

2
Prejunctional and Postjuntional AT_1 Blockade

Several new, non-peptidergic AT_1-selective receptor blockers are now available. A few studies have recently been published in which the potency of these drugs with respect to the attenuation of the facilitation by Ang II of sympathetic neurotransmission was compared (Ohlstein et al. 1997; Shetty and DelGrande 2000; Balt et al. 2001a,b).

Ohlstein et al. reported that, in the pithed rat model, eprosartan (0.3 mg/kg) attenuated stimulation-induced BP responses, whereas valsartan, losartan, and irbesartan did not (Ohlstein et al. 1997). In earlier reports, higher doses of losartan and irbesartan had already been shown to be effective in blocking the facilitating effect of Ang II on noradrenergic neurotransmission in this model (Wong et al. 1992; Christophe et al. 1995). In the isolated rat mesenteric arteries, Balt et al. found the ranking order of the inhibition the facilitating effects of Ang II on stimulation-induced contractile responses to be telmisartan>irbesartan>losar-

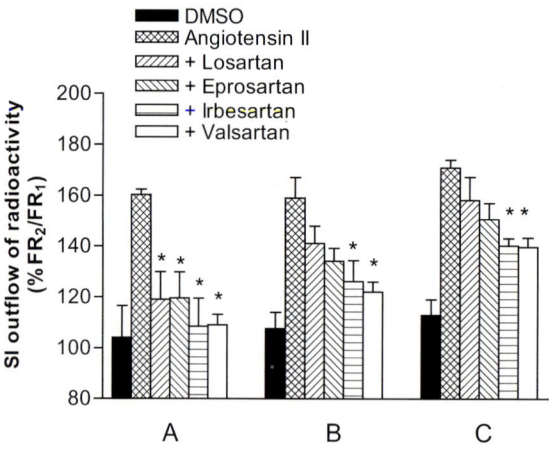

Fig. 5 Effects of losartan, eprosartan, irbesartan, and valsartan (A, 10^{-6} M; B, 10^{-7} M; C, 10^{-8} M) on Ang II (10^{-7} M)-evoked augmentation-induced efflux of radioactive [3H]NA from isolated rat atria. $*p< 0.05$ compared with control Ang II response. (From Shetty and DelGrande 2000)

Fig. 6A–C Effect of AT_1 receptor blockade by irbesartan on: **A** the frequency-response curve induced by electrical stimulation of the spinal cord (T5-L4) $*p<0.05$ compared to control; **B** the dose–response curve induced by intravenously administered noradrenaline; **C** the dose–response curve induced by intravenously administered Ang II in the pithed normotensive rat. From these data it can be concluded that (1) endogenously generated Ang II can enhance noradrenergic neurotransmission in this model and (2) the facilitation by Ang II is mediated by prejunctional AT_1 receptors. In this model, pre- and postjunctional AT_1 blockade can be compared. See also Fig. 7

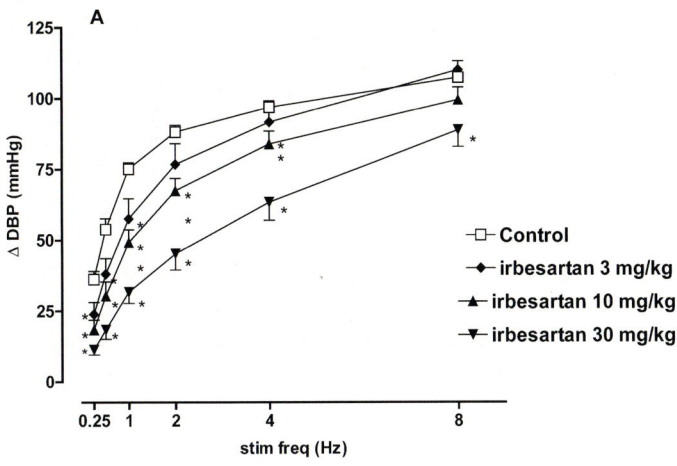
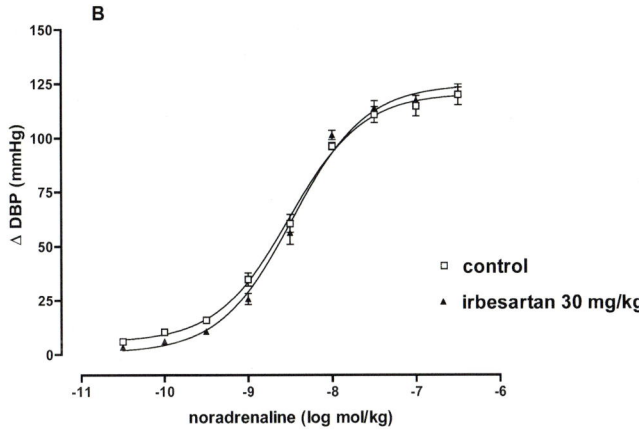
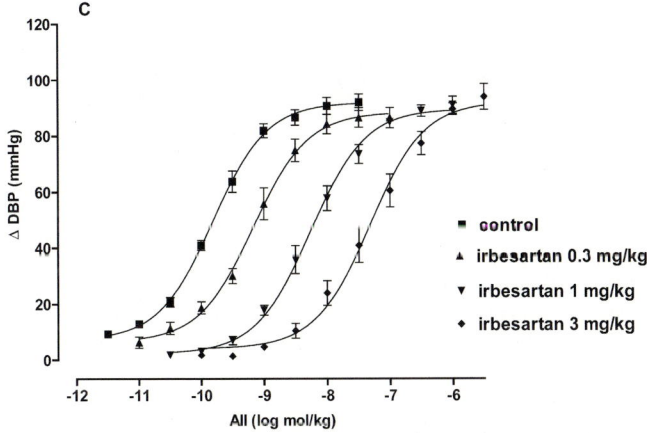

tan (Balt et al. 2001a). Nap et al. found telmisartan more potent than losartan and irbesartan regarding sympatho-inhibition in a model in which stimulation-induced NA release from rabbit aortic rings was measured (Nap et al. 2002). In isolated rat atria, in which stimulation-induced NA release was measured, Shetty and DelGrande reported a ranking order of valsartan=irbesartan>losartan=eprosartan (Shetty and DelGrande 2000) (Fig. 5). So, in contrast to Ohlstein's findings, in Shetty and DelGrande's model eprosartan was not significantly more potent than losartan, regarding sympatho-inhibition.

Differences in subtype between prejunctional AT_1 receptors (on sympathetic nerve terminals) and postsynaptic AT_1 receptors (on vascular smooth muscle cells) have been suggested earlier by Guimarães et al. (2001). In the canine isolated mesenteric artery, these authors saw no effect of losartan on Ang II-induced facilitation in doses which clearly inhibited the postsynaptic receptor (Guimarães et al. 2001). Similar findings were published by Balt et al. (2001). In pithed rat, these authors investigated the effects of a number of AT_1 receptor antagonists on (1) stimulation-induced responses, (2) responses to exogenous NA, and (3) responses to exogenous Ang II (Fig. 6A–C, irbesartan as an example). They found that the ranking order of the AT_1 receptor antagonists regarding blockade of the prejunctional AT_1 receptor differed from the ranking order regarding blockade of direct vasoconstrictor effects of Ang II (Balt et al. 2001b). The ratios between ED_{20} values (as a measure of presynaptic inhibitory dose) and A_2 values (calculated by taking the exponential of the pA_2 values, a measure of the postsynaptic inhibitory dose) differ considerably between the various antagonists (Fig. 7). These findings were confirmed in an in vitro set up: in the rabbit isolated mesenteric artery, eprosartan exhibits sympathoinhibition in

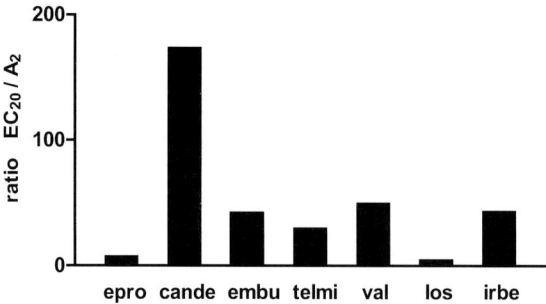

Fig. 7 Ratio between the ED_{20} values (obtained from stimulation experiments; reflect sympatho-inhibitory potency) and A_2 values (obtained from Ang II experiments; reflect inhibitory potency regarding direct effects of Ang II on the vasculature). ED_{20}: dose which at 2 Hz reduced ΔDBP by 20 mmHg. A_2 values were calculated by taking the exponential of the pA_2 values. Note the substantial differences between the AT_1 blockers studied. The presynaptic effect related with impairment of NA release occurs at relatively low doses for eprosartan and losartan when compared with postsynaptic inhibition of vasoconstriction by Ang II. In contrast, there is a substantial difference between these two dosages for candesartan

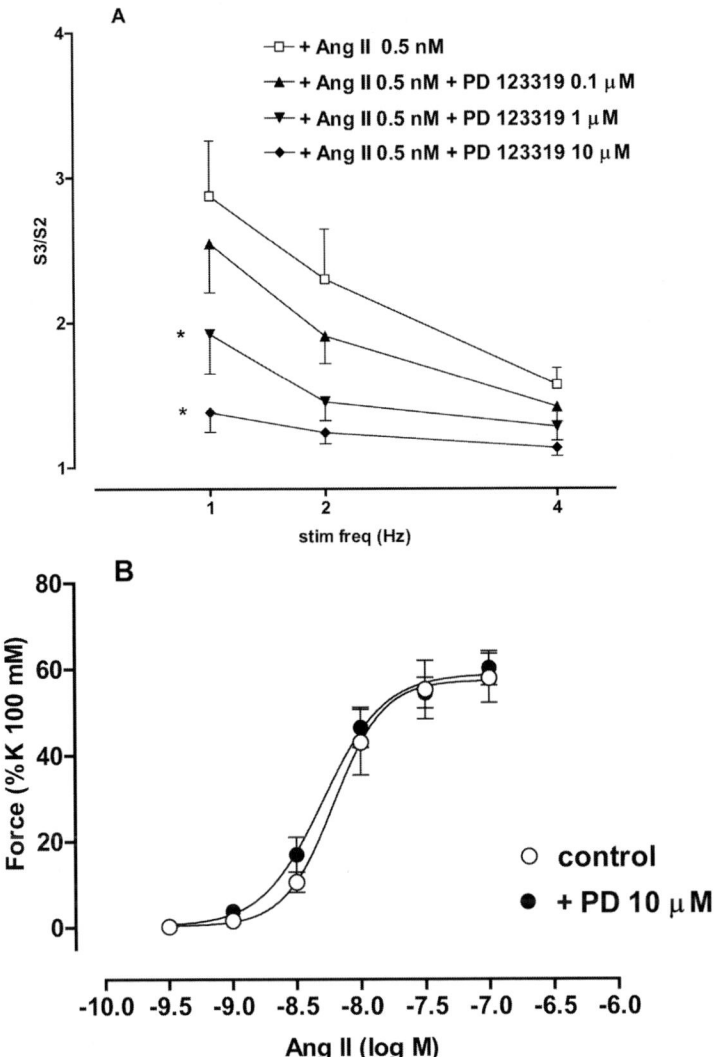

Fig. 8 A Inhibitory effects of PD123319 (0.1–10 μM) on the facilitation by Ang II of stimulation-induced contractions. Ang II (0.5 nM) in the presence or absence of PD123319 was added to the organ bath 2 min prior to the third electrical field stimulation (S3). The ratio between forces induced by S2 and S3 (S3/S2) is shown at the ordinate and stimulation frequencies at the abscissa. Values are given as mean±SEM. *$p<0.05$ at each stimulation frequency compared to responses in presence of Ang II (0.5 nM). **B** PD123319 (10 μM) had no effect on the cumulative concentration–response curve for Ang II in the isolated rabbit mesenteric artery. Values are shown as mean±SEM

concentrations that inhibit AT_1 receptor-mediated vasoconstriction, whereas candesartan does not (Balt et al. 2002).

However, in another study, performed in the pithed rat, the assumption that prejunctional neuronal and postjunctional vascular AT_1 receptors differ could not be confirmed (Dendorfer et al. 2002). In a set-up in which inhibition of Ang II-induced NA release was measured on the one hand and inhibition of acute increases in BP increase on the other, the order of potency for both pre- and postjunctional inhibitory effects was candesartan>eprosartan>EXP 3174 (the active metabolite of losartan)>irbesartan. Although measuring catecholamine release is a more direct means of sympathetic activity compared to measuring contractile force or BP increase, in that study, the investigators relied on a "tyramine"-like effect of Ang II, whereas in other studies a facilitating role of Ang II on electrically stimulation-induced NA release was investigated.

In the isolated rat caudal artery, the facilitation by Ang II of stimulation-induced contractions as well as NA release could be inhibited by PD123319 (0.1 μM), which at this concentration is assumed to block the AT_{1B}-sybtype (Cox et al. 1996). These findings were confirmed in the isolated rabbit mesenteric artery and aorta, in which stimulation-induced contractions and NA efflux was measured. In contrast, PD123319, in concentrations as high as 10 μM, had no effect on direct vasoconstrictor effects of Ang II (Nap et al. 2003; Fig. 8). These findings strongly suggest that the prejunctional AT_1 receptor is of the AT_{1B}-subtype, and the postjunctional AT receptor is not.

However, in humans, the AT_1 receptor is encoded by one gene only, so these findings may not apply to humans. No studies so far have been published in which sympatho-inhibitory properties in humans are compared.

3
Interactions Between the Sympathetic Nervous System and the Renin–Angiotensin System in Hypertension

Both the RAS and sympathetic nervous system have been extensively studied in animal models as well as in human hypertensives. Pharmacological tools targeting various components of both BP regulatory systems have been successfully applied in the treatment of hypertension.

The spontaneously hypertensive rat (SHR) has been used as a model for human essential hypertension. The substantial role of the sympathetic nervous system in the development and maintenance of hypertension in this model is firmly established. Likewise, in human hypertension the etiological role of the sympathetic nervous system has been demonstrated by measurements of regional NA spillover, spectral analysis of heart rate variability, and measurements of sympathetic nerve activity (Esler et al. 1990). The efficacy of ACE inhibitors and AT_1 antagonists in the treatment of hypertension clearly demonstrates the involvement of the RAS in cardiovascular control. In recent years, substantial knowledge has been acquired concerning the role of Ang II in the development of hypertension and hypertensive target organ injury (Weir and Dzau 1999).

The facilitating effect of Ang II on sympathetic neurotransmission in SHR has been shown to be increased compared to their normotensive controls, Wistar-Kyoto (WKY) rats, despite similar or even lower plasma renin activity (PRA) or plasma Ang II levels (Lokhandwala and Eikenburg 1983; Dang et al. 1999). Possibly, an increased density of AT_1 receptors, as reported in several neuronal tissues of the SHR (Raizada et al. 1995) or a decreased neuronal NA uptake (Kawasaki et al. 1982) could be involved in the enhanced facilitation in the SHR. Part of the evidence concerning this issue is derived from experiments with isolated tissues (Collis et al. 1979; Clough et al. 1982; Kawasaki et al. 1982). In the pithed rat model, Clough et al. demonstrated, using an ACE inhibitor, an increased potentiation of sympathetic neurotransmission by Ang II in SHR compared to WKY both at the prejunctional as well as at the postjuntional level (Clough et al. 1982). These authors postulated that such a mechanism may explain the potent hypotensive action of captopril compared to that in the WKY, where captopril does not lower BP. The same was postulated for AT_1 blockade (Jonsson et al. 1993). Indeed, losartan does not lower BP in the WKY (Wong et al. 1990b). In pithed SHR, selective AT_1 receptor antagonists have been shown to attenuate stimulation-induced diastolic (D)BP responses (Häuser et al. 1998; Moreau et al. 1993b,c).

Recently, Raash et al. demonstrated that chronic ACE inhibition increased cardiac NA reuptake in SHR (Raasch et al. 2001) (Fig. 9). Interestingly, in patients with essential hypertension, a reduced presynaptic NA uptake has been described (Rumantir et al. 2000). It is not clear to which extent the sympatho-inhibitory effects of RAS blockade also apply to humans: some studies in hypertensives demonstrated that both ACE-inhibition and AT_1 blockade reduce plasma NA (Weinberger 1982; Moan et al. 1994). In other studies, no such effect was found (Nicholls et al. 1981; Grossman et al. 1994). We already mentioned the

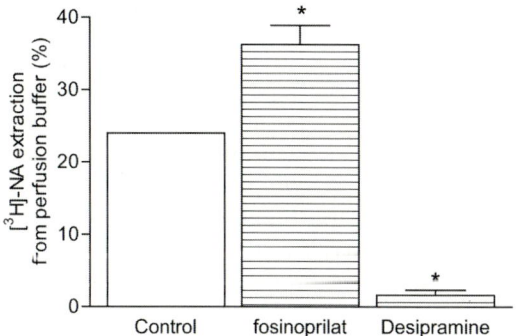

Fig. 9 ACE inhibition by fosinoprilat stimulates uptake of [^{3}H]NA into isolated perfused rat hearts. In contrast, uptake-1 inhibition by desipramine almost abolished [^{3}H]NA-extraction. (From Raasch et al. 2001)

limited value of plasma NA levels. Sakata et al., using an [^{123}I]metaiodobenzylguanidine (MIBG) scintigraphy technique, demonstrated that both chronic ACE-inhibitor therapy as well as AT$_1$ blockade reduced cardiac sympathetic activity (Sakata et al. 2002). In another study, acute losartan treatment blunted the sympathetic stimulatory effect of cold stress in hypertensive subjects on BP and NA concentrations, without changing baseline plasma NA (Rodriguez-Garcia et al. 2000). On the other hand, chronic ACE-inhibition treatment did not significantly affect sympathetic nerve traffic, measured by microneurography, nor plasma NA in hypertensive subjects (Grassi et al. 1998). Struck et al. demonstrated that antihypertensive treatment with amlodipine caused an increase in muscle sympathetic nerve activity, whereas treatment with valsartan did not (Struck et al. 2002). These authors conclude that AT$_1$ blockade not only lowers BP, but also changes the relation between BP and sympathetic outflow, in other words it resets the baroreflex set-point toward lower BP values.

4
Interactions Between the Sympathetic Nervous System and the Renin–Angiotensin System in Congestive Heart Failure

Various forms of congestive heart failure are known to be accompanied by activation of the RAS and the sympathetic nervous system (Mancia 1990). Plasma levels of NA have been shown to be increased and heart rate is usually elevated (Thomas and Marks 1978; Cody et al. 1982). NA spillover from the heart is increased as much as 50-fold in heart failure patients (Esler et al. 1997). In studies in which sympathetic nerve traffic was studied with the microneurography technique, the number of sympathetic bursts per minute was higher in patients with heart failure compared to healthy age-matched subjects (Leimbach et al. 1986). Sympathetic stimulation of renin release results in sodium and water retention, ultimately causing congestion. Indeed, sympatho-excitation, reflected by increased levels of plasma NA, is now seen as a risk factor in various, even asymptomatic stages of heart failure (Rector et al. 1987; Mancia 1990; Benedict et al. 1996).

As early as 1962, the activation of the RAS in heart failure was reported (Davis 1962). In 1978, two groups simultaneously reported that blocking the RAS improves cardiac and systemic haemodynamics (Curtiss et al. 1978; Gavras et al. 1978). It combines afterload reduction with diminished retention of salt and volume. Since then, suppression of the activity/influence of the RAS has shown to have favorable effects in experimental heart failure (Pfeffer et al. 1985; Richer et al. 1999) as well as in clinical trials, such as the CONSENSUS-study. (For review, see Brunner-La Rocca et al. 1999.)

In the treatment of heart failure, β-blockers are now more and more recognized as beneficial drugs (Sharpe 1999). Apart from inhibiting the sympathetic nervous system, these drugs inhibit renin secretion by acting on β-receptors in the juxtaglomerular apparatus in the kidney, and also in the vasculature, as described above. Indeed, in patients with heart failure, treatment with a β-blocker (whether or not in addition to ACE inhibition) reduces Ang II levels (Campbell et al. 2001). Interestingly, in the Val-HEFT trial, a combination of ACE inhibition, AT_1 blockade, and β-blockade, so-called "triple blockade," significantly increased mortality in heart failure patients (Cohn and Tognoni 2001).

In the CONSENSUS trial, ACE-inhibition treatment was associated with a significant reduction of plasma NA levels (Swedberg et al. 1990). In smaller studies, similar findings were reported with AT_1 blockade by some (Gottlieb et al. 1993;

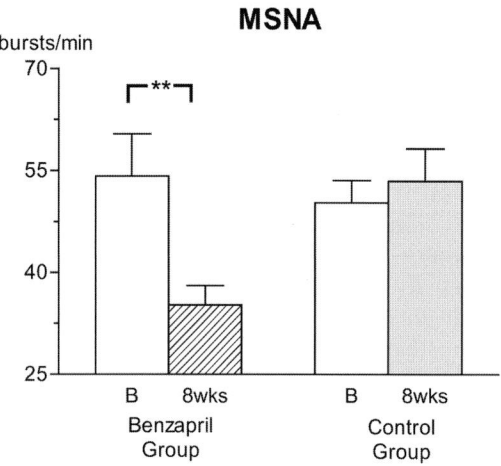

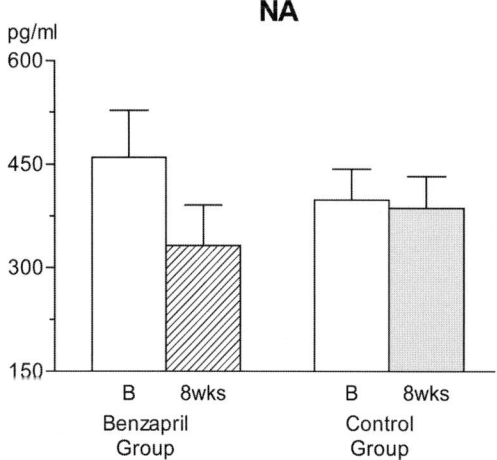

Fig. 10 Bar graphs showing muscle sympathetic nerve activity (MSNA) and plasma NA-values before (*open bars*) and after 8 weeks of benazepril treatment (*hatched bars*) or an 8-week observational period (*shaded bars*) in heart failure treatment. ACE-inhibition significantly reduced MSNA. (From Grassi et al. 1997)

Crozier et al. 1995) but not by all authors (Houghton et al. 1999). In addition, a low dose of an ACE inhibitor was shown to increase parasympathetic activity (Kamen et al. 1997). In patients with heart failure, acute losartan administration increases forearm blood flow by 25%, in contrast to age- and sex-matched controls (Newby et al. 1998). These findings suggest that endogenous Ang II contributes to the maintenance of peripheral resistance. Since there is a sympathetic tone in the forearm under basal conditions, this increased sympathetic tone may be caused by a facilitatory effect of Ang II. In the same study, lower body negative pressure (LBNP) failed to reduce forearm blood flow in heart failure patients, consistent with defective baroreflex control. Additionally, losartan had no effect on the blood flow response to LBNP. Hence, no direct evidence of Ang II-induced facilitation was found. In a pacing-induced heart failure model, chronic AT_1 blockade was shown to reduce renal sympathetic nerve activity, as well as to enhance baroreflex sensitivity (Murakami et al. 1997). In an analogy to this animal model, similar findings were obtained in humans subjected to chronic ACE inhibition (Grassi et al. 1997): in patients with NYHA class II heart failure, treatment with the ACE inhibitor benazepril caused significant reductions in muscle sympathetic nerve activity (see Fig. 10). Plasma NA values were slightly but insignificantly reduced in this study. Additionally, ACE inhibition improved baroreflex restraint on sympathetic drive.

5
Conclusions

In summary, there is an overwhelming body of evidence that the RAS and SNS interact. However, most of the evidence is derived from animal experiments, sometimes far from the clinical setting, using high, unphysiological concentrations of Ang II. Nevertheless, clinical importance of this interaction may follow from the fact that both the RAS and the SNS play a pivotal role in the pathophysiology of hypertension and even more so in heart failure. Consequently, there exists important potential for the positive reciprocal feedback loop between the RAS and the SNS to be involved in the pathophysiology and therapeutic approaches of these conditions.

References

Balt JC, Mathy MJ, Nap A, Pfaffendorf M, van Zwieten PA (2001a) Effect of the AT1-receptor antagonists losartan, irbesartan, and telmisartan on angiotensin II-induced facilitation of sympathetic neurotransmission in the rat mesenteric artery. J Cardiovasc Pharmacol 38:141–148
Balt JC, Mathy MJ, Nap A, Pfaffendorf M, van Zwieten PA (2002) Prejunctional and postjunctional inhibitory actions of eprosartan and candesartan in the isolated rabbit mesenteric artery. J Cardiovasc Pharmacol 40:50–57

Balt JC, Mathy MJ, Pfaffendorf M, van Zwieten PA (2001b) Inhibition of angiotensin II-induced facilitation of sympathetic neurotransmission in the pithed rat: a comparison between losartan, irbesartan, telmisartan, and captopril. J Hypertens 19:465–473

Benedict CR, Shelton B, Johnstone DE, Francis G, Greenberg B, Konstam M, Probstfield JL, Yusuf S (1996) Prognostic significance of plasma norepinephrine in patients with asymptomatic left ventricular dysfunction. SOLVD Investigators. Circulation 94:690–697

Bickerton RK, Buckley JP (1961) Evidence for a central mechanism in angiotensin induced hypertension. Proc Soc Exp Biol 106:834–836

Boke T, Malik KU (1983) Enhancement by locally generated angiotensin II of release of the adrenergic transmitter in the isolated rat kidney. J Pharmacol Exp Ther 226:900–907

Brasch H, Sieroslawski L, Dominiak P (1993) Angiotensin II increases norepinephrine release from atria by acting on angiotensin subtype 1 receptors. Hypertension 22:699–704

Brunner-La Rocca HP, Vaddadi G, Esler MD (1999) Recent insight into therapy of congestive heart failure: focus on ACE inhibition and angiotensin-II antagonism. J Am Coll Cardiol 33:1163–1173

Campbell DJ, Aggarwal A, Esler M, Kaye D (2001) beta-blockers, angiotensin II, and ACE inhibitors in patients with heart failure. Lancet 358:1609–1610

Castren E, Kurihara M, Gutkind JS, Saavedra JM (1987) Specific angiotensin II binding sites in the rat stellate and superior cervical ganglia. Brain Res 422:347–351

Chang PC, Grossman E, Kopin IJ, Goldstein DS, Folio CJ, Holmes C (1995) On the existence of functional angiotensin II receptors on vascular sympathetic nerve terminals in the human forearm. J Hypertens 13:1275–1284

Christophe B, Libon R, Cazaubon C, Nisato D, Manning A, Chatelain P (1995) Effects of irbesartan (SR47436/BMS-186295) on angiotensin II-induced pressor responses in the pithed rat: potential mechanisms of action. Eur J Pharmacol 281:161–171

Clemson B, Gaul L, Gubin SS, Campsey DM, McConville J, Nussberger J, Zelis R (1994) Prejunctional angiotensin II receptors. Facilitation of norepinephrine release in the human forearm. J Clin Invest 93:684–691

Clough DP, Hatton R, Keddie JR, Collis MG (1982) Hypotensive action of captopril in spontaneously hypertensive and normotensive rats. Interference with neurogenic vasoconstriction. Hypertension 4:764–772

Cody RJ, Franklin KW, Kluger J, Laragh JH (1982) Sympathetic responsiveness and plasma norepinephrine during therapy of chronic congestive heart failure with captopril. Am J Med 72:791–797

Cohn JN, Tognoni G (2001) A randomized trial of the angiotensin-receptor blocker valsartan in chronic heart failure. N Engl J Med 345:1667–1675

Collis MG, de May C, Vanhoutte PM (1979) Enhanced release of noradrenaline in the kidney of the young spontaneously hypertensive rat. Clin Sci (Lond) 57 Suppl 5:233s–234s

Collis MG, Keddie JR (1981) Captopril attenuates adrenergic vasoconstriction in rat mesenteric arteries by angiotensin-dependent and -independent mechanisms. Clin Sci 61:281–286

Cox SL, Story DF, Ziogas J (1996) Multiple prejunctional actions of angiotensin II on noradrenergic transmission in the caudal artery of the rat. Br J Pharmacol 119:976–984

Crozier I, Ikram H, Awan N, Cleland J, Stephen N, Dickstein K, Frey M, Young J, Klinger G, Makris L (1995) Losartan in heart failure. Hemodynamic effects and tolerability. Losartan Hemodynamic Study Group. Circulation 91:691–697

Curtiss C, Cohn JN, Vrobel T, Franciosa JA (1978) Role of the renin-angiotensin system in the systemic vasoconstriction of chronic congestive heart failure. Circulation 58:763–770

Dang A, Zheng D, Wang B, Zhang Y, Zhang P, Xu M, Liu G, Liu L (1999) The role of the renin-angiotensin and cardiac sympathetic nervous systems in the development of hypertension and left ventricular hypertrophy in spontaneously hypertensive rats. Hypertens Res 22:217–221

Davis JO (1962) Adrenocortical and renal hormonal function in experimental cardiac failure. Circulation 25:1002–1014

de Jonge A, Knape JT, van Meel JC, Kalkman HO, Wilffert B, Thoolen MJ, Timmermanns PB, van Zwieten PA (1982) Effect of converting enzyme inhibition and angiotensin receptor blockade on the vasoconstriction mediated by alpha 1-and alpha 2- adrenoceptor stimulation in pithed normotensive rats. Naunyn Schmiedebergs Arch Pharmacol 321:309–313

Dendorfer A, Raasch W, Tempel K, Dominiak P (2002) Comparison of the vascular and antiadrenergic activities of four angiotensin II type 1 antagonists in the pithed rat. J Hypertens 20:1151–1156

Duckles SP (1981) Angiotensin II potentiates responses of the rabbit basilar artery to adrenergic nerve stimulation. Life Sci 28:635–640

Endo T, Starke K, Bangerter A, Taube HD (1977) Presynaptic receptor systems on the noradrenergic neurones of the rabbit pulmonary artery. Naunyn Schmiedebergs Arch Pharmacol 296:229–247

Esler M, Kaye D, Lambert G, Esler D, Jennings G (1997) Adrenergic nervous system in heart failure. Am J Cardiol 80:7L-14L

Esler M, Lambert G, Jennings G (1990) Increased regional sympathetic nervous activity in human hypertension: causes and consequences. J Hypertens Suppl 8: S53-S57

Farr WC, Grupp G (1971) Ganglionic stimulation: mechanism of the positive inotropic and chronotropic effects of angiotensin. J Pharmacol Exp Ther 177:48–55

Finch L, Leach DH (1969) The role of the sympathetic nervous system in the cardiovascular responses to angiotensin in the pithed rat. Br J Pharmacol : 481–488

Foucart S, de Champlain J, Nadeau R (1991) Modulation by beta-adrenoceptors and angiotensin II receptors of splanchnic nerve evoked catecholamine release from the adrenal medulla. Can J Physiol Pharmacol 69:1-7

Gavras H, Faxon DP, Berkoben J, Brunner HR, Ryan TJ (1978) Angiotensin converting enzyme inhibition in patients with congestive heart failure. Circulation 58:770–776

Goldsmith SR, Garr M, McLaurin M (1998) Regulation of regional norepinephrine spillover in heart failure: the effect of angiotensin II and beta-adrenergic agonists in the forearm circulation. J Card Fail 4:305–310

Gottlieb SS, Dickstein K, Fleck E, Kostis J, Levine TB, LeJemtel T, DeKock M (1993) Hemodynamic and neurohormonal effects of the angiotensin II antagonist losartan in patients with congestive heart failure. Circulation 88:1602–1609

Grassi G (2001) Renin-angiotensin-sympathetic crosstalks in hypertension: reappraising the relevance of peripheral interactions. J Hypertens 19:1713–1716

Grassi G, Cattaneo BM, Seravalle G, Lanfranchi A, Pozzi M, Morganti A, Carugo S, Mancia G (1997) Effects of chronic ACE inhibition on sympathetic nerve traffic and baroreflex control of circulation in heart failure. Circulation 96:1173–1179

Grassi G, Turri C, Dell'Oro R, Stella ML, Bolla GB, Mancia G (1998) Effect of chronic angiotensin converting enzyme inhibition on sympathetic nerve traffic and baroreflex control of the circulation in essential hypertension. J Hypertens 16:1789–1796

Grossman E, Peleg E, Carroll J, Shamiss A, Rosenthal T (1994) Hemodynamic and humoral effects of the angiotensin II antagonist losartan in essential hypertension. Am J Hypertens 7:1041–1044

Guimarães S, Pinheiro H, Tavares P, Loio A, Moura D (2001) Differential effects of eprosartan and losartan at prejunctional angiotensin II receptors. Naunyn Schmiedebergs Arch Pharmacol 363:509–514

Häuser W, Dendorfer A, Nguyen T, Dominiak P (1998) Effects of the AT1 antagonist HR 720 in comparison to losartan on stimulated sympathetic outflow, blood pressure,

and heart rate in pithed spontaneously hypertensive rats. Kidney Blood Press Res 21:29-35
Houghton AR, Harrison M, Cowley AJ (1999) Haemodynamic, neurohumoral and exercise effects of losartan vs. captopril in chronic heart failure: results of an ELITE trial substudy. Evaluation of Losartan in the Elderly. Eur J Heart Fail 1:385-393
Jones DL (1984) Injections of phentolamine into the anterior hypothalamus-preoptic area of rats blocks both pressor and drinking responses produced by central administration of angiotensin II. Brain Res Bull 13:127-133
Jonsson JR, Smid SD, Frewin DB, Head RJ (1993) Angiotensin II-mediated facilitation of sympathetic neurotransmission in the spontaneously hypertensive rat is not associated with neuronal uptake of the peptide. J Cardiovasc Pharmacol 22:750-753
Kamen PW, Krum H, Tonkin AM (1997) Low-dose but not high-dose captopril increases parasympathetic activity in patients with heart failure. J Cardiovasc Pharmacol 30:7-11
Kawasaki H, Cline WH Jr, Su C (1982) Enhanced angiotensin-mediated facilitation of adrenergic neurotransmission in spontaneously hypertensive rats. J Pharmacol Exp Ther 221:112-116
Kawasaki H, Cline WH, Jr., Su C (1984) Involvement of the vascular renin-angiotensin system in beta adrenergic receptor-mediated facilitation of vascular neurotransmission in spontaneously hypertensive rats. J Pharmacol Exp Ther 231:23-32
Knape JT, van Zwieten PA (1988) Positive chronotropic activity of angiotensin II in the pithed normotensive rat is primarily due to activation of cardiac beta 1- adrenoceptors. Naunyn Schmiedebergs Arch Pharmacol 338:185-190
Knape JTA (1986) Direct and indirect pressor effects of angiotensin II in the pithed normotensive rat. Thesis: 28-40
Laher I, Thompson LP, Gagne L (1990) Protein kinase C as a modulator of response amplification in vascular smooth muscle. Blood Vessels 27:333-340
Leimbach WN Jr, Wallin BG, Victor RG, Aylward PE, Sundlof G, Mark AL (1986) Direct evidence from intraneural recordings for increased central sympathetic outflow in patients with heart failure. Circulation 73:913-919
Lokhandwala MF, Eikenburg DC (1983) Minireview. Presynaptic receptors and alterations in norepinephrine release in spontaneously hypertensive rats. Life Sci 33:1527-1542
Ma X, Abboud FM, Chapleau MW (2001) A novel effect of angiotensin on renal sympathetic nerve activity in mice. J Hypertens 19:609-618
Majewski HK, Musgrave IF (1995) Second messenger pathways in the modulation of neurotransmitter release. Aust N Z J Med 25:817-821
Malik KU, Nasjletti A (1976) Facilitation of adrenergic transmission by locally generated angiotensin II in rat mesenteric arteries. Circ Res 38:26-30
Mancia G (1990) Sympathetic activation in congestive heart failure. Eur Heart J 11 Suppl A:3-11
Mendelsohn FA, Doyle AE, Gray GW (1980) Lack of response of sympathetic nervous system to angiotensin infusion. Lancet 1:492-493
Moan A, Risanger T, Eide I, Kjeldsen SE (1994) The effect of angiotensin II receptor blockade on insulin sensitivity and sympathetic nervous system activity in primary hypertension. Blood Press 3:185-188
Mölderings GJ, Likungu J, Hentrich F, Gothert M (1988) Facilitatory presynaptic angiotensin receptors on the sympathetic nerves of the human saphenous vein and pulmonary artery. Potential involvement in beta-adrenoceptor-mediated facilitation of noradrenaline release. Naunyn Schmiedebergs Arch Pharmacol 338:228-233
Moreau N, Richer C, Vincent MP, Giudicelli JF (1993c) [Interaction between SR 47436, a new angiotensin II antagonist and sympathetic nervous system in pithed SHR rats]. Arch Mal Coeur Vaiss 86:1269-1274
Moreau N, Richer C, Vincent MP, Giudicelli JF (1993b) Sympathoinhibitory effects of losartan in spontaneously hypertensive rats. J Cardiovasc Pharmacol 22:126-134

Moreau N, Richer C, Vincent MP, Giudicelli JF (1993a) Sympathoinhibitory effects of losartan in spontaneously hypertensive rats. J Cardiovasc Pharmacol 22:126–134

Morganti A, Grassi G, Giannattasio C, Bolla G, Turolo L, Saino A, Sala C, Mancia G, Zanchetti A (1989) Effect of angiotensin converting enzyme inhibition on cardiovascular regulation during reflex sympathetic activation in sodium-replete patients with essential hypertension. J Hypertens 7:825–835

Murakami H, Liu JL, Zucker IH (1997) Angiotensin II blockade [corrected] enhances baroreflex control of sympathetic outflow in heart failure. Hypertension 29:564–569

Nakamaru M, Jackson EK, Inagami T (1986) Beta-adrenoceptor-mediated release of angiotensin II from mesenteric arteries. Am J Physiol 250: H144–H148

Nap A, Balt JC, Pfaffendorf M, van Zwieten PA (2002) Sympatholytic properties of several AT1-receptor antagonists in the isolated rabbit thoracic aorta. J Hypertens 20:1821–1828

Nap A, Balt JC, Pfaffendorf M, Zwieten PA (2003) No involvement of the AT2-receptor in angiotensin II-enhanced sympathetic transmission in vitro. J Renin Angiotensin Aldosterone Syst 4:100–105

Newby DE, Goodfield NE, Flapan AD, Boon NA, Fox KA, Webb DJ (1998) Regulation of peripheral vascular tone in patients with heart failure: contribution of angiotensin II. Heart 80:134–140

Nicholls MG, Espiner EA, Miles KD, Zweifler AJ, Julius S (1981) Evidence against an interaction of angiotensin II with the sympathetic nervous system in man. Clin Endocrinol (Oxf) 15:423–430

Ohlstein EH, Brooks DP, Feuerstein GZ, Ruffolo RR, Jr. (1997) Inhibition of sympathetic outflow by the angiotensin II receptor antagonist, eprosartan, but not by losartan, valsartan or irbesartan: relationship to differences in prejunctional angiotensin II receptor blockade. Pharmacology 55:244–251

Panisset JC, Bourdois P (1968) Effect of angiotensin on the response to noradrenaline and sympathetic nerve stimulation, and on 3H-noradrenaline uptake in cat mesenteric blood vessels. Can J Physiol Pharmacol 46:125–131

Peach MJ (1971) Adrenal medullary stimulation induced by angiotensin I, angiotensin II, and analogues. Circ Res 28: Suppl-17

Pfeffer MA, Pfeffer JM, Steinberg C, Finn P (1985) Survival after an experimental myocardial infarction: beneficial effects of long-term therapy with captopril. Circulation 72:406–412

Purdy RE, Weber MA (1988) Angiotensin II amplification of alpha-adrenergic vasoconstriction: role of receptor reserve. Circ Res 63:748–757

Raasch W, Betge S, Dendorfer A, Bartels T, Dominiak P (2001) Angiotensin converting enzyme inhibition improves cardiac neuronal uptake of noradrenaline in spontaneously hypertensive rats. J Hypertens 19:1827–1833

Raizada MK, Lu D, Sumners C (1995) AT1 receptors and angiotensin actions in the brain and neuronal cultures of normotensive and hypertensive rats. Adv Exp Med Biol 377:331–348

Rector TS, Olivari MT, Levine TB, Francis GS, Cohn JN (1987) Predicting survival for an individual with congestive heart failure using the plasma norepinephrine concentration. Am Heart J 114:148–152

Reid IA (1992) Interactions between ANG II, sympathetic nervous system, and baroreceptor reflexes in regulation of blood pressure [editorial]. Am J Physiol 262: E763–E778

Richer C, Domergue V, Vincent MP, Giudicelli JF (1996) Involvement of nitric oxide, but not prostaglandins, in the vascular sympathoinhibitory effects of losartan in the pithed spontaneously hypertensive rat. Br J Pharmacol 117:315–324

Richer C, Fornes P, Cazaubon C, Domergue V, Nisato D, Giudicelli JF (1999) Effects of long-term angiotensin II AT1 receptor blockade on survival, hemodynamics and cardiac remodeling in chronic heart failure in rats. Cardiovasc Res 41:100–108

Rodriguez-Garcia JL, Paule A, Dominguez J, Garcia-Escribano JR, Vazquez M (2000) Changes in plasma norepinephrine and endothelin levels and metabolic profile after AT1-receptor blockade in human hypertension. Am J Cardiol 85:1147–50, A10

Roth RH (1972) Action of angiotensin on adrenergic nerve endings: enhancement of norepinephrine biosynthesis. Fed Proc 31:1358–1364

Rumantir MS, Kaye DM, Jennings GL, Vaz M, Hastings JA, Esler MD (2000) Phenotypic evidence of faulty neuronal norepinephrine reuptake in essential hypertension. Hypertension 36:824–829

Sakata K, Yoshida H, Obayashi K, Ishikawa J, Tamekiyo H, Nawada R, Doi O (2002) Effects of losartan and its combination with quinapril on the cardiac sympathetic nervous system and neurohormonal status in essential hypertension. J Hypertens 20:103–110

Saxena PR (1992) Interaction between the renin-angiotensin-aldosterone and sympathetic nervous systems. J Cardiovasc Pharmacol 19 Suppl 6:S80–S88

Seidelin PH, Collier JG, Struthers AD, Webb DJ (1991) Angiotensin II augments sympathetically mediated arteriolar constriction in man. Clin Sci (Lond) 81:261–266

Sharpe N (1999) Benefit of beta-blockers for heart failure: proven in 1999. Lancet 353:1988–1989

Shetty SS, DelGrande D (2000) Differential inhibition of the prejunctional actions of angiotensin II in rat atria by valsartan, irbesartan, eprosartan, and losartan. J Pharmacol Exp Ther 294:179–186

Stadler T, Veltmar A, Qadri F, Unger T (1992) Angiotensin II evokes noradrenaline release from the paraventricular nucleus in conscious rats. Brain Res 569:117–122

Starke K, Taube HD, Browski E (1977) Presynaptic receptor systems in catecholaminergic transmission. Biochem Pharmacol 26:259–268

Story DF, Ziogas J (1987) Interaction of angiotensin with noradrenergic neuroeffector transmission. Trends Pharmacol Sci 8:269–271

Struck J, Muck P, Trubger D, Handrock R, Weidinger G, Dendorfer A, Dodt C (2002) Effects of selective angiotensin II receptor blockade on sympathetic nerve activity in primary hypertensive subjects. J Hypertens 20:1143–1149

Swedberg K, Eneroth P, Kjekshus J, Wilhelmsen L (1990) Hormones regulating cardiovascular function in patients with severe congestive heart failure and their relation to mortality. Consensus Trial Study Group. Circulation 82:1730–1736

Taddei S, Favilla S, Duranti P, Simonini N, Salvetti A (1991) Vascular renin-angiotensin system and neurotransmission in hypertensive persons. Hypertension 18:266–277

Taher MS, McLain LG, McDonald KM, Schrier RW, Gilbert LK, Aisenbrey GA, McCool AL (1976) Effect of beta adrenergic blockade on renin response to renal nerve stimulation. J Clin Invest 57:459–465

Thomas JA, Marks BH (1978) Plasma norepinephrine in congestive heart failure. Am J Cardiol 41:233–243

Tofovic SP, Pong AS, Jackson EK (1991) Effects of angiotensin subtype 1 and subtype 2 receptor antagonists in normotensive versus hypertensive rats. Hypertension 18:774–782

van Zwieten PA, de Jonge A (1986) Interaction between the adrenergic and renin-angiotensin-aldosterone- systems. Postgrad Med J 62 Suppl 1:23–27

Verhage M, Ghijsen WE, Lopes da Silva FH (1994) Presynaptic plasticity: the regulation of Ca(2+)-dependent transmitter release. Prog Neurobiol 42:539–574

Weinberger MH (1982) Role of sympathetic nervous system activity in the blood pressure response to long-term captopril therapy in severely hypertensive patients. Am J Cardiol 49:1542–1543

Weir MR, Dzau VJ (1999) The renin-angiotensin-aldosterone system: a specific target for hypertension management. Am J Hypertens 12:205S–213S

Wong PC, Bernard R, Timmermans PB (1992) Effect of blocking angiotensin II receptor subtype on rat sympathetic nerve function. Hypertension 19:663–667

Wong PC, Price WA, Chiu AT, Duncia JV, Carini DJ, Wexler RR, Johnson AL, Timmermans PB (1990a) Nonpeptide angiotensin II receptor antagonists. VIII. Characterization of functional antagonism displayed by DuP 753, an orally active antihypertensive agent. J Pharmacol Exp Ther 252:719–725

Wong PC, Price WA, Jr., Chiu AT, Duncia JV, Carini DJ, Wexler RR, Johnson AL, Timmermans PB (1990b) Hypotensive action of DuP 753, an angiotensin II antagonist, in spontaneously hypertensive rats. Nonpeptide angiotensin II receptor antagonists: X. Hypertension 15:459–468

Xiang JZ, Linz W, Becker H, Ganten D, Lang RE, Scholkens B, Unger T (1985) Effects of converting enzyme inhibitors: ramipril and enalapril on peptide action and sympathetic neurotransmission in the isolated heart. Eur J Pharmacol 113:215–223

Yokoyama H, Minatoguchi S, Koshiji M, Uno Y, Kakami M, Nagata C, Ito H, Fujiwara H (1997) Losartan and captopril follow different mechanisms to decrease pressor responses in the pithed rat. Clin Exp Pharmacol Physiol 24:697–705

Ziogas J, Story DF, Rand MJ (1984) Facilitation of noradrenergic transmission by locally generated angiotensin II in guinea-pig isolated atria and in the perfused caudal artery of the rat. Clin Exp Pharmacol Physiol 11:413–418

Part 3
ANG Receptors
AT$_2$

Molecular Aspects of AT$_2$ Receptor

C. Nahmias · C. Boden

Department of Cell Biology, Institut Cochin 22, rue Mechain, 75014 Paris, France
e-mail: nahmias@cochin.inserm.fr

1	Introduction	376
2	Structural Features of the AT$_2$ Receptor	377
2.1	Functional Domains Involved in Ligand Binding	377
2.2	Disulfide Bonds in the AT$_2$ Receptor	379
2.3	Ligand-Independent Activation of the AT$_2$ Receptor	380
2.4	Functional Domains Involved in Intracellular Signaling	380
3	Coupling to G Proteins	381
3.1	Coupling to Gi	381
3.2	Other Interacting Partners of the AT$_2$ Receptor	382
4	Intracellular Signaling Pathways Activated by AT$_2$ Receptor	383
4.1	Cell Proliferation and Apoptosis	383
4.1.1	Protein Tyrosine Phosphatases	384
4.1.2	Inhibition of Protein Kinases	385
4.1.3	Ceramides	385
4.2	Cellular Differentiation	385
4.3	Migration	386
4.4	Neuronal Activation	387
4.5	Renal Function	387
5	Negative Cross-Talk Between AT$_2$ and AT$_1$ Receptors	388
5.1	Opposite Signaling Pathways	388
5.2	Heterodimerization	389
6	AGTR2 Gene Alterations Associated with Human Diseases	389
6.1	Congenital Anomalies of the Kidney and Urinary Tract	390
6.2	X-Linked Mental Retardation	390
References		391

Abstract The octapeptide angiotensin II (Ang II) binds to two subtypes of receptors, AT$_1$ and AT$_2$, that both belong to the superfamily of G protein-coupled receptors (GPCRs). The AT$_1$ subtype is a classical GPCR in terms of coupling and signaling, and appears to mediate all known physiological actions of Ang II. In contrast, the AT$_2$ subtype is an atypical receptor that has remained a puzzle since its discovery in 1989 and molecular cloning in 1993. Over the past 10 years, a number of studies have aimed at elucidating the signaling pathways and functions of the AT_2 subtype. A role for AT_2 receptors has been established during fetal development and in cardiovascular, brain, and renal functions. In most cases, the AT_2 receptor has been shown to counteract the effects of the AT_1 subtype. AT_2 also negatively cross-talks with growth factor receptors and plays a major role in the processes of apoptosis, migration, differentiation, and tissue regeneration. Depending on the cellular model and function examined, the AT_2 receptor activates different signaling pathways, that can be classified into three major types: regulation of protein phosphorylation, activation of phospholipases, and/or regulation of nitric oxide (NO)/cGMP. In the present chapter, we review recent advances on the molecular aspects of the AT_2 receptor: its structural features (functional domains involved in ligand binding and receptor activation), signaling pathways, coupling to G proteins and association with other intracellular partners. We then examine the molecular mechanisms by which AT_2 antagonizes the effects of the AT_1 subtype. Examples of AT_2 receptor gene alterations associated with human diseases such as congenital anomalies of kidney and urinary tract (CAKUT) or X-linked mental retardation, are also discussed.

Keywords G protein-coupled receptor · Structure/function relationship · Signaling pathways · Negative cross-talk · Phosphorylation · Arachidonic acid · Nitric oxide · Cardiovascular · Apoptosis · Migration · Neuronal differentiation · AGTR2 gene alteration

1
Introduction

The first evidence that the octapeptide angiotensin II (Ang II) was able to bind to two distinct receptor subtypes was provided in 1989, in studies showing distinct binding sites for non-peptidic analogs of Ang II in the rat adrenal gland (Chiu et al. 1989a) and different sensitivity of Ang II binding sites to the reducing reagent dithiothreitol (DTT) (Chiu et al. 1989b). These receptor subtypes, first designated as ATA and ATB, are now well-known as type 1 (AT_1) and type 2 (AT_2) receptors. The molecular cloning of cDNAs encoding the AT_1 and AT_2 receptors revealed that both subtypes belong to the superfamily of G protein-coupled receptors (GPCR) organized as single polypeptides that span the cell membrane seven times. AT_1 and AT_2 subtypes thus belong to the same superfamily of receptors and they bind the same ligand; however they exhibit no more than 35% amino acid sequence homology, and they differ in many aspects. The AT_1 subtype appears to mediate most known physiological effects of Ang II and behaves as a classical GPCR with regard to G protein coupling and associated intracellular signaling pathways. In contrast, the AT_2 receptor is an unconventional GPCR; in many cell types, this receptor is not coupled to classical G proteins, its intracellular signaling pathways remain controversial, and its physiological functions are still to be fully characterized. Over the past 10 years, a number of studies have aimed at elucidating the signaling pathways and functions of the

AT_2 subtype. A role for AT_2 receptors has been established in cardiovascular, brain, and renal functions and in the processes of apoptosis and tissue regeneration. In most cases, the AT_2 receptor has been shown to counteract the effects of the AT_1 subtype. In the present chapter, we will focus on the structural features of the AT_2 receptor, the signaling pathways associated with AT_2 receptor activation, and the molecular mechanisms by which AT_2 antagonizes the effects of the AT_1 subtype. Examples of AT_2 receptor gene alterations associated with human diseases will also be discussed.

2
Structural Features of the AT_2 Receptor

2.1
Functional Domains Involved in Ligand Binding

The AT_2 receptor has been distinguished from the AT_1 subtype on the basis of its selective binding to specific synthetic ligands—among them the peptidic CGP 42112 and non-peptidic PD123319 compounds—and its insensitivity to AT_1 receptor antagonists such as losartan. AT_1 and AT_2 receptors share only 35% amino acid sequence identity but they bind Ang II with similar affinity, and the question has been raised whether both receptors bind Ang II through their conserved amino acid residues, or whether they display distinct binding sites to the same ligand.

GPCR binding sites have been shown to differ considerably depending on the type of ligand considered. For instance, binding of β-adrenergic receptors to small molecules like catecholamines involves residues located in the transmembrane regions (tm), whereas receptors for glycoproteins essentially use their large N-terminal portion for ligand binding. In the case of AT_1 receptors, binding to the octapeptide Ang II has been shown to involve amino acid residues located in the tm as well as in the N-terminus.

Initial studies of the AT_2 receptor binding site, using photoaffinity-labeled analogs of Ang II (Servant et al. 1997), revealed that the amino-terminal end of Ang II interacts with the first 30 N-terminal residues of the AT_2 receptor, whereas the carboxyl-terminal end of Ang II interacts with the inner half of the third transmembrane domain (tm3) of the receptor (residues 129–138) (Fig. 1). Further studies of deleted mutants of the AT_2 receptor confirmed the essential role of N-terminal residues in binding to Ang II (Yee et al. 1998). Interestingly, these latter studies also revealed that the N-terminal part of the receptor does not interfere with binding to the synthetic AT_2-selective ligand CGP 42112, indicating distinct binding sites for the natural and synthetic ligands.

The N-terminal portion of the AT_2 receptor contains five potential N-glycosylation sites, at least three of which are utilized (Lazard et al. 1994b). Indeed, the AT_2 receptor is heavily glycosylated; however, enzymatic deglycosylation only has a modest effect on AT_2 receptor binding properties (Servant et al. 1996).

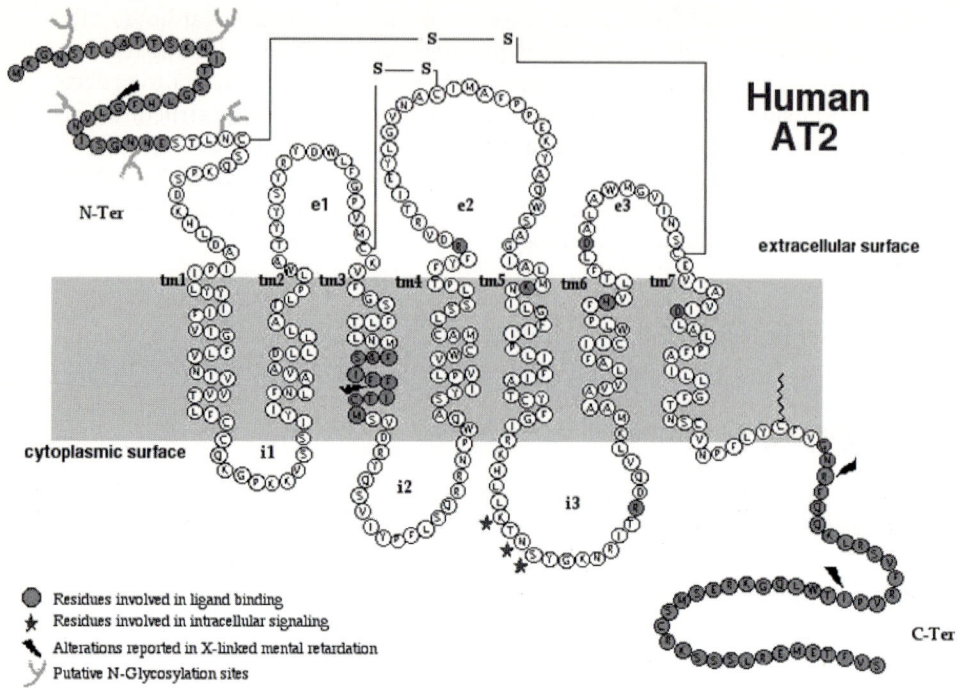

Fig. 1 Structural features of human AT$_2$ receptor

Site-directed mutagenesis indicated that, as for the AT$_1$ receptor, charged residues located in the transmembrane domains of the AT$_2$ receptor play a critical role in binding to Ang II (Fig. 1). Thus, amino acid Lys 215, a conserved residue corresponding to the critical Lys 199 of the rat AT$_1$ receptor in the tm5, is required for efficient binding of AT$_2$ to Ang II (Yee et al. 1997; Pulakat et al. 1998). Replacement of Asp 279 in the tm6 by an alanine also leads to a loss of binding to Ang II (Heerding et al. 1998). Residues Arg 182 located at the junction between e2 and tm4 (Heerding et al. 1997; Kurfis et al. 1999) and His 273 in the tm6 (Turner et al. 1999) are required for high-affinity binding to both Ang II and CGP 42112. In contrast, substitution of Asp 297 to Lys in the tm7 resulted in loss of affinity for Ang II (Heerding et al. 1997; Knowle et al. 2001) but retained partial binding to CGP 42112 (Knowle et al. 2001), further indicating that binding sites for Ang II and CGP 42112 are overlapping but not superimposable.

In addition to the N-terminus and tm regions, intracellular portions of the AT$_2$ receptor, namely the third intracellular loop (i3) and the C-terminal tail, have also been identified as playing a role in ligand binding. This finding was unexpected, as it is generally accepted that cytoplasmic portions of GPCRs are rather involved in intracellular signaling and internalization. Chimeric receptors in which the i3 of AT$_2$ has been replaced by that of AT$_1$ were no longer able to bind Ang II and CGP 42112 (Dittus et al. 1999). Reciprocally, chimeric receptors in which the i3 of AT_1 has been replaced by that of AT_2 showed high-affinity binding to CGP 42112A—but not to PD123319—which was further enhanced in the presence of the AT_2 receptor tm7 domain (Hines et al. 2001b). The role of the C-terminal tail (residues 322–363) in binding and activation of AT_2 was recently examined (Pulakat et al. 2002). A C-terminally deleted mutant of the AT_2 receptor showed reduced affinity to Ang II and unexpectedly showed increased affinity to the peptide analog CGP 42112.

Chimeric receptors in which tm6 and/or tm7 domains of the AT_2 receptor were substituted into the AT_1 receptor revealed functional similarities between the two subtypes of receptors in the domains mediating agonist-dependent activation (Hines et al. 2001a). Further studies combining photo-affinity labeling, site-directed mutagenesis and modeling recently indicated that although poorly homologous in sequence, AT_1 and AT_2 binding sites are superimposable, and revealed that Ang II binds to both receptors in an extended form parallel to the transmembrane regions (Deraet et al. 2002).

2.2
Disulfide Bonds in the AT_2 Receptor

The observation that the AT_2 binding site for a small peptide such as Ang II involves residues located in the N-terminus as well as in the tm6 domain suggested that these regions may be spatially close to each other, possibly via the formation of disulfide bonds. The existence of disulfide bonds in the AT_2 receptor had for a long time remained a puzzle because of the unusual behavior of this receptor in the presence of the reducing reagent DTT. While the AT_1 subtype is inactivated in the presence of DTT, the AT_2 receptor paradoxically shows higher affinity to Ang II in the presence of the reducing reagent (Chiu et al. 1989b). Several cysteines are conserved at similar positions in the sequences of the AT_1 and AT_2 receptors. Site-directed mutagenesis of conserved cysteines in the AT_2 receptor (Feng et al. 2000; Heerding et al. 2001) have confirmed the existence of two disulfide bonds (Cys35–Cys290) and (Cys117–Cys195) that link N-ter to e3 and e1 to e2, respectively, in the same manner as in the AT_1 subtype (Fig. 1). As for the AT_1 subtype, abolishing the (Cys117–Cys195) bond (between e1 and e2) leads to inactivation of the AT_2 receptor. Surprisingly, disruption of the other bond (Cys35–Cys290) that links N-ter to e3, leads to increased binding affinity for the ligands. Analyses of single Cys mutants suggested that a Cys–disulfide bond exchange may occur between the free SH group of Cys(35) or Cys(290) and the (Cys117–Cys195) disulfide bond, resulting in production of an inactive population of receptors through formation of a non-native disulfide bond (Feng et al. 2000). Further analysis of AT_1/AT_2 chimeras suggested to Heerding et al. that the N-ter and e3 regions of the AT_2 receptor may possess latent binding epitopes that are only uncovered after DTT exposure (Heerding et al. 2001).

2.3
Ligand-Independent Activation of the AT_2 Receptor

By comparing the affinity profiles of AT_1 and AT_2 receptors towards a panel of substituted analogs of Ang II, Miura and Karnik suggested that the AT_1 subtype is in a constrained conformation and is activated only when bound to the agonist, whereas the AT_2 receptor is "relaxed," in that no single interaction is critical for binding (Miura and Karnik 1999). In this respect, the native AT_2 receptor resembles the N111G mutant AT_1 receptor, which is constitutively active. These observations led to the interesting hypothesis that activation of the AT_2 receptor may be ligand-independent, a hypothesis consistent with previous findings that replacement of the i3 loop of the AT_2 receptor by that of the AT_1 subtype led to constitutive activation of AT_1 pathways (Wang et al. 1995). It was indeed shown shortly afterwards that AT_2-mediated apoptosis in cultured fibroblasts, epithelial cells, and vascular smooth muscle cells was regulated by overexpression of the receptor rather than by ligand activation (Miura and Karnik 2000). A ligand-independent effect of the AT_2 receptor was also indicated in recent experiments showing heterodimerization between AT_1 and AT_2 receptors, leading to attenuation of the AT_1 signaling pathways (AbdAlla et al. 2001). These authors showed that AT_2-mediated antagonism of AT_1 receptor activation is attenuated by release of AT_1/AT_2 heterodimers but remains unaffected in the presence of the AT_2 receptor antagonist PD123319. Constitutive activation of the AT_2 receptor may, however, depend on the cell type or signaling pathway considered, since other recent studies indicate that AT_2-mediated activation of tyrosine phosphatase SHP-1 in N1E-115 and transfected COS cells is not ligand-independent (Feng et al. 2002).

2.4
Functional Domains Involved in Intracellular Signaling

As for most GPCRs, the third intracellular domain of the AT_2 receptor has been shown to be a major determinant of receptor activation. Intracellular injection into cultured rat neurons of a 22-amino acid peptide corresponding to the i3 loop of the AT_2 receptor elicited increased opening of potassium channels similar to that produced by AT_2 receptor activation (Kang et al. 1995). Transfection of the same i3 peptide into vascular smooth muscle cells resulted in reduction of serum-stimulated DNA synthesis and cell proliferation as well as a decrease in mitogen-activated protein kinase activity, therefore simulating the effects of AT_2 receptor activation (Hayashida et al. 1996). In both studies, the effect of the i3 peptide was blocked by pertussis toxin, indicating the involvement of a Gi protein.

The role of the i3 loop as a major determinant of AT_2-mediated responses was further established in studies showing that chimeric AT_1 receptors in which the i3 loop was replaced by that of the AT_2 were able to transduce AT_2-specific signals, while keeping their AT_1-selective binding properties (Daviet et al. 2001).

Using site-directed mutagenesis, crucial residues (Lys 240, Asn 242, and Ser 243) located in the intermediate portion of the i3 loop have been identified as being involved in AT_2-induced apoptosis, extracellular signal-regulated kinase (ERK) inhibition and SHP-1 activation in PC12 cells (Lehtonen et al. 1999a). Consistently, AT_2 chimeras containing either the i3 or the i2 loop of the AT_1 receptor were no longer able to associate with, nor activate, tyrosine phosphatase SHP-1 (Feng et al. 2002). Interestingly, loss-of-function of the i3 chimera was restored when the C-terminal tail was additionally replaced by that of AT_1, indicating that all three intracellular portions of the AT_2 receptor participate in the coupling to phosphatase SHP-1 (Feng et al. 2002). Studies by Pulakat et al. recently indicated a negative effect of the C-terminal tail in signaling, as AT_2-mediated reduction of cyclic guanosine monophosphate (cGMP) levels in oocytes was further enhanced by C-terminal (322–363) deletion (Pulakat et al. 2002). The same C-terminally deleted mutant was no longer able to interact with ErbB3, a recently identified partner of the AT_2 receptor (see Sect. 3).

3
Coupling to G Proteins

3.1
Coupling to Gi

The AT_2 receptor clearly belongs to the superfamily of GPCRs; however, coupling of this receptor to classical G proteins still remains a matter of debate. Initial studies performed in pheochromocytoma PC12 cells by Bottari showed that the AT_2 receptor is insensitive to analogs of guanosine triphosphate (GTP) (Bottari et al. 1991) and does not activate a GTPase (Brechler et al. 1994), suggesting that the AT_2 receptor is not coupled to heterotrimeric G proteins. Later studies reported the involvement of pertussis toxin-sensitive Gi proteins in AT_2 signaling pathways, leading to opening of potassium channels (Kang et al. 1994; Huang et al. 1995) and apoptosis (Yamada et al. 1996; Horiuchi et al. 1997), and established that AT_2 is able to interact with $G\alpha(i2)$ and $G\alpha(i3)$ in rat fetus extracts (Zhang and Pratt 1996). More recently, Hansen et al. showed in reconstituted systems that AT_2 is able to trigger rapid activation of purified $G\alpha(i)$ and $G\alpha(o)$ but not of $G\alpha(q)$ and $G\alpha(s)$, as measured by radioactive GTPγS binding (Hansen et al. 2000). The ability of AT_2 to couple to Gi, but not to other $G\alpha$ proteins, was also demonstrated in recent studies using chimeric G proteins (Sasamura et al. 2000).

Thus, while most GPCRs function by coupling to a large panel of G proteins sometimes simultaneously in the same cell type, the AT_2 receptor seems to be restricted to coupling to $G\alpha(i)$. However, signaling pathways elicited by the AT_2 receptor in a number of cell types remained insensitive to pertussis toxin, indicating coupling to other G proteins, or to G protein-independent pathways. Two studies have indicated coupling of the AT_2 receptor to other G proteins: in neuroblastoma NG108–15 cells, AT_2 has been shown to inhibit the opening of calci um channels through activation of a non-identified G protein different from Gi (Buisson et al. 1995). In a recent study, Feng et al. have demonstrated an essential role for $G\alpha(s)$ as a scaffold protein in the constitutive association between AT_2 and tyrosine phosphatase SHP-1, illustrating a novel mechanism for GPCR-induced activation of SHP-1 that is independent of $\beta\gamma$ subunits of heterotrimeric G proteins and does not involve activation of $G\alpha(s)$ (Feng et al. 2002).

3.2
Other Interacting Partners of the AT_2 Receptor

Increasing evidence indicates that in addition to coupling to heterotrimeric G proteins, GPCRs also transduce their activation signals via other intracellular regulatory proteins, most of which often interact with the C-termini of GPCRs (Marinissen and Guntkind 2001; Brady and Limbird 2002). This is also the case for the AT_1 receptor, which interacts via its C-terminal tail with a novel protein designated AT_1 receptor-associated protein (ATRAP) (Daviet et al. 1999). Expression of ATRAP in vascular smooth muscle cells results in enhanced AT_1 receptor internalization and inhibition of AT_1-mediated activation of signal transducer and activator of transcription (STAT)3, Akt, and cell proliferation, indicating that ATRAP is a negative modulator of the AT_1 pathway (Cui et al. 2000).

In an attempt to identify novel interacting partners of the AT_2 receptor, Knowle et al. have screened a random peptide library using the whole AT_2 receptor as a bait in the two-hybrid system method. A peptide corresponding to the ATP binding site of ErbB3, a member of the epidermal growth factor (EGF) receptor family, has been isolated. The cytoplasmic tail containing the ATP binding site of ErbB3 has been shown to interact with the i3 loop and C-ter portion of the AT_2 receptor (Knowle et al. 2000; Pulakat et al. 2002). Although the consequence of ErbB3 interaction on AT_2 receptor signaling has not yet been established, this finding is particularly relevant in the light of the major role played by the ErbB family in growth and development (Olayioye et al. 2000) and of the inhibitory effect of AT_2 on EGF receptor (EGFR) signaling (Elbaz et al. 2000; Shibasaki et al. 2001; DePaolis et al. 2002).

A similar approach has been used in our laboratory to identify novel interacting partners of the AT_2 receptor. The 52 C-terminal residues of the AT_2 receptor were used as a bait to screen a mouse fetal cDNA library in the two-hybrid system. One novel sequence, designated AT2 receptor interacting protein (ATIP), was isolated and found to specify a new family of proteins that all interact with the AT_2 receptor, via a conserved coiled-coil domain allowing dimerization (S. Nouet et al., submitted). Transfection of ATIP into eukaryotic cells leads to inhibition of insulin receptor autophosphorylation and growth factor-induced ERK activation, indicating that ATIP simulates AT_2 receptor activation and may be a novel component of growth-inhibitory pathways.

4
Intracellular Signaling Pathways Activated by AT$_2$ Receptor

The AT$_2$ receptor is highly expressed at late stages of fetal development and its expression falls abruptly after birth. In the adult, this receptor is found in specific sites—adrenals, kidney, ovary, brain, vasculature—and is upregulated during the process of tissue regeneration following injury. At the cellular level, most documented AT$_2$-associated signaling pathways are those related to organogenesis and tissue repair, renal function, and neuronal activation. Although differing extensively depending on the cellular model and biological response examined, AT$_2$ signaling pathways can be classified into three major types : (1) regulation of protein phosphorylation, (2) activation of phospholipases, and (3) regulation of nitric oxide (NO)/cGMP (Fig. 2), that have been found to work in concert in some cell types.

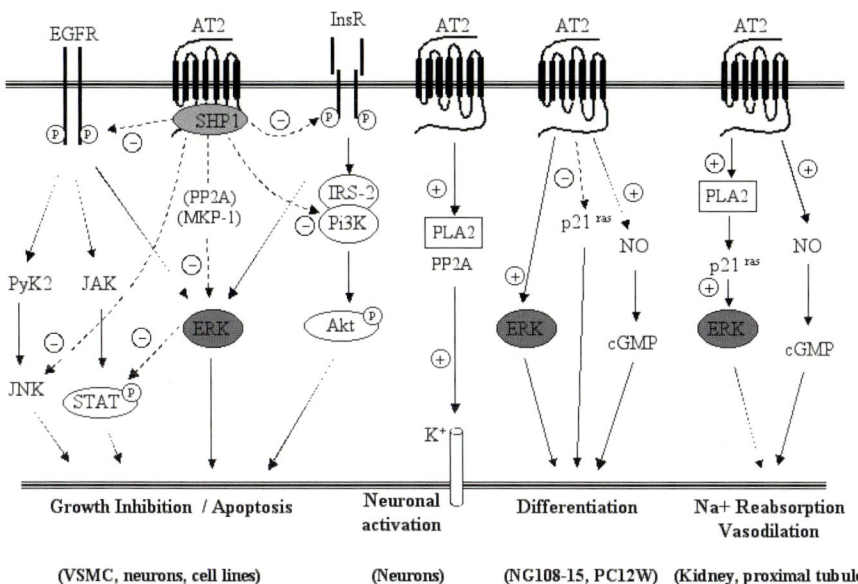

Fig. 2 Major AT$_2$ receptor signaling pathways

4.1
Cell Proliferation and Apoptosis

It is now well established that the AT$_2$ receptor inhibits cell proliferation and promotes apoptosis in vivo and in vitro (for reviews, see Horiuchi et al. 1999b; Nouet and Nahmias 2000; Stoll and Unger 2001). Growth-inhibitory effects of the AT$_2$ receptor are due, at least in part, to negative cross-talk with growth factor receptor tyrosine kinases and mainly involve activation of phosphatases and inhibition of protein kinases.

4.1.1
Protein Tyrosine Phosphatases

Protein tyrosine dephosphorylation and activation of orthovanadate-sensitive protein tyrosine phosphatases were the first cellular responses that were associated with AT_2 receptor activation (Bottari et al. 1992; Brechler et al. 1994; Nahmias et al. 1995). Consistent with a role on growth inhibition, the AT_2 receptor has been shown to rapidly activate the SH2 domain-containing tyrosine phosphatase SHP-1 (Bedecs et al. 1997), a vanadate-sensitive phosphatase that directly interacts with phosphorylated receptor tyrosine kinases and negatively regulates cytokine and growth factor receptor signaling pathways. A number of studies conducted by M. Horiuchi group have established a pivotal role of SHP-1 in AT_2-mediated ERK inactivation and apoptosis (Lehtonen et al. 1999a; Cui et al. 2001, 2002). In vascular smooth muscle cells (VSMC) from AT_2 transgenic mice, AT_2 induces increased association of SHP-1 with EGFR, leading to inhibition of AT_1-mediated EGFR *trans*-activation and activation of ERK (Shibasaki et al. 2001). AT_2 acting through SHP-1 also mediates inhibition of AT_1-induced c-Jun N-terminal kinase (JNK) activation and c-jun expression (Matsubara et al. 2001) as well as inhibition of the insulin-induced PI3 kinase association with IRS-2 and Akt phosphorylation (Cui et al. 2002). Moreover, SHP-1 appears responsible for Janus kinase (JAK)2 dephosphorylation and termination of the AT_1-activated JAK/STAT cascade (Marrero et al. 1998) and has been shown to participate in AT_2-mediated inhibition of AT_1-induced superoxide formation in human umbilical venous endothelial cells (Sohn et al. 2000).

How does the AT_2 receptor activate SHP-1? In some cell types, AT_2-mediated activation of SHP-1 is sensitive to pertussis toxin, indicating coupling to Gi. Although not firmly demonstrated, it may be possible that AT_2 functions like the somatostatin sst2 receptor that constitutively interacts with SHP-1 in a complex including Gi (Lopez et al. 1997). In other cell types however, AT_2-mediated SHP-1 activation is not sensitive to pertussis toxin. A recent study has demonstrated constitutive association of AT_2 with the inactive form of SHP-1 in a complex containing $G\alpha(s)$—but not other $G\alpha$-proteins—independently of $\beta\gamma$ subunits of G proteins. Upon AT_2 receptor stimulation, SHP-1 and Gs still interact but they dissociate from the receptor and SHP-1 becomes activated (Feng et al. 2002).

In addition to SHP-1, two other phosphatases, i.e., protein phosphatase 2A (PP2A) and mitogen-activated protein (MAP) kinase phosphatase (MKP)-1, are involved in AT_2-mediated ERK inactivation and apoptosis. The okadaic acid-sensitive Ser/Thr phosphatase PP2A is involved in inhibition of AT_1-mediated ERK activity in neurons (Huang et al. 1996), and in AT_2-mediated apoptosis in the presence of UV irradiation (Shenoy et al. 1999). The AT_2 receptor also stimulates dual-specificity (Tyr/Thr) MKPs that inactivate ERK by direct dephosphorylation on Tyr and Thr residues. In PC12W cells, AT_2 induces increased expression of MKP-1, leading to inactivation of ERK1 and ERK2 (Yamada et al. 1996) and reduced phosphorylation of the anti-apoptotic protein bcl-2

(Horiuchi et al. 1997). Recently, a role for MKP-3 in Ang II-mediated apoptosis and bcl-2 dephosphorylation and degradation in endothelial cells has also been demonstrated (Rossig et al. 2002). In contrast to SHP-1 whose catalytic activity is increased within minutes, MKP-1 and MKP-3 are transcriptionally activated within hours, suggesting that a relay between different tyrosine phosphatases may be necessary to allow the pro-apoptotic effects of the AT_2 receptor (Nouet and Nahmias 2000).

4.1.2
Inhibition of Protein Kinases

In parallel to tyrosine phosphatase activation, the AT_2 receptor mediates inhibition of protein kinase activity, leading to attenuation of mitogenic signaling cascades. AT_2 interferes at early steps of mitogenic cascades, by inhibiting insulin- and EGF- induced receptor tyrosine kinase autophosphorylation (Elbaz et al. 2000) and EGF receptor *trans*-activation by the AT_1 receptor (Shibasaki et al. 2001; DePaolis et al. 2002). AT_2 also inactivates downstream protein kinases: in addition to its widely described effect on ERK, AT_2 inhibits AT_1-induced JNK activity (Matsubara et al. 2001), and inhibits insulin-induced IRS-2-associated PI3 K and subsequent Akt phosphorylation (Cui et al. 2002). As a consequence of ERK inactivation, AT_2 inhibits the phosphorylation of STAT1, STAT2, and STAT3 induced by several growth factors including AT_1, with no effect on JAK (Horiuchi et al. 1999a). Studies of AT_2 knockout mice have indicated that AT_2 also negatively regulates the phosphorylation of p70S6 kinase in the vasculature—an effect which is not due to inhibition of upstream kinases ERK nor Akt—suggesting a role for AT_2 in the regulation of vascular hypertrophy in vivo (Brede et al. 2001).

4.1.3
Ceramides

In PC12W cells, AT_2 induces generation of ceramides, illustrating another pathway that links AT_2 receptor activation with apoptosis (Gallinat et al. 1999). AT_2-mediated accumulation of ceramide in PC12W cells was shown to be due to de novo biosynthesis of sphingolipids. This pathway was blocked by pertussis toxin and orthovanadate, indicating the involvement of Gi and phosphotyrosine phosphatases in this process (Lehtonen et al. 1999b).

4.2
Cellular Differentiation

Consistent with its role in nerve regeneration (Lucius et al. 1998), the AT_2 receptor induces morphological differentiation and neurite elongation in neuronal cell lines NG108–15 (Laflamme et al. 1996) and PC12W (Meffert et al. 1996), as well as in granular cells from rat cerebellum (Cote et al. 1999). These morphological changes are correlated with increased levels of polymerized tubulin and microtubule-associated proteins Tau and MAP2c, which are involved in neurite extension (Laflamme et al. 1996; Stroth et al. 1998; Cote et al. 1999). In PC12W cells, AT_2 downregulates the expression of MAP1B, in contrast to what is seen during nerve growth factor (NGF)-induced differentiation (Stroth et al. 1998). AT_2 stimulation in PC12W also results in reduced expression levels of the middle-sized neurofilament subunit (NF-M), which is decreased in regenerating neurons and in neuronal cultures undergoing apoptosis (Gallinat et al. 1997).

AT_2 receptor signaling pathways leading to neurite elongation in NG108–15 cells include inhibition of GTP-bound p21ras (Gendron et al. 1998), but surprisingly also include a sustained increase in ERK1 and ERK2 activation that is essential for neurite outgrowth (Gendron et al. 1999). Inhibition of p21ras using the dominant negative mutant RasN17 failed to impair the stimulatory effect of Ang II on ERK, suggesting that p21ras is not an upstream component of the AT_2-activated ERK cascade (Gendron et al. 1999). AT_2 also mediates activation of ERK1 and ERK2 in quiescent PC12W cells (Stroth et al. 2000) and AT_2-mediated neurite extension is blocked by the MEK inhibitor PD098059, confirming the essential role of ERK activation in AT_2-mediated differentiation (Stroth et al. 2000). Taken together, these studies tend to suggest that the net cellular effect of the AT_2 receptor may depend on the growth status of the cell. Thus, in cells cultured in the presence of growth factors, AT_2 inhibits ERK activation, thereby leading to inhibition of cell proliferation, whereas in quiescent cells, AT_2 activates ERK to promote neuronal differentiation and neurite extension.

Another signaling pathway linking AT_2 activation to neuronal differentiation involves increased NO production (Cote et al. 1998). In NG108–15 cells, AT_2 induces a rapid increase in NO synthase (NOS) activity and a subsequent increase in intracellular cGMP levels via a pertussis toxin-sensitive Gi protein (Gendron et al. 2002). AT_2-mediated activation of NO/cGMP is independent of ERK activation, and rather represents a parallel and complementary pathway involved in neurite branching.

A role for AT_2 receptors in the differentiation of VSMC has also been suggested by studies of AT_2 receptor knockout mice, showing reduced expression of calponin and h-caldesmon—two constituents of vascular contractile apparatus—in null mice as compared to wild-type mice (Yamada et al. 1999).

4.3
Migration

Cell migration is an essential step during development and tissue repair. The AT_2 receptor has been reported to induce cell migration in microexplant cultures of rat cerebellum (Cote et al. 1999), and to inhibit the migration of AT_2-transfected VSMC plated on laminin (Chassagne et al. 2002). The decreased migration of AT_2-transfected VSMC was correlated with increased synthesis and binding to fibronectin. Addition of GRGDTP peptide, which prevents cell attachment to fibronectin, reversed the inhibitory effect of AT_2 on cell migration, indicating that VSMC migration is inhibited via synthesis and binding of fibronectin (Chassagne et al. 2002).

In another study conducted on AT_2-transfected VSMC, AT_2 receptor activation was shown to induce a significant increase of collagen synthesis, and this effect was attenuated by treatment with pertussis toxin and Gi antisense oligonucleotides (Mifune et al. 2000). Similar effects were observed in AT_2-transfected mesangial cells, but not in transfected NIH-3T3(AT_2) fibroblasts, further indicating differential effects of AT_2 receptor stimulation in different cellular models (Mifune et al. 2000).

In endothelial cells derived from rat heart (CEC) or bovine aorta (BAEC), AT_2 upregulates the expression of thrombospondin-1 and fibronectin, indicating a role for this receptor in remodeling of endothelial extracellular matrix (Fischer et al. 2001).

4.4
Neuronal Activation

Electrophysiological studies conducted by C. Sumners group have shown that in cultured neurons from newborn rat brain, the AT_2 receptor stimulates a delayed rectifier K^+ current (Martens et al. 1996), leading to increased neuronal firing rate (Zhu et al. 2001). In catecholaminergic neurons in which AT_1 and AT_2 receptors are co-localized, the two subtypes induce opposite effects on K^+ current, indicating functional negative cross-talk between AT_1 and AT_2 in regulating electrophysiological responses of catecholaminergic neurons (Gelband et al. 1997). AT_2-mediated stimulation of K^+ current in rat neuronal cells involves a pertussis toxin-sensitive Gi protein and activation of serine/threonine phosphatase PP2A (Kang et al. 1994; Huang et al. 1995). In these cells, AT_2 receptor activation also leads to increased activity of phospholipase A $(PLA)_2$, generation of arachidonic acid and production of 12-lipoxygenase (12-LO) metabolites (Zhu et al. 1998, 2000). Inhibition of PP2A abolished the stimulatory effects of AT_2 on neuronal delayed-rectifier potassium (K^+) current but did not affect Ang II-stimulated (3H)-AA release, suggesting that PP2A is a distal event in this pathway (Zhu et al. 1998).

Another signaling pathway involving NO/cGMP production has been shown to link the AT_2 receptor to neuronal activation in PC12W and NG108–15 cells. In these cells, AT_2 negatively cross-talks with N-methyl-D-aspartate (NMDA) receptors, leading to attenuation of NMDA-induced increase in cGMP levels and/or NO release (Schelman et al. 1997).

4.5
Renal Function

In the kidney, the AT_2 receptor counteracts the anti-natriuretic and pressor effects mediated by the AT_1 subtype (Siragy et al. 1999). In conditions of elevated Ang II such as sodium depletion, AT_2 mediates a renal vasodilator cascade including generation of bradykinin, NO, and cGMP (Carey et al. 2000), a response partially mediated by neural NOS in addition to other NOS isoforms (Siragy and Carey 1997). A similar AT_2 signaling pathway involving NO and cGMP production has been described in the gastro-intestinal tract (Carey et al. 2000) and AT_2 has also been shown to increase aortic cGMP by a kinin-dependent mechanism in stroke-prone spontaneously hypertensive rats (SHRSPs) (Gohlke et al. 1998).

In isolated microperfused rabbit preglomerular afferent arteriole, activation of the AT_2 receptor causes endothelium-dependent vasodilation, and this effect is mediated by cytochrome P450 epoxygenase metabolites of arachidonic acid (AA) rather than by the NO pathway (Kohagura et al. 2000).

Thus, although triggering different cellular responses, renal and neuronal AT_2 receptors are linked to similar signaling pathways, i.e., regulation of NO/cGMP and AA metabolite production. Extensive studies by J. Douglas group have shown AT_2-mediated activation of PLA2 and generation of arachidonic acid metabolite in rabbit kidney proximal tubule epithelial cells (Jacobs and Douglas 1996; Harwalkar et al. 1998). Of interest, AT_2-mediated activation of PLA2 in these cells leads to activation of MAP kinase (Dulin et al. 1998) and p21ras, via a Shc–Grb2–SOS pathway usually associated with receptor tyrosine kinase (RTK) activation (Jiao et al. 1998).

5
Negative Cross-Talk Between AT_2 and AT_1 Receptors

Although showing distinct patterns of expression, the AT_1 and AT_2 receptors have been shown to be expressed within the same cell type. Their relative contribution to the effects of Ang II is therefore of particular importance in the light of the recent utilization of anti-AT_1 receptor antagonists as potent anti-hypertensive drugs. Functional negative cross-talk between AT_1 and AT_2 receptors on growth and hypertrophy, cardiovascular and neuronal functions has been demonstrated (Masaki et al. 1998; Sumners and Gelband 1998; Horiuchi et al. 1999b; Unger 1999; Stoll and Unger 2001). Obviously, negative cross-talk between AT_1 and AT_2 receptors is a consequence of opposite signaling pathways, but another mechanism, i.e., receptor heterodimerization, has recently been shown to participate in AT_2-mediated antagonism of AT_1 function.

5.1
Opposite Signaling Pathways

The mitogenic AT_1 receptor activates several independent phosphorylation cascades, including MAP kinase (ERK and JNK), JAK/STAT, and PI3-kinase/Akt pathways, which are inactivated following stimulation of the AT_2 receptor. ERK1 and ERK2 seem to be a major target of the AT_1/AT_2 cross-talk, as AT_2 inhibits AT_1-activated ERK in a variety of cellular models through activation of either SHP-1, PP2A, or MKP-1 phosphatases (Nouet and Nahmias 2000). The ERK cascade was originally shown to be activated by receptor tyrosine kinases. As for other GPCRs, AT_1-mediated activation of ERK at least partly results from *trans*-activation of receptor tyrosine kinases, in particular the EGF receptor (Hackel et al. 1999). The observations that AT_2 inhibits autophosphorylation of insulin receptors and EGF receptors (Elbaz et al. 2000; Shibasaki et al. 2001; DePaolis et al. 2002) and that AT_2 interacts with ErbB3, a member of the EGFR family (Knowle et al. 2000), therefore suggest that RTKs may constitute important cross-points between AT_1 and AT_2 signaling pathways at the level of the membrane. Other cross-points, however, have been identified downstream of membrane receptors, since AT_2 mediates dephosphorylation of STAT1, STAT2, and STAT3 with no effect on JAK activity (Horiuchi et al. 1999a), inhibition of JNK with no effect on Pyk2 (Matsubara et al. 2001), and inhibition of PI3-kinase and Akt with no effect on IRS-1 nor on the insulin receptor (Cui et al. 2002).

5.2
Heterodimerization

Increasing evidence indicates that, like other families of membrane receptors, GPCRs are able to self-assemble in the membrane, and that homo- and heterodimerization is a novel mechanism controlling GPCR activation and regulation (Salahpour et al. 2000). Homodimerization of AT_2 receptors in human myometrium has been suggested by cross-linking experiments even before the cloning of the AT_2 cDNA (Lazard et al. 1994b). Recent studies conducted by the U. Quitterer group (AbdAlla et al. 2001) have demonstrated heterodimerization between AT_1 and AT_2 receptors, leading to attenuation of the AT_1 signaling cascade in PC12 cells, fetal fibroblasts, and in human myometrial biopsies. AT_1 antagonism induced by the AT_2 receptor does not involve sequestration of $G\alpha$ proteins, and is rather mediated by direct receptor interaction independently of ligand binding and signaling by the AT_2 receptor. AT_1/AT_2 heterodimerization, being agonist-independent, is mainly regulated by the expression level of AT_2, unraveling the importance of AT_2 receptor transcriptional and post-transcriptional regulation in the AT_1/AT_2 cross-talk. The AT_2 receptor can thus be considered as a natural antagonist of the AT_1 subtype, illustrating a novel mode of negative communication between the two subtypes of Ang II receptors

6
AGTR2 Gene Alterations Associated with Human Diseases

In agreement with the widely described effect of AT_2 receptor in growth, development, and neuronal activation, two studies have reported alterations of the AT_2 receptor gene (AGTR2) that were associated with human diseases: (1) congenital anomalies of the kidney and urinary tract (CAKUT) and (2) X-linked mental retardation.

6.1
Congenital Anomalies of the Kidney and Urinary Tract

One interesting phenotype of AT_2 receptor-null mice includes anomalies of kidney development that strikingly mimic various features of human CAKUT (Nishimura et al. 1999). Establishment of these anomalies in the AT_2 knockout mice is preceded by delayed apoptosis of mesenchymal cells that surround the urinary tract during ontogeny. Detailed analysis of human AGTR2 gene in male Caucasians revealed a significant association between CAKUT and a nucleotide transition (A1332 to G) in the first intron of the gene. Splicing of AT_2 mRNA resulting from the G allele was found to be abnormal—lacking exon 2—and inefficient, suggesting that a single A/G mutation in the first intron may simulate the effect of AT_2 gene knockout (Nishimura et al. 1999).

The human AGTR2 gene is located on chromosome X (Lazard et al. 1994a) and comprises three exons that span more than 5 kb of genomic sequence (Martin and Elton 1995). The entire coding region of the AT_2 receptor gene is contained in the third exon, the first two exons being untranslated. Studies by Warnecke et al. have revealed the presence of enhancer intronic elements in the first intron of AGTR2 that are necessary for efficient transcription of the gene (Warnecke et al. 1999b). In human heart, exon 2 is alternatively spliced and the major AT_2 transcript is that containing exons 1, 2, and 3. Of interest, exon 2 has an inhibitory effect on reporter gene expression, indicating possible regulation of AT_2 receptor expression in vivo by differential splicing (Warnecke et al. 1999a).

Alterations in the AGTR2 gene leading to abnormal transcription of AT_2 mRNA may thus be responsible for developmental anomalies of the kidney. In recent studies, however, no correlation was found between A/G transition in the first intron of AGTR2 and the occurrence of CAKUT in a population of 66 Japanese male patients (Hiraoka et al. 2001) nor between A1332/G transition and the pathogenesis of primary familial vesicoureteral reflux (Yoneda et al. 2002).

6.2
X-Linked Mental Retardation

A recent study has reported alterations in the sequence of the AGTR2 gene in several cases of X-linked mental retardation (MR) (Vervoort et al. 2002). One female patient with a chromosomal (X;7) translocation showed complete extinction of AT_2 receptor mRNA, although the chromosomal breakpoint occurred outside the AGTR2 gene. Further examination of males from families with possible X-linked MR revealed a deletion of one thymidine at nucleotide position 395, causing a frameshift at position Phe133 in the tm3 and resulting in a truncated protein (Fig. 1) in two affected males within the same family and one unrelated patient. In addition, three missense mutations in the coding region of AGTR2: a (Gly21/Val) substitution in the N-ter portion of the AT_2 receptor, and Arg324/Gln and Ile337/Val mutations in the C-terminal portion of the receptor were detected in seven sporadic patients with MR (Fig. 1). Although the functional consequences of these mutations on AT_2 receptor activation and regulation have not yet been examined, it is interesting to note that all mutations are located in regions that have been shown to be involved in ligand binding or signaling (Fig. 1). These findings are consistent with other studies showing (1) expression of the AT_2 receptor mRNA in developing and adult brain (Nuyt et al. 1999), (2) a role for AT_2 in nerve regeneration, and neuronal differentiation and activation, and (3) a possible role for AT_2 in behavior and cognitive functions (Hein et al. 1995; Ichiki et al. 1995; Okuyama et al. 1999; Braszko 2002), further supporting the hypothesis that the AT_2 receptor may play a role in human brain development and/or trophic maintenance of neuronal connections important for learning and memory.

References

AbdAlla S, Lother H, Abdel-tawab A, Quitterer U (2001) The angiotensin II AT_2 receptor is an AT_1 receptor antagonist. J Biol Chem 276:39721–39726

Bedecs K, Elbaz N, Sutren M, Masson M, Susini C, Strosberg A, Nahmias C (1997) Angiotensin II type 2 receptors mediate inhibition of mitogen-activated protein kinase cascade and functional activation of SHP-1 tyrosine phosphatase. Biochem J 325:449–454

Bottari S, Taylor V, King I, Bogdal Y, Whitebread S, Gasparo Md (1991) Angiotensin II AT_2 receptors do not interact with guanine nucleotide binding proteins. Eur J Pharmacol 207:157–163

Bottari S, King I, Reichlin S, Dahlstroem I, Lydon N, Gasparo Md (1992) The angiotensin AT_2 receptor stimulates protein tyrosine phosphatase activity and mediates inhibition of particulate guanylate cyclase. Biochem Biophys Res Commun 183:206–211

Brady A, Limbird L (2002) G protein-coupled receptor interacting proteins: emerging roles in localization and signal transduction. Cell Signal 14:297–309

Braszko J (2002) AT(2) but not AT(1) receptor antagonism abolishes angiotensin II increase of the acquisition of conditioned avoidance responses in rats. Behav Brain Res 131:79–86

Brechler V, Reichlin S, Gasparo MD, Bottari S (1994) Angiotensin II stimulates protein tyrosine phosphatase activity through a G-protein independent mechanism. Receptors Channels 2:89–98

Brede M, Hadamek K, Meinel L, Wiesmann F, Peters J, Engelhardt S, Simm A, Haase A, Lohse M, Hein L (2001) Vascular hypertrophy and increased P70S6 kinase in mice lacking the angiotensin II AT(2) receptor. Circulation 104:2602–2607

Buisson B, Laflamme L, Bottari S, Gasparo Md, Gallo-Payet N, Payet M (1995) A G protein is involved in the angiotensin AT_2 receptor inhibition of the T-type calcium current in non-differentiated NG108-15 cells. J Biol Chem 270:1670–1674

Carey R, Jin X, Wang Z, Siragy H (2000) Nitric oxide: a physiological mediator of the type 2 (AT_2) angiotensin receptor. Acta Physiol Scand 168:65–71

Chassagne C, Adamy C, Ratajczak P, Gingras B, Teiger E, Planus E, Oliviero P, Rappaport L, Samuel J, Meloche S (2002) Angiotensin II AT(2) receptor inhibits smooth muscle cell migration via fibronectin cell production and binding. Am J Physiol Cell Physiol 282:C654–C664

Chiu A, Herblin W, McCall D, Ardecky R, Carini D, Duncia J, Pease L, Wong P, Wexler R, AL AJ (1989a) Identification of angiotensin II receptor subtypes. Biochem Biophys Res Commun 165:196–203

Chiu A, McCall D, Nguyen T, Carini D, Duncia J, Herblin W, Uyeda R, Wong P, Wexler R, Johnson A (1989b) Discrimination of angiotensin II receptor subtypes by dithiothreitol. Eur J Pharmacol 170:117–118

Cote F, Laflamme L, Payet M, Gallo-Payet N (1998) Nitric oxide, a new second messenger involved in the action of angiotensin II on neuronal differentiation of NG108-15 cells. Endocr Res 24:403–407

Cote F, Do T, Laflamme L, Gallo J, Gallo-Payet N (1999) Activation of the AT(2) receptor of angiotensin II induces neurite outgrowth and cell migration in microexplant cultures of the cerebellum. J Biol Chem 274:31686–31692

Cui T, Nakagami H, Iwai M, Takeda Y, Shiuchi T, Tamura K, Daviet L, Horiuchi M (2000) ATRAP, novel AT_1 receptor associated protein, enhances internalization of AT_1 receptor and inhibits vascular smooth muscle cell growth. Biochem Biophys Res Commun 279:938–941

Cui T, Nakagami H, Iwai M, Takeda Y, Shiuchi T, Daviet L, Nahmias C, Horiuchi M (2001) Pivotal role of tyrosine phosphatase SHP-1 in AT_2 receptor-mediated apoptosis in rat fetal vascular smooth muscle cell. Cardiovasc Res 49:863–871

Cui T, Nakagami H, Nahmias C, Shiuchi T, Takeda-Matsubara Y, Li J, Wu L, Iwai M, Horiuchi M (2002) Angiotensin II Subtype 2 Receptor Activation Inhibits Insulin-Induced Phosphoinositide 3-Kinase and Akt and Induces Apoptosis in PC12W Cells. Mol Endocrinol 16:2113–2123

Daviet L, Lehtonen J, Tamura K, Griese D, Horiuchi M, Dzau V (1999) Cloning and characterization of ATRAP, a novel protein that interacts with the angiotensin II type 1 receptor. J Biol Chem 274:17058–17062

Daviet L, Lehtonen J, Hayashida W, Dzau V, Horiuchi M (2001) Intracellular third loops in AT_1 and AT_2 receptors determine subtype specificity. Life Sci 69:509–516

DePaolis P, Porcellini A, Savoia C, Lombardi A, Gigante B, Frati G, Rubattu S, Musumeci B, Volpe M (2002) Functional cross-talk between angiotensin II and epidermal growth factor receptors in NIH3T3 fibroblasts. J Hypertens 20:693–699

Deraet M, Rihakova L, Boucard A, Perodin J, Sauve S, Mathieu A, Guillemette G, Leduc R, Lavigne P, Escher E (2002) Angiotensin II is bound to both receptors AT_1 and AT_2, parallel to the transmembrane domains and in an extended form. Can J Physiol Pharmacol 80:418–425

Dittus J, Cooper S, Obermair G, Pulakat L, Obermeir G (1999) Role of the third intracellular loop of the angiotensin II receptor subtype AT_2 in ligand-receptor interaction. FEBS Lett 445:23–26

Dulin N, Alexander L, Harwalkar S, Falck J, Douglas J (1998) Phospholipase A2-mediated activation of mitogen-activated protein kinase by angiotensin II. Proc Natl Acad Sci USA 95:8098–8102

Elbaz N, Bedecs K, Masson M, Sutren M, Strosberg AD, Nahmias C (2000) Functional trans-inactivation of insulin receptor kinase by growth-inhibitory angiotensin II AT_2 receptor. Mol Endocrinol 14:795–804

Feng Y, Saad Y, Karnik SS (2000) Reversible inactivation of AT(2) angiotensin II receptor from cysteine-disulfide bond exchange. FEBS Lett 484:133–138

Feng Y, Sun Y, Douglas J (2002) $G\beta\gamma$-independent constitutive association of $G\alpha s$ with SHP-1 and angiotensin II receptor AT_2 is essential in AT_2-mediated ITIM-independent activation of SHP-1. Proc Natl Acad Sci USA 99:12049–12054

Fischer J, Stoll M, Hahn A, Unger T (2001) Differential regulation of thrombospondin-1 and fibronectin by angiotensin II receptor subtypes in cultured endothelial cells. Cardiovasc Res 51:784–791

Gallinat S, Csikos T, Meffert S, Herdegen T, Stoll M, Unger T (1997) The angiotensin AT_2 receptor down-regulates neurofilament M in PC12W cells. Neurosci Lett 227:29–32

Gallinat S, Busche S, Schutze S, Kronke M, Unger T (1999) AT_2 receptor stimulation induces generation of ceramides in PC12W cells. FEBS Lett 443:75–79

Gelband C, Zhu M, Lu D, Reagan L, Fluharty S, Posner P, Raizada M, Sumners C (1997) Functional interactions between neuronal AT_1 and AT_2 receptors. Endocrinology 138:2195–2198

Gendron L, Laflamme L, Asselin C, Payet M, Gallo-Payet N (1998) A role for p21ras in the angiotensin II AT_2 receptor transduction pathway. Endocr Res 24:409–412

Gendron L, Laflamme L, Rivard N, Asselin C, Payet M, Gallo-Payet N (1999) Signals from the AT_2 (angiotensin type 2) receptor of angiotensin II inhibit p21ras and activate MAPK (mitogen-activated protein kinase) to induce morphological neuronal differentiation in NG108-15 cells. Mol Endocrinol 13:1615–1626

Gendron L, Cote F, Payet M, Gallo-Payet N (2002) Nitric oxide and cyclic GMP are involved in angiotensin II AT(2) receptor effects on neurite outgrowth in NG108-15 cells. Neuroendocrinology 75:70–81

Gohlke P, Pees C, Unger T (1998) AT_2 receptor stimulation increases aortic cyclic GMP in SHRSP by a kinin-dependent mechanism. Hypertension 31:349–355

Hackel P, Zwick E, Prenzel N, Ullrich A (1999) Epidermal growth factor receptors: critical mediators of multiple receptor pathways. Curr Opin Cell Biol 11:184–189

Hansen J, Servant G, Baranski T, Fujita T, Iiri T, Sheikh S (2000) Functional reconstitution of the angiotensin II type 2 receptor and G(i) activation. Circ Res 87:753–759

Harwalkar S, Chang C, Dulin N, Douglas J (1998) Role of phospholipase A2 isozymes in agonist-mediated signaling in proximal tubular epithelium. Hypertension 31:809–814

Hayashida W, Horiuchi M, Dzau V (1996) Intracellular third loop domain of angiotensin II type-2 receptor. Role in mediating signal transduction and cellular function. J Biol Chem 271:21985–21992

Heerding J, Yee D, Jacobs S, Fluharty S (1997) Mutational analysis of the angiotensin II type 2 receptor: contribution of conserved extracellular amino acids. Regul Pept 72:97–103

Heerding J, Yee D, Krichavsky M, Fluharty S (1998) Mutational analysis of the angiotensin type 2 receptor: contribution of conserved amino acids in the region of the sixth transmembrane domain. Regul Pept 74:113–119

Heerding J, Hines J, Fluharty S, Yee D (2001) Identification and function of disulfide bridges in the extracellular domains of the angiotensin II type 2 receptor. Biochemistry 40:8369–8377

Hein L, Barsh G, Pratt R, Dzau V, Kobilka B (1995) Behavioural and cardiovascular effects of disrupting the angiotensin II type-2 receptor in mice. Nature 377:744–747

Hines J, Fluharty S, Yee D (2001a) Chimeric AT_1/AT_2 receptors reveal functional similarities despite key amino acid dissimilarities in the domains mediating agonist-dependent activation. Biochemistry 40:11251–11260

Hines J, Heerding J, Fluharty S, Yee D (2001b) Identification of angiotensin II type 2 (AT_2) receptor domains mediating high-affinity CGP 42112A binding and receptor activation. J Pharmacol Exp Ther 298:665–673

Hiraoka M, Taniguchi T, Nakai H, Kino M, Okada Y, Tanizawa A, Tsukahara H, Ohshima Y, Muramatsu I, Mayumi M (2001) No evidence for AT_2R gene derangement in human urinary tract anomalies. Kidney Int 59:1244–1249

Horiuchi M, Hayashida W, Kambe T, Yamada T, Dzau V (1997) Angiotensin type 2 receptor dephosphorylates Bcl-2 by activating mitogen-activated protein kinase phosphatase-1 and induces apoptosis. J Biol Chem 272:19022–19026

Horiuchi M, Hayashida W, Akishita M, Tamura K, Daviet L, Lehtonen J, Dzau V (1999a) Stimulation of different subtypes of angiotensin II receptors, AT_1 and AT_2 receptors, regulates STAT activation by negative crosstalk. Circ Res 84:876–882

Horiuchi M, Lehtonen J, Daviet L (1999b) Signaling Mechanism of the AT_2 Angiotensin II Receptor: Crosstalk between AT_1 and AT_2 Receptors in Cell Growth. Trends Endocrinol Metab 10:391–396

Huang X, Richards E, Sumners C (1995) Angiotensin II type 2 receptor-mediated stimulation of protein phosphatase 2A in rat hypothalamic/brainstem neuronal cocultures. J Neurochem 65:2131–2137

Huang X, Richards E, Sumners C (1996) Mitogen-activated protein kinases in rat brain neuronal cultures are activated by angiotensin II type 1 receptors and inhibited by angiotensin II type 2 receptors. J Biol Chem 271:15635–15641

Ichiki T, Labosky P, Shiota C, Okuyama S, Imagawa Y, Fogo A, Niimura F, Ichikawa I, Hogan B, Inagami T (1995) Effects on blood pressure and exploratory behaviour of mice lacking angiotensin II type-2 receptor. Nature 377:748–750

Jacobs L, Douglas J (1996) Angiotensin II type 2 receptor subtype mediates phospholipase A2-dependent signaling in rabbit proximal tubular epithelial cells. Hypertension 28:663–668

Jiao H, Cui X, Torti M, Chang C, Alexander L, Lapetina E, Douglas J (1998) Arachidonic acid mediates angiotensin II effects on p21ras in renal proximal tubular cells via the tyrosine kinase-Shc-Grb2-Sos pathway. Proc Natl Acad Sci USA 95:7417–7421

Kang J, Posner P, Sumners C (1994) Angiotensin II type 2 receptor stimulation of neuronal K+ currents involves an inhibitory GTP binding protein. Am J Physiol 267:1389–1397

Kang J, Richards E, Posner P, Sumners C (1995) Modulation of the delayed rectifier K+ current in neurons by an angiotensin II type 2 receptor fragment. Am J Physiol 268:C278–C282

Knowle D, Ahmed S, Pulakat L (2000) Identification of an interaction between the angiotensin II receptor sub-type AT_2 and the ErbB3 receptor, a member of the epidermal growth factor receptor family. Regul Pept 87:73–82

Knowle D, Kurfis J, Gavini N, Pulakat L (2001) Role of Asp297 of the AT_2 receptor in high-affinity binding to different peptide ligands. Peptides 22:2145–2149

Kohagura K, Endo Y, Ito O, Arima S, Omata K, Ito S (2000) Endogenous nitric oxide and epoxyeicosatrienoic acids modulate angiotensin II-induced constriction in the rabbit afferent arteriole. Acta Physiol Scand 168:107–112

Kurfis J, Knowle DK, Pulakat L (1999) Role of Arg182 in the second extracellular loop of angiotensin II receptor AT_2 in ligand binding. Biochem Biophys Res Commun 263:816–819

Laflamme L, Gasparo M, Gallo J, Payet M, Gallo-Payet N (1996) Angiotensin II induction of neurite outgrowth by AT_2 receptors in NG108-15 cells. Effect counteracted by the AT_1 receptors. J Biol Chem 271:22729–22735

Lazard D, Briend-Sutren M, Villageois P, Mattei M, Strosberg A, Nahmias C (1994a) Molecular characterization and chromosome localization of a human angiotensin II AT_2 receptor gene highly expressed in fetal tissues. Receptors Channels 2:271–280

Lazard D, Villageois P, Briend-Sutren M, Cavaille F, Bottari S, Strosberg A, Nahmias C (1994b) Characterization of a membrane glycoprotein having pharmacological and biochemical properties of an AT_2 angiotensin II receptor from human myometrium. Eur J Biochem 220:919–926

Lehtonen J, Daviet L, Nahmias C, Horiuchi M, Dzau V (1999a) Analysis of functional domains of angiotensin II type 2 receptor involved in apoptosis. Mol Endocrinol 13:1051–1060

Lehtonen J, Horiuchi M, Daviet L, Akishita M, Dzau V (1999b) Activation of the de novo biosynthesis of sphingolipids mediates angiotensin II type 2 receptor-induced apoptosis. J Biol Chem 274:16901–16906

Lopez F, Esteve J, Buscail L, Delesque N, Saint-Laurent N, Theveniau M, Nahmias C, Vaysse N, Susini C (1997) The tyrosine phosphatase SHP-1 associates with the sst2 somatostatin receptor and is an essential component of sst2-mediated inhibitory growth signaling. J Biol Chem 272:24448–24454

Lucius R, Gallinat S, Rosenstiel P, Herdegen T, Sievers J, Unger T (1998) The angiotensin II type 2 (AT$_2$) receptor promotes axonal regeneration in the optic nerve of adult rats. J Exp Med 188:661–670

Marinissen M, Gutkind J (2001) G-protein-coupled receptors and signaling networks: emerging paradigms. Trends Pharmacol Sci 22:368–376

Marrero M, Venema V, Ju H, Eaton D, Venema R (1998) Regulation of angiotensin II-induced JAK2 tyrosine phosphorylation: roles of SHP-1 and SHP-2. Am J Physiol 275:C1216–C1223

Martens J, Wang D, Sumners C, Posner P, Gelband C (1996) Angiotensin II type 2 receptor-mediated regulation of rat neuronal K+ channels. Circ Res 79:302–309

Martin M, Elton T (1995) The sequence and genomic organization of the human type 2 angiotensin II receptor. Biochem Biophys Res Commun 209:554–562

Masaki H, Kurihara T, Yamaki A, Inomata N, Nozawa Y, Mori Y, Murasawa S, Kizima K, Maruyama K, Horiuchi M, Dzau V, Takahashi H, Iwasaka T, Inada M, Matsubara H (1998) Cardiac-specific overexpression of angiotensin II AT$_2$ receptor causes attenuated response to AT$_1$ receptor-mediated pressor and chronotropic effects. J Clin Invest 101:527–535

Matsubara H, Shibasaki Y, Okigaki M, Mori Y, Masaki H, Kosaki A, Tsutsumi Y, Uchiyama Y, Fujiyama S, Nose A, Iba O, Tateishi E, Hasegawa T, Horiuchi M, Nahmias C, Iwasaka T (2001) Effect of angiotensin II type 2 receptor on tyrosine kinase Pyk2 and c-Jun NH2-terminal kinase via SHP-1 tyrosine phosphatase activity: evidence from vascular-targeted transgenic mice of AT$_2$ receptor. Biochem Biophys Res Commun 282:1085–1089

Meffert S, Stoll M, Steckelings U, Bottari S, Unger T (1996) The angiotensin II AT$_2$ receptor inhibits proliferation and promotes differentiation in PC12W cells. Mol Cell Endocrinol 122:59–67

Mifune M, Sasamura H, Shimizu-Hirota R, Miyazaki H, Saruta T (2000) Angiotensin II type 2 receptors stimulate collagen synthesis in cultured vascular smooth muscle cells. Hypertension 36:845–850

Miura S, Karnik S (1999) Angiotensin II type 1 and type 2 receptors bind angiotensin II through different types of epitope recognition. J Hypertens 17:397–404

Miura S, Karnik S (2000) Ligand-independent signals from angiotensin II type 2 receptor induce apoptosis. EMBO J 19:4026–4035

Nahmias C, Cazaubon S, Briend-Sutren M, Lazard D, Villageois P, Strosberg A (1995) Angiotensin II AT$_2$ receptors are functionally coupled to protein tyrosine dephosphorylation in N1E-115 neuroblastoma cells. Biochem J 306:87–92

Nishimura H, Yerkes E, Hohenfellner K, Miyazaki Y, Ma J, Hunley T, Yoshida H, Ichiki T, Threadgill D, Phillips J, Hogan B, Fogo A, Brock J, Inagami T, Ichikawa I (1999) Role of the angiotensin type 2 receptor gene in congenital anomalies of the kidney and urinary tract, CAKUT, of mice and men. Mol Cell 3:1–10

Nouet S, Nahmias C (2000) Signal transduction from the angiotensin II AT$_2$ receptor. Trends Endocrinol Metab 11:1–6

Nuyt A, Lenkei Z, Palkovits M, Corvol P, Llorens-Cortes C (1999) Ontogeny of angiotensin II type 2 receptor mRNA expression in fetal and neonatal rat brain. J Comp Neurol 407:193–206

Okuyama S, Sakagawa T, Chaki S, Imagawa Y, Ichiki T, Inagami T (1999) Anxiety-like behavior in mice lacking the angiotensin II type-2 receptor. Brain Res 821:150–159

Olayioye M, Neve R, Lane H, Hynes N (2000) The ErbB signaling network: receptor heterodimerization in development and cancer. EMBO J 19:3159–3167

Pulakat L, Tadessee A, Dittus J, Gavini N (1998) Role of Lys215 located in the fifth transmembrane domain of the AT$_2$ receptor in ligand-receptor interaction. Regul Pept 73:51–57

Pulakat L, Gray A, Johnson J, Knowle D, Burns V, Gavini N (2002) Role of C-terminal cytoplasmic domain of the AT_2 receptor in ligand binding and signaling. FEBS Lett 524:73–78

Rossig L, Hermann C, Haendeler J, Assmus B, Zeiher A, Dimmeler S (2002) Angiotensin II-induced upregulation of MAP kinase phosphatase-3 mRNA levels mediates endothelial cell apotosis. Basic Res Cardiol 97:1–8

Salahpour A, Angers S, Bouvier M (2000) Functional significance of oligomerization of G-protein-coupled receptors. Trends Endocrinol Metab 11:163–168

Sasamura H, Mifune M, Nakaya H, Amemiya T, Hiraki T, Nishimoto I, Saruta T (2000) Analysis of Galpha protein recognition profiles of angiotensin II receptors using chimeric Galpha proteins. Mol Cell Endocrinol 170:113–121

Schelman W, Kurth J, Berdeaux R, Norby S, Weyhenmeyer J (1997) Angiotensin II type-2 (AT_2) receptor-mediated inhibition of NMDA receptor signaling in neuronal cells. Brain Res Mol Brain Res 48:197–205

Servant G, Dudley D, Escher E, Guillemette G (1996) Analysis of the role of N-glycosylation in cell-surface expression and binding properties of angiotensin II type-2 receptor of rat pheochromocytoma cells. Biochem J 313:297–304

Servant G, Laporte S, Leduc R, Escher E, Guillemette G (1997) Identification of angiotensin II-binding domains in the rat AT_2 receptor with photolabile angiotensin analogs. J Biol Chem 272:8653–8659

Shenoy U, Richards E, Huang X, Sumners C (1999) Angiotensin II type 2 receptor-mediated apoptosis of cultured neurons from newborn rat brain. Endocrinology 140:500–509

Shibasaki Y, Matsubara H, Nozawa Y, Mori Y, Masaki H, Kosaki A, Tsutsumi Y, Uchiyama Y, Fujiyama S, Nose A, Iba O, Tateishi E, Hasegawa T, Horiuchi M, Nahmias C, Iwasaka T (2001) Angiotensin II type 2 receptor inhibits epidermal growth factor receptor transactivation by increasing association of SHP-1 tyrosine phosphatase. Hypertension 38:367–372

Siragy H, Carey R (1997) The subtype 2 (AT_2) angiotensin receptor mediates renal production of nitric oxide in conscious rats. J Clin Invest 100:264–269

Siragy H, Inagami T, Ichiki T, Carey R (1999) Sustained hypersensitivity to angiotensin II and its mechanism in mice lacking the subtype-2 (AT_2) angiotensin receptor. Proc Natl Acad Sci USA 96:6506–6510

Sohn H, Raff U, Hoffmann A, Gloe T, Heermeier K, Galle J, Pohl U (2000) Differential role of angiotensin II receptor subtypes on endothelial superoxide formation. Br J Pharmacol 131:667–672

Stoll M, Unger T (2001) Angiotensin and its AT_2 receptor: new insights into an old system. Regul Pept 99:175–182

Stroth U, Meffert S, Gallinat S, Unger T (1998) Angiotensin II and NGF differentially influence microtubule proteins in PC12W cells: role of the AT_2 receptor. Brain Res Mol Brain Res 53:187–195

Stroth U, Blume A, Mielke K, Unger T (2000) Angiotensin AT(2) receptor stimulates ERK1 and ERK2 in quiescent but inhibits ERK in NGF-stimulated PC12W cells. Brain Res Mol Brain Res 78:175–180

Sumners C, Gelband C (1998) Neuronal ion channel signalling pathways: modulation by angiotensin II. Cell Signal 10:303–311

Turner C, Cooper S, Pulakat L (1999) Role of the His273 located in the sixth transmembrane domain of the angiotensin II receptor subtype AT_2 in ligand-receptor interaction. Biochem Biophys Res Commun 257:704–707

Unger T (1999) The angiotensin type 2 receptor: variations on an enigmatic theme. J Hypertens 17:1775–1786

Vervoort V, Beachem M, Edwards P, Ladd S, Miller K, Mollerat X, Clarkson K, DuPont B, Schwartz C, Stevenson R, Boyd E, Srivastava AK (2002) AGTR2 mutations in X-linked mental retardation. Science 296:2401–2403

Wang C, Jayadev S, Escobedo J (1995) Identification of a domain in the angiotensin II type 1 receptor determining Gq coupling by the use of receptor chimeras. J Biol Chem 270:16677–16682

Warnecke C, Surder D, Curth R, Fleck E, Regitz-Zagrosek V (1999a) Analysis and functional characterization of alternatively spliced angiotensin II type 1 and 2 receptor transcripts in the human heart. J Mol Med 77:718–727

Warnecke C, Willich T, Holzmeister J, Bottari S, Fleck E, Regitz-Zagrosek V (1999b) Efficient transcription of the human angiotensin II type 2 receptor gene requires intronic sequence elements. Biochem J 340:17–24

Yamada H, Akishita M, Ito M, Tamura K, Daviet L, Lehtonen J, Dzau V, Horiuchi M (1999) AT_2 receptor and vascular smooth muscle cell differentiation in vascular development. Hypertension 33:1414–1419

Yamada T, Horiuchi M, Dzau V (1996) Angiotensin II type 2 receptor mediates programmed cell death. Proc Natl Acad Sci USA 93:156–160

Yee D, Kisley L, Heerding J, Fluharty S (1997) Mutation of a conserved fifth transmembrane domain lysine residue (Lys215) attenuates ligand binding in the angiotensin II type 2 receptor. Brain Res Mol Brain Res 51:238–241

Yee D, Heerding J, Krichavsky M, Fluharty S (1998) Role of the amino terminus in ligand binding for the angiotensin II type 2 receptor. Brain Res Mol Brain Res 57:325–329

Yoneda A, Cascio S, Green A, Barton D, Puri P (2002) Angiotensin II type 2 receptor gene is not responsible for familial vesicoureteral reflux. J Urol 168:1138–1141

Zhang J, Pratt R (1996) The AT_2 receptor selectively associates with Gialpha2 and Gialpha3 in the rat fetus. J Biol Chem 271:15026–15033

Zhu M, Gelband C, Moore J, Posner P, Sumners C (1998) Angiotensin II type 2 receptor stimulation of neuronal delayed-rectifier potassium current involves phospholipase A2 and arachidonic acid. J Neurosci 18:679–686

Zhu M, Natarajan R, Nadler J, Moore J, Gelband C, Sumners C (2000) Angiotensin II increases neuronal delayed rectifier K(+) current: role of 12-lipoxygenase metabolites of arachidonic acid. J Neurophysiol 84:2494–2501

Zhu M, Sumners C, Gelband C, Posner P (2001) Chronotropic effect of angiotensin II via type 2 receptors in rat brain neurons. J Neurophysiol 85:2177–2183

AT$_2$ Receptor of Angiotensin II and Cellular Differentiation

N. Gallo-Payet[1] · L. Gendron[1] · E. Chamoux[1] · M. D. Payet[2]

[1] Service of Endocrinology, Department of Medicine, Faculty of Medicine,
University of Sherbrooke, Sherbrooke, QC, J1H 5N4, Canada
e-mail: Nicole.Gallo-Payet@USherbrooke.ca

[2] Department of Physiology and Biophysics, Faculty of Medicine,
University of Sherbrooke, Sherbrooke, QC, J1H 5N4, Canada

1	AT$_2$ Receptor Expression in Fetal Tissues	400
2	AT$_2$ Receptor and Induction of Neuronal Differentiation	400
2.1	Neurite Outgrowth	401
2.2	Cell Migration	403
2.3	Neuronal Excitability	403
2.4	Mechanisms Involved in AT$_2$-Induced Neurite Outgrowth	404
2.5	Conclusion for the Involvement of AT$_2$ in Neuronal Differentiation	408
3	Differentiation of Steroidogenic Tissues	409
3.1	AT$_2$ Receptors in the Adrenal Gland	409
3.2	Differentiation of Ovarian Granulosa Cells	412
4	AT$_2$ Receptors in Chromaffin Cells	413
5	Differentiation of Smooth Muscle Cells	414
6	Differentiation of Adipocytes	414
7	Conclusion and Perspectives	415
	References	416

Abstract Since its discovery, the AT$_2$ receptor of angiotensin II has been one of the most controversial G protein-coupled receptors. The AT$_2$ receptor is widely distributed in the fetus, but in most tissues, its expression is dramatically diminished few hours after birth. These observations have led to the hypothesis that this receptor may play an important role during fetal development. During the last decade, many studies have been conducted to elucidate the role of the AT$_2$ receptor in many different tissues and cell lines. Apart from a well described action in cell apoptosis, one of the major roles attributed to the AT$_2$ receptor of angiotensin II is its involvement in cellular differentiation. The AT$_2$ receptor is involved in differentiation of many tissues. For example, in cells from neuronal origin, activation of the AT$_2$ receptor was shown to induce neurite outgrowth and elongation, to modulate neuronal excitability, and to promote cellular migration. In steroidogenic tissues, the AT$_2$ receptor is associated with the development of the human fetal adrenal gland, where it induces apoptosis and probably cell migration, as well as of granulosa ovarian cells, where it promoted ovulation and oocyte maturation. The AT_2 receptor is also transiently expressed in smooth muscle cells where it may play a role in vasculogenesis, and in adipocytes where it induces production of prostacyclin (PGI_2) involved in preadipocyte differentiation. This chapter describes in detail how the AT_2 receptor of angiotensin II acts on cellular differentiation, with a particular emphasis on neuronal differentiation. Indeed, signaling mechanisms involved in the AT_2 effects are various and can be classified as atypical. For instance, AT_2 receptor activation promotes mitogen-activated protein kinase (MAPK) cascade through a $p21^{ras}$-independent pathway, and stimulates, in a parallel way, nNOS/cGMP/PKG signaling; both are essential to promote neurite outgrowth.

Keywords AT_2 receptor · Angiotensin · Cellular differentiation · Signaling mechanisms · MAPK cascade · Neurite outgrowth · Fetal adrenal · NG108-15 cells

1
AT_2 Receptor Expression in Fetal Tissues

One of the most remarkable features of the AT_2 receptor is its high level of expression in many fetal tissues (Shanmugam et al. 1995; Tanaka et al. 1995; Breault et al. 1996; Schütz et al. 1996), including the brain (Grady et al. 1991; Millan et al. 1991; Tsutsumi et al. 1991) (see chapter by Nishimura, this volume), suggesting an involvement of the AT_2 receptor in fetal development. Indeed, involvement of the AT_2 receptor has been documented in different models of differentiation such as morphological neuronal differentiation, steroidogenesis in gonads and in adrenal gland, contractility in smooth muscle cells, and prostaglandin production in adipocytes. The most important features of these events will be reviewed in this chapter.

2
AT_2 Receptor and Induction of Neuronal Differentiation

Due to its high level of expression in the neonatal brain and in some area of the adult brain involved in cognition and behavior, it was hypothesized that the AT_2 receptor was important in neuronal development and in the formation and/or maintenance of neuronal connections. Both the development and differentiation of cells of neuronal origin have been extensively studied and were shown to involve four different, essential stages of cellular evolution. These processes include cell growth, cell migration, neurite outgrowth, and synaptogenesis, all of which are controlled by the extracellular environment (hormones, growth factors, neurotransmitters, the extracellular matrix, cell adhesion molecules, and electrical activity).

2.1
Neurite Outgrowth

Over the last decade, evidence has been obtained that binding of angiotensin II (Ang II) to its AT_2 receptor modulates at least one of the above-mentioned criteria for neuronal differentiation, the induction of neurite outgrowth. Ang II-induced morphological changes have been studied in NG108-15 cells (a neuroblastoma–glioma hybrid cell line) and PC12 W cells (cell line derived from rat pheochromocytoma), two cell lines which express only the AT_2 receptor, or in primary cultures of rat cerebellar granule cells, in which the two types of receptors, AT_1 and AT_2 are present (Côté et al. 1999) (Fig. 1E, F and Fig. 3A). Three-day treatment of non-differentiated NG108-15 cells (Laflamme et al. 1996) or PC12 W cells (Meffert et al. 1996) with 100 nM Ang II induces morphological differentiation as characterized by the outgrowth of neurites (Fig. 1A–D).

Neurite extension is initiated at the growth cone and involves several biochemical steps directed towards promoting the assembly of tubulin monomers into microtubules necessary to support the growing neurites (Laferriere et al. 1997). Several molecules play crucial roles in regulating neurite outgrowth and differentiation. Among these are the microtubule-associated proteins (MAPs), which include MAP1B, MAP2, and tau (Matus 1988; Smith 1994). These proteins promote tubulin polymerization, stabilize microtubules, and exhibit embryonic and adult isoforms, whose differential expression during brain development correlates with the maturation of neuronal circuitry. For example, MAP2 and tau bind to distinct populations of microtubules in adult neurons: MAP2 to somatodendritic microtubules and tau to axonal microtubules (Matus et al. 1981; Binder et al. 1984). In NG108-15 cells, neurite outgrowth induced by Ang II stimulation of the AT_2 receptor is correlated with an increase in the level of polymerized tubulin and in the levels of the microtubule-associated protein, MAP2c (Laflamme et al. 1996). Mediation by the AT_2 receptor may be inferred since these cells contain only AT_2 receptors; these effects are mimicked by CGP 42112 (an AT_2 receptor agonist); they are not suppressed by addition of DUP 753 (an AT_1 receptor antagonist) but are abolished by co-incubation with PD 123319 (an AT_2 receptor antagonist). In rat pheochromocytoma PC12 W cells, similar neurite elongation was observed upon Ang II stimulation, and was also associated with an increase in MAP levels and/or association with microtubules, in particular MAP2 and MAP1B (Meffert et al. 1996) but a decrease in the level of neurofilament middle molecular weight subunit, NF-M (Gallinat et al. 1997; Stroth et al. 1998).

The effects of Ang II, through the AT_2 receptor, on neurite elongation have also been observed in more physiological models, such as postnatal retinal explants (Lucius et al. 1998) and microexplant cultures of cerebellum (which contain both AT_1 and AT_2 receptors) (Côté et al. 1999). In the latter, Ang II treatment, through the AT_2 receptor localized in neurons, strongly increased βIII-tubulin polymerization, a neuron-specific tubulin isoform (Matus 1988), and increased expression and association of phosphorylated forms of MAP2 (isoforms a, b and c) and tau (Côté et al. 1999). These findings confirm that AT_2 receptor

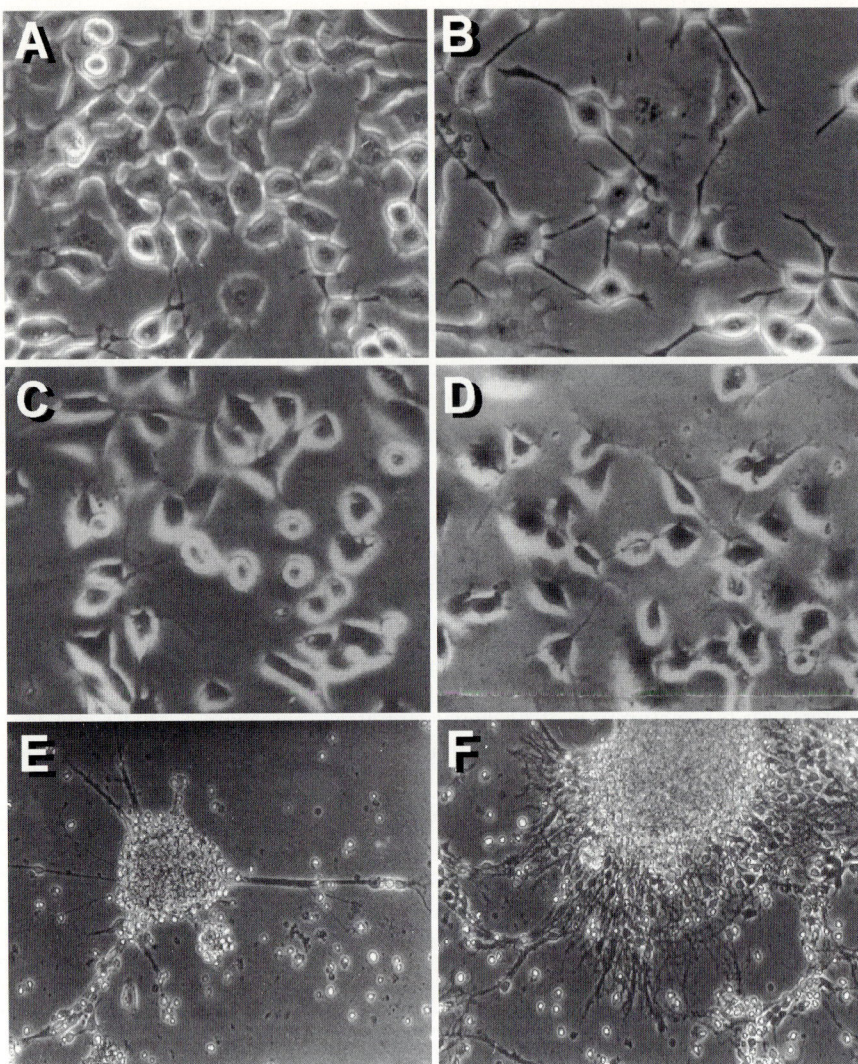

Fig. 1A–F Phase-contrast morphology of angiotensin II-treated cells of neuronal origin. Cells were cultured in media containing 10% fetal calf serum (FCS) for the NG108-15 cells (**A, B**), 0.5% FCS for the PC12 W cells (**C, D**) or neurobasal medium supplemented with B27 for cerebellar microexplants (**E, F**). Cells were cultured in the absence (**A, C, E**) or in the presence of 100 nM Ang II (**B, D, F**) over 3 days for the NG108-15 cells, 5 days for the PC12 W cells and 4 days for cerebellar microexplants. (Reproduced with permission from Gendron et al. 2002; (**A, B**); Stroth et al. 1998 (**C, D**); and Côté et al. 1999 (**E, F**)

stimulation activates all the components involved in the process of neurite elongation. Several developmental studies indicate that increases in tau and MAP expression precede that of tubulin, suggesting that the primary effect of the AT_2 receptor activation may be on tau and MAP2 rather than on tubulin itself.

Thus, alone, AT_2 receptor activation promotes neurite outgrowth and specific phosphorylation/association of MAPs with tubulin. However, in conditions where the AT_1 receptor is also present, such as in dibutyryl cAMP (dbcAMP)-treated NG108-15 cells [a well-known differentiating factor for NG108-15 cells (Hamprecht et al. 1985; Beaman-Hall et al. 1993; Laflamme et al. 1996)], or in cerebellum (Côté et al. 1999), activation of AT_1 receptor inhibits the effect produced by the AT_2 receptor. These observations further illustrate the well-described negative cross-talk interaction between the two types of Ang II receptors (for reviews see Inagami et al. 1999; Unger 1999; Stoll et al. 2001).

2.2
Cell Migration

In addition to elongation, Ang II application in cerebellar microexplants induces a marked cell migration observed at the center of the microexplant moving towards the periphery (Côté et al. 1999) (Fig. 1 E, F). Similar cell migration is observed during AT_2 receptor-induced regeneration of postnatal retinal microexplants (Lucius et al. 1998). These effects were more pronounced in cells treated with Ang II and DUP 753 or in cells treated with 10 nM of CGP 42112. Moreover, incubation with Ang II and PD 123319 blocked the AT_2 receptor-mediated effect. The study by Côté et al. (1999) confirms the hypothesis raised by Jöhren et al. (1998) that the high expression of AT_2 receptors in the inferior olivary–cerebellar pathway may be associated with a role for the AT_2 receptor in neuronal plasticity and cerebellar development. Indeed, neuronal development and differentiation of the cerebellum involve several steps including proliferation (in the ventricular zone), migration (through the ventricular zone to the cortical zone), and finally either neurite extension or apoptosis, once the cells have reached their specific destination (Cambray-Deakin et al. 1987; Komuro et al. 1998). In the cerebellar microexplants, where both neuronal and glial cells are present, AT_2 receptor activation induces not only neurite outgrowth, a process associated with morphological differentiation, but interestingly, cell migration as well. However, although extensively described for other systems (Yamada et al. 1996; Shenoy et al. 1999), the AT_2 receptors did not induce apoptosis in these cerebellar microexplants. Because the decision between apoptosis versus survival depends on the specific combination of local factors, the effect of Ang II observed in microexplant cultures may be due to interaction of AT_2 receptor signalization with factors locally produced by the mixed population of cells present in the microexplants, such as BDNF (brain-derived neurotrophic factor), NT-3 (neurotrophin 3) (Segal et al. 1995) or PACAP (pituitary adenylyl cyclase activating peptide) (Basille et al. 1994; Cavallaro et al. 1996).

2.3
Neuronal Excitability

Neuronal differentiation is characterized by a differential expression of ionic channels resulting in a change in the electrical activity. The T-type Ca^{2+} channel is involved in protein synthesis, growth, and proliferation and is predominantly present at earlier stages and disappears later (Yaari et al. 1987; McCobb et al. 1989; Kostyuk et al. 1993; Schmid et al. 1999). Activation of the AT_2 receptor by Ang II or CGP 42112 reduced the amplitude of the T-type current in non-differentiated NG108-15 cells by an as-yet-unknown mechanism which involved a phosphotyrosine phosphatase and a pertussis toxin (PTX)-insensitive G protein (Buisson et al. 1992; Buisson et al. 1995). Considering the antiproliferative effect of mibefradil, a selective antagonist of the T-type Ca^{2+} channel (Schmitt et al. 1995), it is tempting to postulate that the AT_2-induced blockage of the T-type current could be involved in Ang II-dependent NG108-15 differentiation. The transient K^+ current I_A and the delayed-rectifier K^+ current I_{Kv} are enhanced by AT_2 receptor activation (Kang et al. 1993). This effect on K^+ current results in the shortening of both action potential duration and refractory period, leading to an increased firing rate (Zhu et al. 2001). Increased membrane excitability is one of the fundamental characterizations of neuronal development during embryonic and postnatal ages (Fitzgerald 1987; Gao et al. 1998), particularly for the establishment of synaptic transmission (Gao et al. 1998). It is interesting to note that the decrease in the T-type current amplitude and enhancement of K^+ currents' amplitude, both occurring during differentiation, are mimicked by activation of the AT_2 receptor (for more details, see chapter by Diez in the next volume).

An inhibition of the visual response to flash stimulation was reported in rat superior colliculus (Merabet et al. 1997). This inhibition was mainly attributed to AT_1 receptor but a slight effect was also observed with CGP 42112. Furthermore, activation of the AT_2 receptor depresses glutamate depolarization and excitatory postsynaptic potentials in locus coeruleus (Xiong et al. 1994). However, AT_2 receptors have an excitatory effect on inferior olivary neurons (Ambühl et al. 1992). All together, these results indicate that the AT_2 receptor is able to modulate several different types of currents, but it is not yet known if one common or several signaling pathways are used.

During differentiation, microtubules are characterized by phases of rapid elongation and shortening. This dynamic instability is regulated by phosphorylation–dephosphorylation of microtubule-associated proteins (Maccioni et al. 1995; Sanchez et al. 2000), by environmental cues (Komuro et al. 1998) often targeted at the growth cone level for the axon (Letourneau 1996; Williamson et al. 1996), by calcium transients and electrical activity (Kater et al. 1991; Gomez et al. 2000; Spitzer et al. 2000). All these phenomena are stimulated by the AT_2 receptor.

2.4
Mechanisms Involved in AT_2-Induced Neurite Outgrowth

As indicated above, the association/dissociation of MAPs to microtubules is rapid and dynamic and is dependent on MAPs' phosphorylation states. Each MAP can be phosphorylated on multiple Ser/Thr and Tyr residues by the activation of different protein kinases such as glycogen synthase kinase (GSK3β), p42/p44^{mapk}, cyclin-dependent protein kinases (CDKs), cAMP- or calcium-dependent protein kinases (respectively, PKA and PKC) or dephosphorylated by the activation of different phosphatases (PP1, PP2A) (Sanchez et al. 2000). Moreover, MAP-2 is the main substrate for ERK1 and ERK2 while tau is phosphorylated by several kinases, including proline-directed serine/threonine protein kinase (Avila et al. 1994). Over the past 5 years, substantial amounts of data have been obtained, using both biochemical and morphological approaches to determine how Ang II could induce such differentiation processes.

Activation of the AT$_2$ receptor has been shown to modulate many signaling events including the activation/inhibition of kinases and phosphatases (Horiuchi et al. 1999; Nishimura et al. 1999; Unger 1999) (see chapter by Balt and Pfaffendorf, this volume). In NG108-15 cells cultured in the presence of 10% fetal bovine serum, Ang II (or CGP42112) application induced a time-dependent modulation of tyrosine phosphorylation of several proteins. In particular, Ang II induced a sustained increase in the activities of p42 and p44 MAPK (p42^{mapk} and p44^{mapk}). This delayed-but-sustained activation of p42/p44^{mapk} was shown to be essential for Ang II to promote neurite outgrowth and elongation. Accordingly, when cells were treated with a specific inhibitor for MEK1 (PD98059, 10 µM), the Ang II-induced p42/p44^{mapk} activity and neurite outgrowth were completely abolished (Gendron et al. 1999). Supporting these observations, Stroth and coworkers (2000) then demonstrated that the Ang II-induced neuronal differentiation of PC12 W cells was also dependent on p42/p44^{mapk} activation. Again, in this model, PD 98059 was sufficient in blocking Ang II's effects on neurite outgrowth and on p42/p44^{mapk} activation (Stroth et al. 2000). Thus, p42/p44^{mapk} plays an important role in the regulation of the mechanisms involved in Ang II-induced neuronal differentiation.

Gendron et al. (1999) demonstrated that, during the course of p42/p44^{mapk} activation by Ang II, p21^{ras} activation was inhibited. Reinforcing these observations, cells transfected with the dominant negative form of the Ras protein, RasN17, exhibit an increased basal state of p42/p44^{mapk} activation, which remains sensitive to Ang II stimulation. These results indicate that p21^{ras} inhibition alone activates p42/p44^{mapk} and that Ang II stimulation further increases this activity, by a p21^{ras}-independent mechanism. Among the alternative pathways reported to activate p42/p44^{mapk} independently of p21^{ras}, the nitric oxide (NO) signaling cascade and the Rap1/B-Raf signaling cassette are the most extensively described. The former is known to be activated by the AT$_2$ receptor, both in the kidney (Siragy et al. 1996; Siragy et al. 1997; Carey et al. 2000) and in NG108-15 cells (Schelman et al. 1997; Côté et al. 1998). Gendron et al. (2002) have shown that the increase in cyclic guanosine monophosphate (cGMP) was consecutive to a rapid increase in NO synthase (NOS) activity induced by dephosphorylation of neuronal NOS (nNOS). This increased activity did not involve modification in expression of nNOS. This effect, specifically stimulated by Ang II via the AT$_2$ receptor, is mediated by a PTX-sensitive Gi protein. However, cGMP is not involved in Ang II-induced activation of p42/p44^{mapk}. Indeed, the blockade of any protein from this signaling cascade (NOS), soluble guanylyl cyclase (sGC), or cGMP-dependent protein kinase (PKG) failed to interfere with the effects of Ang II on the p42/p44^{mapk} cascade of signaling (Gendron et al. 2002). In summary, these investigators stated that the activation of p42/p44^{mapk} cascade is independent of NO production, at least in NG108-15 cells.

As recent observations obtained in developing brain or in neuronal cell lines have indicated that NO could be involved in morphological neuronal differentiation (Peunova et al. 1995; Poluha et al. 1997; Sheehy et al. 1997) and in migration of neurons (Wright et al. 1998), this same group further investigated if NO could be involved in the AT$_2$ receptor mechanism of action in the induction of morphological neuronal differentiation of NG108-15 cells. Daily application of dbcGMP was sufficient to induce neurite outgrowth, suggesting that NO/sGC/cGMP pathway is involved in the AT$_2$-mediated differentiation. In agreement with these observations, inhibition of sGC or PKG (LY83583 or methylene blue and KT5823, respectively) impaired Ang II effects on neurite elongation. However, fine morphological comparison between Ang II- and dbcGMP-stimulated NG108-15 cells revealed that cGMP itself was responsible for neurite branching and filopodia formation, while PKG was necessary for elongation. Indeed, preincubation of the NG108-15 cells with KT5823 blocked Ang II-induced elongation, but did not interfere with branching and filopodia formation. Such observations corroborate results obtained by Phung et al. (1999) that cGMP and PKG have distinct effects on differentiation of PC12 cells. Together, these results indicate that p42/p44^{mapk}, NO/cGMP, and PKG have independent but complementary effects in the induction of dynamic neurite outgrowth and neurite branching. Axonal branches may be regulated by a cGMP-dependent pathway. Such collateral branches are important in the establishment and refinement of neuronal connections during development (Gallo et al. 2000).

What remains unclear at this point is the initial events linking AT$_2$ receptor activation and the increase in p42/p44^{mapk} activity. In a recent study (Gendron et al. 2003), we demonstrated that the activation of the AT$_2$ receptor rapidly, but transiently, activated the Rap1/B-Raf complex of signaling proteins. In RapN17- and Rap1GAP-transfected cells, the effects induced by Ang II were abolished, demonstrating that activation of these proteins was responsible for the observed p42/p44^{mapk} phosphorylation and for morphological differentiation. In several models—such as PC12 cells—cAMP and cAMP-activated proteins are required for nerve growth factor (NGF) activation of Rap1, as well as for p42/p44^{mapk} (York et al. 1998). To assess whether cAMP was involved in the activation of Rap1/B-Raf and neuronal differentiation induced by Ang II, NG108-15 cells were treated with stimulators or inhibitors of the cAMP pathway. However, dbcAMP and forskolin did not stimulate Rap1 nor p42/p44^{mapk} activities. Furthermore, addition of H-89, an inhibitor of protein kinase A, or Rp-8-Br-cAMPS, an inactive cAMP analog, failed to impair p42/p44^{mapk} activity and neurite outgrowth induced by Ang II. These observations clearly indicate that cAMP, a well-known stimulus of neuronal differentiation, do not participate in the AT$_2$ receptor signaling pathways in the NG108-15 cells. Therefore, the AT$_2$ receptor of Ang II ac-

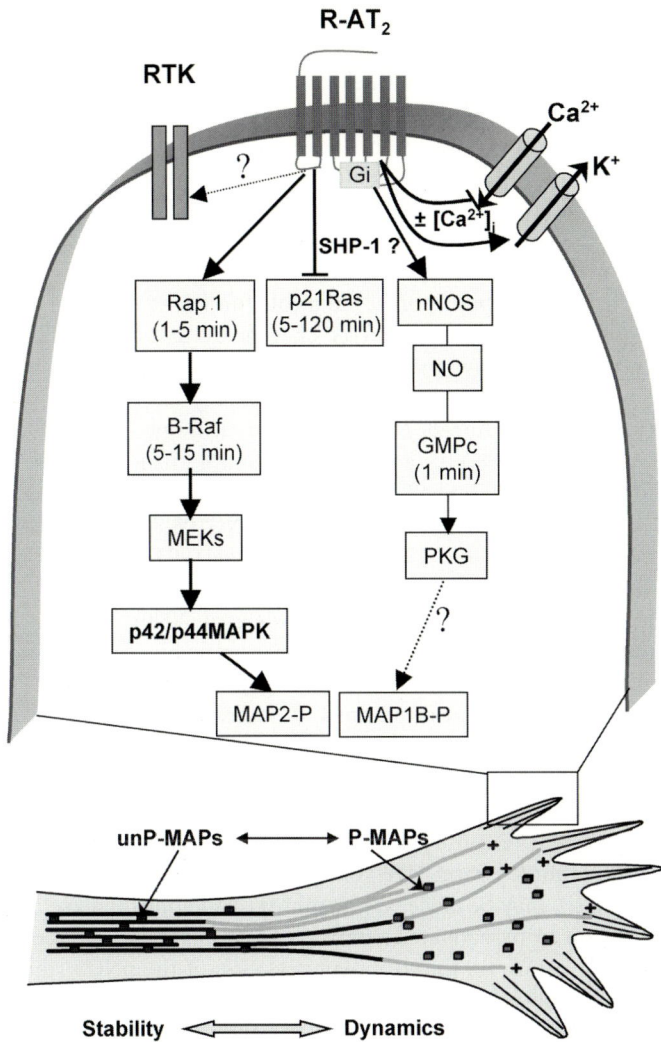

Fig. 2 Schematic representation of the AT_2 receptor signaling mechanisms involved in neurite elongation. The Ang II AT_2 receptor induction of neurite outgrowth and elongation involves at least two pathways. After binding of Ang II, the activated AT_2 receptor rapidly inactivates $p21^{ras}$ but stimulates the Rap1-B-Raf pathway, leading to a delayed $p42/p44^{mapk}$ phosphorylation. A second pathway involving the NO/sGC/cGMP signaling cascade is also required for observable neurite outgrowth. In addition, AT_2 receptor modulates intracellular calcium concentration ($[Ca^{2+}]_i$) by inhibiting Ca^{2+} channels and activating K^+ channels. Together, these parallel pathways could modulate gene expression and the phosphorylation state of different microtubule-associated proteins (MAP2, MAP1B) to control microtubule stability/dynamics responsible for neurite elongation. *nNOS*, neuronal NOS (nitric oxide synthase); *NO*, nitric oxide; *sGC*, soluble guanylyl cyclase; *MEK*, mitogen-activated protein kinase kinase; *unP-MAPs*, unphosphorylated MAPs; *P-MAPs*, phosphorylated MAPs. (The *lower panel* is adapted from Gordon-Weeks 1991)

tivates the signaling modules of Rap1/B-Raf and p42/p44^{mapk} via a cAMP-independent pathway to induce morphological differentiation of NG108-15 cells. Thus, how Rap1 could be activated in this model remains unknown.

Several questions still remain to be answered in regards to the AT$_2$ receptor. However, in agreement with other reported results, some hypotheses could be raised. For example, the effect of Ang II on neuronal differentiation and on stimulation of p42/p44^{mapk} activity are clearly observed when cells are cultured in the presence of serum, suggesting that a growth factor-promoting effect is required in the signaling mechanism of the AT$_2$ receptor. It is also not yet clear how NOS is dephosphorylated by Ang II; however, a role for SHP-1 could be envisaged (Bedecs et al. 1997; Feng et al. 2002) (see chapter by Balt and Pfaffendorf, this volume). Another alternative may be through interaction between the AT$_2$ receptor and the bradykinin receptor. Indeed, it is already known that bradykinin mediates AT$_2$ receptor-induced NO production (Siragy et al. 1996; Gohlke et al. 1998; Searles et al. 1999). In addition, heterodimerization between the bradykinin (B2) receptor and the AT$_1$ receptors has been demonstrated (AbdAlla et al. 2000). Similar effects could be envisaged in NG108-15 cells, where BK receptors are expressed (Chiang et al. 1989; McIntyre et al. 1993). In addition, the AT$_2$ subtype has been shown to interplay with members of the EGF-receptors family or with insulin receptors (Elbaz et al. 2000; Knowle et al. 2000).

All these observations concerning AT$_2$ receptor signaling involved in neurite outgrowth are summarized in Fig. 2. After binding of Ang II, the activated AT$_2$ receptor induces a sustained activation of p42/p44^{mapk} cascade and an increase in NO and cGMP content. The initial events, at least in NG108-15 cells, involve inactivation of p21^{ras} activity (5–120 min), but a rapid activation of Rap1 (1–5 min). Activated Rap1 then enhances the activity of B-Raf (5–15 min) that in turn stimulates p42/p44^{mapk} phosphorylation (30–60 min). Finally, the return of p42/p44^{mapk} phosphorylation to basal level, occurring much later (seen after 120 min of Ang II treatment), may be under a phosphotyrosine phosphatase activity such as SHP-1, shown to be activated after the AT$_2$ receptor stimulation. In parallel, the initial effect of AT$_2$ on the dephosphorylation of the nNOS (thus activation of nNOS) may be subsequent to the activation of the phosphotyrosine phosphatase SHP-1. All these events could occur at the end of the dendrite (or in the growth cone for the axon), where elongation, retraction, and pathfinding are initiated. Together, p42/p44^{mapk}, cGMP, and PKG could regulate gene expression and modulation of the phosphorylation states of different microtubule-associated proteins (MAPs) such as MAP2, tau, and MAP1b.

2.5
Conclusion for the Involvement of AT$_2$ in Neuronal Differentiation

In conclusion, the results reported to date indicate that AT$_2$ receptor activation promotes and/or accelerates all the processes involved in morphological differentiation since, its stimulation (1) increases polymerization of the βIII-tubulin, the specific neuronal isoform, (2) increases tau expression, thereby increasing axonal outgrowth, (3) increases MAP2c and MAP1B expression and phosphorylation, promoting polymerization of unstable microtubules, (4) decreases expression of NF-M protein, affecting stability to favor plasticity, (5) stimulates cell migration and, finally, (6) modulates membrane excitability. In addition, AT_2 effects are antagonized by the AT_1 receptor, indicating that AT_1 and AT_2 receptors have opposite actions on neuronal differentiation.

These observations clearly demonstrate the involvement of the AT_2 receptor in the fine tuning of neuronal differentiation and cell migration, two very important events occurring during brain development. Thus, both AT_1 and AT_2 receptor types may participate, together with other environmental growth factors, adhesion molecules and the components of the extracellular matrix, in the differentiation of brain areas expressing these receptors.

Lessons from knockout mice or from neurological disorders reinforce the idea that AT_2 receptor may be important for neuronal development. Indeed, perturbations in exploratory behavior and locomotor activity have been observed (Hein et al. 1995; Ichiki et al. 1995), as well as an anxiety-like behavior (Okuyama et al. 1999). In addition, a decrease in the expression of the AT_2 receptor is observed in areas of the adult brain, implicated in the development of neurological disorders such as Alzheimer disease, Huntington disease, or Parkinson disease (caudate nucleus, putamen and substantia nigra, temporal cortex) (Ge et al. 1996). Finally, recent observations indicate that *AGTR2* mutations are correlated with mental retardation (Vervoort et al. 2002). Together, such observations indicate inappropriate neuronal differentiation and plasticity (in adult) when AT_2 is absent or genetically modified.

3
Differentiation of Steroidogenic Tissues

3.1
AT_2 Receptors in the Adrenal Gland

Among the peripheral tissues which express AT_2 receptors are the adrenal glands in rats, ovine, and human (Aguilera et al. 1994; Shanmugam et al. 1995; Wintour et al. 1998). In the human fetal adrenal gland, immunocytochemical studies revealed that AT_2 receptor labeling is predominant in the fetal zone (identified using P450C17 as a fetal zone marker) and in a population of cells having a radial disposition, which could correspond to the future chromaffin cells, as attested by chromogranin A and dopamine β-hydroxylase reactivities (Breault et al. 1996) (Fig. 3B). Labeling was also present around the central medulla vein.

The human fetal adrenal gland is morphologically and functionally different from both the adult adrenal gland and fetal gland of non-primates. Morphological evidence indicates that the human fetal adrenal gland is a very dynamic organ, in which proliferating cells located at the periphery migrate, then differentiate, and finally undergo senescence in the central part of the gland. The func-

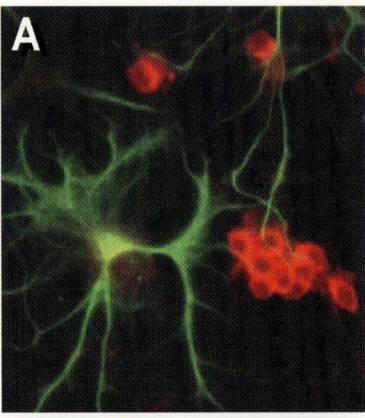
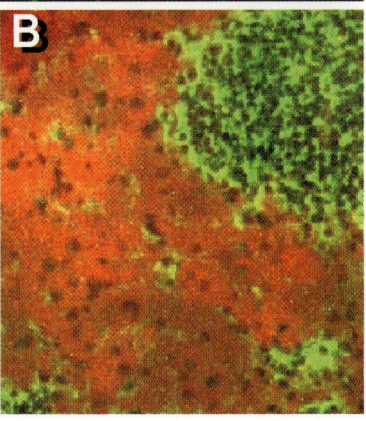

Fig. 3A, B Immunofluorescence localization of AT$_2$ receptor in cerebellar granule cells and in human fetal adrenal gland. Primary cultures of cerebellar microexplants or frozen sections of an 18-week-old human fetal adrenal gland were processed for immunofluorescence labeling using a primary antibody directed against the AT$_2$ receptor, and visualized after incubation with a secondary anti-rabbit-coupled antibody. In cerebellar microexplants, AT$_2$ receptor (labeled with rhodamine, *red*) was identified in neurons but not in astrocytes (labeled with anti-glial fibrillary acidic protein labeled with fluorescein, *green*; **A**). In the fetal adrenal gland, AT$_2$ receptor (labeled with fluorescein, *green*) was identified in fetal cells and in islets of chromaffin cells (*upper right*). Visualization of cellular composition has been achieved using Evans's blue solution (*orange*; **B**). The anti-AT$_2$ receptor antibody was a gift from Dr. Ian Bird, Department of Obstetrics and Gynecology, University of Wisconsin, Madison, USA

tional role of the fetal adrenal gland is to produce large amounts of DHEA/DHEAS (dihydroepiandrosterone and its sulfated derivative) (from the fetal zone). In contrast, cortisol secretion is absent during the first 15 weeks of gestation, and appears in a zone called transitional zone, visible between the fetal and definitive zones (Mesiano et al. 1997).

Data by Chamoux et al. (1999) indicate that the high expression of AT$_2$ receptors in the fetal zone may be correlated with the high level of apoptosis present in this area of the adrenal gland during development. Indeed, in human fetal adrenal cells cultured for 24 h, Ang II, via the AT$_2$ receptor, induced DNA fragmentation and cleavage of the DNA repair enzyme, poly-(ADP-ribose) polymerase (PARP), thus corroborating numerous studies indicating that the AT$_2$ receptor can induce cell apoptosis (see previous chapter, by Nahmias and Boden, this volume). Furthermore, stimulation with Ang II or CGP 42112 strongly modified the actin network, inducing membrane blebbing and a complete disappearance of the stress fiber network. At that time, actin was localized exclusively along the plasma membrane, with labeling predominant at the base of the bleb formation.

The same group has also shown that the AT_2 receptor overlaps the fibronectin expression pattern in the human fetal adrenal gland (Chamoux et al. 2001), whereas laminin is expressed only at the gland periphery, where the AT_2 receptor is absent, while collagen IV is expressed throughout the fetal gland. Because it is now well accepted that extracellular matrix can induce intracellular signaling or interact with hormonal or growth factor transduction pathways leading to specific cell behaviors such as proliferation, migration, apoptosis, or gene expression (Aplin et al. 1999; Giancotti et al. 1999), the authors examined Ang II effects on cells cultured on various matrices. Altogether, their results indicate that the extracellular matrix modulates cell behavior and hormonal responsiveness in the human fetal adrenal gland. In particular, collagen IV favors steroid secretion in response to adrenocorticotropin (ACTH) and Ang II stimulation, laminin alone enhances proliferation, and fibronectin favors cell behavior associated with fetal zone cells by enhancing apoptosis and ACTH responsiveness in terms of DHEA/DHEAS secretion.

Interaction between Ang II and matrices is indeed well known (see chapter by Schiffrin, following volume). Chassagne et al. recently confirmed a positive interaction between AT_2 and fibronectin: in this study, stimulation of the AT_2 receptor stopped migration of smooth muscle cells by stimulating the synthesis of fibronectin as well as increasing binding to this matrix (Chassagne et al. 2002). Fisher et al. indicated that activation of the AT_2 receptor enhanced the expression of fibronectin mRNA in endothelial cells (Fischer et al. 2001). These data indicate a strong correlation between fibronectin and AT_2-receptor mediated effects of Ang II in terms of inhibition of proliferation and induced cell apoptosis.

Interestingly, the synergistic effect of the AT_2-receptor/collagen IV on steroid secretion seems to be specific to cytochrome P450C17 activity, since no increase was observed for DHEAS production. This observation is of interest, since DHEA-sulfotransferase activity usually parallels that of P450C17 (Parker et al. 1999). However, in vitro, the effects of collagen IV on Ang II responsiveness are not due to an increase in mRNA expression of the enzyme, but rather to a specific hormonal/environmental regulation of the sulfotransferase and/or lyase activity. In addition, Ang II abrogated expression of the 3-β-hydroxysteroid dehydrogenase (3β-HSD) mRNA. These results may be compared with the in vivo physiology: P450C17 is largely expressed in the fetal zone, but absent in the periphery, where collagen and laminin are both present. In contrast, 3β-HSD is present at the periphery where fibronectin is less present. Taken together, these data highlight the intricate regulation of 3β-HSD onset and P450C17 maintenance in the human fetal adrenal gland, and suggest that AT_2 receptor, together with fibronectin- or collagen-induced events, may influence activity of these enzymes. Interestingly, the Gα_{i1-2} and Gα_{i3} proteins also exhibit a mirror distribution, with Gα_{i3} being localized in fetal cells, as the AT_2 receptor. These results are summarized in Fig. 4, outlining the interactions between hormonal and environmental factors in controlling human fetal adrenal development.

Fibronectin is known to favor proliferation when recognized by its high-affinity integrin receptor $\alpha5\beta1$. No typical receptor for fibronectin is expressed in

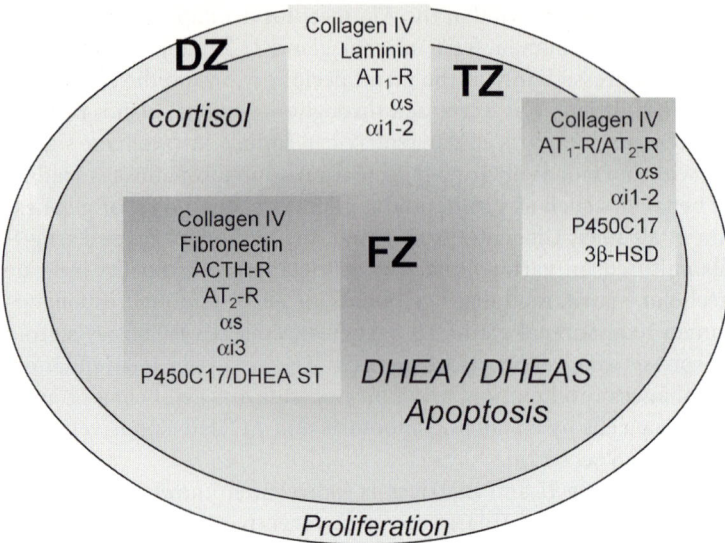

Fig. 4 Summary of the differential distribution of the angiotensin II receptors (AT$_1$ and AT$_2$), extracellular matrix components and heterotrimeric G proteins in the human fetal adrenal gland during the second trimester of gestation. Correlation is presented with known steroid production. Collagen IV and Gs are distributed throughout the fetal gland, while laminin is predominant in the definitive zone. Fibronectin is predominant in the fetal zone, as are the AT$_2$ receptor and the Gαi3 protein. *DZ*, definitive zone; *FZ*, fetal zone; *TZ*, transitory zone. (Reproduced with the permission of Chamoux et al. 2002)

the human fetal adrenal gland, but the $\alpha3\beta1$ integrin, which can serve as a moderate-affinity receptor is present (Kuhn et al. 1994; Chamoux et al. 2001). Recent studies have shown that $\alpha3\beta1$ disrupts cell–cell adhesion (DiPersio et al. 2000; Kawano et al. 2001), which may be correlated with the migration observed in the fetal zone. The absence of $\alpha5\beta1$ may also explain why these cells are non-proliferative and are subject to apoptosis. These observations reinforce the notion that extracellular matrix and their receptors integrins can modulate hormone responsiveness. The AT$_2$ receptor, which induces various physiological effects, could be particularly sensitive to such interactions.

3.2
Differentiation of Ovarian Granulosa Cells

Autoradiographic studies revealed the intense localization of Ang II receptors present in the granulosa layers of ovaries in rat (Pucell et al. 1991), rabbit (Yoshimura et al. 1996), and human (Johnson et al. 1997). AT$_2$ receptors are predominantly located in granulosa cells layers of the preovulatory follicles, whereas AT$_1$ receptors are more concentrated in the thecal cell layers and stroma. Ang II, through AT$_2$ receptor activation, induced ovulation and oocyte maturation, characterized by meiotic maturation of ovulated ova and follicular oocytes in the absence of gonadotropin. AT_2 receptor stimulation also stimulated production of prostaglandins (PGE_2 and $PGE_{2\alpha}$) and estradiol, but not progesterone (Yoshimura et al. 1996). This observation suggests an AT_2 receptor-specific effect on aromatase activity (Yoshimura et al. 1996) or on the passive conversion of Ang II-induced thecal androgen production to granulosa cells estrogen (Pucell et al. 1991). In addition, Johnson et al. (1997), using human granulosa cells, showed that the decrease in progesterone production by activation of the AT_2 receptor was mediated through inhibition of 3β-HSD activity (but did not affect protein expression). In another study, AT_2 receptor expression was discernable only on granulosa cells present in follicle, attaining the tertiary stage of atresia, and representing the advanced process of ovulation and follicle disorganization (Obermuller et al. 1998). At this stage of maturation, the neoexpression of AT_2 receptors counteracts follicle-stimulating hormone (FSH)-stimulated survival signals and induces cell death (Kotani et al. 1999).

These data indicate that AT_2, but not AT_1 receptors for Ang II, play an important role during follicular development and in the ovulatory process. Moreover, as in the adrenal gland, AT_2 receptor activation induces specific steroid production, stimulating pathways involving P450C17 enzyme activity, inhibiting those involving 3β-HSD, and inducing apoptosis as well.

4
AT_2 Receptors in Chromaffin Cells

As mentioned earlier, AT_2 receptors are found in fetal adrenal gland, not only on adrenocortical cells, but also on chromaffin cells (Fig. 3B), as attested by co-labeling with chromogranin A and dopamine β-hydroxylase reactivities (Breault et al. 1996). Even if not yet formally demonstrated, a role for the AT_2 receptor in cell migration could be hypothesized, based on observations made in neuronal cells (Lucius et al. 1998; Côté et al. 1999). Indeed, pheochromoblasts originating from the neural crest begin migration throughout the fetal cortex as early as 6 weeks of pregnancy, and progressively colonize the center of the gland leading to the formation of the medulla. Paracrine action of steroids from the fetal cells induce their progressive differentiation into chromaffin cells.

In contrast to most tissues, expression of the AT_2 receptor in chromaffin cells persists in the adult (Martineau et al. 1999; Takekoshi et al. 2000). However, the role of the AT_2 receptor on catecholamine secretion is not yet clearly established. Martineau et al. (1999) found that AT_2 receptor stimulation increases both epinephrine and norepinephrine secretion. Initially, Takekoshi et al. (2001), found an inhibitory effect of the AT_2 receptor; however, more recently, the same group found that CGP42112, as well Ang II, stimulated catecholamine synthesis, through a process involving cGMP-dependent regulation of calcium influx through voltage-dependent Ca^{2+} channels (Takekoshi et al. 2001). The overall proposed hypothesis is that AT_2 stimulation decreases cGMP production, resulting in a reduction of PKG activity which leads to an increase of Ca^{2+} mobilization through voltage-dependent channels, stimulating catecholamine release.

However, the specific target of action, as well as the net effect of AT_2 receptor stimulation on catecholamine secretion, still remains unclear. Because the two types of Ang II receptors are present on chromaffin cells (Ishii et al. 2001), an interaction between the transduction pathways activated by both AT_1 and AT_2 receptors could be envisaged when attempting to explain Ang II action.

5
Differentiation of Smooth Muscle Cells

Yamada et al. (1999) have shown that the AT_2 receptor is transiently expressed during late gestation in the fetal vasculature and that this expression rapidly declines after birth. The authors examined the expression of various differentiation markers for vascular smooth muscle cells in the aorta of wild-type and of AT_2 receptor-null mice. Although there is no change in the expression of α-smooth muscle actin, mRNA levels of calponin and of the high-molecular weight caldesmon (h-caldesmon) are lower in the AT_2 receptor knockout mice than in control, wild-type mice. Their data demonstrated a correlation between time-dependent expression of the AT_2 receptor with that of h-caldesmon and calponin. All three proteins are absent during early development (E13–E15), their expression increases between E18–E20, then declines after 1 day in the neonate. These data suggest that the AT_2 receptor plays a role in the phenotypic differentiation of smooth muscle cells, by acting on cytoskeletal-associated proteins, rather than on actin itself. These proteins, h-caldesmon and calponin, bind actin and inhibit actomyosin ATPase, thus inhibiting contractility. Taken together, the delayed expression of these smooth muscle cells markers suggests that the AT_2 receptor may play some role in vasculogenesis.

6
Differentiation of Adipocytes

Adipose tissue contains differentiated and undifferentiated cells. Preadipocytes contain both AT_1 and AT_2 receptors (Darimont et al. 1994; Mallow et al. 2000). In addition, the AT_2 receptor expression is high in preadipocytes, but is completely lost after differentiation, suggesting a role for AT_2 during cell differentiation (Mallow et al. 2000). Using Ob1771 preadipocyte clonal cell line, Darimont et al. (1994) have shown that Ang II stimulation of AT_2 receptors stimulated the production of prostacyclin (PGI_2), the major metabolite of arachidonic acid in these cells. Through paracrine action, PGI_2 released by these cells is then able to participate in the differentiation of other undifferentiated preadipocytes (Darimont et al. 1994).

7
Conclusion and Perspectives

Altogether, these studies demonstrate that the AT_2 receptors (1) are involved in the modulation of apoptosis and migration, (2) affect expression of specific differentiating markers, such as MAPs in neurons, and (3) affect P450C17 expression in adrenals or caldesmon in smooth muscle cells. Even if initial studies conducted using models of knockout mice for the AT_2 receptor did not reveal major changes in development, more recent studies indicate several perturbations, in addition to the regulation of blood pressure. A role of AT_2 receptor in migration may appear as an important one, since knockout mice for AT_2 receptor present congenital anomalies of the kidney and urinary tract (CAKUT), both in mice and in human, presenting anomalies of the *AGTR2* gene (Nishimura et al. 1999; Miyazaki et al. 2001).

Despite numerous studies, little is known about the initial events associated with the AT_2 receptor, and the overall mechanism of action remains incompletely understood. For many of the cell models described in this chapter, stimulation of the AT_2 receptor induced modifications in cytoskeleton organization. Much of the data in the literature indicate that most of the proteins involved in intracellular signaling cascades are linked to microfilaments or microtubules (Schmidt et al. 1998; Schlaepfer et al. 1999; Nikolic 2002), among which are the heterotrimeric G proteins (Ibarrondo et al. 1995; Côté et al. 1997), the phosphatases SHP-1 (Brumell et al. 1997; Kim et al. 1999), and SHP-2 (Schoenwaelder et al. 2000; Xu et al. 2001), and the tyrosine kinases Src or Pyk2. Of note, a recent study showed that the AT_2 receptor inhibited the kinase Pyk2 (Matsubara et al. 2001). It is now well accepted that extracellular matrix can induce intracellular cell signals or interact with hormonal or growth factor transduction pathways leading to specific cell behaviors such as proliferation, migration, apoptosis, or gene expression (Aplin et al. 1999; Giancotti et al. 1999). Accumulating evidence suggests that the AT_2 receptor activates different signaling pathways, which may be attributed to interactions or associations with cytoskeleton-associated proteins, which together are linked to specific integrins; all are coordinated to imprint specific cell responses according to the nature of the extracellular matrix components.

Acknowledgements. We thank the current members of our laboratories for comments and also the past members, who over the years have contributed to many of the different findings regarding the AT_2 receptor action, in particular Liette Laflamme, Bruno Buisson, Frédéric Côté, Jean-François Oligny, and our technicians, Lyne Bilodeau and Lucie Chouinard. This work was supported by grants from the Canadian Institutes of Health Research to NGP and MDP (MOP 37912 and MOP 37891). Nicole Gallo-Payet is holder of a Canada Research Chair in Endocrinology of the Adrenal Gland.

References

AbdAlla S, Lother H, Quitterer U (2000) AT_1-receptor heterodimers show enhanced G-protein activation and altered receptor sequestration. Nature 407:94–98

Aguilera G, Kapur S, Feuillan P, Sunar-Akbasak B, Bathia AJ (1994) Developmental changes in angiotensin II receptor subtypes and AT_1 receptor mRNA in rat kidney. Kidney Int 46:973–979

Ambühl P, Felix D, Imboden H, Khosla MC, Ferrario CM (1992) Effects of angiotensin II and its selective antagonists on inferior olivary neurones. Regul Pep 41:19–26

Aplin AE, Juliano RL (1999) Integrin and cytoskeletal regulation of growth factor signaling to the MAP kinase pathway. J Cell Sci 112:695–706

Avila J, Dominguez J, Diaz-Nido J (1994) Regulation of microtubule dynamics by microtubule-associated protein expression and phosphorylation during neuronal development. Int J Dev Biol 38:13–25

Basille M, Gonzalez B, Fournier A, Vaudry H (1994) Ontogeny of pituitary adenylate cyclase-activating polypeptide) (PACAP) receptors in the rat cerebellum: a quantitative autoradiographic study. Dev Brain Res 82:81–89

Beaman-Hall CM, Vallano ML (1993) Distinct mode of microtubule-associated proteins expression in the neuroblastoma, glioma cell line 108CC15/NG108-15. J Neurobiol 24:1500–1516

Bedecs K, Elbaz N, Sutren M, Masson M, Susini C, Strosberg AD, Nahmias C (1997) Angiotensin II type 2 receptors mediate inhibition of mitogen-activated protein kinase cascade and functional activation of SHP-1 tyrosine phosphatase. Biochem J 325:449–454

Binder LI, Kim H, Caceres A, Payne MR, Rebhun LI (1984) Heterogeneity of microtubule-associated protein 2 during rat brain development. Proc Natl Acad Sci USA 81:5613–5617

Breault L, Lehoux J-G, Gallo-Payet N (1996) The angiotensin receptor AT_2 is present throughout the human fetal adrenal gland of the second trimester gestation. J Clin Endocrinol Metab 81:3914–3922

Brumell JH, Chan CK, Butler J, Borregaard N, Siminovitch KA, Grinstein S, Downey GP (1997) Regulation of Src homology 2-containing tyrosine phosphatase 1 during activation of human neutrophils. Role of protein kinase C. J Biol Chem 272:875–882

Buisson B, Bottari SP, De Gasparo M, Gallo-Payet N, Payet M-D (1992) The angiotensin AT_2 receptor modulates T-type calcium current in non-differentiated NG108-15 cells. FEBS Lett 309:161–164

Buisson B, Laflamme L, Bottari SP, De Gasparo M, Gallo-Payet N, Payet M-D (1995) A G protein is involved in the angiotensin AT_2 receptor inhibition of the T-type calcium current in non-differentiated NG 108-15 cells. J Biol Chem 270:1670–1674

Cambray-Deakin M, Morgan A, Burgoyne R (1987) Sequential appearance of cytoskeletal components during the early stages of neurite outgrowth from cerebellar granule cells in vitro. Dev Brain Res 37:197–207

Carey R, Jin X, Wang Z, Siragy H (2000) Nitric oxide: a physiological mediator of the type 2 (AT2) angiotensin receptor. Acta Physiol Scand 168:65–71

Cavallaro S, Copani A, D'Agata V, Musco S, Petralia S, Ventra C, Stivala F, Travali S, Canonico P (1996) Pituitary adenylate cyclase activating polypeptide prevents apoptosis in cultured cerebellar granule neurons. Mol Pharmacol 50:60–66

Chamoux E, Bolduc L, Lehoux JG, Gallo-Payet N (2001) Identification of extracellular matrix components and their integrin receptors in the human fetal adrenal gland. J Clin Endocrinol Metab 86:2090–2098

Chamoux E, Breault L, Lehoux J-G Gallo-Payet N (1999) Involvement of AT_2 receptor of angiotensin II in apoptosis during human fetal adrenal gland development. J Clin Endocrinol Metab 84:4722–4730

Chamoux E, Narcy A, Lehoux A, Gallo-Payet N (2002) Fibronectin, laminin and collagen IV as modulators of cell behavior during adrenal gland development in the human fetus. J Clin Endocrinol Metab 87:1819–1828

Chassagne C, Adamy C, Ratajczak P, Gingras B, Teiger E, Planus E, Oliviero P, Rappaport L, Samuel JL, Meloche S (2002) Angiotensin II AT_2 receptor inhibits smooth muscle cell migration via fibronectin cell production and binding. Am J Physiol Cell Physiol 282: C654–C664

Chiang CF, Hauser G (1989) Effects of bradykinin, GTP gamma S, R59022 and N-ethylmaleimide on inositol phosphate production in NG108-15 cells. Biochem Biophys Res Commun 165:175–182

Côté F, Do T, Laflamme L, Gallo J, Gallo-Payet N (1999) Activation of the AT_2 receptor of angiotensin II induces neurite outgrowth and cell migration in microexplant cultures of the cerebellum. J Biol Chem 274:31686–31692

Côté F, Laflamme L, Payet MD, Gallo-Payet N (1998) Nitric oxide, A new second messenger involved in the action of angiotensin II on neuronal differentiation of NG 108-15 cells. Endocrine Res 24:403–407

Côté M, Payet M-D Dufour M-N Guillon G, Gallo-Payet N (1997) Association of the G protein α_q/α_{11}-subunit with cytoskeleton in adrenal glomerulosa cells: role in receptor-effector coupling. Endocrinology 138:3299–3307

Darimont C, Vassaux G, Ailhaud G, Negrel R (1994) Differentiation of preadipose cells: paracrine role of prostacyclin upon stimulation of adipose cells by angiotensin-II. Endocrinology 135:2030–2036

DiPersio CM, van der Neut R, Georges-Labouesse E, Kreidberg JA, Sonnenberg A, Hynes RO (2000) alpha3beta1 and alpha6beta4 integrin receptors for laminin-5 are not essential for epidermal morphogenesis and homeostasis during skin development. J Cell Sci 113:3051–3062

Elbaz N, Bedecs K, Masson M, Sutren M, Strosberg AD, Nahmias C (2000) Functional trans-inactivation of insulin receptor kinase by growth- inhibitory angiotensin II AT_2 receptor. Mol Endocrinol 14:795–804

Feng YH, Sun Y, Douglas JG (2002) Gbeta gamma -independent constitutive association of $G\alpha s$ with SHP- 1 and angiotensin II receptor AT_2 is essential in AT_2-mediated ITIM- independent activation of SHP-1. Proc Natl Acad Sci USA 99:12049–12054

Fischer JW, Stoll M, Hahn AW, Unger T (2001) Differential regulation of thrombospondin-1 and fibronectin by angiotensin II receptor subtypes in cultured endothelial cells. Cardiovasc Res 51:784–791

Fitzgerald M (1987) Spontaneous and evoked activity of fetal primary afferents in vivo. Nature 326:603–605

Gallinat S, Csikos T, Meffert S, Herdegen T, Stoll M, Unger T (1997) The angiotensin AT_2 receptor down-regulates neurofilament M in PC12 W cells. Neurosci Lett 9:29–32

Gallo G, Letourneau P (2000) Neurotrophins and the dynamic regulation of the neuronal cytoskeleton. J Neurobiol 44:159–173

Gao BX, Ziskind-Conhaim L (1998) Development of ionic currents underlying changes in action potential waveforms in rat spinal motoneurons. J Neurophysiol 80:3047–3061

Ge J, Barnes NM (1996) Alterations in angiotensin AT_1 and AT_2 receptor subtype levels in brain regions from patients with neurodegenerative disorders. Eur J Pharmacol 297:299–306

Gendron L, Côté F, Payet M, Gallo-Payet N (2002) Nitric oxide and cyclic GMP are involved in angiotensin II AT_2 receptor effects on neurite outgrowth in NG108-15 cells. Neuroendocrinology 75:70–81

Gendron L, Laflamme L, Rivard N, Asselin C, Payet M, Gallo-Payet N (1999) Signals from the AT_2 receptor of angiotensin II inhibit p21^{ras} and activate MAPK to induce morphological neuronal differentiation in NG108-15 cells. Mol Endocrinol 13:1615–1626

Gendron L, Oligny JF Payet MD Gallo-Payet N (2003). Cyclic AMP-independent involvement of Rap1/B-Raf in the angiotensin II AT$_2$ receptor signaling pathway in NG108-15 cells. J Biol Chem 278:3606–3614

Giancotti F, Ruoslahti E (1999) Integrin signaling. Science 285:1028–1032

Gohlke P, Pees C, Unger T (1998) AT$_2$ receptor stimulation increases aortic cyclic GMP in SHRSP by a kinin-dependent mechanism. Hypertension 31:349–355

Gomez TM, Spitzer NC (2000) Regulation of growth cone behavior by calcium: new dynamics to earlier perspectives. J Neurobiol 44:174–183

Gordon-Weeks P (1991) Microtubule organization in growth cones. Biochem Soc Trans 19:1080–1085

Grady EF, Sechi L, Griffin C, Schambelan M, Kalinyak J (1991) Expression of AT$_2$ receptors in the developing rat fetus. J Clin Invest 88:921–933

Hamprecht B, Glaser T, Reiser G, Bayer E, Propst F (1985) Culture and characteristics of hormone-responsive neuroblastoma X glioma hybrid cells. Methods Enzymol 109:316–341

Hein L, Barsh G, Pratt R, Dzau V, Kobilka B (1995) Behavioural and cardiovascular effects of disrupting the angiotensin II type-2 receptor in mice. Nature 377:744–747

Horiuchi M, Akishita M, Dzau V (1999) Recent progress in angiotensin II type 2 receptor research in the cardiovascular system. Hypertension 33:613–321

Ibarrondo J, Joubert D, Dufour M-N Cohen-Solal A, Homburger V, Jard S, Guillon G (1995) Close association of the α subunits of G$_q$ and G$_{11}$ G proteins with actin filaments in WRK$_1$ cells: Relation to G protein-mediated phospholipase C activation. Proc Natl Acad Sci USA 92:8413–8417

Ichiki T, Labosky PA, Shiota C, Okuyama S, Imagawa Y, Fogo A, Niimura F, Ichikawa I, Hogan B LM, Inagami T (1995) Effects on blood pressure and exploratory behaviour of mice lacking angiotensin II type-2 receptor. Nature (London) 377:748–750

Inagami T, Kambayashi Y, Ichiki T, Tsuzuki S, Eguchi S, Yamakawa T (1999) Angiotensin receptors: molecular biology and signalling. Clin Exp Pharmacol Physiol 26:544–549

Ishii K, Takekoshi K, Shibuya S, Kawakami Y, Isobe K, Nakai T (2001) Angiotensin subtype-2 receptor (AT$_2$) negatively regulates subtype-1 receptor (AT$_1$) in signal transduction pathways in cultured porcine adrenal medullary chromaffin cells. J Hypertens 19:1991–1999

Johnson MC, Vega M, Vantman D, Troncoso JL, Devoto L (1997) Regulatory role of angiotensin II on progesterone production by cultured human granulosa cells. Expression of angiotensin II type-2 receptor. Mol Hum Reprod 3:663–668

Jöhren O, Häuser W, Saavedra J (1998) Chemical lesion of the inferior olive reduces [125I]Sarcosine1-angiotensin II binding to AT$_2$ receptors in the cerebellar cortex of young rats. Brain Res 793:176–186

Kang J, Sumners C, Posner P (1993) Angiotensin II type 2 receptor-modulated changes in potassium currents in cultured neurons. Am J Physiol 265:C607–C616

Kater SB, Mills LR (1991) Regulation of growth cone behavior by calcium. J Neurosci 11:891–899

Kawano K, Kantak SS, Murai M, Yao CC, Kramer RH (2001) Integrin alpha3beta1 engagement disrupts intercellular adhesion. Exp Cell Res 262:180–196

Kim CH, Qu CK, Hangoc G, Cooper S, Anzai N, Feng GS, Broxmeyer HE (1999) Abnormal chemokine-induced responses of immature and mature hematopoietic cells from motheaten mice implicate the protein tyrosine phosphatase SHP-1 in chemokine responses. J Exp Med 190:681–690

Knowle D, Ahmed S, Pulakat L (2000) Identification of an interaction between the angiotensin II receptor sub-type AT$_2$ and the ErbB3 receptor, a member of the epidermal growth factor receptor family. Regul Pept 87:73–82

Komuro H, Rakic P (1998) Orchestration of neuronal migration by activity of ion channels, neurotransmitter receptors, and intracellular Ca^{2+} fluctuations. J Neurobiol 37:110–130

Kostyuk P, Pronchuk N, Savchenko A, Verkhratsky A (1993) Calcium currents in aged rat dorsal root ganglion neurones. J Physiol 461:467–483

Kotani E, Sugimoto M, Kamata H, Fujii N, Saitoh M, Usuki S, Kubo T, Song K, Miyazaki M, Murakami K, Miyazaki H (1999) Biological roles of angiotensin II via its type 2 receptor during rat follicle atresia. Am J Physiol 276: E25–33

Kuhn K, Eble J (1994) The structural bases of integrin-ligand interactions. Trends Cell Biol 4:256–261

Laferriere N, MacRae T, Brown D (1997) Tubulin synthesis and assembly in differentiating neurons. Biochem Cell Biol 75:103–117

Laflamme L, De Gasparo M, Gallo J-M Payet MD, Gallo-Payet N (1996) Angiotensin II induction of neurite outgrowth by AT_2 receptors in NG108-15 cells. Effect counteracted by the AT_1 receptors. J Biol Chem 271:22729–22735

Letourneau P (1996) The cytoskeleton in nerve growth cone motility and axonal pathfinding. Perspect Dev Neurobiol 4:111–123

Lucius R, Gallinat S, Rosenstiel P, Herdegen T, Sievers J, Unger T (1998) The angiotensin II type 2 (AT_2) receptor promotes axonal regeneration in the optic nerve of adult rats. J Exp Med 188:661–670

Maccioni R, Cambiazo V (1995) Role of microtubule-associated proteins in the control of microtubule assembly. Physiol Rev 75:835–864

Mallow H, Trindl A, Loffler G (2000) Production of angiotensin II receptors type one (AT_1) and type two (AT_2) during the differentiation of 3T3-L1 preadipocytes. Horm Metab Res 32:500–503

Martineau D, Lamouche S, Briand R, Yamaguchi N (1999) Functional involvement of angiotensin AT_2 receptor in adrenal catecholamine secretion in vivo. Can J Physiol Pharmacol 77:367–374

Matsubara H, Shibasaki Y, Okigaki M, Mori Y, Masaki H, Kosaki A, Tsutsumi Y, Uchiyama Y, Fujiyama S, Nose A, Iba O, Tateishi E, Hasegawa T, Horiuchi M, Nahmias C, Iwasaka T (2001) Effect of angiotensin II type 2 receptor on tyrosine kinase Pyk2 and c-Jun NH2-terminal kinase via SHP-1 tyrosine phosphatase activity: evidence from vascular-targeted transgenic mice of AT_2 receptor. Biochem Biophys Res Commun 282:1085–1091

Matus A (1988) Microtubule-associated proteins: their potential role in determining neuronal morphology. Ann Rev Neurosci 11:29–44

Matus A, Bernhardt R, Hugh-Jones T (1981) High molecular weight microtubule-associated proteins are preferentially associated with dendritic microtubules in brain. Proc Natl Acad Sci USA 78:3010–3014

McCobb DP, Best PM, Beam KG (1989) Development alters the expression of calcium currents in chick limb motoneurons. Neuron 2:1633–1643

McIntyre P, Phillips E, Skidmore E, Brown M, Webb M (1993) Cloned murine bradykinin receptor exhibits a mixed B1 and B2 pharmacological selectivity. Mol Pharmacol 44:346–355

Meffert S, Stoll M, Steckelings UM, Bottari SP, Unger T (1996) The angiotensin II AT_2 receptor inhibits proliferation and promotes differentiation in PC12 W cells. Mol Cell Endocrinol 122:59–67

Merabet L, de Gasparo M, Casanova C (1997) Dose-dependent inhibitory effects of angiotensin II on visual responses of the rat superior colliculus: AT_1 and AT_2 receptor contributions. Neuropeptides 31:469–481

Mesiano S, Jaffe RB (1997) Developmental and functional biology of the primate fetal adrenal cortex. Endocr Rev 18:378–403

Millan MA, Kiss A, Aguilera G (1991) Developmental changes in brain angiotensin II receptors in the rat. Peptides 12:723–737

Miyazaki Y, Ichikawa I (2001) Role of the angiotensin receptor in the development of the mammalian kidney and urinary tract. Comp Biochem Physiol A Mol Integr Physiol 128:89–97

Nikolic M (2002) The role of Rho GTPases and associated kinases in regulating neurite outgrowth. Int J Biochem Cell Biol 34:731–745

Nishimura H, Yerkes E, Hohenfellner K, Miyazaki Y, Ma J, Hunley TE, Yoshida H, Ichiki T, Threadgill D, Phillips JA, 3rd Hogan BM, Fogo A, Brock JW, 3rd Inagami T, Ichikawa I (1999) Role of the angiotensin type 2 receptor gene in congenital anomalies of the kidney and urinary tract, CAKUT, of mice and men. Mol Cell 3:1–10

Obermuller N, Schlamp D, Hoffmann S, Gentili M, Inagami T, Gretz N, Weigel M (1998) Localization of the mRNA for the angiotensin II receptor subtype 2 (AT_2) in follicular granulosa cells of the rat ovary by nonradioactive in situ hybridization. J Histochem Cytochem 46:865–870

Okuyama S, Sakagawa T, Chaki S, Imagawa Y, Ichiki T, Inagami T (1999) Anxiety-like behavior in mice lacking the angiotensin II type-2 receptor. Brain Res 821:150–159

Parker CR, Jr. Stankovic AM, Goland RS (1999) Corticotropin-releasing hormone stimulates steroidogenesis in cultured human adrenal cells. Mol Cell Endocrinol 155:19–25

Peunova N, Enikolopov G (1995) Nitric oxide triggers a switch to growth arrest during during differentiation of neural cells. Nature 375:68–73

Phung Y, Bekker J, Hallmark O, Black S (1999) Both neuronal NO synthase and nitric oxide are required for PC12 cell differentiation: a cGMP independent pathway. Brain Res Mol Brain Res 64:165–178

Poluha W, Schonhoff CM, Harrington KS, Lachyankar M, Crosbie N, Buesco D, Ross A (1997) A novel, nerve growth factor-activated pathway involving nitric oxide, p53, and p21WAF1 regulates neuronal differentiation of PC12 cells. J Biol Chem 272:24002–24007

Pucell AG, Hodges JC, Sen I, Bumpus FM, Husain A (1991) Biochemical properties of the ovarian granulosa cell type 2-angiotensin II receptor. Endocrinology 128:1947–1959

Sanchez C, Diaz-Nido J, Avila J (2000) Phosphorylation of microtubule-associated protein 2 (MAP2) and its relevance for the regulation of the neuronal cytoskeleton function. Prog Neurobiol 61:133–168

Schelman WR, Kurth JL, Berdeaux RL, Norby SW, Weyhenmeyer JA (1997) Angiotensin II type-2 (AT_2) receptor-mediated inhibition of NMDA receptor signalling in neuronal cells. Brain Res Mol Brain Res 48:197–205

Schlaepfer DD, Hauck CR, Sieg DJ (1999) Signaling through focal adhesion kinase. Prog Biophys Mol Biol 71:435–478

Schmid S, Guenther E (1999) Voltage-activated calcium currents in rat retinal ganglion cells in situ: changes during prenatal and postnatal development. J Neurosci 19:3486–3494

Schmidt A, Hall M (1998) Signaling to the actin cytoskeleton. Annu Rev Cell Dev Biol 14:305–338

Schmitt R, Clozel JP, Iberg N, Buhler FR (1995) Mibefradil prevents neointima formation after vascular injury in rats. Possible role of the blockade of the T-type voltage-operated calcium channel. Arterioscler Thromb Vasc Biol 15:1161–1165

Schoenwaelder SM, Petch LA, Williamson D, Shen R, Feng GS, Burridge K (2000) The protein tyrosine phosphatase Shp-2 regulates RhoA activity. Curr Biol 10:1523–1526

Schütz S, Le Moullec JM, Corvol P, Gasc J-M (1996) Early expression of all the components of the renin-angiotensin-system in human development. Am J Pathol 149:2067–2079

Searles CD, Harrison DG (1999) The interaction of nitric oxide, bradykinin, and the angiotensin II type 2 receptor: lessons learned from transgenic mice. J Clin Invest 104:1013–1014

Segal R, Pomeroy S, Stiles C (1995) Axonal growth and fasciculation linked to differential expression of BDNF and NT3 receptors in developing cerebellar granule cells. J Neurosci 15:4970–4981

Shanmugam S, Z.G. Lenkei J.-M. Gasc P.L. Corvol Llorens-Cortes. CM (1995) Ontogeny of angiotensin II type 2 (AT_2) receptor mRNA in the rat. Kidney Int 47:1095–1100

Sheehy A, Phung Y, Riemer R, Black S (1997) Growth factor induction of nitric oxide synthase in rat pheochromocytoma cells. Brain Res Mol Brain Res 52:71–77

Shenoy U, Richards E, Huang X, Sumners C (1999) Angiotensin II type 2 receptor-mediated apoptosis of cultured neurons from newborn rat brain. Endocrinology 140:500–509

Siragy H, Carey R (1996) The subtype-2 (AT_2) angiotensin receptor regulates renal cyclic guanosine 3′,5′-monophosphate and AT_1 receptor-mediated prostaglandin E2 production in conscious rats. J Clin Invest 97:1978–1982

Siragy HM, Carey RM (1997) The subtype-2 (AT_2) angiotensin receptor mediates renal production of nitric oxide in conscious rats. J Clin Invest 100:264–269

Siragy HM, Jaffa AA, Margolius HS, Carey RM (1996) Renin-angiotensin system modulates renal bradykinin production. Am J Physiol 271:R1090–R1095

Smith C L (1994) Cytoskeletal movements and substrate interactions during initiation of neurite outgrowth by sympathetic neurons in vitro. J Neurosci 14:384–398

Spitzer NC, Lautermilch NJ, Smith RD, Gomez TM (2000) Coding of neuronal differentiation by calcium transients. Bioessays 22:811–817

Stoll M, Unger T (2001) Angiotensin and its AT_2 receptor: new insights into an old system. Regul Pept 99:175–182

Stroth U, Blume A, Mielke K, Unger T (2000) Angiotensin AT_2 receptor stimulates ERK1 and ERK2 in quiescent but inhibits ERK in NGF-stimulated PC12 W cells. Brain Res Mol Brain Res 78:175–180

Stroth U, Meffert S, Gallinat S, Unger T (1998) Angiotensin II and NGF differentially influence microtubule proteins in PC12 W cells: role of the AT_2 receptor. Mol Brain Res 53:187–195

Takekoshi K, Ishii K, Isobe K, Nanmoku T, Kawakami Y, Nakai T (2000) Angiotensin-II subtype 2 receptor agonist (CGP-42112) inhibits catecholamine biosynthesis in cultured porcine adrenal medullary chromaffin cells. Biochem Biophys Res Commun 272:544–550

Takekoshi K, Ishii K, Kawakami Y, Isobe K, Nakai T (2001) Activation of angiotensin II subtype 2 receptor induces catecholamine release in an extracellular Ca^{2+}-dependent manner through a decrease of cyclic guanosine 3′,5′-monophosphate production in cultured porcine adrenal medullary chromaffin Cells. Endocrinology 142:3075–3086

Tanaka M, Ohnishi J, Ozawa Y, Sugimoto M, Usuki S, Naruse M, Murakami K, Miyazaki H (1995) Characterization of angiotensin II receptor type 2 during differentiation and apoptosis of rat ovarian cultured granulosa cells. Biochem Biophys Res Commun 207:593–598

Tsutsumi K, Strömberg C. Viswanathan M. Saavedra J.M. (1991) Angiotensin-II receptors subtypes in fetal tissues of the rat: autoradiography, guanine nucleotide sensitivity and association with phosphoinositide hydrolysis. Endocrinology 129:1075–1082

Unger T (1999) The angiotensin type 2 receptor: variations on an enigmatic theme. J Hypertens 17:1775–17786

Vervoort VS, Beachem MA, Edwards PS, Ladd S, Miller KE, de Mollerat X, Clarkson K, DuPont B, Schwartz CE, Stevenson RE, Boyd E, Srivastava AK (2002) AGTR2 mutations in X-linked mental retardation. Science 296:2401–2403

Williamson T, Gordon-Weeks P, Schachner M, Taylor J (1996) Microtubule reorganization is obligatory for growth cone turning. Proc Natl Acad Sci USA 93:15221–15226

Wintour EM, Alcorn D, Albiston A, Boon WC, Butkus A, Earnest L, Moritz K, Shandley L (1998) The renin-angiotensin system and the development of the kidney and adrenal in sheep. Clin Exp Pharmacol Physiol Suppl 25: S97–S100

Wright JW, Schwinof KM, Snyder MA, Copenhaver PF (1998) A delayed role for nitric oxide-sensitive guanylate cyclases in a migratory population of embryonic neurons. Dev Biol 204:15–33

Xiong H, Marshall K (1994) Angiotensin II depresses glutamate depolarizations and excitatory postsynaptic potentials in locus coeruleus through angiotensin II subtype 2 receptors. Neuroscience 62:163–175

Xu F, Zhao R, Peng Y, Guerrah A, Zhao ZJ (2001) Association of tyrosine phosphatase SHP-2 with F-actin at low cell densities. J Biol Chem 276:29479–29484

Yaari Y, Hamon B, Lux HD (1987) Development of two types of calcium channels in cultured mammalian hippocampal neurons. Science 235:680–682

Yamada H, Akishita M, Ito M, Tamura K, Daviet L, Lehtonen JY, Dzau VJ, Horiuchi M (1999) AT_2 receptor and vascular smooth muscle cell differentiation in vascular development. Hypertension 33:1414–1419

Yamada T, Horiuchi M, Pratt R, Dzau V (1996) Angiotensin II type 2 receptor mediates programmed cell death. Proc Natl Acad Sci USA 93:156–160

York R, Yao H, Dillon T, Ellig C, Eckert S, McCleskey E, Stork P (1998) Rap1 mediates sustained MAP kinase activation induced by nerve growth factor. Nature 392:622–626

Yoshimura Y, Karube M, Aoki H, Oda T, Koyama N, Nagai A, Akimoto Y, Hirano H, Nakamura Y (1996) Angiotensin II induces ovulation and oocyte maturation in rabbit ovaries via the AT_2 receptor subtype. Endocrinology 137:1204–1211

Zhu M, Sumners C, Gelband CH, Posner P (2001) Chronotropic effect of angiotensin II via type 2 receptors in rat brain neurons. J Neurophysiol 85:2177–2183

AT_2 Renal Aspects

E. M. Abdel-Rahman · H. M. Siragy

Department of Medicine, University of Virginia Health System, Box 801409, Charlottesville, VA 22908, USA
e-mail: hms7a@virginia.edu

1	Introduction	424
2	AT_2 Development and Structural Features	425
3	AT_2 Signaling	426
4	AT_2R Localization	427
5	AT_1/AT_2 Interactions	428
6	AT_2R/NO Interaction	429
7	AT_2R and the Kidney	430
7.1	AT_2R Localization in the Kidney	431
7.2	AT_2R Actions and the Kidney	433
7.2.1	Pressure Natriuresis and Diuresis	433
7.2.2	Effects on Blood Vessels	434
7.2.3	Blood Pressure Regulation	435
7.2.4	Apoptosis	436
7.2.5	Antiproliferative Effects	437
7.2.6	Glomerular Actions	438
7.2.7	Diabetes Mellitus	438
7.2.8	Congenital Anomalies of the Kidney and Urinary Tract	439
8	Conclusion	439
	References	440

Abstract Renin–angiotensin system (RAS) plays an important role in the regulation of body fluids and electrolyte homeostasis and blood pressure control. Angiotensin II (Ang II), the major effector hormone of the RAS, functions mainly through stimulation of two angiotensin receptors subtypes, AT_1R and AT_2R. The well-known actions of Ang II have been attributed mainly to activation of various signal-transduction pathways modulated by AT_1R. Further research has led to the identification of the angiotensin subtype-2 (AT_2R) that is clearly distinct from AT_1R in its structure, tissue-specific expression, and signaling mechanisms. In this chapter, we review the different aspects of AT_2R including its development and structural features, its signaling pathways, and its localization in different tissues. The main focus of this review is on the AT_2 receptor locations and actions within the kidneys, and the different ways that AT_1R interacts with AT_2R. While on one hand AT_1R activation stimulates vasoconstriction, vascular cell hypertrophy and hyperplasia, sodium retention, stimulation of reactive oxygen species (ROS) production, and induction of inflammatory, thrombotic, and fibrotic processes, and under some conditions can lead to apoptosis in some cell types, on the other hand, AT_2R has vasodilatory, antigrowth, and apoptotic actions that opposes AT_1R actions. In the kidneys, AT_2R promotes vasodilation and natriuresis and plays a role in blood pressure (BP)-lowering effect of the AT_1R blockers. This vasodilatory action of AT_2R has been shown to be through bradykinin (BK)-/nitric oxide (NO)-dependent pathways. Besides reviewing the potential physiological actions of AT_2R in the kidneys, we also shed some light on the potential roles played by AT_2R in several disease processes affecting the kidneys, such as glomerular diseases, diabetes mellitus, and in congenital anomalies of the kidney and urinary tract.

Keywords AT_2 receptor · AT_1 receptor · Kidney · Nitric Oxide · Bradykinin

1
Introduction

Renin–angiotensin system (RAS) plays an important role in the regulation of body fluids and electrolyte homeostasis and BP control. Although RAS was originally regarded as a circulating system, most of its components are localized in tissues, indicating the existence of a local tissue RAS as well (Dzau 1988; Navar 1999). All the components of RAS, including renin, the substrate angiotensinogen, angiotensins, the enzymes involved in the synthesis and degradation of angiotensins, as well as angiotensin receptors are present in the kidney.

Angiotensin II (Ang II), the major effector hormone of the RAS, functions mainly through stimulation of two angiotensin receptors subtypes, AT_1R and AT_2R. These cellular receptors have been cloned, pharmacologically characterized, and localized at various tissues. The angiotensin subtype-1-receptor (AT_1R) is widely distributed throughout the body and mainly in the vasculature, heart, adrenal glands, kidneys, and nervous and endocrine systems.

The well-known actions of Ang II have been attributed mainly to activation of various signal–transduction pathways modulated by AT_1R. However, the discovery of the highly selective, peptide and non- peptide ligands such as CGP42112A and PD123319 has led to the identification of the angiotensin subtype-2 (AT_2R) (Timmermans et al. 1993). AT_2R is clearly distinct from AT_1R in its structure, tissue-specific expression, and signaling mechanisms.

The Ang II receptors can be distinguished according to inhibition by specific antagonists. AT_1R is selectively antagonized by biphenylimidazoles such as losartan, whereas tetrahydroimidazopyridines specifically inhibits AT_2R (Ardaillou 1999). AT_2R is also identified by being selectively activated by CGP42112A. This is a hexapeptide analog of Ang II, which at low dose may act as an AT_2R agonist, but may also inhibit the AT_2R and AT_1R at higher concentrations (Criscione et al. 1990). The availability of the radiolabeled forms of the highly selective AT_2R ligands (Whitebread et al. 1991; Heemskerk et al. 1993) made direct localization of AT_2R a practical possibility.

Drugs that antagonize AT_1R cause marked elevation of the levels of plasma Ang II, which in turn activates the AT_2R. With the increasing usage of AT_1R blockers, elucidation and understanding of AT_2R-mediated physiological actions will have important pharmacotherapeutic implications.

2
AT$_2$ Development and Structural Features

The AT_2R is a seven-transmembrane receptor comprising 363 amino acids, with a molecular mass of approximately 41 kDa. It has low amino acids sequence homology (around 32%–34%) with AT_1R (Kambayashi et al. 1993; Mu Koyama et al. 1993; Table 1).

The AT_2R protein consists of five potential N-glycosylation sites in the extracellular N-terminal domain and 14 cysteine residues. The second intracellular loop consists of a potential protein kinase C phosphorylation site (Griendling et al. 1996). The third intracellular loop is essential for AT_2R signal transduction via mitogen-activated protein (MAP) kinase inactivation (Lehtonen et al. 1997). The AT_2R expression is species-dependent. In general, humans have a much higher level of expression of AT_2R than rodents. In humans, AT_2R expression can be equal to, or even exceed that of AT_1R, but the functional significance of the AT_2R in humans remains unknown (Carey et al. 2000). The gene for AT_1R is located on chromosome 3 and exists as a single copy of the X-chromosome in

Table 1 General properties of AT$_2$

Structure	363 amino acids
	7 transmembrane
Molecular mass	41 kDa
Chromosomal localization	X-chromosome
Selective ligands	PD123319, PD123177, CGP42112-A
Signal transduction	Activation of phosphatase
	Inhibition of kinases
	Closing T-type Ca^{++} channel
	Opening delayed rectifier K^+ channel
	Stimulation of cGMP/NO
	Activation of phospholipase A_2
	Activation of STAT-1
	Generation of ceramides
Action	Pressure diuresis and natriuresis
	Vasodilatation
	Apoptosis
	Blood pressure reduction
	??? Protective role in D.M. complication
	??? Role in glomerular disease

???, There is some information to support a protective role but not enough to accept it as a fact.

human, rat, and mouse. The genes encoding the AT_2R are localized at xq22-q2 in human (Koike et al. 1994), chromosome xq3 in rat (Tissir et al. 1995), and chromosome x in mouse (Hein et al. 1995), with 99% amino acid residue sequence identity between rat and mouse, and 72% amino acid residue sequence identity between rat and human. While in some species AT_1R is expressed in two subtypes, genomic analyses have indicated that there is only one subtype to the AT_2R gene family (Mukoyama et al. 1993; Ichiki et al. 1995). While a 4.5-kb AT_2 cDNA was cloned from rat PC12 W cells and rat fetuses (Kambayashi et al. 1993), an AT_2 genomic 4.4-kb DNA fragment was cloned from mice (Ichiki et al. 1994). The AT_2R genomic DNA of all three species, humans, rats, and mice, consists of three exons with an uninterrupted coding region being confined in the third exon (Tsuzuki et al. 1994). This confinement of the entire coding system in one exon allowed the AT_2 coding sequence, or fragments containing more extensive sequence, to be cloned.

The expression of the AT_2R gene is dependent on growth state. The AT_2R gene is expressed ubiquitously at very high level in the fetus but then declines soon after birth in many, but not all, tissues. While it decreases to undetectable levels in the skin, the decline stops at certain levels in the heart and adrenal gland. Moreover, while AT_2R is expressed in uterine myometrium under nonpregnant conditions, it declines during pregnancy to return to nonpregnant levels after partition (de Gasparo et al. 2000). Furthermore, in special cells such as PC12 W cells (Kijima et al.1995), R3T3 cells (Horiuchi et al.1995), or mesangial cells (Goto et al. 1997), AT_2R expression is markedly increased when a confluent quiescent state is reached.

Several conditions were shown to modulate the expression of AT_2R. AT_2R expression is increased in adults after tissue injury, myocardial infarction, vascular injury, heart failure, and after sodium depletion (Siragy and Bedigian 1999). In contrast, Ang II, norepinephrine, insulin-like growth factor, basic fibroblast growth factor, and transforming growth factor beta (TGF) all decrease AT_2R expression (Carey et al. 2000b). Several cell lines express the AT_2R, but not the AT_1R, such as PC12 W, R3T3, and some lines of N1E115.

3
AT_2 Signaling

Ang II mediates its effects by acting directly through its receptors and via crosstalk with intracellular signaling cascades of other vasoactive agents, growth factors, and cytokines. Although the exact signaling pathways and the functional roles of AT_2R are not fully elucidated, these receptors may antagonize, under physiological conditions, the AT_1R-mediated actions (Ciuffo et al. 1998; Yamada et al. 1998).

Several signal transduction elements for AT_2R have been described. The growth-inhibitory effects of the AT_2R are reported to be at least partly mediated by the activation of protein tyrosine phosphatase, which results in the inhibition of AT_1R-activated mitogen-activated protein MAP kinase (Horiuchi et al. 1999).

In neuronal cells cultured from neonatal rat hypothalamus and brain stem, serine/threonine phosphatase 2A (PP2A) activation and subsequent extracellular signal-regulated kinase (ERK) inactivation through AT_2R have been reported (Huang et al. 1996). In the heart of transgenic mice overexpressing cardiac AT_2R, ERK activity was noted to be decreased, suggesting that ERK inactivation by the AT_2R has a physiological role in vivo (Masaki et al. 1998). Other signaling pathways, such as closing of a T-type Ca^{++} channel (Buisson et al. 1995), opening a delayed rectifier potassium channel (Kang et al. 1995), elevation in cyclic guanosine monophosphate (cGMP), stimulation of particulate guanylate cyclase, and stimulation of NO production have been proposed as signal transduction elements for AT_2R (Siragy and Carey 1996; Tsutsumi et al. 1999; Siragy et al. 2000). In vivo studies in the kidney have shown that Ang II can activate phospholipase A2 via AT_2R, resulting in the release of arachidonic acid and its metabolites, which in turn can activate Ras kinase and MAP kinase. (Dulin et al. 1998; Jiao et al. 1998).

Seebach et al. explored the signal transducer and activator of transcription (STAT) pathway as a possible mechanism of Ang II signaling in fetal human mesangial cells. These studies showed that AT_2R mediated the Ang II activation of STAT1, which may be an important signaling pathway for Ang II-induced cellular responses (Seebach et al. 2001). Another signaling pathway involving apoptosis through generation of ceramides was suggested by Gallinat et al. (1999).

4
AT₂R Localization

AT2R is expressed in abundance in mesenchymal tissues of the developing fetus, such as the uterus, the adrenal medulla, and specific brain regions, suggesting neuronal and developmental roles for the AT_2R (Matsubara 1998). However, in almost all tissues there is a rapid regression to low levels or even disappearance of expression of the AT_2R in the early postnatal period. Arce et al. showed high expression of AT_2R in 2-week-old rat hindbrains, which was absent in the adult rat, suggesting a probable role of these receptors in neuronal development (Arce et al. 2001). In adults, AT_2R were demonstrated in rat kidney (Zhuo 1992; Aguilera 1994; Ozono et al. 1997), rat heart (Sechi et al. 1992; Wang Z-Q et al. 1998), rat brain (Tsutsumi and Saavedra 1991), and rat adrenal glands (Shanmugam 1995). Other tissues that showed AT_2R expression were uterine myometrium, ovary, and pancreas (Matsubara 1998).

Using binding studies and quantitative autoradiography, AT_2 expression in adult Sprague-Dawley (SD) rat aorta, has been localized through out the aortic wall (Viswanathan et al. 1991). Similar results were demonstrated in Wistar-Kyoto (WKY) and spontaneous hypertensive (SHR) rats (Song et al. 1995). Both AT_1R and AT_2R were expressed in endothelial cells derived from SHR coronary arteries in an 80:20 ratio. (Stoll et al. 1995).

Immunohistochemistry staining showed that rat AT_2 expression varies with vessel size. It is more abundant in endothelial and vascular smooth muscle cells

(VSMC) in micro-vessels than in large vessels (Nora et al. 1998). Tsutsumi et al. failed to identify AT_2R expression in the endothelium of aorta from wild-type (WT) or transgenic mice overexpressing the AT_2R (Tsutsumi et al. 1999). Taken together, these results suggest that demonstration of the AT_2R in endothelial cells depends on species, size of vessels, and experimental conditions.

AT_2R is also expressed in both medial and adventitial layers of the rat mesenteric vasculature with predominant localization in the adventitial layer (Bonnet et al. 2001). In the heart, immunostaining for AT_2R was positive in the myocardium and coronary vessels throughout the ventricles and atria of neonatal and young rat hearts. The AT_2R expression was significantly more in the neonatal rat heart as compared to young rat heart (Wang et al. 1998).

5
AT_1/AT_2 Interactions

The interaction between receptor subtypes may be important to the physiological response to the agonist. Integrated responses to Ang II are the result of combined AT_1R- and AT_2R-mediated actions (Berry et al. 2001). It has been postulated that Ang II effects at the AT_1R are opposed by actions at the AT_2R (Carey et al. 2000b). Furthermore, it has been viewed that infusion of Ang II with an AT_2R antagonist represents an alternative model of AT_1R stimulation (Diep et al. 1999). Thus the concept of "crosstalk" was proposed where one Ang II subtype receptor will affect the other.

AT_1R activation stimulates vasoconstriction, vascular cell hypertrophy and hyperplasia, sodium retention (Berry et al. 2001), stimulation of reactive oxygen species (ROS) production (Berry et al. 2000), and induction of inflammatory (Muller et al. 2000), thrombotic (Vaughan et al. 1995) and fibrotic processes (Boffa et al. 1999), and under some conditions can lead to apoptosis in some cell types (Li et al. 1999). In contrast, AT_2R has vasodilatory, antigrowth, and apoptotic actions that oppose AT_1R actions (Berry et al. 2001).

Several organs and tissues showed manifestations of the crosstalk between AT_1R and AT_2R. In the heart, these receptors seem to exert opposite effects in terms of cardiovascular hemodynamics and cell growth (Xoriuchi et al. 1999). In neuronal cells, Laflamme et al. illustrated negative crosstalk interaction between AT_1R and AT_2R receptors (Laflamme et al. 1996). Tsutsumi et al. demonstrated that AT_1R and AT_2R play opposite roles in aortic vascular smooth muscles (VSM). In transgenic mice overexpressing AT_2R, chronic infusion of Ang II completely abolished the AT_1R- mediated pressor effects. (Tsutsumi et al. 1999) In the mesenteric artery of SD rats, Bonnet et al. demonstrated that AT_1R plays a role in the up-regulation of the AT_2R following Ang II infusion (Bonnet et al. 2001).

In the kidneys, the AT_1R is shown to induce vasoconstriction and sodium retention, whereas the AT_2R is postulated to promote vasodilation and natriuresis (Haithcock et al. 1999; Siragy and Carey 1999; Siragy et al. 1999). AT_2R inhibition prevents the hypotensive effects of AT_1R blockade. Thus the AT_2R seems to mediate the depressor responses to Ang II (Inagami et al. 1999; Siragy and Carey 1999; Tanaka et al. 1999). AT_2R has been shown to induce vasodilation through bradykinin (BK)-/nitric oxide (NO)-dependent pathways (Siragy and Carey 1999; Siragy et al. 2000). It has been suggested that AT_2R may play a role in the BP-lowering effect of AT_1R blockers.

Several studies examined the underlying mechanism(s) of the crosstalk occurring between AT_1 and AT_2 receptors. Hunley et al. showed that AT_2 decreases angiotensin converting enzyme (ACE) activity tonically, which by decreasing the formation of Ang II, attenuates AT_1R-mediated actions (Hunley et al. 2000). Moreover, the AT_2R appears to counterbalance the AT_1R by increasing the production of BK, NO, and cGMP, thus mediating vasodilation and BP-lowering actions.

6
AT_2R/NO Interaction

NO is synthesized from the amino acid L-arginine in endothelial as well as other cell types by the action of NO synthase (NOS) (Moncada et al. 1991). In the kidney, the NOS isoenzymes are distributed in close proximity to the sites of the RAS components, which may explain the interaction between these two systems (Millatt et al. 1999). Intrarenal NO helps to maintain the normally low renal vascular resistance, being responsible for up to one-third of the renal blood flow (Navar et al. 1996). It participates in the maintenance of renal perfusion and glomerular filtration in the normal kidney by its vasodilatory effect, mainly on the afferent arterioles of the superficial glomeruli (Kone 1997; Kone and Baylis 1997). NO also plays an important role in controlling tubular function. It acts as a natriuretic factor, by directly inhibiting Na^+ reabsorption in the proximal and collecting tubules, by inhibiting Na^+/H^+ exchange and Na^+-K^+-ATPase (Majid and Navar 1997). Other actions that involve NO include controlling cell growth, apoptosis, and inflammation (Gross and Wolin 1995).

The stimulation of AT_2R is associated with increased generation of BK (Siragy et al. 1996, 1998), NO (Siragy et al. 1998; Carey et al. 2000a), and cGMP (Siragy and Carey 1996), all of which have vasodilatory properties (Carey et al. 2000b). A vasotonic role for BK production during AT_2R activation has been reported in several animal models such as in the stroke-prone spontaneously hypertensive rats (SHR-SP) aorta (Gohlke et al. 1998) and in rats with two-kidney, one figure-eight wrap hypertension (Siragy and Carey 1999). Seyedi et al. studied coronary microvessels and large coronary arteries obtained from normal dogs. Nitrite, a product of NO metabolism, was increased following Ang II infusion, an effect blocked by BK B_2R antagonist, AT_1R antagonist, and AT_2R antagonist (Seyedi et al. 1995). These effects suggested a role for local kinin formation in the coronary vessels, and that AT_1R and AT_2R may play a role in mediating the release of nitrite. Golke et al. demonstrated that the attenuation in Ang II-induced pressor response in SHR-SP rats by AT_1R blockade was associated with a rise in aortic cGMP concentration. BK-B2R antagonist or L-nitroarginine methyl ester (L-NAME) (Golke et al. 1998) inhibited this increase in cGMP concentration. This study suggests that AT_2R activation during AT_1R blockade is associated with an increase in BK production, which in turn stimulates NO generation and can cause a vasodilatory effect. Bucher et al. described another interaction between AT_2R and NO. In this study SD rats were injected with either lipopolysaccharide or lipoteichoic acid to stimulate gram-negative and gram-positive sepsis, respectively. In both models of sepsis, AT_2R expression in the adrenal gland was downregulated and NOS II expression was induced. This downregulation of AT_2R was prevented by blocking of NO synthesis (Bucher et al. 2001). This study points to another important aspect of the AT_2R–NO interaction, where AT_2R could play an important role in the pathogenesis of septic shock that is NO-dependent.

Siragy and Carey demonstrated that Ang II stimulates cGMP production in the kidney, which could be inhibited by AT_2R blockade or NOS inhibitor (Siragy and Carey 1996, 1997a). These data demonstrated that AT_2R activation leads to an increase in renal NO production. They further demonstrated that the physiological action of Ang II in the kidney, such as the release of cGMP by the AT_2R, is potentiated during stressful conditions such as sodium depletion in which the RAS is stimulated. Reports from the same laboratory also demonstrated that the increase in Ang II, in response to AT_1R blocker valsartan, stimulated AT_2R, which mediates a BK and NO cascade (Siragy et al. 2000). Furthermore, it was shown that the AT_2-null mouse has low basal levels of renal BK and renal cGMP, an index of NO production, with no change in these levels following Ang II infusion or dietary salt restriction (Siragy et al. 1999). Israel et al. showed that systemic administration of AT_1R blocker, losartan, not only inhibited the pressor response induced by mild footshocks but also resulted in vasodepression. When AT_2R was blocked by PD123319 in combination with losartan, the vasodepressor response was eliminated (Israel et al. 2000). These data suggested that the vasodepressor response to footshocks in the presence of AT_1R antagonist is triggered by activation of AT_2R. This vasodepressor response was also blunted by BK, NO, and prostaglandin (PG) inhibitors, suggesting the involvement of BK, NO, and PG in the depressor response triggered by AT_2R activation in that model. Taken together, these results demonstrate that AT_2R is necessary for normal physiological responses of BK and NO, and that the renal NO production by AT_2R activation can be a major contributing factor in the regulation and control of multiple renal functions.

7
AT$_2$R and the Kidney

The kidney serves as one of the major target tissues that mediate the physiological actions of Ang II. Ang II has multiple effects on renal function including modulation of renal blood flow, glomerular filtration rate (GFR), tubular epithelial transport, renin release, and cellular growth (Matsubara 1998). Ang II levels in renal interstitial fluid (RIF) were reported to be 1,000 times higher than systemic plasma levels, suggesting the importance of Ang II in the local regulation of renal function (Siragy et al. 1995). Furthermore, Nishiyama et al. confirmed that RIF concentrations of Ang I and Ang II are substantially higher than the corresponding plasma concentrations (Nishiyama et al. 2002). The high RIF concentrations of Ang II were not responsive to acute ACE inhibition or volume expansion, suggesting the compartmentalization and independent regulation of RIF Ang II.

AT_2R has specific effects on renal function. The detection of increased incidence of congenital anomalies of the kidney and the urinary tract in AT_2R-deficient mice, suggests a role for AT_2R in genitourinary development (Guron and Friberg 2000). Among the renal effects of the AT_2R stimulation are: mediation of renal interstitial production of NO, BK, and cGMP; counter-regulatory protective role against AT_1R-mediated anti natriuretic and pressor actions of Ang II; and enhancing the activity of 9-ketoreductase enzyme leading to conversion of prostaglandin E2 (PGE2) to prostaglandin F2 alpha (PGF2 alpha) (Siragy and Carey 1996, 1997b).

7.1
AT_2R Localization in the Kidney

A paucity of information exists about the localization of the AT_2R in the kidney. Autoradiography coupled with competitive binding studies have been used to characterize the distribution of Ang II receptors subtypes in renal tissues. By using these techniques, the distribution of AT_1R and AT_2R subtypes showed remarkable species-differences (Table 2).

Table 2 AT_2 localization in the kidney

Species	Age	Location
Mouse	Fetal	Mesenchymal cells of mesonephros surrounding the mesonephric tubules
Rat	Fetal	Mesenchymal cells surrounding the epithelial structure and ureteric buds, immature glomeruli, primitive cortical and medullary tubules, interstitial cells
	Newborn	Glomeruli, proximal and distal tubules, blood vessel
	Young adult	Glomeruli and tubules. Re-expressed with sodium depletion in glomerular mesangial cells, tubules and mesenchyma, interstitial cells of outer cortex
	Mature adult	Glomeruli and tubules. Enhanced by sodium depletion in glomeruli, interstitial cells, and adventitia of preglomerular arcuate and interlobular arteries.
Rabbit		Fibrous sheath around the kidney
Rhesus monkey		JGA, vasculature of renal cortex
Human	Fetal	Mesenchymal cells adjacent to the Stalk of ureter epithelium, superficial cortex, between collecting ducts
	Children	Interlobular arterial
	Adult	Glomeruli and tubules, large cortical preglomerular vessels, interlobular arteries, tubulo-interstitium

Kakuchi et al. reported that AT_2R is intensely expressed in the mesenchymal cells of the mesonephros surrounding the mesonephric tubules of fetal mouse kidney (Kakuchi et al. .1995). In the rat fetal kidney, AT_2R is heavily expressed in immature glomeruli, cortical and medullary tubules and interstitial cells, but within few days of birth, AT_2R expression diminishes rapidly, and is observed only at low levels in adult kidney (Ozono et al. 1997; Carey et al. 2000b).

Ozono et al. published an eloquent study using immunohistochemical staining to detect AT_2R gene in fetal (day 14 and 19 of fetal life), newborn (day 1 postpartum) and adult rat kidney (4 weeks and 3 months old). The study showed heavy immunohistochemical staining for the AT_2 receptor in the undifferentiated mesenchymal cells surrounding the epithelial structures and in the ureteric buds in the F14 fetus. In the F19 fetus, AT_2 receptor signal was detected in the S-shaped glomeruli, primitive tubules, and mesenchymal tissue. The study further showed positive staining in glomeruli, proximal and distal renal tubule cells, and blood vessels of the newborn rat kidney at D1. In young adult (4-week-old) rats on normal sodium intake, renal staining for the AT_2 receptor was markedly reduced but remained detectable in the glomeruli and tubules. Sodium depletion increased AT_2R expression in young adult rats, in the glomeruli, tubules, and mesenchyma. The glomerular staining was localized to mesangial cells and also was present in the interstitial cells of the outer cortex. Similar results were shown in the mature adult (3-month-old) rats. On normal sodium intake, there was some glomerular and tubule expression of AT_2R protein signal that was enhanced during sodium depletion predominantly in the glomeruli and in the interstitial cells of the outer cortex (Ozono et al. 1997).

The status of renal AT_2R in the adult rat kidney has been controversial. Initial studies (Gibson et al. 1991) suggested that there was no gene expression for AT_2R in adult kidney. However, using immunohistochemical techniques and autoradiography, it has been shown that AT_2R is present in low levels in the glomerular epithelial cells, cortical tubules, and interstitial cells of glomeruli and tubules of rat kidney. (Ozono et al. 1997; Wang et al. 1999). Miyata et al., using immunohistochemistry and reverse transcription-polymerase chain reaction (RT-PCR) found the AT_2R mRNA and protein expression throughout the rat kidney, but little expression was found in the glomerulus and the medullary thick ascending limbs of Henle (Miyata et al.1999). Sharma et al. demonstrated the presence of AT_1R and AT_2R in rat glomerular epithelial cells, and proposed that Ang II and its interaction with AT_1R and AT_2R on glomerular epithelial cells play an important part in the regulation of glomerular function (Sharma et al. 1998).

AT_2R and its mRNA are most prominent in the fetal and newborn mammalian kidney. In contrast, the expression of AT_2R in the adult mammalian kidney has been reported to be very low and localized mainly in the glomerular and mesangial cells (Goto et al. 1997; Ozono et al. 1997), or adventitia of the preglomerular arcuate and interlobular arteries (Zhou et al. 1996). Autoradiography revealed that in rabbit, the fibrous sheath around the kidney contains AT_2R binding sites (Herblin et al. 1991), and that in rhesus monkeys it is present in juxta-glomerular apparatus (JGA) and in the vasculature of the renal cortex (Gibson et al. 1991).

In the human fetal kidney, the presence of AT_2Rm RNA was detected in undifferentiated mesenchymal cells adjacent to the stalk of the ureter epithelium, in the superficial cortex near nephrogenic areas, and in the area between collecting ducts (Grone et al. 1992). In adult human kidney, AT_2 mRNA was localized in the medial layers of interlobular arteries (Matsubara et al. 1998). AT_2R was also found in large cortical preglomerular vessels and the tubulo-interstitium (Grone et al. 1992; Goldfarb et al. 1994; Zhou et al. 1996). The presence of AT_2R in the interlobular arteries was also shown in children. Vinswanathan et al. examined kidneys obtained at autopsy from five children (aged 1 month to 2 years) during the first 24 h after death. Ang II receptor subtypes were localized by quantitative autoradiography. While renal glomeruli expressed exclusively AT_1R, 80% of the Ang II receptors in the interlobular artery belonged to AT_2R (Vinswanathan et al. 2000). Immunohistochemistry of the human adult kidney with the use of a polyclonal antibody to the AT_2R revealed a positive signal in the glomerular epithelial elements, tubules, and large-order renal blood vessel (Carey et al. 2000b).

7.2
AT_2R Actions and the Kidney

Ang II takes part in a number of regulatory events in the kidney mainly the control of renal vascular resistance, glomerular filtration, and tubular epithelial transport. These renal actions of Ang II are mediated by AT_1R, which predominates in glomeruli, JGA, and proximal tubules in adult human kidneys. However, the AT_2R also seems to have some specific effects on renal functions. AT_2R is postulated to play a role in pressure natriuresis, vascular and BP regulation, apoptosis and antiproliferation, glomerular action, and certain pathologic renal conditions.

7.2.1
Pressure Natriuresis and Diuresis

Controversial data exist regarding the role of AT_2R in pressure natriuresis and diuresis. While most studies proposed that AT_2R enhance both pressure natriuresis and diuresis (Siragy et al. 1999; Gross et al. 2000), other studies did not support this proposal (Lo et al. 1995; Madrid et al. 1997). Siragy et al. suggested that AT_2R stimulation can physiologically increase pressure natriuresis and proposed that the effect is mediated by BK/NO cascade. In AT_2-null mice, chronic infusion of sub-pressor doses of Ang II for 1 week increased the systolic BP (SBP) and reduced urinary sodium excretion ($U_{NA}V$), whereas neither SBP nor $U_{NA}V$ changed in WT mice. Furthermore, in WT mice, dietary sodium restriction or Ang II infusion increased RIF BK and cGMP, with no changes noted in the AT_2-null mice. These results suggest that the absence of AT_2R leads to sodium retention and hypertension in the presence of physiologically increased

Ang II (Siragy et al. 1999). In agreement with Siragy et al. was the study by Gross and colleagues. Under the same renal perfusion pressure (RPP), Gross et al. noted that the control WT mice excreted threefold the quantity of sodium and water compared to the AT_2 null mice, suggesting that AT_2R plays a role in enhancing pressure natriuresis and diuresis (Gross et al. 2000). Haithcoch et al. showed that AT_2R is linked to inhibition of bicarbonate reabsorption in the proximal tubule, an effect that opposes AT_1R-mediated facilitation of sodium and bicarbonate reabsorption and suggests a role for AT_2R in natriuresis. These natriuretic and diuretic responses were not related to hemodynamic changes (Haithcoch et al. 1999)

7.2.2
Effects on Blood Vessels

Ang II directly influences blood vessels structure and function. The atrophic and proliferative effects of Ang II on the vasculature are modulated by both AT_1R and AT_2R. Tsutsumi et al. demonstrated that AT_1R and AT_2R play opposite roles in aortic VSM. In transgenic mice with overexpressed AT_2, chronic infusion of Ang II completely abolished the AT_1R-mediated pressor effects, suggesting an AT_2-mediated vasodilation (Tsutsumi et al. 1999). Sheuer et al. and Munzenmaier et al. showed data confirming the vasodilatory actions of AT_2R. Sheuer et al. demonstrated that AT_2 mediates the depressor phase of the biphasic BP response to Ang II in mature rats (Sheuer et al. 1993), while Munzenmaier and coworkers showed the pressor action of Ang II to be enhanced with chronic treatment with AT_2R antagonist PD123319 in SD rats (Munzenmaier et al. 1996). In agreement with these data is the observation that the pressor actions of Ang II are further enhanced in the AT_2-null mice model (Hein et al. 1995; Ichicki et al. 1995). Furthermore, Ichiki et al. and Siragy et al. demonstrated that AT_2R-null mice have a higher sensitivity to BP and exert enhanced pressor actions of Ang II (Ichicki et al. 1995; Siragy et al. 1999).

To delineate the mechanisms of the AT_2R vasodilatory effects, Tsutsumi et al. chronically infused Ang II into mice overexpressing the AT_2R. In these studies, Ang II completely abolished the AT_1-mediated pressor effect, which was blocked by inhibitors of bradykinin type 2 receptor (icatibant) and NO synthase (L-NAME). The vasoconstrictive response of Ang II was enhanced by AT_2R blockade, suggesting that AT_2 antagonizes the AT_1-mediated vasoconstrictive actions in aortic VSMC through the stimulation of bradykinin–NO–cGMP cascade (Tsutsumi et al. 1999). The proposed mechanism of stimulation of this cascade is that AT_2 causes intracellular acidic changes through inhibition of amiloride-sensitive Na^+/H^+ exchanger activity. The increase in cellular acidity stimulates the activity of kininogenases that in turn increase formation of bradykinin (Tsutsumi et al. 1999). Arima and Ito proposed another mechanism while studying rabbit afferent arteriole (Af-Art). They showed that AT_1R blockade abolished the Ang II-induced vasoconstriction. Further infusion of Ang II in the setting of AT_1R blockade caused dose-dependent dilation, which was abolished by

AT$_2$R blockade, suggesting an AT$_2$R-mediated vasodilation. While the dilation was not affected by inhibiting the synthesis of NO or prostaglandin (PG), it was completely abolished by either disrupting the endothelium or inhibiting the synthesis of epoxyeicosatrienoic acid (EET), a cytochrome P-450 (CYP) epoxygenase metabolite of arachidonic acid that is a potent vasodilator in the Af-Art. Thus, AT$_2$R induces endothelium-dependent and EET-mediated vasodilation, which is independent of NO release (Arima and Ito 2000). Using the same model of rabbit Af-Art, Kohagura et al. induced vasoconstriction by inhibiting NOS or EET using L-NAME or miconazole respectively. The addition of Ang II enhanced further this vasoconstriction. When the experiment was repeated with the addition of AT$_2$R blocker PD123319, only L-NAME augmented the vasoconstrictive activity of Ang II, while EET inhibitor, miconazole had no effect. These studies suggest an additional mechanism whereby AT$_2$R mediates vasodilation via coupling to EET (Kohagura et al. 2000).

Contrary to the proposed vasodilator effect of AT$_2$R, Croft et al. showed that Ang II increases CYP monooxygenase in preglomerular microvessels (PGMVs) of SD rats, which in turn cause an increase in the vasoconstrictor 20-hydroxyeicosatrienoic acid (HETE) by two- to threefolds. This effect was blocked by AT$_2$R blockade (PD123319) and phospholipase C (PLC) inhibitor (U-73122). These data suggested that AT$_2$R-PLC effector unit is associated with synthesis of a vasoconstrictor product in PGMVs (Croft et al. 2000).

7.2.3
Blood Pressure Regulation

Several studies suggest that AT$_2$R plays a role in BP control, counterbalancing the AT$_1$R pressor action and causing a reduction in Ang II-induced hypertension. AT$_2$R has been suggested to have a direct vasodilatory action that can cause reduction in BP. Furthermore, AT$_2$R through its pressure natriuresis and diuresis actions can achieve similar BP reduction effects. The malfunction of pressure natriuresis and diuresis leads to arterial hypertension.

In a series of studies (Siragy and Carey 1996, 1997a, 1999; Siragy et al. 1996), Siragy et al. provided evidence for a vasodilatory cascade mediated by AT$_2$R. These studies showed that Ang II, via AT$_2$R, triggers a cascade of increase tissue BK, synthesis of NO, and cGMP, and ultimately leading to vasodilation and reduction of the increased BP associated with Ang II.

The protective role of the renal AT$_2$R was further investigated in a 2-kidney, 1 figure-8 wrap model (Siragy and Carey 1999). Siragy et al. observed that BK, NO, and cGMP were all higher in the intact kidney than in the wrapped kidney. While AT$_1$R blockade increased BK, NO, and cGMP further in the intact kidney, AT$_2$R blockade decreased these products in both kidneys. They concluded that Ang II triggers the production of BK in the kidney via the stimulation of AT$_2$R, which then releases NO and cGMP to produce counter-regulatory vasodilation in this model. The study also suggests that the hypotensive response associated with AT$_1$R blockade is partially mediated by the AT$_2$R. To further expand on the role that AT_2R plays in BP regulation, Carey et al. infused the selective AT_2R agonist CGP42112A, and showed that it decreased BP. They further showed that this BP-lowering response was blocked by either AT_2R blockade PD123319 or the NOS inhibitor L-NAME, suggesting that the AT_2R induces a vasodilator response mediated by NO (Carey et al. 2001).

The role of AT_2R in BP regulation was studied further in AT_2-null mice (Hein et al. 1995; Ichiki et al. 1995; Siragy et al. 1999). At base line in the absence of AT_2R, AT_2-null mice have slightly elevated BP. This mouse model demonstrated pressor hypersensitivity to Ang II compared with WT controls. When the RAS was activated by dietary sodium restriction or Ang II infusion, the AT_2-null mice did not increase BK, NO, nor cGMP in contrast to the WT control mice (Siragy et al. 1999). These data support the hypothesis that AT_2R plays a role in lowering BP through the vasodilatory actions of BK/NO/cGMP.

Additional studies using AT_1R blockers (Matrougui et al. 1999) alone or in combination with AT_2R agonist (Barber et al. 1999) or in combination with salt restriction (Siragy et al. 2000) confirmed the vasodilator actions of AT_2R stimulation and its role in BP regulation.

7.2.4
Apoptosis

Apoptosis is characterized by a series of morphological events such as shrinkage of the cell, condensation of chromatin, fragmentation into apoptotic bodies, and rapid phagocytosis by neighboring cells. It plays a crucial role in the normal development as well as in the pathophysiology of a variety of tissues that have undergone injury or damage. Among many other factors, the balance between pro- and anti-apoptotic proteins determines whether a cell will undergo programmed cell death or survive.

AT_2R stimulation has been demonstrated to induce apoptosis in a number of cells in correlation with extensive cell death in the developing fetus. AT_2R induces apoptosis in cells of neuronal origin, such as PC12 W (Yamada et al. 1996) and cultured neurons from newborn rat (Shenoy et al. 1999), rat ovarian granulosa cells (Tanaka et al. 1995), fibroblast from mouse embryo (Li et al. 1998), and endothelial cells from human (Dimmeler et al. 1997). AT_2R was also shown to mediate vascular mass regression by stimulating smooth muscle cells apoptosis (Tea et al. 2000). In a study by Suzuki et al., vascular injury was induced by polyethylene cuff placement around the left femoral artery of AT_2-null mice. They found that the number of apoptotic cells decreased in this mouse model, pointing to a proapoptotic effect of AT_2R on VSMC in the process of neointimal formation after vascular injury (Suzuki et al. 2002).

In the kidney, it has been suggested that the balance between cell proliferation and apoptosis plays an important role in the pathophysiology of various renal diseases (Kovacs and Gomba 1998; Thomas et al. 1998; Truong et al. 1998). Ang II was shown to induce apoptosis in renal tubular cells. This was demonstrated by the immunohistochemical techniques and confirmed by electron microscopy. Both AT₁R and AT₂R antagonists attenuated this effect of Ang II on apoptosis. In a kidney model of rat with unilateral ureteral obstruction, AT₂R blockade suppressed tubular cell apoptosis (Morrissey and Klahr 1999). Using the same model of unilateral ureteral obstruction in male WT and AT₂R-deficient mice, Ma et al. investigated the role of Ang II in renal remodeling. They observed that the obstructed kidney from the AT₂R knock-out mice showed significantly fewer apoptotic cells compared to WT control mice, supporting a role for AT₂R in apoptotic cell death (Ma et al. 1998).

Ceramide, an intracellular lipid second messenger and one of the most hydrophobic molecules in mammalian cells, has been implicated as an important mediator of apoptosis (Pushkareva et al. 1995). Gallinat et al. demonstrated that Ang II increased ceramide levels during apoptosis in PC12 W cells, and that this action was completely abolished by co-incubation with an AT₂R antagonist, confirming that apoptosis is mediated by AT₂R (Gallinat et al. 1999). Miura and Karnik further showed that induction of apoptosis could be a constitutive function of the AT₂R, and proposed that overexpression of the AT₂R itself may act as a signal for apoptosis that did not require Ang II (Miura and Karnik 2000).

7.2.5
Antiproliferative Effects

As mentioned before, the balance between cell proliferation and apoptosis plays an important role in the pathophysiology of various diseases (Kovacs and Gomba 1998; Thomas et al. 1998; Truong et al. 1998). Ang II was noted to be able to induce both cellular proliferation (Bunkenburg et al. 1992; Kunert-Radek et al. 1994) and apoptosis (Anaka et al. 1995; Kajstura et al. 1997) in renal cystic diseases, renal fibrosis, and obstructive uropathy. While the proliferative properties of Ang II were considered to be mediated by AT₁R, the AT₂R was considered to promote apoptosis (Nakajima et al. 1995; Yamada et al. 1996). The opposite was also suggested where AT₁R mediates apoptosis (Croft et al. 2000) and AT₂R mediates proliferation (Levy et al. 1996). To establish a mechanistic pathway through which Ang II exerts its growth promoting effect, Li et al. infused Ang II to WKY rats with AT₁R blocker or AT₂R blocker for 21 days. They noted that growth of the heart, aorta, coronary, renal, mesenteric, and femoral arteries are mediated by the AT₁R, with little evidence of a role of AT₂R (Li et al.1998). This result raised the question of whether AT₂R subtype has growth-inhibitory and proapoptotic properties or no role in growth at all. Stoll et al. demonstrated an antiproliferative effect of Ang II on coronary endothelial cells that could be blocked by the AT₂R blocker PD123319 (Stoll et al. 1995). These results suggest that the antiproliferative effect of Ang II was mediated via AT₂R. In agreement with this observation were Brogelli et al. and Suzuki et colleagues. Using an in vitro organ culture model, Brogelli et al. demonstrated that the growth of VSMC of male Wistar rat aorta was inhibited by AT₁R blockade, yet increased by the AT₂R antagonist. This data demonstrated that AT₂R-mediated inhibition of VSMC proliferation in adult vessels (Brogelli et al.2002). Suzuki et al. demonstrated that neointimal formation as well as DNA synthesis in VSMC was exaggerated in AT_2-null mice following vascular injury, suggesting an antiproliferative effects of AT_2R (Suzuki et al. 2002).

In the kidney, Goto et al. examined AT_2 expression using culture mesangial cells (MC) from normotensive WKY and from SHR-SP rats. MC from SHR-SP, whose proliferative activity was much higher than WKY, showed only AT_1R subtype. The lower expression of AT_2R in MC from SHR-SP may suggest its involvement in the MC higher proliferative activity and possibly in the development of renal disorders (Goto et al. 1997).

7.2.6
Glomerular Actions

AT_2R seems to play a role in regulation of glomerular blood flow (GBF). Selective activation of AT_2R using afferent arteriole microperfusion technique was shown to cause endothelium-dependent vasodilation via CYP, possibly by releasing vasodilatory EET (Kohagura et al. 2000). On the other hand, AT_2R was shown to be able to do the opposite by increasing synthesis of the vasoconstrictor product of arachidonic acid metabolite, 20 HETE (Croft et al. 2000).

In a human study, Mifune et al. examined kidney samples from 21 patients with or without glomerular lesions. The glomerular lesions included IgA nephropathy, minimal change disease, and membranous glomerulonephritis. While mild–moderate immunohistochemical staining of AT_2R was seen in the blood vessels, weaker staining was found in the glomeruli. These results were observed in both the normal and diseased kidney samples, suggesting a role for the AT_2R in both normal and diseased human kidneys (Mifune et al. 2001).

7.2.7
Diabetes Mellitus

Diabetes mellitus (DM) has grown into an epidemic. Several studies pointed toward a significant role of the RAS activation in DM. Clinical studies demonstrated a role for interrupting the RAS at the ACE level and/or at the AT_1R levels, using ACEI and AT_1R blockade, in delaying progression of diabetic nephropathy (Lewis et al. 1993, 2001; Parving et al. 2001). In spite of that, the source of intrarenal Ang II formation remains unclear. Several studies have shown that renal renin mRNA and activity, proximal tubule angiotensinogen expression, and proximal tubule cell mRNA expression are increased in the diabetic model (Anderson et al. 1993; Wang TT et al. 1998; Zimpelmann et al. 2000). Once Ang II is formed in the kidney, it exerts most of its effects as vasoconstriction, increased glomerular capillary pressure, and mechanical stretch-induced glomerular injury via AT_1R (Yuan et al. 1990; Akai et al. 1994). Recently, Wehbi et al. studied the role of intrarenal AT_2R in the progression of diabetic nephropathy in streptozotocin (STZ)-induced diabetic SD rats. The study demonstrated by RT-PCR that early diabetes had no significant effect on glomerular mRNA expression of renin, angiotensinogen, or ACE. While the non-glycosylated AT_1R expression was increased in the diabetic rats, there was a significant decrease in glomerular AT_2R protein expression. Furthermore, AT_2R expression was decreased in all kidney regions in early diabetes. These data suggest that alterations in the balance of kidney AT_1R and AT_2R expression may contribute to Ang II-mediated glomerular injury in progressive diabetic nephropathy (Wehbi et al. 2001).

Bonnet et al. further confirmed that AT_2R was reduced in long-term diabetic SHR. In a study where they assessed gene and protein expression of the AT_1R and the AT_2R using RT-PCR, immunohistochemistry, and autoradiography in STZ-induced diabetes in SHR and WKY, they showed that both AT_1R and AT_2R mRNA levels in the kidney, glomerular, and tubular–interstitial staining for both AT_1R and AT_2R, and their binding was reduced (Bonnet et al. 2002). These data suggest a role for the AT_2R in the pathogenesis of renal injury, including DM.

7.2.8
Congenital Anomalies of the Kidney and Urinary Tract

The RAS plays a role in cell proliferation and differentiation, and organogenesis, especially nephrogenesis (Wolf and Neilson 1996). AT_2R is more expressed during development and appears to affect tissue remodelling through apoptosis and plays a role in nephrogenesis.

Congenital anomalies of the kidney and urinary tract account for more than 50% of cases of abdominal mass found in neonates, and involve 0.5% of all pregnancies (Scott et al. 1988). Studies of the mice and human AT_2R gene showed an association between being AT_2R null and congenital anomalies of the kidney and urinary tract (Nishimura et al. 1999). The AT_2R is postulated to be involved in the development of the ureteric lumen through apoptotic resorption of cells. Thus, the occurrence of a mutation of the AT_2R was shown to occur more frequently in individuals with pelvi-ureteric junction obstruction as well as lower ureter malformation (Nishimura et al. 1997). Hohenfellner et al., while confirming a role for AT_2R for the normal development of the ureter, could not find a relationship between AT_2R and vesico-ureteral reflux (Hohenfellner et al. 1999).

8
Conclusion

While our knowledge of AT_1R actions in the kidney is better delineated, most of the data regarding the role of AT_2R in normal physiological and in pathological situations remain unclear. A vast number of basic and clinical studies over the past few decades have confirmed a major role for drugs that interrupt the RAS at the ACE or the AT_1R levels in the management of several diseases including hypertension, diabetic nephropathy, congestive heart failure, and some of the glomerular diseases.

The possible protective role that AT_2R may play in some of these diseases is intriguing. Further studies aiming at better understanding of the physiologic and pathophysiologic roles of AT_2R and evaluating the potential beneficial actions of drugs that work as agonists for this receptor is much needed.

Acknowledgements. This study was supported by grants HL-47669 and HL-57503 (H.M.S.) from the National Institutes of Health. H.M.S. was the recipient of Research Career Development Award K04-HL-03006 from the National Institutes of Health.

References

Aguilera G, Kapur S, Feuillan P, Sunar-Akbasak B, Bathia AJ (1994) Developmental changes in angiotensin II receptor subtypes and AT1 receptor mRNA in rat kidney. Kidney Int 46:973–979
Akai Y, Homma T, Burns KD, Yasuda T, Badr KF, Harris RC (1994) Mechanical stretch/relaxation of cultured rat mesangial cells induces protooncogenes and cyclooxygenase. Am J Physiol 267:C482–C490
Anderson S, Jung FF, Ingelfinger JR (1993) Renal renin-angiotensin system in diabetes: functional, immunohistochemical, and molecular biological correlations. Am J Physiol 265:F477–F486
Arce ME, Sanchez S, Seltzer A, Ciuffo GM (2001) Autoradiographic localization of angiotensin II receptors in developing rat cerebellum and brainstem. Regul Pept 99:53–60
Ardaillou R (1999) Angiotensin II receptors. J Am Soc Nephrol 10 Suppl 11:S30–S39
Arima S, Ito S (2000) Angiotensin II type 2 receptors in the kidney: evidence for endothelial-cell-mediated renal vasodilatation. Nephrol Dial Transplant 15:448–451
Barber MN, Sampey DB, Widdop RE (1999) AT (2) receptor stimulation enhances antihypertensive effect of AT (1) receptor antagonist in hypertensive rats. Hypertension 34:1112–1116
Berry C, Hamilton CA, Brosnan MJ, Magill FG, Berg GA, McMurray JJ, Dominiczak AF (2000) Investigation into the sources of superoxide in human blood vessels: angiotensin II increases superoxide production in human internal mammary arteries. Circulation 101:2206–2212
Berry C, Touyz R, Dominiczak AF, Webb RC, Johns DG (2001) Angiotensin receptors: signaling, vascular pathophysiology, and interactions with ceramide. Am J Physiol Heart Circ Physiol 281:H2337–H2365
Boffa JJ, Tharaux PL, Placier S, Ardaillou R, Dussaule JC, Chatziantoniou C (1999) Angiotensin II activates collagen type I gene in the renal vasculature of transgenic mice during inhibition of nitric oxide synthesis: evidence for an endothelin-mediated mechanism. Circulation 100:1901–1908
Bonnet F, Cooper ME, Carey RM, Casley D, Cao Z (2001) Vascular expression of angiotensin type 2 receptor in the adult rat: influence of angiotensin II infusion. J Hypertens 19:1075–1081
Bonnet F, Candido R, Carey, RM Casley D, Russo LM, Osicka TM, Cooper ME, Cao Z (2002) Renal expression of angiotensin receptors in long-term diabetes and the effects of angiotensin type 1 receptor blockade. J Hypertens 20:1615–1624
Brogelli L, Parenti A, Ledda F (2002) Inhibition of vascular smooth muscle cell growth by angiotensin type 2 receptor stimulating for in vitro organ culture model. J Cardio Pharm 39:739–745
Bucher M, Hobbhahn J, Kurtz A (2001) Nitric oxide-dependent down-regulation of angiotensin II type 2 receptors during experimental sepsis. Crit Care Med 29:1750–1755

Buisson B, Laflamme L, Bottari SP, de Gasparo M, Gallo-Payet N, Payet MD (1995) A G protein is involved in the angiotensin AT2 receptor inhibition of the T-type calcium current in non-differentiated NG108-15 cells. J Biol Chem 270:1670–1674

Bunkenburg B, van A, Rogg H, Wood JM (1992) Receptor-mediated effects of angiotensin II on growth of vascular smooth muscle cells from spontaneously hypertensive rats. Hypertension 20:746–754

Carey RM, Jin XH, Wang ZQ, Siragy HM (2000a) Nitric oxide: a physiological mediator of the type 2 (AT_2) angiotensin receptor. Acta Physiol Scand 168:65–71

Carey RM, Wang ZQ, Siragy HM (2000b) Role of the angiotensin type 2 receptor in the regulation of blood pressure and renal function. Hypertension 35(1 Pt 2):155–163

Carey RM, Howell NL, Jin XH, Siragy HM (2001) Angiotensin type 2 receptor-mediated hypotension in angiotensin type-1 receptor-blocked rats. Hypertension 38:1272–1277

Ciuffo GM, Alvarez SE, Fuentes LB (1998) Angiotensin II receptors induce tyrosine dephosphorylation in rat fetal membranes. Regul Pept 74:129–135

Criscione L, Thomann H, Whitebread S, de Gasparo M, Buhlmayer P, Herold P, Ostermayer F, Kamber B (1990) Binding characteristics and vascular effects of various angiotensin II antagonists. J Cardiovasc Pharmacol 16 Suppl 4:S56–S59

Croft KD, McGiff JC, Sanchez-Mendoza A, Carroll MA (2000) Angiotensin II releases 20-HETE from rat renal microvessels. Am J Physiol Renal Physiol 279:F544–F551

de Gasparo M, Catt KJ, Inagami T, Wright JW, Unger T (2000) International union of pharmacology. XXIII. The angiotensin II receptors. Pharmacol Rev 52:415–472

Diep QN, Li JS, Schiffrin EL (1999) In vivo study of AT (1) and AT (2) angiotensin receptors in apoptosis in rat blood vessels. Hypertension 34:617–624

Dimmeler S, Rippmann V, Weiland U, Haendeler J, Zeiher AM (1997) Angiotensin II induces apoptosis of human endothelial cells. Protective effect of nitric oxide. Circ Res 81:970–976

Dulin NO, Alexander LD, Harwalkar S, Falck JR, Douglas JG (1998) Phospholipase A2-mediated activation of mitogen-activated protein kinase by angiotensin II. Proc Natl Acad USA 95:8098–8102

Dzau VJ (1988) Tissue renin-angiotensin system: physiologic and pharmacologic implications. Introduction. Circulation 77:I1-I3

Gallinat S, Busche S, Schutze S, Kronke M, Unger T (1999) AT2 receptor stimulation induces generation of ceramides in PC12 W cells. FEBS Lett 443:75–79

Gibson RE, Thorpe HH, Cartwright ME, Frank JD Schorn TW, Bunting PB, Siegl PK (1991) Angiotensin II receptor subtypes in renal cortex of rats and rhesus monkeys. Am J Physiol 261:F512–F518

Gohlke P, Pees, C, Unger, T (1998) AT2 receptor stimulation increases aortic cyclic GMP in SHRSP by a kinin-dependent mechanism. Hypertension 31:349–355

Goldfarb DA, Diz DI, Tubbs RR, Ferrario CM, Novick AC (1994) Angiotensin II receptor subtypes in the human renal cortex and renal cell carcinoma. J Urol 151:208–213

Goto M, Mukoyama M, Suga S, Matsumoto T, Nakagawa M, Ishibashi R, Kasahara M, Sugawara A, Tanaka I, Nakao K (1997) Growth-dependent induction of angiotensin II type 2 receptor in rat mesangial cells. Hypertension 30:358–362

Griendling KK, Minieri CA, Ollerenshaw JD, Alexander RW (1994) Angiotensin II stimulates NADH and NADPH oxidase activity in cultured vascular smooth muscle cells. Circ Res 74:1141–8

Grone HJ, Simon M, Fuchs E (1992) Autoradiographic characterization of angiotensin receptor subtypes in fetal and adult human kidney. Am J Physiol 262:F326–F331

Gross SS, Wolin MS (1995) Nitric oxide: pathophysiological mechanisms. Ann Rev Physiol 57:737–769

Gross V, Schunck WH, Honeck H, Milia AF, Kargel E, Walther T, Bader M, Inagami T, Schneider W, Luft FC (2000) Inhibition of pressure natriuresis in mice lacking the AT2 receptor. Kidney Int 57:191–202

Guron G, Friberg P (2000) An intact renin-angiotensin system is a prerequisite for normal renal development. J Hypertens 18:123-137

Haithcock D, Jiao H, Cui XL, Hopfer U, Douglas JG (1999) Renal proximal tubular AT2 receptor: signaling and transport. J Am Soc Nephrol 10 Suppl 11:S69-S74

Heemskerk FM, Zorad S, Seltzer A, Saavedra JM (1993) Characterization of brain angiotensin II AT2 receptor subtype using [125I] CGP 42112A. Neuro Rep 4:103-105

Hein L, Barsh GS, Pratt RE, Dzau VJ, Kobilka BK (1995) Behavioural and cardiovascular effects of disrupting the angiotensin II type-2 receptor in mice. Nature 377:744-747

Herblin WF, Diamond SM, Timmermans PB (1991) Localization of angiotensin II receptor subtypes in the rabbit adrenal and kidney. Peptides 12:581-584

Hohenfellner K, Hunley TE, Yerkes E, Habermehl P, Hohenfellner R, Kon V (1999) Angiotensin II, type 2 receptor in the development of vesico-ureteric reflux. BJU Int 83:318-322

Horiuchi M, Koike G, Yamada T, Mukoyama M, Nakajima M, Dzau VJ (1995) The growth-dependent expression of angiotensin II type 2 receptor is regulated by transcription factors interferon regulatory factor-1 and -2. J Biol Chem 270:20225-20230

Horiuchi M, Akishita M, Dzau VJ (1999) Recent progress in angiotensin II type 2 receptor research in the cardiovascular system. Hypertension 33:613-621

Huang XC, Richards EM, Sumners C (1996) Mitogen-activated protein kinases in rat brain neuronal cultures are activated by angiotensin II type 1 receptors and inhibited by angiotensin II type 2 receptors. J Biol Chem 271:15635-15641

Hunley TE, Iwasaki S, Homma T, Kon V (1995) Nitric oxide and endothelin in pathophysiological settings. Pediatr Nephrol 9:235-244

Hunley TE, Tamura M, Stoneking BJ, Nishimura H, Ichiki, T, Inagami, T, Kon V (2000) The angiotensin type II receptor tonically inhibits angiotensin-converting enzyme in AT2 null mutant mice. Kidney Int 57: 570-577

Ichiki T, Herold CL, Kambayashi Y, Bardhan S, Inagami T (1994) Cloning of the cDNA and the genomic DNA of the mouse angiotensin II type 2 receptor. Biochim Biophys Acta 19 1189:247-250

Ichiki T, Labosky PA, Shiota C, Okuyama S, Imagawa Y, Fogo A, Niimura F, Ichikawa I, Hogan BL, Inagami T (1995) Effects on blood pressure and exploratory behaviour of mice lacking angiotensin II type-2 receptor. Nature 377:748-750

Inagami T, Eguchi S, Numaguchi K, Motley ED, Tang H, Matsumoto T, Yamakawa T (1999) Cross-talk between angiotensin II receptors and the tyrosine kinases and phosphatases. J Am Soc Nephrol 10:S57-S61

Israel A, Cierco M, Sosa B (2000) Angiotensin AT (2) receptors mediate vasodepressor response to footshock in rats. Role of kinins, nitric oxide and prostaglandins. Eur J Pharmacol 394:103-108

Jiao H, Cui XL, Torti M, Chang CH, Alexander LD, Lapetina EG, Douglas JG (1998) Arachidonic acid mediates angiotensin II effects on p21ras in renal proximal tubular cells via the tyrosine kinase-Shc-Grb2-Sos pathway. Proc Natl Acad USA 95:7417-7421

Kajstura J, Cigola E, Malhotra A, Li P, Cheng W, Meggs LG, Anversa P (1997) Angiotensin II induces apoptosis of adult ventricular myocytes in vitro. J Mol Cell Cardiol 29:859-870

Kakuchi J, Ichiki T, Kiyama S, Hogan BL, Fogo A, Inagami T, Ichikawa I (1995) Developmental expression of renal angiotensin II receptor genes in the mouse. Kidney International 47:140-147

Kambayashi Y, Bardhan S, Takahashi K, Tsuzuki S, Inui H, Hamakubo T, Inagami T (1993) Molecular cloning of a novel angiotensin II receptor isoform involved in phosphotyrosine phosphatase inhibition. J Biol Chem 268:24543-24546

Kang J, Richards EM, Posner P, Sumners C (1995) Modulation of the delayed rectifier K+ current in neurons by an angiotensin II type 2 receptor fragment. Am J Physiol 268:C278-C282

Kijima K, Matsubara H, Murasawa S, Maruyama K, Mori Y, Inada M (1995) Gene transcription of angiotensin II type 2 receptor is repressed by growth factors and glucocorticoids in PC12 cells. Biochem Biophys Res Commun 216:359–366

Kohagura K, Endo Y, Ito O, Arima S, Omata K, Ito S (2000) Endogenous nitric oxide and epoxyeicosatrienoic acids modulate angiotensin II-induced constriction in the rabbit afferent arteriole. Acta Physiol Scand 168:107–112

Koike G, Horiuchi M, Yamada T, Szpirer C, Jacob HJ, Dzau VJ (1994) Human type 2 angiotensin II receptor gene: cloned, mapped to the X chromosome, and its mRNA is expressed in the human lung. Biochem Biophys Res Commun 203:1842–1850

Kone BC (1997) Nitric oxide in renal health and disease. Am J Kidney Dis 30:311–333

Kone BC, Baylis C (1997) Biosynthesis and homeostatic roles of nitric oxide in the normal kidney. Am J Physiol 272:F561–F578

Kovacs J, Gomba S (1998) Analysis of the role of apoptosis and cell proliferation in renal cystic disorders. Kidney Blood Press Res 21:325–328

Kunert-Radek J, Stepien H, Komorowski J, Pawlikowski M (1994) Stimulatory effect of angiotensin II on the proliferation of mouse spleen lymphocytes in vitro is mediated via both types of angiotensin II receptors. Biochem Biophys Res Commun 198:1034–1039

Laflamme L, Gasparo M, Gallo JM, Payet MD, Gallo-Payet N (1996) Angiotensin II induction of neurite outgrowth by AT_2 receptors in NG108-15 cells. Effect counteracted by the AT_1 receptors. J Biol Chem 271:22729–22735

Lehtonen JY, Hayashida W, Horiuchi M, Dzau VJ (1997) Angiotensin II receptor subtype specific effects are mediated by intracellular third loops: chimeric receptor approach. Circulation 96:1478

Levy BI, Benessiano J, Henrion D, Caputo L, Heymes C, Duriez M, Poitevin P, Samuel JL (1996) Chronic blockade of AT2-subtype receptors prevents the effect of angiotensin II on the rat vascular structure. J Clin Invest 98:418–425

Lewis EJ, Hunsicker LG, Bain RP, Rohde RD (1993) The effect of angiotensin-converting-enzyme inhibition on diabetic nephropathy. The Collaborative Study Group. N Engl J Med 329:1456–1462

Lewis EJ, Hunsicker LG, Clarke WR, Berl T, Pohl MA, Lewis JB, Ritz E, Atkins RC, Rohde R, Raz I, Collaborative S (2001) Renoprotective effect of the angiotensin-receptor antagonist irbesartan in patients with nephropathy due to type 2 diabetes. N Engl J Med 20; 345:851–860

Li D, Yang B, Philips MI, Mehta JL (1999) Proapoptotic effects of ANG II in human coronary artery endothelial cells: role of AT1 receptor and PKC activation. Am J Physiol 276:H786–H792

Li JS, Touyz RM, Schiffrin EL (1998) Effects of AT1 and AT2 angiotensin receptor antagonists in angiotensin II-infused rats. Hypertension 31:487–492

Li W, Ye Y, Fu B, Wang J, Yu L, Ichiki T, Inagami T, Ichikawa I, Chen X (1998) Genetic deletion of AT2 receptor antagonizes angiotensin II-induced apoptosis in fibroblasts of the mouse embryo. Biochem Biophys Res Commun 250:72–76

Lo M, Liu KL, Lantelme P, Sassard J (1995) Subtype 2 of angiotensin II receptors controls pressure-natriuresis in rats. J Clin Invest 95:1394–1397

Ma J, Nishimura H, Fogo A, Kon V, Inagami T, Ichikawa I (1998) Accelerated fibrosis and collagen deposition develop in the renal interstitium of angiotensin type 2 receptor null mutant mice during ureteral obstruction. Kidney International 53:937–944

Madrid MI, Garcia-Salom M, Tornel J, de Gasparo M, Fenoy FJ (1997) Effect of interactions between nitric oxide and angiotensin II on pressure diuresis and natriuresis. Am J Physiol 273:R1676–R1682

Majid DS, Navar LG (1997) Nitric oxide in the mediation of pressure natriuresis. Clin Exp Pharmacol Physiol 24:595–599

Masaki H, Kurihara T, Yamaki A, Inomata N, Nozawa Y, Mori Y, Murasawa S, Kizima K, Maruyama K, Horiuchi M, Dzau VJ, Takahashi H, Iwasaka T, Inada M, Matsubara H

(1998) Cardiac-specific overexpression of angiotensin II AT2 receptor causes attenuated response to AT1 receptor-mediated pressor and chronotropic effects. J Clin Invest 101:527–535

Matrougui K, Loufrani L, Heymes C, Levy BI, Henrion D (1999) Activation of AT (2) receptors by endogenous angiotensin II is involved in flow-induced dilation in rat resistance arteries. Hypertension 34:659–665

Matsubara H (1998) Pathophysiological role of angiotensin II type 2 receptor in cardiovascular and renal diseases. Circ Res 83:1182–1191

Matsubara H, Sugaya T, Murasawa S, Nozawa Y, Mori Y, Masaki H, Maruyama K, Tsutumi Y, Shibasaki Y, Moriguchi Y, Tanaka Y, Iwasaka T, Inada M (1998) Tissue-specific expression of human angiotensin II AT1 and AT2 receptors and cellular localization of subtype mRNAs in adult human renal cortex using in situ hybridization. Nephron 80:25–34

Mifune M, Sasamura H, Nakazato Y, Yamaji Y, Oshima N, Saruta T (2001) Examination of angiotensin II type 1 and type 2 receptor expression in human kidneys by immunohistochemistry. Clin Exp Hypertens (New York) 23:257–266

Millatt LJ, Abdel-Rahman EM, Siragy HM (1999) Angiotensin II and nitric oxide: a question of balance. Reg Pept 81:1–10

Miura S, Karnik SS (2000) Ligand-independent signals from angiotensin II type 2 receptor induce apoptosis. EMBO J 19:4026–4035

Miyata N, Park F, Li XF, Cowley AWJ (1999) Distribution of angiotensin AT1 and AT2 receptor subtypes in the rat kidney. Am J Physiol 277:F437–F446

Moncada S, Palmer RM, Higgs EA (1991) Nitric oxide: physiology, pathophysiology, and pharmacology. Pharmacological Reviews 43:109–142

Morrissey JJ, Klahr S (1999) Effect of AT2 receptor blockade on the pathogenesis of renal fibrosis. Am J Physiol 276:F39–F45

Mukoyama M, Nakajima M, Horiuchi M, Sasamura H, Pratt RE, Dzau VJ (1993) Expression cloning of type 2 angiotensin II receptor reveals a unique class of seven-transmembrane receptors. J Biol Chem 268:24539–24542

Muller DN, Dechend R, Mervaala EM, Park JK, Schmidt F, Fiebeler A, Theuer J, Breu V, Ganten D, Haller H, Luft FC (2000) NF-kappa B inhibition ameliorates angiotensin II-induced inflammatory damage in rats. Hypertension 35:193–201

Munzenmaier DH, Greene AS (1996) Opposing actions of angiotensin II on microvascular growth and arterial blood pressure. Hypertension 27:760–765

Nakajima M, Hutchinson HG, Fujinaga M, Hayashida W, Morishita R, Zhang L, Horiuchi M, Pratt RE, Dzau VJ (1995) The angiotensin II type 2 (AT2) receptor antagonizes the growth effects of the AT1 receptor: gain-of-function study using gene transfer. Proc Natl Acad USA 92:10663–10667

Navar LG, Inscho EW, Majid SA, Imig JD, Harrison-Bernard LM, Mitchell KD (1996) Paracrine regulation of the renal microcirculation. Physiol Rev 76:425–536

Navar LG, Imig JG, Zou L, Wang CT (1997) Intrarenal production of angiotensin II Semin Nephrol 17:412–422

Nishimura H, Yerkes E, Schulman M (1997) The angiotensin type 2 receptor null mutant mice: a model of the diverse spectrum of congenital urinary tract anomalies in humans. J Am Soc Nephrol 8:364A

Nishimura H, Yerkes, E, Hohenfellner, K, Miyazaki Y, Ma J, Hunley TE, Yoshida H, Ichiki T, Threadgill D, Phillips JA, Hogan BM, Fogo A, Brock JW, Inagami T, Ichikawa I (1999) Role of the angiotensin type 2 receptor gene in congenital anomalies of the kidney and urinary tract, CAKUT, of mice and men. Mol Cell 3:1–10

Nishiyama A, Seth DM, Navar LG (2002) Renal interstitial fluid concentrations of angiotensins I and II in anesthetized rats. Hypertension 39:129–134

Nora EH, Munzenmaier DH, Hansen-Smith FM, Lombard JH, Greene AS (1998) Localization of the ANG II type 2 receptor in the microcirculation of skeletal muscle. Am J Physiol 275:H1395–H1403

Ozono R, Wang ZQ, Moore AF, Inagami T, Siragy HM, Carey RM (1997) Expression of the subtype 2 angiotensin (AT2) receptor protein in rat kidney. Hypertension 30:1238-1246

Parving HH, Lehnert H, Brochner-Mortensen J, Gomis R, Andersen S, Arner P (2001) The effect of irbesartan on the development of diabetic nephropathy in patients with type 2 diabetes. N Engl J Med 20; 345:870-878

Pushkareva M, Obeid LM, Hannun YA (1995) Ceramide: an endogenous regulator of apoptosis and growth suppression. Immunol Today 16:294-297

Scheuer DA, Perrone MH (1993) Angiotensin type 2 receptors mediate depressor phase of biphasic pressure response to angiotensin. Am J Physiol 264:R917-R923

Scott JE, Renwick M (1988) Antenatal diagnosis of congenital abnormalities in the urinary tract. Results from the Northern Region Fetal Abnormality Survey. Br J Urol 62:295-300

Sechi LA, Griffin CA, Grady EF, Kalinyak JE, Schambelan M (1992) Characterization of angiotensin II receptor subtypes in rat heart. Circ Res 71:1482-1489

Seebach FA, Welte T, Fu XY, Block LH, Kashgarian M (2001) Differential activation of the STAT pathway by angiotensin II via angiotensin type 1 and type 2 receptors in cultured human fetal mesangial cells. Experimental & Molecular Pathology 70:265-273

Seyedi N, Xu X, Nasjletti A, Hintze TH (1995) Coronary kinin generation mediates nitric oxide release after angiotensin receptor stimulation. Hypertension 26:164-170

Shanmugam S, Llorens-Cortes C, Clauser E, Corvol P, Gasc JM (1995) Expression of angiotensin II AT2 receptor mRNA during development of rat kidney and adrenal gland. Am J Physiol 268:F922-F930

Sharma M, Sharma R, Greene AS, McCarthy ET, Savin VJ (1998) Documentation of angiotensin II receptors in glomerular epithelial cells. Am J Physiol 274:F623-F627

Shenoy UV, Richards EM, Huang XC, Sumners C (1999) Angiotensin II type 2 receptor-mediated apoptosis of cultured neurons from newborn rat brain. Endocrinology 140:500-509

Siragy HM, Bedigian M (1999) Mechanism of action of angiotensin-receptor blocking agents. Curr Hypertens Rep 1:289-295

Siragy HM, Carey RM (1996) The subtype-2 (AT2) angiotensin receptor regulates renal cyclic guanosine 3', 5'-monophosphate and AT1 receptor-mediated prostaglandin E2 production in conscious rats. J Clin Invest 97:1978-1982

Siragy HM, Carey RM (1997a) The subtype 2 (AT2) angiotensin receptor mediates renal production of nitric oxide in conscious rats. J Clin Invest 100:264-269

Siragy HM, Carey RM (1997b) The subtype 2 angiotensin receptor regulates renal prostaglandin F2 alpha formation in conscious rats. Am J Physiol 273:R1103-R1107

Siragy HM, Carey RM (1999) Protective role of the angiotensin AT2 receptor in a renal wrap hypertension model. Hypertension 33:1237-1242

Siragy HM, Howell NL, Ragsdale NV, Carey RM (1995) Renal interstitial fluid angiotensin. Modulation by anesthesia, epinephrine, sodium depletion, and renin inhibition. Hypertension 25:1021-1024

Siragy HM, Jaffa AA, Margolius HS, Carey RM (1996) Renin-angiotensin system modulates renal bradykinin production. Am J Physiol 271:R1090-R1095

Siragy HM, de Gasparo M, Carey RM (1998) Angiotensin AT_1 receptor (AT_1R) blockade mediates a renal bradykinin (BK)-nitric oxide (NO) cascade. Hypertension 32:157

Siragy HM, Inagami T, Ichiki T, Carey RM (1999) Sustained hypersensitivity to angiotensin II and its mechanism in mice lacking the subtype-2 (AT2) angiotensin receptor. Proc Natl Acad USA 96:6506-6510

Siragy HM, de Gasparo M, Carey RM (2000) Angiotensin type 2 receptor mediates valsartan-induced hypotension in conscious rats. Hypertension 35:1074-1077

Song Y et al (1995) Mapping of angiotensin II receptor subtypes in peripheral tissues of spontaneous hypertensive rats by in vitro autoradiography. Clin Exp Pharmacol Physiol 22:S17-S19

Stoll M, Steckelings UM, Paul M, Bottari SP, Metzger R, Unger T (1995) The angiotensin AT2-receptor mediates inhibition of cell proliferation in coronary endothelial cells. J Clin Invest 95:651–657

Suzuki J, Iwai M, Nakagami H, Wu L, Chen R, Sugaya T, Hamada M, Hiwada K, Horiuchi M (2002) Role of angiotensin II-regulated apoptosis through distinct AT_1 and AT_2 receptors in neointimal formation. Circulation 106:847–853

Tanaka M, Ohnishi J, Ozawa Y, Sugimoto M, Usuki S, Naruse M, Murakami K, Miyazaki H (1995) Characterization of angiotensin II receptor type 2 during differentiation and apoptosis of rat ovarian cultured granulosa cells. Biochem Biophys Res Commun 207:593–598

Tea BS, Der S, Touyz RM, Hamet P, deBlois D (2000) Proapoptotic and growth-inhibitory role of angiotensin II type 2 receptor in vascular smooth muscle cells of spontaneously hypertensive rats in vivo. Hypertension 35:1069–1073

Thomas GL, Yang B, Wagner BE, Savill J, El N (1998) Cellular apoptosis and proliferation in experimental renal fibrosis. Nephrol Dial Transpl 13:2216–2226

Timmermans PB, Wong PC, Chiu AT, Herblin WF, Benfield P, Carini DJ, Lee RJ, Wexler RR, Saye JA, Smith, RD (1993) Angiotensin II receptors and angiotensin II receptor antagonists. Pharmacol Rev 45:205–251

Tissir F, Riviere M, Guo DF, Tsuzuki S, Inagami T, Levan G, Szpirer J, Szpirer C (1995) Localization of the genes encoding the three rat angiotensin II receptors, Agtr1a, Agtr1b, Agtr2, and the human AGTR2 receptor respectively to rat chromosomes 17q12, 2q24 and Xq34, and the human Xq22. Cytogenetics & Cell Genetics 71:77–80

Truong LD, Sheikh-Hamad D, Chakraborty S, Suki WN (1998) Cell apoptosis and proliferation in obstructive uropathy. Semin Nephrol 18:641–651

Tsutsumi K, Saavedra JM (1991) Characterization and development of angiotensin II receptor subtypes (AT_1 and AT_2) in rat brain. Am J Physiol 261:R209–R216

Tsutsumi Y, Matsubara H, Masaki H, Kurihara H, Murasawa S, Takai S, Miyazaki M, Nozawa Y, Ozono R, Nakagawa K, Miwa T, Kawada N, Mori Y, Shibasaki Y, Tanaka Y, Fujiyama S, Koyama Y, Fujiyama A, Takahashi H, Iwasaka T (1999) Angiotensin II type 2 receptor overexpression activates the vascular kinin system and causes vasodilation. J Clin Invest 104:925–935

Tsuzuki S, Ichiki T, Nakakubo H, Kitami Y, Guo DF, Shirai H, Inagami T (1994) Molecular cloning and expression of the gene encoding human angiotensin II type 2 receptor. Biochem Biophys Res Commun 200:1449–1454

Vaughan, DE, Lazos SA, Tong K (1995) Angiotensin II regulates the expression of plasminogen-activator inhibitor-1 in cultured endothelial cells: a potential link between the renin-angiotensin system and thrombosis. J Clin Invest 95:995–1001

Viswanathan M, Tsutsumi K, Correa FM, Saavedra JM (1991) Changes in expression of angiotensin receptor subtypes in the rat aorta during development. Biochem Biophys Res Commun 179:1361–1367

Viswanathan M, Selby DM, Ray PE (2000) Expression of renal and vascular angiotensin II receptor subtypes in children. Pediatr Nephrol 14:1030–1036

Wang TT, Wu XH, Zhang SL, Chan JS (1998) Effect of glucose on the expression of the angiotensinogen gene in opossum kidney cells. Kidney International 53:312–319

Wang ZQ, Moore AF, Ozono R, Siragy HM, Carey RM (1998) Immunolocalization of subtype 2 angiotensin II (AT2) receptor protein in rat heart. Hypertension 32:78–83

Wang ZQ, Millatt LJ, Heiderstadt NT, Siragy HM, Johns RA, Carey RM (1999) Differential regulation of renal angiotensin subtype AT1A and AT2 receptor protein in rats with angiotensin-dependent hypertension. Hypertension 33:96–101

Wehbi GJ, Zimpelmann J, Carey RM, Levine DZ, Burns KD (2001) Early streptozotocin-diabetes mellitus downregulates rat kidney AT2 receptors. Am J Physiol Renal Physiol 280:F254–F265

Whitebread SE, Taylor V, Bottari SP, Kamber B, de Gasparo M (1991) Radioiodinated CGP 42112A: a novel high affinity and highly selective ligand for the characterization of angiotensin AT2 receptors. Biochem Biophys Res Commun 181:1365–1371

Wolf G, Neilson EG (1996) From converting enzyme inhibition to angiotensin II receptor blockade: new insight on angiotensin II receptor subtypes in the kidney. Experimental Nephrology 4 Suppl 1:8–19

Xoriuchi M, Hamai M, Cui TX, Iwa M, Minokoshi Y (1999) Cross talk between angiotensin II type 1 and type 2 receptors: cellular mechanism of angiotensin type 2 receptor-mediated cell growth inhibition. Hypertens Res Clin Exp 22:67–74

Yamada T, Horiuchi M, Dzau VJ (1996) Angiotensin II type 2 receptor mediates programmed cell death. Proc Natl Acad USA 93:156–160

Yamada T, Akishita M, Pollman MJ, Gibbons GH, Dzau VJ, Horiuchi M (1998) Angiotensin II type 2 receptor mediates vascular smooth muscle cell apoptosis and antagonizes angiotensin II type 1 receptor action: an in vitro gene transfer study. Life Sci 63:L289–L295

Yuan BH, Robinette JB, Conger JD (1990) Effect of angiotensin II and norepinephrine on isolated rat afferent and efferent arterioles. Am J Physiol 258:F741-F750

Zhuo J, Song K, Harris PJ, Mendelsohn FA (1992) In vitro autoradiography reveals predominantly AT1 angiotensin II receptors in rat kidney. Renal Physiol Biochem 15:231–239

Zhuo J, Dean R, MacGregor D, Alcorn D, Mendelsohn FA (1996) Presence of angiotensin II AT2 receptor binding sites in the adventitia of human kidney vasculature. Clin Exp Pharmacol Physiol Suppl 3:S147–S154

Zimpelmann J, Kumar D, Levine DZ, Wehbi G, Imig JD, Navar LG, Burns KD (2000) Early diabetes mellitus stimulates proximal tubule renin mRNA expression in the rat. Kidney Int 58:2320–2330

AT$_2$ Function and Target Genes

C. Wruck · M. Stoll · T. Unger

Institute of Pharmacology and Toxicology, Charité, Humboldt-University Berlin, Dorotheenstr. 94, 10117 Berlin, Germany
e-mail: thomas.unger@charite.de

1	Introduction	450
2	Function	453
2.1	Growth and Development	453
2.1.1	Development of the Nephron and Aorta	454
2.1.2	Development of the Brain and Cognitive Function	455
2.2	Tissue Regeneration and Pathophysiological Aspects	457
2.2.1	Nerve Regeneration	458
2.2.2	Remodeling of the Heart After Injury	458
2.2.3	Renal Physiological and Pathophysiological Function	463
3	Target Genes	464
4	Summary and Conclusion	468
	References	469

Abstract Angiotensin II (ANG II), the biologically active component of renin–angiotensin system (RAS), acts through two receptor subtypes, the AT$_1$ and the AT$_2$ receptors. All classic physiological effects of ANG II, such as vasoconstriction, aldosterone and vasopressin release, sodium and water retention, as well as ANG II-mediated growth, are mediated by the AT$_1$ receptor. Recent investigations have established a role for the AT$_2$ receptor in cardiovascular, brain, and renal function as well as in the modulation of various biological processes involved in development, cell differentiation, and tissue repair. Furthermore, binding of ANG II to AT$_2$ receptors inhibits, in certain cells, proliferation, mediates differentiation in neural cell lines, and even induces apoptosis. Interestingly, one study suggests that AT$_2$ receptor-mediated apoptosis may be independent of the ligand ANG II and requires only AT$_2$-receptor expression on a specific target cell. As all these effects of AT$_2$ receptor activation are the opposite of those mediated through AT$_1$ receptors, a ying–yang hypothesis has been proposed in which AT$_1$ and AT$_2$ receptors have principally contrary functions. This review summarizes new insights in the function of the AT$_2$ receptor and its affected target genes.

Keywords Angiotensin II · AT_2 receptor · Function · Target gene · Angiotensin receptors

1
Introduction

The peptide hormone angiotensin II (ANG II) exerts a broad field of physiological and pathophysiological actions on the cardiovascular and nervous system, fetal development, and in response to injuries. In 1989, two independent groups of investigators provided pharmacological evidence for the existence of two major subtypes of ANG II receptor (Chiu et al. 1989; Whitebread et al. 1989). These subtypes were subsequently named the AT_1 and AT_2 receptors. There may be even more ANG II receptor types, but they are not yet cloned. The AT_1 receptor is responsible for mediating many of the well-known stimulatory physiological actions of ANG II, including vasoconstriction, secretion of aldosterone, and renal sodium reabsorption. The AT_1 receptor can activate G protein-coupled and -uncoupled signal transduction pathways, such as down-regulation of adenylate cyclase, activation of phospholipase A2, and protein kinase C, stimulation of the Janus kinase/STAT (signal transducer and activator of transcription) cascade, and opening of calcium channels. The AT_1 receptor is the predominant subtype in the adult and is responsible for the most physiological effects of ANG II (de Gasparo et al. 2000).

The AT_2 receptor shares only about 33% amino acid sequence identity with its AT_1 counterpart (Nakajima et al. 1993). The lowest homology between these receptors exists between their third intracellular loops (ICL). The third ICL is regarded as one of the major regions of the G protein-coupled receptor that is responsible for their signal transduction. Delivery of synthetic third ICL peptides of the AT_2 receptor into vascular smooth muscle cells (VSMC) results in decreased DNA synthesis, diminished cell proliferation, and a reduction in mitogen-activated protein (MAP) kinase activity, similar to the effects of ANG II acting via the AT_2 receptor (Hayashida et al. 1996a). Moreover, in neurons cultured from newborn rat hypothalamus and brain stem, intracellular injection of a 22-amino acid peptide (PEP-22) corresponding to the putative third ICL of the AT_2 receptor elicited an increase in delayed rectifier K^+ current (Kv) similar to that obtained with ANG II acting via AT_2 receptors in these cells (Kang et al. 1996). Using a chimeric receptor in which the third ICL of the AT_2 receptor was replaced with that of the AT_1 receptor, Dittus et al. observed a loss in affinity to [^{125}I-Sar^1-Ile^8]ANG II and ^{125}I-CGP42112A (Dittus et al. 1999). Collectively these data support the hypothesis that the third ICL of the AT_2 receptor is as closely linked to intracellular signaling as to the determination of ligand-binding properties.

Regardless of the low homology of the AT_1 and AT_2 receptors, ANG II exhibits a similar affinity to both receptor subtypes. Discrimination of these subtypes became possible with the development of highly selective receptor antagonists (Timmermans et al. 1993). AT_1 receptors exhibit a low affinity to tetrahydroimidazolepyridines, such as PD123177 and PD123319, and a high affinity for a class of compounds called "sartans" which, from the chemical point of view, belong to the biphenylimidazoles. On the other hand, AT_2 receptors show a very low affinity for biphenylimidazoles and a high affinity for tetrahydroimidazolepyridines. Moreover, compared with AT_1 receptors, AT_2 receptors also display a high affinity for the peptide CGP42112A, which is a partial agonist (Brechler et al. 1993).

The exposure of G protein-coupled receptors (GPCRs) to agonists often results in a rapid attenuation of receptor responsiveness. This process, termed desensitization, is the consequence of a combination of different mechanisms including the uncoupling of the receptor from heterotrimeric G proteins, the internalization of cell surface receptors to intracellular membranous compartments, and the down-regulation of the total cellular receptors due to reduced receptor mRNA and protein synthesis, as well as both the lysosomal and plasma membrane degradation of pre-existing receptors (Ferguson et al. 2001). In contrast to this concept, the AT_2 receptor neither undergoes internalization after agonist binding (Hein et al. 1997) nor down-regulation of mRNA or protein level. Contrariwise, the AT_2 receptor density on the cell surface is adjusted by self up-regulation mechanisms on transcriptional and translational levels by the receptor itself (Shibata et al. 1997; Li et al. 1999).

The AT_2 receptor gene (AGTR2) is located on the X chromosome. The gene comprises three exons with the entire coding region in the third exon (Ichiki et al. 1995a). There is no intron in the coding region, thus excluding multiple forms of the AT_2 receptor derived by alternative splicing.

The majority of studies performed on AT_2 receptor regulation have been in vitro cell culture experiments. In the course of these experiments, it became evident early on that the cell culture conditions can affect the expression of AT_2 receptors. For example, in the fibroblast cell line R3T3, which is known to express AT_2 but not AT_1 receptors, the number of AT_2 binding sites depends greatly on the cell density (Dudley et al. 1993). In actively growing R3T3 cells, the expression of AT_2 receptors is very low, but their expression is substantially increased when the cells become confluent. This regulation of AT_2 receptors seems to occur at multiple levels, involving translational and/or posttranslational as well as transcriptional control (Camp et al. 1995). The fact that serum dramatically influences the expression of AT_2 receptors in cultured cells led to the idea that growth factors may have a role in the regulation of these receptors, and this is the case. The promoter activity of the AT_2 receptor gene is indeed modulated by several factors. To be specific, the AT_2 receptor is consistently down-regulated by factors which are known to induce cell proliferation, such as basic fibroblast growth factor (bFGF), epidermal growth factor (EGF), nerve growth factor (NGF), platelet-derived growth factor (PDGF), and ANG II via AT_1 receptor (Ichiki et al. 1995a; Kizima et al. 1996; Li et al. 1998, 1999). In contrast, AT_2 receptor expression is up-regulated during serum deprivation and cell contact inhibition as well as by factors like insulin, insulin-like growth factor, the inflammatory cytokine interleukin 1b, estrogen, and ANG II via the AT_2 receptor itself

(Ichiki et al. 1995b; Kambayashi et al. 1996; Shibata et al. 1997; Armando et al. 2002; Stoll et al. 2002). It should also be pointed out that the effects of growth factors or serum on AT_2 receptor expression in PC12W and R3T3 cells are dependent on the respective cell passage (Li et al. 1998), indicating another level of complexity in the regulation of AT_2 receptor expression in these cells. In primary neuronal cultures, short-term incubations with NGF resulted in increased AT_2 receptor mRNA levels, whereas long-term incubations with this factor generated the opposite effect (Huang et al. 1997). The finding that the effects of certain factors on the AT_2 receptor expression are not necessarily identical in different cell types emphasizes the importance of thoroughly discriminating data gathered in each situation.

In vivo, AT_2 receptors are abundantly expressed in fetal organs but are present in distinct tissues in the adult organism, including brain, adrenal glands, uterine myometrium, ovarian follicles, kidney, and heart (Matsubara et al. 1998). In the brain, the receptor is expressed in brain stem, several thalamic nuclei, lateral septum, and amygdala (Tsutsumi and Saavedra 1991). Interestingly, the AT_2 receptor is re-expressed under pathological conditions such as congestive heart failure, renal failure, or following incidents like skin lesions, vascular injury, myocardial infarction, and lesions of the nervous system like brain ischemia, brain lesions, sciatic, or optic nerve transection (Unger 1999).

The signal transduction pathways for the AT_2 receptor still remain enigmatic. Although the AT_2 receptor possesses all features of a G protein-coupled receptor, the coupling of the receptor to G proteins is still a matter of controversy. The insensitivity of the AT_2 receptor to guanosine triphosphate (GTP)γS raised the assumption that this receptor is not a member of the G protein-coupled receptor family (Bottari et al. 1991). Evidence that linked the AT_2 receptor to G proteins came from studies in neuronal cells demonstrating that the receptor modulates K^+ channel activity by the G protein G_i (Kang et al. 1994). It turned out that at least three phosphatases lead to the growth-inhibitory signals of the AT_2 receptor: mitogen-activated protein kinase phosphatase (MKP)-1, protein phosphatase 2A (PP2A) and SH2 domain-containing phosphatase (SHP)-1 (Nouet and Nahmias 2000). Moreover, stimulation of the AT_2 receptor in certain cell types leads to changes in cellular cGMP levels and nitric oxide (NO) production (Sumners et al. 1991; Siragy and Carey 1997; Gohlke et al. 1998). In renal tubular epithelial cells and in cultured neurons, the receptor is also linked to a membrane-associated phospholipase A2 (PP2A) pathway which involves the G protein p21ras. This pathway can be blocked by the inhibition of the PP2A (Zhu et al. 1998). Another fatty acid used by the AT_2 receptor as a second messenger is the sphingolipid ceramide. In PC12W cells, a rat pheochromocytoma cell line, long-term activation of the AT_2 receptor increases the synthesis of ceramide and may lead via this pathway to the AT_2 receptor-mediated apoptosis (Gallinat et al. 1999; Lehtonen et al. 1999).

Unexpectedly, studies using AT_2 receptor knockout mice show that the AT2 receptor appears to have a physiological role in blood pressure control as well as behavior modulation (Hein et al. 1995; Ichiki et al. 1995c; Sakagawa et al.

2000). These knockout models are part of a separate section of this volume (i.e., see Brede and Hein) and discussed there in more detail.

As most of these effects of AT_2 receptor activation are opposing those mediated by AT_1 receptor activation, a yin–yang hypothesis has been proposed in which the AT_1 and AT_2 receptors have principally opposed functions (Unger 1999; Gallinat et al. 2000; Stoll and Unger 2001; Kaschina and Unger 2003).

2
Function

ANG II has a number of well-known physiological effects, including stimulation of drinking behavior, increases in blood pressure, modulation of baroreceptor function, and stimulation of vasopressin release. However, the majority of early studies have indicated that they are AT_1 receptor-mediated effects and have failed to recognize any role of AT_2 receptors (Hogarty et al. 1992; Qadri et al. 1993). One notable exception involves studies showing that administration of the AT_2 receptor antagonist PD123177 potentiates the stimulatory effects of ANG II on drinking behavior and vasopressin secretion (Höhle et al. 1996). This is one example of a situation in which AT_2 receptors appear to oppose AT_1 receptor-mediated effects.

The AT_2 receptor is expressed at a very high level in numerous fetal tissues (Shanmugam and Sandberg 1996). In contrast, its expression is restricted to certain organs and brain areas in the adult. The AT_2 receptor is re-expressed in the context of cardiac and vascular injury (Nakajima et al. 1995) and nerve crush (Gallinat et al. 1998; Lucius et al. 1998) as well as during wound healing (Kimura et al. 1992) or kidney obstruction (Morrissey and Klahr 1999). These observations led to the hypothesis that the AT_2 receptor exerts its main functions during the processes of cellular development and reorganization. Therefore, in this chapter, we will focus on the role of the AT_2 receptor during ontogenesis and in tissue regeneration.

2.1
Growth and Development

During the process of ontogenesis, the AT_2 receptor is expressed in an abundant and transient manner. In fetal tissues, the AT_2 receptor emerges on embryonic days 11–13 (E11–E13) and reaches its maximal level on E19. The AT_2 receptor is localized in tissues derived from embryonic mesoderm such as the striated and smooth muscle, undifferentiated mesenchymal tissues surrounding the developing cartilage, diaphragm, tongue, connective tissue, bronchi, blood vessels, ligaments, and submucosal layer of the stomach and intestine, and the choroid surrounding the retina and dermis. The AT_2 receptor expression then rapidly declines in after birth to lower or undetectable levels (Table 1; Shanmugam and Sandberg 1996). In comparison, the AT_1 receptor is expressed less in fetal organs such as the aorta, kidney, adrenal gland, liver, and lungs (Grady et al.

Table 1 Tissue distribution of the AT_2 receptor and ontogenic change

Tissue	Fetus	Newborn	3 week	8 week	Reference(s)
Adrenal					
Cortex	+	+	+	+	Shanmugam et al. 1995
Medulla		+	+	+	Feuillan et al. 1993
Ovary					
Follicular granulosa	+	+		+	Pucell et al. 1991
Kidney					
Cortex	+++	+			Shanmugam et al. 1995
Medulla (outer strip)	+	++			Shanmugam et al. 1995; Kakuchi et al. 1995
Heart	±				Shanmugam et al. 1995
Uterus					
Myometrium					
Blood vessels			+		
Heart	±				Shanmugam et al. 1995
Aorta	+			+	Feuillan et al. 1997; Shanmugam et al. 1995
Pancreas	+				Chappell et al. 1992
Trachea	+				Shanmugam et al. 1995
Stomach	+				Shanmugam et al. 1995
Mesenchyme	+				
Skin	+++	±			Grady et al. 1991; Shanmugam et al. 1995; Feuillan et al. 1993
Tongue	+++	±			Grady et al. 1991; Feuillan et al. 1993
Skeletal muscle	+				Feuillan et al. 1993

1991) which are the tissues where ANG II is known to play a physiological role during adult life.

2.1.1
Development of the Nephron and Aorta

In the fetal kidney on the embryonic days E12–E16, the AT_2 receptor is expressed in mesenchymal cells of differentiating cortex and medulla, which surround the glomeruli in the cortex and the tubular tissues in the medulla (Kakuchi et al. 1995). These mesenchymal cells will finally be replaced by tubular tissue after their apoptosis. On E14, the AT_2 receptor emerges in the interstitial mesenchym of the kidney, but not in the glomeruli or the S-body, whereas the AT_1 receptor is richly expressed in the preglomeruli and S-bodies. On E16, AT_2 receptor mRNA is seen in the renal capsule and inner medulla where it is prominent along the papillary duct and between the collecting ducts. In this context, the apoptosis of the AT_2 receptor-expressing cells in fetal tissue development is an attractive concept. On the other hand, mice with targeted deletions of genes for angiotensinogen or angiotensin-converting enzyme (ACE) revealed more severe renal abnormalities compared to AT_1 or AT_2 receptor knockout animals with milder abnormalities of the kidney (Hein 1998). These findings suggest that AT_1 and AT_2 receptors are both involved in the developmental process of the nephron and that ANG II growth-stimulatory effects, though AT_1 receptors may be adjusted by AT_2 receptor-mediated apoptosis and growth inhibition. Therefore, changes in AT_2 receptor signaling may lead to changes in the sensitive balance between growth stimulation and inhibition and to alterations in tubular formation.

In the fetal aorta, AT_2 receptor expression is undetectable in early stages (E15) but rises to very high levels during late embryonic development (E16–E21) whereupon it decreases 2 weeks after birth (D22) (Viswanathan et al. 1991). A similar pattern of AT_2 receptor expression during development was also observed in other cardio-pulmonary tissues. By in utero application of the AT_2 receptor antagonist PD123319, the role of the AT_2 receptor in the development of the fetal aorta could be examined. The drug was administered for 3 days prior to tissue harvesting and the rates of DNA synthesis in the fetal aortae in utero were measured. During early embryonic development (E15), when the AT_2 receptor expression is low and the aortic DNA synthesis rates are near maximum, PD123319 has no effect on DNA synthesis. Between E16 and E21, the growth rates in the fetal aorta decline as the AT_2 receptor is expressed. PD123319 treatment attenuates significantly this reduction in aortic DNA synthesis (Nakajima et al. 1995). These results suggest that the AT_2 receptor mediates an antigrowth effect on the aorta in vivo which may be causally related to the delayed vasculogenesis observed in AT_2 receptor knockout mice (Yamada et al. 1998).

2.1.2
Development of the Brain and Cognitive Function

Although circumventricular organs of the brain respond to circulating ANG II, an endogenous brain renin–angiotensin system can generate ANG II in tissues protected from circulating ANG II by the blood–brain barrier. ANG II generated by this system reacts with angiotensin receptors within the brain. The brain angiotensin receptors have been studied and reviewed extensively (Gehlert et al. 1991a,b; Rowe et al. 1992; Saavedra 1992; Song et al. 1992; Höhle et al. 1995; Lenkei et al. 1996). An expression of the AT_2 receptor in fetal (E18) brain tissues was reported in several areas: inferior olive, paratrigeminal nucleus, and hypoglossal nucleus (Table 2). Saavedra determined AT_1 and AT_2 receptor expression in various regions of the brain, and the changes of the expression depending on age (Saavedra 1992). Although AT_1 receptor expression did not show marked age dependence, the AT_2 receptor showed a significant decrease from 2 to 8 weeks of age. Using competitive radioligand binding the distribution of ANG II receptor types in adult rat brain nuclei were investigated (Obermüller et al. 1991). Whereas midbrain and brain stem contained AT_1 and AT_2 receptors in

Table 2 Distribution of the AT_2 receptor in the brain: effect of age (fmol/mg protein, mean±SEM)

	2-week-old	8-week-old
Regions containing only the AT_2 receptor		
Persistent AT_2 receptor with age		
Lateral septal nucleus	58±6	18±3
Ventral thalamic nuclei	101±8	24±3
Mediodorsal thalamic nucleus	165±11	38±13
Locus ceruleus	289±19	98±13
Principal sensory trigeminal nucleus	75±6	15±4
Parasolitary nucleus	220±15	57±14
Inferior olive	1328±61	181±32
Medial amygdaloid nucleus	159±8	94±9
Medial geniculate nucleus	338±24	71±6
Transient expression of the AT_2 receptor		
Anterior pretectal nucleus	53±8	ND
Nucleus of the optic tract	101±13	ND
Ventral tegmental area	101±11	ND
Posteodorsal tegmental nucleus	110±21	ND
Hypoglossal nucleus	141±11	ND
Central medial and paracentral thalamic nucleus	202±14	ND
Laterodorsal thalamic nucleus	110±10	ND
Oculomotor nucleus	98±13	ND
Regions containing both AT_1 and AT_2 receptors		
Persistent AT_2 receptor with age		
Superior colliculus	145±5	65±8
Cingulate cortex	19±4	8±4
Transient expression of the AT_2 receptor		
Cerebellar cortex	59±6	ND

ND, not determined.

comparable concentrations, AT_1 receptors were by far predominant in several hypothalamic nuclei, although a limited amount of AT_2 receptors was detectable in most of them. Although minor differences may exist among these studies, they share the following observations: AT_2 receptor expression is consistently high in the cerebellar nuclei, inferior olive, and locus coeruleus in the brain stem, which is rich in noradrenergic neurons. In contrast to the AT_1 receptor, which participates in various central cardiovascular functions and is expressed in the hypothalamus and in brain stem nuclei (nucleus of the solitary tract, dorsal motor nucleus of vagus at a low level), the AT_2 receptor is much less present in distinct hypothalamic and brain stem nuclei associated with the regulation of cardiovascular functions. The presence of the AT_2 receptor in the dorsal motor nucleus of vagus is not completely settled.

The AT_2 receptor is exclusively or predominantly expressed in two limbic systems, the lateral septal nucleus, and the medial amygdaloid nucleus. This distribution of the AT_2 receptor remains unchanged during development and in the adult animal; however, the density of the AT_2 receptors shows a marked decline after birth (Millan et al. 1991; Tsutsumi and Saavedra 1991). The fact that the

AT$_2$ receptor is much more highly expressed in fetal brain tissues compared to adult suggests that the AT$_2$ receptor may modulate auditory, visual, motor, and limbic formation during the process of development. In vitro studies also support that assumption, e.g., the long-term treatment of the AT$_2$ receptor with ANG II induces two major morphological changes in 3-day-old rat cerebellar cells: increased elongation of neurites and cell migration from the edge of the microexplant toward the periphery. These are two important processes in the organization of the various layers of the cerebellum (Cote et al. 1999).

The role of the AT$_2$ receptor in brain development may explain the altered behavior of AT$_2$ receptor knockout mice. These mice have been reported to show an anxiety-like and markedly reduced exploratory behavior (Hein et al. 1995a; Ichiki et al. 1995c), and the pain threshold was significantly lower in AT$_2$ receptor knockout compared to findings in wild-type mice. Furthermore, AT$_2$ receptor knockout mice show a greater stimulation of dipsogenesis after water deprivation (Hein et al. 1995a). In the passive avoidance task and cold injury, no differences were found between wild-type and AT$_2$-deficient mice (Sakagawa et al. 2000).

Another genetic study points to a role for the AT$_2$ receptor in brain development and cognitive function in human. This study shows that AT$_2$ receptor expression was missing in a female patient with mental retardation (MR) who had a balanced X;7 chromosomal translocation. Additionally, 8 of 590 unrelated male patients with MR were found to have mutations in the AT$_2$ receptor (Vervoort et al. 2002). Thus it appears that the AT$_2$ receptor in human seems to be involved in the development of cognitive function and learning behavior.

Although the distribution of AT$_2$ receptors in the brain is well known, their effects are still not clear. As outlined above, brain AT$_2$ receptors may play a role in cognitive functions and certain types of behavior, such as exploration or drinking (Hein et al. 1995a). On the other hand, they may also antagonize the central effects of angiotensin peptides in osmoregulation mediated via AT$_1$ receptors (Höhle et al. 1995, 1996). Taken together, the data suggest the AT$_2$ receptor regulates the differentiation of various precursor cells and enables them to assume their final tasks. A possibly important role of AT$_2$ receptors in neuroregeneration and neuroprotection will be dealt with below.

2.2
Tissue Regeneration and Pathophysiological Aspects

AT$_2$ receptor expression is dramatically altered in a number of tissues after experimental- or disease-induced damage, and there is much evidence that AT$_2$ receptors contribute to the cellular processes that follow injury situations. The role of AT$_2$ receptors in pathological situations will be reviewed in the following paragraphs, with particular focus on their function in tissue regeneration.

2.2.1
Nerve Regeneration

In vitro treatment of undifferentiated, NGF pre-stimulated NG108-15 or PC12W cells, which are used as in vitro models of sympathetic neurons, with ANG II induces growth arrest and morphological differentiation to neuronal cells via a Rap1/B-Raf-pathway. This effect can be blocked by the selective AT_2 receptor antagonist PD123177 (Laflamme et al. 1996; Meffert et al. 1996; Stroth et al. 1998; Gendron et al. 2002). During this differentiation process the cytoskeleton of PC12W cells undergoes substantial morphological changes, pointing to an involvement of the AT_2 receptor in axonal regeneration (Fig. 1). These morphological changes are paralleled by alterations in the expression patterns of different cytoskeletal proteins, such as neurofilament-M (NF-M) subunit (Gallinat et al. 1997) and microtubule-associated proteins (Stroth et al. 1998), pointing to an involvement of the AT_2 receptor in axonal regeneration. As demonstrated by Gallinat et al., a lasting and pronounced increase in the AT_2 receptor gene expression after sciatic nerve transection occurs in both sciatic nerves as well as in dorsal root ganglia neurons (Gallinat et al. 1998). Together with the observation that Schwann cells express AT_2 receptors (Bleuel et al. 1995), this specific AT_2 receptor expression pattern after peripheral nerve injury points to a role of these cells in neuroregenerative events.

Both in vitro and in vivo, the AT_2 receptor promotes axonal elongation of postnatal retinal explants and dorsal root ganglia neurons and, in adult rats, axonal regeneration of retinal ganglion cells after optic nerve crush. It has been demonstrated that AT_2 receptor stimulation directly contributes to the process of axonal regeneration in the CNS of adult rats. Intravitreous application of ANG II-soaked collagen foams significantly improved the axonal regeneration of retinal ganglion cells after optic nerve crush. Co-treatment with the AT_2 receptor antagonist PD123177, but not the AT_1 receptor antagonist losartan, entirely abolished the ANG II-induced axonal regeneration, providing for the first time evidence for receptor-specific neurotrophic actions of ANG II via the AT_2 receptor (Figs. 2 and 3; Lucius et al. 1998).

Thus, the AT_2 receptor plays a role in neuronal cell differentiation and nerve regeneration via regulation of the neural cytoskeleton of affected neurons (for review, Rosenstiel et al. 2002).

2.2.2
Remodeling of the Heart After Injury

Pharmacological evidence indicates that most of the known effects of ANG II in adult cardiovascular tissues are due to the AT_1 receptor predominantly expressed in the adult heart. However, the relative proportion of the AT_1 and AT_2 receptor density changes during myocardial infarction (MI) (Fig. 4). AT_2 receptor density was shown to be increased in experimental MI 1 day after infarction in the infarcted area. Additionally, the AT_2 receptor was further upregulated

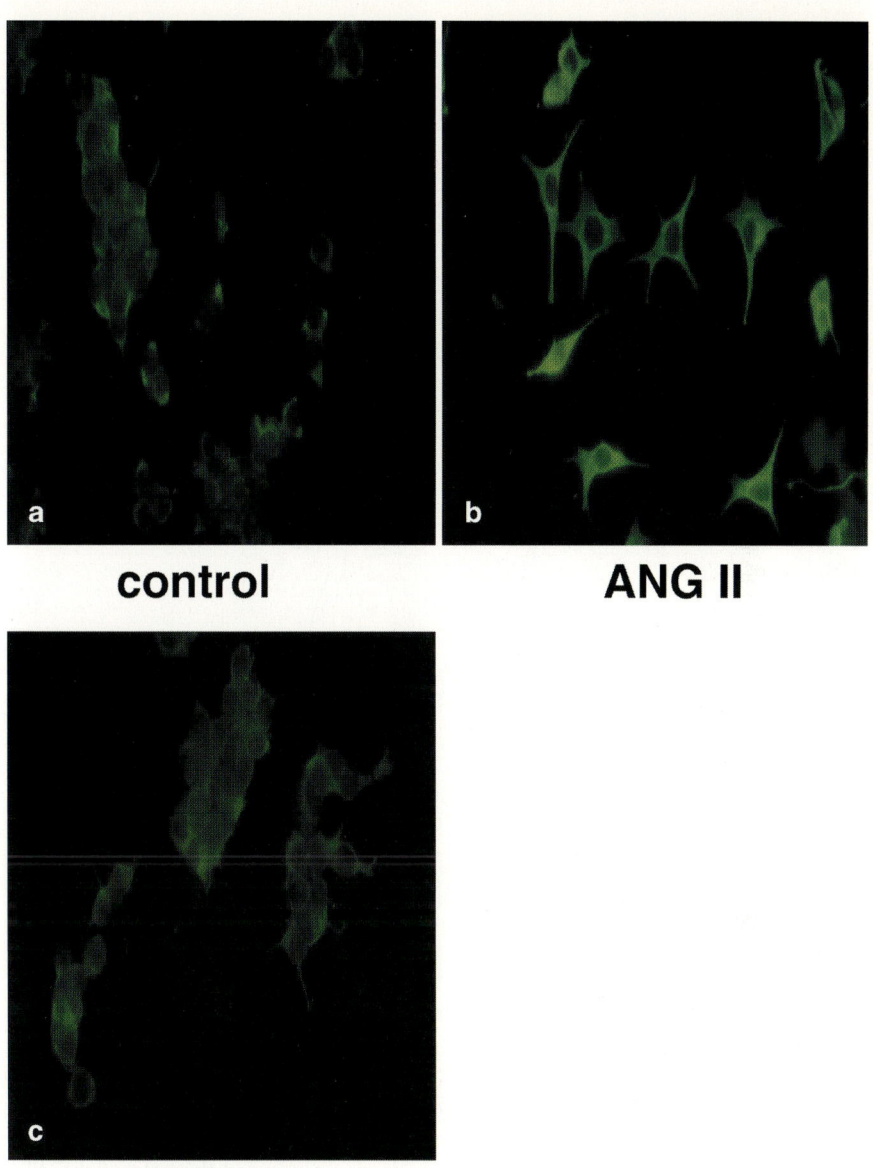

Fig. 1a–c Immunofluorescence microscopy analysis of the effect of ANG II on β-tubulin in PC12W cells.
a Control; **b** ANG II; **c** ANG II+

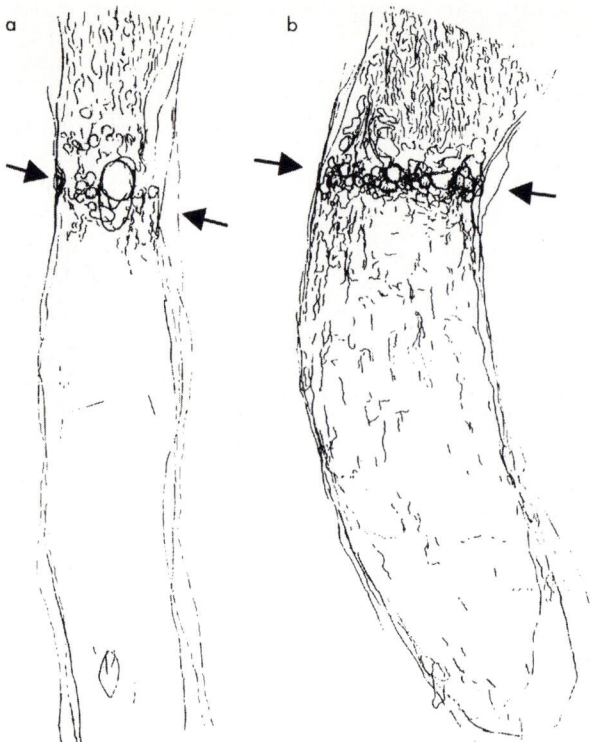

Fig. 2a, b Camera-lucida projections from serial sections of paraffin-embedded and GAP-43-stained optic nerves. The lesion site is demarcated by *two arrows*. In control animals (**a**), only few fibers grew over the lesion, whereas in animals receiving the ANG II gel foam (**b**), numerous axons crossed the lesion site and regenerated over a distance of several millimeters. ANG II increased the number of GAP-43-positive axons in the proximal optic nerve stump compared with controls

7 days post infarction in both the infarcted and the noninfarcted area (Nio et al. 1995; Busche et al. 2000). These findings resemble the hypertrophied heart, where the ratio of AT_2 to AT_1 receptor densities is also increased (Lopez et al. 1995).

Most studies using AT_2 receptor-deficient mice suggested that AT_2 receptors inhibit hypertrophy and vascular fibrosis (Akishita et al. 2000b; Brede et al. 2001; Sandmann et al. 2001a). A recent study demonstrated the role of the AT_2 receptor in wound healing after MI using an in vivo intervention study in AT_2 receptor knockout mice with MI. Mice lacking the AT_2 receptor revealed a decreased collagen deposition, followed by cardiac rupture. The impaired wound healing caused a significantly greater mortality rate of the AT_2 receptor knockout mice compared to wild-type mice (Ichihara et al. 2002). Experiments using cultured cell systems showed that experimental overexpression of the AT_2 receptor in VSMC (vascular smooth muscle cells) induced collagen expression (Mifune at al. 2000). Further in vitro studies demonstrated that the AT_2 receptor in-

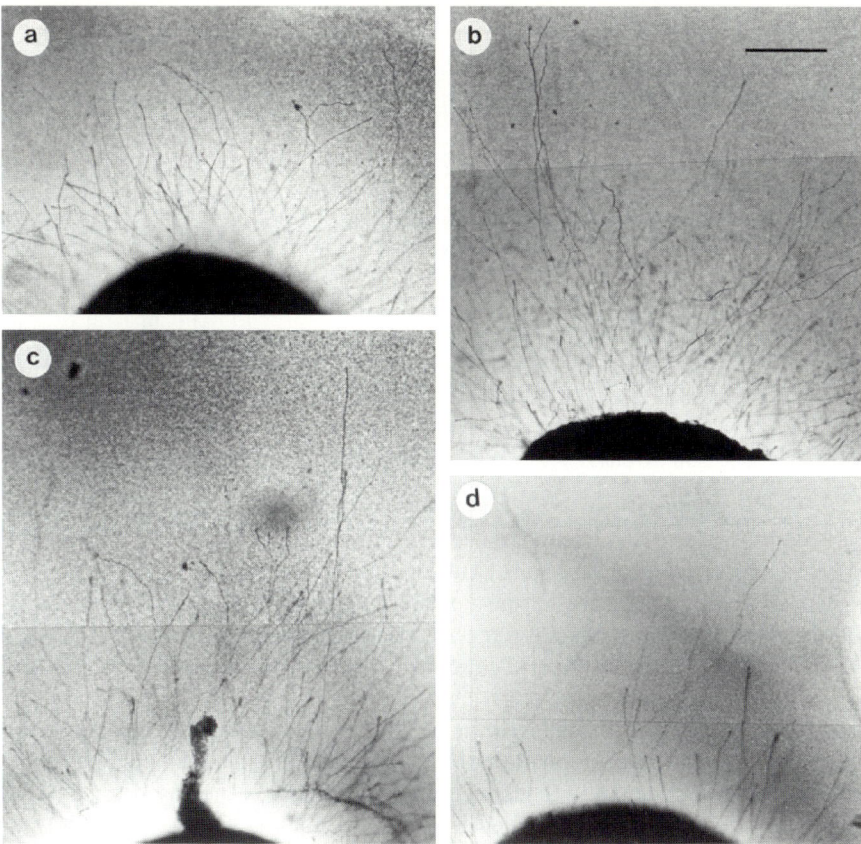

Fig. 3. a Retinal explant (postnatal day 11) in control cultures after 3 days in vitro. Many regenerating, but short neurites werde dectected. **b** Addition of BDNF (10 ng/ml) enhanced axonal growth up to 160±3%. **c** In the present of ANG II (10^6 M), the number and length of neurites was increased to comparable amounts (162±4%), whereas **d** the AT_2 receptor antagonist, PD123177 (10^5 M), blocked axonal regrowth (102±3%). On the other hand, the AT_1 receptor-antagonist, losartan (10^5 M) did not influence the ANG II-induced axonal elongation (data not shown). Bar=100 μm

hibitied the proliferation of rat coronary endothelial cells and transfected VSMCs (Stoll et al. 1995; Fischer et al. 2001) and prevented ANG II-induced growth of cultured neonatal rat myocytes (Booz and Baker 1996). The AT_2 receptor appears also to control the transcriptional regulation of the Na^+-H^+-exchanger NBC-1 in the ischemic myocardium and contributes to the control of pH regulation in the diseased cardiac tissue (Sandmann et al. 2001b). Therefore, the AT_2 receptor counteracts acidification in the affected areas through cytoplasmic alkalinization and thereby reduces apoptosis and inflammation, both consequences of acidification.

Considering the ying–yang hypothesis, where AT_2 receptor functions are the opposite of those mediated through the AT_1 receptor, the up-regulation of the

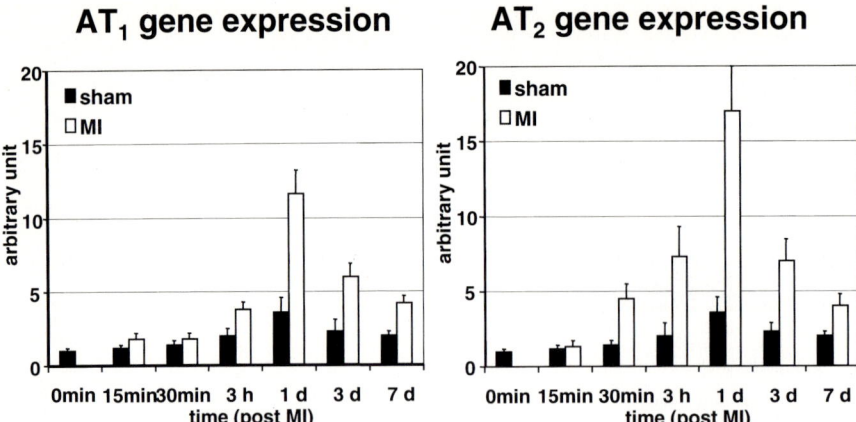

Fig. 4 Regulation of AT receptors following myocardial infarction

AT$_2$ receptor in the diseased heart leads to interesting pharmacological consequences of AT$_1$ receptor antagonist treatment. AT$_1$ receptor antagonists cause a reduction in blood pressure, cardiac and vascular hypertrophy, proteinuria, and glomerular sclerosis. By blocking the AT$_1$ receptor, the circulating ANG II preferentially binds to the cardiac AT$_2$ receptor. There is evidence that the AT$_2$ receptor is activated in failing hearts, especially in cardiac fibroblasts. Liu et al. reported that the cardioprotective function of AT$_1$ receptor antagonists is mainly exerted by the selective stimulation of AT$_2$ receptor, which is mediated partly by the kinin/NO system. ANG II-induced cardiac fibrosis is increased by chronic inhibition of NO synthase, and generation of coronary kinin mediates NO release after ANG II receptor stimulation (Liu et al. 1997). Moderate endothelial NO synthase (eNos) activity and NO generation is cardio protective because NO reversibly depresses mitochondrial respiration and hence decreases basal myocardial oxygen consumption (Xie et al. 1996). Interestingly, Ritter et al. showed that AT$_2$ receptor stimulation with ANG II increases eNOS protein expression 3.3-fold in rat cardiomyocytes (Ritter et al. 2003; see Sect. 11). AT$_2$ receptor-mediated activation of the kinin/NO system is also involved in pressure natriuresis and diuresis of the kidneys. Kijima et al. reported that the AT$_2$ receptor in hypertrophic myocytes at least partially exerts an inhibitory effect on AT$_1$ receptor-mediated positive chronotropic or hypertrophic actions by showing up-regulation of AT$_2$ receptors in stretch-induced myocyte hypertrophy (Kijima et al. 1996). Taken together, these results show that the AT$_2$ receptor plays an important role in cardiovascular recovery in various ways (Table 3; Matsubara 1998; Sandmann and Unger 2002).

Table 3 Pathophysiological roles of AT_2 receptors in cardiovascular systems

Effect	Cell/tissue	Reference(s)
Growth inhibition	VSMC	Nakajima et al. 1995
	Endothelial cell	Stoll et al.1995
	Neonatal rat myocytes Cardiomyocyte	Fischer et al. 2001
	Cardiac fibroblast	Booz and Baker 1996
		Van Kesteren et al. 1997
		Ohkubo et al. 1997
Apoptosis	Endothelial cell	Dimmeler et al. 1997
	VSMC	Yamada et al. 1998
	Cardiomyocyte	Leri et al. 1998
		Goldenberg et al. 2001
Differentiation	VSMC	Yamada et al. 1999
Decrease in cellular matrix	Heart	Ohkubo et al. 1997
		Fischer et al. 2001
		Shimizu-Hirota et al. 2001
		Chassagne et al. 2002
pH regulation	Heart	Kohout et al. 1995
		Sandmann et al. 2001b
Vasodilatation	Glomerular afferent arteriole	Arima et al. 1997
NO production	Coronaryartery and microvessel	Seyedi et al. 1995
	Kidney	Siragy et al. 1996
	Aorta	Gohlke et al. 1998
	Heart	Ritter et al. 2002

2.2.3
Renal Physiological and Pathophysiological Function

The AT_1 receptor is clearly expressed in the glomeruli and in the inner stripe of the outer medulla. Glomerular AT_1 receptors may control the hemodynamic function of the kidney. The proportion of the AT_2 receptor in the kidney cortex is less than 10% of the AT_1 receptor in the rat and rabbit, and roughly 55% in monkey (Chang and Lotti 1991). Nevertheless, the AT_2 receptor influences renal function by controlling pressure natriuresis (Lo et al. 1995). In sodium-depleted rats it is involved in decreasing cGMP production by a bradykinin-dependent mechanism (Siragy and Carey 1996; Chung und Unger 1998).

Morrissey and Klahr determined the effects of the AT_2 receptor antagonist PD123319 on pathophysiological processes in the kidneys of rats undergoing unilateral ureteral obstruction. The authors showed that blockade of the AT_2 receptor flattens the increase in interstitial volume and the collagen IV matrix score of the ureteral-obstructed kidney. The AT_2 receptor antagonist inhibited apoptosis of tubular cells, α-smooth muscle actin expression within the interstitium, and p53 expression in the ureteral-obstructed kidney. These findings suggest that the AT_2 receptor exerts an antifibrotic effect under pathophysiological conditions, again an effect that is opposite to the profibrotic effects of the AT_1 receptor (Morrissey and Klahr 1999). In support of this hypothesis, an in vitro study shows that ANG II induces growth inhibition and apoptosis in the cultured proximal tubule cell line LLC-PK1 overexpressing of AT_2 receptors (Zimpelmann and Burns 2001). Taken together, these reports demonstrate that the AT_2 receptor also plays a pivotal role in renal remodeling following injury.

3
Target Genes

As demonstrated above, the AT_2 receptor is mainly involved in cellular development and tissue rearrangement after injury. In order to elucidate the expression profile of AT_2 receptor target genes involved in these processes, various studies have been performed. Recently, two studies using differential display performed in cell-culture systems discovered some of these target genes. Wolf and co-workers investigated the alteration in gene expression in PC12W cells after AT_2 receptor stimulation. They were able to show that ANG II suppresses the mRNA expression of SM-20, a growth factor-responsive gene. The SM-20 transcription was also reduced by ANG II acting on AT_2 receptors in rat glomerular endothelial cells that express both AT_1 and AT_2 receptors (Wolf et al. 2002). These effects could be antagonized by PD123177, but not by losartan, a selective AT_1 antagonist. SM-20 is associated with mitochondria and is involved in neurotrophin withdrawal-mediated apoptosis through a caspase-dependent mechanism (Wax et al. 1994). Its down-regulation via AT_2 receptors may set the course from apoptosis to differentiation of the affected cells.

Stoll and co-workers identified two additional target genes of the AT_2 receptor. One of them was the transcription factor Zfhep, a zinc finger homeodomain enhancer-binding protein which is implicated in various developmental processes. The expression patterns of Zfhep in vivo and in a P19 cell model of neurogenesis suggest that Zfhep may play a role in the differentiation of neural cells (Yen et al. 2001). In quiescent PC12W cells, Zfhep mRNA was time-dependently induced by ANG II, an effect that could be blocked by the selective AT_2 antagonist PD123177. These observations were supported by data from a coronary endothelial cell model. In this model, Zfhep mRNA expression was also induced after AT_2 receptor stimulation. Interestingly, this Zfhep induction was only observed when the AT_1 receptor was blocked by losartan. This observation suggests that a negative regulatory influence of AT_1 receptors on AT_2 induced Zfhep mRNA expression. Zfhep is present in both hyperphosphorylated and hypophosphorylated forms. It is primarily phosphorylated on serine and threonine residues and dephosphorylation enables the translocation from the cytosol into the nucleus of the transcription factor. This dephosphorylation could by accomplished by PP2A (Costantino et al. 2002), a phosphatase which is activated via AT_2 receptors. In this way the AT_2 receptor may modulate the differentiation process in neuronal cells.

The second gene identified by Stoll et al. was the AT_2 receptor itself (Stoll et al. 2002), an observation that is consistent with an older study using cultured rat cortical cells (Shibata et al.1997). This self up-regulation is exceptional because the majority of receptors are down-regulated after stimulation, preventing an "overload" of the receptor signal by a positive feedback loop. But this observation can be reconciled with the general unique nature of the AT_2 receptor signaling, including the lack of ligand-stimulated receptor internalization and desensitization of the receptor (Mukoyama et al. 1993; Mukoyama et al. 1995). These findings led to the assumption of the existence of an unknown endogenous factor which, like the AT_2 receptor ligands, seems to prevent AT_2 receptor degradation (Csikos et al. 1998). Miura and Karnik were first to report that the stimulation of the AT_2 receptor via ANG II is not obligatory for its signaling and also that the receptor expression level is crucial. They proposed that an overexpression of the AT_2 receptor itself is a receptor-intrinsic signal for apoptosis that does not require ANG II (Miura and Karnik 2000).

Since then, several publications have confirmed this concept of an intrinsic activity of the AT_2 receptor. Another interesting finding is that the AT_2 receptor binds directly to the AT_1 receptor and thereby antagonizes the function of the AT_1 receptor. This inhibitory effect of the AT_2 receptor is independent of the stimulation status. This report is notable since it is the first identified interaction between two G protein-coupled receptors, where one receptor acts like an antagonist (AbdAlla et al. 2001).

All these data fit to two recently published studies which show that overexpression of the AT_2 receptor significantly alters the expression profile of cells, given the receptor density crosses the critical threshold. In VSMC, overexpressed AT_2 receptor leads to decreased expression of AT_1 and transforming growth factor (TGF)-β type I receptors (Su et al. 2002). In the same cell line, over-expressed AT_2 receptor upregulates bradykinin and inducible NO in the absence of ANG II. The bradykinin B2 receptor antagonist HOE-140 and the NO synthase inhibitor N^{ω}-nitro-L-arginine methylester (L-NAME) inhibit the decrease in AT_1 receptor expression via the overexpression of AT_2 receptor in VSMCs, whereas L-arginine enhances a decrease of AT_1 receptor expression (Jin et al. 2002). By this, the AT_2 receptor antagonizes the AT_1 receptor via a bradykinin B2 receptor/NO pathway without the requirement of ANG II (Fig. 5). These observations are consistent with the presumed AT_2 receptor-mediated vasodilator signaling cascade that includes bradykinin, nitric oxide, and cyclic guanosine 5-monophosphate (see Sect. 9).

Another gene that is negatively regulated by AT_2 receptor overexpression is the TGF-β I receptor. Conversely, AT_1 receptor stimulation is accompanied by an up-regulated TGF-β 1 gene expression. TGF-β 1 is a potent regulator of VSMC proliferation, migration, and extracellular matrix synthesis, components which partially contribute to the pathogenesis of cardiovascular end organ damage. Furthermore, the AT_1 receptor stimulates collagen production in VSMCs via a paracrine/autocrine loop of TGF-β (Ford et al. 1999). The suppression of the AT_1 receptor as well as the TGF-β I receptor appear to have a potential benefit in preventing pathological myocardial fibrosis.

To investigate the molecular events that follow the initiation of neurite outgrowth in PC12W cells via AT_2 receptor stimulation, Gallinat and co-workers

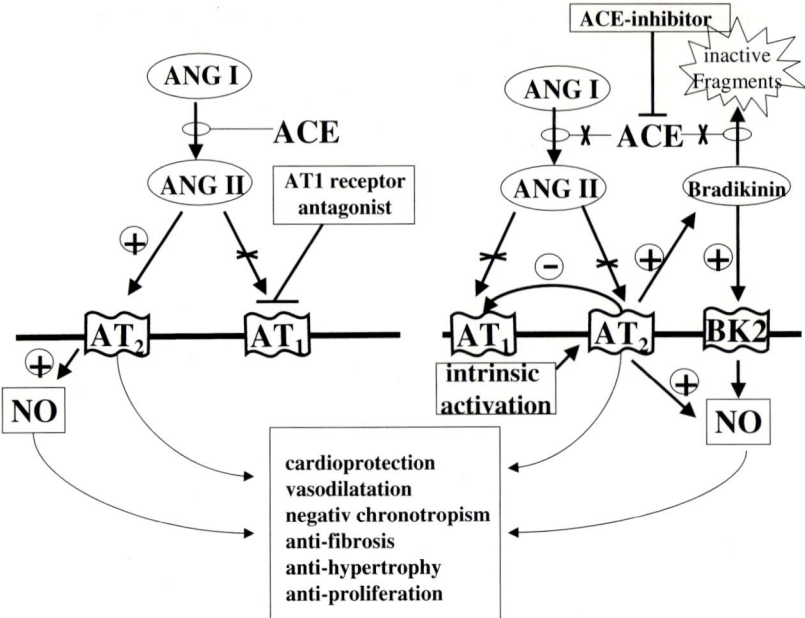

Fig. 5 Influence of AT_1 receptor antagonist and ACE-inhibitor on the AT_2 receptor effects

determined the expression pattern of the middle-sized neurofilament subunit (NF-M). They showed that the protein and mRNA levels of NF-M was not only down-regulated by ANG II via AT_2 receptor directly, but also by counteracting the NGF-mediated NF-M up-regulation. In view of previous findings of decreased NF levels in regenerating neurons and in neuronal cultures undergoing apoptosis, this observation suggests a role of AT_2 receptors in either of these processes (Gallinat et al. 1997).

As outlined above, the AT_2 receptor is re-expressed and plays a role in tissue remodeling. Change in the expression profile of the concerned areas is one possible option of the cells to implement this reconstruction. The AT_2 receptor appears to take a direct and active part in the remodeling of the extracellular matrix of the injured regions. Fischer et al. showed that stimulation of the AT_2 receptor in VSMCs, retrovirally transfected with the AT_2 receptor, resulted in a significant increase in thrombospondin 1 synthesis, which was abolished by an AT_2 receptor antagonist and attenuated by pretreatment with pertussis toxin, suggesting a G protein-dependent pathway (Fischer et al 2001). Using the same cell model, Mifune and co-workers were able to show that the AT_2 receptor stimulation increases collagen synthesis also via a G protein-mediated mechanism (Mifune at al. 2000).

In the myocardium, acidosis depresses the contractility of cardiac myocytes by affecting virtually every step in excitation–contraction coupling. Hence, in order to preserve a proper function of the heart, cardiac myocytes possess pH-regulating transporters that maintain the pH within very narrow limits. Ion exchange in myocardial cells represents a major mechanism for pH regulation during normal physiological processes, but particularly during ischaemia and early reperfusion. The AT_2 receptor, being upregulated itself within the infarcted area, upregulates and activates the Na^+-$HCO3$ symporter NBC-1 in the ischemic myocardium and thereby contributes to the control of pH in cardiac tissue (Kohout et al. 1995; Sandmann et al. 2001b). Since acidification is also a feature of apoptosis and inflammation, the AT_2 receptor acts to obviate this process by preventing acidification through cytoplasmic alkalinization.

A recent study using in vivo and in vitro experiments determined the influence of the AT_2 receptor on myocardial eNOS (NOS-III) protein expression. AT_2 receptor stimulation with ANG II increases the eNOS protein expression 3.3-fold in rat cardiomyocytes. This augmentation could be blocked by PD123319, whereas inhibition of the AT_1 receptor did not reduce ANG II-mediated eNOS protein expression. The modulatory effects of the AT_2 receptor on eNOS expression was confirmed in vivo using an AT_2 receptor knockout mouse. Mice lacking AT_2 receptors express significantly less eNOS protein in the myocardium then wild-type mice (Ritter et al. 2003). As outlined above, AT_2 receptor stimulation releases NO (see Sect. 9). This has been shown for coronary arteries and the aorta (Seyedi et al. 1995; Golhke et al. 1998). Moderate eNOS activity and NO generation is considered to be cardioprotective, because NO reversibly depresses mitochondrial respiration and hence decreases basal myocardial oxygen consumption (Xie et al. 1996; Liu et al. 1997).

Many inflammatory renal diseases are connected with glomerular influx of monocytes/macrophages (M/M). The signals guiding the M/M into the glomerulus are cytokines and chemokines. In vitro ANG II stimulation of the AT_2 receptor induces mRNA and protein expression of the chemokine RANTES in cultured rat glomerular endothelial cells, a chemokine that is chemotactic for monocytes. In an in vivo study it was shown that intraperitoneal infusion of ANG II into naive rats for 4 days significantly stimulated glomerular RANTES mRNA and protein expression compared to control. Immunohistochemistry revealed induction of RANTES protein mainly in glomerular endothelial cells and small capillaries. Under medical conditions the AT_2 receptor may have a proinflammatory implication mediated through its target gene RANTES (Wolf et al. 1997).

In summary, the functions of the AT_2 receptor in development and regeneration imply a wide range of target genes (Table 4) that are controlled via ANG II stimulation and AT_2 receptor density. The suppression of proliferation signals by the AT_2 receptor via activation of various phosphatases is mediated via down-regulation of growth-mediating receptors as well as up-regulation of the receptor itself, thereby producing a positive feedback loop. During its function in cell differentiation, the AT_2 receptor regulates miscellaneous genes in order to reorganize precursor cells during maturation and enable them to assume their finale tasks.

Table 4 Target genes modulated by the AT_2 receptor

Target gene	Regulation	Reference
AT_1 receptor	Down	Su et al. 2002
AT_2 receptor	Up	Stoll et al. 2002
Bradykinin	Up	Jin et al. 2002
Collagen	Up	Mifune et al. 2000
eNOS	Up	Ritter et al. 2002
NBC-1 Na^+-$HCO3$ symporter	Up	Sandmann et al. 2001b
Thrombospondin 1	Up	Fischer et al. 2001
RANTES	Up	Wolf et al. 1997
SM-20	Down	Wolf et al. 2002
TGF-β type I receptor	Down	Su et al. 2002
Zfhep	Up	Stoll et al. 2002

4 Summary and Conclusion

The AT_2 receptor signals and functions in unexpected ways compared to the "classical" AT_1 receptor. Moreover, some of the actions of the AT_2 receptor are even directly opposing those of the AT_1 receptor, especially concerning the growth- and differentiation-modulating actions after ANG II stimulation. The modulatory or suppressive function of the AT_2 receptor often counteracts stimulatory signals. This requires measuring reduction in activities, which is technically more difficult than measurement of increased activity from a low level. This may have been the reason the AT_2 receptor has sometimes escaped the attention of investigators.

The signaling mechanisms of the AT_2 receptor are diverse, and only a few of them have yet been characterized reasonably well. The unique nature of the AT_2 receptor includes the lack of desensitization and internalization after ligand binding and the intrinsic activity of the receptor in itself. AT_2 receptor expression in fetal tissues is widespread. However, in several tissues the receptor disappears after birth or—under in vitro conditions—when cells are transferred in the presence of serum and growth factors. On the other hand, a sometimes dramatic up-regulation of the AT_2 receptor does occur after tissue injuries. Especially during heart failure its expression is significantly increased. Pharmacologically relevant is the fact that AT_2 receptor stimulation inhibits cardiac fibroblast growth and extracellular matrix formation and exerts a negative chronotropic effect, indicating that AT_2 receptor stimulation has a cardioprotective effect (Fig. 5). It appears that the AT_2 receptor accepts the role as a modulator of biological programs in embryonic development and cellular processes following injuries, particularly by altering the expression profile of its target genes.

References

AbdAlla S, Lother H, Abdel-tawab AM, Quitterer U (2001) The angiotensin II AT2 receptor is an AT1 receptor antagonist. J Biol Chem 276(43):39721–39726

Akishita M, Iwai M, Wu L, Zhang L, Ouchi Y, Dzau VJ, Horiuchi M (2000) Inhibitory effect of angiotensin II type 2 receptor on coronary arterial remodeling after aortic banding in mice. Circulation 102:1684–1689

Arima S, Endo Y, Yaoita H, Omata K, Ogawa S, Tsunoda K, Abe M, Takeuchi K, Abe K, Ito S (1997) Possible role of P-450 metabolite of arachidonic acid in vasodilator mechanism of angiotensin II type 2 receptor in the isolated microperfused rabbit afferent arteriole. J Clin Invest 100(11):2816–2823

Armando I, Jezova M, Juorio AV, Terron JA, Falcon-Neri A, Semino-Mora C, Imboden H, Saavedra JM (2002) Estrogen upregulates renal angiotensin II AT(2) receptors. Am J Physiol Renal Physiol 283(5):F934–F943

Bleuel A, de Gasparo M, Whitebread S, Puttner I, Monard D (1995) Regulation of protease nexin-1 expression in cultured Schwann cells is mediated by angiotensin II receptors. J Neurosci 15:750–761

Booz GW, Baker KM (1996) Role of type 1 and type 2 angiotensin receptors in angiotensin II-induced cardiomyocyte hypertrophy. Hypertension 28(4):635–640

Bottari SP, Taylor V, King IN, Bogdal Y, Whitebread S, de Gasparo M (1991) Angiotensin II AT2 receptors do not interact with guanine nucleotide binding proteins. Eur J Pharmacol 207(2):157–63

Brechler VP, Jones W, Levens NR, de Gasparo M, Bottari SP (1993) Agonistic and antagonistic properties of angiotensin analogs at the AT_2 receptor in PC12W cells. Regul Pept 44:207–213

Brede M, Hadamek K, Meinel L, Wiesmann F, Peters J, Engelhardt S, Simm A, Haase A, Lohse MJ, Hein L (2001) Vascular hypertrophy and increased P70S6 kinase in mice lacking the angiotensin II AT(2) receptor. Circulation 104:2602–2607

Busche S, Gallinat S, Bohle RM, Reinecke A, Seebeck J, Franke F, Fink L, Zhu M, Sumners C, Unger T (2000) Expression of Angiotensin AT1 and AT2 Receptors in Adult Rat Cardiomyocytes after Myocardial Infarction : A Single-Cell Reverse Transcriptase-Polymerase Chain Reaction Study. Am J Pathol 157(2):605–611

Camp HS, Dudley DT (1995) Modulation of angiotensin II receptor (AT_2) mRNA levels in R3T3 cells. Receptor 5:123–132

Chassagne C, Adamy C, Ratajczak P, Gingras B, Teiger E, Planus E, Oliviero P, Rappaport L, Samuel JL, Meloche S (2002) Angiotensin II AT(2) receptor inhibits smooth muscle cell migration via fibronectin cell production and binding. Am J Physiol Cell Physiol 282:C654–C664

Chang RS, Lotti VJ (1991) Angiotensin receptor subtypes in rat, rabbit and monkey tissues: relative distribution and species dependency. Life Sci 49 (20):1485–1490

Chiu AT, Herblin WF, McCall DE, Ardecky RJ, Carini DJ, Duncia JV, Pease LJ, Wong PC, Wexler RR, Johnson AL, et al (1989) Identification of angiotensin II receptor subtypes. Biochem Biophys Res Commun 165(1):196–203

Chung O, Unger T (1998) Unopposed stimulation of the angiotensin AT_2 receptor in the Kidney. Nephrol Dial Transplant 13:537–540

Costantino ME, Stearman RP, Smith GE, Darling DS (2002) Cell-specific phosphorylation of Zfhep transcription factor. Biochem Biophys Res Commun 296(2):368–373

Cote F, Do TH, Laflamme L, Gallo JM, Gallo-Payet N (1999) Activation of the AT(2) receptor of angiotensin II induces neurite outgrowth and cell migration in microexplant cultures of the cerebellum. J Biol Chem 274:31686–31692

Csikos T, Balmforth AJ, Grojec M, Gohlke P, Culman J, Unger T (1998) Angiotensin AT2 receptor degradation is prevented by ligand occupation. Biochem Biophys Res Commun 243(1):142–147 de Gasparo M, Catt KJ, Inagami T, Wright JW, Unger T (2000) International union of pharmacology. XXIII. The angiotensin II receptors. Pharmacol Rev 52(3):415–472

Dimmeler S, Rippmann V, Weiland U, Haendeler J, Zeiher AM (1997) Angiotensin II induces apoptosis of human endothelial cells: protective effect of nitric oxide. Circ Res 81:970–976

Dittus J, Cooper S, Obermeir G, Pulakat L (1999) Role of the third intracellular loop of the angiotensin II receptor subtype AT_2 in ligand-receptor interaction. FEBS Lett 445:23–26

Dudley DT and Summerfelt RM (1993) Regulated expression of angiotensin II (AT_2) binding sites in R3T3 cells. Regul Pept 44:199–206

Ferguson SSG (2001) Evolving Concepts in G Protein-Coupled Receptor Endocytosis: The Role in Receptor Desensitization and Signaling. Pharmacol Rev 53:1-24

Fischer JW, Stoll M, Hahn AW, Unger T (2001) Differential regulation of thrombospondin-1 and fibronectin by angiotensin II receptor subtypes in cultured endothelial cells. Cardiovasc Res 51(4):784–791

Ford CM, Li S, Pickering JG (1999) Angiotensin II stimulates collagen synthesis in human vascular smooth muscle cells. Involvement of the AT(1) receptor, transforming growth factor-beta, and tyrosine phosphorylation. Arterioscler Thromb Vasc Biol 19(8):1843–1851

Gallinat S, Csikos T, Meffert S, Herdegen T, Stoll M, Unger T (1997) The angiotensin AT2 receptor down-regulates neurofilament M in PC12W cells. Neurosci Lett 227(1):29–32

Gallinat S, Yu M, Dorst A, Unger T, Herdegen T (1998) Sciatic nerve transection evokes lasting up-regulation of angiotensin AT2 and AT1 receptor mRNA in adult rat dorsal root ganglia and sciatic nerves. Brain Res Mol Brain Res 57(1):111–122

Gallinat S, Busche S, Schutze S, Kronke M, Unger T (1999) AT2 receptor stimulation induces generation of ceramides in PC12W cells. FEBS Lett 443(1):75–79

Gallinat S, Busche S, Raizada MK, Sumners C (2000) The angiotensin II type 2 receptors: an enigma with multiple variations. Am J Physiol 278:E357–E374

Gehlert DR, Gackenheimer SL, Schober DA (1991a) Angiotensin II receptor subtypes in rat brain: Dithiotherital inhibits ligand binding to AII-1 and enhances binding to AII-2. Brain Res 546:161–165

Gehlert DR, Gackenheimer SL, Schober DA (1991b) Autoradiographic localization of subtypes of angiotensin II antagonist binding in the rat brain. Neuroscience 44:501–514

Gehlert DR, Speth RC, Wamsley JK (1986) Quantitative autoradiography of angiotensin II receptors in SHR brain. Peptides 19867:1021–1027

Gendron L, Oligny JF, Payet MD, Gallo-Payet N (to be published) Cyclic AMP-independent involvement of Rap1/B-Raf in the angiotensin II AT2 receptor signaling pathway in NG108-15 cells. J Biol Chem

Gohlke P, Pees C, Unger T (1998) AT_2 receptor stimulation increases aortic cyclic GMP in SHRSP by a kinin-dependent mechanism. Hypertension 81:349–355

Goldenberg I, Grossman E, Jacobson KA, Shneyvays V, Shainberg A (2001) Angiotensin II-induced apoptosis in rat cardiomyocyte culture: a possible role of AT1 and AT2 receptors. J Hypertens 19(9):1681–1689

Grady EF, Sechi LA, Griffin CA, Schambelan M, Kalinyak JE (1991) Expression of AT2 receptors in the developing rat fetus. J Clin Invest 88:921–933

Hayashida W, Horiuchi M, Dzau V J (1996a) Intracellular third loop domain of angiotensin II type-2 receptor. Role in mediating signal transduction and cellular function. J Biol Chem 271:21985–21992

Hayashida W, Horiuchi M, Grandchamp J, Dzau VJ (1996b) Antagonistic action of angiotensin II type-1 and type-2 receptors on apoptosis in cultured neonatal rat ventricular myocytes. Hypertension 28:535

Hein L (1998) Genetic deletion and overexpression of angiotensin II receptors. J Mol Med 76:756–763

Hein L, Barsk GS, Pratt RE, Dzau VJ, Kobilka BK (1995) Behavioral and cardiovascular effects of disruption the angiotensin II type-2 receptor gene in mice. Nature 377:744–747

Hein L, Meinel L, Pratt RE, Dzau VJ, Kobilka BK (1997) Intracellular trafficking of angiotensin II and its AT1 and AT2 receptors: evidence for selective sorting of receptor and ligand. Mol Endocrinol 11(9):1266–1277

Höhle S, Blume A, Lebrun C, Culman J, Unger T (1995) Angiotensin receptors in the brain. Pharmacol Toxicol 77:306–315

Höhle S, Culman J, Boser M, Qadri F, Unger T (1996) Effect of angiotensin AT_2 and muscarinic receptor blockade on osmotically induced vasopressin release. Eur J Pharmacol 300:119–123

Hogarty DC, Speakman EA, Puig V, Phillips MI (1992) The role of angiotensin, AT_1 and AT_2 receptors in the pressor, drinking and vasopressin responses to central angiotensin. Brain Res 586:289–294

Horiuchi M, Yamada T, Hayashida W, Dzau VJ (1997) Interferon regulatory factor-1 upregulates angiotensin II type 2 receptor and induces apoptosis. J Biol Chem 272(18):11952–11958

Huang XC, Shenoy UV, Richards EM, Sumners C (1997) Modulation of angiotensin II type 2 receptor mRNA in rat hypothalamus and brainstem neuronal cultures by growth factors. Brain Res. Mol Brain Res 47:229–236

Ichihara S, Senbonmatsu T, Price E Jr, Ichiki T, Gaffney FA, Inagami T (2002) Targeted deletion of angiotensin II type 2 receptor caused cardiac rupture after acute myocardial infarction. Circulation 22;106 (17):2244–2249

Ichiki T, Inagami T (1995a) Expression, genomic organization, and transcription of the mouse angiotensin II type 2 receptor gene. Circ Res 5:693–700

Ichiki T, Kambayashi Y, Inagami T (1995b) Multiple growth factors modulate mRNA expression of angiotensin II type-2 receptor in R3T3 cells. Circ Res 77(6):1070–1076

Ichiki T, Labosky PA, Shiota C, Okuyama S, Imagawa Y, Fogo A, Niimura F, Ichikawa I, Hogan BLM, Inagami T (1995c) Effects on blood pressure and reduced exploratory behavior in mice lacking angiotensin II type 2 receptor. Nature 377:748–750

Jin XQ, Fukuda N, Su JZ, Lai YM, Suzuki R, Tahira Y, Takagi H, Ikeda Y, Kanmatsuse K, Miyazaki H (2002) Angiotensin II type 2 receptor gene transfer downregulates angiotensin II type 1a receptor in vascular smooth muscle cells. Hypertension 39(5):1021–1027

Kambayashi Y, Nagata K, Ichiki T, Inagami T (1996) Insulin and insulin-like growth factors induce expression of angiotensin type-2 receptor in vascular-smooth-muscle cells. Eur J Biochem 239(3):558–565

Kang J, Posner P, Sumners C (1994) Angiotensin II type 2 receptor stimulation of neuronal K+ currents involves an inhibitory GTP binding protein. Am J Physiol 267 (5 Pt 1):C1389–C1397

Kang J, Richards EM, Posner P, Sumners C (1995) Modulation of the delayed rectifier K^+ current in neurons by an angiotensin II type 2 receptor fragment. Am J Physiol Cell Physiol 268:C278–C282

Kaschina E, Unger T (2003) Angiotensin AT1/AT2 receptors: regulation, signalling and function. Blood Press 12:70–88

Kijima K, Matsubara H, Komuro I, Yazaki Y, Inada M (1996) Mechanical stretch induces enhanced expression of angiotensin II receptors in neonatal rat cardiac myocytes. Circ Res 79:887–897

Kimura B, Sumners C, Phillips MI (1992) Changes in skin angiotensin II receptors in rats during wound healing. Biochem Biophys Res Commun 187:1083–1909

Kizima K, Matsubara H, Murasawa S, Maruvama K, Ohkubo N, Mori Y, Inada M (1996) Regulation of angiotensin II type 2 receptor gene by the protein kinase C-calcium pathway. Hypertension 27:529–534

Kohout TA, Rogers TB (1995) Angiotensin II activates the Na+/HCO3- symport through a phosphoinositide-independent mechanism in cardiac cells. J Biol Chem 270(35):20432–20438

Laflamme L, de Gasparo M, Gallo JM, Payet MD, Gallopayet N (1996) Angiotensin II induction of neurite outgrowth by AT_2 receptors in NG108-15 cells—effect counteracted by the AT_1 receptors. J Biol Chem 271:22729–22735

Lehtonen JY, Horiuchi M, Daviet L, Akishita M, Dzau VJ (1999) Activation of the de novo biosynthesis of sphingolipids mediates angiotensin II type 2 receptor-induced apoptosis. J Biol Chem 274(24):16901–16906

Leri A, Claudio PP, Li Q, Wang XW, Reiss K, Wang SG, Malhotra A, Kajstura J, Anversa P (1998) Stretch-mediated release of angiotensin II induces myocyte apoptosis by activating p53 that enhances the local renin-angiotensin system and decreases the Bcl-2-to-Bax protein ratio in the cell. J Clin Invest 101:1326–1342

Lenkei Z, Palkovits M, Corvol P, Llorens-Cortes C (1996) Distribution of angiotensin II type-2 receptor (AT_2) mRNA expression in the adult rat brain. J Comp Neurol 373:322–339

Li JY, Avallet O, Berthelon MC, Langlois D, Saez JM (1998) Effects of growth factors on cell proliferation and angiotensin II type 2 receptor number and mRNA in PC12W and R3T3 cells. Mol Cell Endocrinol 139(1–2):61–69

Li JY, Avallet O, Berthelon MC, Langlois D, Saez JM (1999) Transcriptional and translational regulation of angiotensin II type 2 receptor by angiotensin II and growth factors. Endocrinology 140 (11):4988–4994

Liu YH, Yang XP, Sharov VG, Nass O, Sabbah HN, Peterson E, Carretero OA (1997) Effects of angiotensin-converting enzyme inhibitors and angiotensin II type 1 receptor antagonists in rat with heart failure. Role of kinins and angiotensin II type 2 receptors. J Clin Invest 99:1926–1935

Lo, M., Liu KL, Lanteime P, Sassard J (1995) Subtype 2 of angiotensin II receptors controls pressure-natriuresis in rats. J Clin Invest 95:1394–1397

Lopez JJ, Lorell BH, Ingelfinger JR, Weinberg EO, Schunkert H, Diamant D, Tang SS (1994) Distribution and function of cardiac angiotensin AT1- and AT2-receptor subtypes in hypertrophied rat hearts. Am J Physiol 267:H844–H852

Lucius R, Gallinat S, Rosenstiehl P, Herdegen T, Sievers J, Unger T (1998) The angiotensin II type 2 (AT_2) receptor promotes axonal regeneration in the optic nerve of adult rats. J Exp Med 188:661–670

Matsubara H (1998) Pathophysiological role of angiotensin II type 2 receptor in cardiovascular and renal diseases. Circ Res 83(12):1182–1191

Matsubara H, Sugaya T, Murasawa S, Nozawa Y, Mori Y, Masaki H, Maruyama K, Tsutumi Y, Shibasaki Y, Moriguchi Y, Tanaka Y, Iwasaka T, Inada M (1998) Tissue-specific expression of human angiotensin II AT1 and AT2 receptors and cellular localization of subtype mRNAs in adult human renal cortex using in situ hybridization. Nephron 80(1):25–34

Meffert S, Stoll M, Steckelings UM, Bottari SP, Unger T (1996) The angiotensin AT_2 receptor inhibits proliferation and promotes differentiation in PC12W cells. Mol Cell Endocrinol 122:59–67

Mifune M, Sasamura H, Shimizu-Hirota R, Miyazaki H, Saruta T (2000) Angiotensin II type 2 receptors stimulate collagen synthesis in cultured vascular smooth muscle cells. Hypertension 36:845–50

Millan M, Jacobowitz DM, Aguilera G, Catt KJ (1991) Differential distribution of AT1 and AT2 angiotensin II receptor subtypes in the rat brain during development. Proc Natl Acad Sci USA 88:11440–11444

Miura S and Karnik SS (2000) Ligand-independent signals from angiotensin II type 2 receptor induce apoptosis. EMBO J 19:4026–4035

Morrissey JJ, Klahr S (1999) Effect of AT_2 receptor blockade on the pathogenesis of renal fibrosis Am J Physiol 276:F39–F45

Mukoyama M, Nakajima M, Horiuchi M, Sasamura H, Pratt RE, Dzau VJ (1993) Expression cloning of type 2 angiotensin II receptor reveals a unique class of seven-transmembrane receptors. J Biol Chem 268(33):24539–24542

Mukoyama M, Horiuchi M, Nakajima M, Pratt RE, Dzau VJ (1995) Characterization of a rat type 2 angiotensin II receptor stably expressed in 293 cells. Mol Cell Endocrinol 112(1):61–68

Murasawa S, Matsubara H, Kijima K, Maruyama K, Ohkubo N, Mori Y, Iwasaka T, Inada M (1996) Down-regulation by cAMP of angiotensin II type 2 receptor gene expression in PC12 cells. Hypertens Res 19(4):271–279

Nakajima M, Mukoyama M, Pratt RE, Horiuchi M, Dzau VJ (1993) Cloning of cDNA and analysis of the gene for mouse angiotensin II type 2 receptor. Biochem Biophys Res Commun 197(2):393–399

Nakajima M, Hutchinson HG, Fujinaga M, Hayashida W, Morishita R, Zhang L, Horichi M, Pratt R, Dzau VJ (1995) The angiotensin II (AT2) receptor antagonizes the growth effects of the AT1 receptor: Gain-of-function study using gene transfer. Proc Natl Acad Sci USA 92:10663–10667

Nio Y, Matsubara H, Murasawa S, Kanasaki M, Inada M (1995) Regulation of gene transcription of angiotensin II receptor subtypes in myocardial infarction. J Clin Invest 95:46–54

Nouet S, Nahmias C (2000) Signal transduction from the angiotensin II AT2 receptor. Trends Endocrinol Metab 11(1):1–6

Obermüller N, Unger T, Culman J, Gohlke P, de Gasparo M, Bottari SP (1991) Distribution of angiotensin II receptor subtypes in rat brain nuclei. Neuro Sci Lett 132:11–15

Ohkubo N, Matsubara H, Nozawa Y, Mori Y, Murasawa S, Kijima K, Maruyama K, Masaki H, Tsutumi Y, Shibazaki Y, Iwasaka T, Inada M (1997) Angiotensin type 2 receptors are reexpressed by cardiac fibroblasts from failing myopathic hamster hearts and inhibit cell growth and fibrillar collagen metabolism. Circulation 96:3954–3962

Okuyama S, Sakagawa T, Inagami T (1999) Role of the angiotensin II type-2 receptor in the mouse central nervous system. Jpn J Pharmacol Nov;81(3):259–263

Qadri F, Culman J, Veltmar A, Maas K, Rascher W, Unger T (1993) Angiotensin II-induced vasopressin release is mediated through alpha-1 adrenoceptors and angiotensin II AT_1 receptors in the supraoptic nucleus. J Pharmacol Exp Ther 267:567–574

Ritter O, Schuh K, Brede M, Rothlein N, Burkard N, Hein L, Neyses L (2003) AT2 receptor activation regulates myocardial eNOS expression via the calcineurin-NF-AT pathway. FASEB J 17:283–285

Rosenstiel P, Gallinat S, Arlt A, Unger T, Sievers J, Lucius R (2002) Angiotensin AT2 receptor ligands: do they have potential as future treatments for neurological disease? CNS Drugs 16(3):145–153

Rowe BP, Grovwe DL, Saylor DL, Speth RC (1990) Angiotensin II receptor subtypes in the rat brain. Eur J Pharmacol 186:339–342

Rowe BP, Saylor DL, Speth RC (1992) Analysis of angiotensin II receptor subtypes in individual rat brain nuclei. Neuroendocrinology 55: 563–573

Saavedra JM (1992) Brain and pituitary angiotensin. Endocr Rev 13:324–380

Saavedra JM (1999) Emerging features of brain angiotensin receptors. Regul Pept 30;85(1):31–45

Sakagawa T, Okuyama S, Kawashima N, Hozumi S, Nakagawasai O, Tadano T, Kisara K, Ichiki T, Inagami T (2000) Pain threshold, learning and formation of brain edema in mice lacking the angiotensin II type 2 receptor. Life Sci 67(21):2577–2585

Sandmann S, Yu M, Unger T (2001a) Transcriptional and translational regulation of calpain in the rat heart after myocardial infarction-effects of AT(1) and AT(2) receptor antagonists and ACE inhibitor. Br J Pharmacol Feb 132(3):767–777

Sandmann S, Yu M, Kaschina E, Blume A, Bouzinova E, Aalkjaer C, Unger T (2001b) Differential effects of angiotensin AT1 and AT2 receptors on the expression, translation and function of the Na+-H+ exchanger and Na+-HCO3-symporter in the rat heart after myocardial infarction. J Am Coll Cardiol 37(8):2154-2165

Sandmann S, Unger T (2002) Pathophysiological and clinical implications of AT(1)/AT(2) angiotensin II receptors in heart failure and coronary and renal failure. Drugs 62 Spec No 1:43-52

Seyedi N, Xu X, Nasjletti A, Hintze TH (1995) Coronary kinin generation mediates nitric oxide release after angiotensin receptor stimulation. Hypertension 26(1):164-170

Shanmugam S, Llorens-Cortes C, Clauser E, Corvol P, Gasc JM (1995)Expression of angiotensin II AT2 receptor mRNA during development of rat kidney and adrenal gland. Am J Physiol 268:F922-F930

Shanmugam S, Sandberg K (1996) Ontogeny of angiotensin II receptors. Cell Biol Int 20(3):169-176

Shimizu-Hirota R, Sasamura H, Mifune M, Nakaya H, Kuroda M, Hayashi M, Saruta T (2001) Regulation of vascular proteoglycan synthesis by angiotensin II type 1 and type 2 receptors. J Am Soc Nephrol 12:2609-2615

Shibata K, Makino I, Shibaguchi H, Niwa M, Katsuragi T, Furukawa T (1997) Up-regulation of angiotensin type 2 receptor mRNA by angiotensin II in rat cortical cells. Biochem Biophys Res Commun 239(2):633-637

Siragy HM, Carey RM (1996) The subtype-2 (AT_2) angiotensin receptor regulates renal cyclic guanosine 3',5'-monophosphate and AT_1 receptor-mediated prostaglandin E_2 production in conscious rats. J Clin Invest 97:1978-1982

Siragy HM, Carey RM (1997) The subtype 2 (AT2) angiotensin receptor mediates renal production of nitric oxide in conscious rats. J Clin Invest 100(2):264-269

Siragy HM, Carey RM (2001) Angiotensin type 2 receptors: potential importance in the regulation of blood pressure. Curr Opin Nephrol Hypertens 10(1):99-103

Song K, Allen AM, Paxinos G, Mendelsohn F (1992) Mapping of angiotensin II receptor subtype heterogeneity in rat brain. J Comp Neurol 316:467-484

Steckelings UM, Artuc M, Paul M, Stoll M, Henz BM (1996) Angiotensin II stimulates proliferation of primary human keratinocytes via a non-AT1, non-AT2 angiotensin receptor. Biochem Biophys Res Commun 229(1):329-333

Stoll M, Steckelings UM, Paul M, Bottari SP, Metzger R, Unger T (1995) The angiotensin AT_2-receptor mediates inhibition of cell proliferation in coronary endothelial cells. J Clin Invest 95:651-657

Stoll M, Unger T (2001) Angiotensin and its AT2 receptor: new insights into an old system. Regul Pept 99(2-3):175-182

Stoll M, Hahn AW, Jonas U, Zhao Y, Schieffer B, Fischer JW, Unger T (2002) Identification of a zinc finger homoeodomain enhancer protein after AT(2) receptor stimulation by differential mRNA display. Arterioscler Thromb Vasc Biol 22(2):231-237

Stroth U, Meffert S, Gallinat S, Unger T (1998) Angiotensin II and NGF differentially influence microtubule proteins in PC12W cells-role of the AT_2 receptor. Mol Brain Res 53:187-195

Su JZ, Fukuda N, Jin XQ, Lai YM, Suzuki R, Tahira Y, Takagi H, Ikeda Y, Kanmatsuse K, Miyazaki H (2002) Effect of AT(2) Receptor on expression of AT(1) and TGF-beta Receptors in VSMCs from SHR. Hypertension 40(6):853-858

Timmermans PBMW, Wong PC, Chiu AT, Herblin WF, Benfield P, Carin DJ, Lee RJ, Wexler RR, Saye JA, Smith RD (1993) Angiotensin II receptors and angiotensin II receptor antagonists. Pharmacol Rev 45:205-251

Tsutsumi K, Saavedra JM (1992) Heterogeneity of angiotensin II AT2 receptors in the rat brain. Mol Pharmacol 41:290-297

Unger T (1999) The angiotensin type 2 receptor: variations on an enigmatic theme. J Hypertens 17:1775-1786

Vervoort VS, Beachem MA, Edwards PS, Ladd S, Miller KE, de Mollerat X, Clarkson K, DuPont B, Schwartz CE, Stevenson RE, Boyd E, Srivastava AK (2002) AGTR2 mutations in X-linked mental retardation. Science 296:2401-2403 van Kesteren CA, van Heugten HA, Lamers JM, Saxena PR, Schalekamp MA, Danser AH (1997) Angiotensin II-mediated growth and antigrowth effects in cultured neonatal rat cardiac myocytes and fibroblasts. J Mol Cell Cardiol 29(8):2147–2157

Viswanathan M, Tsutsumi K, Correa FM, Saavedra JM (1991) Changes in expression of angiotensin receptor subtypes in the rat aorta during development. Biochem Biophys Res Commun 179:1361–1367

Wax SD, Rosenfield C L, Taubman MB (1994) Identification of a novel growth factor-responsive gene in vascular smooth muscle cells. J Biol Chem 269:13041–13047

Whitebread S, Mele M, Kamber B, de Gasparo M (1989) Preliminary biochemical characterization of two angiotensin II receptor subtypes. Biochem Biophys Res Commun163 (1):284–291

Wolf G, Harendza S, Schroeder R, Wenzel U, Zahner G, Butzmann U, Freeman RS, Stahl RA (2002) Angiotensin II's antiproliferative effects mediated through AT2-receptors depend on down-regulation of SM-20. Lab Invest 82(10):1305–1317

Wolf G, Ziyadeh FN, Thaiss F, Tomaszewski J, Caron RJ, Wenzel U, Zahner G, Helmchen U, Stahl RA (1997) Angiotensin II stimulates expression of the chemokine RANTES in rat glomerular endothelial cells. Role of the angiotensin type 2 receptor. J Clin Invest 100(5):1047–1058

Xie YW, Shen W, Zhao G, Xu X, Wolin MS, Hintze TH (1996) Role of endothelium-derived nitric oxide in the modulation of canine myocardial mitochondrial respiration in vitro. Implications for the development of heart failure. Circ Res 79(3):381–387

Yamada H, Akishita M, Lto M, Tamura K, David L, Lehtonen JYA, Dzau VJ, Horiuchi M (1999) AT2 receptor and vascular smooth muscle cell differentiation in vascular development. Hypertension 33:1414–1419

Yamada T, Akishita M, Pollman MJ, Gibbons GH, Dzau VJ, Horiuchi M (1998) Angiotensin II type 2 receptor mediates vascular smooth muscle cell apoptosis and antagonizes angiotensin II type 1 receptor action: an in vitro gene transfer study. Life Sci 63(19):PL289–PL295

Yen G, Croci A, Dowling A, Zhang S, Zoeller RT, Darling DS (2001) Developmental and functional evidence of a role for Zfhep in neural cell development. Brain Res Mol Brain Res 96(1–2):59–67

Zimpelmann J, Burns KD (2001) Angiotensin II AT_2 receptors inhibit growth responses in proximal tubule cells. Am J Physiol Renal Physiol 281(2):F300–F308

Zhu M, Gelband CH, Moore JM, Posner P, Sumners C (1998) Angiotensin II type 2 receptor stimulation of neuronal delayed-rectifier potassium current involves phospholipase A_2 and arachidonic acid. J Neurosci 18:679–686

Angiotensin-(1-7). Its Contribution to Arterial Pressure Control Mechanisms

C. M. Ferrario · D. B. Averill · K. B. Brosnihan · M. C. Chappell · D. I. Diz · P. E. Gallagher · E. A. Tallant

Hypertension and Vascular Disease Center, Wake Forest University School of Medicine, Winston-Salem, NC, 27157 USA
e-mail: cferrari@wfubmc.edu

1	Introduction	478
2	**Production and Metabolism of Angiotensin-(1-7)**	478
2.1	Pathways of Angiotensin-(1-7) Formation and Degradation	479
2.2	Immunocytochemical Findings	481
3	**Physiological Actions of Angiotensin-(1-7)**	484
3.1	Vascular Mechanisms	486
3.1.1	Species and Regional Bed Differences in Angiotensin-(1-7) Vasodilator Responses	486
3.1.2	Interactions of Angiotensin-(1-7) and Bradykinin in the Vasculature	488
3.2	Brain/Neural Mechanisms	489
3.3	Angiotensin-(1-7) in the Kidneys	494
3.4	Angiotensin-(1-7) and Trophic Mechanisms	496
4	**Mechanisms of Signal Transduction by Angiotensin-(1-7)**	498
4.1	Angiotensin-(1-7) Receptor(s)	501
5	**Role of Angiotensin-(1-7) in Human Hypertension**	504
6	**Summary**	505
	References	506

Abstract A formidable body of evidence accumulated to date reveals the pleiotropic actions of the renin–angiotensin system and its role in cardiovascular medicine. The characterization of the biological actions of angiotensin-(1-7) expanded knowledge of the endogenous control mechanisms regulating the actions of angiotensin II and revealed alternate enzymatic pathways for the processing of angiotensin I and angiotensin II. The following chapter reviews the tissue-specific pathways for formation and metabolism of this member of the renin–angiotensin system as well as its localization within brain, renal, and cardiac tissue. The unique actions of angiotensin-(1-7) in general oppose those of angiotensin II, and the preponderance of data identifies the regulation of the balance between the two peptides as important to overall cardiovascular health. We emphasize the physiological actions of the peptide in vascular, neural, and renal systems and provide a comprehensive review of its trophic and signaling mechanisms. The current status of the receptors mediating the actions of the peptide and the potential for non-receptor mechanisms are also presented. Finally, we discuss the significance of angiotensin-(1-7) in human disease, based on evidence emerging from recent studies suggesting that deficits in angiotensin-(1-7) are associated with hypertension in specific subpopulations of patients.

Keywords Blood pressure · Angiotensin-(1-7) · Hypertension · Kidney · Heart · Vasculature · Brain

1
Introduction

With the publication of the first edition of *Angiotensin* 30 years ago (Ryan 1974), it appeared that the role of the renin–angiotensin system in the regulation of arterial pressure was cast in stone. In retrospect, this was not the case. The formidable body of evidence accumulated to date has provided a more complete understanding of the pleiotropic actions of the system, its role in cardiovascular medicine, and the critical role of the angiotensin peptide system in heart failure (Pitt 2002), hypertension (Dahlof et al. 2002), diabetes (Brenner et al. 2001; Dahlof et al. 2002), atherosclerosis (Strawn et al. 2000), and end-stage renal disease (Brenner et al. 2001; Lewis et al. 2001).

The advent of new technologies and the synthesis of selective antagonists to subtype 1 and 2 angiotensin receptors provided the opportunity to explore the molecular mechanisms at a more complex level. A re-examination of the putative role of angiotensin fragments other than angiotensin II (Ang II) was one of the outcomes of these endeavors. The characterization of the biological actions of angiotensin-(1-7) [Ang-(1-7)] expanded knowledge of the endogenous control mechanisms regulating the actions of Ang II and uncovered alternate enzymatic pathways for the processing of angiotensin I (Ang I) and Ang II. Until recently, the dogma was that the carboxyl-terminus of the Ang II molecule was required for binding to its receptor. This idea, advanced by Bumpus and associates (1961), required revision with the discovery that Ang-(1-7), a fragment where the C-terminal end of the Ang II molecule was cleaved at the eigth position, had pharmacological activity. A comprehensive review of the lessons learned from studying the biological role of Ang-(1-7) in both animals and humans is provided below.

2
Production and Metabolism of Angiotensin-(1-7)

The newer concept, that the biochemical mechanisms for the processing of Ang I into smaller biologically active forms, departed significantly from the original concept, which implicated angiotensin converting enzyme (ACE) and aminopeptidases as the two critical enzymes for synthesis and degradation of angiotensin peptides.

2.1
Pathways of Angiotensin-(1-7) Formation and Degradation

Illustrated in Fig. 1 are the major bioactive components of the renin–angiotensin system following conversion of angiotensinogen to the decapeptide Ang I. The peptide cascade diverges with the processing of Ang I to Ang II and Ang-(1-7) yielding products with different carboxy termini and contrasting biological actions. Ang II generation occurs primarily by the hydrolysis of the Phe^8-His^9 bond of Ang I by ACE (EC 3.4.15.1) (Corvol et al. 1995; Deddish et al. 1998). Numerous studies demonstrated that the mast cell-derived serine protease chymase also generates Ang II from Ang I or the ACE-resistant analog $[Pro^{11},Ala^{12}]$-Ang I (Urata et al. 1990; Balcells et al. 1996). However, the lack of direct evidence that chymase contributes to endogenous production of Ang II questions the relevance of this pathway. Ang II is metabolized further either at its amino terminus by aminopeptidases, internally at the Tyr^4-Ile^5 bond by endopeptidases, or at the carboxy terminus by carboxypeptidases. Several enzymes expressing carboxypeptidase-like activity, including prolyl oligopeptidase and prolyl carboxypeptidase, hydrolyze the Pro^7-Phe^8 bond to generate Ang-(1-7) (Welches et al. 1991; Welches et al. 1993). Both enzymes are primarily cytosolic with prolyl carboxypeptidase localized to lysosomes. Their peptide substrates generally require a carboxy terminal proline residue. Shariat-Madel et al. (2002) demonstrated a novel activity of prolyl carboxypeptidase that processes prokallikrein to the active form of the enzyme in human endothelial cells. The ability of prolyl carboxypeptidase to activate kallikrein as well as to form Ang-(1-7) is of particular interest in lieu of the positive interaction between Ang-(1-7) and bradykinin in vascular function (see below).

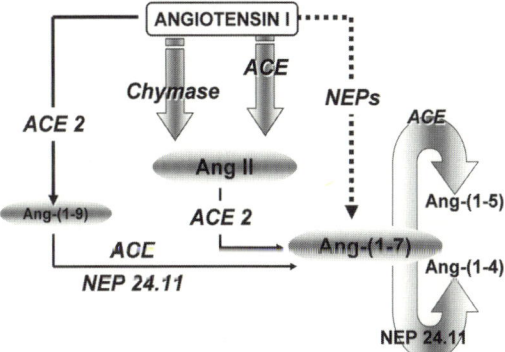

Fig. 1 Biochemical pathways involved in the synthesis and degradation of angiotensin-(1-7). *ACE*, angiotensin converting enzyme; *ACE2*, angiotensin converting enzyme 2 (Crackower et al. 2002); *NEPs*, neutral endopeptidases 24.11, 24.26, and 24.15

Recent studies identified a third enzymatic candidate that processes Ang II to Ang-(1-7). Although originally termed ACE2 due to its sequence homology with ACE, this enzyme does not cleave Ang I to Ang II nor is it sensitive to ACE inhibitors (Donoghue et al. 2000). ACE2 is a carboxypeptidase that exhibits a high catalytic efficiency for the generation of Ang-(1-7) from Ang II—almost 500-fold greater than that for the conversion of Ang I to Ang-(1-9) and 10- to 600-fold higher than that of prolyl oligopeptidase and prolyl carboxypeptidase, respectively, to form Ang-(1-7). From an array of over 120 peptides, only dynorphin A and apelin 13 were hydrolyzed by ACE2 with comparable kinetics to the conversion of Ang II to Ang-(1-7) (Vickers et al. 2002). ACE2 is a type I membrane protein that is primarily localized to the heart (myocytes), kidney (endothelium and tubular elements), and testes (Donoghue et al. 2000). Although the physiological role of ACE2 in the formation of Ang-(1-7) is unknown, ACE2 knockout mice have altered heart function that may be associated with changes in the cardiac renin–angiotensin system (Crackower et al. 2002). Moreover, ACE2 shows a parallel distribution with Ang-(1-7) in kidney tubules and is increased in the spontaneously hypertensive rat (SHR) following combined ACE–neprilysin blockade with omapatrilat (Chappell et al. 2002).

Ang-(1-7) is formed directly from Ang I or Ang-(1-9) by multiple endopeptidases that cleave the Pro^7-Phe^8 bond, completely bypassing the formation of Ang II (Gafford et al. 1983; Stephenson and Kenny 1987; Welches et al. 1993). In the circulation, this pathway is mediated principally by the metalloendopeptidase neprilysin (EC 3.4.24.11) (Campbell et al. 1998; Iyer et al. 1998b; Chappell et al. 2000). Similar to ACE and ACE2, neprilysin is a membrane-bound enzyme located on the luminal side of the endothelium with access to circulating peptides. Although the kidney, particularly the brush border of the proximal tubules, exhibits the highest concentration of neprilysin, significant activity is found in extra-renal sites including the lung, aorta, mesenteric artery, endothelial cells, and neutrophils (Erdos et al. 1989; Tamburini et al. 1989; Llorens-Cortes et al. 1992; Soleilhac et al. 1992; Graf et al. 1995; Tharaux et al. 1997). Neprilysin inhibitors significantly reduce circulating levels of Ang-(1-7), particularly in the presence of ACE inhibition; however, they do not completely abolish its production, suggesting the contribution of other enzymatic pathways (Anastasopoulos et al. 1998; Iyer et al. 1998b; Chappell et al. 2000; Ferrario et al. 2002a). The extent to which other endopeptidases form endogenous Ang-(1-7) from Ang I is not adequately defined at the present time nor have complete kinetic analyses been obtained. Endopeptidases that process Ang I to Ang-(1-7) in vitro include endothelin-converting enzyme in human endothelial cells (Pirro et al. 2001), thimet oligopeptidase (EC 3.4.24.15) in rat vascular smooth muscle cells (Chappell et al. 1994, 2000) as well as prolyl oligopeptidase in neuronal cells, brain tissue, and the perfused hindlimb (Santos et al. 1988; Welches et al. 1991).

As the pathways for the formation of Ang-(1-7) were defined, the route for the metabolism of the peptide was investigated. We and others showed that Ang-(1-7) is hydrolyzed at the Ile^5-His^6 bond to form angiotensin-(1-5) and the dipeptide His-Pro by ACE (Chappell et al. 1998; Deddish et al. 1998; Allred et al. 2000). Although somatic ACE contains two active sites (N and C domains), Ang-(1-7) and the hematopoietic fragment Acetyl-Ser-Asp-Lys-Pro are the only two known substrates hydrolyzed exclusively by the N-domain of human ACE (Deddish et al. 1998). The favorable kinetic constants (K_m=0.8 μM, kcat/K_m of 2,200 mM^{-1} sec^{-1} for canine ACE) suggest an in vivo role for ACE in the metabolism of Ang-(1-7). The K_m of ACE for Ang-(1-7) is surpassed only by that of the potent vasodilatory peptide bradykinin. Similar to kinins, the ACE-dependent metabolism may account for the remarkably short half-life of Ang-(1-7) in the circulation ($t_{1/2}$ of 9 s versus 60 s for Ang II) (Yamada et al. 1998). Moreover, chronic treatment with the ACE inhibitor lisinopril prolonged the half-life of infused Ang-(1-7) approximately six-fold in the SHR while the AT_1 antagonist losartan had no effect (Yamada et al. 1998). The marked increase in circulating levels of Ang-(1-7) following ACE inhibition most likely reflects increased synthesis (due to higher Ang I levels) as well as reduced metabolism. In the circulation, ACE constitutes a nexus in this pathway as it may balance the magnitude of the pressor and proliferative properties of Ang II and the depressor and antiproliferative actions of Ang-(1-7) and bradykinin. However, in the kidney, both ACE and neprilysin are involved in the metabolism of Ang-(1-7). In salt-sensitive hypertensives and the SHR, chronic administration of the combined ACE/neprilysin inhibitor omapatrilat resulted in a significant elevation in the urinary excretion of Ang-(1-7) (Ferrario et al. 2002a,b). In isolated brush border membranes from rat, we demonstrated that neprilysin hydrolyzed the Tyr^4-Ile^5 bond of Ang-(1-7) to form Ang-(1-4) and the tripeptide Ile-His-Pro (Allred et al. 2000). Moreover, addition of omapatrilat increased the $t_{1/2}$ of Ang-(1-7) in urine approximately eight-fold (5.5 min to 34 min) (Chappell et al. 2001; Ferrario et al. 2002a). Thus, in tissues such as the kidney with high concentrations of neprilysin, this endopeptidase may primarily contribute to Ang-(1-7) metabolism.

2.2
Immunocytochemical Findings

Selective antibodies to Ang-(1-7) reveal some of the sites at which the heptapeptide may act in a true paracrine or autocrine form. The earliest observations were obtained in rat brain where it was first shown that Ang-(1-7) modulated the secretion of vasopressin (Schiavone et al. 1988, 1990; Moriguchi et al. 1994) and the central control of the baroreflex system (Campagnole-Santos et al. 1989). Localization of immunoreactive (ir-) Ang-(1-7) within the rat hypothalamus (Block et al. 1988) gave impetus to more detailed studies of Ang-(1-7) distribution in rat brain. Ang-(1-7) immunolabeling was detected in neurons in the nucleus circularis and in the suprachiasmatic, supraoptic, and paraventricular nuclei (PVN) of [mRen-2]27 transgenic hypertensive rats (Krob et al. 1998). Interestingly, robust immunolabeling was present in the posterior magnocellular division of the PVN (Fig. 2, top panel), while double labeling experiments provided direct evidence that vasopressin and Ang-(1-7) co-localized within in-

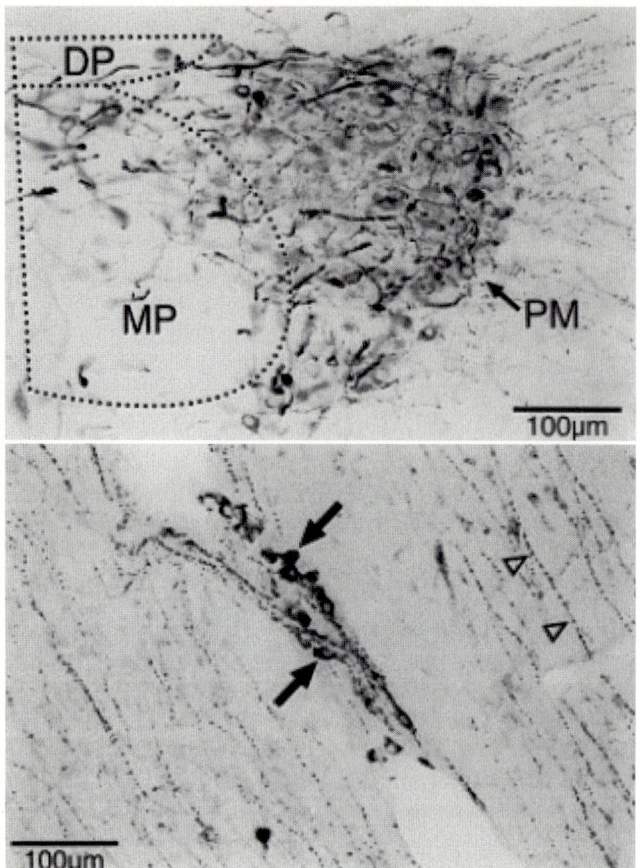

Fig. 2 *Top panel*: Ang-(1-7) immunoreactive labeling in the posterior paraventricular nucleus in a [mRen-2]27 transgenic hypertensive rat. Labeled neurons are primarily found in the dorsal and lateral parts of the posterior magnocellular division. *DP*, dorsal component of the paraventricular nuclei; *MP* and *PM*, medial and posterior components of the paraventricular nuclei. *Bottom panel*: Ang-(1-7) immunoreactive neurons associated with a blood vessel in the hypothalamus immediately anterior to nucleus circularis. The vessel lumen, which is visible on *top* and *bottom*, is obscured by labeled neurons (*black arrows*) and processes. *Arrowheads at the right* point to a beaded axon that is typical of the axons showing ir-Ang-(1-7). (Reprinted by permission from Krob et al. 1998)

dividual neurons projecting to the median eminence, neurohypophysis, and both PVN and supraoptic nuclei (Krob et al. 1998). The anatomical distribution of Ang-(1-7) neurons is in keeping with the seminal observation of Schiavone et al. (1988) showing Ang-(1-7)-mediated release of vasopressin from isolated rat hypothalamic-hypophysial explants and the finding that central Ang-(1-7) infusions stimulated vasopressin secretion in normal rats (Moriguchi et al. 1994). Furthermore, the presence of ir-Ang-(1-7) in neuron-like cells surrounding ce-

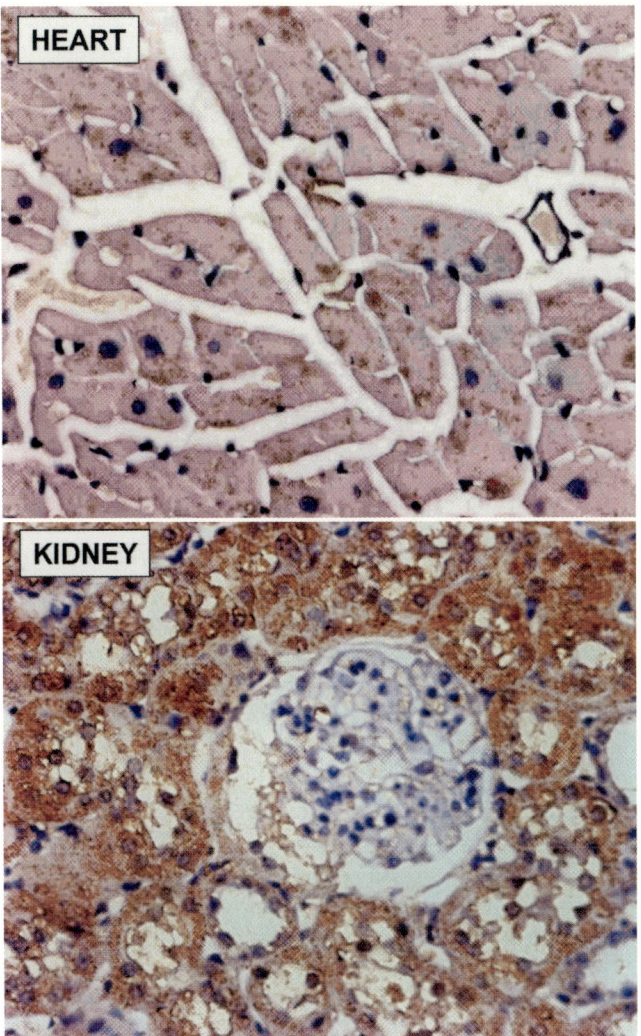

Fig. 3 Localization of immunoreactive Ang-(1-7) products in cardiac myocytes (*top*) and kidneys (*bottom*) provide additional evidence for paracrine functions of the peptide. Dense immunoreactive products positive for Ang-(1-7) are present in cardiac myocytes of the left ventricle from a normotensive Lewis rat. In the kidney, immunoreactive Ang-(1-7) is localized in proximal and distal tubules but completely absent in the glomeruli. (From studies reported by Ferrario; Ferrario 2002; Ferrario et al. 2002a)

rebral arterioles (Fig. 2, lower panel) is consistent with the observation that Ang-(1-7) dilates cerebral vessels (Meng and Busija 1993).

Recent studies demonstrated intense ir-Ang-(1-7) in rat cardiac (Ferrario 2002) and renal (Ferrario et al. 2002a) tissues. In the heart, ir-Ang-(1-7) was found only in cardiac myocytes (Fig. 3, upper panel) (Ferrario 2002). Moreover, we also found that Ang-(1-7) staining was more intense in myocytes found within the region of the penumbra surrounding the infarct zone (Ferrario 2002). These findings correlate with previous observations showing the presence of Ang-(1-7) in the venous effluent from the canine coronary sinus (Santos et al. 1990) and within the interstitial fluid collected from microdialysis probes placed in canine left ventricle (Wei et al. 2002).

As discussed below, Ang-(1-7) has important actions in the control of glomerulotubular function. That these actions may be mediated, at least in part, by local production of Ang-(1-7) in renal tubules gained credence with the demonstration of intense ir-Ang-(1-7) in renal proximal tubules and the ascending loop of Henle in the rat (Ferrario et al. 2002a). The lower panel of Fig. 3 shows that Ang-(1-7) immunostaining is absent in renal glomeruli but present in the cytoplasm of renal proximal convoluted tubules and the thick ascending limbs of loops of Henle (Ferrario et al. 2002a).

3
Physiological Actions of Angiotensin-(1-7)

The pithed rat served as the first biological model to characterize the effects of Ang-(1-7) on blood pressure and vascular tone (Benter et al. 1993). In this animal model, the absence of autonomic nervous system function achieved through spinal cord destruction provided a tool to determine the vasoactive properties of Ang-(1-7) in the systemic circulation. Injection of the peptide into a peripheral vein in the pithed rat produced a biphasic response in blood pressure that was blocked by the non-selective Ang II antagonist [Sar^{1}Thr^8]-Ang II. The initial short-lasting pressor component was blocked by losartan which also attenuated the duration, but not the magnitude of the depressor component. The entire hypotensive response was blocked by indomethacin. These early studies, confirmed by others, showed that the depressor component of the response may be variably influenced by the species studied, whether animals are anesthetized or non-anesthetized, and the dose of the peptide employed (Santos et al. 2000). With one exception (Kono et al. 1986), a vasodilator response was recorded in pigs (Porsti et al. 1994), dogs (Brosnihan et al. 1996), and rabbits (Ren et al. 2002). In general, the vasodilator effect was largely intact after administration of either AT_1 or AT_2 receptor blockers, a finding that supports the possibility that the vascular effects of Ang-(1-7) are mediated by a receptor subtype that is not sensitive to either AT_1 or AT_2 receptor blockade.

Vasodilators such as bradykinin, prostaglandins, and nitric oxide (NO) constitute a system of autacoid regulators acting to oppose the pressor molecules such as Ang II, catecholamines, and endothelin. With the possible exceptions of bradykinin and atrial natriuretic peptide, none of these factors shares biochemical pathways for either formation or degradation of the biologically active products. In contrast, enzymes which form and degrade angiotensin peptides characteristically share common substrates and biochemical pathways. The intertwining of these pathways suggested that the vasodilator properties of Ang-(1-7) could represent a mechanism within the renin–angiotensin system to op-

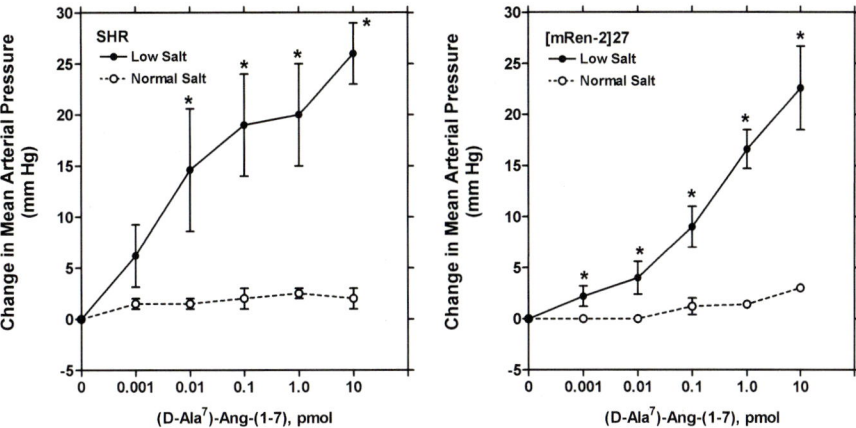

Fig. 4 Comparative effect of intravenous infusions of [d-Ala⁷]Ang-(1-7) on the blood pressure of conscious SHR (*left panel*) and [mRen-2]27 hypertensive transgenic rats maintained in either a normal (*open symbols, dashed lines*) or reduced intake of sodium (*filled symbols, solid lines*) 12 days after initiation of the dietary regimen. (Reprinted by kind permission from Iyer et al. 2000a)

pose the pressor and trophic actions of Ang II. This concept was first evaluated in two-kidney one-clip hypertensive dogs, as the elevated blood pressure in this model of experimental hypertension is clearly dependent on increased circulating levels of Ang II. Nakamoto et al. (1995) showed that the vasodilator component of the blood pressure response due to intravenous injections of Ang-(1-7) was augmented in both untreated and L-arginine-treated two kidney, one-clip hypertensive dogs but significantly attenuated in hypertensive dogs fed the NO synthase inhibitor.

In these studies (Nakamoto et al. 1995), the magnitude of the vasodilator effects of Ang-(1-7) was consistently dependent on whether the renin–angiotensin system was activated. Ferrario et al. (1997) thus proposed that Ang-(1-7) may serve as an intrinsic negative regulator of the pressor actions of Ang II. The concept was explored further in studies in which dietary salt depletion was used to assess the contribution of the renin–angiotensin system to the regulation of arterial pressure (Iyer et al. 2000a). During salt depletion, renin–angiotensin system activation fails to elevate blood pressure, presumably due to blood volume contraction and downregulation of AT₁ receptors. However, recent experiments implicate a role for Ang-(1-7) in opposing the actions of Ang II. Fig. 4 shows the dose-dependent effects of administering a selective antagonist of Ang-(1-7) (Santos et al. 1994)—[D-Ala⁷]Ang-(1-7)—to SHR and [mRen-2]27 transgenic hypertensive rats maintained on a low-salt diet for 11 days (Iyer et al. 2000a). Blockade of endogenous Ang-(1-7) activity caused dose-dependent increases in the arterial pressure of salt-restricted rats, suggesting that the heptapeptide prevented the increased levels of Ang II from raising blood pressure. The pharmacological effects of the selective Ang-(1-7) receptor antagonist were observed in the presence of increased plasma levels of renin and Ang II and duplicated by administering an Ang-(1-7) antibody (Iyer et al. 1998b). These data provide a mechanistic understanding for the finding that stimulation of the renin–angiotensin system by salt depletion is not accompanied by increases in arterial pressure.

A tonic role of Ang-(1-7) in contributing to blood pressure regulation was also demonstrated in hypertensive rats in which the elevated blood pressure was normalized by chronic oral administration of an ACE inhibitor or an Ang II receptor blocker. In both situations, control of arterial blood pressure by either agent was reversed in part by acute administration of a selective Ang-(1-7) antibody (Iyer et al. 1998a) or inhibition of neprilysin (Iyer et al. 1998b). In these treated hypertensive rats, the pressor effects of Ang-(1-7) blockade could not be ascribed to activation of AT_2 receptors or a direct effect of bradykinin (Iyer et al. 1998a,b).

3.1
Vascular Mechanisms

3.1.1
Species and Regional Bed Differences in Angiotensin-(1-7) Vasodilator Responses

The vascular-related actions of Ang-(1-7) are species-, region-, and mediator-specific. In porcine and canine coronary arteries, Porsti et al. (1994) and Brosnihan et al. (1996) reported that Ang-(1-7) produced a dose-dependent dilation that was dependent upon an intact endothelium. The vasodilator response was mediated by a non-AT_1/non-AT_2 angiotensin receptor that stimulated the endothelium-dependent release of NO and bradykinin, but not prostaglandins (Porsti et al. 1994; Brosnihan et al. 1996). The confirmatory nature of these two studies strongly supports the characterization of Ang-(1-7) receptors in canine coronary endothelium (Ferrario et al. 1997). In contrast, Gorelik et al. (1998) reported no vasodilator response to Ang-(1-7) in porcine coronary arteries. It is difficult to explain the results of Gorelik et al. (1998) since the preparation used (species, pre-constricting agent, and regional bed) was similar to those reported by Porsti et al. (1994). In piglet pial arterioles (Meng and Busija 1993), Ang-(1-7) and Ang II produced mild vasodilation at concentrations of 10^{-5} and 10^{-4} M, as assessed with intravital microscopy. Intravenous administration of indomethacin blocked the dilator responses. The results of these studies may reflect a lack of pre-constriction of the pial preparation as well as differences in regional beds. Furthermore, in canine coronary vessels, Ang II and Ang-(1-7) at equivalent concentration ranges produced diametrically opposite changes in the contractile state of coronary artery rings (Fig. 5) (Brosnihan et al. 1996). Weaker agonistic vasoconstrictor and vasodilator actions were elicited by Ang-(3-8) and Ang-(3-7), respectively.

Intraarterial injections of Ang-(1-7) in feline superior mesenteric vascular beds produced vasodilation at 1-, 3-, and 10-µg doses and vasoconstriction at 30 and 100 µg (Osei et al. 1993). In the hindquarter vascular bed, intraarterial in-

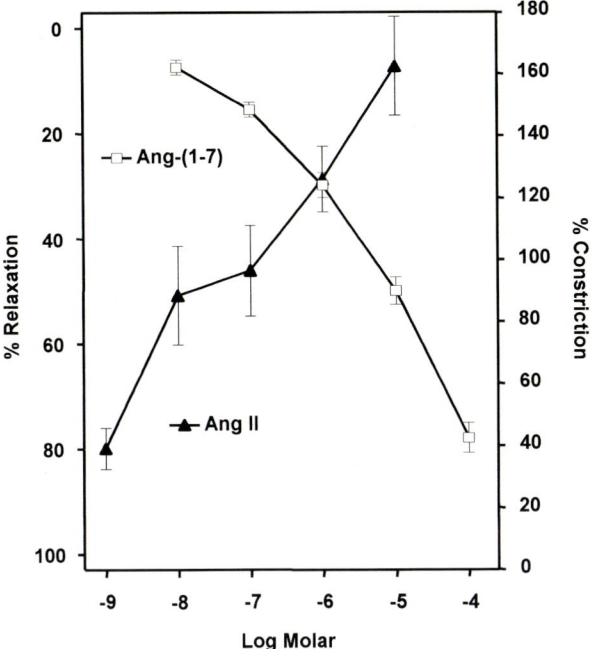

Fig. 5 Contrasting effects of Ang II and Ang-(1-7) on isolated coronary artery rings obtained from dogs illustrates the comparative effects of the peptides on vascular tension. (Adapted from studies reported in Brosnihan et al. 1998)

jection of 1 µg produced an increase in perfusion pressure, whereas a decrease in perfusion pressure was found at 10- and 30-µg doses. While the pressor effects of Ang-(1-7) were prevented by blockade of AT_1 receptors with losartan, the depressor responses to lower doses of the peptide were mediated by release of NO (Osei et al. 1993).

In Wistar rat mesenteric vessels, topical applications of Ang-(1-7) elicited 4% and 10% vasodilation at 100 and 1000 pmol, respectively (Oliveira et al. 1999). The vasodilator response was reduced but not abolished by pretreatment with either N^{ω}-nitro-L-arginine methylester (L-NAME) or indomethacin but was unaffected by the bradykinin B_2 receptor antagonist HOE 140. [D-Ala^7]Ang-(1-7) completely blocked the Ang-(1-7) response, whereas losartan was without effect. Le Tran and Forster (Forster and le Tran 1997) showed that Ang-(1-7) dilated pre-contracted rat aorta (ED_{50} of 2.3 nM) in an endothelial-dependent manner. Pre-constriction was obtained with 20 mM KCl, which produced a submaximal contraction. The degree of pre-constriction and the agent used appear to be critical factors in assessing the vasodilator actions of Ang-(1-7). Ang-(1-7) caused the vasodilation of norepinephrine-preconstricted rabbit renal afferent arterial vessels with an ED_{50} of 2 nM (Ren et al. 2002). The response was mediated by NO and was selectively blocked by [D-Ala^7]Ang-(1-7). Ang-(1-7) caused the release of prostanoids including prostacyclin from cultured rabbit aortic vascular smooth muscles cells (Muthalif et al. 1998), rat and porcine aortic vascular smooth muscle cells (Jaiswal et al. 1993a,b), and endothelial cells (Jaiswal et al. 1992). It is thus unclear why Ang-(1-7)-stimulated release of prostacyclin does not cause vasodilation in in vitro vascular preparations that are endothelium-denuded. In humans, Ueda et al. (2000) showed that Ang-(1-7) alone had no significant effect on forearm blood flow at concentrations from 1 to 2,000 pmol/min; at greater concentrations (0.1–40 nmol/min) they observed a reduction in forearm blood flow. However, the antagonism of Ang II-mediated vasoconstriction of human forearm resistance vessels by nanomolar concentrations of Ang-(1-7) observed by Ueda et al. (2000) may result from activation of an $AT_{(1-7)}$ receptor and the release of compensatory vasodilators. Wilsdorf et al. (Wilsdorf et al. 2001) also found no vasodilator response to Ang-(1-7) in human forearm. Finally, Davie and McMurray (1999) reported no vasodilator but a slight vasoconstrictor response in patients with heart failure and treated with ACE inhibitors.

A recent innovative study demonstrated the vasodilator effect of Ang-(1-7) in mature skin and sponge-induced neovasculature in mice (Machado et al. 2002). Blood flow increased in the presence of Ang-(1-7) and was blocked by [D-Ala^7] Ang-(1-7), NO synthase inhibitors, or indomethacin. Treatment with an AT_2 receptor antagonist failed to alter the growth of newly formed vessels but prevented the vasodilation in the pre-existing vessels. These studies suggest that Ang-(1-7) may contribute to newly formed vascular bed formation during wound healing, chronic inflammatory processes, and tumor growth and development.

3.1.2
Interactions of Angiotensin-(1-7) and Bradykinin in the Vasculature

The interactions of Ang-(1-7) and bradykinin were first demonstrated by Paula et al. (1995, 1999), who showed that Ang-(1-7) potentiates the hypotensive effect of bradykinin in conscious normotensive and hypertensive rats. The response was attenuated by pretreatment with indomethacin and exaggerated with ACE inhibition. The potentiation of bradykinin by Ang-(1-7) was also shown in isolated dog and pig coronary arteries (Li et al. 1997; Gorelik et al. 1998), rat mesenteric microvessels (Oliveira et al. 1999; Fernandes et al. 2001), isolated rat heart, and human forearm brachial arteries (Ueda et al. 2001). The specificity of the response was demonstrated by showing that Ang-(1-7) did not affect the vasodilator responses of acetylcholine or sodium nitroprusside (Li et al. 1997; Oliveira et al. 1999) and neither Ang I nor Ang II potentiated the bradykinin response (Li et al. 1997). The potentiating effect of Ang-(1-7) was blocked by a NO synthase inhibitor in canine and porcine coronary arteries (Schappert 1996; Gorelik et al. 1998), rat mesenteric vessels (Oliveira et al. 1999), and human forearm brachial arteries (Ueda et al. 2001) while indomethacin inhibited the response in the isolated heart preparation (Almeida et al. 2000), and in mesenteric microvessels of Wistar (Oliveira et al. 1999) and SHR rats (Fernandes et al.

2001). In contrast, endothelium-derived hyperpolarizing factor participated in the response in SHR microvessels (Fernandes et al. 2001). In addition, [D-Ala^7]Ang-(1-7), which is not a substrate or inhibitor of ACE, blocks the Ang-(1-7) potentiation of the responses to bradykinin in both Wistar rats (Almeida et al. 2000) and SHRs (Fernandes et al. 2001). These results suggest that Ang-(1-7) potentiates bradykinin through the release of prostaglandins, NO or endothelium-derived hyperpolarizing factor (EDHF), depending upon the species and the vascular bed, and is mediated by an AT$_{(1-7)}$ receptor.

Alternatively, Ang-(1-7) may potentiate bradykinin responses through its ability to inhibit ACE activity. Micromolar concentrations of Ang-(1-7) inhibit ACE activity, through interactions at the C-terminal domain (Chappell et al. 1998; Deddish et al. 1998). However, the K_m for Ang-(1-7) is similar to that of bradykinin, suggesting that competition for the enzyme would occur even at modestly elevated levels of either peptide. Indeed, ACE inhibitors increase levels of both peptides. ACE inhibitors potentiate responses to bradykinin, by inhibiting bradykinin B$_2$ receptor desensitization (Minshall et al. 1997; Benzing et al. 1999). This resensitization is thought to occur through protein-protein interactions between the bradykinin receptor and ACE. In porcine coronary arteries, bradykinin-induced vasodilation was potentiated by captopril, quinaprilat, or Ang-(1-7) (Tom et al. 2001). The mechanisms for the interaction between Ang-(1-7) and bradykinin in the dilation of blood vessels and the regulation of blood pressure are complex and require further investigation.

3.2
Brain/Neural Mechanisms

The first report of a biological function for Ang-(1-7) indicated that it was as potent as Ang II in releasing vasopressin from isolated hypothalami-pituitary explants (Schiavone et al. 1988). Since that observation, a wide variety of actions were reported for the peptide in the central nervous system (CNS). Unlike the counter-regulatory actions described thus far, many actions of Ang-(1-7) in the CNS mimic those of Ang II at a cellular level. However, when Ang-(1-7) is evaluated by effects exerted through integrative regulatory systems, the heptapeptide generally opposes the actions of Ang II. Ang-(1-7) is present in brain tissues such as the hypothalamus and medulla oblongata at concentrations equivalent to or greater than those of Ang II (Chappell et al. 1989). The distribution of the peptide in neurons and pathways of the medulla oblongata, PVN (Fig. 2), median eminence and pituitary-hypophysial portal system (Block et al. 1988; Lawrence et al. 1992, Krob et al. 1998) and its co-distribution with nicotinamide adenine dinucleotide phosphate, reduced diaphorase (NADPH) (Calka and Block 1993) and vasopressin (Block et al. 1988) underscore the importance of this peptide as a critical component of the mechanisms that participate in fluid balance and osmoregulation.

Hypothalamus. Ang-(1-7) stimulated release of vasopressin from isolated explants (Schiavone et al. 1988) and from hypothalamic sites in the whole animal (Moriguchi et al. 1994; Qadri et al. 1998). This is consistent with excitatory actions on paraventricular neurons (Felix et al. 1991). However, plasma vasopressin levels were not different in these studies, suggesting that the peptide acted as a local modulator of tissue vasopressin release. Although the release of vasopressin from isolated organ explants by Ang-(1-7) was comparable to Ang II, plasma release of vasopressin in response to paraventricular injections of the peptide was less than that of Ang II following injection (Qadri et al. 1998). Moreover, the effects in whole animals were not accompanied by the increase in blood pressure or drinking normally seen with Ang II (Moriguchi et al. 1994; Qadri et al. 1998). In the hypothalamus of [mRen-2]27 transgenic hypertensive rat, both peptides were elevated to levels 10- to 20-fold higher than in normotensive Hannover Sprague Dawley (SD) rats (Senanayake et al. 1994). Although no direct effects of the peptide on blood pressure were observed, Moriguchi et al. (1995b) clearly showed that in [mRen-2]27 transgenic hypertensive rats, Ang II and Ang-(1-7) produced opposing actions on blood pressure through central mechanisms. Intracerebroventricular administration of an Ang II antibody lowered pressure, while an antibody that selectively bound to Ang-(1-7) raised pressure.

The overall effect of exogenously administered Ang-(1-7) on plasma vasopressin may depend upon the fluid balance state of the animal and the balance between endogenous levels of Ang II and Ang-(1-7). Long-term infusion of Ang-(1-7) in Wistar-Kyoto (WKY) or SD rats tended to increase plasma vasopressin levels (Benter et al. 1995b). In contrast, Ang-(1-7) treatment of SHRs with elevated levels of plasma vasopressin lowered plasma vasopressin (Benter et al. 1995b). These data, which interestingly mimicked the effects of ACE inhibition (Thibonnier et al. 1981), showed that effects of Ang-(1-7) were dependent upon the state of the animal and that Ang-(1-7) acts as a modulator rather than a direct secretagogue. Similarly, in [mRen-2]27 transgenic hypertensive rats with elevated hypothalamic tissue levels of both Ang II and Ang-(1-7) (Senanayake et al. 1994), the release of vasopressin (Moriguchi et al. 1994) or substance P (Diz et al. 1997) in response to Ang-(1-7) was diminished.

The failure of Ang-(1-7) to stimulate thirst or increase blood pressure while modulating vasopressin release provided some of the first indications that Ang-(1-7) and Ang II must be acting through separate receptors and transmitter systems or pathways. Indeed, patterns of release for norepinephrine, dopamine, and substance P were distinctly different for the two peptides. Ang II released norepinephrine, dopamine, and substance P from hypothalamic slices, whereas Ang-(1-7) primarily released substance P, although delayed effects on dopamine and norepinephrine were observed (Diz and Pirro 1992). Ang-(1-7) may inhibit potassium-stimulated release of norepinephrine from hypothalamic slices (Diz and Pirro 1992; Gironacci et al. 2000) through a mechanism mediated by NO (Gironacci et al. 2000). Ang-(1-7) shows a remarkable co-distribution with NADPH diaphorase, suggesting that NO may play a role in the effects of Ang-

(1-7) in the brain (Calka and Block 1993). However, studies in glial cell cultures support a role for the peptide in release of other neuroactive substances, such as prostaglandins (Jaiswal et al. 1991b; Tallant et al. 1991b).

The failure of Ang-(1-7) to evoke release of the monoamines in a similar time frame as Ang II may be the mechanism for the absence of drinking and pressor responses to the heptapeptide (Kawabe et al. 1986). However, since tachykinins participate in Ang II-mediated release of vasopressin (Perfumi et al. 1988; Massi et al. 1991a,b), the similar effects of the two angiotensin peptides on substance P may explain their concerted ability to evoke release of vasopressin. Intracerebroventricular injection of Ang-(1-7) induced *c-fos* expression primarily in the organum vasculosum of the lamina terminalis and median preoptic nucleus, whereas enhanced *c-fos* expression by Ang II was observed in the PVN and subfornical organ as well as the organum vasculosum of the lamina terminalis and median preoptic nucleus, providing additional evidence that the two peptides may act through different pathways (Mahon et al. 1995).

Rowe et al. (1995) showed that Ang-(1-7) has moderate affinity for the AT_1 receptor (~100 nM) at several brain sites including the PVN. The affinity of neuronal AT_1 receptors for Ang-(1-7) is greater than that reported for the peptide at peripheral AT_1 sites (>1 µM) although lower than Ang II. Qadri et al. (1998) showed that actions of Ang-(1-7) in the PVN required higher doses of Ang-(1-7) than Ang II, and actions of both peptides were blocked by AT_1 receptor antagonists. The AT_2 antagonist PD123177 also blocked the response to Ang-(1-7) (Qadri et al. 1998). In fact, Ang-(1-7) stimulated dopamine release only after AT_1 receptor blockade (Pawlak et al. 2001). Electrophysiological studies of Ang-(1-7) in the PVN showed that neuronal responses to Ang-(1-7) were blocked by [D-Ala^7]Ang-(1-7) as well as by AT_1 and AT_2 antagonists (Felix et al. 1991; Ambuhl et al. 1992) but that blockade by AT_2 antagonists required higher doses, suggesting an interaction with AT_1 receptors. Substance P release (Diz and Pirro 1992) and dopamine release (Pawlak et al. 2001) in response to Ang-(1-7) were attenuated in hypothalamic slices by [D-Ala^7]Ang-(1-7) and AT_2 antagonists as well as [D-Ala^7]Ang-(1-7) blocked the release of substance P or catecholamines in response to Ang II (Diz and Pirro 1992; Pawlak et al. 2001). Critical evaluation of these findings clearly support the interpretation that the receptor activated by Ang-(1-7) does not show the same pharmacologic selectivity as the classic AT_1 or AT_2 receptors or the non-AT_1, non-AT_2 receptor present in the peripheral vasculature, as discussed elsewhere in the chapter.

Medulla Oblongata. Ang-(1-7) elicited excitatory neuronal actions in the nucleus of the solitary tract (nTS) and the dorsal motor nucleus of the vagus (dmnX) (Barnes et al. 1990). In both nuclei, the percentage of cells excited was comparable to that of Ang II. However, some cells responded to both peptides or only one peptide, suggesting that separate receptors mediated the observed actions. Microinjection of low doses of Ang-(1-7) into the dorsomedial medulla evoked depressor and bradycardic responses, whereas higher doses of the peptide produced a biphasic response consisting of an initial increase in blood pressure followed by a decrease in blood pressure (Campagnole-Santos et al. 1989). These responses appeared similar to those evoked by injection of Ang II into the nTS (Diz et al. 1984). However, subsequent studies showed that unilateral sino-aortic denervation potentiated the cardiovascular responses to ipsilateral injection of Ang-(1-7), whereas responses to Ang II injection were not affected (Campagnole-Santos et al. 1990). Studies also demonstrated that Ang II may act at nerve terminals in the nTS as well as at intrinsic neurons of the nTS (Qu et al. 1996; Diz et al. 2001). Thus, Ang II has both pre- and post-synaptic sites of actions in the dorsomedial medulla and Ang II has excitatory actions on cells of the nodose ganglion (Widdop et al. 1992). In contrast, Ang-(1-7) is devoid of actions on cells in the nodose ganglion (Widdop et al. 1992). The site of action of Ang-(1-7) (pre- or post-synaptic) awaits further studies.

The response patterns to Ang-(1-7) in the medulla oblongata are dependent on the cardiovascular state of the animal, similar to the actions at forebrain sites. For example, the depressor/bradycardic responses to Ang-(1-7) injected into the nTS were greater in the SHR. In contrast, microinjection of Ang-(1-7) in the nTS of ASrAogen rats (angiotensinogen antisense linked to a glial fibrillary acidic protein) caused smaller depressor responses than those produced by injection of the peptide in the nTS of Hannover SD rats (Couto et al. 2002). Thus, the decrease in tissue levels of angiotensin peptides in the brain stem of ASrAogen rats (Schinke et al. 1999; Monti et al. 2001) and/or the relatively lower level of arterial pressure in these rats altered the cardiovascular response to Ang-(1-7). The opposite effects observed in the SHR may reflect the importance of resting levels of pressure on the response.

In terms of the baroreceptor reflex, the effects of Ang II and Ang-(1-7) are in direct opposition. Ang II attenuated, whereas Ang-(1-7) augmented, baroreceptor reflex control of heart rate. Ang-(1-7) given by either intracerebroventricular administration (Campagnole-Santos et al. 1992; Oliveira et al. 1996) or long-term peripheral infusion (Benter et al. 1995a) also increased the gain of the baroreceptor reflex control of heart rate in SHR and normotensive rats. This effect was mediated at sites in the nTS (Chaves et al. 2000; Diz and Westwood 2000), where a tonic action of the peptide contributed to the resting level of reflex function in normotensive rats (Diz and Westwood 2000). Actions of Ang-(1-7) in the dorsal medulla were apparently influenced by hypertension and by anti-hypertensive treatments. In SHRs with an impaired baroreceptor reflex, the role of endogenous Ang-(1-7) was enhanced, since the injection of [D-Ala^7]Ang-(1-7) inhibited the reflex to a greater extent than that produced by injection of the Ang-(1-7) antagonist in the nTS of Wistar rats (Chaves et al. 2000). In [mRen-2]27 transgenic hypertensive rats with lower tissue levels of Ang-(1-7), [D-Ala^7]Ang-(1-7) had no effect on the reduced baroreflex sensitivity (Senanayake et al. 1994; Diz and Westwood 2000). In ASrAogen animals (Couto et al. 2002), Ang-(1-7) no longer facilitated the reflex and, surprisingly, inhibited baroreflex sensitivity. When ACE inhibitors were given to SHRs (Heringer-Walther et al. 2001) or rats with renal hypertension (Britto et al. 1997), the ability of ACE inhibition to improve the baroreceptor reflex was mediated, in part, by endogenous

Ang-(1-7). The role of Ang-(1-7) in this beneficial effect of ACE inhibition was substantiated by the finding that [D-Ala^7]Ang-(1-7) reversed the enhanced gain in the reflex seen with this type of treatment. These findings underscore that Ang II and Ang-(1-7) exist in a balance to set the sensitivity or gain of the baroreceptor reflex at the level of the nTS.

In the ventrolateral medulla (VLM), Ang-(1-7) like Ang II has actions that mimic excitatory neurotransmitters in the caudal and rostral cell groups involved in blood pressure control. In the caudal VLM, Ang-(1-7) lowered arterial pressure by activation of pathways providing tonic inhibition to the pressor rostral VLM (Potts et al. 2000a,b). In the rostral VLM, the peptide excited pathways leading to activation of the sympathetic nervous system (Fontes et al. 1994; Lima et al. 1996; Fontes et al. 1997). These actions mimic those observed when Ang II or glutamate was injected into the caudal or rostral VLM (Averill and Diz 1999; Diz et al. 2002). The effects of Ang-(1-7) at the rostral VLM are mediated by both vasopressin and the sympathetic nervous system (Oliveira et al. 1998). However, the increases in blood pressure in response to Ang-(1-7) injections into the rostral VLM were enhanced after hemorrhage, while those to Ang II were not (Lima et al. 1999). In animals with reduced levels of Ang II (Schinke et al. 1999; Monti et al. 2001) and presumably Ang-(1-7), the ASrAogen animals, there was no role of endogenous Ang-(1-7), since [D-Ala^7]Ang-(1-7) did not decrease pressure observed in a SD control animal (Baltatu et al. 2001). In [mRen-2]27 transgenic animals with higher levels of Ang II and lower levels of Ang-(1-7) than in SD rats (Senanayake et al. 1994), [D-Ala^7]Ang-(1-7) caused a greater fall in pressure than in SD rats, suggesting a greater contribution of the Ang-(1-7) to the elevated pressure in the rostral VLM (Fontes et al. 2000). Taken together with the effects of Ang-(1-7) on the baroreceptor reflex in the dorsal medulla, the data suggest that, much like Ang II (Weinstock and Gorodetsky 1995), the overall influence of Ang-(1-7) depends upon its site of action, the relative balance of Ang II and Ang-(1-7) endogenously present, the prevailing level of arterial pressure and heart rate, and the sensitivity of the reflex.

We observed no effect of Ang-(1-7) on basal release of serotonin and substance P in slices prepared from the medulla oblongata (Diz and Pirro 1992). However, Ang-(1-7) caused a significant release in the presence of a potassium stimulus (Diz and Pirro 1992). Since Ang-(1-7) facilitated, whereas Ang II attenuated, the gain of the baroreceptor reflex control of heart rate, this pattern of transmitter release is particularly relevant given that serotonin and substance P are known to enhance reflex function. The relationship between these changes in transmitters and the actions of Ang II or Ang-(1-7) in the ventral medulla are not known. However, both peptides may act directly upon neurons of the caudal and rostral VLM and not through release of additional transmitters at those sites similar to other excitatory neurotransmitters (Li and Guyenet 1995, 1996). Ang II acts in the rostral VLM to enhance the release of glutamate (Moriguchi et al. 1995a; Yamada et al. 1995; Zhu et al. 1998), suggesting a presynaptic effect for Ang II. We do not know if Ang-(1-7) shares a similar mechanism of action.

As indicated above, Ang-(1-7) has moderate affinity for the AT_1 receptor in the nTS (~100 nM) (Rowe et al. 1995). We reported that losartan competed for Ang II binding in the nTS and dmnX of the dog, suggesting AT_1 sites predominated in the dorsal medulla (Diz and Ferrario 1996). However, in the more rostral aspects of these nuclei, these losartan-sensitive binding sites were 10- to 100-fold more sensitive to Ang-(1-7) and the AT_2 antagonist PD123177. Couto et al. (1998) reported a smaller response to Ang II in the rostral nTS of the rat, whereas Ang-(1-7) acted similarly at both sites. Importantly, the actions of the Ang-(1-7) were blocked by [D-Ala^7]Ang-(1-7) (Santos et al. 1994). In contrast, Potts et al. (2000b) found that the actions of Ang-(1-7) in the ventrolateral medulla required higher doses than Ang II and actions of both peptides were blocked by AT_1 receptor antagonists (either losartan or candesartan). [D-Ala^7]Ang-(1-7) also blocked the actions of both Ang II and Ang-(1-7) in this study (Potts et al. 2000b). Thus, receptors responding to Ang-(1-7) in the medulla oblongata are also sensitive to AT_1 and AT_2 antagonists; however, in contrast to the hypothalamus, there is less of an effect of the AT_2 antagonists in this brain region. There remains a great deal of controversy about the nature of the receptors mediating responses to angiotensin peptides in the ventrolateral medulla. The sarcosine-substituted peptide antagonists have actions that do not appear related to either Ang II, Ang-(1-7), or other angiotensin peptides (Potts et al. 2000a) and the specificity of the actions of losartan are questionable (Averill et al. 1994).

3.3
Angiotensin-(1-7) in the Kidneys

The kidney is a critical target site for the actions of the renin–angiotensin system. The ability of Ang II to promote the retention of sodium and water through its potent vasoconstrictor effects, reduction in the filtration coefficient, and stimulation of various transporters in the proximal epithelium are well-characterized (Ichikawa and Harris 1991). In contrast, infusion of Ang-(1-7) into the renal artery stimulates a marked diuresis and natriuresis in both the rat and dog, as well as an increase in glomerular filtration rate (GFR) (DelliPizzi et al. 1994; Heyne et al. 1995; Hilchey and Bell-Quilley 1995; Handa et al. 1996; Vallon et al. 1998; Heller et al. 2000). The diuretic and natriuretic actions of Ang-(1-7) are independent of either AT_1 or AT_2 receptors, but are attenuated by the [D-Ala^7]Ang-(1-7) antagonist. At concentrations 1,000-fold higher than that of Ang II, Ang-(1-7) did not antagonize Ang II-dependent vasoconstriction in the rat kidney (Handa et al. 1996). Moreover, in the rabbit, low nanomolar doses of Ang-(1-7) induced vasodilation of pre-constricted afferent arterioles that was attenuated solely by [D-Ala^7]Ang-(1-7) (Ren et al. 2002). These studies suggest that the renal actions of Ang-(1-7) encompass both tubular and vascular sites that are sensitive to blockade of a non-AT_1/AT_2 receptor. Ang-(1-7) or its metabolite Ang-(3-7) are potent inhibitors of Na^+,K^+-ATPase activity in isolated convoluted proximal tubules and the renal cortex (Handa et al. 1996; Lopez et al.

1998; Handa 1999). In cultured rabbit renal tubular epithelial cells, Ang-(1-7) stimulated transcellular flux of sodium which was associated with activation of phospholipase A_2 (Andreatta-Van Leyen et al. 1993). Furthermore, inhibition of sodium transport by Ang I was markedly potentiated by captopril—strong evidence for the bio-activation to Ang-(1-7) in the proximal tubules of the kidney. Indeed, the natriuretic and diuretic actions of Ang-(1-7) in the perfused kidney were associated with increased levels of prostacyclin that were attenuated by the cyclooxygenase inhibitor indomethacin (Hilchey and Bell-Quilley 1995). Furthermore, low picomolar doses of Ang-(1-7) stimulated phosphatidylcholine incorporation in the renal cortex, a mechanism that may provide a substrate for arachidonic acid synthesis (Gironacci et al. 2002). Recently, we showed that the Ang-(1-7)-dependent inhibition of oubain-sensitive rubidium (Rb^{86}) influx was blocked by both [D-Ala^7]Ang-(1-7) and inhibitors of the cytochrome P450 (CYP450) system in isolated tubules (Chappell 2002). These data suggest that Ang-(1-7) may selectively activate the proximal epithelial CYP450 system to augment sodium excretion, in contrast to the Ang II-dependent stimulation of the vascular CYP450 system to produce potent vasoconstrictors (McGiff et al. 1993; Harder et al. 1995; Roman and Alonso-Galicia 1999). Our studies with the mixed vasopeptidase inhibitor omapatrilat in the SHR model also support the renal actions of Ang-(1-7) described above. The pronounced diuresis accompanying the chronic administration of omapatrilat was associated with a large increase in urinary excretion of Ang-(1-7) and enhanced immunocytochemical staining of the peptide in the kidney (Ferrario et al. 2002a).

Other reports suggest that the renal actions of Ang-(1-7) are somewhat comparable to those of Ang II. For example, in a perfused preparation of straight proximal tubules, peritubular application of Ang-(1-7) displayed biphasic effects on bicarbonate transport that were essentially identical to Ang II (Garcia and Garvin 1994). A low concentration (10^{-12} M) of Ang-(1-7) stimulated transport, while higher levels (10^{-8} M) inhibited absorption, probably by altering the Na$^+$/H$^+$ exchanger. At this site, the biphasic actions of Ang-(1-7) were completely blocked by losartan, and partially attenuated by the AT_2 antagonist PD123319 (Garcia and Garvin 1994). Intratubular application of Ang-(1-7) stimulated transport in the loop of Henle, but did not affect reabsorption in either the proximal or distal tubule (Vallon et al. 1997). In this case, the lack of an effect may result from the high luminal concentrations of Ang-(1-7) in the proximal or distal segments of the tubule. Indeed, luminal application of an AT_1 antagonist or ACE inhibitor attenuates basal reabsorption in the proximal tubule, suggesting that endogenous levels of Ang II are sufficient to facilitate water and sodium transport (Quan and Baum 1996). In contrast to these actions, Santos and colleagues (Santos and Baracho 1992) found that subcutaneous administration of Ang-(1-7) promoted an anti-diuretic action in water-loaded Wistar rats, an effect blocked by both losartan and [D-Ala^7]Ang-(1-7), but unaffected by an AT_2 antagonist or a vasopressin V_2 antagonist (Santos et al. 1996; Baracho et al. 1998). Studies to date have not determined whether other AT_1 antagonists also inhibit this response, nor have the cellular mechanisms been defined. Ang-(1-7) stimulates water transport in collecting duct tubules, suggesting that these terminal epithelial cells may be the site of action for Ang-(1-7) (Santos et al. 1996). Borges et al. (2002) also investigated the influence of Ang-(1-7) on water absorption in the jejunum and reported similar antidiuretic actions that were blocked by [D-Ala^7]Ang-(1-7). However, in the presence of losartan, the same dose of Ang-(1-7) potently inhibited water reabsorption. Finally, a series of studies reported that Ang-(1-7) alone stimulated ouabain-insensitive Na-ATPase activity in the ovine renal cortex, but inhibited Ang II-dependent stimulation of this ATPase through a [D-Ala^7]Ang-(1-7)-sensitive site (Caruso-Neves et al. 2000; Lara et al. 2002). Although Burgelova et al. (2002) showed that intrarenal administration of Ang-(1-7) produced natriuresis, the peptide also blocked the anti-natriuretic actions of Ang II. In this case, Ang-(1-7) did not attenuate the reduction in GFR or increased renal perfusion pressure associated with Ang II. Furthermore, infusion of the [D-Ala^7]Ang-(1-7) alone produced a comparable response to that of Ang II (Burgelova et al. 2002). Importantly, these latter studies emphasize the complex array of actions of Ang-(1-7) and suggest that the overall state of the renin–angiotensin system as well as the dose, route of administration, and site(s) of the nephron exposed to Ang-(1-7) clearly influence the physiological actions of the peptide in the kidney.

3.4
Angiotensin-(1-7) and Trophic Mechanisms

Early observations of responses to Ang-(1-7) included its ability to regulate cellular growth by either preventing mitogen-stimulated proliferation or stimulating the growth of quiescent cells. We and others showed that nanomolar concentrations of Ang-(1-7) inhibited mitogen-stimulated growth of cultured rat thoracic aortic vascular smooth muscle cells (VSMCs) through activation of a [D-Ala^7]Ang-(1-7)-sensitive receptor (Freeman et al. 1996; Zeng et al. 2001). Ang-(1-7) also has antiproliferative properties in vivo, reducing neointimal formation following balloon catheter injury to the rat carotid artery (Strawn et al. 1999). These results demonstrate that Ang-(1-7) inhibits smooth muscle cell growth following vascular injury and could be used to reduce restenosis following angioplasty and stent placement. In similar studies in neonatal rat cardiomyocytes, Ang-(1-7) reduced Ang II-mediated hypertrophy by preventing the increase in protein synthesis and cell surface area which was insensitive to either AT$_1$ or AT$_2$ receptor antagonists, suggesting that Ang-(1-7) inhibits myocyte hypertrophy through activation of a novel receptor (Zeng et al. 2000). These studies are in agreement with a recent report by Loot et al. (2002) demonstrating that infusion of Ang-(1-7) reduced myocyte surface area in infarcted rat hearts following coronary artery ligation. Cardiac tissue contains an endogenous renin–angiotensin system (Dzau and Re 1994), and infusion of Ang I into the canine left ventricle results in an 800-fold elevation in interstitial Ang-(1-7) (Wei et al. 2002). In addition, we recently showed significant amounts of immunoreactive Ang-(1-7) in rat myocardium (Ferrario 2002). These results suggest a potential role for Ang-(1-7) in the regulation of myocyte growth and an impact on cardiac dynamics.

Ang-(1-7) also regulates the formation of new capillary blood vessels. Machado et al. (2000) demonstrated an antiangiogenic effect of Ang-(1-7) using a mouse sponge model. These results suggest a potential therapeutic use for Ang-(1-7) in angiogenesis-dependent diseases such as tumor growth. We recently showed that Ang-(1-7) inhibited DNA synthesis in human lung and breast cancer cells, effects mediated by a non-AT_1, non-AT_2 angiotensin peptide receptor sensitive to [D-Ala^7]Ang-(1-7) (Tallant et al. 2001). Collectively, these results suggest a potential role for Ang-(1-7) as a chemopreventive and chemotherapeutic agent in human cancers through a reduction in cancer cell growth and angiogenesis.

In contrast to the antiproliferative responses to Ang-(1-7), the heptapeptide stimulates the growth of human skin and cardiac fibroblasts. Nickenig et al. (1997) reported that nanomolar concentrations of both Ang II and Ang-(1-7) stimulated DNA synthesis in cultured human skin fibroblasts. Although the proliferative response to Ang II was partially blocked by an AT_1 receptor antagonist, the stimulatory effect of Ang-(1-7) was insensitive to either an AT_1 or an AT_2 receptor antagonist. Fleck and colleagues (Neuss et al. 1994, 1996) also showed that Ang II and Ang-(1-7) stimulated the growth of cardiac fibroblasts isolated from the left ventricle of explanted end-stage failing hearts. The proliferative responses to Ang II and Ang-(1-7) had EC_{50}s in the nanomolar range and were not prevented by AT_1 or AT_2 receptor antagonists. These results suggest that human fibroblasts contain a novel angiotensin peptide receptor that binds both Ang II and Ang-(1-7) and is coupled to the stimulation of cell growth. Ang II and Ang-(1-7) were also equipotent in increasing DNA synthesis in neonatal rat cortical astrocytes and human astrocytoma cells, responses insensitive to either AT_1 or AT_2 receptor antagonists (Fogarty et al. 2002). However, micromolar concentrations of Ang II and Ang-(1-7) were required to stimulate astrocyte proliferation, leading to questions about the physiological role of these responses. The ability of Ang-(1-7) to either inhibit or stimulate cell growth is dependent upon the type of cell, suggesting that the $AT_{(1-7)}$ receptor may activate unique signaling pathways in different types of cells.

Finally, roles for Ang-(1-7) in hematopoietic recovery following radiation injury and in dermal repair and wound healing were recently suggested by Rodgers et al. (2001, 2002). Treatment of irradiated mice with either Ang II or large doses of Ang-(1-7) accelerated hematopoietic recovery by increasing the number of hematopoietic progenitors in the bone marrow and in the blood. Although losartan tended to block the response to Ang II and not to Ang-(1-7), experiments with additional angiotensin receptor antagonists are required to identify the receptor mediating these responses (Rodgers et al. 2002). Ang II and Ang-(1-7) also accelerated dermal repair in four different models of wound healing, including excision wounds and thermal injury and increased keratinocyte proliferation (Rodgers et al. 2001). Since Ang II and Ang-(1-7) were equipotent in stimulating hematopoietic progenitor formation and accelerating dermal repair but Ang-(1-7) does not increase blood pressure, Ang-(1-7) may have important therapeutic effects following myelosuppressive irradiation and dermal injury.

4
Mechanisms of Signal Transduction by Angiotensin-(1-7)

Prostanoid Production by Ang-(1-7). We showed that Ang-(1-7) stimulates the release of prostaglandin E_2 (PGE_2) and prostacyclin from C6 glioma cells, human astrocytes, porcine aortic endothelial cells, both porcine and rat aortic VSMCs and rabbit vas deferens (Trachte et al. 1990; Jaiswal et al. 1991a,b, 1992, 1993a,b; Tallant et al. 1991b). In cultured rabbit renal tubular epithelial cells, Ang-(1-7) activated phospholipase A_2 to release arachidonic acid (Andreatta Van Leyen et al. 1993). An antisense oligonucleotide to the cytosolic phospholipase A_2 blocked arachidonic acid release by Ang-(1-7) in rabbit VSMCs, and Ang-(1-7) stimulated translocation of the cytosolic phospholipase A_2 to the nuclear membrane, demonstrating that Ang-(1-7) activates phospholipase A_2 to release arachidonic acid for prostanoid production (Muthalif et al. 1998). In addition, Ang-(1-7) stimulation of arachidonic acid release in rabbit VSMCs was dependent upon the sequential activation of calmodulin-dependent protein kinase II and mitogen-activated protein kinase prior to activation of the cytosolic phospholipase A_2. Thus, Ang-(1-7) stimulates phospholipase A_2 activity to release arachidonic acid for prostanoid production. In contrast to Ang II, Ang-(1-7) does not activate phospholipase C in vascular smooth muscle or endothelial cells, astrocytes, or mesangial cells (Tallant et al. 1991b; Jaiswal et al. 1993a; Chansel et al. 2001; Heitsch et al. 2001; Tallant and Higson 1997) or phospholipase D in VSMCs (Freeman and Tallant 1994).

Many physiological responses to Ang-(1-7) are dependent upon prostanoid production, based upon studies using the cyclooxygenase inhibitor indomethacin. Indomethacin blocked the Ang-(1-7)-mediated inhibition of VSMC growth by Ang-(1-7) (Tallant and Clark 2003), the reduced incidence and duration of arrhythmias in rat hearts (Ferreira et al. 2001), the vasodilation of piglet pial arteries (Meng and Busija 1993), and the depressor component in the pithed rat (Benter et al. 1993). In SHRs treated with lisinopril, losartan, or both, an antibody against Ang-(1-7), an inhibitor of Ang-(1-7) formation, or [D-Ala^7]Ang-(1-7) caused an increase in blood pressure (Iyer et al. 2000b). Indomethacin caused a similar increase in blood pressure, and the increase in blood pressure by [D-Ala^7]Ang-(1-7) was prevented by pretreatment with the cyclooxygenase inhibitor. These results suggest that vasodilatory prostaglandins mediate many of the vascular effects of Ang-(1-7).

In Dahl salt-sensitive rats, the Ang-(1-7)-mediated reduction in mean arterial pressure was associated with an increase in prostacyclin and a reduction in thromboxane A_2 levels (Bayorh et al. 2002). Since thromboxanes are vasoconstrictive, the decrease in thromboxane production by Ang-(1-7) may contribute to its hypotensive response. Treatment of two-kidney, one-clip (2K-1C) hypertensive rats with Ang-(1-7) reduced thrombus weight following vena cava occlusion (Kucharewicz et al. 2000). Ang-(1-7) also reduced collagen adhesion to platelets isolated from 2K-1C rats. These antithrombotic effects could be due to increased production of prostaglandins or reduced thromboxane synthesis by Ang-(1-7), since prostaglandins are known anti-thrombotic mediators and thromboxanes induce platelet aggregation. Since Ang-(1-7) also increased plasminogen activator inhibitor (PAI)-1 and tissue plasminogen activator (TPA) release from human endothelial cells (HUVECs) (Yoshida et al. 2002), the potential role for Ang-(1-7) as an antithrombotic agent requires further investigation.

Prostaglandins also participate in many of the renal effect of Ang-(1-7). Ang-(1-7) stimulated arachidonic acid release from renal tubular epithelial cells (Andreatta-Van Leyen et al. 1993). Infusion of Ang-(1-7) into isolated rat kidneys caused an increase in prostacyclin in both the urine and the perfusate; pretreatment with indomethacin blocked the increase in prostacyclin as well as the diuretic effects (Hilchey and Bell-Quilley 1995). Ang-(1-7) infusion into SHRs caused diuresis and natriuresis during the first 3 days of infusion, concomitant with increased urinary prostaglandins measured on day 2 of the infusion (Benter et al. 1995b). In agreement with these studies, patients chronically treated with the ACE inhibitor captopril had increased plasma concentrations of Ang-(1-7) and prostacyclin with no significant effect on plasma concentrations of Ang II (Luque et al. 1996). These results suggest that Ang-(1-7) and prostacyclin may contribute to the antihypertensive effects of chronic therapy with captopril.

Nitric Oxide Production by Ang-(1-7). The Ang-(1-7)-mediated vasodilation of canine and porcine coronary arteries, isolated feline mesenteric and hindquarter vascular beds, canine middle cerebral arteries, and rabbit afferent arterioles were attenuated by NO synthase inhibitors (Osei et al. 1993; Porsti et al. 1994; Brosnihan et al. 1996; Feterik et al. 2001; Ren et al. 2002). Ang-(1-7) reduced norepinephrine release from the rat hypothalamus, an additional response which was blocked by NO synthase inhibitors (Gironacci et al. 2000). In addition, the antiangiogenic effects of Ang-(1-7) in the mouse sponge model was mediated by NO (Machado et al. 2001). Since the effects of Ang-(1-7) on the canine, porcine, and rabbit arteries as well as in the rat hypothalamus were also blocked by the bradykinin B_2 receptor antagonist HOE 140, Ang-(1-7) may stimulate the production of vasoactive kinins to release NO.

The reduction in mean arterial pressure of Dahl salt-sensitive rats by Ang-(1-7) was associated with an enhanced release of prostacyclin, as described above, as well as enhanced NO production (Bayorh et al. 2002). Ang-(1-7) also caused the release of NO from aortic rings isolated from Dahl salt-sensitive rats on a low- or high-salt diet. In addition, Ang (1-7) stimulated the production of aortic cyclic guanosine monophosphate (cGMP). These results suggest that the release of NO by Ang-(1-7) stimulates the production of cGMP to induce vascular relaxation.

Stimulation of NO synthase by Ang-(1-7) was measured in isolated canine vessels and bovine aortic endothelial cells. In isolated canine microvessels or coronary arteries, Ang-(1-7) caused nitrite release but only at a concentration of 10 μM (Seyedi et al. 1995). The effect of Ang-(1-7) was blocked by HOE 140 or losartan and may result from activation of AT_1 receptors at this concentration. Ang-(1-7)-mediated NO release from bovine aortic endothelial cells was only observed at micromolar concentrations of the peptide, was blocked by HOE 140, and was partially attenuated by [D-Ala^7]Ang-(1-7), losartan, or PD123177 (Heitsch et al. 2001). The requirement of micromolar concentrations of Ang-(1-7) and blockade of NO production by various receptor antagonists may result from activation of $AT_{(1-7)}$, AT_1, AT_2, or B_2 receptors by the peptide or through inhibition of ACE activity.

Effects of Ang-(1-7) at the AT1 Receptor. Pharmacological concentrations of Ang-(1-7) bind to the AT_1 receptor and either stimulate AT_1-like responses or antagonize Ang II responses at this receptor. Chansel et al. (2001) showed that Ang-(1-7) prevented the Ang II-mediated increase in Ca^{2+} in rat mesangial cells. However, the effect of Ang-(1-7) occurred at micromolar concentrations of the heptapeptide which competed at the AT_1 receptor in these cells. Ang-(1-7) also constricted rat renal microvessels at micromolar concentrations (van Rodijnen et al. 2002). Consistent with these findings, we showed that micromolar concentrations of Ang-(1-7) reduced Ang II binding to the AT_1 receptor and Ang II-mediated phospholipase C activity in rat aortic VSMCs and CHO cells transfected with the AT_{1A} receptor (Clark et al. 2001a,b). Micromolar concentrations of Ang-(1-7) compete for binding to the AT_1 receptor in VSMCs and CHO-AT_{1A} cells and AT_1 receptor antagonists blocked downregulation of the AT_1 receptor by Ang-(1-7), demonstrating that Ang-(1-7) interacts with the AT_1 receptor to reduce Ang II-mediated responses. We also showed that Ang-(1-7) downregulates the AT_1 receptor in kidney slices (Clark et al. 2003). This effect was blocked by [D-Ala^7]Ang-(1-7) and indomethacin, suggesting that it occurred through activation of an $AT_{(1-7)}$ receptor. Since ACE inhibitors increase Ang-(1-7) levels (Kohara et al. 1993), the decreases in AT_1 receptor density seen in brain areas such as the nTS or PVN of SHR treated with ACE inhibitors (Diz et al. 1992, 1994) and hypothalamic or medullary neuronal cells in culture (Wilson et al. 1988) may reflect actions of Ang-(1-7) at the AT_1 receptor.

Competition for the AT_1 receptor by pharmacological concentrations of Ang-(1-7) may be responsible for blockade of contractile responses to Ang II in rabbit aortic rings by Ang-(1-7), which required 1–30 µM Ang-(1-7) (Mahon et al. 1994) and human internal mammary arteries, in response to 10 µM Ang-(1-7) (Roks et al. 1999). However, this latter study also addressed the role of Ang-(1-7) as an ACE inhibitor to reduce Ang I contractions in human atrial tissue. They demonstrated that Ang-(1-7) had an IC_{50} value of 3 µM as an ACE inhibitor and acted in a noncompetitive/mixed way. In functional studies, Ang-(1-7) inhibited the contraction to Ang I and [Pro^{10}]Ang I, a substrate solely converted by ACE, but not to [Pro^{11}-D-Ala^{12}] Ang I, an ACE-resistant substrate. Thus, Ang-(1-7) could also modulate responses at the AT_1 receptor through inhibition of ACE; however, the concentrations required for ACE inhibition by Ang-(1-7) do not occur under physiological conditions. In contrast to inhibition of Ang II responses at micromolar concentrations, the antagonism of Ang II-mediated vasoconstriction of human forearm resistance vessels by nanomolar concentrations of Ang-(1-7) observed by Ueda et al. (2000) may result from activation of an $AT_{(1-7)}$ receptor and the release of compensatory vasodilators.

Although Ang-(1-7) acutely downregulates the AT_1 receptor, Neves et al. (2000) showed that a 24-h treatment with Ang-(1-7) upregulated AT_1 mRNA in VSMCs isolated from the Akron strain of WKY and SHR rats. This suggests that chronic exposure to Ang-(1-7) may alter the expression of the AT_1 receptor in a strain-specific manner.

4.1
Angiotensin-(1-7) Receptor(s)

Recognition of the physiological responses to Ang-(1-7) occurred concurrently with the identification of subtype-selective ligands for different molecular forms of Ang II receptors. AT_1 angiotensin receptors are pharmacologically defined by their selectivity for the prototypical ligand losartan and similar antagonists (valsartan, L-158,809, etc.), while AT_2 receptors show selectivity for the antagonist PD123177 and PD123319 as well as the agonist CGP42112 (de Gasparo et al. 1995). Characterization of physiological or cellular responses to Ang-(1-7) was therefore accompanied by attempts to define the receptor mediating these responses. Ang-(1-7) is a poor competitor (IC_{50}s in the micromolar range) at pharmacologically defined AT_1 receptors in VSMC (Jaiswal et al. 1993a,b) or AT_2 receptors in differentiated NG108-15 cells, pancreatic cells, and human myometrium (Tallant et al. 1991a; Chappell et al. 1995; Bouley et al. 1998). [D-Ala^7]Ang-(1-7), a modified form of Ang-(1-7) in which proline at position 7 is replaced with D-alanine, was designed as a selective antagonist for the Ang-(1-7) receptor (Santos et al. 1994). In initial studies, [D-Ala^7]Ang-(1-7) blocked hemodynamic and renal responses to Ang-(1-7), did not compete for binding of ^{125}I-Ang II to rat adrenal AT_1 or AT_2 receptors, and did not block pressor or contractile responses to Ang II, demonstrating selectivity for Ang-(1-7) (Santos et al. 1994).

We identified an Ang-(1-7) binding site on bovine aortic endothelial cells (BAEC) which was competed by [D-Ala^7]Ang-(1-7) and [Sar^1-Ile^8]Ang II, but not by losartan or PD123319 (Tallant et al. 1997). In agreement with this identification, treatment of BAEC with Ang-(1-7) stimulated NO release, an effect blocked by [D-Ala^7]Ang-(1-7) (Heitsch et al. 2001). A similar ^{125}I-Ang-(1-7) binding site, sensitive to Ang-(1-7) and [D-Ala^7]Ang-(1-7), was visualized in the endothelium of canine coronary artery rings (Ferrario et al. 1997), consistent with functional effects of Ang-(1-7) in endothelium-intact canine and porcine coronary arteries (Xiang et al. 1985; Porsti et al. 1994; Brosnihan et al. 1996). We recently identified a similar binding site for Ang-(1-7) in the rat vasculature that is sensitive to both [D-Ala^7]Ang-(1-7) and [Sar^1-Thr^8]Ang II, but not to losartan or PD123319 (Iyer et al. 2000b). Ang-(1-7) inhibited the mitogen-stimulated growth of rat VSMCs which was blocked by [D-Ala^7]Ang-(1-7) and sarcosine analogs of Ang II but not by losartan and PD123177 (Freeman et al. 1996; Tallant et al. 1999; Min et al.

2000). Additionally, the specific receptor antagonist [D-Ala^7]Ang-(1-7) blocked the Ang-(1-7)-induced inhibition of angiogenesis in the mouse sponge model (Machado et al. 2001). The identification of high-affinity binding sites for Ang-(1-7) that are selectively competed by [D-Ala^7]Ang- (1-7) agrees with the ability of this ligand to block functional responses at these sites.

A multitude of physiological responses to Ang-(1-7) are mediated by a specific AT$_{(1-7)}$ receptor that is selectively blocked by [D-Ala^7]Ang-(1-7) or the sarcosine analogs of Ang II, but not by pharmacologically-defined AT$_1$ or AT$_2$ receptor antagonists. Studies in the pithed rat (Benter et al. 1993) provided the first evidence for a hypotensive response to Ang-(1-7) that was mediated by a non-AT$_1$, non-AT$_2$ receptor. Moreover, treatment with an Ang-(1-7) antibody, the non-selective angiotensin peptide antagonist [Sar^1-Thr^8]Ang II, or a neprilysin inhibitor to prevent Ang-(1-7) formation caused a rapid increase in blood pressure in SHRs chronically treated with an ACE inhibitor and an AT$_1$ receptor antagonist (Iyer et al. 1998a,b). These effects were not blocked by the AT$_2$ receptor antagonist PD123319. Similar effects of the Ang-(1-7) antibody, a neprilysin inhibitor, or [D-Ala^7]Ang-(1-7) were observed in salt-depleted SHRs or [mRen-2]27 hypertensive rats (Iyer et al. 2000a). In agreement, Ang-(1-7) infusion decreased blood pressure in SHRs on day 4 and 5 which was blocked by [D-Ala^7]Ang-(1-7) (Widdop et al. 1999). Bayorh et al. (2002) recently showed that Ang-(1-7) also caused a depressor response in salt-sensitive Dahl rats which was specifically blocked by [D-Ala^7]Ang-(1-7); Ang-(1-7) infusion caused an increased in the release of prostacyclin and NO and reduced thromboxane A$_2$ levels, all of which were prevented by pretreatment with the selective AT$_{(1-7)}$ receptor antagonist. Collectively, these results demonstrate that the hypotensive response to Ang-(1-7) was mediated by a non-AT$_1$, non-AT$_2$ [D-Ala^7]-Ang-(1-7)-sensitive receptor. Other studies reviewed above are also consistent with the existence of an AT$_{(1-7)}$ receptor in brain (Ambuhl et al. 1994; Fontes et al. 1997; Chaves et al. 2000; Fontes et al. 2000) and renal tissue (Vallon et al. 1998; Gironacci et al. 2002; Lara et al. 2002). Taken together, these data provide overwhelming evidence that the actions of Ang-(1-7) are mediated by a non-AT$_1$, non-AT$_2$ receptor. We refer to this receptor as the AT$_{(1-7)}$ receptor, in accordance with the guidelines established by the International Union of Pharmacology Nomenclature Subcommittee for Angiotensin Receptors (Bumpus et al. 1991; de Gasparo et al. 1995). The unique AT$_{(1-7)}$ receptor is defined by its sensitivity to Ang-(1-7), its antagonism by [Sar^1-Thr^8]Ang II and [D-Ala^7]Ang-(1-7), and its lack of response to losartan or PD123319, either functionally or in competition for binding.

Recently, Santos et al. (2002) reported the G protein-coupled receptor *mas* as a functional AT$_{(1-7)}$ receptor. Ang-(1-7) bound with high affinity to either CHO or HEK cells transfected with the *mas* gene. [D-Ala^7]Ang-(1-7) blocked the heptapeptide binding to the transfected CHO cells; however, the AT$_1$ or AT$_2$ receptor antagonists had no effect. The results suggest that *mas* serves as a selective Ang-(1-7) binding site. *Mas* is predominately expressed in the testis and the hippocampus and amygdala of the mammalian forebrain with minimal but detectable levels in the rodent heart and kidney (Metzger et al. 1995). This tissue distribution differs from many previous reports of Ang-(1-7) binding and functional responses, suggesting the possibility of additional receptor subtypes.

Conversely, some of the renal effects of Ang-(1-7) implicate a losartan-sensitive receptor (Garcia and Garvin 1994). Ang-(1-7) inhibited sodium transport in the rat proximal tubule, an effect that was partially blocked by losartan but completely inhibited by [Sar^1-Thr^8]Ang II (Handa et al. 1996). In the basolateral membranes of porcine renal proximal tubules, Ang-(1-7)-stimulated Na$^+$-ATPase activity was reversed by losartan, but not by PD123319 or [D-Ala^7]Ang-(1-7) (Caruso-Neves et al. 2000). In contrast to the diuretic and natriuretic effects of Ang-(1-7), the heptapeptide has anti-diuretic effects in water-loaded rats that were also specifically blocked by both [D-Ala^7]Ang-(1-7) and losartan, but not by AT$_2$ receptor antagonists (Baracho et al. 1998). In support of losartan-sensitive responses to Ang-(1-7) in the kidney, Ang-(1-7) competed with high affinity (K_i=8 nM) for an ^{125}I-Ang II binding site in rat renal cortex that was totally displaced by losartan (Gironacci et al. 1999). Likewise, microinjection of [D-Ala^7]Ang-(1-7) or losartan into the ventrolateral medulla inhibited both the pressor and depressor responses to Ang-(1-7) (Potts et al. 2000b). Although Ang-(1-7)-stimulated prostaglandin release from C6 glioma cells was also blocked by the AT$_1$ antagonist, micromolar concentrations of the antagonist were required, suggesting that this may be a non-selective effect of the antagonist (Jaiswal et al. 1991a). Collectively, these results suggest that some of the Ang-(1-7)-stimulated responses in the kidney and brain are partially mediated by a receptor that is sensitive to losartan and may be sensitive to other AT$_1$ receptor antagonists, but retains sensitivity to [D-Ala^7]Ang-(1-7).

There are also several reports of responses to Ang-(1-7) that are sensitive to AT$_2$ receptor antagonists. Ang-(1-7) stimulated arachidonic acid release from rabbit VSMCs which was partially blocked by [D-Ala^7]Ang-(1-7) or PD123319. However, the combination of [D-Ala^7]Ang-(1-7) and PD123319 completely blocked the response to Ang-(1-7), suggesting that the heptapeptide activates both a [D-Ala^7]Ang-(1-7)-sensitive and a PD123319-sensitive receptor on rabbit VSMCs (Muthalif et al. 1998). Ang-(1-7) inhibition of norepinephrine release from the rat hypothalamus was blocked by PD123319 and by [D-Ala^7]Ang-(1-7), but not by losartan. These results suggest that the Ang-(1-7) inhibition of noradrenergic neurotransmission is mediated by a receptor that is sensitive to both [D-Ala^7]Ang-(1-7) and PD123319 (Gironacci et al. 2000) or that the two receptors act in series. In cultured human astrocytes, the Ang-(1-7)-mediated release of both PGE$_2$ and prostacyclin was blocked by CGP42112A; however, other AT$_2$ receptor antagonists were not available at the time of these studies. Since CGP42112A is now recognized as an agonist for the AT$_2$ receptor, further studies are required to determine whether this response is mediated by an AT$_2$ receptor (Jaiswal et al. 1991b).

Identification of an AT$_{(1-7)}$ receptor is further confounded by reports of responses to Ang-(1-7) that are sensitive to both AT$_1$ and AT$_2$ receptor antagonists. Iontophoresis of Ang-(1-7) excited neurons in the paraventricular nucleus. Although both AT$_1$ and AT$_2$ receptor antagonists blocked the responses to Ang-

(1-7), AT_1 receptor antagonists were more potent than AT_2 receptor antagonists (Ambuhl et al. 1992). Ang-(1-7) caused the release of norepinephrine from the rat atria, which was inhibited by both losartan and PD123319 (Gironacci et al. 1994). Additionally, Ang-(1-7), at concentrations from 1 to 10 μM, stimulated NO and superoxide release from bovine aortic endothelial cells. These responses were partially blocked by [D-Ala^7]Ang-(1-7), losartan, or PD123177 (Heitsch et al. 2001). However, since the K_d for binding of Ang-(1-7) to the $AT_{(1-7)}$ receptor in bovine aortic endothelial cells was in the nanomolar range (Tallant et al. 1997), the observed responses may be due to interactions with ACE or resensitization of the B_2 receptor on endothelial cells. In porcine vascular smooth muscle cells, Ang-(1-7)-mediated prostanoid release was not inhibited by losartan but was partially inhibited by L-158,809, EXP3174, or PD123319 (Jaiswal et al. 1993b).

Finally, Handa (1999) reported that Ang-(1-7) recognizes the AT_4 receptor in renal proximal tubules and Madin-Darby canine kidney (MDCK) collecting duct cells. This interaction requires the metabolism of Ang-(1-7) to Ang-(3-7), a high-affinity agonist that subsequently binds to the AT_4 receptor (Handa 2000a,b). However, the AT_4 receptor was identified as the insulin-regulated aminopeptidase, suggesting a substrate–enzyme interaction rather than an agonist–receptor pathway (Albiston et al. 2001).

[D-Ala^7]Ang-(1-7) exhibited agonistic properties in two recent reports. Gironacci et al. (2002) showed that [D-Ala^7]Ang-(1-7) blocked Ang-(1-7)-stimulated phosphatidylcholine synthesis in rat renal cortical slices. However, [D-Ala^7]Ang-(1-7) itself stimulated phosphatidylcholine synthesis. Additionally, Ang-(1-7) as well as [D-Ala^7]Ang-(1-7) decreased the release of PAI-1 and TPA in human umbilical vein endothelial cells (Yoshida et al. 2002). These results suggest that [D-Ala^7]Ang-(1-7) may also act as a partial agonist in some circumstances.

5
Role of Angiotensin-(1-7) in Human Hypertension

A limited number of studies in humans provides a foundation for Ang-(1-7) as another mechanism implicated in cardiovascular function and blood pressure dysregulation. The vasodilator effects of Ang-(1-7) contribute to the antihypertensive effects of ACE inhibitors (Ferrario et al. 1998) and the vasopeptidase inhibitor omapatrilat (Ferrario et al. 2002b). These two studies provided evidence that salt-sensitive hypertensive patients may have lower excretion rates of Ang-(1-7), a finding that we interpreted as a deficit in Ang-(1-7) synthesis or activity as a contributor to the syndrome of low-renin hypertension. The bulk of the physiological, pharmacological, and molecular evidence gathered to date makes this field a fruitful point for further investigation.

Recent studies demonstrated that circulating and urinary levels of Ang-(1-7) are increased in normal human pregnancy. Valdes et al. (2001) showed that urinary Ang II and Ang-(1-7) increased throughout gestation. By the 35th week of gestation, both urinary Ang II and Ang-(1-7) reached their highest levels, which were 16- and 20-fold higher than that obtained during the menstrual cycle. In addition, plasma levels of Ang I, Ang II, and Ang-(1-7) were increased in normotensive pregnant women during the third trimester of pregnancy (Merrill et al. 2002). These changes were accompanied by increased levels of other components of the renin–angiotensin system, including plasma renin activity and aldosterone. In age—and race—matched third-trimester preeclampsia subjects, plasma levels of Ang-(1-7) were significantly decreased compared with normal pregnant subjects. Other components of the renin–angiotensin system, with the exception of ACE, were reduced in preeclamptic subjects; only plasma Ang II remained elevated in preeclamptic subjects as compared to non-pregnant subjects. There was an inverse association between circulating Ang-(1-7) and both systolic ($r=-0.51$, $p<0.004$) and diastolic ($r=-0.44$, $p<0.02$) blood pressure, supporting a role of Ang-(1-7) in the control of blood pressure. This study suggests that Ang-(1-7) may contribute to the elevated blood pressure in preeclamptic women. Furthermore, these data are consistent with the negative correlation between urinary Ang-(1-7) and blood pressure previously reported in normotensive and hypertensive subjects (Ferrario et al. 1998).

6
Summary

Hypertension research has provided fundamental insights into the regulation of cardiovascular function and expanded knowledge of the biochemical mechanisms by which hormones and locally generated autacoids regulate tissue perfusion and blood distribution to maintain homeostasis. Research on Ang-(1-7) has expanded knowledge of the biochemistry of the renin–angiotensin system and its interaction with molecules participating in the processes of inflammation, growth, and vascular tone. The unique actions of Ang-(1-7) in generally opposing the pleiotropic effects of Ang II in cardiovascular function led us to propose that the heptapeptide represents a negative feedback mechanism controlling the agonistic actions of Ang II in tissue function and body fluid control mechanisms. The demonstration that ACE is the primary enzyme responsible for Ang-(1-7) inactivation provides a mechanistic point of balance between synthesis of Ang II and Ang-(1-7) degradation. The half-life of Ang-(1-7) in blood and tissue, similar to that of bradykinin and far shorter than the half-life of circulating Ang II, supports the role of this hormone as a vasodilator factor. Modulatory effects of Ang-(1-7) described in neural and central neuroendocrine mechanisms regulating blood pressure and fluid volumes are strong, as is the physiological and pharmacological evidence that Ang-(1-7) contributes to the glomerulotubular regulation of sodium and, perhaps, water excretion.

While there is overwhelming evidence that Ang-(1-7) binds to a unique, non-AT_1, non-AT_2 receptor, additional studies suggest that the molecular interactions of Ang-(1-7) with its cellular receptor are complex. Data accumulated thus far indicate that the heptapeptide also may bind to additional receptors that are sensitive to AT_1 and/or AT_2 receptor antagonists as well as blocked by [D-Ala^7] Ang-(1-7). Alternatively, Ang-(1-7) may bind to a unique receptor that physically interacts with the pharmacologically defined AT_1 and AT_2 receptors, resulting in a protein complex with combinatorial characteristics. Such a heterodimer is formed between the AT_1 and bradykinin B_2 receptors, resulting in a complex with altered G protein signaling and receptor endocytosis (AbdAlla et al. 2000). Cloning of the unique non-AT_1, non-AT_2 receptor as well as further studies directed toward protein–protein interactions between the $AT_{(1-7)}$ receptor and other membrane receptors are needed to elucidate the precise molecular mechanisms involved in Ang-(1-7)-mediated responses.

Acknowledgements. This work is supported in part by the following grants: HL 51951, HL-56973, from the National Heart, Lung and Blood Institute of the National Institutes of Health; HD 42631 from the National Institute of Child Health and Human Development of National Institutes of Health; and Grants 0355741U and 0151521U from The American Heart Association.

References

AbdAlla S, Lother H, Quitterer U (2000) AT_1-receptor heterodimers show enhanced G-protein activation and altered receptor sequestration. Nature 407:94–98

Albiston AL, McDowall SG, Matsacos D, Sim P, Clune E, Mustafa T, Lee J, Mendelsohn FAO, Simpson RJ, Connolly LM, Chai SY (2001) Evidence that the angiotensin IV (AT4) receptor is the enzyme insulin-regulated aminopeptidase. J Biol Chem 276:48623–48626

Allred AJ, Diz DI, Ferrario CM, Chappell, MC (2000) Pathways for angiotensin-(1-7) metabolism in pulmonary and renal tissues. Am J Physiol 279:F841–F850

Almeida AP, Frabregas BC, Madureira MM, Santos RJ, Campagnole-Santos MJ, Santos RA (2000) Angiotensin-(1-7) potentiates the coronary vasodilatory effect of bradykinin in the isolated rat heart. Braz J Med Biol Res 33:709–713

Ambuhl P, Felix D, Imboden H, Khosla MC, Ferrario CM (1992) Effects of angiotensin analogues and angiotensin receptor antagonists on paraventricular neurones. Regul Pept 38:111–120

Ambuhl P, Felix D, Khosla MC (1994) (7-D-ALA)-Angiotensin-(1-7): selective antagonism of angiotensin-(1-7) in the rat para ventricular nucleus. Brain Res 35:289–291

Anastasopoulos F, Leung R, Kladis A, James GM, Briscoe TA, Gorski TP, Campbell DJ (1998) Marked difference between angiotensin-converting enzyme and neutral endopeptidase inhibition in vivo by a dual inhibitor of both enzymes. J Pharmacol Exp Ther 284:799–805

Andreatta-Van Leyen S, Romero MF, Khosla MC, Douglas JG (1993) Modulation of phospholipase A2 activity and sodium transport by angiotensin-(1-7). Kidney Int 44:932–936

Averill DB, Diz DI (1999) Angiotensin peptides and baroreflex control of sympathetic outflow: pathways and mechanisms of the medulla oblongata. Brain Res Bull 51:119–128

Averill DB, Tsuchihashi T, Khosla MC, Ferrario CM (1994) Losartan, nonpeptide angiotensin II-type 1 (AT_1) receptor antagonist, attenuates pressor and sympathoexcitatory response evoked by angiotensin II and L-glutamate in rostral ventrolateral medulla. Brain Res 665:245–252

Balcells E, Meng QC, Hageman GR, Palmer RW, Durand JN, Dell'Italia LJ (1996) Angiotensin II formation in dog heart is mediated by different pathways in vivo and in vitro. Am J Physiol 40:H417–H421

Baltatu O, Fontes MA, Campagnole-Santos MJ, Caligiorni S, Ganten D, Santos RA, Bader M (2001) Alterations of the renin–angiotensin system at the RVLM of transgenic rats with low brain angiotensinogen. Am J Physiol Regul Integr Comp Physiol 280:R428–R433

Baracho NC, Simoes-e-Silva A, Khosla MC, Santos RA (1998) Effect of selective angiotensin antagonists on the antidiuresis produced by angiotensin-(1-7) in water-loaded rats. Braz J Med Biol Res 31:1221–1227

Barnes KL, Knowles WD, Ferrario CM (1990) Angiotensin II and angiotensin-(1-7) excite neurons in the canine medulla in vitro. Brain Res Bull 24:275–280

Bayorh MA, Eatman D, Walton M, Socci RR, Thierry-Palmer M, Emmett N (2002) A-779 attenuates angiotensin-(1-7) depressor response in salt-induced hypertensive rats. Peptides 23:57–64

Benter IF, Diz DI, Ferrario CM (1993) Cardiovascular actions of angiotensin-(1-7). Peptides 14:679–684

Benter IF, Diz DI, Ferrario CM (1995a) Pressor and reflex sensitivity is altered in spontaneously hypertensive rats treated with angiotensin-(1-7). Hypertension 26:1138–1144

Benter IF, Ferrario CM, Morris M, Diz DI (1995b) Antihypertensive actions of angiotensin-(I-7) in spontaneously hypertensive rats. Am J Physiol 269:H313–H319

Benzing T, Fleming I, Blaukat A, Müller-Esterl W, Busse R (1999) Angiotensin-converting enzyme inhibitor ramiprilat interferes with the sequestration of the B2 kinin receptor within the plasma membrane of native endothelial cells. Circulation 99:2034–2040

Block CH, Santos RAS, Brosnihan KB, Ferrario CM (1988) Immunocytochemical localization of angiotensin-(1-7) in the rat forebrain. Peptides 9:1395–1401

Borges EL, Cabral BM, Braga AA, Neves MJ, Santos RA, Rogana E (2002) Effect of angiotensin-(1-7) on jejunal absorption of water in rats. Peptides 23:51–56

Bouley R, Perodin J, Plante H, Rihakova L, Bernier SG, Maletinska L, Guillemette G, Escher E (1998) N- and C-terminal structure-activity study of angiotensin II on the angiotensin AT2 receptor. Eur J Pharmacol 343:323–331

Brenner BM, Cooper ME, de Z, Keane WF, Mitch WE, Parving HH, Remuzzi G, Snapinn SM, Zhang Z, Shahinfar S, RENAAL Study Investigators (2001) Effects of losartan on renal and cardiovascular outcomes in patients with type 2 diabetes and nephropathy. N Engl J Med 345:861–869

Britto RR, Santos RAS, Fagundes-Moura CR, Khosla MC, Campangnole-Santos MJ (1997) Role of angiotensin-(1-7) in the modulation of the baroreflex in renovascular hypertensive rats. Hypertension 30:549–556

Brosnihan KB, Li P, Ferrario CM (1996) Angiotensin-(1-7) dilates canine coronary arteries through kinins and nitric oxide. Hypertension 27:523–528

Brosnihan KB, Li P, Tallant EA, Ferrario CM (1998) Angiotensin-(1-7): a novel vasodilator of the coronary circulation. Biol Res 31:227–234

Bumpus FM, Catt KJ, Chiu AT, DeGasparo M, Goodfriend T, Husain A, Peach MJ, Taylor DG Jr, Timmermans PBWM (1991) Nomenclature for angiotensin receptors: A report of the Nonmenclature Committee for the Council for High Blood Pressure Research. Hypertension 17:720–721

Bumpus FM, Khairallah PA, Arakawa K, Page III, Smeby RR (1961) The relationship of structure to pressor and oxytocic actions of isoleucine angiotensin octapeptide and various analogues. Biochim Biophys Acta 46:38–44

Burgelova M, Kramer HJ, Teplan V, Velickova G, Vitko S, Heller J, Maly J:Cervenka L (2002) Intrarenal infusion of angiotensin-(1-7) modulates renal functional responses to exogenous angiotensin II in the rat. Kidney Blood Press Res 25:202–210

Calka J, Block CH (1993) Angiotensin-(1-7) and nitric oxide synthase in the hypothalamo-neurohypophysial system. Brain Res Bull 30:677–685

Campagnole-Santos MJ, Diz DI, Ferrario CM (1990) Actions of angiotensin peptides after partial denervation of the solitary tract nucleus. Hypertension (Suppl I) 15:I-34-I-39

Campagnole-Santos MJ, Diz DI, Santos RAS, Khosla MC, Brosnihan KB, Ferrario CM (1989) Cardiovascular effects of angiotensin-(1-7) injected into the dorsal medulla of rats. Am J Physiol 257:H324–H329

Campagnole-Santos MJ, Heringer SB, Batista EN, Khosla MC, Santos RAS (1992) Differential baroreceptor reflex modulation by centrally infused angiotensin peptides. Am J Physiol 263:R89–R94

Campbell DJ, Anastasopoulos F, Duncan AM, James GM, Kladis A, Briscoe TA (1998) Effects of neutral endopeptidase inhibition and combined angiotensin converting enzyme and neutral endopeptidase inhibition on angiotensin and bradykinin peptides in rats. J Pharmacol Exp Ther 287:567–577

Caruso-Neves C, Lara LS, Rangel LB, Grossi AL, Lopes AG (2000) Angiotensin-(1-7) modulates the ouabain-insensitive Na^+-ATPase activity from basolateral membrane of the proximal tubule. Biochim Biophys Acta 1467:189–197

Chansel D, Vandermeersch S, Oko A, Curat C, Ardaillou R (2001) Effects of angiotensin IV and angiotensin-(1-7) on basal and angiotensin II-stimulated cytosolic Ca2+ in mesangial cells. Eur J Pharmacol 414:165–175

Chappell MC (2002) Inhibition of Na^+, K^+-ATPase activity by angiotensin-(1-7) in rat proximal tubules is attenuated by cytochrome P450 blockade. FASEB 16:A496

Chappell MC, Allred AJ, Ferrario CM (2001) Pathways of angiotensin-(1-7) metabolism in the kidney. Nephrol Dial Transplant 16:22–26

Chappell MC, Brosnihan KB, Diz DI, Ferrario CM (1989) Identification of angiotensin-(1-7) in rat brain: evidence for differential processing of angiotensin peptides. J Biol Chem 264:16518–16523

Chappell MC, Gomez,MN, Pirro NT, Ferrario CM (2000) Release of angiotensin-(1-7) from the rat hindlimb: influence of angiotensin-converting enzyme inhibition. Hypertension 35:348–352

Chappell MC, Jacobsen DW, Tallant EA (1995) Characterization of angiotensin II receptor subtypes in pancreatic acinar AR42 J cells. Peptides 16:741–747

Chappell MC, Jung F, Gallagher PE, Averill DB, Crackower MA, Penninger JM, Ferrario CM (2002) Omapatrilat treatment is associated with increased ACE-2 and angiotensin-(1-7) in spontaneously hypertensive rats. Hypertension 40:409

Chappell MC, Pirro NT, Sykes A, Ferrario CM (1998) Metabolism of angiotensin-(1-7) by angiotensin converting enzyme. Hypertension 31:362–36

Chappell MC, Tallant EA, Brosnihan KB, Ferrario CM (1994) Conversion of angiotensin I to angiotensin-(1-7) by thimet oligopeptidase (EC 3.4.24.15) in vascular smooth muscle cells. J Vasc Med Biol 5:129–137

Chaves GZ, Caligiorne SM, Santos RAS, Khosla MC, Campagnole-Santos MJ (2000) Modulation of the baroreflex control of heart rate by angiotensin-(1-7) at the nucleus tractus solitarii of normotensive and spontaneously hypertensive rats. J Hypertens 18:1841–1848

Clark MA, Tallant EA, Diz DI (2001a) Downregulation of the AT_{1A} receptor by pharmacologic concentrations of angiotensin-(1-7). J Cardiovasc Pharmacol 37:437–448

Clark MA, Tommasi EN, Bosch SM, Tallant EA, Diz DI (2003) Angiotensin-(1-7) reduces renal angiotensin II receptors through a cyclooxygenase dependent pathway. J Cardiovasc Pharmacol (in press)

Clark MA, Diz DI, Tallant EA (2001b) Angiotensin-(1-7) downregulates the angiotensin II type 1 receptor in vascular smooth muscle cells. Hypertension 37:1141–1146

Corvol P, Michaud A, Soubrier F, WilliamsTA (1995) Recent advances in knowledge of the structure and function of the angiotensin I converting enzyme. J Hypertens 13:S3–S10

Couto AS, Baltatu O, Santos RA, Ganten D, Bader M, Campagnole-Santos MJ (2002) Differential effects of angiotensin II and angiotensin-(1-7) at the nucleus tractus solitarii of transgenic rats with low brain angiotensinogen. J Hypertens 20:919–925

Couto AS, Santos RAS, Campagnole-Santos MJ, DeMinas Gerais UF, Horizonte B, Brazil MG (1998) Cardiovascular effects produced by angiotensin II and angiotensin-(1-7) in different areas of the nucleus tractus solitary. J Hypertens 16 [Suppl 2]:S169

Crackower MA, Sarao R, Oudit GY, Yagil C, Kozieradzki I, Scanga SE, Oliveira-dos-Santo AJ, da Costa J, Zhang L, Pei Y, Scholey J, Bray MR, Ferrario CM, Backx PH, Manoukian AS, Chappell MC, Yagil Y, Penninger JM (2002) Angiotensin-converting enzyme 2 is an essential regulator of heart function. Nature 417:822–828

Dahlof B, Devereux RB, Kjeldsen SE, Julius S, Beevers G, deFaire U, Fyhrquist F, Ibsen H, Kristiansson K, Lederballe-Pedersen O, Lindholm LH, Nieminen MS, Omvik P, Oparil S, Wedel H (2002) Cardiovascular morbidity and mortality in the losartan intervention for endpoint reduction in hypertension study (LIFE): a randomised trial against atenolol. Lancet 359:995–1003

Davie AP, McMurray JJ (1999) Effect of angiotensin-(1-7) and bradykinin in patients with heart failure treated with an ACE inhibitor. Hypertension 34:457–460

de Gasparo M, Husain A, Alexander W, Catt KJ, Chiu AT, Drew M, Goodfriend T, Harding JW, Inagami T, Timmermans PBMWM (1995) Proposed update of angiotensin receptor nomenclature. Hypertension 25:924–927

Deddish PA, Marcic B, Jackman HL, Wang HZ, Skidgel RA, Erdos EG (1998) N-domain specific substrate and C-domain inhibitors of angiotensin converting enzyme. Hypertension 31:912–917

DelliPizzi A, Hilchey SD, Bell-Quilley CP (1994) Natriuretic action of angiotensin (1-7). Br J Pharmacol 111:1–3

Diz DI, Barnes KL, Ferrario CM (1984) Hypotensive actions of microinjections of angiotensin II into the dorsal motor nucleus of the vagus. J Hypertens (Suppl 3) 2:53–56

Diz DI, Falgui B, Bosch SM, Westwood BM, Kent J, Ganten D, Ferrario CM (1997) Hypothalamic substance P release: Attenuated angiotensin responses in mRen2(27) transgenic rats. Hypertension 29:510–513

Diz DI, Ferrario CM (1996) Angiotensin receptor heterogeneity in dorsal medulla oblongata as defined by angiotensin-(1-7). In: Raizada MR, Phillips MI, Summers KM (eds) Recent advances in cellular and molecular aspects of angiotensin receptors. Advances in experimental biology and medicine. Plenum Press, New York, pp 225–235

Diz DI, Jessup JA, Westwood BM, Bosch SM, Vinsant S, Gallagher PE, Averill DB (2002) Angiotensin peptides as neurotransmitters/neuromodulators in the dorsomedial medulla. Clin Exp Pharmacol Physiol 29:473–482

Diz DI, Kohara K, Ferrario C M (1992) Normalization of angiotensin (Ang) II receptors in the dorsal medulla oblongata of spontaneously hypertensive rats (SHR) follows converting enzyme inhibition and increases in plasma angiotensin-(1-7) concentrations. Am J Hypertens 5:16A

Diz DI, Moriguchi A, Bosch SM, Ganten D, Ferrario CM (1994) Angiotensin receptor regulation in paraventricular nucleus (PVN) differs in SHR and transgenic (mREN2) 27 rats. Am J Hypertens 7(4):30A

Diz DI, Pirro N (1992) Differential actions on angiotensin II and angiotensin-(1-7) on transmitter release. Hypertension 19:II-41–II-48

Diz DI, Westwood B, Averill DB (2001) AT(1) antisense distinguishes receptors mediating angiotensin II actions in solitary tract nucleus. Hypertension 37:1292–1297

Diz DI, Westwood BM (2000) Deficiency of endogenous angiotensin-(1-7) in the nucleus tractus solitarii of (mREN2)27 transgenic rats may account for diminished baroreceptor reflex function. Hypertension 36[4]:681

Donoghue M, Hsieh F, Baronas E, Godbout K, Gosselin M, Stagliano N, Donovan M, Woolf B, Robison K, Jeyaseelan R, Breitbart RE, Acton S (2000) A novel angiotensin-converting enzyme-related carboxypeptidase (ACE2) converts angiotensin I to angiotensin 1-9. Circ Res 87:E1-E9

Dzau VJ, Re R (1994) Tissue angiotensin system in cardiovascular medicine. Circulation 89:493-498

Erdos EG, Wagner B, Harbury CB, Painter RG, Skidgel RA, Fa XG (1989) Down-regulation and inactivation of neutral endopeptidase 24.11 (enkephalinase) in human neutrophils. J Biol Chem 264:14519-14523

Felix D, Khosla MC, Barnes KL, Imboden H, Montani B, Ferrario CM (1991) Neurophysiological responses to angiotensin-(1-7). Hypertension 17:1111-1114

Fernandes L, Fortes ZB, Nigro D Tostes RC, Santos RA, Carvalho MHC (2001) Potentiation of bradykinin by angiotensin-(1-7) on arterioles of spontaneously hypertensive rats studied in vivo. Hypertension 37:703-709

Ferrario CM (2002) Does angiotensin-(1-7) contribute to cardiac adaptation and preservation of endothelial function in heart failure? Circulation 105:1523-1525

Ferrario CM, Averill DB, Brosnihan KB, Chappell MC, Iskandar SS, Dean RH, Diz DI (2002a) Vasopeptidase inhibition and Ang-(1-7) in the spontaneously hypertensive rat. Kidney Int 62:1349-1357

Ferrario CM, Chappell MC, Tallant EA, Brosnihan KB, Diz DI (1997) Counterregulatory actions of angiotensin-(1-7). Hypertension 30:535-541

Ferrario CM, Martell N, Yunis C, Flack JM, Chappell MC, Brosnihan,KB, Dean RH, Fernandez A, Novikov S, Pinillas C, Luque M (1998) Characterization of angiotensin-(1-7) in the urine of normal and essential hypertensive subjects. Am J Hypertens 11:137-146

Ferrario CM, Smith RD, Brosnihan KB, Chappell MC, Campese VM, Vesterqvist O, Liao W, Ruddy MC, Grim CE (2002b) Effects of omapatrilat on the renin angiotensin system in salt sensitive hypertension. Am J Hypertens 15:557-564

Ferreira AJ, Santos RA, Almeida AP (2001) Angiotensin-(1-7): cardioprotective effect in myocardial ischemia/reperfusion. Hypertension 38:665-668

Feterik K, Smith L, Katusic ZS (2001) Angiotensin-(1-7) causes endothelium-dependent relaxation in canine middle cerebral artery. Brain Res 873:75-82

Fogarty DJ, Sanchez-Gomez MV, Matute C (2002) Multiple angiotensin receptor subtypes in normal and tumor astrocytes in vitro. Glia 39:304-313

Fontes MA, Baltatu O, Caligiorne SM, Campagnole-Santos MJ, Ganten D, Bader M, Santos RA (2000) Angiotensin peptides acting at rostral ventrolateral medulla contribute to hypertension of TGR(mREN2)27 rats. Physiol Genom 2:137-142

Fontes MA, Pinge MC, Naves V, Campagnole-Santos MJ, Lopes OU, Khosla MC, Santos RA (1997) Cardiovascular effects produced by microinjection of angiotensins and angiotensin antagonists into the ventrolateral medulla of freely moving rats. Brain Res 750:305-310

Fontes MA, Silva LC, Campagnole-Santos MJ, Khosla MC, Guertzenstein PG, Santos RA (1994) Evidence that angiotensin-(1-7) plays a role in the central control of blood pressure at the ventro-lateral medulla acting through specific receptors. Brain Res 665:175-180

Forster C, le Tran Y (1997) Angiotensin-(1-7) and the rat aorta: modulation by the endothelium. J Cardiovasc Pharmacol 30:676-682

Freeman EJ, Chisolm GM, Ferrario CM, Tallant EA (1996) Angiotensin-(1-7) inhibits vascular smooth muscle cell growth. Hypertension 28:104-108

Freeman EJ, Tallant EA (1994) Vascular smooth-muscle cells contain AT_1 angiotensin receptors coupled to phospholipase D activation. Biochem J (Great Britain) 304:543-548

Gafford JT, Skidgel RA, Erdos EG, Hersh LB (1983) Human kidney enkephalinase a neutral metalloendopeptidase that cleaves active peptides. Biochemistry 22:3265-3271

Garcia NH, Garvin JL (1994) Angiotensin 1-7 has a biphasic effect on fluid absorption in the proximal straight tubule. J Am Soc Nephrol 5:1133-1138

Gironacci MM, Adler-Graschinsky E, Pena C, Enero MA (1994) Effects of angiotensin II and angiotensin-(1-7) on the release of [^{3}H] norepinephrine from rat atria. Hypertension 24:457–460

Gironacci MM, Fernandez-Tome MC, Speziale E, Sterin-Speziale N, Pena C (2002) Enhancement of phosphatidylcholine biosynthesis by angiotensin-(1-7) in the rat renal cortex. Biochem Pharmacol 63:507–514

Gironacci MM, Coba MP, Pena C (1999) Angiotensin-(1-7) binds at the type 1 angiotensin II receptors in rat renal cortex. Regul Pept 84:51–54

Gironacci MM, Vatta M, Rodriguez-Fermepin M, Fernandez BE, Pena C (2000) Angiotensin-(1-7) reduces norepinephrine release through a nitric oxide mechanism in rat hypothalamus. Hypertension 35:1248–1252

Gorelik G, Carbini LA, Scicli AG (1998) Angiotensin 1-7 induces bradykinin-mediated relaxation in porcine coronary artery. J Pharmacol Exp Therap 286:403–410

Graf K, Koehne P, Grafe M, Zhang M, Auch-Schwalk W, Fleck E (1995) Regulation and differential expression of neutral endopeptidase 24.11 in human endothelial cells. Hypertension 26:230–235

Handa RK (1999) Angiotensin-(1-7) can interact with the rat proximal tubule AT(4) receptor system. Am J Physiol 277:F75–F83

Handa RK (2000a) Binding and signaling of angiotensin-(1-7) in bovine kidney epithelial cells involves the AT(4) receptor. Peptides 21:729–736

Handa RK (2000b) Metabolism alters the selectivity of angiotensin-(1-7) receptor ligands for angiotensin receptors. J Am Soc Nephrol 11:1377–1386

Handa RK, Ferrario CM, Strandhoy JW (1996) Renal actions of angiotensin-(1-7) in vivo and in vitro studies. Am J Physiol 270:F141–F147

Harder DR, Campbell WB, Roman RJ (1995) Role of cytochrome P-450 enzymes and metabolites of arachidonic acid in the control of vascular tone. J Vas Res 32:79–92

Heitsch H, Brovkovych S, Malinski T, Wiemer G (2001) Angiotensin-(1-7)-stimulated nitric oxide and superoxide release from endothelial cells. Hypertension 37:72–76

Heller J, Kramer HJ, Maly J, Cervenka L, Horacek V (2000) Effect of intrarenal infusion of angiotensin-(1-7) in the dog. Kidney Blood Press Res 23:89–94

Heringer-Walther S, Batista EN, Walther T, Khosla MC, Santos RA, Campagnole-Santos MJ (2001) Baroreflex improvement in SHR after ACE inhibition involves angiotensin-(1-7). Hypertension 37:1309–1314

Heyne N, Beer W, Muhlbauer B, Osswald H (1995) Renal response to angiotensin (1-7) in anesthetized rats. Kidney Int 47:975–976

Hilchey SD, Bell-Quilley CP (1995) Association between the natriuretic action of angiotensin-(1-7) and selective stimulation of renal prostaglandin I_2 release. Hypertension 25:1238–1244

Ichikawa I, Harris RC (1991) Angiotensin actions in the kidney: Renewed insight into the old hormone. Kidney Int 40:583–596

Iyer SN, Averill DB, Chappell MC, Yamada K, Jones AG, Ferrario CM (2000a) Contribution of angiotensin-(1-7) to blood pressure regulation in salt-depleted hypertensive rats. Hypertension 36:417–422

Iyer SN., Chappell MC, Averill DB, Diz DI, Ferrario CM (1998a) Vasodepressor actions of angiotensin-(1-7) unmasked during combined treatment with lisinopril and losartan. Hypertension 31:699–705

Iyer SN, Ferrario CM, Chappell MC (1998b) Angiotensin-(1-7) contributes to the antihypertensive effects of blockade of the renin–angiotensin system. Hypertension 31:356–361

Iyer SN, Yamada K, Diz DI, Ferrario CM, Chappell MC (2000b) Evidence that prostaglandins mediate the antihypertensive actions of angiotensin-(1-7) during chronic blockade of the renin–angiotensin system. J Cardiovasc Pharmacol 36:109–117

Jaiswal N Diz DI, Chappell MC, Khosla MC, Ferrario CM (1992) Stimulation of endothelial cell prostaglandin production by angiotensin peptides: Characteristics of receptors. Hypertension (Suppl II) 19:49–55

Jaiswal N, Diz DI, Tallant EA, Khosla MC, Ferrario CM (1991a) Characterization of angiotensin receptors mediating prostaglandin synthesis in C6 glioma cells. Am J Physiol 260:R1000–R1006

Jaiswal N, Jaiswal RK, Tallant EA, Diz DI, Ferrario CM (1993a) Alterations in prostaglandin production in spontaneously hypertensive rat smooth muscle cells. Hypertension 21:900–905

Jaiswal N, Tallant EA, Diz DI, Khosla MC, Ferrario CM (1991b) Subtype 2 angiotensin receptors mediate prostaglandin synthesis in human astrocytes. Hypertension 17:1115–1120

Jaiswal N, Tallant EA, Jaiswal RK, Diz DI, Ferrario CM (1993b) Differential regulation of prostaglandin synthesis by angiotensin peptides in porcine aortic smooth muscle cells: Subtypes of angiotensin receptors involved. J Pharmacol Exp Therap 265:664–673

Kawabe H, Brosnihan KB, Diz DI, Ferrario CM (1986) Role of brain dopamine in centrally evoked angiotensin II responses in conscious rats. Hypertension (Suppl I) 8:I-84–I-89

Kohara K, Brosnihan KB., Ferrario CM (1993) Angiotensin-(1-7) in the spontaneously hypertensive rat. Peptides 14:883–891

Kono T, Taniguchi A, Imura H, Oseko F, Khosla MC (1986) Biological activities of angiotensin II-(1-6)-hexapeptide and angiotensin II-(1-7)-heptapeptide in man. Life Sci 38:1515–1519

Krob HA, Vinsant SL, Ferrario CM, Friedman DP (1998) Angiotensin-(1-7) immunoreactivity in the hypothalamus of the (mRen-2^d)27 transgenic rat. Brain Res 798:36–45

Kucharewicz I, Chabielska E, Pawlak D, Matys T, Rolkowski R, Buczko W (2000) The antithrombotic effect of angiotensin-(1-7) closely resembles that of losartan. J Renin–angiotensin-Aldosterone Syst 1:268–272

Lara LS, Bica RB, Sena SL, Correa JS, Marques-Fernandes MF, Lopes AG, Caruso-Neves C (2002) Angiotensin-(1-7) reverts the stimulatory effect of angiotensin II on the proximal tubule Na(+)-ATPase activity via a A779-sensitive receptor. Regul Pept 103:17–22

Lawrence AC, Clark IJ, Campbell DJ (1992) Increased angiotensin-(1-7) in hypophysial-portal plasma of conscious sheep. Neuroendocrinology 55:105–114

Lewis EJ, Hunsicker LG, Clarke WR, Berl T, Pohl MA, Lewis JB, Ritz E, Atkins RC, Rohde R, Raz I (2001) Renoprotective effect of the angiotensin-receptor antagonist irbesartan in patients with nephropathy due to type 2 diabetes. N Engl J Med 345:851–860

Li P, Chappell MC, Ferrario CM, Brosnihan KB (1997) Angiotensin-(1-7) augments bradykinin-induced vasodilation by competing with ACE and releasing nitric oxide. Hypertension 29:394–400

Li YW, Guyenet PG (1995) Neuronal excitation by angiotensin II in the rostral ventrolateral medulla of the rat in vitro. Am J Physiol 268:R272–R277

Li YW, Guyenet PG (1996) Angiotensin II decreases a resting K+ conductance in rat bulbospinal neurons of the C1 area. Circ Res 78:274–282

Lima DX, Campagnole-Santos MJ, Fontes MAP, Khosla MC, Santos RAS (1999) Haemorrhage increases the pressor effect of angiotensin-(1-7) but not of angiotensin II at the rat rostral ventrolateral medulla. J Hypertens 17:1152

Lima DX, Fontes MA, Oliveira RC, Campagnole-Santos MJ, Khosla MC, Santos RA (1996) Pressor action of angiotensin I at the ventrolateral medulla: effect of selective angiotensin blockade. Immunopharmacology 33:305–307

Llorens-Cortes C, Huang H, Vicart P, Gasc JM, Paulin D, Corvol P (1992) Identification and characterization of neutral endopeptidase in endothelial cells from venous or arterial origins. J Biol Chem 267:14012–14018

Loot AE, Roks AJM, Henning RH, Tio RA, Suurmeijer AHH, Boomsma F, vanGilst WH (2002) Angiotensin-(1-7) attenuates the development of heart failure after myocardial infarction in rats. Circulation 105:1548–1550

Lopez O, Gironacci M, Rodriguez D, Pena C (1998) Effect of angiotensin-(1-7) on ATPase activities in several tissues. Regul Pept 77:135–139

Luque M, Martin P, Martell N, Fernandez C, Brosnihan KB, Ferrario CM (1996) Effects of captopril related to increased levels of prostacyclin and angiotensin-(1-7) in essential hypertension. J Hypertens 14:799–805

Machado RD, Ferreira MA, Belo AV, Santos RA, Andrade SP (2002) Vasodilator effect of angiotensin-(1-7) in mature and sponge-induced neovasculature. Regul Pept 107:105–113

Machado RD, Santos RA, Andrade SP (2000) Opposing actions of angiotensins on angiogenesis. Life Sci 66:67–76

Machado RD, Santos RA, Andrade SP (2001) Mechanisms of angiotensin-(1-7)-induced inhibition of angiogenesis. Am J Physiol Regul Integr Comp Physiol 280:R994–R1000

Mahon JM, Allen M, Herbert J, Fitzsimons JT (1995) The association of thirst, sodium appetite and vasopressin release with c-fos expression in the forebrain of the rat after intracerebroventricular injection of angiotensin II, angiotensin-(1-7) or carbachol. Neurosci Lett 69:199–208

Mahon JM, Carr RD, Nicol AK, Henderson IW (1994) Angiotensin-(1-7) is an antagonist at the type 1 angiotensin II receptor. J Hypertens 12:1377–1381

Massi M, Polidori G, Perfumi M, Gentili L, DeCaro G (1991a) Tachykinin receptor subtypes involved in the cental effects of tachykinins on water and salt intake. Brain Res 26:155–160

Massi M, Saija A, Polidori C, Perfumi M, Gentili L, Costa G, De Caro G (1991b) The hypothalamic paraventricular nucleus is a site of action for the central effect of tachykinins on plasma vasopressin. Brain Res Bull 26:149–154

McGiff JC, Quilley CP, Carroll MA (1993) The contribution of cytochrome P450-dependent arachidonate metabolites to integrated renal function. Steroids 58:573–579

Meng W, Busija DW (1993) Comparative effects of angiotensin-(1-7) and angiotensin II on piglet pial arterioles. Stroke 24:2041–2045

Merrill DC, Karoly M, Chen K, Ferrario CM, Brosnihan KB (2002) Angiotensin-(1-7) in normal and preeclamptic pregnancy. Endocrine 18:239–245

Metzger R, Bader M, Ludwig T, Berberich C, Bunnemann B, Ganten D (1995) Expression of the mouse and rat mas proto-oncogene in the brain and peripheral tissues. FEBS Lett 357:27–32

Min L, Sim MK, Xu XG (2000) Effects of des-aspartate-angiotensin I on angiotensin II-induced incorporation of phenylalanine and thymidine in cultured rat cardiomyocytes and aortic smooth muscle cells. Regul Pept 95:93–97

Minshall RD, Tan F, Nakamura F, Rabito SF, Becker RP, Marcic B, Erdos EG (1997) Potentiation of the actions of bradykinin by angiotensin I-converting enzyme inhibitors: The role of expressed human bradykinin B_2 receptors and angiotensin I-converting enzyme in CHO cells. Circ Res 81:848–856

Monti J, Schinke M, Bohm M, Ganten D, Bader M, Bricca G (2001) Glial angiotensinogen regulates brain angiotensin II receptors in transgenic rats TGR (ASrAOGEN). Am J Physiol Regul Integr Comp Physiol 280:R233–R240

Moriguchi A, Ferrario CM., Brosnihan KB, Ganten D, Morris M (1994) Differential regulation of central vasopressin in transgenic rats harboring the mouse Ren-2 gene. Am J Physiol 267:R786–R791

Moriguchi A, Mikami H, Otsuka A, Katahira K, Kohara K, Ogihara T (1995a) Amino acids in the medulla oblongata contribute to baroreflex modulation by angiotensin II. Brain Res Bull 36:85–89

Moriguchi A, Tallant EA, Matsumura K, Reilly TM, Walton H, Ganten D, Ferrario CM (1995b) Opposing actions of angiotensin-(1-7) and angiotensin II in the brain of transgenic hypertensive rats. Hypertension 25:1260–1265

Muthalif MM, Benter IF, Uddin MR, Harper JL, Malik KU (1998) Signal transduction mechanisms involved in angiotensin-(1-7)-stimulated arachidonic acid release and prostanoid synthesis in rabbit aortic smooth muscle cells. J Pharm Exp Ther 284:388–398

Nakamoto H, Ferrario CM, Fuller SB, Robaczwski DL, Winicov E, Dean RH (1995) Angiotensin-(1-7) and nitric oxide interaction in renovascular hypertension. Hypertension 25:796–802

Neuss M, Regitz-Zagrosek V, Hildebrandt A, Fleck E (1994) Human cardiac fibroblasts express an angiotensin receptor with unusual binding characteristics which is coupled to cellular proliferation. Biochem Biophys Res Commun 204:1334–1339

Neuss M, Regitz-Zagrosek V, Hildebrandt A, Fleck E (1996) Isolation and characterisation of human cardiac fibroblasts from explanted adult hearts. Cell Tissue Res 286:145–153

Neves LA, Santos RA, Khosla MC, Milsted A (2000) Angiotensin-(1-7) regulates the levels of angiotensin II receptor subtype AT1 mRNA differentially in a strain-specific fashion. Regul Pept 95:99–107

Nickenig G, Geisen G, Vetter H, Sachinidis A (1997) Characterization of angiotensin receptors on human skin fibroblasts. J Mol Med 75:217–222

Oliveira DR, Santos RAS, Santos GFP, Khosla MC, Campagnole-Santos MJ (1996) Changes in the baroreflex control of heart rate produced by central infusion of selective angiotensin antagonists in hypertensive rats. Hypertension 27:1284–1290

Oliveira MA, Fortes ZB, Santos RA, Kosla MC, De C (1999) Synergistic effect of angiotensin-(1-7) on bradykinin arteriolar dilation in vivo. Peptides 20:1195–1201

Oliveira RC, Campagnole-Santos MJ, Santos RAS, DeMinas Gerais UF, Horizonte B, Brazil MG (1998) The pressor effect of angiotensin-(1-7) at the rostral ventrolateral medulla is multimediated. J Hypertens 16 [Suppl 2]:S129

Osei SY, Ahima RS, Minkes RK, Weaver JP, Khosla MC, Kadowitz PJ (1993) Differential responses to angiotensin-(1-7) in the feline mesenteric and hindquarters vascular beds. Eur J Pharmacol 234:35–42

Paula RD, Lima CV, Britto RR, Campagnole-Santos MJ, Khosla MC, Santos RAS (1999) Potentiation of the hypotensive effect of bradykinin by angiotensin-(1-7)-related peptides. Peptides 20:493–500

Paula RD, Lima CV, Khosla MC, Santos RAS (1995) Angiotensin-(1-7) potentiates the hypotensive effect of bradykinin in conscious rats. Hypertension 26:1154–1159

Pawlak R, Napiorkowska-Pawlak D, Takada Y, Urano T, Nagai N, Ihara H, Takada A (2001) The differential effect of angiotensin II and angiotensin 1-7 on norepinephrine, epinephrine, and dopamine concentrations in rat hypothalamus: the involvement of angiotensin receptors. Brain Res Bull 54:689–694

Perfumi M, Sajia A, Costa G, Massi M, Polidori C (1988) Vasopressin release induced by intracranial injection of eledoisin is mediated by central angiotensin II. Pharmacol Res Commun 20:811–826

Pirro NT, Ferrario CM, Chappell MC (2001) Neprilysin and endothelin converting enzyme contribute to the processing of Ang I and Ang-(1-9) to Ang-(1-7) in human endothelial cells. FASEB J 15[5]:A778

Pitt B (2002) Clinical trials of angiotensin receptor blockers in heart failure: what do we know and what will we learn? Am J Hypertens 15:22S–27S

Porsti I, Bara AT, Busse R, Hecker M (1994) Release of nitric oxide by angiotensin-(1-7) from porcine coronary endothelium: implications for a novel angiotensin receptor. Br J Pharmacol 111:652–654

Potts PD, Allen AM, Horiuchi J, Dampney RA (2000a) Does angiotensin II have a significant tonic action on cardiovascular neurons in the rostral and caudal VLM? Am J Physiol Regul Integr Comp Physiol 279:R1392–R1402

Potts PD, Horiuchi J, Coleman MJ, Dampney RA (2000b) The cardiovascular effects of angiotensin-(1-7) in the rostral and caudal ventrolateral medulla of the rabbit. Brain Res 877:58–64

Qadri F, Wolf A, Waldmann T, Rascher W, Unger T (1998) Sensitivity of hypothalamic paraventricular nucleus to C- and N-terminal angiotensin fragments—vasopressin release and drinking. J Neuroendocrinol 10:275–281

Qu L, McQueeney AJ, Barnes KL (1996) Presynaptic or postsynaptic location of receptors for angiotensin II and substance P in the medial solitary tract nucleus. J Neurophysiol 75:2220–2228

Quan A, Baum M (1996) Endogenous production of angiotensin II modulates rat proximal tubule transport. J Clin Invest 97:2878–2882

Ren Y, Garvin JL, Carretero OA (2002) Vasodilator action of angiotensin-(1-7) on isolated rabbit afferent arterioles. Hypertension 39:799–802

Rodgers K, Xiong S, Felix J, Roda N, Espinoza T, Maldonado S, Dizerega G (2001) Development of angiotensin (1-7) as an agent to accelerate dermal repair. Wound Repair Regen 9:238–247

Rodgers KE, Xiong S, DiZerega GS (2002) Accelerated recovery from irradiation injury by angiotensin peptides. Cancer Chemother Pharmacol 49:403–411

Roks AJ, van Geel PP, Pinto YM, Buikema H, Henning RH, De Zeeuw D, van Gilst WH (1999) Angiotensin-(1-7) is a modulator of the human renin–angiotensin system. Hypertension 34:296–301

Roman RJ, Alonso-Galicia M (1999) P450-Eicosanoids: A novel signaling pathway regulating renal function. News Physiol Sci 14:238–242

Rowe BP, Saylor DL, Speth RC, Absher DR (1995) Angiotensin-(1-7) binding at angiotensin II receptors in the rat brain. Regulatory Pept 56:139–146

Ryan JW (1974) The fate of angiotensin II. In: Page IH, Bumpus FM (eds) Handbook of experimental pharmacology. Springer-Verlag, Berlin, pp 81–110

Santos RA, Simoes e Silva AC, Maric C, Silva DMR, Machado RP, de Buhr I, Heringer-Walter S, Pinheiro VB, Lopes Mt, Bader M, Mendes EP, Lemos VS, Campagnole-Santos MJ, Schultheiss HP, Speth R, Walther T (2003) Angiotensin-(1-7) is an endogenous ligand for the G protein-coupled receptor Mas. Proc Natl Acad Sci USA 100(14):8258–8263

Santos RAS, Baracho NCV (1992) Angiotensin-(1-7) is a potent antidiuretic peptide in rats. Braz J Med Biol Res 25:651–654

Santos RAS, Brosnihan KB, Chappell MC, Pesquero J, Chernicky CL, Greene LJ, Ferrario CM (1988) Converting enzyme activity and angiotensin metabolism in the dog brainstem. Hypertension (Suppl I) 11:153–157

Santos RAS, Brum JM, Brosnihan KB, Ferrario CM (1990) Renin–angiotensin system during acute myocardial ischemia in dogs. Hypertension (Suppl I) 15:I-121–I-127

Santos RAS, Campagnole-Santos MJ, Andrade SP (2000) Angiotensin-(1-7): an update. Regul Pept 91:45–62

Santos RAS, Campagnole-Santos MJ, Baracho NCV, Fontes MAP, Silva LCS, Neves LAA, Oliveira DR, Caligiorne SM, Rodrigues ARV, Gropen C Jr, Carvalho WS, Silva ACSE, Khosla MC (1994) Characterization of a new angiotensin antagonist selective for angiotensin-(1-7): evidence that the actions of angiotensin -(1-7) are mediated by specific angiotensin receptors. Brain Res Bull 35:293–398

Santos RAS, Silva ACS, Magaldi AJ, Khosla MC, Ceasr KR, Passaglio KT, Baracho NCV (1996) Evidence for a physiological role of angiotensin-(1-7) in the control of hydroelectrolyte balance. Hypertension 27:875–884

Schappert SM (1996) National ambulatory medical care survey: 1994 summary. Advance Data, Vital Health Stat, Hyattsville, MD, National Center for Health Statistics, 273, PHS 96-1250:1-20

Schiavone MT, Khosla MC, Ferrario CM (1990) Angiotensin-[1-7]: evidence for novel actions in the brain. J Cardiovasc Pharmacol 16 (Suppl.4):S19-S24

Schiavone MT, Santos RAS, Brosnihan KB, Khosla MC, Ferrario CM (1988) Release of vasopressin from the rat hypothalamo-neurohypophysial system by angiotensin-(1-7) heptapeptide. Proc Natl Acad Sci USA 85:4095-4098

Schinke M, Baltatu O, Bohm M, Peters J, Rascher W, Bricca G, Lippoldt A, Ganten D, Bader M (1999) Blood pressure reduction and diabetes insipidus in transgenic rats deficient in brain angiotensinogen. Proc Natl Acad Sci USA 96:3975-3980

Senanayake PD, Moriguchi A, Kumagai H, Ganten D, Ferrario CM, Brosnihan KB (1994) Increased expression of angiotensin peptides in the brain of transgenic hypertensive rats. Peptides 15:919-926

Seyedi N, Xu X, Nasjletti A, Hintze TH (1995) Coronary kinin generation mediates nitric oxide release after angiotensin receptor stimulation. Hypertension 26:164-170

Shariat-Madar Z, Mahdi F, Schmaier AH (2002) Identification and characterization of prolylcarboxypeptidase as an endothelial cell prekallikrein activator. J Biol Chem 277:17962-17969

Soleilhac JM, Lucas E, Beaumont A, Turcaud S, Michel JB, Ficheux D, Fournie-Zaluski MC, Roques BP (1992) A 94-kDa protein, identified as neutral endopeptidase-24.11, can inactive atrial natriuretic peptide in the vascular endothelium. Mol Pharmacol 41:609-614

Stephenson SL, Kenny AJ (1987) Metabolism of neuropeptides. Biochem J (Great Britain) 241:237-247

Strawn WB, Chappell MC, Dean RH, Kivlighn S, Ferrario CM (2000) Inhibition of early atherogenesis by losartan in monkeys with diet-induced hypercholesterolemia. Circulation 101:1586-1593

Strawn WB, Ferrario CM, Tallant EA (1999) Angiotensin-(1-7) reduces smooth muscle growth after vascular injury. Hypertension 33:207-211

Tallant EA, Clark MA (2003) Molecular mechanisms of inhibition of vascular growth by angiotensin-(1-7). Hypertension 42:574-579

Tallant EA, Higson JT (1997) Angiotensin II activates distinct signal transduction pathways in astrocytes isolated from neonatal rat brain. Glia 19:333-342

Tallant EA, Diz DI, Ferrario CM (1999) Antiproliferative actions of angiotensin-(1-7) in vascular smooth muscle. Hypertension 34:950-957

Tallant EA, Diz DI, Khosla MC, Ferrario CM (1991a) Identification and regulation of angiotensin II receptor subtypes on NG108-15 cells. Hypertension 17:1135-1143

Tallant EA, Jaiswal N, Diz DI, Ferrario CM (1991b) Human astrocytes contain two distinct angiotensin receptor subtypes. Hypertension 18:32-39

Tallant EA, Landrum MH, Gallagher PE (2001) Attenuation of human breast and lung cancer cell growth by angiotensin-(1-7). FASEB J 15[5]:A778

Tallant EA, Lu X, Weiss RB, Chappell MC, Ferrario CM (1997) Bovine aortic endothelial cells contain an angiotensin-(1-7) receptor. Hypertension 29:388-392

Tamburini PP, Koehn JA, Gilligan JP, Charles D, Palmesino RA, Sharif R, McMartin C, Erion MD, Miller MJS (1989) Rat vascular tissue contains a neutral endopeptidase capable of degrading atrial natriuretic peptide. J Pharm Exp Ther 251:956-961

Tharaux PL, Stefanski A, Ledoux S, Soleilhac JM, Ardaillou R, Dussaule JC (1997) EGF and TGF-β regulate neutral endopeptidase expression in renal vascular smooth muscle cells. Am J Physiol 272:C1836-C1843

Thibonnier M, Soto ME, Menard J, Aldegir JC (1981) Reduction of plasma and urinary vasopressin during treatment of severe hypertension by captopril. Eur J Clin Invests II:449-453

Tom B, deVries R, Saxena PR, Danser AH (2001) Bradykinin potentiation by angiotensin-(1-7) and ACE inhibitors correlates with ACE C- and N-domain blockade. Hypertension 38:95–99

Trachte GJ, Ferrario CM, Khosla MC (1990) Selective blockade of angiotensin responses in the rabbit isolated vas deferens by angiotensin receptor antagonists. J Pharmacol Exp Therap 255:929–934

Ueda S, Masumori-Maemoto S, Wada A, Ishii M, Brosnihan KB, Umemura S (2001) Angiotensin(1-7) potentiates bradykinin-induced vasodilatation in man. J Hypertens 19:2001–2009

Ueda S, Masumpori-Maemoto S, Ashino K, Nagahara T, Gotoh E, Umemura S, Ishii M (2000) Angiotensin-(1-7) attenuates vasoconstriction evoked by angiotensin II but not by noradrenaline in man. Hypertension 35:998–1001

Urata H, Kinoshita A, Misono KS, Bumpus FM, Husain A (1990) Identification of a highly specific chymase as the major angiotensin II-forming enzyme in the human heart. J Biol Chem 265:22348–22357

Valdes G, Germain AM, Corthorn J, Berrios C, Foradori AC, Ferrario CM, Brosnihan KB (2001) Urinary vasodilator and vasoconstrictor angiotensins during menstrual cycle, pregnancy, and lactation. Endocrine 16:117–122

Vallon V, Heyne N, Richter K, Khosla MC, Fechter K (1998) [7-D-ALA]-Angiotensin 1-7 blocks renal actions of angiotensin 1-7 in the anesthetized rat. J Cardiovasc Pharmacol 32:164–167

Vallon V, Richter K, Heyne N, Osswald H (1997) Effect of intratubular application of angiotensin 1-7 on nephron function. Kidney Blood Press Res 20:233–239

van Rodijnen WF, van Lambalgen TA, Tangelder GJ, van Dokkum RP, Provoost AP, ter Wee PM (2002) Reduced reactivity of renal microvessels to pressure and angiotensin II in fawn-hooded rats. Hypertension 39:111–115

Vickers C, Hales P, Kaushik V, Dick L, Gavin J, Tang J, Godbout K, Parsons T, Baronas E, Hsieh F, Acton S, Patane M, Nichols A, Tummino P (2002) Hydrolysis of biological peptides by human angiotensin-converting enzyme-related carboxypeptidase. J Biol Chem 277:14838–14843

Wei CC, Ferrario CM, Brosnihan KB, Farrell DM, Bradley WE, Jaffa AA, Dell'Italia LJ (2002) Angiotensin peptides modulate bradykinin levels in the interstitium of the dog heart in vivo. J Pharmacol Exp Ther 300:324–329

Weinstock M, Gorodetsky E (1995) Comparison of the effects of angiotensin II, losartan, and enalapril on baroreflex control of heart rate in conscious rabbits. J Cardiovasc Pharmacol 25:501–507

Welches WR, Brosnihan KB, Ferrario CM (1993) A comparison of the properties, and enzymatic activity of three angiotensin processing enzymes: angiotensin converting enzyme, prolyl endopeptidase and neutral endopeptidase 24.11. Life Sci 52:1461–1480

Welches WR, Santos RAS, Chappell MC, Brosnihan KB, Greene LJ, Ferrario CM (1991) Evidence that prolyl endopeptidase participates in the processing of brain angiotensin. J Hypertens 9:631–638

Widdop RE, Krstew E, Jarrott B (1992) Electrophysiological responses of angiotensin peptides on the rat isolated nodose ganglion. Clin Exp Hypertens A, 14:597–613

Widdop RE, Sampey DB, Jarrott B (1999) Cardiovascular effects of angiotensin-(1-7) in conscious spontaneously hypertensive rats. Hypertension 34:964–968

Wilsdorf T, Gainer JV, Murphey LJ, Vaughan DE, Brown NJ (2001) Angiotensin-(1-7) does not affect vasodilator or TPA responses to bradykinin in human forearm. Hypertension 37:1136–1140

WilsonKM, Magargal W, Berecek KH (1988) Long-term captopril treatment. Angiotensin II receptors and responses. Hypertension (Suppl I) 11:I-148–I-152

Xiang J, Linz W, Becker H, Ganten D, Lang RE, Scholkens B, Unger T (1985) Effects of converting enzyme inhibitors: ramipril and enalapril on peptide action and sympathetic neurotransmission in the isolated heart. Eur J Pharmacol 113:215–223

Yamada K, Iyer SN, Chappell MC, Ganten D, Ferrario CM (1998) Converting enzyme determines the plasma clearance of angiotensin-(1-7). Hypertension 98:496–502

Yamada K, Moriguchi A, Mikami H, Okuda N, Higaki J, Ogihara T (1995) The effect of central amino acid neurotransmitters on the antihypertensive response to angiotensin blockade in spontaneous hypertension 4226. J Hypertens 13:1624–1630

Yoshida M, Naito Y, Urano T, Takada A, Takada Y (2002) L-158,809 and (D-Ala(7))-angiotensin I/II (1-7) decrease PAI-1 release from human umbilical vein endothelial cells. Thromb Res 105:531–536

Zeng W, Hong MA, Wang L, et al (2001) The role of angiotensin-(1-7) in the proliferative response of vascular smooth muscle cells induced by endothelin-1. Chin J Geriatr Cardiovasc Cerebrovasc Dis 3:107–109

Zeng W, Ma H, Lu W, et al (2000) The role of angiotensin-(1-7) in myocardial cell hypertrophy induced by angiotensin II. Chin J Cardiol 28:460–463

Zhu DN, Moriguchi A, Mikami H, Higaki J, Ogihara T (1998) Central amino acids mediate cardiovascular response to angiotensin II in the rat. Brain Res Bull 45:189–197

Angiotensin AT$_4$ Receptor

S. Y. Chai · F. A. O. Mendelsohn · J. Lee · T. Mustafa · S. G. McDowall · A. L. Albiston

Howard Florey Institute of Experimental Physiology and Medicine,
The University of Melbourne, Parkville, Victoria 3010, Australia
e-mail: sychai@hfi.unimelb.edu.au

1	Definition of the AT$_4$ Receptor	520
2	Ligands of the AT$_4$ Receptor	520
3	Distribution of the AT$_4$ Receptor	521
3.1	Cardiovascular	522
3.2	Renal	522
3.3	Brain	523
4	Effects of Ang IV and Other AT$_4$ Receptor Ligands in the CNS	523
5	Effects of Ang IV and Other AT$_4$ Receptor Ligands in Peripheral Tissues	525
5.1	Vasculature	525
5.2	Heart	526
5.3	Renal	526
6	Biochemical Properties of the AT$_4$ Receptor	527
7	Identification of the AT$_4$ Receptor	528
7.1	Purification	528
7.2	Proof of Concept: The AT$_4$ Receptor Is IRAP	528
8	Insulin-Regulated Aminopeptidase	530
8.1	Biochemical Properties of IRAP	530
8.2	Tissue Distribution of IRAP	531
8.3	Trafficking of GLUT4/IRAP Vesicles	532
8.4	AT$_4$ Ligands	532
	References	533

Abstract Although angiotensin IV (Ang IV) was thought initially to be an inactive product of angiotensin II (Ang II) degradation, the hexapeptide was subsequently shown to markedly enhance learning and memory in normal rodents and reverse memory deficits observed in animal models of amnesia. These central nervous system effects of Ang IV are mediated by binding to a specific site known as the AT$_4$ receptor which is found in appreciable levels throughout the brain and concentrated particularly in regions involved in cognition. The AT$_4$ receptor has a broad distribution and is found in a range of tissues including the adrenal gland, kidney, lung and heart. In addition to Ang IV, the peptide LVV-haemorphin-7 binds with high affinity to the AT_4 receptor and has been demonstrated to mediate the same effects as Ang IV. In the kidney Ang IV has been demonstrated to mediate a number of effects including increasing renal cortical blood flow and decreasing Na^+ transport in isolated renal proximal tubules. Biochemical studies define the AT_4 receptor as a single transmembrane glycoprotein, 165 kDa in size. This field of research was redefined by the identification the AT_4 receptor as the transmembrane enzyme insulin-regulated membrane aminopeptidase (IRAP). Insulin-regulated aminopeptidase is a type II integral membrane-spanning protein belonging to the M1 family of aminopeptidases. Ang IV has been demonstrated to be a potent inhibitor of IRAP enzymatic activity.

Keywords Aminopeptidase · Angiotensin · AT_4 receptor · Memory · Haemorphin · Insulin-regulated aminopeptidase

1
Definition of the AT_4 Receptor

Angiotensin IV (Ang IV), the 3–8 fragment of angiotensin II (Ang II), is formed by consecutive actions of aminopeptidase A and N on Ang II (Zini et al. 1996). This hexapeptide was initially thought to be inactive because it does not elicit the classical effects of Ang II on blood pressure and fluid balance. However, in 1992 a specific high-affinity binding site was discovered for Ang IV in bovine adrenal membranes (Swanson et al. 1992). This binding site is pharmacologically distinct from angiotensin AT_1 and AT_2 receptors and binds Ang II with only micromolar affinity. This binding site was named the angiotensin AT_4 receptor by an International Union of Pharmacology (IUPHAR) nomenclature committee (de Gasparo et al. 1995).

Ang IV exhibits a high affinity for the AT_4 receptor in bovine adrenal cortical membranes with a K_D of between 0.2 and 0.6 nM and a B_{max} of between 0.5 and 2.9 pmol/mg protein (Swanson et al. 1992; Bernier et al. 1994). The binding of Ang IV to the AT_4 receptor in bovine adrenal cortical membranes is saturable and reversible.

2
Ligands of the AT_4 Receptor

The amino acids of Ang IV critical for binding to the AT_4 receptor have been identified (Sardinia et al. 1993, 1994; Krishnan et al. 1999). The presence of an amino-terminal valine or, more precisely, a primary α-amine in the L-amino acid conformation in position 1, is important for binding to the AT_4 receptor. Glycine substitutions at positions 1, 2 or 3 of Ang IV greatly reduce affinity for the AT_4 receptor, whereas substitutions at positions 4, 5 or 6 of Ang IV have little effect (Sardinia et al. 1993). N-terminal elongation of Ang IV to Ang III and

Ang II results in a marked reduction in affinity, whilst the C-terminal extensions (Ang I 3–9, Ang I 3–10) do not alter binding affinity (Sardinia et al. 1993). Thus, the N-terminal amino acids of the Ang IV peptide are critical for receptor binding, whilst the C-terminal portion is less critical.

Two N-terminally modified analogues of the Ang IV, namely Nle^1-Ang IV and norleucinal Ang IV, exhibit higher affinities for the AT$_4$ receptor site from bovine adrenal membranes, with K_i of 3.6×10^{-12} and 1.8×10^{-10} M respectively (Sardinia et al. 1994). These peptides were thought to be more resistant to degradation by aminopeptidases and were found to be "agonists" at the AT$_4$ receptor site, eliciting vasodilatation and facilitation of memory as discussed in the following sections. Replacement of the Ile^3 with Val^3 and the amide bonds connecting Val^1 and Tyr^2, Val^3 and His^4 with methylene peptide bond isosteres [Vψ(CH2-NH2)YVψ(Ch2-NH2)HPF] results in a non-peptide partial antagonist of the AT$_4$ receptor named divalinal Ang IV (Krebs et al. 1996). Divalinal Ang IV bound to the AT$_4$ receptor from bovine adrenal membranes with a K_D of 0.4 nM (Krebs et al. 1996).

In addition to Ang IV and its analogues, we isolated a decapeptide, LVVYPWTQRF, from sheep cerebral cortex that binds with high affinity to the AT$_4$ receptor (Moeller et al. 1997). This peptide is identical to an internal sequence of β,δ,γ and ε globin and has been called LVV-haemorphin-7 because it was derived from haemoglobin and exhibited weak opioid activity in the guinea-pig ileum (Garreau et al. 1995). In bovine adrenal membranes, LVV-haemorphin-7 competes for ^{125}I-Ang IV binding with an IC$_{50}$ of 35 nM, while Ang IV competes for the same binding site with an IC$_{50}$ of 2.3 nM (Moeller et al. 1997). The binding of ^{125}I-LVV-haemorphin-7 and ^{125}I-Ang IV on adjacent sections of the sheep hindbrain exhibit identical patterns (Moeller et al. 1997), suggesting that both of the peptides bind to the same site, the AT$_4$ receptor. In contrast to Ang IV, LVV-haemorphin-7 is more resistant to degradation in vitro (Moeller et al. 1999a).

Deletion of the last four amino acids from the carboxy terminus or N-terminal truncation of Leu^1 does not alter binding of LVV-haemorphin-7 to the AT$_4$ receptor. In fact, removal of Val^2 from the N-terminus results in a tenfold increase in its affinity for the receptor site. By contrast, substitution of Tyr^4 and Trp^6 residues with alanine results in a tenfold decrease in affinity with respect to LVV-haemorphin-7. However, the replacement of Pro^5 and Thr^7 with alanine has little effect on the affinity of the decapeptide for the AT$_4$ receptor.

3
Distribution of the AT$_4$ Receptor

Although the AT$_4$ receptor was first identified and characterised in bovine adrenal membranes, where it occurs in high levels in the zona glomerulosa (Krebs et al. 1996), the receptor has since been detected in many tissues. These include heart, spleen, colon, kidney and brain of the guinea-pig, human prostate and bladder, rhesus monkey bladder and rat kidney (reviewed in Wright et al. 1995).

3.1
Cardiovascular

In the guinea-pig heart, the AT_4 receptors were found in cardiomyocytes, blood vessels, epicardium and endocardium (Hanesworth et al. 1993) and in the rabbit in cardiac fibroblasts (Wang et al. 1995b). The AT_4 receptor has also been detected in bovine coronary (Hall et al. 1995), bovine aortic (Bernier et al. 1995) and porcine aortic (Riva and Galzin 1996) endothelial cells, and in bovine aortic vascular smooth muscle cells (Hall et al. 1993). In rabbit carotid arteries, we found the AT_4 receptor distributed most abundantly in the smooth muscle layer and in vaso vasorum of the adventitia. Much lower levels are detected in the endothelium (Moeller et al. 1999b). Moreover, we found that the AT_4 receptor was upregulated in the media and the thickened neointima following endothelial denudation (Moeller et al. 1999b).

3.2
Renal

In the rat kidney, the AT_4 receptors have been reported in outer stripe of the medulla (Harding et al. 1994; Coleman et al. 1998; Handa et al. 1998) and at lower levels in the cortex, including the proximal and distal tubules. However, we find a different pattern of distribution of the AT_4 receptors in sheep, guinea-pig, rabbit, mouse and human kidney (A.L. Albiston, unpublished results). In our studies, high levels of the AT_4 receptors occur in the medullary rays and also in the inner medulla. Moderate levels of the receptor are also detected in the proximal and distal tubules, over the glomeruli and in the renal vasculature (Fig. 1). In WKY rats fed a high salt diet (8%), the AT_4 receptors in the kidney were reported to increase by 28% (Grove and Deschepper 1999).

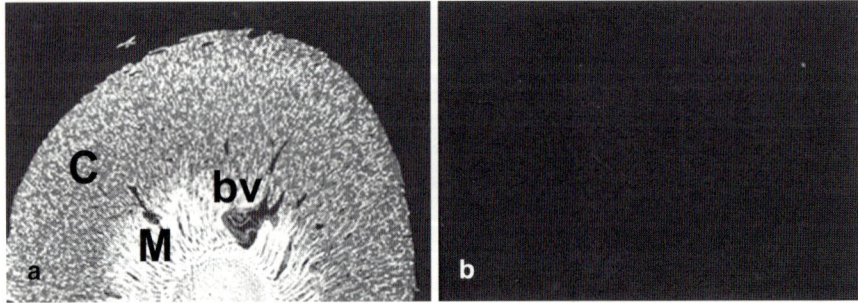

Fig. 1a, b Receptor autoradiography using [^{125}I]Nle^1-Ang IV on 10-μm adjacent ovine kidney sections. Incubations were carried out as described previously. (Chai et al. 2000). The binding sites are displayed as white regions against the black background. Total: [^{125}I]Nle1-Ang IV binding, (**a**) non-specific : [^{125}I]Nle1-Ang IV binding in the presence of 1 μM Ang IV (**b**). *C*, cortex; *M*, medulla; *bv*, blood vessel

3.3
Brain

Angiotensin AT_4 receptors, as detected by $[^{125}I]$Ang IV or $[^{125}I]Nle^1$-Ang IV binding, are distributed widely in the brains of the guinea-pig (Miller-Wing et al. 1993), rat (Roberts et al. 1995), sheep (Moeller et al. 1995), macaque monkey (Moeller et al. 1996a) and human (Chai et al. 2000). The distribution of the AT_4 receptors is highly conserved throughout these species. The AT_4 receptors are found in high concentrations in the basal nucleus of Meynert, in the pyramidal cell layer of CA1 to CA3 of Ammon's horn of the hippocampus, and throughout the neocortex in layers IV/V. This pattern of AT_4 receptor distribution closely resembles that of cholinergic neurons and their projections and its abundance in brain regions involved in cognitive processing is consistent with the memory-enhancing properties of the AT_4 receptor agonists. High levels of binding also occur in most brain regions involved in motor control including the motor cortex, ventral lateral thalamic nucleus, cerebellum and the cranial nerve motor nuclei and spinal cord ventral horn motor neurons. Many sensory regions, and some hypothalamic regions, also contained moderate levels of the AT_4 receptor, suggesting a more widespread role for the receptor in the brain.

The AT_4 receptor has also been demonstrated in primary cultures of embryonic rat neurons and neonatal rat astrocytes where the receptor was found to be abundant with B_{max}s of 1.7 and 1.4 pmol/mg protein respectively (Greenland et al. 1996).

4
Effects of Ang IV and Other AT_4 Receptor Ligands in the CNS

The first report of central effects of Ang IV was by Braszko and colleagues in 1988 who infused Ang IV (1 nmol) into the lateral cerebral ventricles of the rat and observed improved acquisition of conditioned avoidance responses and improved recall in the passive avoidance paradigm (Braszko et al. 1988). They also reported stimulation of exploratory locomotor activity and enhanced stereotypy with apomorphine or amphetamine (Braszko et al. 1988).

Following the discovery of the AT_4 receptor in discrete regions of the guinea-pig brain, particularly in the cerebral cortex and hippocampus, the effect of intracerebroventricular infusion of Ang IV on the passive avoidance paradigm was reinvestigated (Wright et al. 1993). A single dose of Ang IV (100 and 1,000 pmol) given before the final conditioning trial improved performance thereby enhancing memory retention (Wright et al. 1993). Ang IV (10, 100 and 1,000 pmol), given prior to retesting on day 2, was found to enhance retrieval of memory (Wright et al. 1993). In contrast to the earlier study by Braszko, no significant increase was observed in exploratory behaviour in the open field test (Wright et al. 1995).

In a spatial learning paradigm, the Morris water maze, chronic infusion of Nle^1-Ang IV (0.1–0.5 nmol/h for 6 days) facilitated the rate of acquisition of this task for the first 3 days (Wright et al. 1999). In the same study, chronic infusion of the "antagonist" divalinal Ang IV (0.5–5 nmol/h for 6 days) interfered with the acquisition process on days 3–6 (Wright et al. 1999). In our study using the Barnes circular maze, another spatial learning paradigm, a single dose of Nle^1-Ang IV (100 or 1,000 pmol) or the structurally unrelated AT$_4$ ligand, LVV-haemorphin-7 (1,000 pmol), given intracerebroventricularly dramatically accelerated spatial learning by up to 3 days compared to control rats. The concurrent delivery of 10 nmol of the antagonist divalinal Ang IV with 100 pmol of Nle^1-Ang IV prevented the increased rate of acquisition mediated by theAT$_4$ receptor.

The role of AT$_4$ receptors in memory was reported even in invertebrates, since both Ang IV and Ang II facilitated long-term memory of an escape response in the crab *Chasmagnathus* (Delorenzi et al. 1997). However, they later concluded that Ang II, but not Ang IV, is the endogenous mediator of long-term memory in this species (Delorenzi and Maldonado 1999).

In addition to its memory-enhancing effects in normal animals, AT$_4$ receptor ligands are effective in various models of amnesia. Bolus i.c.v. injection of 1 nmol Nle^1-Ang IV prior to the testing reverses scopolamine-induced amnesia in rats, demonstrated in the Morris water maze paradigm (Pederson et al. 1998). Subsequently it was demonstrated that i.c.v. injection of Nle^1-Ang IV dose-dependently (0.1–1 nmol) reversed scopolamine-induced amnesia in the same paradigm, an effect blocked by the AT$_4$ receptor antagonist, Nle^1,Leual^3-Ang IV [1 nmol (Pederson et al. 2001)]. Similarly, we find that i.c.v. injection of the AT$_4$ receptor ligand, LVV-haemorphin-7 (100 pmol), reverses scopolamine-induced amnesia in a passive-avoidance paradigm (S.Y. Chai et al., unpublished results). In a second rat model of amnesia, a single i.c.v. injection of the AT$_4$ receptor agonist, norleucinal (1 nmol), 5 min before the training trial each day, reversed the deficit in spatial learning induced by bilateral perforant pathway knife-cuts in the rat (Wright et al. 1999).

Recent studies demonstrate a role for AT$_4$ receptors in long-term potentiation (LTP) (Kramar et al. 2001; Wayner et al. 2001), a cellular basis for learning and memory. Application of Nle^1-Ang IV (1 μM) enhances synaptic transmission during low-frequency test pulses (0.1 Hz), and increases tetanus-induced LTP by 63% with respect to control in CA1 region of the rat hippocampus in vitro. Paired stimulation before and during infusion of Nle^1-Ang IV had no effect on paired-pulse facilitation, suggesting that the Nle^1-Ang IV-induced increase in synaptic transmission and LTP is likely to be a post-synaptic event. Similarly, administration of Ang IV (0.8 μm) enhanced LTP in the rat dentate gyrus in vivo (Wayner et al. 2001). A complex time-related effect was observed with a maximum at 5 min, a return to normal LTP at 30 min, a minimum below normal at 90 min, and a return to normal at 120 min. The Ang IV-induced enhancement of LTP was attenuated in the presence of divalinal-Ang IV (5 μM).

In cultured embryonic chicken sympathetic neurons Ang IV inhibited neurite outgrowth in a dose-dependent manner (Moeller et al. 1996b; Reed et al. 1996). This effect was blocked by the Ang IV analogues WSU 4042, Nle^1-Y-I-amide or Nle^1-Ang IV, but not by the Ang II antagonists, [Sar^1, Ile^8]Ang II or CGP 42112, indicating an AT_4 receptor-mediated effect (Moeller et al. 1996b). Other compounds, such as acetylcholine, have been shown to reduce neurite outgrowth of embryonic chicken ciliary neurons (Pugh and Berg 1994).

Based on the distribution of the AT_4 receptor in the brain, and involvement of the cholinergic system in hippocampal-dependent memory acquisition, retention and retrieval, we investigated the effect of AT_4 receptor ligands on acetylcholine release from hippocampal slices. Ang IV and LVV-haemorphin-7 both potentiated depolarisation-induced [^{3}H]-acetylcholine release from the rat hippocampus in a concentration-dependent manner with the maximal effective dose (0.1 μM) of each inducing increases of 45% and 96% above control, respectively (Lee et al. 2001). Potentiation of release by both agonists was attenuated by the AT_4 receptor antagonist, 1 μM divalinal-Ang IV. Ang IV-induced potentiation was not affected by AT_1 and AT_2 receptor antagonists (Lee et al. 2001). AT_4 receptors therefore can potentiate depolarisation-induced release of acetylcholine from hippocampal neurons and may be one mechanism by which AT_4 receptor ligands enhance cognition.

There is considerable evidence to support a role for AT_4 receptors in mitogenesis (reviewed in Mustafa 2002). Work from this laboratory demonstrated that the AT_4 receptor ligand, LVV-haemorphin-7, stimulates DNA synthesis, in the neuronal/glial hybrid cell line, NG108-15 (Moeller et al. 1999a). However, treatment with Ang IV alone had no effect on DNA synthesis, and co-incubation of the cells with excess Ang IV (1 μM) inhibited LVV-haemorphin-7-stimulated DNA synthesis. In contrast, we also demonstrated that both Ang IV and LVV-haemorphin-7 (100–1.0 nM) stimulated DNA synthesis in the human neuroblastoma cell line SK-N-MC by up to 72% and 81% above control levels (Mustafa et al. 2001). These trophic effects may play a role in the AT_4 receptor-mediated reversal of memory deficits resulting from bilateral perforant pathway lesion (see above).

5
Effects of Ang IV and Other AT_4 Receptor Ligands in Peripheral Tissues

5.1
Vasculature

Intra-arterial infusions of either Ang IV or Nle^1- Ang IV at 100 pmol/min into the carotid artery increased cerebral blood flow in anaesthetised rats, an effect blocked by pretreatment with N^ω-nitro-L-arginine methylester (L-NAME), suggesting the involvement of nitric oxide (Kramar et al. 1998). Intravenous infusions of Ang IV (1 μg/kg/min) also reversed acute cerebral blood flow reduction after subarachnoid haemorrhage (Naveri et al. 1994).

Ang IV dose-dependently stimulated plasminogen activator inhibitor-1 (PAI-1) expression in cultured bovine aortic endothelial cells (Kerins et al. 1995) and Ang IV at 1 μM increased PAI-1 mRNA and protein levels in human coronary artery endothelial cells (Mehta et al. 2002). The actions of Ang IV on the plasminogen activator system were inhibited by divalinal Ang IV, indicating an AT_4 receptor-mediated effect. In porcine pulmonary aortic endothelial cells, Ang IV stimulated nitric oxide release (Hill-Kapturczak et al. 1999) and increased activity of endothelial cell nitric oxide synthase (Patel et al. 1998). Ang IV also mobilised intracellular calcium and increased release from intracellular stores in porcine pulmonary endothelial cells (Patel et al. 1999; Chen et al. 2000). These findings suggest that the Ang IV-mediated vasorelaxation results from mobilisation of calcium and release of nitric oxide from the endothelium.

5.2
Heart

The AT_4 receptor has been characterised in guinea-pig and rabbit heart membranes (Hanesworth et al. 1993) and in rabbit cardiac fibroblasts (Wang et al. 1995b) where Ang IV stimulates DNA and RNA synthesis (Wang et al. 1995a). This contrasts with an earlier report that Ang IV (at 1 μM) inhibited the effect of Ang II to increase protein synthesis in chick myocytes whilst having no effect alone (Baker and Aceto 1990). In the isolated rabbit heart preparation, Nle^1-Ang IV (100 pM) was reported to modulate left ventricular systolic function by reducing pressure-generating capability and increasing the sensitivity of pressure development to volume change. The AT_4 receptor agonist also reduced left ventricular ejection capability and accelerated relaxation (Slinker et al. 1999). Treatment with 100 pM Nle^1-Ang IV also decreased *c-fos* and *egr-1* mRNA expression in the mechanically loaded, isolated rabbit heart (Yang et al. 1997).

5.3
Renal

In the anaesthetised rat, intra-renal arterial infusion of Ang IV or Nle^1-Ang IV (0.1–100 pmol/min) produced a dose-dependent increase in renal cortical blood flow without altering systemic blood pressure (Coleman et al. 1998). Increased renal cortical blood flow and increased sodium excretion following intra-renal arterial infusion of Ang IV occurred with no change in mean arterial pressure, glomerular filtration rate or urinary volume (Hamilton et al. 2001). These effects were blocked by pretreatment with divalinal Ang IV (Coleman et al. 1998; Hamilton et al. 2001). In contrast, another study in the anaesthetised rats reported no vasodilation after intra-renal infusion of Ang IV (10, 100 and 1,000 pmol/min) or LVV-haemorphin-7 (10, 100) (Fitzgerald et al. 1999).

In view of the abundance of AT_4 receptors in renal proximal tubules (Handa et al. 1998), the effects of Ang IV have been investigated in a number of proximal tubular epithelial cell lines. In isolated rat proximal tubules, Ang IV dose-dependently (0.1 pM–1 μM) decreased transcellular Na^+ transport as measured by proximal tubule oxygen consumption rates (Handa et al. 1998). The effect of 1 pM Ang IV on Na^+ transport was mimicked by 1 pM Nle^1- Ang IV and blocked by 1 μM divalinal Ang IV (Handa et al. 1998). In the HK2 human immortalised proximal tubular epithelial cells, Ang IV stimulated PAI-1 expression (Gesualdo et al. 1999). In the same cell line, Ang IV induced a concentration-dependent increase in intracellular sodium and activation of extracellular signal regulated kinase (ERK)-2 and p38 (Handa 2001). The intracellular calcium response to Ang IV was biphasic, increasing from 10^{-10} to 10^{-8} then decreasing at 10^{-7} and 10^{-6} M (Handa 2001). Surprisingly, divalinal Ang IV also increased intracellular sodium and calcium levels and ERK-2 and p38 phosphorylation (Handa 2001). Ang IV also dose-dependently stimulated tyrosine phosphorylation of p125-focal adhesion kinase and p68-paxillin in the LLC-PK1/Cl4 porcine proximal tubule epithelial cells (Chen et al. 2001).

In the distal tubular-like epithelial cells, the Mardin-Darby bovine kidney (MDBK) epithelial cell line Ang IV (0.1 and 10 nM) increased intracellular calcium concentration (Handa et al. 1999), a response which was no longer evident at 1 μM Ang IV. Interestingly, in this cell line, as in the HK-2 cells, divalinal Ang IV also produced the same effect as Ang IV (Handa et al. 1999).

Angiotensin AT_4 receptors have also been detected in a human collecting duct cell line, where 0.1 μM Ang IV stimulated adenosine monophosphate (cAMP) production in the presence of forskolin (Czekalski et al. 1996).

6
Biochemical Properties of the AT_4 Receptor

The biochemical properties of the AT_4 receptor have been characterised in bovine adrenal and aortic endothelial cells using photosensitive analogues of Ang IV (Bernier et al. 1998; Zhang et al. 1998). In bovine aortic endothelial cells, the AT_4 receptor was found to be a glycoprotein of 186 kDa, with N-linked carbohydrate moieties which accounted for 30% of the mass of the receptor (Bernier et al. 1998). Mild trypsin treatment of intact endothelial cell membranes produced the release of a high molecular weight fragment of the AT_4 receptor of approximately 177 kDa into the supernatant, suggesting that the receptor comprises a large extracellular domain (Bernier et al. 1998). A smaller protein was reported in the bovine adrenal membranes with a molecular weight of 165 kDa containing approximately 20% N-linked carbohydrates (Zhang et al. 1998). The receptor occurred as a dimer, with the ligand-binding subunit of 165 kDa and a smaller subunit of approximately 70 kDa (Zhang et al. 1998).

Structural analysis of the AT_4 receptor in different bovine tissues suggests the presence of a brain-specific isoform of the receptor. In contrast to the AT_4 receptor in other peripheral tissues, the receptor in the hippocampus was smaller (150 kDa) and did not occur as a dimer (Zhang et al. 1999). The difference in size between AT_4 receptor from the hippocampal and the other sources could be attributed to differential glycosylation of the proteins. There were also minor differences between the hippocampal AT_4 receptor and the receptor from peripheral sources that were uncovered by endopeptidase C fingerprinting, although these were poorly characterised (Zhang et al. 1999).

In both the bovine aortic endothelial and rat neuronal NG108-15 cells, the cell surface AT_4 receptor is rapidly internalised upon binding of [^{125}I]Ang IV (Briand et al. 1999; Moeller et al. 1999a) whereas binding of [^{125}I]divalinal Ang IV did not result in internalisation of the receptor (Briand et al. 1999).

7
Identification of the AT_4 Receptor

7.1
Purification

In order to identify the AT_4 receptor, our group purified the receptor from bovine adrenal membranes, an abundant source (B_{max}=3 nmol/mg protein) of this receptor (Wright et al. 1995; Zhang et al. 1998). The AT_4 receptor was cross-linked to a photoactivatable analogue of Ang IV (Mustafa et al. 2001) and solubilised. After ion exchange chromatography and sodium dodecyl sulphate polyacrylamide gel electrophoresis (SDS-PAGE), a silver-stained protein band co-migrating with the radioactive band was excised and digested with trypsin. Tryptic peptides were subjected to capillary-column reversed phase-high performance liquid chromatography, coupled to an electrospray ionisation ion trap mass spectrometer for peptide sequencing, as described (Simpson 2000).

Of three peptides identified, one was 95% identical to amino acid sequence of human oxytocinase, the homologue of rat IRAP. IRAP was the most likely candidate since its size (Keller et al. 1995) and tissue distribution (Keller et al. 1995) closely resemble those of the AT_4 receptor, and the enzyme had been demonstrated to bind Ang IV (Herbst et al. 1997).

7.2
Proof of Concept: The AT_4 Receptor Is IRAP

In order to demonstrate that the "AT_4 receptor" is in fact IRAP, HEK 293T cells were transfected with the full-length cDNA for human IRAP (pCI-IRAP), and analysed for the biochemical and pharmacological properties of "AT_4 receptor" (Albiston et al. 2001).

First, in cells transfected with IRAP, the specific binding of ^{125}I-Nle^1-Ang IV was 30- to 40-fold higher than binding to membranes from cells transfected with empty vector. Unlabelled Ang IV and LVV-haemorphin-7 competed for the binding of ^{125}I-Nle^1-Ang IV with IC_{50} values of 32 nM and 140 nM, respectively, which are in close agreement with the IC_{50} values for the endogenous "AT_4 receptor" (Mustafa et al. 2001). This indicates that the pharmacological profile of IRAP was identical to that of the AT_4 receptor (Albiston et al. 2001).

Second, membranes from transfected cells were cross-linked with [^{125}I]Nle^1-BzPhe^6-Gly^7-Ang IV and resolved by SDS-PAGE. In cells transfected with pCI-IRAP, but not with empty vector, a major radiolabelled band of 165 kDa and a minor band of greater than 250 kDa were observed under non-reducing conditions. Thus, the molecular weight of IRAP was identical to that of the AT_4 receptor observed in previous studies (Bernier et al. 1998; Briand et al. 1998; Mustafa et al. 2001). Both bands were absent in the presence of 10 μM unlabelled Ang IV, confirming the specific interaction of the photoactivatable Ang IV analogue with the "AT_4 receptor" expressed from the IRAP cDNA (Albiston et al. 2001).

Third, we compared the distribution of "AT_4 receptors", detected by ^{125}I-Nle^1-Ang IV binding, with the distribution of IRAP mRNA and IRAP immunoreactivity in the mouse brain (Albiston et al. 2001). In line with previous studies (Miller-Wing et al. 1993; Moeller et al. 1996a; Chai et al. 2000), ^{125}I-Nle^1-Ang IV binding sites were widely distributed and occurred in high abundance in the medial septum, in the pyramidal cell layer of CA1 to CA3 region of the hippocampus, and throughout the neocortex in the mouse brain. High levels of binding were also found in brain regions involved in motor control, such as the cerebellum. Using in situ hybridisation histochemistry and immunohistochemistry, the distribution of IRAP mRNA and immunoreactivity closely paralleled

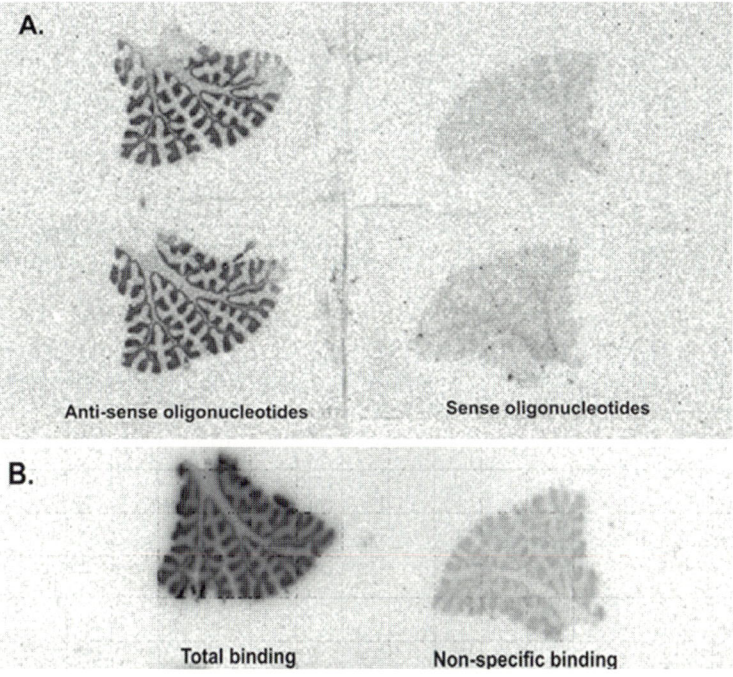

Fig. 2A, B Parallel distribution of [^{125}I]Nle-Ang IV binding and IRAP gene expression in human cerebellar cortex. **A** Distribution of IRAP mRNA in human cerebellar sections determined by in situ hybridisation histochemistry, using antisense and sense oligonucleotide sequences derived from the human IRAP cDNA sequence. The antisense sequences used include nucleotides 1136–1109, nucleotides 1407–1377 and nucleotides 2587–2560 of the human IRAP sequence (Human Genbank Acc.U62768). **B** The adjacent sections of the cerebellum were labelled with [^{125}I]Nle-Ang IV as previously described. (Chai et al. 2000)

^{125}I-Nle^1-Ang IV binding in the mouse brain (Albiston et al. 2001). Specifically, IRAP mRNA and protein was localised in regions such as the medial septum, hippocampus, neocortex and cerebellum. In the CA1–CA3 region of the hippocampus, IRAP-positive immunoreactive cells co-localised with NeuN staining in pyramidal neurons, but not in glial cells, as revealed by glial fibrillary acidic protein (GFAP) staining (Albiston et al. 2001). We also find a parallel distribution of IRAP mRNA and ^{125}I-Nle^1-Ang IV binding in the human brain (Fig. 2).

8
Insulin-Regulated Aminopeptidase

8.1
Biochemical Properties of IRAP

IRAP is a 916 amino acid protein with a calculated molecular mass of 104,575 Da. The predicted 105-kDa molecular mass for IRAP is less than the apparent 165-kDa size of the protein on SDS-PAGE, which is due to glycosylation of the protein. IRAP is a type II integral membrane-spanning protein such that the C-terminal domain containing the catalytic site is extracellular followed by a single 22 amino acid potential transmembrane domain, and a short 109 amino acid hydrophilic N-terminal segment, which projects into the cytoplasm (Keller et al. 1995). The N-terminal cytoplasmic domain contains two dileucine motifs, which are preceded by highly acidic regions; these motifs are involved in vesicular trafficking (Fig. 3) (Johnson et al. 2001).

The large extracellular region of IRAP contains the catalytic domain consisting of zinc binding [HEXXH-(18–64X)-E] and exopeptidase (GAMEN) motifs, making it a member of the M1 family of zinc metallopeptidases (Fig. 3) (Laustsen et al. 1997, 2001; Rasmussen et al. 2000). IRAP degrades a range of peptides in vitro including oxytocin, arginine vasopressin, Ang II, Ang III, somatostatin, met-enkephalin, dynorphin A (1–8), neurokinin A, neuromedin B and CCK-8 (26–33; Tsujimoto et al. 1992; Herbst et al. 1997; Matsumoto et al. 2000). In hydrolysing fully active hormones into inactive peptides or alternatively processing precursors into active peptides, IRAP may regulate in vivo peptide levels and their specific actions.

IRAP was initially cloned from rat adipocytes and skeletal muscle (Keller et al. 1995), where it is found predominantly in vesicles containing the insulin-regulated glucose transporter (GLUT4) (Kandror et al. 1994; Keller et al. 1995). This enzyme is also known as placental leucine aminopeptidase (P-LAP), or oxytocinase (OTase), because it was also cloned from a human placental cDNA library as the peptidase involved in the degradation of oxytocin (Rogi et al. 1996; Laustsen et al. 1997). Analysis of the amino acid sequence revealed that human P-LAP/oxytocinase shares 85% identity with the rat IRAP sequence (Keller et al. 1995), indicating that IRAP is the rat homologue of human P-LAP.

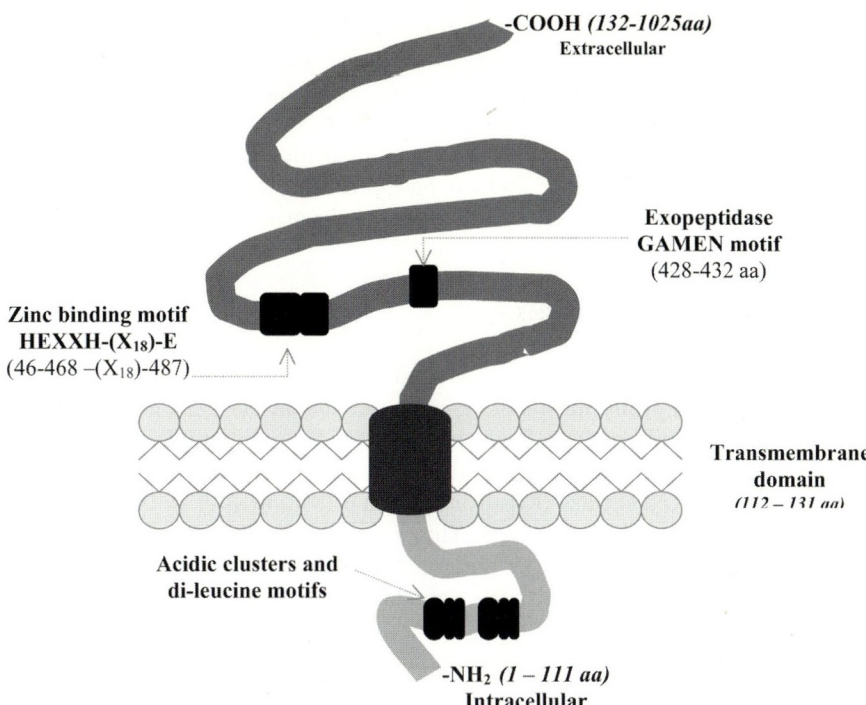

Fig. 3 Schematic diagram representing the structure and plasma membrane topology of IRAP. The large extracellular catalytic domain of IRAP (131–1025 aa) contains the Zn^{2+} binding (HEXXH-E) and exopeptidase motif (GAMEN) that is conserved in all M1 zinc metallopeptidases family members. The short intracellular domain (1–112 aa) contains di-leucine motifs in an acidic region

8.2
Tissue Distribution of IRAP

In addition to its characterisation in classic insulin responsive tissues (adipocytes, skeletal muscle and cardiomyocytes), IRAP has been identified in a range of other tissues. IRAP was first described in syncytiotrophoblasts of the placenta and was predicted to be the primary site of synthesis and release of the enzyme into the maternal circulation during pregnancy (Small and Watkins 1975). Following cloning of IRAP, the enzyme was also detected by both Western and Northern blot analysis in the heart, skeletal muscle, placenta, brain and pancreas (Keller et al. 1995; Rogi et al. 1996; Laustsen et al. 1997). Immunohistochemical localisation of IRAP protein has subsequently been demonstrated not only present in the placenta (Nagasaka et al. 1997) but also in other reproductive tissues such as human umbilical vascular endothelial cells (Nakamura et al. 2000) and endometrial epithelial cells (Toda et al. 2002).

Earlier studies reported immunohistochemical localisation of IRAP in neuronal cells but did not describe the brain region from which the sections were taken (Nagasaka et al. 1997). Later studies extended these findings, localising the enzyme in neurons of the rat cerebral cortex, basal ganglion, and cerebral cortex (Matsumoto et al. 2001). Localisation of IRAP at a subcellular level in neurons revealed granular and punctate binding throughout the cell body and dendritic spines, suggesting localisation of IRAP in cytoplasmic vesicles (Nagasaka et al. 1997; Matsumoto et al. 2001). The rat pheochromocytoma cell line PC12 endogenously expresses a high level of IRAP in the absence of GLUT4.

8.3
Trafficking of GLUT4/IRAP Vesicles

In the basal state, the majority of IRAP and GLUT4 are co-localised intracellularly in specialised post-endosomal vesicles, and also in the trans-golgi network in adipocytes (Kandror 1999; Pessin et al. 1999; Holman and Sandoval 2001; Simpson et al. 2001). During insulin stimulation, both GLUT4 and IRAP rapidly translocate to the cell surface (Kandror and Pilch 1994; Ross et al. 1996) and are inserted into the plasma membrane. This process, in part, resembles synaptic vesicle trafficking and docking in the regulation of neurotransmitter release in the brain. IRAP and GLUT4 are both trafficked by the same specialised insulin-regulated pathway in adipocytes, and a similar trafficking mechanism has also been reported for skeletal muscle (Sumitani et al. 1997; Zhou et al. 2000). Other than insulin, treatment with peptides such as endothelin and oxytocin has also been shown to stimulate the translocation of IRAP in adipocytes and endothelial cells, respectively (Wu-Wong et al. 1999; Nakamura et al. 2000). The sub-cellular localisation of IRAP in non-insulin responsive tissues is poorly defined, as are the factors that may mediate its translocation to the plasma membrane in these cells. The physiological relevance of the translocation of IRAP to the cell surface remains to be elucidated.

8.4
AT_4 Ligands

The aminopeptidase activity of IRAP can be measured by the hydrolysis of a synthetic fluorescent substrate, L-leucine-β-naphthylamide (K_m=31 μM). We have used this system to investigate the effect of AT_4 ligands on IRAP catalytic activity. As stated in Sect. 16, transient transfection of HEK293T cells with pCI-IRAP results in a roughly 30- to 40-fold increase of IRAP expression compared to cells transfected with the vector alone, as assessed by [^{125}I]Nle^1-Ang IV binding to cell membranes. For evaluation of effects on IRAP catalytic activity, HEK293T-solubilised membranes were incubated with 10 μM of the substrate, L-leucine-β-naphthylamide in the absence and presence of AT_4 ligands (Albiston et al. 2001). Addition of increasing concentrations (0.1–10 μM) of the AT_4 receptor ligands Ang IV and LVV-H7 resulted in a marked dose-dependent decrease in catalytic activity. These findings indicate that AT_4 receptor ligands bind with high affinity to IRAP and dose-dependently inhibit catalytic activity.

To determine whether the AT_4 ligands are potential substrates of IRAP, Ang IV was incubated with solubilised, crude membranes from HEK293T-transfected cells and peptide degradation determined by HPLC separation with on-line mass spectrometric detector analysis. There was no difference in the degradation profile between membranes from IRAP-transfected and vector-only-transfected cells, indicating that Ang IV is not a substrate for IRAP (R.A. Lew et al., unpublished results).

We therefore propose that AT_4 ligands may, at least in part, mediate their effects by acting as inhibitors of IRAP. The AT_4 ligands are structurally similar to a number of substrates for IRAP, being small peptides of less than 10 amino acids in length and with at least two aromatic residues. Therefore, it is likely that the AT_4 ligands are substrate analogue inhibitors of IRAP, binding at the catalytic site. If this is the case, we would predict that all AT_4 ligands are competitive inhibitors of IRAP.

References

Albiston AL, McDowall SG, Matsacos D, Sim P, Clune E, Mustafa T, Lee J, Mendelsohn FA, Simpson RJ, Connolly LM, Chai SY (2001) Evidence that the angiotensin IV (AT4) receptor is the enzyme insulin regulated aminopeptidase. J Biol Chem 13:13

Baker MF, Aceto FJ (1990) Angiotensin II stimulation of protein synthesis and cell growth in chick heart cells. Am J Physiol 259 (Heart Cir Physiol 28):H610–H618

Bernier SG, Bellemare JM, Escher E, Guillemette G (1998) Characterization of AT4 receptor from bovine aortic endothelium with photosensitive analogues of angiotensin IV. Biochemistry 37:4280–4287

Bernier SG, Fournier A, Guillemette G (1994) A specific binding site recognizing a fragment of angiotensin II in bovine adrenal cortex membranes. Eur J Pharmacol 271:55–63

Bernier SG, Servant G, Boudreau M, Fournier A, Guillemette G (1995) Characterization of a binding site for angiotensin IV on bovine aortic endothelial cells. Eur J Pharmacol 291:191–200

Braszko JJ, Kupryszewski G, Witczuk B, Wisniewski K (1988) Angiotensin II-(3-8)-hexapeptide affects motor activity, performance of passive avoidance and a conditioned avoidance response in rats. Neuroscience 27:777–783

Briand SI, Bellemare JM, Bernier SG, Guillemette G (1998) Study on the functionality and molecular properties of the AT4 receptor. Endocr Res 24:315–323

Briand SI, Neugebauer W, Guillemette G (1999) Agonist-dependent AT(4) receptor internalization in bovine aortic endothelial cells. J Cell Biochem 75:587–597

Chai SY, Bastias MA, Clune EF, Matsacos DJ, Mustafa T, Lee JH, McDowall SG, Mendelsohn FA, Paxinos G, Albiston AL (2000) Distribution of angiotensin IV binding sites (AT(4) receptor) in the human forebrain, midbrain and pons as visualised by in vitro receptor autoradiography. J Chem Neuroanat 20:339–348

Chen JK, Zimpelmann J, Harris RC, Burns KD (2001) Angiotensin IV induces tyrosine phosphorylation of focal adhesion kinase and paxillin in proximal tubule cells. Am J Physiol Renal Physiol 280:F980–F988

Chen S, Patel JM, Block ER (2000) Angiotensin IV-mediated pulmonary artery vasorelaxation is due to endothelial intracellular calcium release. Am J Physiol Lung Cell Mol Physiol 279:L849–L8456

Coleman JK, Krebs LT, Hamilton TA, Ong B, Lawrence KA, Sardinia MF, Harding JW, Wright JW (1998) Autoradiographic identification of kidney angiotensin IV binding sites and angiotensin IV-induced renal cortical blood flow changes in rats. Peptides 19:269–277

Czekalski S, Chansel D, Vandermeersch S, Ronco P, Ardaillou R (1996) Evidence for angiotensin IV receptors in human collecting duct cells. Kidney Int 50:1125–1131

de Gasparo M, Husain A, Alexander W, Catt KJ, Chiu AT, Drew M, Goodfriend T, Harding JW, Inagami T, Timmermans PB (1995) Proposed update of angiotensin receptor nomenclature. Hypertension 25:924–927

Delorenzi A, Locatelli F, Romano A, Nahmod V, Maldonado H (1997) Angiotensin II (3–8) induces long-term memory improvement in the crab Chasmagnathus. Neurosci Lett 226:143–146

Delorenzi A, Maldonado H (1999) Memory enhancement by the angiotensinergic system in the crab Chasmagnathus is mediated by endogenous angiotensin II. Neurosci Lett 266:1–4

Fitzgerald SM, Evans RG, Bergstrom G, Anderson WP (1999) Renal hemodynamic responses to intrarenal infusion of ligands for the putative angiotensin IV receptor in anesthetized rats. J Cardiovasc Pharmacol 34:206–211

Garreau I, Zhao Q, Pejoan C, Cupo A, Piot JM (1995) VV-hemorphin-7 and LVV-hemorphin-7 released during in vitro peptic hemoglobin hydrolysis are morphinomimetic peptides. Neuropeptides 28:243–250

Gesualdo L, Ranieri E, Monno R, Rossiello MR, Colucci M, Semeraro N, Grandaliano G, Schena FP, Ursi M, Cerullo G (1999) Angiotensin IV stimulates plasminogen activator inhibitor-1 expression in proximal tubular epithelial cells. Kidney Int 56:461–470

Greenland K, Wyse B, Sernia C (1996) Identification and characterization of angiotensinIV binding sites in rat neurone and astrocyte cell cultures. J Neuroendocrinol 8:687–693

Grove KL, Deschepper CF (1999) High salt intake differentially regulates kidney angiotensin IV AT4 receptors in Wistar-Kyoto and spontaneously hypertensive rats. Life Sci 64:1811–1818

Hall KL, Hanesworth JM, Ball AE, Felgenhauer GP, Hosick HL, Harding JW (1993) Identification and characterisation of a novel angiotensin binding site in cultured vascular smooth muscle cells that is specific for the hexapeptide (3–8) fragment of angiotensin II, angiotensin IV. Regul Pept 44:225–232

Hall KL, Venkateswaran S, Hanesworth JM, Schelling ME, Harding JW (1995) Characterization of a functional angiotensin IV receptor on coronary microvascular endothelial cells. Regul Pept 58:107–115

Hamilton TA, Handa RK, Harding JW, Wright JW (2001) A role for the angiotensin IV/AT4 system in mediating natriuresis in the rat. Peptides 22:935–944

Handa RK (2001) Characterization and signaling of the AT(4) receptor in human proximal tubule epithelial (HK-2) cells. J Am Soc Nephrol 12:440–449

Handa RK, Harding JW, Simasko SM (1999) Characterization and function of the bovine kidney epithelial angiotensin receptor subtype 4 using angiotensin IV and divalinal angiotensin IV as receptor ligands. J Pharmacol Exp Ther 291:1242–1249

Handa RK, Krebs LT, Harding JW, Handa SE (1998) Angiotensin IV AT4-receptor system in the rat kidney. Am J Physiol 274:F290–F299

Hanesworth JM, Sardinia MF, Krebs LT, Hall KL, Harding JW (1993) Elucidation of a specific binding site for angiotensin II(3–8), angiotensin IV, in mammalian heart membranes. J Pharmacol Exp Ther 266:1036–1042

Harding JW, Wright JW, Swanson GN, Hanesworth JM, Krebs LT (1994) AT4 receptors: specificity and distribution. Kidney Int 46:1510–1512

Herbst JJ, Ross SA, Scott HM, Bobin SA, Morris NJ, Lienhard GE, Keller SR (1997) Insulin stimulates cell surface aminopeptidase activity toward vasopressin in adipocytes. Am J Physiol 272:E600–E606

Hill-Kapturczak N, Kapturczak MH, Block ER, Patel JM, Malinski T, Madsen KM, Tisher CC (1999) Angiotensin II-stimulated nitric oxide release from porcine pulmonary endothelium is mediated by angiotensin IV. J Am Soc Nephrol 10:481–491

Holman GD, Sandoval IV (2001) Moving the insulin-regulated glucose transporter GLUT4 into and out of storage. Trends Cell Biol 11:173–179

Johnson AO, Lampson MA, McGraw TE (2001) A di-leucine sequence and a cluster of acidic amino acids are required for dynamic retention in the endosomal recycling compartment of fibroblasts. Mol Biol Cell 12:367–381

Kandror KV (1999) Insulin regulation of protein traffic in rat adipose cells. J Biol Chem 274:25210–25217

Kandror KV, Pilch PF (1994) gp160, a tissue-specific marker for insulin-activated glucose transport. Proc Natl Acad Sci USA 91:8017–8021

Kandror KV, Yu L, Pilch PF (1994) The major protein of GLUT4-containing vesicles, gp160, has aminopeptidase activity. J Biol Chem 269:30777–30780

Keller SR, Scott HM, Mastick CC, Aebersold R, Lienhard GE (1995) Cloning and characterization of a novel insulin-regulated membrane aminopeptidase from Glut4 vesicles. J Biol Chem 270:23612–23618

Kerins DM, Hao Q, Vaughan DE (1995) Angiotensin induction of PAI-1 expression in endothelial cells is mediated by the hexapeptide angiotensin IV. J Clin Invest 96:2515–2520

Kramar EA, Armstrong DL, Ikeda S, Wayner MJ, Harding JW, Wright JW (2001) The effects of angiotensin IV analogs on long-term potentiation within the CA1 region of the hippocampus in vitro. Brain Res 897:114–121

Kramar EA, Krishnan R, Harding JW, Wright JW (1998) Role of nitric oxide in angiotensin IV-induced increases in cerebral blood flow. Regul Pept 74:185–192

Krebs LT, Kramar EA, Hanesworth JM, Sardinia MF, Ball AE, Wright JW, Harding JW (1996) Characterization of the binding properties and physiological action of divalinal-angiotensin IV, a putative AT4 receptor antagonist. Regul Pept 67:123–130

Krishnan R, Hanesworth JM, Wright JW, Harding JW (1999) Structure-binding studies of the adrenal AT4 receptor: analysis of position two- and three-modified angiotensin IV analogs. Peptides 20:915–920

Laustsen PG, Rasmussen TE, Petersen K, Pedraza-Diaz S, Moestrup SK, Gliemann J, Sottrup-Jensen L, Kristensen T (1997) The complete amino acid sequence of human placental oxytocinase. Biochim Biophys Acta 1352:1–7

Laustsen PG, Vang S, Kristensen T (2001) Mutational analysis of the active site of human insulin-regulated aminopeptidase. Eur J Biochem 268:98–104

Lee J, Chai SY, Mendelsohn FA, Morris MJ, Allen AM (2001) Potentiation of cholinergic transmission in the rat hippocampus by angiotensin IV and LVV-hemorphin-7. Neuropharmacology 40:618–623

Matsumoto H, Nagasaka T, Hattori A, Rogi T, Tsuruoka N, Mizutani S, Tsujimoto M (2001) Expression of placental leucine aminopeptidase/oxytocinase in neuronal cells and its action on neuronal peptides. Eur J Biochem 268:3259–3266

Matsumoto H, Rogi T, Yamashiro K, Kodama S, Tsuruoka N, Hattori A, Takio K, Mizutani S, Tsujimoto M (2000) Characterization of a recombinant soluble form of human placental leucine aminopeptidase/oxytocinase expressed in Chinese hamster ovary cells. Eur J Biochem 267:46–52

Mehta J, Li DY, Yang H, Raizada MK (2002) Angiotensin II and IV stimulate expression and release of plasminogen activator inhibitor-1 in cultured human coronary artery endothelial cells. J Cardiovasc Pharmacol 39:789–794

Miller-Wing AV, Hanesworth JM, Sardinia MF, Hall KL, Wright JW, Speth RC, Grove KL, Harding JW (1993) Central angiotensin IV binding sites: distribution and specificity in guinea pig brain. J Pharmacol Exp Ther 266:1718–1726

Moeller I, Albiston AL, Lew RA, Mendelsohn FA, Chai SY (1999a) A globin fragment, LVV-hemorphin-7, induces [3H]thymidine incorporation in a neuronal cell line via the AT4 receptor. J Neurochem 73:301–308

Moeller I, Chai SY, Oldfield BJ, McKinley MJ, Casley D, Mendelsohn FA (1995) Localization of angiotensin IV binding sites to motor and sensory neurons in the sheep spinal cord and hindbrain. Brain Res 701:301–306

Moeller I, Clune EF, Fennessy PA, Bingley JA, Albiston AL, Mendelsohn FA, Chai SY (1999b) Up regulation of AT4 receptor levels in carotid arteries following balloon injury. Regul Pept 83:25–30

Moeller I, Lew RA, Mendelsohn FA, Smith AI, Brennan ME, Tetaz TJ, Chai SY (1997) The globin fragment LVV-hemorphin-7 is an endogenous ligand for the AT4 receptor in the brain. J Neurochem 68:2530–2537

Moeller I, Paxinos G, Mendelsohn FA, Aldred GP, Casley D, Chai SY (1996a) Distribution of AT4 receptors in the Macaca fascicularis brain. Brain Res 712:307–324

Moeller I, Small DH, Reed G, Harding JW, Mendelsohn FA, Chai SY (1996b) Angiotensin IV inhibits neurite outgrowth in cultured embryonic chicken sympathetic neurones. Brain Res 725:61–66

Mustafa T, Chai SY, Mendelsohn FA, Moeller I, Albiston AL (2001) Characterization of the AT(4) receptor in a human neuroblastoma cell line (SK-N-MC). J Neurochem 76:1679–1687

Nagasaka T, Nomura S, Okamura M, Tsujimoto M, Nakazato H, Oiso Y, Nakashima N, Mizutani S (1997) Immunohistochemical localization of placental leucine aminopeptidase/oxytocinase in normal human placental, fetal and adult tissues. Reprod Fertil Dev 9:747–753

Nakamura H, Itakuara A, Okamura M, Ito M, Iwase A, Nakanishi Y, Okada M, Nagasaka T, Mizutani S (2000) Oxytocin stimulates the translocation of oxytocinase of human vascular endothelial cells via activation of oxytocin receptors. Endocrinology 141:4481–4485

Naveri L, Stromberg C, Saavedra JM (1994) Angiotensin IV reverses the acute cerebral blood flow reduction after experimental subarachnoid hemorrhage in the rat. J Cereb Blood Flow Metab 14:1096–1099

Patel JM, Li YD, Zhang J, Gelband CH, Raizada MK, Block ER (1999) Increased expression of calreticulin is linked to ANG IV-mediated activation of lung endothelial NOS. Am J Physiol 277:L794–L801

Patel JM, Martens JR, Li YD, Gelband CH, Raizada MK, Block ER (1998) Angiotensin IV receptor-mediated activation of lung endothelial NOS is associated with vasorelaxation. Am J Physiol 275:L1061–L1068

Pederson ES, Harding JW, Wright JW (1998) Attenuation of scopolamine-induced spatial learning impairments by an angiotensin IV analog. Regul Pept 74:97–103

Pederson ES, Krishnan R, Harding JW, Wright JW (2001) A role for the angiotensin AT4 receptor subtype in overcoming scopolamine-induced spatial memory deficits. Regul Pept 102:147–156

Pessin JE, Thurmond DC, Elmendorf JS, Coker KJ, Okada S (1999) Molecular basis of insulin-stimulated GLUT4 vesicle trafficking. J Biol Chem 274:2593–2596

Pugh PC, Berg DK (1994) Neuronal acetylcholine receptors that bind alpha-bungarotoxin mediate neurite retraction in a calcium-dependent manner. J Neurosci 14:889–896

Rasmussen TE, Pedraza-Diaz S, Hardre R, Laustsen PG, Carrion AG, Kristensen T (2000) Structure of the human oxytocinase/insulin-regulated aminopeptidase gene and localization to chromosome 5q21. Eur J Biochem 267:2297–2306

Reed G, Moeller I, Mendelsohn FA, Small DH (1996) A novel action of angiotensin peptides in inhibiting neurite outgrowth from isolated chick sympathetic neurons in culture. Neurosci Lett 210:209–212

Riva L, Galzin AM (1996) Pharmacological characterization of a specific binding site for angiotensin IV in cultured porcine aortic endothelial cells. Eur J Pharmacol 305:193–199

Roberts KA, Krebs LT, Kramar EA, Shaffer MJ, Harding JW, Wright JW (1995) Autoradiographic identification of brain angiotensin IV binding sites and differential c-Fos expression following intracerebroventricular injection of angiotensin II and IV in rats. Brain Res 682:13–21

Rogi T, Tsujimoto M, Nakazato H, Mizutani S, Tomoda Y (1996) Human placental leucine aminopeptidase/oxytocinase. A new member of type II membrane-spanning zinc metallopeptidase family. J Biol Chem 271:56–61

Ross SA, Scott HM, Morris NJ, Leung WY, Mao F, Lienhard GE, Keller SR (1996) Characterization of the insulin-regulated membrane aminopeptidase in 3T3-L1 adipocytes. J Biol Chem 271:3328–3332

Sardinia MF, Hanesworth JM, Krebs LT, Harding JW (1993) AT4 receptor binding characteristics: D-amino acid- and glycine- substituted peptides. Peptides 14:949–954

Sardinia MF, Hanesworth JM, Krishnan F, Harding JW (1994) AT4 receptor structure-binding relationship: N-terminal-modified angiotensin IV analogues. Peptides 15:1399–1406

Simpson F, Whitehead JP, James DE (2001) GLUT4—at the cross roads between membrane trafficking and signal transduction. Traffic 2:2–11

Simpson RJ CL, Eddes JS, Pereira JJ, Moritz RL, Reid GE (2000) Proteomic analysis of the human colon carcinoma cell line (LIM 1215): development of a membrane protein database. Electrophoresis 21:1707–1732

Slinker BK, Wu Y, Brennan AJ, Campbell KB, Harding JW (1999) Angiotensin IV has mixed effects on left ventricle systolic function and speeds relaxation. Cardiovasc Res 42:660–669

Small CW, Watkins WB (1975) Oxytocinase-immunohistochemical demonstration in the immature and term human placenta. Cell Tissue Res 162:531–539

Sumitani S, Ramlal T, Somwar R, Keller SR, Klip A (1997) Insulin regulation and selective segregation with glucose transporter-4 of the membrane aminopeptidase vp165 in rat skeletal muscle cells. Endocrinology 138:1029–1034

Swanson GN, Hanesworth JM, Sardinia MF, Coleman JK, Wright JW, Hall KL, Miller-Wing AV, Stobb JW, Cook VI, Harding EC, et al (1992) Discovery of a distinct binding site for angiotensin II (3-8), a putative angiotensin IV receptor. Regul Pept 40:409–419

Toda S, Ando H, Nagasaka T, Tsukahara S, Nomura M, Kotani Y, Nomura S, Kikkawa F, Tsujimoto M, Mizutani S (2002) Existence of placental leucine aminopeptidase/oxytocinase/insulin- regulated membrane aminopeptidase in human endometrial epithelial cells. J Clin Endocrinol Metab 87:1384–1389

Tsujimoto M, Mizutani S, Adachi H, Kimura M, Nakazato H, Tomoda Y (1992) Identification of human placental leucine aminopeptidase as oxytocinase. Arch Biochem Biophys 292:388–392

Wang L, Eberhard M, Erne P (1995a) Stimulation of DNA and RNA synthesis in cultured rabbit cardiac fibroblasts by angiotensin IV. Clin Sci (Colch) 88:557–562

Wang L, Eberhard M, Kohler F, Erne P (1995b) A specific binding site for angiotensin II(3–8), angiotensin IV, in rabbit cardiac fibroblasts. J Recept Signal Transduct Res 15:517–527

Wayner MJ, Armstrong DL, Phelix CF, Wright JW, Harding JW (2001) Angiotensin IV enhances LTP in rat dentate gyrus in vivo. Peptides 22:1403–1414

Wright JW, Krebs LT, Stobb JW, Harding JW (1995) The angiotensin IV system: functional implications. Front Neuroendocrinol 16:23–52

Wright JW, Miller-Wing AV, Shaffer MJ, Higginson C, Wright DE, Hanesworth JM, Harding JW (1993) Angiotensin II(3–8) (ANG IV) hippocampal binding: potential role in the facilitation of memory. Brain Res Bull 32:497–502

Wright JW, Stubley L, Pederson ES, Kramar EA, Hanesworth JM, Harding JW (1999) Contributions of the brain angiotensin IV-AT4 receptor subtype system to spatial learning. J Neurosci 19:3952–3961

Wu-Wong JR, Berg CE, Wang J, Chiou WJ, Fissel B (1999) Endothelin stimulates glucose uptake and GLUT4 translocation via activation of endothelin ETA receptor in 3T3-L1 adipocytes. J Biol Chem 274:8103–8110

Yang Q, Hanesworth JM, Harding JW, Slinker BK (1997) The AT4 receptor agonist [Nle1]-angiotensin IV reduces mechanically induced immediate-early gene expression in the isolated rabbit heart. Regul Pept 71:175–183

Zhang JH, Hanesworth JM, Sardinia MF, Alt JA, Wright JW, Harding JW (1999) Structural analysis of angiotensin IV receptor (AT4) from selected bovine tissues. J Pharmacol Exp Ther 289:1075–1083

Zhang JH, Stobb JW, Hanesworth JM, Sardinia MF, Harding JW (1998) Characterization and purification of the bovine adrenal angiotensin IV receptor (AT4) using [125I]benzoylphenylalanine-angiotensin IV as a specific photolabel. J Pharmacol Exp Ther 287:416–424

Zhou M, Vallega G, Kandror KV, Pilch PF (2000) Insulin-mediated translocation of GLUT-4-containing vesicles is preserved in denervated muscles. Am J Physiol Endocrinol Metab 278:E1019–E1026

Zini S, Fournie-Zaluski MC, Chauvel E, Roques BP, Corvol P, Llorens-Cortes C (1996) Identification of metabolic pathways of brain angiotensin II and III using specific aminopeptidase inhibitors: predominant role of angiotensin III in the control of vasopressin release. Proc Natl Acad Sci USA 93:11968–11973

Subject Index

accelerated
- aging   71, 79
- cellular turnover   73

ACE, see angiotensin-converting enzyme

[^{3}H]-acetylcholine release   525
Acetyl-Ser-Asp-Lys-Pro   481
acquisition   523
ACTH   47, 55
activation   300, 376
activator protein (AP)-1   150
active conformation   287
adenoviruses   255
adenylate cyclase   337
adequate oxygen delivery   100
adipocytes   532
adipose tissue   178
adrenal   135
- gland   51, 409
- medulla   121, 355
- steroid   47
adrenergic nerve endings   45
β-adrenergic system   56
adrenoceptor
- α-adrenoceptor antagonist   45
- β-adrenoceptor   43
affinity purification   113
age   19, 78
aglomerular toadfish   44
agonist   279
AGT, see angiotensinogen
AGTR2   390
albuminuria   162

aldosterone   46, 154, 450
- hypertension   157
allostery   279
alternative splicing   272
aminopeptidase   532
aminopeptidase A   195
amphibians   38
amygdala   452
Ang, see angiotensin
angiogenesis   100, 497
angiopoietin-2   104
angiotensin (AT, Ang)   477
- activity   58
- ^{125}I-Ang I and II infusion   133
- (1–7)   300, 477, 504
- - [D-Ala$_7$]Ang-(1–7)   485, 492, 495, 501, 503, 506
- - plasma levels   505
- - receptor   497
- fragments   478
- I (AT$_1$)   38, 51, 137
- - antagonists   88
- - AT$_1$/AT$_2$ cross-talk   389
- - converting enzyme   181
- - homolog receptor   38, 47
- - mRNA   50, 52
- - receptor   8, 100, 135, 185, 207, 216, 297, 501
- - receptor expression   319
- - receptor regulation   321
- - receptor-dependent endocytosis   133, 134
- - selective antagonist   54

– IA
– – receptor   237
– – receptor gene   271
– IB
– – receptor   362
– – receptor gene   272
– II (AT$_2$)   38, 51, 73, 186, 232, 237, 297, 399
– – [^{125}I]-labelled angiotensin II   116
– – [Sar$_1$Thr$_8$]-Ang II   484
– – antagonist   495
– – binding   270
– – binding site   276
– – fragments   299
– – gene-null mutant mice   54
– – hypertension   153
– – mRNA   50
– – receptor   8, 100, 106, 112, 115, 185, 189, 207, 218, 240, 299, 399, 457, 501
– – receptor blocker   486
– – receptor expression   425
– – receptor localization   427
– – receptor subtypes   112
– – selective antagonist   54
– – signaling   426
– – smooth muscle cell proliferation   101
– III   195, 299
– IV (AT$_4$)   299, 519, 521
– – receptor   300, 504, 520
– native   36
– peptides   7
– receptor   45
– – antagonist   38
angiotensin-converting enzyme (ACE)   6, 43, 50, 81, 101, 214, 478, 504, 505
– 2   194, 207, 480
– AS-ODN   258
– gene   207
– inhibitor   253, 488
– upregulation   139
angiotensinogen (AGT)   5, 36, 174, 212, 232
– gene
– – deletion   53
– – null-mutant mice   53
– knockout mice   53
– mRNA   48
– antisense   492
antagonist   279, 303
– slow dissociating   306

anti-natriuretic   387
antisense oligonucleotides (AS-ODN)   254
antithrombotic agent   499
apoptosis   54, 71, 106, 383, 436, 437, 454
– stimuli   84
aquaporin   60
arachidonic acid   388, 495
Arg$_{167}$ 310
L-arginine   429
arrestin   285
– β-arrestin   286, 289
arterial pressure   478
arteriovenous differences   131
Asn$_{111}$ 311
aspartly protease   36
ASrAogen rats   492
association studies   183
astrocytes   120, 503
astrocytic angiotensinogen   119
AT, see angiotensin
atherosclerosis   327, 478
ATRAP   283
atrial natriuretic peptide   10, 19
atrophy   77
autoregulation of GFR   14

balloon catheter   496
baroreceptor   43
– reflex   492
baroreflex   238, 252
– sensitivity   492
basal nucleus of Meynert   523
bcl-2   384
– expression   106
binding   376, 501
– pocket   275
biochemical mechanisms   478
biomarker   81
biphenylimidazoles   451
biphenyltetrazole   305
blood
– pressure   31, 120, 209, 252, 257, 504
– – regulation   435
– – regulatory substance   73
– vessels   434
– volume   31, 55
blood–brain barrier   455
BODE   116, 122

Subject Index

BP 55
- regulation 52
bradykinin 6, 434, 465, 481, 505
- B$_2$ receptor 489
brain 118, 218, 503
- angiotensin 119
- stem nuclei 456
Braun-Menendez 5
Brown-Séquard, Charles-Édouard 5

Ca^{2+} channel agonist 44
CAKUT 389, 390
calcium 282
candesartan 303, 362
captopril 46, 101
- treatment 55
carboxyl group 308
cardiac 24
- fibrosis suppression 88
- hypertrophy 216, 233, 239
- myocytes 483
- remodeling 158
cardiomyocytes 158
cardiovascular
- function 388, 505
- medicine 477
- remodeling 151, 155
cAT$_1$ mRNA 47
catecholamine 491
- release 45
cathepsin D 36
cation channels 81
cell
- cycle control genes 77
- growth 32
- mesangial 137
- proliferation 383
- surface 138
central nervous system (CNS) 353, 489
ceramide 385, 437
cerebellar microexplants 403
c-fos 491
cGMP 45
cGMP-dependent protein kinase 406
CGP42112 377
- A 59
chaperones 78
chicken (cAT$_1$) 41
chimeric receptors 379

chloramphenicol acetyltransferase 232
CHO-hAT$_1$ 308
chromaffin tissues 45
chromatin fragments 83
chromosomal mapping 79
chromosome X 390
chymase 139, 240
clathrin-coated
- pits 286
- vesicle 289
coagulation 14
colchicine 123
colocalization 121
competitive antagonists 305
conformation 302
congestive heart failure 152, 364
constitutive activation 302, 380
contractility 14
controlling renin release 43
coronary
- artery 501
- heart disease 178
- vessels 486
cortex 432
crosstalk 428
C-terminal tail 378
cumulative
- cell death rate 86
- hemorrhage 43
cyclic guanosine monophosphate
  (cGMP) 120, 465, 499
Cyr61 105, 106
cytochrome P450 (CYP450) 236, 495
cytoplasmic alkalinization 467
cytoskeleton 458
cytosolic Ca^{2+} 41, 44

de novo synthesis 132
deletion 182
dephosphorylation 384, 464
desensitization 284, 286
diabetes mellitus 438
diabetic nephropathy 178, 438
differentiation 54, 386, 399
diffusion 136
dileucine motifs 530
dipsogenesis 46, 457
disulfide
- bonds 379

– bridge   273
diuresis   433, 434, 494
DNA
– fragmentation   74
– synthesis   525
domestic fowl   45
dorsal motor nucleus of the vagus (dmnX)   491
down-regulation   284
drinking   258, 490
– behavior   453
– rate   46
dynamin   289

effects   387
EGFR   103, 384
ELAINE trial   116, 118
elasmobranch   35
elevated $(Na^+)_i/(K^+)_i$ ratio   85
embryonic development   48
endocrine system   13
endoglin   103
end-organ damage   188
endothelial
– dysfunction   26
– nitric oxide synthase (eNOS)   103, 106, 462
endothelin (ET)   150, 163
– endothelin-converting enzyme   480
endothelium
– endothelium-dependent relaxation   45
– endothelium-derived hyperpolarizing factor   489
enhancer   179
epidermal growth factor (EGF)   318
epoxyeicosatrienoic acid (EET)   435
eprosartan   358, 360
ErbB3   382
ERK   384
– 1   405
– 2   405, 527
erythropoietin   14
estrogen   328
– deficiency   329
euryhaline teleosts   48
evolution   33, 59
– process   61
expression cDNA libraries   271

extracellular
– fluid volume   13, 19
– matrix   387
eye   135

Fak   342
familial resemblances   187
fat tissue   240
fetal
– development   48, 383
– growth   54
– – restriction   53
– kidney   49, 57
– life   33, 52
– zone   409
fibroblast growth factor (FGF)   318
fibronectin   411
filtration fraction   13
fluid-mineral balance   52
fluorescein isothiocyanate   118

gain-of-function mutants   281, 286
β-galactosidase   81, 232
gene
– cassette   79
– expression   71
– targeting   52
– therapy   256, 259
genotype concordance   79
GFP transgene expression   57
glioblastoma multiforme   124
glomerular
– antidiuresis   47
– blood flow (GBF)   438
– diuresis   47
– filtration   13
– – rate (GFR)   430
glycogen   14
glycoprotein   274
Goldblatt, Harry   5, 22
GPCR   274
G protein   337, 380
– coupling   338
– G protein-coupled receptor (GPCR)   376, 451
– receptor   284
Gq/11 protein   282
green fluorescent protein   232

Subject Index

GRK 285
growth
– factor 52
– inhibition 88
– – pathway 382
growth-promoting
– action 59
– effect 59
GTPγS 282
guanosine triphosphate (GTP) 337, 452
guanylyl cyclase 406

Haeckel's hypothesis 33, 56
hematocrit 14
haplotypes 181
heart
– failure 24, 162, 235, 478
– size 77
heat stress proteins 77
α-helix 338
heterodimerization 380, 506
high-affinity 502
high-molecular-weight AGT 179
hindlimb ischemia 101
hippocampus 523
holocephalians 35
homodimerization 389
Hox genes 50
human
– angiotensinogen 233
– gene 271
– study 504
hydrogen peroxide 326
3-hydroxy-3-methylglutaryl 327
18-hydroxydehydrogenase activity 46
hypercholesterolemia 327, 329
hyperinsulinemia 329
hyperplasia 74
hypertension 71, 152, 176, 178, 234, 478, 505
– essential 23
– low-renin essential hypertension 20
– one kidney-one clip (1K1C) 23
– pregnancy-induced 178
– two kidney-one clip (2K1C) 23
– volume-dependent 24
hypertrophy 12, 77, 151, 258, 388, 460, 496
– remodeling 157

– secondary
hypothalamus 489, 503

immunocytochemistry 113
immunohistochemistry 427, 432
inactivating mutations 279
increased
– aorta 77
– telomere restriction fragments 77
indomethacin 484
inositol (1,4,5) triphosphate 282
insertion 182
insulin-regulated glucose transporter 530
insurmountable 303
internalization 287, 288, 308
interstitial fluid 138
intracellular
– angiotensin generation 137
– loop 378
– signaling 84
inverse agonism 279, 281, 311
ion
– carrier 83
– transport inhibitor 83
IRAP 528
– substrates 533
irbesartan 358, 360
isoproterenol 43
– renin release 44

JAK 343, 385
– JAK2/STAT1 pathway 283
juxtaglomerular (JG)
– apparatus 9, 34, 57
– area 57
– cells 8

K$^+$ current 387
kidney 49, 77, 238, 241, 428, 430, 503
– abnormalities 455
– AT$_2$
– – actions 433
– – localization 431
– congenital anomalies 439
– cortex 463
– kidney-derived renin 133

- mesonephric 50
- metanephric 50
kinases 285
Krüppel-like zinc-finger transcription factor (Klf5) 105, 106

lacZ reporter gene 242
LAK2 tyrosine kinase 341
Langendorff hearts 131
lar trafficking 530
large arteries 157
left ventricular hypertrophy 158
linkage disequilibrium 175, 182
loop of Henle 495
losartan 59, 119, 236, 358, 360, 430
- binding site 278
luciferase 232
LVV-haemorphin-7 520
Lys^{199} 310

M6P/IGFII receptor 139
macula densa 9, 35
maldevelopment
- of the kidney 54
- renal structure 53
mannose 6-phosphate insulin-like growth factor II 136
MAP2 401
mas oncogene 275
maternal dietary protein restriction 54
mean circulation filling pressure 19
medulla 51, 115, 463
- oblongata 489
- ventrolateral 493
medullary rays 522
membrane hyperpolarization 81
memory 519
- retention 523
mental retardation 390, 457
mesonephros 49
meta-analysis 177
miconazole 435
microinjection 230
microvascular angiogenesis 102
migration 386, 403
mineralcorticoids 46
minicistron 273
mitochondria 135

mitogen-activated protein (MAP) 425
- kinase (MAPK) 340, 344, 498
mitotic misregulation 79
MKP-1 384
model of osteoporosis 81
molecular cloning 300
monoamines 491
monovalent ion handling 81
MOT-2 expression 79
mRNA binding protein 326
mutagenesis 301
myocardial
- AT receptors 38
- infarction 152, 162, 458
myocardium 466
myocytes 462
α-myosin heavy-chain promoter 239

Na
- regulation 55
- retaining effect of Ang II 47
- transport 47
Na$^+$
- H$^+$ exchanger 81
- K$^+$ pump 81
- K$^+$, Cl$^-$ cotransport 60, 81
- K$^+$-ATPase 494
- Na^+_i elevation 79
- transport 526
NAD(P)H oxidasis 21, 25
natriuresis 434, 494
necrosis 85
negative feedback control 55
neocortex 523
neointima 220
- formation 496
neonatal hyperplasia 71
nephrectomy 131, 133, 241
nephrogenesis 439
nephron 496
neprilysin 480
nerve regeneration 391, 458
neural crest cell 51
neurite 401
- extension 386
- outgrowth 399, 524
neuroblastoma NG108-15 cells 381
neurohypophysis 121
neuron 401, 530

neuronal connections 391
newborn 50
NG108-15 cells 403
N-glycosylation 377, 425
nitric oxide (NO) 405, 406, 429, 465, 484, 525
– synthase 101, 405, 499
L-nitroarginine methyl ester (L-NAME) 429, 430, 487
nitrocellulose 123
Nle$_1$-Ang IV 521
N-methyl-D-aspartate 120
NO/cGMP 387
nodose ganglion 492
non-AT$_1$ receptor 502, 505
non-AT$_2$
– [D-Ala$_7$]Ang-(1–7)-sensitive receptor 502
– receptor 486, 502, 505
nonmammalian species 33
nonpeptide 303
– AT$_1$ antagonists 277
norepinephrine 155, 504
N-terminal residues 377
nuclear
– factor-1 150
– polyploidy 86
– ribonucleoprotein 322

oligonucleotides 255
ontogenesis 453
ontogeny 32, 56, 60, 390
organ
– mass reduction 87
– remodeling 83
organogenesis 48
organum vasculosum laminae terminalis 257
orthostatic hypotension 14
osmoregulation 489
ovary 135
oxidase 159
oxidative stress 86, 159, 160
oxygen radicals 25
oxytocinase 528

p21$_{ras}$ 405
p38 phosphorylation 527

p42$_{mapk}$ 405
p44$_{mapk}$ 405
p130Cas 343
Page 5
papaverine-induced hypotension, plasma cortisol 46
paraventricular nucleus (PVN) 117, 481
partial agonists 281
Paxillin 343
PD123177 464
PD123319 59, 102, 362, 377, 455
PDGFR 103
perinatal programming 54
peroxynitrite 25
pharmacogenetics 195
phenotype 79, 81
– concordance 79
pheochromocytoma 381
phosphatase 2A 384
phosphatidylcholine 495
phospholipase 336, 337, 495
– C (PLC) 498
– – β 282
– – γ1 283
phosphorylation 286, 326, 340
phylogeny 32, 56, 60
PI3 K 385
pial arterioles 486
pituitary-hypophysial portal system 489
PKC 285
plasma
– ACE 182
– renin activity 132, 152
plasmid vectors 255
plasminogen activator inhibitor-1 (PAI-1) 525
platelet
– aggregation 499
– platelet-derived growth factor (PDGF) 318
PLC 341
polymorphism 175
polyploidy 74
postjunctional vascular AT$_1$ receptor 362
potassium 427
potentiation, long-term 524
PRA 43, 55
pre-activation 302
pregnancy 504
prejunctional 362

pressure
- mechanism 87
- natriuresis 16, 433
- - curve 16
primitive bony fish 35
progeria 79
programmed cell death 71
proliferation 496
- imbalance 73
prolyl
- carboxypeptidase 479
- oligopeptidase 479
promoter 176
prorenin 135, 242
- binding 137
- receptor 136
- uptake 137
prostaglandin 435
- E_2 (PGE_2) 498
prostanoid 498, 504
protein synthesis 526
proto-oncogenes 52
proximal tubules 495
purified antibodies 114
Pyk2 342
pyramidal cell layer 529

quantitative trait locus 75
quinapril 101

rab5 288
radioligand binding 305
RANTES study 467
rat hindquarter 131
recapitulation 56
recombinant inbred strains 76
regional clearance 132
regulatory protein 382
renin 179, 241, 259
- actions of angiotensin 47
- afferent arterial vessels 487
- binding 137
- body fluid feedback mechanism 16
- development 54, 209
- fibrosis 233
- gene 8, 234
- glomeruli 484
- growth 54

- interstitial fluid (RIF) 430
- low-renin 180
- membrane-bound 138
- mRNA 49, 57
- proximal tubules 484, 503, 504
- receptor 136
- Ren-1_C 8, 213
- Ren-1_D 213
- Ren-2 213
- renin-angiotensin system (RAS) 31, 76, 112, 232, 477, 505
- - local (or tissue) 131
- renin-binding proteins 136
- renin-binding receptors 136
- - genes 174
- renin-containing juxtaglomerular cell 49
- renin-like enzyme 57
- renin-secretory cell 57, 58
- secretion 44
- sympathetic nerve 9, 19
- transgene expression 57
- uptake 137
renovascular hypertension 22
replicative senescence 79
resensitization 287
resistance arteries 156
retrieval of memory 523
retroviruses 255
reverse transcription-polymerase chain reaction (RT-PCR) 432
right atrial pressure 19

salt
- appetite 13
- conservation 11
- depletion 485
- sensitivity 19
saralasin 277
sarile 303
scaffolding 286, 382
scopolamine-induced amnesia 524
senescence pathway 72
sequence identity 274
serotonin 493
serum-deprivation 84
signal
- cascade 82
- sequence 273

- transduction 311, 427
single nucleotide polymorphism 175
sino-aortic denervation 492
site-directed mutagenesis 275
small
- artery 157
- G protein 344
sodium 494, 505
- diet 11
- excretion 526
- retention 428
solitary tract (nTS) 491
sphingolipid 385
spontaneously hypertensive rat (SHR) 74, 253, 362, 427
sporadic dominant mutation 81
Src family kinase 342
STAT 343
steroidogenic pathways 46
streptozotocin 438
subcellular distribution 135
subfornical organ 46
substance P 490
superoxide
- production 78, 161
- radical 20
surmountable 303
SV40 T antigen 231
sympathetic
- ganglia 355
- nerve terminals 355
- nervous system 351, 493
syncytiotrophoblasts 531

tachykinin 491
target gene 464
telomerase activity 75
telomere length 81
testis 135
tetradecapeptide 119
tetrahydroimidazolepyridines 450, 451
TGR
- (mREN2)27 235
- ASrAOGEN 235, 237, 238
thick ascending limbs of loops of Henle 484
thimet oligopeptidase 480
thirst 13, 490
thrombin 103

thromboxane A_2 498
thymidine incorporation 42
Tigerstedt, Robert A. 5
tissue
- angiotensin II 130
- - content 135
- mesenchymal 51
- regeneration 457
- repair 386
transactivation 103
transgenic 153
- animals 493
- hypertensive rat
- mammals 230
- mice 207, 234, 236, 253
- [mRen-2]27 481
- rat 234
trans-golgi network 532
transmembrane
- domain 273, 378
- glycoprotein 273
trophic mechanisms 496
tryptophan hydroxylase 238
T-type Ca^{2+} channel 403
tubulin 401
tubuloglomerular feedback 9, 44
turkey (tAT_1) 41
two state model 307
two-hybrid system 382
two-kidney one-clip hypertensive dogs 485
tyrosine
- kinase 283, 340
- phosphatase SHP-1 381
- phosphorylation 340, 527

urinary tract 439
urine-concentrating ability 55

valsartan 358
vasa recta 13
vascular
- fibrosis 460
- growth 58
- hypertrophy 24
- reactivity 155
- regression 88
- remodeling 159

- smooth muscle cells (VSMC) 78, 150, 322, 324, 496, 503
- tone 12
vascularization 53
vasculogenesis 100
vasoconstriction 155, 494
vasodilation 388, 486, 526
vasopressin 13, 119, 120, 236, 258, 481, 490
- secretion 453
VEGF 102, 106
vessel
- mesentric 487
viral vector 255
volume regulation 55

water 505
- absorption 496
wound healing 497

ying-yang hypothesis 461

zebrafish AGT 36
Zfhep 464
zinc binding 530
zona
- fasciculata 52
- glomerulosa 51, 115, 521
- reticularis 52

Printing: Saladruck, Berlin
Binding: Stein+Lehmann, Berlin